This item must be returned or renewed by the last date shown below. The loan period may be shortened if it is reserved by another reader. A fine will be due if it is not returned on time.

DATE OF RETURN

2 8 SEP 1998

1 1 JAN 1999

06 DEC 1999

0 7 JUN 2002

1 9 DEC 2005

UNIVERSITY COLLEGE LONDON

Gower Street London WC1E 6BT

LF4D

WITHDRAWN

RLO323.7.8.97

Time Frequency and Wavelets
in Biomedical Signal Processing

Time Frequency and Wavelets in Biomedical Signal Processing

Edited by

Metin Akay
*Thayer School of Engineering,
Dartmouth College*

IEEE Press Series in Biomedical Engineering
Metin Akay, *Series Editor*

Endorsed by the IEEE Engineering in Medicine and Biology Society

The Institute of Electrical and Electronics Engineers, Inc., New York

UNIVERSITY
COLLEGE
LONDON

To my wife, Yasemin, and son, Altug,

for their infinite support, patience, and love.

Printed in the United States of America

10   9   8   7   6   5   4   3   2   1

ISBN 0-7803-1147-7

IEEE Order Number: PC5619

Library of Congress Cataloging-in-Publication Data

Time frequency and wavelets in biomedical signal processing / edited
  by Metin Akay.
     p.   cm. — (IEEE Press series in biomedical engineering)
  Includes bibliographical references and index.
  ISBN 0-7803-1147-7 (cloth)
  1. Signal processing. 2. Wavelets (Mathematics) 3. Biomedical
engineering. I. Akay, Metin. II. IEEE Engineering in Medicine and
Biology Society. III. Series.
R857.S47T54   1997
610'.28—dc21   97-19866
                                                     CIP

Contents

Contents

xv

PART IV WAVELETS, NEURAL NETWORKS, AND FRACTALS 643

List of Contributors

Metin Akay
Thayer School of Engineering,
Dartmouth College, New Hampshire

J. J. Bellanger
Laboratoire Traitement du Signal et
de l'Image, INSERM, Université de
Rennes, France

K. J. Blinowska
Laboratory of Medical Physics,
Warsaw University, Poland

Tom Brotherton
Orincon Corporation

G. Carrault
Laboratoire Traitement du Signal et
de l'Image, INSERM, Université de
Rennes, France

Jiande Z. Chen
Institute for Healthcare Research,
Baptist Medical Center, Oklahoma

Robert A. Clark
Department of Radiology, College
of Medicine, University of South
Florida, Florida

Laurence P. Clarke
Department of Radiology, College
of Medicine, University of South
Florida, Florida

Ronald R. Coifman
Department of Mathematics,
Yale University, Connecticut

Dominique Devedeux
Département Génie Biologique,
Université de Technologie de
Compiègne, France

Hartmut Dickhaus
Department of Medical Informatics,
Universität Heidelberg/
Fachhochschule-Heilbronn, Germany

Jacques Duchêne
Université de Technologie de Troyes

Pierre Duhamel
Télécom Paris, France

P. J. Durka
Laboratory of Medical Physics,
Warsaw University, Poland

Russell Fischer
Biomedical Engineering Department,
Rutgers University, New Jersey

Patrick Flandrin
Laboratoire de Physique, École
Normale Supérieure de Lyon, France

Dennis M. Healy, Jr.
Department of Mathematics
and Computer Science, Dartmouth
College, New Hampshire

Hartmut Heinrich
Department of Medical Informatics,
Universität Heidelberg/
Fachhochschule-Heilbronn, Germany

C. Heneghan
Department of Electrical, Computer
and Systems Engineering, Boston
University, Massachusetts

Bruce K. T. Ho
Department of Radiological Sciences,
University of California at Los
Angeles, California

Doug Jones
Coordinated Science Lab,
University of Illinois, Illinois

Maria Kallergi
Department of Radiology, College
of Medicine, University of South
Florida, Florida

Mohsine Karrakchou
INRS Telecommunications, Canada

S. M. Khanna
Department of Otolaryngology,
Columbia College of Physicians
and Surgeons, New York

Jerome Knopp
Department of Electrical and
Computer Engineering, University
of Missouri, Missouri

Kunikazu Kobayashi
Department of Computer Science
and Systems Engineering, Yamaguchi
University, Japan

Iztok Koren
Department of Electrical and
Computer Engineering, University
of Florida, Florida

Murat Kunt
Laboratoire de traitement des
signaux, École Polytechnique
Fédérale de Lausanne, France

Andrew Laine
Computer and Information Science
and Engineering Department,
University of Florida, Florida

Zhiyue Lin
Department of Medicine, University
of Kansas Medical Center, Kansas

Marco Ma
Department of Computer Sciences,
University of California at Los
Angeles, California

Armando Manduca
Department of Physiology
and Biophysics, Mayo Clinic
and Foundation

S. Lawrence Marple, Jr.
Orincon Corporation

Mustafa Matalgah
Department of Electrical
and Computer Engineering,
University of Missouri, Missouri

Salah Mawagdeh
Faculty of Medicine, Jordan
University of Science and
Technology, Jordan

Claudia Mello
Biomedical Engineering Department,
Rutgers University, New Jersey

Tim Olson
Department of Mathematics,
Dartmouth College, New Hampshire

Ramesh K. Panwar
Department of Computer Sciences,
University of California at Los
Angeles, California

Wei Qian
Department of Radiology, College
of Medicine, University of South
Florida, Florida

Walter B. Richardson
Department of Mathematics,
University of Texas, Texas

Olivier Rioul
Télécom Paris, France

Janet C. Rutledge
Electrical Engineering and Computer
Science Department, McCormick

School of Engineering and Applied
Science, Northwestern University,
Illinois

Berkman Sahiner
Department of Electrical Engineering
and Computer Science, University
of Michigan, Michigan

Pongskorn Saiptech
Department of Radiological Sciences,
University of California at Los
Angeles, California

Sergio Schuler
Computer and Information Science
and Engineering Department,
University of Florida, Florida

Robert J. Sclabassi
Laboratory for Computational
Neuroscience, Department of
Neurological Surgery, University
of Pittsburgh, Pennsylvania

L. Senhadji
Laboratoire Traitement du Signal et
de l'Image, INSERM, Université de
Rennes, France

Mingui Sun
Laboratory for Computational
Neuroscience, Department of
Neurological Surgery, University
of Pittsburgh, Pennsylvania

M. C. Teich
Department of Electrical, Computer
Systems Engineering, Boston
University, Massachusetts

Priya Venugopal
Department of Radiology, College
of Medicine, University of South
Florida, Florida

Douglas W. Warner
Department of Mathematics
and Computer Science, Dartmouth
College, New Hampshire

John B. Weaver
Department of Diagnostic
Radiology, Dartmouth – Hitchcock
Medical Center, New Hampshire

Mladen Victor Wickerhauser
Department of Mathematics,
Washington University, Missouri

William J. Williams
Department of Electrical Engineering
and Computer Science, University
of Michigan, Michigan

Andrew E. Yagle
Department of Electrical Engineering
and Computer Science, University
of Michigan, Michigan

Qinghua Zhang
IRISA/INRIA, Campus de Beaulieu,
France

Preface

A signal can be considered to be stationary if its statistical characteristics are NOT changing with time. Stationary signals can be analyzed using classical Fourier transform methods in which the signal can be expanded on the orthogonal basis functions (sin and cosine waves). However, most biomedical signals are nonstationary and have highly complex time-frequency characteristics. In practice, the stationary condition for the nonstationary signals can be satisfied by dividing the signal into blocks of short segments in which the signal segment can be assumed to be stationary. This method, called the short time Fourier transform (STFT) was proposed by Gabor in 1946. However, the problem with the STFT is the length of the desired segment. Choosing a short analysis window may cause poor frequency resolution. On the other hand, a long analysis window may improve the frequency resolution but compromises the assumption of stationary within the window.

To overcome these difficulties with the STFT, several time-frequency analysis methods including the Gabor representation, Wigner-Ville Distribution, Binomial transform, Choi-Williams, Reduced Interference Distribution methods etc. have been proposed. An alternative way to analyze the nonstationary biomedical signals is the wavelet transform which expands the signal onto the basis functions. The basis functions can be constructed bu dilation, contractions and shifts of a unique function called the **wavelet prototype**. The wavelet method act as a **mathematical microscope** in which we can observe different parts of the signal by just adjusting the focus. In practice, it is not necessary to the wavelet transform to have continuous frequency (scale) parameters, to allow fast numerical implementations, the scale can be varied only along the dyadic sequences. Therefore, the wavelet transform has a very good time resolution at the high frequencies and good frequency resolution at the low frequencies. Because of a number of theoretical a well as practical contributions made on various aspects of the WT's, the subject is growing rapidly.

This edited book will cover the introduction to the time-frequency and wavelet transform methods the applications of the wavelet transforms to the biological signals including the EEG, respiratory, auditory, and evoked potential response signals etc and medical images.

PART I

Time-Frequency Analysis Methods with Biomedical Applications

In this part, we will focus on the basics of time-frequency analysis methods and their biomedical applications.

Chapter 1 by Williams will be devoted to the review of recent advances in time-frequency analysis methods, including the reduced interference distribution methods, the time-frequency analysis methods based on adaptive kernels, and the fast algorithms. Chapter 2 by Williams will discuss the biomedical applications of time-frequency methods described in chapter 1. The biomedical examples include the electrophysiological signals and epilepsy—the event-related potentials. In addition, the time-frequency analyses of animal sounds as well as muscle, and heart sound signal analyses are included in detail. Chapter 3 by Marple et al. reviews the adaptive quadratic time-frequency, wavelet, and model-based representation methods and their applications to Doppler ultrasound echoes from cardiac structures. Chapter 4 by Dickhaus and Heinrich presents the application of time-frequency methods to describe the complex behavior of cardiac late potentials in ECGs. Chapter 5 by Duchêne and Devedeux discusses the application of time-frequency methods to uterine EMG characterization to detect preterm delivery risk. Chapter 6 by Lin and Chen summarizes the short-time Fourier transform, the time-frequency analysis method based on the exponential distribution, and the adaptive method in analyzing the electrogastrograph signals to describe the gastric myoelectrical activity. The last chapter in this part is chapter 7 by Mello and Akay and it discusses the traditional analysis methods and the general classes of representations. Biomedical applications of these methods are cited.

Chapter 1

Recent Advances in Time-Frequency Representations: Some Theoretical Foundations*

William J. Williams

1.1. INTRODUCTION

The Fourier transform has been of great value in many areas of engineering and science. However, signals of practical interest often do not conform to the requirements of realistic application of Fourier principles. The approach works best when the signal of interest is composed of a number of discrete frequency components so that time is not a specific issue (e.g., a constant frequency sinusoid) or, somewhat paradoxically, when the signal exists for a very short time so that its time of occurrence is considered to be known (e.g., an impulse function). Much of what we are taught implies that signals that cannot be satisfactorily represented in these ways are somehow suspect and must be forced into the mold or abandoned.

It has been quite difficult to satisfactorily handle nonstationary signals such as chirps using conceptualizations based on stationarity. The spectrogram represents an attempt to apply the Fourier transform for a short-time analysis window, within which it is hoped that the signal behaves reasonably according to the requirements of stationarity. Many real-world signals, particularly biological signals, do not conform to these requirements. By moving the analysis window along the signal, one hopes to track and capture the variations of the signal spectrum as a function of time. The well-known spectrogram is an example of such an approach. The spectrogram has

*This research was supported in part by grants from the Rackham School of Graduate Studies, the Office of Naval Research, ONR contract no. N00014-89-J-1723, the National Science Foundation, NSF Grant BCS 9110571, and a Biomedical Research Support Grant from the Office of Vice President for Research through the National Institutes of Health.

many useful properties including a well-developed general theory. It has been used with great success for many years and has provided many useful insights into biological phenomena, particularly speech. The spectrogram often presents serious difficulties when used to analyze rapidly varying signals, however. If the analysis window is made short enough to capture rapid changes in the signal, it becomes impossible to resolve frequency components of the signal which are close in frequency during the analysis window duration. On the other hand, if the time window is made long enough to permit good frequency resolution, it is difficult to determine where, in time, the various frequency components act. There are many assumptions in conventional engineering analysis which allow us to view signals from an idealized viewpoint. The Fourier transform is defined to be

$$X(\omega) = F[x(t)] = \int_{-\infty}^{\infty} x(t)e^{-j\omega t}dt \tag{1-1}$$

and its inverse,

$$x(t) = F^{-1}[X(\omega)] = \frac{1}{2\pi}\int_{-\infty}^{\infty} X(\omega)e^{j\omega t}d\omega \tag{1-2}$$

This very familiar transform is certainly well known to the reader. However, one seldom questions the integral limits. Everyone knows that it is not possible to obtain $x(t)$ in a practical sense. How could one know $x(t)$ for all time? Likewise, it is impossible to know, in a practical sense, what $X(\omega)$ is for all frequencies. If we have a function which expresses $x(t)$ or $X(\omega)$, then there is no problem. However, we may often neglect to apply this thinking when dealing with real-world signals. The Fourier transform essentially implies that one does not need to worry about time after the transform is applied. Time has been integrated out of the picture. All one cares about is the frequency content of the signal. There is no attention to *when* the signal components of different frequencies act. Likewise, when the inverse transform is obtained, one is supposed to have no interest in the frequency of the various components of the signal $x(t)$. It is tacitly assumed that the frequency components of the signal are eternal and not changing with time. This is the basis of the Fourier series, which is a weighted sum of sine and cosine terms. Figure 1-1 illustrates this. There are three components with different frequencies present in $x(t)$. The magnitude of $X(\omega)$ is also shown. Next, these same components are windowed in time and are combined to provide a sequential combination rather than a simultaneous combination of these frequency components. The results of this experiment are shown in Fig. 1-2.*

One can see that there is no evidence of the difference in the time action of the sinusoidal components in the transform domain. Certainly, the simultaneous sinewaves yield a sharper spectrum because they are longer in duration. Where and when they act is unclear from the spectrum, however. This justifies the need for joint time-frequency representations (TFR). The spectrogram has long been a useful tool in time-frequency (t-f) analysis. The basic idea behind the spectrogram is to assume that the signal is stationary or quasi-stationary over a limited time window. This time

---

*The actual computation was done using a 512-point FFT and the sample rate assumed was 1 Hz.

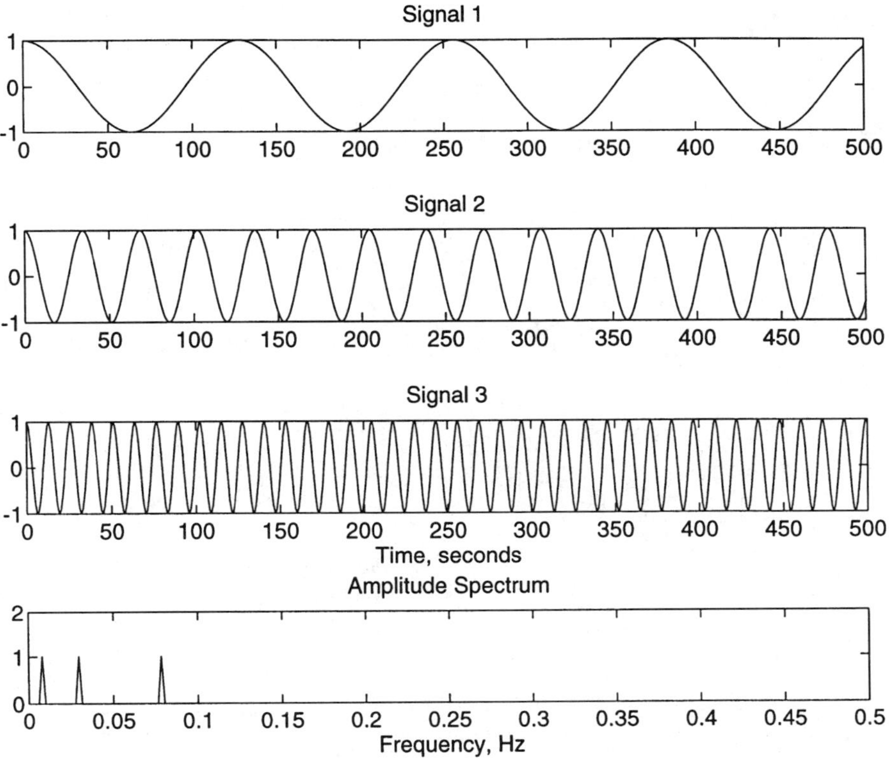

Figure 1-1   Three sinewaves simultaneous in time, and the amplitude spectrum of their sum.

window is moved along the signal and a time-indexed spectrum is computed. The continuous formulation is as follows from the short-time Fourier transform, or STFT:

$$\text{STFT}_x(t, \omega) = \int x(\tau)h(\tau - t)e^{-j\omega\tau}\,d\tau \tag{1-3}$$

and then,

$$\text{SP}_x(t, \omega) = |\text{STFT}_x(t, \omega)|^2 \tag{1-4}$$

where $h(t)$ is the window function. The spectrogram suffers from a window trade-off condition which is often known as the uncertainty principle. Long time windows provide good frequency resolution, but poor time resolution. Short time windows provide good time resolution, but poor frequency resolution. One must make a choice.

The spectrogram has been a very useful tool in time-frequency analysis. However, it has several serious liabilities and limitations that we will cover in detail in this chapter. More recent time-frequency developments have provided useful and interesting alternatives to spectrograms.

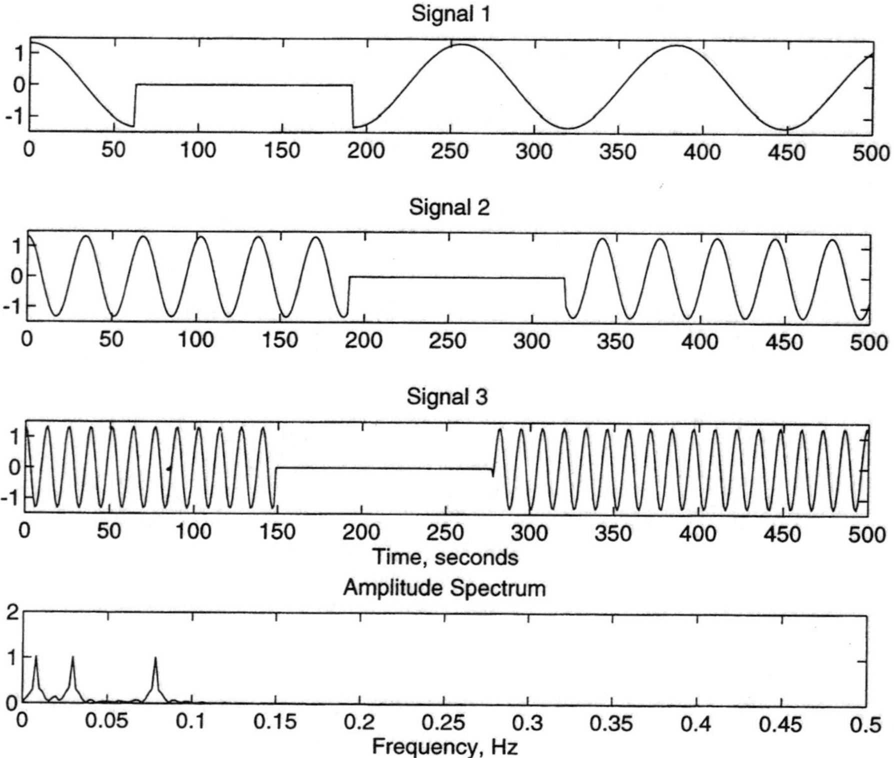

Figure 1-2.   The signals from Fig. 1-1 with gaps and the amplitude spectrum of their sum.

The Wigner distribution (WD) has been employed as an alternative to overcome the liabilities and limitations of the spectrogram. The WD was first introduced in the context of quantum mechanics [1] and revived for signal analysis by Ville [2]. The WD has many important and interesting properties [3–5]. It provides a high-resolution representation in time and in frequency for a nonstationary signal such as a chirp. In addition, the WD has the important property of satisfying the time and frequency marginals in terms of the instantaneous power in time and energy spectrum in frequency. However, its energy distribution is not nonnegative and it often possesses severe cross-terms, or interference terms, between components in different t-f regions, potentially leading to confusion and misinterpretation. An excellent discussion on the geometry of interferences has been provided by Hlawatsch and Flandrin [6–8].

Both the spectrogram and the WD are members of Cohen's class of distributions [9]. Cohen has provided a consistent set of definitions for a desirable set of t-f distributions, which has been of great value in guiding and clarifying efforts in this area of research. Cohen's class of distributions is defined to be

$$C_x(t, \omega; \phi) = \int \int \int e^{j(-\theta t - \tau\theta + \omega u)} \phi(\theta, \tau) x(u + \tau/2) x^*(u - \tau/2) \, du \, d\tau \, d\theta \qquad (1\text{-}5)$$

where $x(t)$ is the time signal, $x^*(t)$ is its complex conjugate, and $\phi(\theta, \tau)$ is the **kernel** of the distribution.*

A recent comprehensive review by Cohen [10] provides an excellent overview of time-frequency distributions and recent results using them. This paper addresses a specific subset of t-f distributions belonging to Cohen's class. These are the time-shift and frequency-shift invariant t-f distributions. For these distributions, a time shift in the signal is reflected as an equivalent time shift in the t-f distribution, and a shift in the frequency of the signal is reflected as an equivalent frequency shift in the t-f distribution.† The spectrogram, the WD, and the reduced interference distribution (RID) all have this property. Different distributions can be obtained by selecting different kernel functions in Cohen's class. Boashash has compared the performances of several time-frequency distributions in terms of resolution [11]. Desirable properties of a distribution and associated kernal requirements have been extensively investigated by Claasen and Mecklenbräuker [3–5].

The Wigner distribution is in general expressed as

$$W_x(t, \omega) = \int x\left(t + \frac{\tau}{2}\right)x^*\left(t - \frac{\tau}{2}\right)e^{-j\omega\tau}\,d\tau \tag{1-6}$$

or, in its dual form, as

$$W_x(t, \omega) = \frac{1}{2\pi}\int X^*\left(\omega + \frac{\theta}{2}\right)X\left(\omega - \frac{\theta}{2}\right)e^{-j\theta t}\,d\theta \tag{1-7}$$

The Wigner distribution often provides high time- and frequency-resolution results for simple monocomponent signals. However, if $x(t) = a(t) + b(t)$, then the Wigner distribution consists of four components, $W_{aa}(t, \omega) + W_{ab}(t, \omega) + W_{ba}(t, \omega) + W_{bb}(t, \omega)$. If, due to symmetry, $W_{ab}(t, \omega)$ and $W_{ba}(t, \omega)$ combine, then an interference term, or cross-term, which has twice the amplitude of $W_{aa}(t, \omega)$ and $W_{bb}(t, \omega)$, results if $W_{aa}(t, \omega)$ and $W_{bb}(t, \omega)$ are equal in amplitude.

There is another classical distribution, the Rihaczek, or Margenau–Hill, distribution [12]. Its form is

$$R_x(t, \omega) = \frac{1}{\sqrt{2\pi}}x(t)X^*(\omega)e^{-j\omega t} \tag{1-8}$$

This is an interesting result, since it treats the TFR as a separable product of the signal and its Fourier transform. However, it has serious cross-term problems and satisfies few desirable properties. It does have a strong support property, to be discussed later.

More recently, Choi and Williams introduced a new distribution having an exponential-type kernel [13], which they called the exponential distribution, or ED. This new distribution overcomes several drawbacks of the spectrogram and WD; this distribution provides high resolution with suppressed interferences [13, 14]. It has been called the Choi–Williams distribution, first by Cohen and

---

*The range of integrals is from $-\infty$ to ∞ throughout this paper unless otherwise indicated.

†Some prefer to call this covariance rather than invariance, since the distribution moves in accordance with shifts in time and frequency. The shape is invariant, however.

subsequently by a number of other investigators. We prefer to refer to this specific example as the ED and the general class of reduced interference distributions as RIDs.

Another new time-frequency distribution has received a lot of attention in recent years. This is the cone kernel distribution, or the ZAM distribution, introduced by Zhao, Atlas, and Marks [15]. The ZAM is spectrogram-like in some aspects, but it overcomes several of the liabilities of the spectrogram and offers high resolution along with sharp time delineation and good frequency resolution of segmented sinewaves.

Time-frequency distributions (TFDs) have been so termed due to their similarities and analogies to probabilistic concepts. Some prefer to call them time-frequency representations (TFRs) to highlight the fact that they are not really distributions in the probabilistic sense. In this work, we will use the term TFRs in general, though "distribution" will be retained when referring to specific members of Cohen's class. There are a number of TFRs which have recently arisen or evolved and are based on the elements of one or more of the TFRs just mentioned. These TFRs will be discussed later in this chapter. One particularly useful method of viewing TFRs will be discussed next. This is the reduced interference distribution, or RID. This chapter is organized around the RID concept, since this is the frame of reference held by this writer. The large literature on spectrograms and Wigner distributions will not be comprehensively addressed, nor will some of the exciting recent developments in time-frequency analysis be covered. The focus will be on concepts developed over the past ten years which have been applied with at least some success in biological signal analysis and have thus withstood the test of time to this point. Applications of these approaches will be covered in another chapter.

1.2. THE REDUCED INTERFERENCE DISTRIBUTION

1.2.1 Ambiguity Function Relationships

The key to understanding t-f relationships and manipulations is a thorough understanding of the ambiguity domain. Let $X(\omega)$ be the Fourier transform (FT) of the signal $x(t)$; let $R_x(t, \tau)$ be the *instantaneous autocorrelation* of a complex signal $x(t)$, defined as

$$R_x(t, \tau) = x(t + \tau/2)x^*(t - \tau/2) \tag{1-9}$$

where f^* denotes the complex conjugate of f. The Wigner distribution of $x(t)$ is defined as the FT of $R_f(t, \tau)$ with respect to the lag variable τ:

$$W_x(t, \omega) = F_\tau[x(t + \tau/2)x^*(t - \tau/2)] = F_\tau[R_x(t, \tau)] \tag{1-10}$$

Similarly, but with a different physical meaning, the symmetrical ambiguity function (AF) is defined as the inverse Fourier transform (IFT) of $R_x(t, \tau)$ with respect to the first variable:

$$A_x(\theta, \tau) = F_t^{-1}[x(t + \tau/2)x^*(t - \tau/2)] = F_t^{-1}[R_x(t, \tau)] \tag{1-11}$$

Thus $W_x(t, \omega)$ and $A_x(\theta\tau)$ are related by the two-dimensional (2-D) FT:

$$W_x(t, \omega) = \int\int A_x(\theta, \tau)e^{-j(t\theta+\omega\tau)}d\theta d\tau \qquad (1\text{-}12)$$

These relationships may be combined with Eq. 1-1 to show that $C_x(t,f;\phi)$ may be found by

$$C_x(t, \omega, \phi) = \int\int \phi(\theta, \tau)A_x(\theta, \tau)e^{-j(t\theta+\omega\tau)}d\theta d\tau \qquad (1\text{-}13)$$

Thus while the Wigner distribution may be found from the symmetric ambiguity function by means of a double Fourier transform, any member of Cohen's class of distributions may be found by first multiplying the kernel, $\phi(\theta, \tau)$, by the symmetric ambiguity function and then carrying out the double Fourier transform. The generalized ambiguity function, $\phi(\theta, \tau)A_x(\theta, \tau)$ [16], is a key concept in t-f which aids one in clearly seeing the effect of the kernel in determining $C_x(t, \omega; \phi)$. A test signal for evaluating some of the properties of time-frequency distributions is introduced at this point. The test signal consists of two sinusoidal segments of differing time and frequency placement. This signal is shown in Fig. 1-3. The Wigner distribution and the ambiguity function of the two sinusoids displaced in time and frequency are shown in Fig. 1-4.

Thus if the Wigner kernel is multiplied by the ambiguity function, the ambiguity function is not altered. The Wigner time-frequency result is shown in Fig. 1-5.

It can be shown [10] that the kernel of the spectrogram is the ambiguity function of the time window itself. Since the time window is Gaussian in this case, the kernel is a 2-D Gaussian function of ω and τ. Figure 1-6 shows the spectrogram kernel and the result of its effect on the ambiguity function.

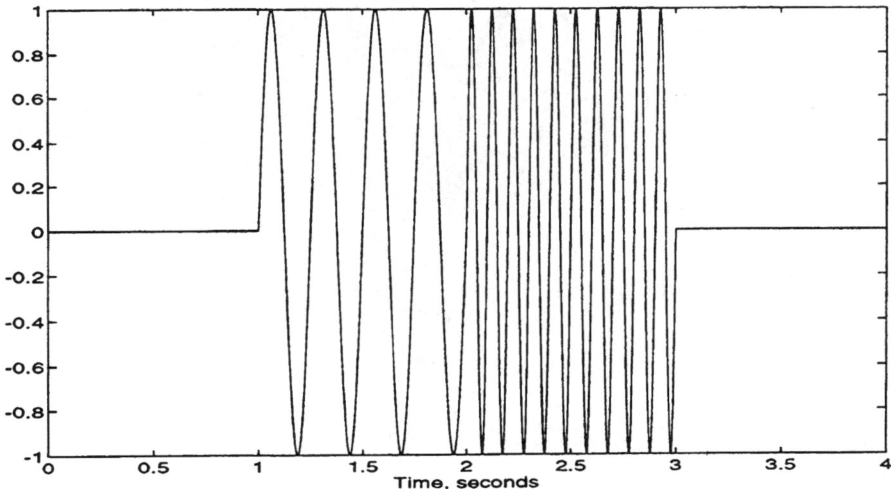

Figure 1-3   Test signal used to evaluate the time-frequency distributions.

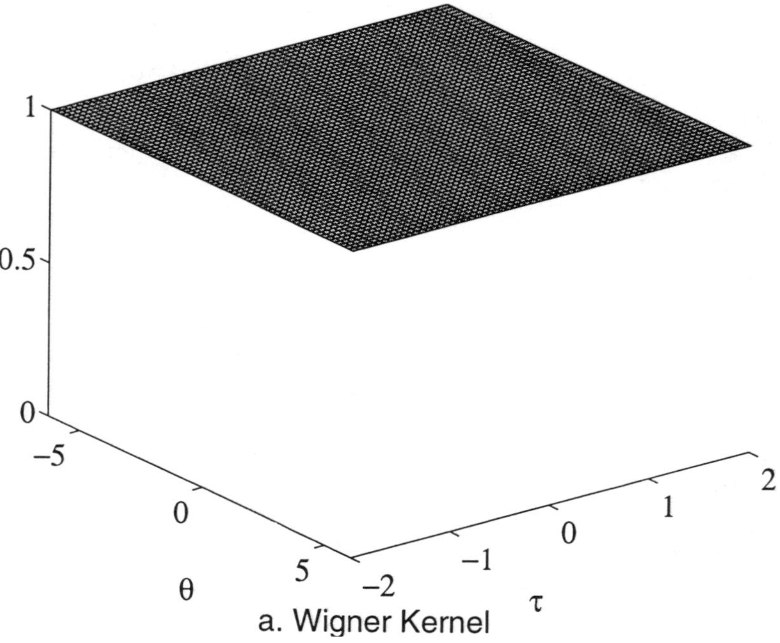

a. Wigner Kernel

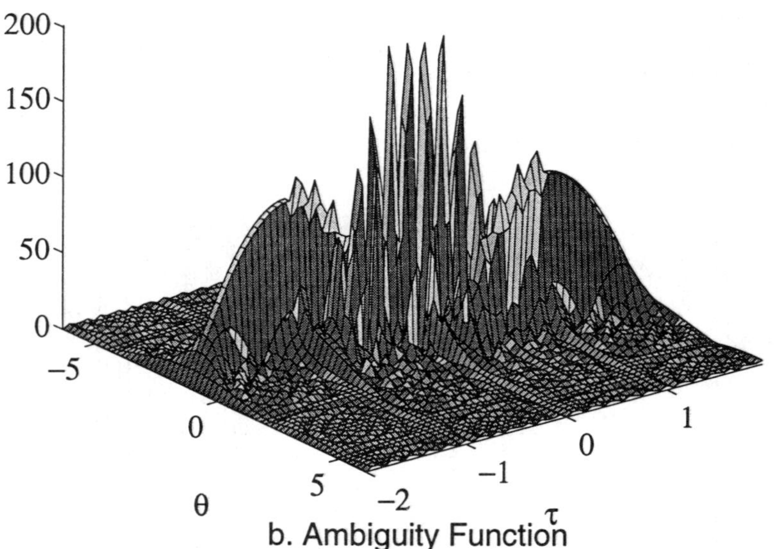

b. Ambiguity Function

Figure 1-4   (a) Wigner distribution kernel and (b) the ambiguity function.

It can readily be seen that the spectrogram kernel radically alters the ambiguity function. This has a marked effect on the time-frequency distribution as well, as shown in Fig. 1-6. The spectrogram kernel filters the ambiguity function in the low-pass region. The resulting spectrogram is shown in Fig. 1-7.

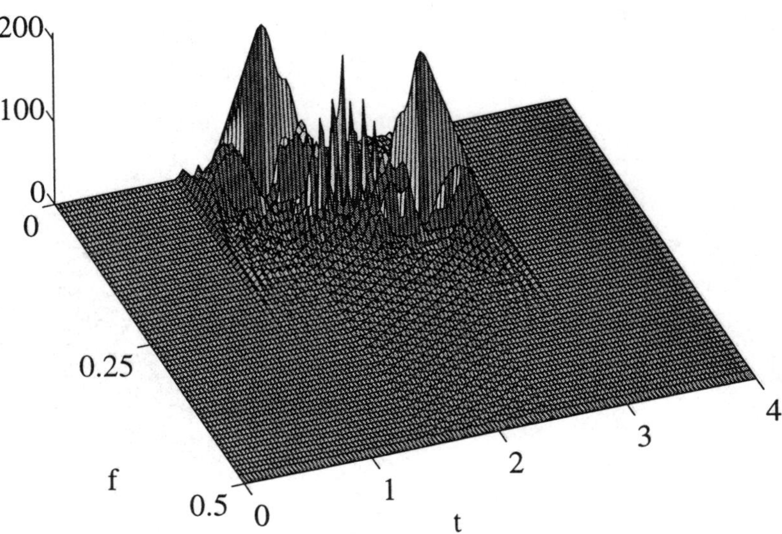

Figure 1-5 Wigner distribution time-frequency result.

The kernel for the WD is unity, so the generalized ambiguity function is identical to the ambiguity function, and its t-f representation (the double Fourier transform) preserves both the auto-terms and the cross-terms. The kernels of the spectrogram and the RID emphasize the auto-terms and deemphasize the cross-terms, but in very different ways.

1.2.2 The Exponential Distribution

The spectrogram and the WD both have properties that are valuable under certain conditions. The ED is an attempt [13] to improve on the WD. It has a kernel, $\phi(\theta, \tau) = \exp[-\theta^2\tau^2/\sigma]$, and it proves to be quite effective in suppressing the interferences while retaining high resolution. Its kernel is similar to Fig. 1-8.* Its performance has been compared to those of the spectrogram and the WD in a variety of environments [10,16]. The σ parameter may be varied over a range of values to obtain different trade-offs between cross-term suppression and high auto-term t-f resolution. In fact, as σ becomes very large, the ED kernel approaches the WD kernel. This provides the best resolution, but the cross-terms become large and approach WD cross-terms in size. Unfortunately, however, in a strict sense, this distribution violates the support properties, but does satisfy them with small error. This is not a very important practical issue, since a window can easily be imposed when the t, τ form of the ED kernel is convolved in time with the local autocorrelation prior to Fourier transforming with respect to τ to obtain the ED (t, ω) form of

---

*The particular parameters of the kernels and the test signals along with the results in this section have been chosen to bring out the important attributes of each approach as opposed to providing a head-to-head quantitative comparison.

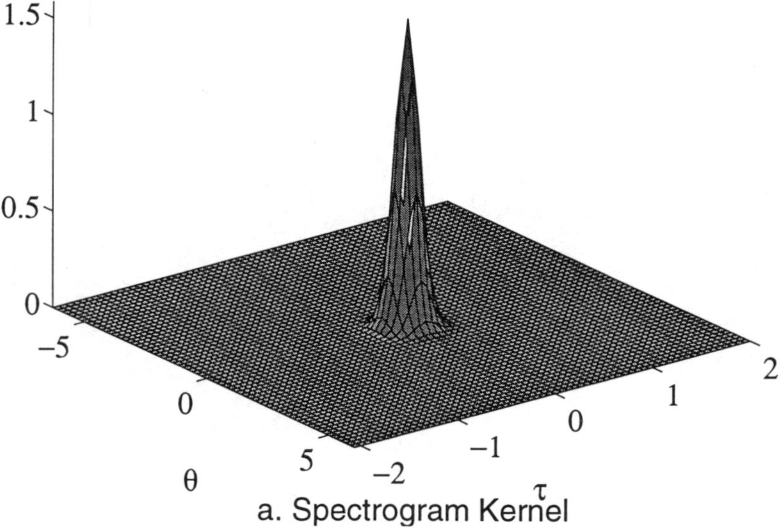

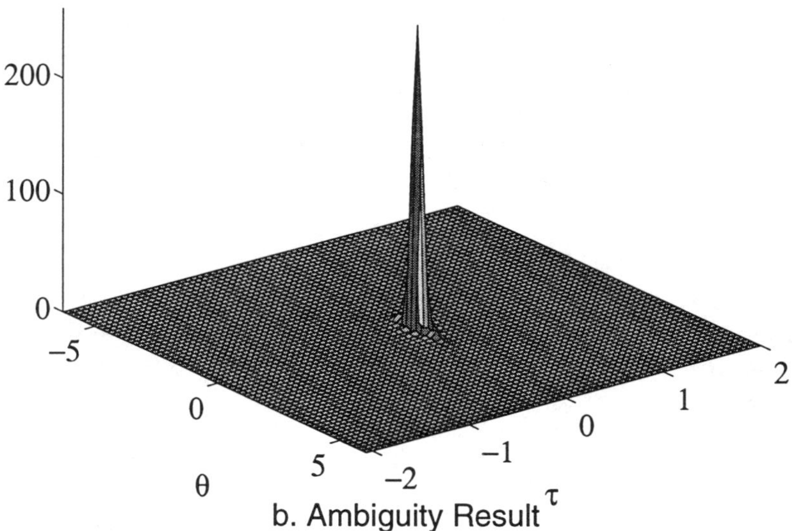

Figure 1-6   (a) Spectrogram kernel and (b) the resulting altered ambiguity function.

the distribution. This insures that the support properties are exactly satisfied. The windowed ED RID ambiguity plane results are shown in Fig. 1-8. It can be seen that the RID kernel captures the central portion of the ambiguity function and rejects the outlying cross-terms.

The RID kernel keeps much more of the ambiguity function. The offending interference terms are essentially excluded here. The resulting time-frequency distribution is shown in Fig. 1-9.

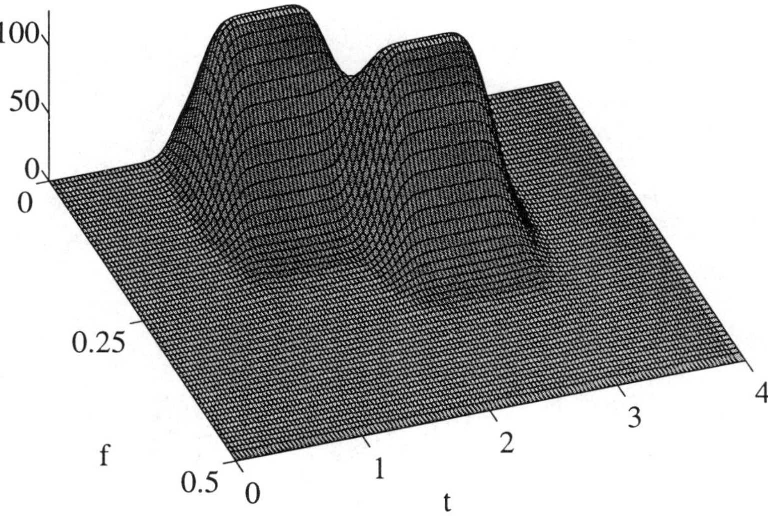

Figure 1-7   Spectrogram time-frequency result.

1.2.3 Zhao–Atlas–Marks

The Zhao–Atlas–Marks (ZAM) [15] distribution, or cone kernel distribution, had a distinctively different motivation than the RID. The formulation was motivated by the phenomenon of lateral inhibition in the auditory system. The ZAM kernel adheres to the requirement that guarantees that the time-support property is met. Its kernel is presented in Table 1-1, but, unlike the Wigner distribution, the spectrogram, and the RID, the ambiguity plane provides a generally confusing picture of how the kernel manifests itself in producing good results. The original form of the ZAM simply enforced the time support property on the local autocorrelation. Its formulation is

$$C_{\mathrm{ZAM}}(t, \omega) = \int \int_{t-\frac{|\tau|}{2}}^{t+\frac{|\tau|}{2}} x\left(u + \frac{1}{2}\tau\right) x^*\left(u - \frac{1}{2}\tau\right) \exp^{-j\omega\tau} \, du\,d\tau \qquad (1\text{-}14)$$

The ZAM result for the test signal is shown in Fig. 1-10. A thorough analysis of the ZAM has been provided by Oh and Marks [17].

1.2.4 Kernel Selection for RID

A more formal description of RIDs is appropriate at this point. Requirements for the RID and the RID's properties are quite similar to the WD. Once these properties are laid out, it will be possible to compare and contrast different distributions with much greater ease. The properties of the WD are investigated in [3–5, 10, 12]. RID requirements and properties will be discussed in comparison with the WD. The unity value of the WD kernel guarantees the desirable properties of the WD.

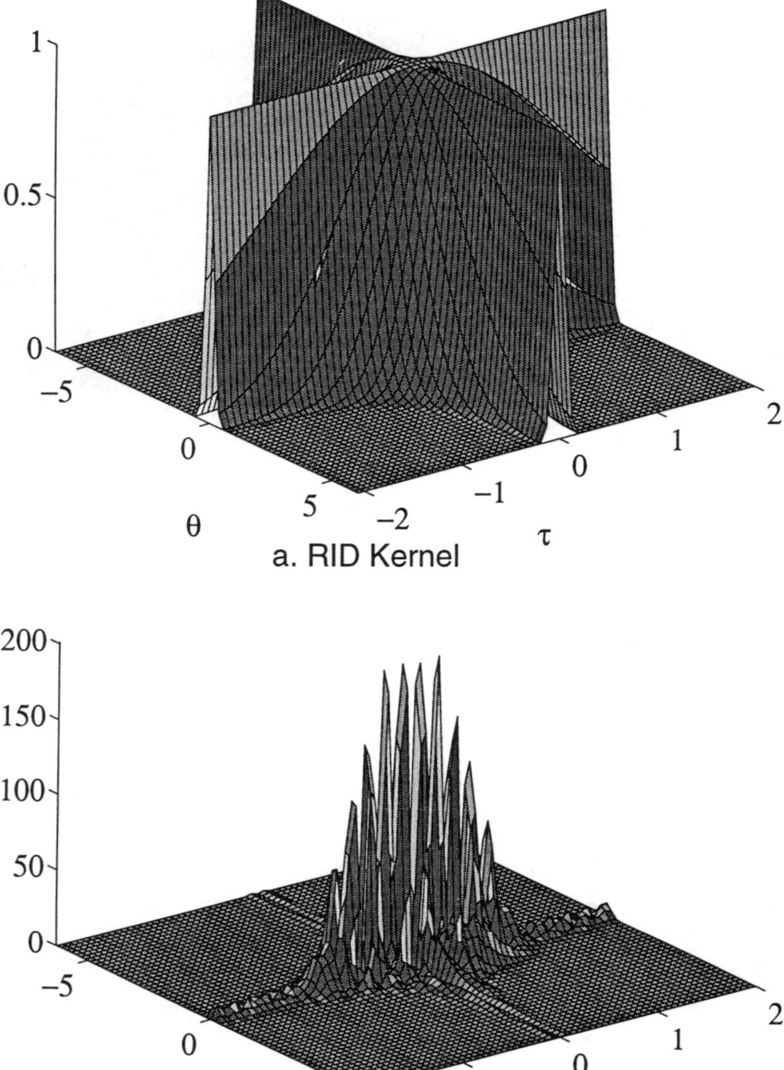

Figure 1-8 (a) RID kernel and (b) the resulting altered ambiguity function.

However, it is not necessary to require the kernel to be unity for all ω in order to maintain most of its desirable properties. It is sufficient to insure that the kernel is unity along $\omega = 0$ and $\tau = 0$ and that the kernel is such that $\phi^*(\theta, \tau) = \psi(-\theta, -\tau)$, the latter property insuring realness. The RID kernel is cross-shaped and acts as a low-pass filter in both θ and τ. Returning to Fig. 1-7, one can see that the spectrogram suffers from poor auto-term resolution, whereas the WD and the RID exhibit good

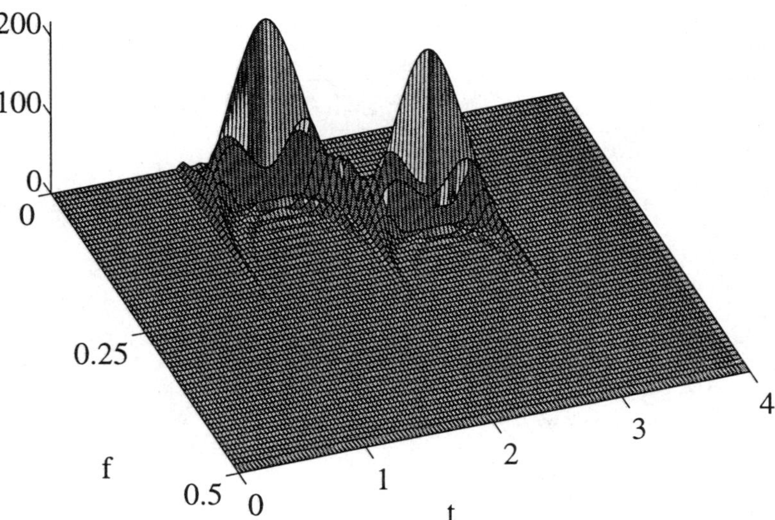

Figure 1-9   RID time-frequency result.

resolution and support properties (Figs. 1-5 and 1-9). However, the WD also exhibits interference terms. The spectrogram has the virtue of suppressing cross-terms as does the RID and has the further advantage of being nonnegative, which is not the case for the WD and the RID. The RID possesses almost all of the desirable properties of the WD except for its unitary property, $|\phi(\theta, \tau)| = 1$ for all θ, τ.

It can be seen that the ZAM produces a nice result, resolving the sinewave segments well in time and in frequency. In contrast to the RID, the ZAM, in general, places the interference terms at approximately the same time-frequency locations as the auto-terms for such signals. There are some clear differences in RID and ZAM that should be taken into consideration, however. These differences will be discussed further as the tools for understanding time-frequency distributions are further developed. It is quite desirable for a distribution to possess the time and frequency support property. This property insures that the distribution does not extend beyond the support of the signal in time or the support of its Fourier transform in frequency. One can see in Fig. 1-7 that the spectrogram violates this property rather badly. The time and frequency support property may be maintained for the RID by insuring that

$$\psi(t, \tau) = \int \phi(\theta, \tau)e^{-j\theta t}d\theta = 0 \quad \text{if} \quad |\tau| < 2|t|. \tag{1-15}$$

This forms a "cone-shaped" region in t, τ. The WD obviously satisfies this support property since the Fourier transform of unity is an impulse function, clearly staying within the t, τ limits. The form of the kernel in θ, ω is also cone-shaped, insuring the frequency support property. Zhao, Atlas, and Marks [15] suggest a cone-shaped kernel for nonstationary signal analysis, but further restrictions which insure a number of desirable RID properties are missing. The ED can be brought into the RID requirements by imposing an RID window as suggested earlier. The RID is not

TABLE 1-1: Properties of Time-Frequency Distributions (P) and Their Requirements (R)

Property
P0. nonnegativity: $C_x(t,\omega;\phi) \geq 0 \ \forall t, \omega$
R0. $\phi(\theta,\tau)$ is the ambiguity function of some function $w(t)$.
P1. realness: $C_x(t,\omega;\phi) \in R$
R1. $\phi(\theta,\tau) = \phi^*(-\theta,-\tau)$
P2. time shift: $g(t) = x(t - t_0) \Rightarrow C_g(t,\omega;\phi) = C_x(t - t_0, \omega;\phi)$
R2. $\phi(\theta,\tau)$ does not depend on t.
P3. frequency shift: $g(t) = x(t)e^{j\omega_0 t} \Rightarrow C_g(t,\omega;\phi) = C_x(t,\omega - \omega_0;\phi)$
R3. $\phi(\theta,\tau)$ does not depend on ω.
P4. time marginal: $\frac{1}{2\pi}\int C_x(t,\omega)d\omega = x(t)x^*(t)$
R4. $\phi(\theta,0) = 1 \forall \theta$
P5. frequency marginal: $\int C_x(t,\omega;\phi)dt = X(\omega)X^*(\omega)$
R5. $\phi(0,\tau) = 1 \forall \tau$
P6. instantaneous frequency: $\dfrac{\int \omega C_x(t,\omega;\phi)d\omega}{\int C_x(t,\omega;\phi)d\omega} = \omega_i(t)$
R6. R4 and $\frac{\partial\phi(\theta,\tau)}{\partial\tau}\big|_{\tau=0} = 0 \ \forall\theta$
P7. group delay: $\dfrac{\int t C_x(t,\omega;\phi)dt}{\int C_x(t,\omega;\phi)dt} = t_g(\omega)$
R7. R5 and $\frac{\partial\phi(\theta,\tau)}{\partial\theta}\big|_{\theta=0} = 0 \ \forall\tau$
P8. time support: $x(t) = 0$ for $|t| > t_c \Rightarrow C_x(t,\omega;\phi) = 0$ for $|t| > t_c$
R8. $\psi(t,\tau) \triangleq \int \phi(\theta,\tau)e^{-j\theta t}d\theta = 0$ for $|\tau| < 2|t|$
P9. frequency support: $x(\omega) = 0$ for $|\omega| > \omega_c \Rightarrow C_x(t,\omega;\phi) = 0$ for $|\omega| > \omega_c$
R9. $\Psi(\theta,\omega) \triangleq \int \phi(\theta,\tau)e^{j\omega\tau}d\tau = 0$ for $|\theta| < 2|\omega|$
P10. Reduced Interference
R10. $\phi(\theta,\tau)$ is a 2-D low-pass filter type.
P11. Scale Invariance
R11. $\phi(\theta,\tau)$ is a product kernel.

a totally new distribution since the Born–Jordon kernel [10], $\phi(\theta,\tau) = \mathrm{sinc}(\theta\tau)$, meets all of the RID requirements.

The windowed-ED and the Born–Jordan (aka Cohen's Born–Jordan) distributions are members of the RID class of distributions. That the Born–Jordan distribution was a member of the RID class was discovered when the RID was defined [14,18]. The RID is a very general concept which can be used to design a large number of distributions with desirable characteristics.

1.2.5 Design Procedures for Effective RID Kernels

There is much more that can be done in terms of kernel design. It is possible to bring much of the work that has been done on windows and digital filters to bear in designing effective RID kernels [18]. We propose the following approach for designing RID kernels.

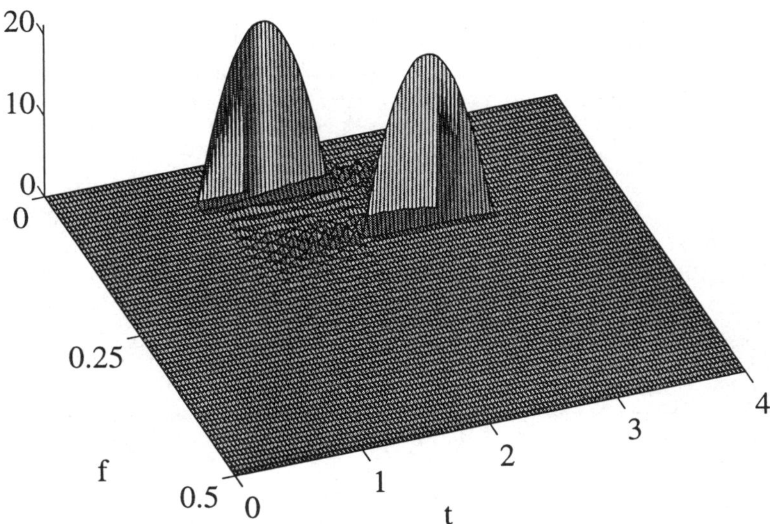

Figure 1-10   ZAM time-frequency result.

1. Design a primitive real-valued function $h(t)$ that satisfies the following:

R1: $h(t)$ has unit area, i.e., $\int h(t)dt = 1$.
R2: $h(t)$ is a symmetrical function of time, i.e., $h(-t) = h(t)$.
R3: $h(t)$ is time-limited on [-1/2, 1/2], i.e., $h(t) = 0$ for $|t| > 1/2$.
R4: $h(t)$ tapers smoothly toward both ends so that its frequency response has little high-frequency content.* That is, $|H(\theta)| \ll 1$ for $|\theta| \gg 0$, where $H(\theta)$ is the FT of $h(t)$.

2. Take the FT of $h(t)$, i.e.,

$$H(\theta) = \int h(t)e^{-j\theta t}dt$$

3. Replace θ by $\theta\tau$ in $H(\theta)$

The primitive function, $h(t)$, may be considered to be a window or impulse response of a filter. Thus a substantial theoretical framework may be easily adapted to RID kernel design.

The RID has the following integral expression:

$$\text{RID}_x(t, \omega; h) = \int\int \frac{1}{|\tau|}h\left(\frac{u-t}{\tau}\right)x(u+\tau/2)x^*(u-\tau/2)e^{-j\tau\omega}du\,d\tau \tag{1-16}$$

For computation, the generalized autocorrelation function is

$$R_x'(t, \tau; h) = \int \frac{1}{|\tau|}h\left(\frac{u-t}{\tau}\right)x(u+\tau/2)x^*(u-\tau/2)du \tag{1-17}$$

*It may be desirable to design in bandstop and bandpass regions for some special cases.

TABLE 1-2: Comparison of Various Time-Frequency Distributions in Meeting Desirable Properties

Distribution	$\phi(\theta,\tau)$	P0	P1	P2	P3	P4	P5	P6	P7	P8	P9	P10	P11
Wigner	1		X	X	X	X	X	X	X	X	X		X
Rihaczek	$e^{j\theta\tau/2}$			X	X	X	X			X	X		X
Re{Rihaczek}	$\cos(\theta\tau/2)$		X	X	X	X	X	X	X	X	X		X
Exponential (ED)	$e^{-\theta^2\tau^2/2\sigma}$		X	X	X	X	X	X	X			X	X
Spectrogram	$A_w(\theta,\tau)$ of a window $w(t)$	X	X	X	X	X						X	
Born–Jordan*	$\frac{\sin(\theta\tau/2)}{\theta\tau/2}$		X	X	X	X	X	X	X	X	X	X	X
Windowed-ED	$e^{-v^2/2\sigma} * W(v)\big|_{v=\theta\tau}$		X	X	X	X	X	X	X	X	X	X	X
Cone (ZAM)*	$g(\tau)|\tau|\frac{\sin(a\theta\tau)}{a\theta\tau}$		X	X	X	X				X		X	

This is sometimes called Cohen's Born–Jordan distribution, since he suggested its form in his 1966 paper. This distribution is essentially an RID.

*The distribution with $a = \frac{1}{2}$ and $g(\tau) = 1$ was originally reported.

and

$$\mathrm{RID}_x(t, \omega; h) = \int R_x'(t, \tau; h)e^{-j\tau\omega}d\tau \tag{1-18}$$

is carried out.

1.2.6 Limitations of RID

One can find signals that will not be effectively handled by the RID, for example, a chirp. If the symmetrical ambiguity function of the chirp falls on a 45-degree diagonal line, then it will not intersect well with the RID kernel. In other situations, cross-terms will not always fall far away from the θ, τ axes. If a cross-term falls on either the θ or τ axis, it will not be suppressed very much. So, the RID is not a panacea for all problems. Kernels should be examined carefully in terms of the signals at hand and kernel design should be optimized to the problem at hand. One may wish to relinquish some desirable properties of t-f distributions in order to meet a specific goal, as will be discussed in a following section. The RID approach seems to have a number of advantages compared to the WD and the spectrogram. However, neither the RID or any other distribution is capable of ideal performance under all situations. Custom kernels may be required in some situations. However, on the balance, the RID seems to perform quite well in a number of real-world situations, particularly with biological signals, which are often compact in both time and frequency. Several aspects of RIDs in comparison with other distributions may be of interest to the reader.

Positivity. The RID is not a nonnegative distribution, as is the spectrogram. However, we have observed that in virtually all practical cases, the RID is more nonnegative than the WD, as should be the case from theoretical considerations. This is because the WD cross-terms often exhibit quite negative values. The RID reduces negativity as a consequence of reducing the size of the cross-terms. One evidently cannot have a manifestly positive energy valued time-frequency distribution with a fixed kernel for all signals and still retain many of the desirable properties of time-frequency distributions. The negative energy values cannot be given a conventional physical interpretation, but they are required for other good attributes. A judgment must be made according to the benefit obtained by relaxing the positivity condition. The Cohen–Posch class of distributions to be discussed in a following section retains this valuable energy positivity. That such distributions can be used in practical situations has been demonstrated fairly convincingly [19].

Cross-Terms.   In general, cross-terms cannot be completely abolished. They even exist in the spectrogram, albeit often in a hidden form [20]. When two signal components are closely spaced in time or frequency, then cross-terms will become rather prominent. In fact, if two signal components should overlap exactly, cross-terms must exist in order to yield the proper energy values for the combined signals. On some occasions, the cross-terms may be of value in reflecting the relationships between signal components.

Smoothing, Smoothed WDs.   There have been a number of attempts to eliminate cross-terms "after the fact." First, the WD is computed. Next smoothing, filtering, masking operations are carried out to reduce or eliminate the cross-terms. On occasion it may be desirable to smooth the WD in an optimal manner [21]; however, this type of smoothing is entirely different from the RID approach. According to Cohen's viewpoint, it is misleading to refer to the RID as a smoothed WD [10]. In fact, according to Cohen, a general relationship between distributions can be derived so that one may be derived from the other by means of appropriate convolution. Thus there is no master distribution and any particular distribution in Cohen's class may be transformed into any other distribution in the class, in theory. One might start with any one of several distributions and build a theoretical time-frequency framework. Other distributions in Cohen's class might then be considered to be derived from this "master" distribution.

Moyal Formula.   This relation, first shown by Moyal [22] reveals an interesting relationship between two signals and the overlap of their Wigner distributions. It is

$$\left| \int s_1(t)s_2^*(t)dt \right|^2 = \int \int W_1(t, \omega)W_2(t, \omega)dtd\omega \qquad (1\text{-}19)$$

The requirement for this property is for $|\phi(\theta, \tau)|^2$ to be 1 for all θ, τ. There are an infinite number of distributions which obey the Moyal formula. The RID does not.

The unitary property is convenient and simplifies the mathematics in some derivations, but is not required since an alternative, albeit more complicated, expression is available [10]. As Cohen points out, "some have made Moyal's formula a requirement for a distribution, but it is not clear why that should be so." Janssen [23] has suggested that it has a certain appeal in quantum mechanics but is "perhaps not necessary for signal analysis." Cohen notes that it is not really used in quantum mechanics either. One may trade the unitary property for better distribution characteristics at the expense of more complex mathematics for some aspects of the problem (such as signal synthesis from the distribution).

Use of the Analytic Signal.   The analytic form of the signal is almost universally used in t-f studies. Using the analytic form eliminates cross-terms between positive and negative frequency components of the signal. However, for certain low-frequency signals, there may be undue smoothing of the low-frequency time components of the t-f representation due to the frequency domain window implied by the discrete form of the distribution [4]. In that case, the nonanalytic form of the signal should be considered. Use of the nonanalytic form of the signal for RID should be less troublesome than would be the case for the WD due to the suppression of the cross-terms between positive and negative frequency components in the RID.

The Uncertainty Principle.   Some people think that the RID results violate the uncertainty principle. In quantum mechanics, it is known that, for a particle, the position and momentum of the particle cannot both be known with certainty. In signal processing, there is no intrinsic requirement to adhere to this idea. Cohen [10] convincingly makes the point that the uncertainty principle has no bearing on the question of joint distribution and relates to the product of the standard deviations of *marginals*. This is not to say that the question of the relationship of time and frequency marginals with the t-f distribution is a trivial question. In fact, it is a matter of some subtlety which deserves a fuller exposition beyond the brief treatment given here. There are a number of "uncertainty principles," each being based on a different conceptual viewpoint. Cohen has suggested several such "uncertainty principles" in a recent work [24].

1.3. ADDITIONAL DISTRIBUTIONS WITH DESIGNED OR ADAPTIVE KERNELS

Quite a few new time-frequency distributions have appeared recently. These new distributions often follow the basic themes already outlined in this paper, but are usually designed to improve the cross-term suppression while retaining auto-term resolution. In general, these new distributions are signal-dependent to a lesser or greater degree. The fixed kernel distributions are covered first.

1.3.1 Fixed Kernel Designs

The ED has been generalized in order to maneuver the kernel so as to avoid cross-terms [25–28]. In general, the product kernel requirement is relaxed in order to gain flexibility in placing the attenuation regions of the kernel where the cross-terms lie while retaining low attenuation in the auto-term regions. These are signal-dependent kernels in a sense, because the design is based on a specific signal or signal type. This may be a valuable approach to eliminate cross-terms. However, there are two cautions to be observed. First, the valuable product kernel requirement for the strict RID is relaxed. The valuable property of scale invariance is thereby lost or compromised. In addition, it may be difficult to apply these kernels. They are designed and conceptualized in the ambiguity plane context. In practice, kernels are often most successfully applied in the local autocorrelation domain.

Guo, Durand, and Lee have adhered to the basic RID principles in the development of their Bessel kernel [29, 30]. The Bessel function is used as the primitive in this development. The resulting distribution is shown to have some advantage in the studies of the femoral artery via Doppler ultrasound techniques. They recommend a form of the Bessel kernel wherein the time support constraint is relaxed in order to improve performance in a noisy environment.

There are several reports which provide a comparison of various TFRs, particularly the spectrogram, several forms of the WD (time-smoothed and frequency-smoothed), the ED, the RID, and the ZAM [31–33]. All of these are valuable in building up insights concerning the use of these various approaches. However, in general, synthetic signals are used to make the points and the results tend to be somewhat biased according to the set of signals utilized. There is not yet a truly objective quantitative means of comparing various approaches for a specific set of real-world signals, though considerable progress has been made in that direction.

1.3.2 Distributions with Adaptive Kernels

There have been several attempts to adapt kernels to optimize resolution and suppress cross-terms. A few examples will be mentioned which adhere to the basic principles of kernel design and Cohen's class. Baraniuk and Jones [34] developed the first truly adaptive distribution from this standpoint. First, they compute the generalized ambiguity function, which is the product of the ambiguity function and the kernel. A functional of the kernel, which is the double integration of the squared magnitude of the generalized ambiguity function over (τ, θ), is formed. This functional is then maximized under the constraints that the kernel tapers off and is nonincreasing radially. A further constraint is placed on the squared magnitude of the area under the kernel surface. The area is kept below some threshold value. One can also impose the constraints suggested in Table 1-1 such that the time and frequency marginals are retained. The performance of this approach is often quite nice, particularly for chirps which are oriented along a 45-degree line in the ambiguity plane. Insisting on the marginal constraints degrades performance a bit, however. Baraniuk and Jones provide some impressive examples of the effectiveness of this distribution for bat sound analysis.

It is possible to retain the RID properties and gain the advantages of the adaptive approach as well. We have used the basic idea of adapting the RID primitive, $h(t)$ under an information-like constraint [35, 36]. The idea is to minimize the uncertainty of the resulting distribution. The information measure used is based on an adaptation of Rényi information [37], to be discussed later in this chapter. The RID is important in this context, since it is *information-invariant*, which means that the information measure does not change with scale, time shift, or frequency shift if the RID is properly normalized. Thus the RID is information-invariant under scale. That means that the optimum kernel for a given signal will be information-invariant for all time-shifted, frequency-shifted, and scaled versions of the signal. All distributions in Cohen's class which have kernels that do not change with time or frequency will be information-invariant to time and frequency shift, but not all are invariant under scale changes, as indicated in Table 1-2. Some examples of the adaptive RID results will be presented in a following section.

Loughlin, Pitton, and Atlas [19] have developed another very interesting adaptive time frequency distribution which they call the minimum cross-entropy (MCE) distribution. Cohen and Posch [16] have provided a means for deriving distributions in Cohen's class that are manifestly positive. The kernels for this subclass of Cohen's class are signal-dependent. Nevertheless, there is a great deal of flexibility in choosing these kernels. The MCE is one way. The MCE can be formulated with positivity constraints (P0 in Table 1-1) and constraints such that the proper time and frequency marginals are retained (P4 and P5 in Table 1-1). The cross-entropy between an initial estimate of the TFD and the desired TFD is minimized in this approach, using iterative methods. Additional constraints such as proper group delay and instantaneous frequency (P6 and P7 in Table 1-1) can also possibly be included. Time and frequency support (P8 and P9 in Table 1-1) are guaranteed by the marginal constraints. There is actually a stronger form of time and frequency support present. This "strong finite support" property guarantees that the distribution is zero everywhere the time and frequency marginals are zero. The MCE can also demonstrate that a scale-invariant, and hence information-invariant, distribution does not necessarily require a product kernel. The MCE concept is a very attractive idea. The results are sometimes difficult to interpret, however, and take some explanation and training of the eye to appreciate, as is the case with many time-frequency distributions.

Wood and Barry [38–40] have reported on adaptive distributions based on Radon transform concepts. These ideas seem to have considerable merit and may be useful in a variety of areas.

Recently, an adaptive ZAM has been reported [41]. It has been successfully applied to wood thrush songs. A single parameter is adapted to provide the results in this case. It appears that this approach has considerable merit in providing an effective entry level tool for adaptive kernel t-f analysis.

There are several criticisms that might commonly be leveled at the approaches outlined in this section. First, the distributions are no longer bilinear or quadratic. Therefore, it is difficult to interpret the results in terms of energy. Second, the time required to compute these distributions is usually quite a bit more than is the case with a fixed kernel distribution. Sometimes the result may not seem to really be worth it. Finally, it will probably be difficult to achieve real-time application of the technique to most data, since the kernel must be continuously updated in order to keep pace with changes in the signal in most cases. Of course, one possibility is to "freeze" the kernel at some desirable compromise that deals with the signal reasonably well most of the time.

1.3.3 Some Adaptive RID Results

An illustration of some adaptive RID results may prove to be helpful. Adaptive kernel approaches have been touched upon briefly. TFRs can range from the WD, where everything is allowed to pass through and manifest itself, to a highly adaptive or signal-based design. The WD has problems with cross-terms and noise. If one designs or adapts a kernel to match the signal very well, then even nonsignal examples may be enhanced and modified to look like the signal. Enforcing various constraints helps matters a bit and may help to prevent the TFR from making something from nothing. The RID has a fairly large number of constraints and also enjoys many desirable properties. There is quite a lot of flexibility left in the kernel, however. The primitive, $h(t)$, may be adapted within the RID constraints to some criterion. We have chosen to use Rényi information as the measure to be minimized. The idea is to minimize the uncertainty. Thus $h(t)$ can be adapted to achieve this. Rényi information of orders 2 and 3 have been used [35,36]. In practice, the problem is formulated in the discrete form and the TFD is normalized using the total volume of the deviation of the TFD with respect to zero.

It is appealing to consider the application of information concepts to obtain an objective measure of resolution in TFDs. It is not possible, in general, to use Shannon information measures on TFDs because many interesting TFDs exhibit negative values. Rényi information provides an escape from this difficulty.

The general definition for Rényi information of order α for a TFR of signal $x(t)$ is

$$R_x^\alpha = \frac{1}{1-\alpha} \log_2 \int_{-\infty}^{\infty} \int_{-\infty}^{\infty} \mathrm{RID}_x^\alpha(t, \omega) dt d\omega \qquad (1\text{-}20)$$

Here, we assume that RID_x has been normalized in some way to provide a unit volume. Shannon information results for $\alpha = 1$. However, Shannon information is not appropriate for most TFRs due to the negative energy values. Rényi information effectively sidesteps the problem of negative energy values in TFRs. It is important that the result of the double integration be positive since it forms the log argument. A proof for this property has been offered [42]. Rényi information of order 3 has been investigated for TFDs and found to have interesting and useful properties [37]. Information should function as an uncertainty measure and thus indicate that there are a small number of clearly resolved signal components in a TFR representation (small information value) as opposed to a less well resolved TFR (large information value). It is to this end that we propose to use Rényi information as an objective criterion for well-resolved TFRs in an adaptive kernel algorithm.

There are several different information measures that can be used. We suggest the following simple adaptation of Rényi information in the discrete form:

$$U_x = -\log_2 \sum_k \sum_n |\mathrm{RID}_x|^2(k, n) \qquad (1\text{-}21)$$

Here, RID_x has been normalized to a unit volume with respect to the zero energy plane. This is, in fact, very similar to the "sharpness criterion" used by Baraniuk and Jones [34] in their adaptive kernel scheme. But, now, we can see that it has some basis in terms of an information measure. It is appealing from several standpoints, which space does not permit discussing here. Other measures have been investigated as well in our studies, including weighted sums of Rényi information of different orders.

The result for three short tones buried in 0 dB noise is shown in Fig. 1-11. The kernel stabilizes after about ten iterations. Uncertainty is decreased by about one bit from the WD starting point, as shown in Fig. 1-11(b). Fig. 1-11(c) shows the WD result. Fig. 1-11(d) shows the optimum kernel result. The primitive function is shown

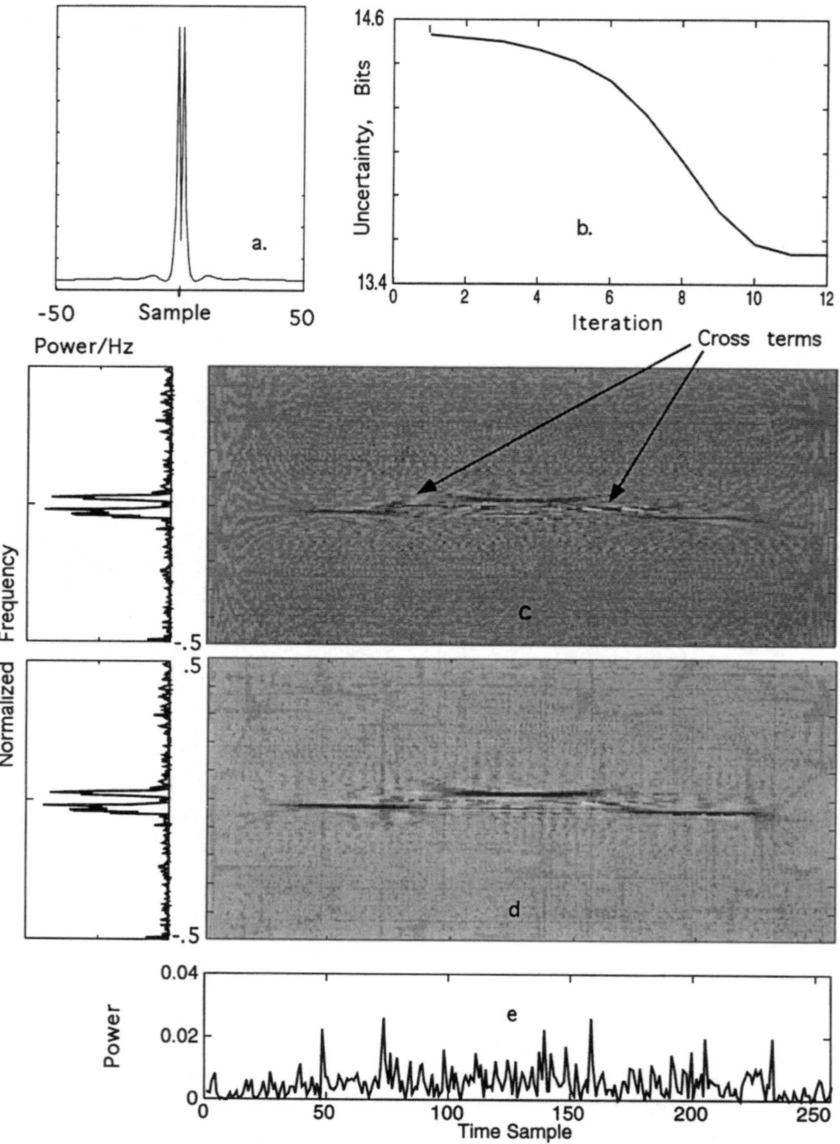

Figure 1-11   RID optimum kernel results for three short tones in 0 dB noise: (a) optimum RID kernel; (b) uncertainty vs. iteration; (c) WD result; (d) optimum kernel result; (e) instantaneous power. (From [36]).

in Fig. 1-11(a). It is clear that neither the instantaneous power [Fig. 1-11(e)] nor the energy spectra, shown along the margins of the TFDs, either singly or together, give a clear idea of the t-f placement of the tones. However, both the WD and the RID show the tone locations. The WD suffers from cross-terms and noise. The result for the optimum kernel RID shows the location of the tones quite clearly. The tones are more clearly delineated than for the WD and the cross-terms are much less evident.

It is interesting to note that the primitive function [Fig. 1-11(a)] has an impulsive dip at the origin and some oscillation close to the origin. The dip represents a negative WD-like contribution. The primitive function has been "tuned" to the situation at hand. The negativity of the RID result is small. The negative WD contribution produced by the dip at the origin actually cancels out cross-terms produced by what would be the primitive function lacking the dip. Other examples have been produced by this approach, which show that the RID is capable of somewhat more than would be suggested by the fixed kernel versions of it. It would be interesting to compare the adaptive RID and the MCE, since both insist on satisfying the marginals and both are positive (MCE) or almost positive (adaptive RID).

The RID is attractive because of its ease of implementation once the primitive function is known. All forms of the kernel are automatically then known, particularly the (t, τ) form used to convolve with the local autocorrelation [18], as mentioned previously in this chapter. The RID is information- and scale-invariant. This means that if the kernel is optimized for a particular $s(t)$, then the kernel is also optimum for all $s[a(t - t_o)] \exp(j\omega_o t)$—that is, all scaled (a), time-shifted (t_o), and frequency-shifted (ω_o) versions of the signal in theory. It is easy to construct a simple proof of information invariance for this. These ideas remain to be explored more fully. It is interesting to find that one can retain the good properties of the RID and yet adapt it well to the situation at hand.

1.4. NOISE CONSIDERATIONS

The effects of noise must be considered in many applications. The Wigner distribution provides notoriously poor results in the presence of noise [10, 12]. The WD essentially sprays noise all over the t-f plane. There are cross-terms between auto-terms and the noise terms, and cross-terms between noise terms and other noise terms. The cone-shape provides the time support property for the kernels and distributions thereby obtained. These kernels are "pinched in" at the $\tau = 0$ value and spread out for increasing magnitudes of τ. Hearon and Amin [43] have shown that such kernels are not good choices for low variance results. Even though the ZAM provides an *unbiased* estimate of the noise [17], Hearon and Amin show that it is quite noisy. In fact, Cohen's Born–Jordan distribution is the minimum variance distribution for white noise. Being a member of the RID class, it is the minimum variance RID. The spectrogram kernel is quite spread out at $\tau = 0$, so it provides a fairly low variance result. However, this is at a considerable price. The WD kernel is very narrow for all τ in the local autocorrelation domain. It is $\delta(\tau)$, an impulse function.

The more time samples a kernel can gather for a particular τ, the better the noise properties, in general. Thus the ED kernel, not explicitly enforcing time support, performs better under noise than does the windowed ED, which meets all of the RID requirements. Smaller values of σ yield better noise results, but compromise time support properties more.

1.5. DISCRETE FORMULATIONS AND FAST ALGORITHMS

Time-frequency distributions are presented in a continuous form for theoretical development and discussion of properties. However, one usually wishes to utilize a discrete form of the distribution for computational convenience using a digital computer. Claasen and Mecklenbräuker [4] developed discrete forms of the Wigner distribution,

$$W_x(n, \omega) = 2 \sum_{k=-\infty}^{\infty} e^{-j2\omega k} x(n + k) x^*(n - k) \tag{1-22}$$

If discrete values of the local autocorrelation, $R_x(n, m)$, were available for all integers n and m, then it would be logical to express the discrete form of the Wigner distribution as the discrete-time Fourier transform (DTFT) of $R_x(n, m)$, or

$$W_x(n, \omega) = \sum_{m=-\infty}^{\infty} e^{-j\omega m} R_x(n, m) \tag{1-23}$$

Note that as k takes on the values $0,1,2, \ldots$, the discrete local autocorrelations $R_x(n, 0) = x(n)x^*(n)$, $R_x(n, 2) = x(n + 1)x^*(n - 1)$, $R_x(n, 4) = x(n + 2)x^*(n - 2)$ are evaluated. Discrete values two samples apart are thus correlated. Local autocorrelation values for odd integer spacings are not available, so the discrete form of the Wigner distribution is formed from the evenly spaced correlation values. This means that the local autocorrelation is undersampled by a factor of 2 compared to the sequence $x(n)$, and aliasing may occur in the discrete WD if $x(n)$ were not sampled at twice the Nyquist frequency for a real valued $x(n)$. The analytic form of the signal presents no problem with aliasing, however, since only half the period of the DTFT spectrum is occupied. If aliasing is a problem with the discrete WD and not with the original sequence $x(n)$, then additional points may be interpolated for $x(n)$ to fill in the missing correlation values required to form the discrete WD.

1.5.1 Discrete Realizations

Except for potential aliasing problems, the discrete form of the Wigner distribution enjoys many of the desirable properties of the continuous form and suffers from similar limitations. Practical computation requires a finite-length sequence of $x(n)$ values. A discrete time-discrete frequency version of the WD may be expressed as a discrete-time windowed version of the infinite-length sequence form [4], with interpolation of the odd indexed values of the local autocorrelation if required. The resulting discrete distribution is termed the "pseudo-Wigner distribution (PWD)." The PWD may be computed by efficient means involving FFTs. Cohen's review and book [10, 12] mentions several such efficient computational algorithms.

There are several computational algorithms available for the ED. Barry has suggested an efficient one [44]. Sometimes one runs into trouble when directly converting continuous forms of TFDs into discrete TFRs. Note, for example, that RID computation may prove to be troublesome when $|\tau| = 0$ (Eqs. 1-18 and 1-19). Usually, this problem can be overcome by evaluating the limiting result at such a point. In the case of the RID, the t, τ form of the kernel is an impulse function for $\tau = 0$, so one simply uses this fact to obtain the correct results. Convolving the impulse function with the local autocorrelation simply yields the local autocorrelation. So, the final result is that of a WD for the troublesome value of τ. Due to such problems, it is often desirable to formulate the discrete form from basic principles rather than as an approximation of a continuous form.

Requirements for discrete forms of the RID are similar to those of the discrete WD. The discrete RID may be formed by

$$\text{RID}_x(n, \omega) = \sum_{m=-\infty}^{\infty} R_x(n, m) *_n \psi(n, m)e^{-j\omega m} \qquad (1\text{-}24)$$

where

$$\phi(m, \omega) = \sum_{n=-\infty}^{\infty} \psi(n, m)e^{-j\omega n} \qquad (1\text{-}25)$$

is the discrete RID kernel.

The discrete RID may thus be conveniently formed by obtaining the local autocorrelation $R_x(n, m)$, convolving it with $\psi(n, m)$ along n, and DTFTing the result with respect to m. A partially discrete form of the ED called the running-windowed exponential distribution (RWED) has been suggested [13]. The RWED retains the continuous form in the ED and the limiting forms (when the kernel approaches an impulse function) must be evaluated as special cases. A fully discrete form of the kernel is more desirable. A very convenient discrete RID kernel has been discovered based on the binomial distribution [14]. The form of the kernel is

$$\psi(n, m) = \frac{1}{2^m} \binom{m}{k} \delta(n - m + k) \qquad (1\text{-}26)$$

The correlation shift index, m, is assumed to take on the values $-\infty \ldots -1, 0, 1, \ldots, \infty$ and the time shift index, n, is assumed to take on the values $-\infty, \ldots, -1, -.5, 0, .5, 1, \ldots, \infty$. It can be shown that the signal structure of the discrete local autocorrelation and the discrete form of the kernel can be easily formulated to include the half-integers [14, 45]. Figure 1-12 illustrates the characteristics of the discrete binomial kernel.

It can easily be seen that $\psi(n, m)$ meets the discrete equivalent of the time support requirement for RID. That is,

$$\psi(n, m) = 0 \text{ if } |m| < 2|n| \qquad (1\text{-}27)$$

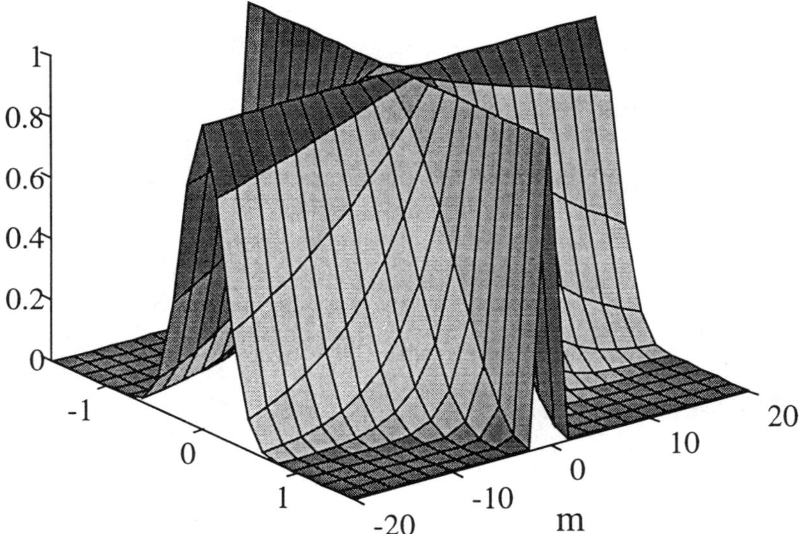

Figure 1-12   Binomial kernel in the ambiguity domain.

The DTFT of $\psi(n, m)$ yields

$$\phi(\theta, m) = \frac{1}{2^m}(e^{j\pi\theta} + e^{-j\pi\theta})^{|m|} = \cos^{|m|}(\pi\theta) \tag{1-28}$$

Figure 1-12 shows that this kernel exhibits the required RID characteristics in the ambiguity domain. The kernel has the required low-pass 2-D filter character over a period of 1 Hz (assuming the sequence $x(n)$ was obtained by sampling at a rate of one sample per second). This result assumes that $\psi(n, m)$ is available for all integer values of m. This form of the kernel is shown in Fig. 1-13.

The z-transform of $\psi(n, m)$ yields a form similar to that obtained by using the DTFT, but a digital filter realization is more easily recognized in this form. In terms of z,

$$g_{zt}(z, m) = \left(\frac{z + z^{-1}}{2}\right)^{|m|} \tag{1-29}$$

This result suggests that $\psi(n, m)$ should be filtered differently for each m. Consider that for the kth value of m, $\psi(n, k)$ is a time sequence which is filtered by a noncausal finite impulse response (FIR) filter of the form $[(z + z^{-1})/2]^{|k|}$. This filter may be formed by cascading k sections of a noncausal FIR filter of the elemental form $[(z + z^{-1})/2]$. The computational requirements for these elemental FIR filter sections are quite modest, involving only the summation of a one sample advance and a one sample delay followed by a division of the sum by 2. It may be convenient to accomplish the division by 2 by means of a right-shifting of bits in some realizations. The noncausal form of the filter is not required, but was used for purposes of exposition. The entire filter structure may be made causal by adding an

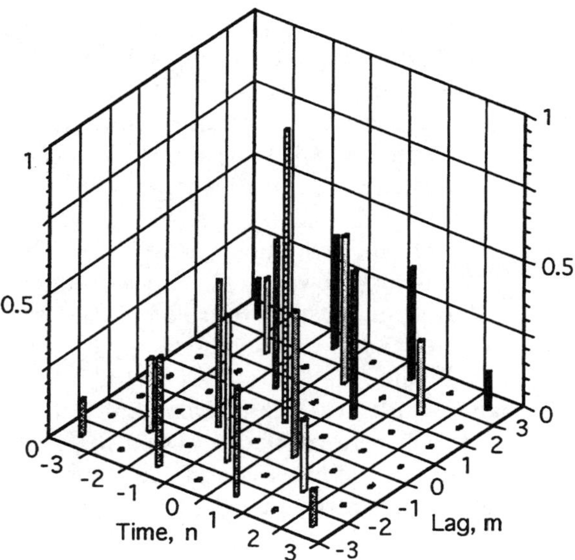

Figure 1-13 Binomial kernel in the local autocorrelation domain.

appropriate common delay to all of the FIR filters such that the FIR filters for all m are causal. This may be attractive for possible real-time computation of discrete RIDs. The DTFT mentioned in these discussions would be replaced by the FFT for fast, recursive RID computation in practice.

There are no doubt many ways of realizing discrete RIDs. Amin [46] has suggested applying singular value decomposition (SVD) to the full-rank description of a kernel matrix in the time and lag variables. The aim is to find a rank-one kernel estimator which is closest to the full-rank kernel description so that the kernel may be more simply expressed, and thus decrease computational complexity. Amin approximated the ED kernel using this technique and obtained fairly nice-looking t-f plots for pairs of chirps. However, the kernel was not well approximated by the rank-one estimate. The question remains as to how much the desirable properties of the distribution are eroded by this approximation. Increasing the rank of the approximation quickly improves the kernel representation. However, this also increases the computational complexity. Practical RID computation in many real-world situations may require parallel VLSI realizations or dedicated digital signal processing (DSP) modules.

1.5.2 Binomial Time-Frequency Distribution Results

The binomial distribution (BD) has been applied in a number of areas. Some of these applications will be covered in another chapter of this volume. Its results are similar to the ED, but it is fully discrete in its formulation and application. It is not an approximation of a continuous form and it may be computed efficiently and recursively. It is possible to compute TFRs of arbitrarily large time series using the BD. A comparison of the spectrogram, WD, and BD may be helpful at this point. A more ambitious test signal is used in this case. The signal consists of a rising chirp which rises from midfrequency to high frequency and a frequency-modulated signal which also falls slowly in frequency with time. This will be termed the "chirp-warble" for purposes of discussion.

The spectrogram result is shown in Fig. 1-14. One can readily see that the chirp is evident, but rather smeared in frequency as one might expect when using a 512-point window. The warble is a continuous smear with little evident fine structure.

The WD result is shown in Fig. 1-15. A 512-point analysis window was used for the WD. The resolution for the WD is dramatically better than is the case for the spectrogram, however. One can see that the chirp is very well resolved and the frequency-modulated "warble" is also well resolved. However, there is a very prominent band of cross-term activity between the chirp and the warble.

Finally, the BD result is shown in Fig. 1-16. The BD result is quite clear. There is very little evidence of cross-term activity. One can see some faint vertical lines extending between the chirp and the warble. These are the cross-terms. They are spread out and hence considerably attenuated in amplitude. Note that there is some cross-term activity nested within the warble. These are the short horizontal pulses interspersed between the loops of the sinusoid. Such cross-terms are sometimes

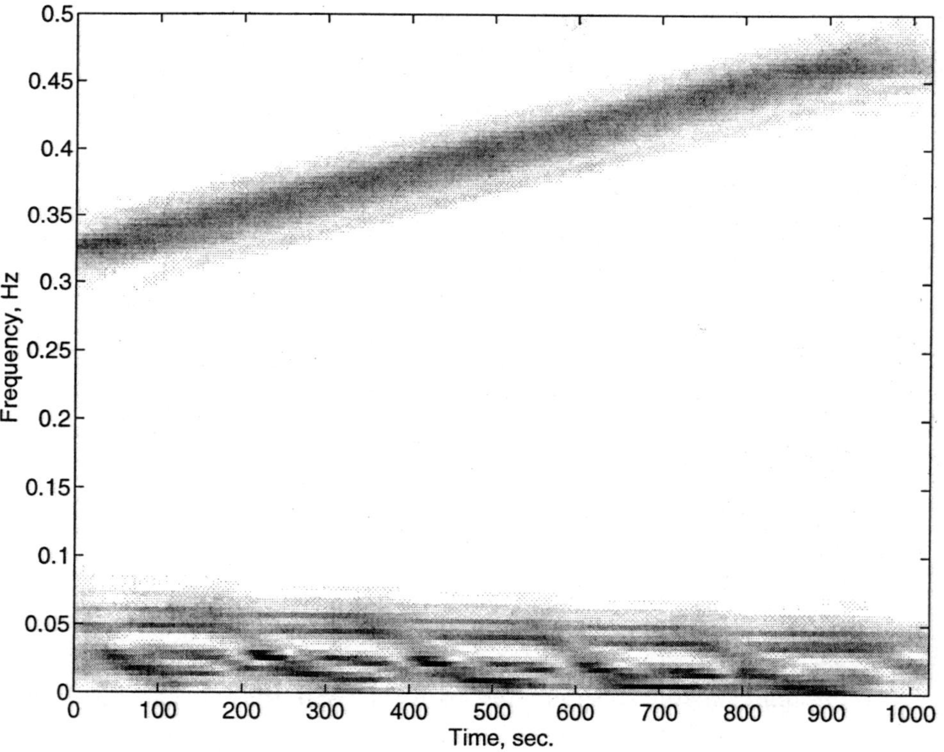

Figure 1-14 Spectrogram result for the chirp-warble using a 512-point window.

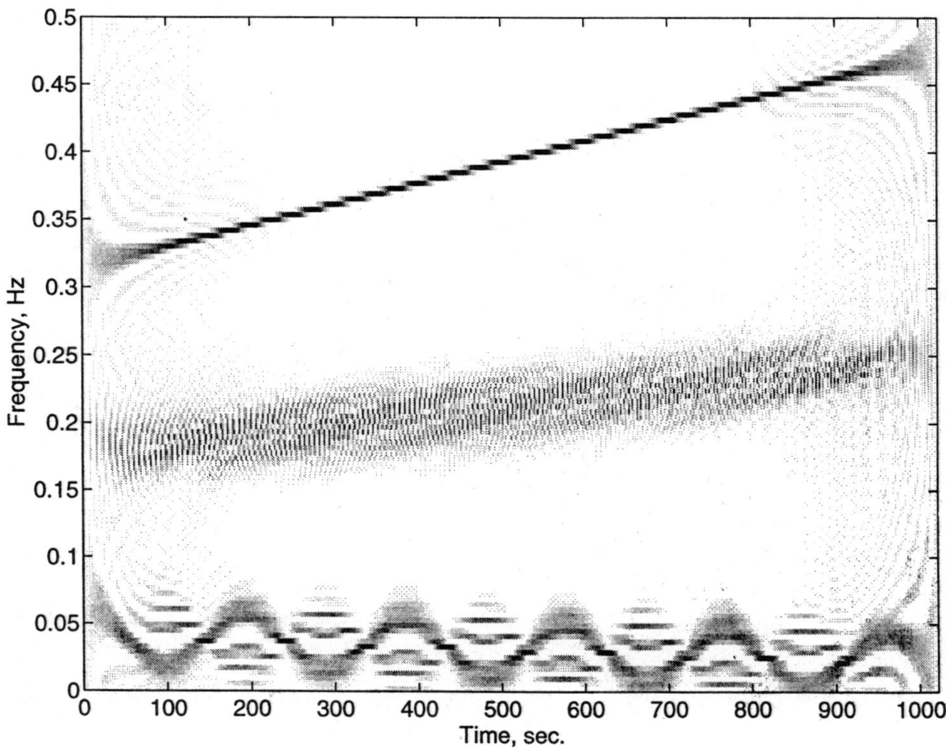

Figure 1-15   WD result for the chirp-warble.

called *self-terms* because they are produced within what is considered to be a single component. However, there is another interpretation. The warble signal is nearly periodic. Thus these self-terms are actually hints of the harmonic structure, or FM sidebands, of the frequency-modulated sinusoid.

1.5.3 Fast Algorithms Using Spectrogram Decompositions

One may also approach TFR discrete realization by means of spectrogram decomposition [46–49]. The approach for developing discrete TFRs discussed to this point is sometimes termed the "outer product approach." The "inner product approach" is an alternative. Development of these concepts is beyond the scope of this paper, but it may be helpful for the reader to have a brief outline of the idea. It can be shown that distributions from Cohen's class can be expressed as a linear weighted sum of spectrograms.

Shift-invariant bilinear discrete TFRs are specified by a discrete kernel, and can be rewritten in the inner product form of

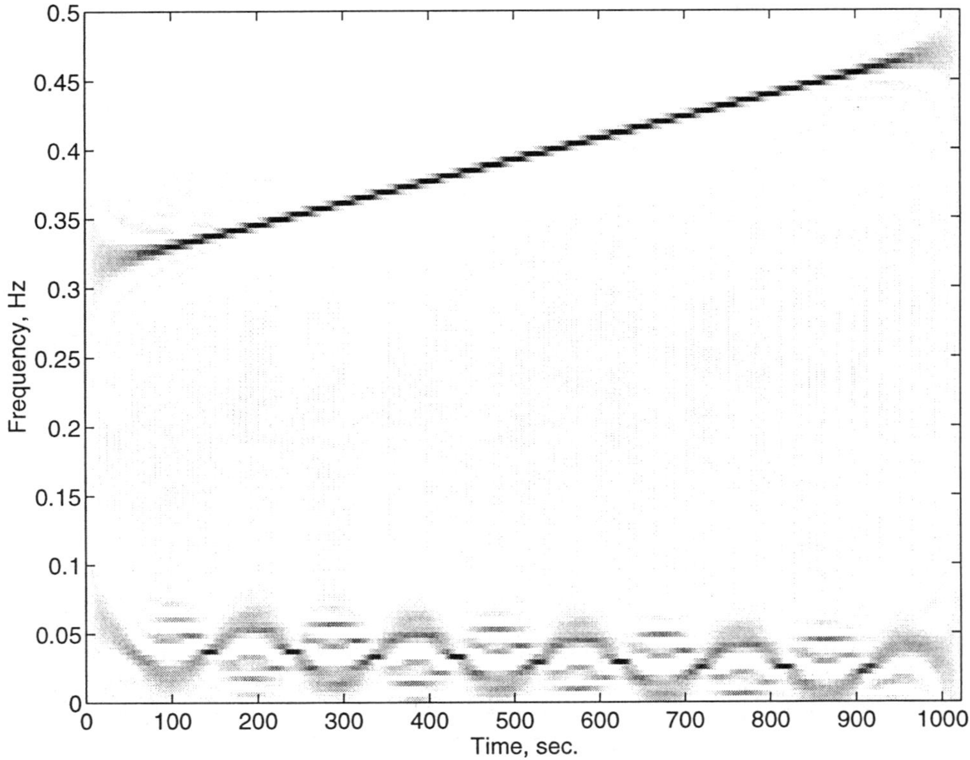

Figure 1-16   BD result for the chirp-warble.

$$\text{TFR}_x(n, w; \psi) = \sum_{n_1}\sum_{n_2}[x(n + n_1)e^{-j\omega(n+n_1)}]\psi\left(-\frac{n_1 + n_2}{2}, n_1 - n_2\right)[x(n + n_2)e^{-j\omega(n+n_2)}]^*$$

$$= \langle\bar{\psi}\bar{S}_{-n}\bar{M}_{-\omega}z, \bar{S}_{-n}\bar{M}_{-\omega}z\rangle \tag{1-30}$$

where $\bar{S}_{-n}$ and $\bar{M}_{-\omega}$ are, respectively, the time- and frequency-shift operators on ℓ_2, the space of finite-energy discrete-time signals, and $\bar{\psi}$ is a bounded linear operator on ℓ_2.

The spectral representation of $\bar{\psi}$ may be used to express the TFR as a weighted sum of spectrograms, or "projectograms." If the kernel is associated with a bounded, self-adjoint linear operator, then the kernel may be decomposed by an eigendecomposition such that one can represent the TFR as being composed of a finite series of spectrograms. The orthonormal windows forming the spectrograms are the eigenfunctions of the decomposition. The eigenvalues of the decomposition provide the weights for summing the set of spectrograms. The viewpoint may be taken that the projections of the signal on the eigenvectors of the kernel decomposition are then time- and frequency-shifted by the time- and frequency-shift operators, yielding, essentially, the STFT. The magnitude-squared STFT is the "projectogram," or spectrogram, associated with that particular window. Expressed mathematically, without proof,

$$\text{TFR}_x(n, \omega; \psi) = \sum_{k=1}^{N} \lambda_k \| P_k \bar{S}_{-n} \bar{M}_{-\omega} z \|^2$$

$$= \sum_{k=1}^{N} \lambda_k \left| \sum_{n_1} x(n + n_1) e^{-j\omega(n+n_1)} e_k^*(n_1) \right|^2 \tag{1-31}$$

Here, e_k is the eigenvector/window and P_k is the signal projection on that window, and λ_k is the eigenvalue for the particular k. One can recognize that $\sum_{n_1} x(n + n_1) e^{-j\omega(n+n_1)} e_k^*(n_1)$ is the STFT of the signal.

Generally, good TFR representation is possible using only a fraction of the windows associated with the largest eigenvalues. The spectrogram itself, of course, requires only one term, since it has only one window in its decomposition. The BD has been found to be represented very well by only about 17 windows. The eigenvalues for the WD all have magnitudes of 1, so even a great many terms may offer a poor representation. If all of the spectrograms are computed in parallel, then the time for computation of the TFR is the same as the time to compute one spectrogram.

O'Hair and Suter have suggested using a Zak transform decimation method [50], which speeds up spectrogram computation considerably, as a means of speeding up general TFR computation, since the individual spectrograms of the spectrogram decomposition can all be computed more rapidly.

1.6. TIME-VARYING FILTERING AND SYNTHESIS

Once the time-frequency result is obtained, one may wish to find the way back to the signal which produced the t-f result. The WD offers the possibility of reconstructing or synthesizing the signal [12]. We repeat the formulation for the WD as

$$W_x(t, \omega) = \int x\left(t + \frac{\tau}{2}\right) x^*\left(t - \frac{\tau}{2}\right) e^{-j\omega\tau} d\tau \tag{1-32}$$

This implies that

$$x\left(t + \frac{\tau}{2}\right) x^*\left(t - \frac{\tau}{2}\right) = \frac{1}{2\pi} \int W_x(t, \omega) e^{j\tau\omega} d\omega \tag{1-33}$$

which, with $t = \frac{\tau}{2}$ and $k = 1/x^*(0)$ and setting $\tau = t$, gives

$$x(t) = \frac{k}{2\pi} \int W_x(t/2, \omega) e^{j\omega t} d\omega \tag{1-34}$$

Thus $x(t)$ is recaptured, except for an arbitrary scale and phase factor k. By normalization, one can determine the scale factor. The *absolute* phase cannot be recovered, however, since we have multiplied the signal by its complex conjugate in forming the WD.

This result and the notion of the WD as an energy localization over the time-frequency plane enables one to conduct "time-varying filtering" via multiplying the WD by a weighting function of time and frequency, i.e.,

$$W_x'(t, \omega) = W_x(t, \omega)\Gamma(t, \omega) \tag{1-35}$$

The modified WD, however, may not be "valid." In other words, there may not exist a time function that produces the modified WD in contrast to the unfiltered WD, which, of course, does represent a valid $x(t)$. Therefore, time-frequency synthesis is often embedded in a least squares (LS) framework [51]. The objective is to find the signal $y(t)$ that best approximates the modified WD in terms of squared error, namely, $y(t)$ that minimizes the mean square error (MSE).

The MSE is the minimization of

$$\epsilon = \int\int |W_y(t, \omega) - W_x'(t, \omega)|^2 dt\, d\omega$$
$$= \int\int \left| \int y(t + \tau/2)y^*(t - \tau/2)e^{-j\omega\tau}d\tau - W_x(t, \omega)\Gamma(t, \omega) \right|^2 dt\, d\omega \tag{1-36}$$

where

$$R_x'(t, \tau) = \frac{1}{2\pi}\int W_x'(t, \omega)e^{j\omega\tau}d\omega \tag{1-37}$$

and

$$R_x''(u, v) = R_x'\left(\frac{u + v}{2}, u - v\right) \tag{1-38}$$

$R_x''(u, v)$ can be decomposed into

$$R_x''(u, v) = \sum_i \lambda_i\phi_i(u)\phi_i^*(v) \tag{1-39}$$

where $\{\lambda_i\}$ are real eigenvalues and the orthonormal basis $\{\phi_i(t)\}$ is the set of eigenfunctions.

Assuming λ_1 is the maximum positive eigenvalue and $\phi_1(t)$ is the associated eigenfunction, the solution $y(t)$ is obtained as

$$y(t) = \sqrt{\lambda_1}\phi_1(t)$$

Since $y(t)e^{j\alpha}$ produces exactly the same WD for arbitrary α, one may wish to find the phase α that makes the synthesized waveform "best" resemble the original waveform $x(t)$, or minimize

$$\epsilon = \int |y(t)e^{j\alpha} - x(t)|^2 dt \tag{1-40}$$

Thus

$$e^{j\alpha} = \frac{\int x(t)y^*(t)dt}{|\int x(t)y^*(t)dt|} = \frac{\int x(t)\phi_1^*(t)dt}{|\int x(t)\phi_1^*(t)dt|} \tag{1-41}$$

The WD-based method described here is highly nonlinear. Next, we discuss some methods that use linear transformations of the input signal. The STFT of $x(t)$, using a sliding window $h(t)$, is defined as

$$\text{STFT}_x(t, \omega) = \int x(u)h^*(t - u)e^{-j\omega u}\,du \tag{1-42}$$

The objective is to reconstruct the signal $y(t)$ from the modified STFT

$$\text{STFT}'_x(t, \omega) = \text{STFT}_x(t, \omega)\Gamma(t, \omega) \tag{1-43}$$

Then,

$$y(t) = \frac{1}{2\pi E_h} \int \int \text{STFT}'_x(u, \omega)h(u - t)e^{j\omega t}\,du\,d\omega \tag{1-44}$$

where E_h, the energy of $h(t)$ is usually normalized; i.e.,

$$E_h = \int |h(t)|^2 dt = 1 \tag{1-45}$$

provides a reasonable solution in the sense that it minimizes the MSE.

$$\epsilon = \int \int |\text{STFT}'_x(t, \omega) - \text{STFT}_y(t, \omega)|^2 dt d\omega \tag{1-46}$$

The LS reconstruction method can be rewritten as linear time-varying (LTV) filtering:

$$\begin{aligned}
y(t) &= \frac{1}{2\pi E_h} \int \int \text{STFT}_x(u, \omega)\Gamma(u, \omega)h(u - t)e^{j\omega t}\,du\,d\omega \\
&= \frac{1}{2\pi E_h} \int \int \int f(v)h^*(u - v)e^{-j\omega v}\,dv \int \Gamma(u, \omega)h(u - t)e^{j\omega t}\,du\,d\omega \\
&= \int f(v)\frac{1}{2\pi E_h} \int h^*(u - v) \int \Gamma(u, \omega)e^{j\omega(t-v)}\,d\omega\, h(u - t)du\,dv \\
&= \int f(v)q(t, v)dv \tag{1-47}
\end{aligned}$$

where

$$q(t, v) = \frac{1}{2\pi E_h} \int h^*(u - v)\gamma(u, t - v)h(u - t)du \tag{1-48}$$

with

$$\gamma(u, t) = \int \Gamma(u, \omega)e^{j\omega t}\,d\omega \tag{1-49}$$

The phase information is retained using this method.

For practical implementation of the WD-based time-varying filtering and signal synthesis, a discrete-time sequence, $\{x(n)\}$, of finite length N, and the discrete-time WD (DTWD) have been used:

$$W_x(n, f) = 2\sum_k e^{-j2k\omega}x(n + k)x^*(n - k) \tag{1-50}$$

Let x be the data vector:

$$x = [x(1), x(2), \cdots, x(N)]^T \tag{1-51}$$

We propose using the extended discrete-time Wigner distribution (EDTWD) [51], which was originally introduced to resolve the aliasing problem of the DTWD:

$$W_x^e(n \in X \cup X', \omega) = \sum_{n \pm k/2 \in X} e^{-j\omega k} x(n + k/2)x^*(n - k/2) \tag{1-52}$$

where $X = \{n : n$ is an integer$\}$ and $X' = \{n + .5 : n \in X\}$. The overall procedure of time-varying filtering and signal synthesis we propose here is then:

1. Given $\{x(n)\}$, compute the EDTWD as previously defined.
2. Apply a time-varying filter:

$$W_x'(n, \omega) = W_x(n, \omega)\Gamma(n, \omega), n \in X \cup X' \tag{1-53}$$

3. Take the inverse Fourier transform:

$$p(n, m) = \frac{1}{2\pi} \int_{-\pi/2}^{\pi/2} W_f'(n, \omega)e^{j\omega m} d\omega, \quad n \pm \frac{m}{2} \in X \tag{1-54}$$

4. Construct a modified product table matrix $\mathbf{Q}$, of size $N \times N$, where each entry q_{ij} is obtained by

$$q_{ij} = p\left(\frac{i+j}{2}, i-j\right) \tag{1-55}$$

5. Solve the following matrix equation,

$$(\mathbf{Q} + \mathbf{Q}^{*T})\mathbf{y} = 2\|\mathbf{y}\|^2\mathbf{y} = \lambda_{\max}\mathbf{y} \tag{1-56}$$

by finding the maximum eigenvalue $\lambda_{\max}$ and associated eigenvector $\mathbf{y}$ of $(\mathbf{Q} + \mathbf{Q}^{*T})$.

6. Scale the eigenvector $\mathbf{y}$ such that $2\|\mathbf{y}\|^2 = \lambda_{\max}$.
7. Correct the phase of the reconstructed signal, if necessary:

$$\mathbf{y}_{\text{opt}} = \mathbf{y}e^{j\alpha} = \mathbf{y}\frac{\mathbf{x}^T\mathbf{y}^*}{|\mathbf{x}^T\mathbf{y}^*|} \tag{1-57}$$

This method differs from that proposed by Boudreaux-Bartels and Parks [52] in that two eigenvector-eigenvalue solutions are obtained in their method, one for the even-indexed and one for the odd-indexed correlation values in the product matrix. This requires two phase correction computations and the odd and even results are combined.

When signals are reasonably separated in time-frequency, then a simple mask which picks out one t-f component will produce a nice result. This is true even if there are interference terms from other signals within the mask. Intuitively, this seems reasonable, since the interference terms are probably cast into other eigenvectors of the eigendecomposition and are thus not selected when the principal eigenvector, containing the desired synthesized signal, is selected. However, when t-f components are close, this approach can be disappointing, since interference terms within the mask may be large and contribute to the principal eigenvector. More sophisticated methods are available which are potentially able to improve synthesis under these conditions [53]. Methods based on Radon transform concepts also appear to have quite a lot of promise (Wood and Barry [38–40]).

Rather than use the WD as the exploratory tool for finding t-f components, it is better to use a TFR that more clearly reveals the components, such as the RID. Then, once the filter or mask support is determined from this, it can readily be applied to the same t-f region of the WD.

Very little use of time-varying filtering and synthesis has been made in biological signal analysis. An exception is the work by Wood and Barry [39], to be discussed more fully in another chapter in this volume. It appears to have a promising future in this application area, however. There are situations where the desired signal and an interfering signal cannot be separated in time (by time windowing or gating) or in frequency (by bandpass filtering). If the signals are cotemporal and cospectral, then the approaches described in this section can be of help. An example is brought to mind in terms of EEG signal processing. A muscle artifact seen via electromyography (EMG), is a common contaminant of EEGs. Even so, an appropriate t-f mask or filter may be able to extract the desired signal because the EMG and EEG will not overlap in many regions of the TFR, even though they do overlap in some regions.

1.7. ANALYSIS WINDOW COMPARISONS: WAVELETS AND COHEN'S CLASS

This chapter has been mainly concerned with Cohen's class of distributions, particularly the newer and more effective distributions with interference reduction properties. However, the wavelet transform has attracted quite a lot of attention and more wondrous properties have been attributed to it than it actually possesses. The wavelet transform is actually similar to the STFT [54], but it may be considered to have a window that changes shape with frequency. The spectrogram, the RID, the ZAM, and many other TFRs in Cohen's class possess analysis windows that stay fixed in shape with frequency. Figure 1-17 illustrates the analysis window shapes for several techniques. In Fig. 1-17(a) one sees the spectrogram window.* It is assumed that the reader is familiar with the wavelet transform either through prior exposure or by means of other chapters in this volume.

Both the wavelet transform and the STFT are linear transforms. TFRs from Cohen's class are bilinear because the integrand involving the signal is the product of two differently shifted versions of the signal. The spectrogram results when the STFT magnitude is squared. Similarly, the scalogram [55] results when the wavelet transform magnitude is squared. The scalogram is nonnegative, but it is equipped with cross-terms just like members of Cohen's class [56]. There are other ways of viewing the relationships between wavelet-based energy distributions and Cohen's class. In fact, if one is willing to change the kernel in Cohen's class with frequency, one can obtain scalogram results and many other novel results as well, perhaps particularly

---

*All window shapes in Fig. 1-17 are composed of squares or rectangles for purposes of illustration. In truth, they would be more ellipsoid in shape, tapering down from their maximum values at the center of the window.

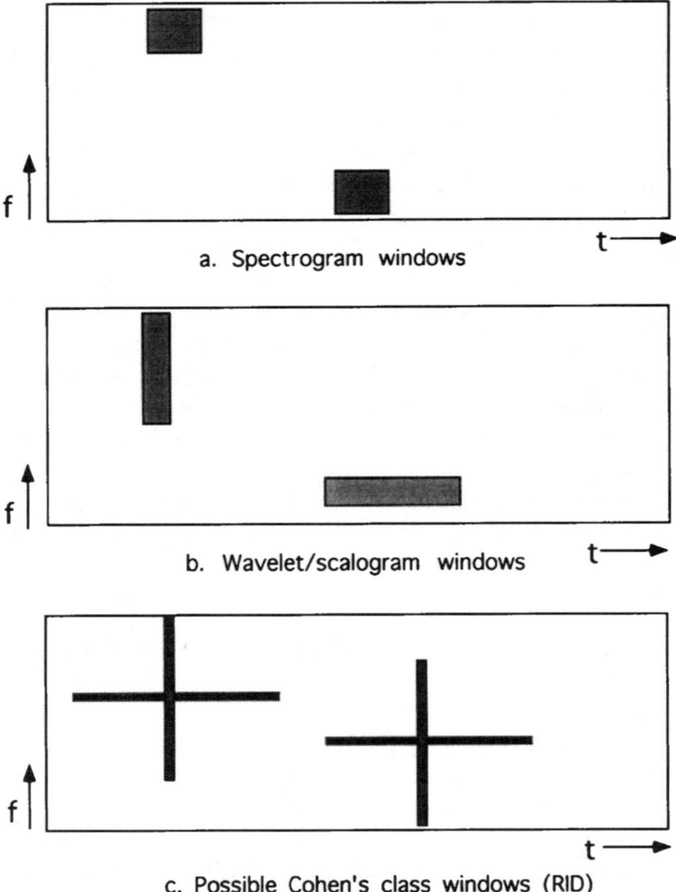

Figure 1-17 A comparison of windows for various analysis schemes: (a) spectrogram; (b) wavelet; (c) RID. Note that the spectrogram and RID windows do not change shape with frequency, whereas the wavelet window does.

designed to capture the t-f components of the WD, since Cohen's class results can be considered from the standpoint that the kernel is a window in time-frequency that is convolved in time and frequency in order to produce the particular distribution of interest [57]. Application in the ambiguity plane is by multiplying the kernel. The WD is a double Fourier transform away from the ambiguity function, so the kernel is the double Fourier transform of the ambiguity plane form and it is now *convolved* in time and frequency. Some kernels, such as the RID kernel, offer high resolution in both time and frequency due to their shape, but have the liability of some negativity as well. One must pay some prices, it seems, and no present technique can solve all problems.

The wavelet transform will work well when t-f components are narrow in frequency and prolonged in time for low frequencies, but narrow in time and broad in frequency for high frequencies, relatively speaking. On occasion, one observes t-f components that are far from that arrangement, perhaps even reversing roles with shorter duration components at low frequencies and long duration components at high frequencies. Then, a kernel for Cohen's class that adjusts properly could be designed, perhaps widening in frequency for high frequencies and narrowing in frequency for low frequencies.

There are even more possibilities. Techniques based on the Radon transform may provide even more diversity in terms of the way the t-f plane is handled. Wood and Barry [38–40] have devised some interesting approaches based on this idea that seem to have very useful biological signal applications.

1.8. CONCLUSIONS

The RID and other new time-frequency concepts, including wavelet transforms, have the potential of revealing many new phenomena and changing the way we view and think about signals. These are only beginnings. New and improved approaches are being reported regularly. This chapter has only touched on the extensive literature now available. An attempt was made to discuss techniques that have been applied with some success on biological and biomedical problems or techniques that appear to have considerable promise in such applications. One should not become attached to a particular TFR or approach, but learn the principles behind them so that well-reasoned choices can be made when confronted with an application. Another chapter in this volume will provide greater insight into the effectiveness of time-frequency distributions in biological and biomedical applications.

ACKNOWLEDGMENTS

It is impossible to mention by name all of the people who have contributed to the work reported in this chapter. Graduate students, research collaborators, and time-frequency colleagues who have provided many hours of lively discussions which have shaped this work are most gratefully acknowledged and thanked.

REFERENCES

[1] E. Wigner, "On the quantum correction for thermodynamic equilibrium," *Phys. Rev.*, vol. 40, pp. 749–759, 1932.

[2] J. Ville, "Theorie et applications de la notion de signal analytique," *Cables et Transmissions*, vol. 20A, pp. 61–74, 1948.

[3] T. C. M. Claasen and W. F. G. Mecklenbräuker, "The Wigner distribution—A tool for time-frequency signal analysis—Part I: Continuous-time signals," *Philips J. Res.*, vol. 35,pp. 217–250, 1980.

[4] T. C. M. Claasen and W. F. G. Mecklenbräuker, "The Wigner distribution—A tool for time-frequency signal analysis—Part II: Discrete time signals," *Philips J. Res.*, vol. 35, pp. 276–300, 1980.

[5] T. C. M. Claasen and W. F. G. Mecklenbräuker, "The Wigner distribution—A tool for time-frequency signal analysis—Part III: Relations with other time-frequency signal transformations," *Philips J. Res.*, vol. 35, pp. 372–389, 1980.

[6] P. Flandrin and F. Hlawatsch, "Signal representations, geometry and catastrophes in signal processing." In *Mathematics in Signal Processing*. T. Durrani, J. Abbiss, J. Hudson, R. Madan, J. McWhirter, and T. Moore (eds.) pp. 3–14, Oxford: Clarendon Press, 1987.

[7] P. Flandrin, "Some features of time-frequency representations of multi-component signals," *IEEE Int. Conf. Acous., Speech, Sig. Proc.*, vol. 41B, pp. 4.1–4.4, 1984.

[8] F. Hlawatsch, "Interference terms in the Wigner distribution," in *Digital Signal Processing—84*, V. Cappellini and A. Constantinides (ed.) pp. 363–367, Amsterdam: North-Holland, 1984.

[9] L. Cohen, "Generalized phase-space distribution functions," *J. Math. Phys.*, vol. 7, pp. 781–786, 1966.

[10] L. Cohen, "Time-frequency distributions—A review," *Proc. IEEE*, vol. 77, pp. 941–981, 1989.

[11] B. Boashash, "Time-frequency signal analysis," In *Advances in Spectrum Analysis and Array Processing*, S. Haykin (ed.) Englewood Cliffs, NJ: Prentice Hall 1991.

[12] L. Cohen, *Time-Frequency Signal Analysis*. Englewood Cliffs, NJ: Prentice Hall, 1995.

[13] H. I. Choi and W. J. Williams, "Improved time-frequency representation of multicomponent signals using exponential kernels," *IEEE Trans. Acoust., Speech, Signal Proc.*, vol. ASSP-37, no. 6, pp. 862–871, 1989.

[14] W. J. Williams and J. Jeong, "Reduced interference time-frequency distributions." In *Time-Frequency Signal Analysis: Methods and Applications*, B. Boashash, (ed.) Melbourne: Longman and Cheshire, 1992.

[15] Y. Zhao, L. E. Atlas, and R. Marks," The use of cone shaped kernels for generalized time-frequency representations of nonstationary signals," *IEEE Trans. Acoust., Speech, Signal Proc.*, vol. 38, pp. 1084–1091, 1990.

[16] L. Cohen and T. E. Posch, "Generalized ambiguity function," *IEEE Int. Conf. Acoust., Speech, Sig. Proc.*, pp. 1033–1036, 1985.

[17] S. Oh and R. J. Marks II, "Some properties of the generalized time-frequency representation with cone shaped kernel," *IEEE Trans. Signal Proc.*, vol. 40, pp. 1735–1745, 1992.

[18] J. Jeong and W. J. Williams, "Kernel design for reduced interference distributions," *IEEE Trans. Signal Proc.*, vol. 40, pp. 402–412, 1992.

[19] P. J. Loughlin, J. W. Pitton, and L. E. Atlas, "Construction of positive time-frequency distributions," *IEEE Trans. Signal Proc.*, vol. 42, pp. 2697–2705, 1994.

[20] J Jeong and W. J. Williams, "Mechanism of the cross terms in the spectrogram," *IEEE Trans. Signal Proc.*, vol. 40, pp. 2608–2613, 1992.

[21] J. C. Andrieux, M. R. Feix, G. Mourgues, P. Bertrand, B. V. T. Izrar, and V. T. Nguyen, "Optimum smoothing of the Wigner–Ville Distribution," *IEEE Trans. Acoust., Speech, Signal Proc.*, vol. ASSP-35, pp. 764–769, 1987.

[22] J. E. Moyal, "Quantum mechanics as a statistical theory," *Proc. Cambridge Phil. Soc.*, vol. 45, pp. 99–124, 1949.

[23] A. J. E. M. Janssen, "On the locus and spread of pseudo-density functions in the time-frequency plane," *Philips J. Res.*, vol. 37, pp. 79–110, 1982.

[24] L. Cohen, "The uncertainty principles of the short-time Fourier transform," *SPIE Advanced Signal Processing Algorithms*, vol. 2563, pp. 80–90, 1995.

[25] E. J. Diethorn, "The generalized exponential time-frequency distribution," *ATT Bell Laboratories Tech. Memo*, 1992.

[26] E. J. Diethorn, "The generalized exponential time-frequency distribution," *IEEE Trans. Signal Proc.*, vol. 42, pp. 1028–1037, 1994.

[27] A. Papandreou and G. F. Boudreaux-Bartels, "Distributions for time-frequency analysis: A generalization of Choi–Williams and the Butterworth distribution," *IEEE Trans. Signal Proc.*, vol. 41, pp. 463–472, 1993.

[28] E. J. Zalubas and M. G. Amin, "The design of signal dependent time-frequency kernels by the McClellan transformation," *Proc. IEEE-SP Int. Symp. Time-Frequency and Time-Scale Anal.*, pp. 91–94, 1992.

[29] Z. Guo, L.-G. Durand, and H. C. Lee, "Comparison of time-frequency distribution techniques for analysis of simulated Doppler ultrasound signals of the femoral artery," *IEEE Trans Biomed. Eng.*, vol. 41, pp. 332–342, 1994.

[30] Z. Guo, L.-G. Durand, and H. C. Lee, "The time-frequency distributions of nonstationary signals based on a Bessel kernel," *IEEE Trans. Signal Proc.*, vol. 42, pp. 1700–1707, 1994.

[31] F. Hlawatsch and G. F. Boudreaux-Bartels, "Linear and quadratic time-frequency signal representations," *IEEE Signal Proc. Mag.*, vol. 9, pp. 21–67, 1992.

[32] D. L. Jones and T. W. Parks, "A resolution comparison of several time-frequency representations," *IEEE Trans. Signal Proc.*, vol. 40, pp. 413–420, 1992.

[33] P. J. Loughlin, J. W. Pitton, and L. E. Atlas, "Bilinear time-frequency representations: New insights and properties," *IEEE Trans. Signal Proc.*, vol. 41, pp. 750–767, 1993.

[34] R. G. Baraniuk and D. J. Jones, "Signal-dependent time-frequency representation: optimal kernel design," *IEEE Trans. Signal Proc.*, vol. 41, pp. 1589–1602, 1993.

[35] T. H. Sang and W. J. Williams, "Rényi information and signal-dependent kernel design," *Proc. IEEE Int. Conf. Acoust., Speech, Signal Proc.*, vol. 2, pp. 997–1000, 1995.

[36] W. J. Williams and T. H. Sang, "Adaptive RID kernels which minimize time-frequency uncertainty," *IEEE Int. Symp. Time-Frequency and Time-Scale Anal.*, pp. 96–99, October, 1994.

[37] W. J. Williams, M. J. Brown, and A. O. Hero III, "Uncertainty, information and time-frequency distributions," *SPIE Advanced Signal Processing Algorithms, Architectures and Implementations II*, vol. 1566, pp. 144–156, 1991.

[38] J. C. Wood and D. T. Barry, "Linear signal synthesis using the Radon–Wigner transform," *IEEE Trans. Signal Proc.*, vol. 42, pp. 2105–2111, 1994.

[39] J. C. Wood and D. T. Barry, "Tomographic time-frequency analysis and its application toward time-varying filtering and adaptive kernel design for multi-component linear-FM signals," *IEEE Trans. Signal Proc.*, vol. 42, pp. 2094–2104, 1994.

[40] J. C. Wood and D. T. Barry, "Radon transformation of time-frequency distributions for analysis of multicomponent signals," *IEEE Trans. Signal Proc.*, vol. 42, pp. 3166–3177, 1994.

[41] R. N. Czerwinski and D. L. Jones, "Adaptive cone-kernel time-frequency analysis," *IEEE Trans. Signal Proc.*, vol. 43, pp. 1715–1718, 1995.

[42] P. Flandrin, R. G. Baranuik, and O. Michel, "Time-frequency complexity and information," *Proc. IEEE ICASSP-94*, vol. 3, pp. 329–332, 1994.

[43] S. Hearon and M. G. Amin, "Minimum variance time-frequency distribution kernels," *IEEE Trans. Signal Proc.*, vol. 43, pp. 1258–1262, 1995.

[44] D. T. Barry, "Fast calculation of the Choi-Williams distribution," *IEEE Trans. Signal Proc.*, vol. 40, pp. 450–455, 1992.

[45] J. Jeong and W. J. Williams, "Alias-free generalized discrete-time time-frequency distributions," *IEEE Trans. Signal Proc.*, vol. 40, pp. 2757–2765, 1992.

[46] M. G. Amin, "Spectral decomposition of the time-frequency distribution kernels," *IEEE Trans. Signal Proc.*, vol. 42, pp. 1156–1166, 1994.

[47] G. S. Cunningham and W. J. Williams, "Kernel decompositions of time-frequency distributions," *IEEE Trans. Signal Proc.*, vol. 42, pp. 1425–1442, 1994.

[48] G. S. Cunningham and W. J. Williams, "Fast implementations of discrete time-frequency distributions," *IEEE Trans. Signal Proc.*, vol. 42, pp. 1496–1508, 1994.

[49] L. B. White, "Transition kernels for bilinear time-frequency signal representations," *IEEE Trans. Acoust., Speech Signal Proc.*, vol. 39, pp. 542–544, 1991.

[50] J. R. O'Hair and B. W. Suter, "The Zak transform and decimated time-frequency distributions," *IEEE Trans. Signal Proc.*, vol. 44, pp. 1099-1110, 1996.

[51] J. Jeong and W. J. Williams, "Time-varying filtering and signal synthesis," In *Time-Frequency Signal Analysis*. B. Boashash (ed.) Melbourne: Longman and Cheshire, 1992.

[52] G. F. Boudreaux-Bartels and T. W. Parks, "Time-varying filtering and signal estimation using Wigner distribution synthesis techniques," *IEEE Trans. Acoust., Speech, Signal Proc.*, vol. 34, pp. 442–451, 1986.

[53] F. Hlawatsch and W. Kozek, "Time-frequency projection filters and time-frequency signal expansions," *IEEE Trans. Signal Proc.*, vol. 42, pp. 3321–3334, 1994.

[54] I. Daubechies, "The wavelet transform, time-frequency localization and signal analysis," *IEEE Trans. Informat. Theory*, vol. 36, pp. 961–1005, 1990.

[55] O. Rioul and P. Flandrin, "Time-scale energy distributions: A general class extending wavelet transforms," *IEEE Trans. Signal Proc.*, vol. 40, pp. 1746–1757, 1992.

[56] S. Kadambe and G. F. Boudreaux-Bartels, "A comparison of the existence of 'cross terms' in the Wigner distribution and the squared magnitude of the wavelet transform and the short time Fourier transform," *IEEE Trans. Signal Proc.*, vol. 40, pp. 2498–2517, 1992.

[57] J. Jeong and W. J. Williams, "Variable windowed spectrograms: connecting Cohen's class and the wavelet transform," *IEEE ASSP Workshop on Spectrum Estimation and Modeling*, pp. 270–273, 1990.

Chapter 2

Biological Applications
and Interpretations of
Time-Frequency Signal Analysis*

William J. Williams

2.1. INTRODUCTION

Great progress has been made in applying linear time-invariant techniques in signal processing. In all such cases the deterministic portion of the signal is tacitly assumed to be composed of complex exponentials which are the solutions to linear time-invariant (LTI) differential equations. These assumptions are implicit or explicit and are valid enough to yield good results when the signals involved result from systems produced by engineering design. Many signals, particularly those of biological origin, do not comply with these assumptions, however. The newly emerging techniques of time-frequency (t-f) analysis can provide new insights into the nature of biological signals. This chapter describes some results using reduced interference distributions (RIDs) and other high-resolution time-frequency distributions (TFDs) in the analysis of several biosignals. Several examples serve to demonstrate that new results and insights can be obtained using these new techniques. However, just as in all pursuits, the methods must be applied properly and interpreted properly in order to obtain the most meaningful results.

The conventional usage of the Laplace transform and the Fourier transform is largely driven by the fact that the eigenfunctions of linear time-invariant differential equations have solutions which are of the form Ce^{st}, where s is complex, t is time,

*This research was supported in part by the Office of Naval Research, ONR contract no. N00014-90-J-1654, the National Science Foundation, NSF grant no. BCS-9110571, and a Biomedical Research Support Grant from the Office of Vice President for Research through the National Institutes of Health.

and C is an arbitrary constant determined by the initial conditions and/or the driving function. These transforms have the important shift and convolution properties which greatly aid in the manipulation of equations and the practical computation of results. These transforms essentially change differential equations into algebraic equations, which then allows the powerful machinery of linear algebra to be used. The z transform provides the same type of algebraic simplification for linear shift-invariant difference equations. Discrete computational tools such as the fast Fourier transform (FFT) then follow.

A sinusoidal signal which exhibits an increasing (or decreasing) frequency with time is often called a chirp. Such signals can result from systems which are linear but exhibit time-varying system parameters. The differential equations describing these systems are not linear time-invariant, but exhibit time-varying coefficients and can properly be called linear time-varying systems. There are, of course, many examples of time-varying systems that exhibit quite complicated responses. In general, the solutions for even simple time-varying systems are fairly intractable unless a clever transformation can be found which will provide a time-invariant form.

It has been quite difficult to analyze signals with frequency content that varies rapidly with time by using conceptualizations based on time-invariance. The spectrogram represents an attempt to apply the Fourier transform (FT) for a short-time analysis window, within which it is hoped that the system being analyzed behaves reasonably according to the requirements of time-invariance. By moving the analysis window along the signal, one hopes to capture and track the variations of the signal spectrum as a function of time. The spectrogram has many useful properties, including a well-developed general theory [1,2]. The spectrogram often presents serious difficulties when it is used to analyze rapidly varying signals, however. If the analysis window is made short enough to capture rapid changes in the signal, it becomes impossible to resolve signal components which are close in frequency within the analysis window duration. Watkins [3] has provided an enlightening view of the limitations of the spectrogram when it is used to analyze marine mammal sounds.

The Wigner distribution (WD) has been employed as an alternative to overcome some of the shortcomings of the spectrogram. The WD was first introduced in the context of quantum mechanics [4] and revived for signal analysis by Ville [5]. The WD has many important and interesting properties [6–8]. It provides a high-resolution representation in time and in frequency. In addition, the WD has the important property of satisfying the time and frequency marginals in terms of the instantaneous power in time and energy spectrum in frequency. While the WD provides unambiguous high-resolution t-f representations of nonstationary monocomponent signals, such as chirps, its representations of multicomponent signals often are not useful. These representations contain interference (cross-terms) in different regions of the t-f plane, resulting from interactions between signal components. Cross-terms are sometimes a hindrance to interpretation since they carry redundant information and may obscure primary features of the signal. An excellent discussion on the geometry of interferences has been provided by Hlawatsch [9] and Flandrin [10].

A great deal of progress has been made in minimizing the liabilities of the spectrogram and the WD in recent years. One such approach is the reduced interference distribution, or RID, approach [11] . Other new approaches have also been suggested. This chaper aims to provide a better understanding of the RID approach, how it differs from other approaches, how to interpret results, and how to properly apply the technique. No attempt will be made to review the entire field. Recent developments using RIDs and other effective cross-term-suppressing distributions are the main focus of this review. Results obtained using WDs and spectrograms will be brought in as appropriate to provide contrast with the newer and generally more effective TFDs. In addition to the RID, the cone kernel distribution (CKD) (or ZAM distribution, developed by Zhao, Atlas, and Marks [12]) will be discussed, as will adaptive kernel distribution concepts pioneered by Baraniuk and Jones [13].

These approaches have already been discussed more fully in Chapter 1, addressing the theories used in the present chapter, which aims to provide a discussion of applications. Much of what is reported in this chapter comes from a paper from a special issue of the *IEEE Proceedings* on applications of time-frequency analysis [14] and from recent articles by our group on this subject area.

2.2. COHEN'S CLASS OF DISTRIBUTIONS

Both the spectrogram and the WD are members of Cohen's class of distributions (see [15]). Although the focus of that paper was on quantum mechanics, the concept has provided a consistent set of definitions for t-f distributions which has been of great value in guiding and clarifying efforts in this area of research. Cohen's class of distributions is defined as

$$C_x(t, \omega; \phi) = \iiint e^{j(-\theta t - \tau v + \omega u)} \phi(\theta, \tau) x(u + \tau/2) x^*(u - \tau/2) du d\tau dv \qquad (2\text{-}1)$$

where $x(t)$ is the time signal, $x^*(t)$ is its complex conjugate, and $\phi(\theta, \tau)$ is the *kernel* of the distribution.* Desirable properties of a distribution and associated kernel requirements have been extensively investigated by Claasen and Mecklenbräuker [6–8]. Further, a comprehensive review [1] and a recent book [2] by Cohen provide an excellent overview of t-f distributions and results obtained using them. A recent tutorial review may also be useful to the reader [16], where some biological applications are discussed.

A specific subset of t-f distributions belonging to Cohen's class are of considerable interest. These are the time-shift and frequency-shift invariant t-f distributions. For these distributions, a time shift in the signal is reflected as an equivalent time shift in the t-f distribution, and a shift in the frequency of the signal is reflected as an equivalent frequency shift in the t-f distribution. The spectrogram, the WD, the RID, and the ZAM all have this property. Although the spectrogram has been used for many years in a variety of areas, this chapter will not review that vast literature, but concentrate on more recent developments that have found some use in biological signal analysis. Indeed, some of the developments have been strongly motivated by biological problems.

---

*The range of integrals is from $-\infty$ to ∞ throughout this paper.

2.2.1 Electrophysiological Signals and Epilepsy

Some applications of time-frequency analysis.   The spectrogram has been used extensively in the analysis of brain electrical potentials (BEPs), primarily in the characterization of background electroencephalograms (EEGs) [17–19]. Over the years, a number of alternate t-f representations have been proposed for BEPs, none of which, however, has found wide acceptance. Kawabata employed the instantaneous power spectrum defined by Page [20], a measure of the rate of change of the energy spectrum, to study dynamic variations in the EEG during photic alpha blocking [21] and performance of mental tasks [22]. De Weerd and Kap [23] compared t-f representations of event-related potentials (ERPs) obtained by a nonuniform filter bank and the complex energy distribution function (CEDF) introduced by Rihazcek [24]. They preferred the use of the filter bank, citing difficulty in interpreting results obtained by the CEDF. The applicability of the WD for the representation of ERPs was examined by Morgan and Gevins [25] and the cross-WD was employed in the study of neonatal EEGs by Scher et al. [26]. Though the WD has been found useful in other fields, it has had limited application in the analysis of BEPs.

RID applications in epilepsy.   In this chapter we will report on RID analyses of electrocorticograms (ECoGs) recorded in instances of temporal lobe epilepsy and EEGs recorded in Lennox–Gastaux epilepsy.* These signals have a time-varying spectral content that can change dramatically in a very short time (e.g., during seizures). It will be shown in this chapter that the RID provides a new viewpoint on biosignals encountered in epilepsy. The RID does not only provide a rather striking visual presentation of the t-f content of these signals [27], it also provides strong insights into the underlying system characteristics responsible for the character of these signals.

The RID has been used extensively in investigating biological signals, particularly biopotentials. EEGs are biopotentials obtained using scalp electrodes. ECoGs are biopotentials obtained by placing electrodes directly on the exposed surface (cortex) of the brain. The latter is done as an investigatory procedure just prior to epilepsy abatement surgery. Figure 2-1 compares the results obtained using the WD, RID, and spectrogram for an "ictal discharge" which occurs during a seizure and is recorded using ECoG data [27,28]. Time-frequency techniques discussed in this chapter have been used in the EEG setting (particularly the spectrogram) for several years. RID approaches have only recently been reported, however. The reader should refer to the citations given in this chapter for a more comprehensive discussion of these applications.

One can readily see in Figure 2-1 that the spectrogram smears the t-f result considerably. The WD provides good resolution of the major components of the signal, but exhibits rather severe cross-term activity, potentially confusing the interpretation of the results. The RID clearly reveals the three components of the signal

---

*ECoGs are BEPs obtained directly from the cortical surface. The recording electrodes are typically placed on grids or strips with regular interelectrode spacings.

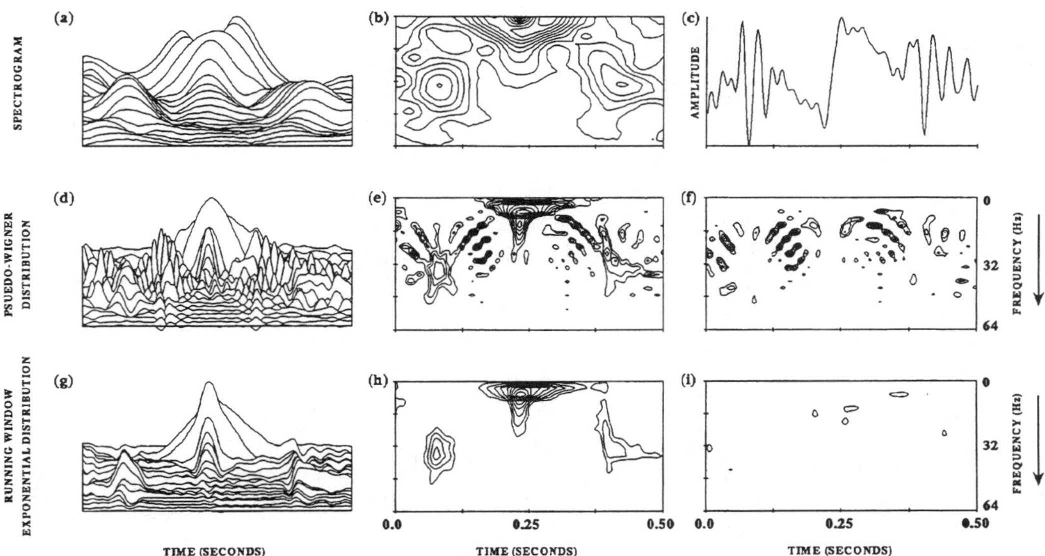

Figure 2-1 A comparison of the spectrogram, WD, and RID for a 500-ms (250 samples) segment of seizure ECoG. The time series is shown in (c). Parts (a), (d), and (g) are surface plots of the spectrogram, WD, and RID representations, respectively. Parts (b), (e), and (h) are positive contour plots of the representations shown in (a), (d), and (g), respectively. Parts (f) and (i) are negative contour plots of the representations shown in (d) and (g), respectively. The ordinate (frequency) axis is reserved to allow an unobstructed view of the t-f plane. The spectrogram was calculated using a 160-ms von Hamm window, shifted one data point between successive evaluations. The WD and RID ($\sigma = 1.0$) were computed as 160-ms estimates of the analytic form of the signal [27,28].

with very little cross-term interference. It is also interesting to note that the RID exhibits very little negativity compared to the WD in this case. This is because the WD cross-terms can actually be twice the size of the auto-terms [1,2]. The RID, on the other hand, flattens the cross-terms. Negative values of TFDs are produced by cross-terms. When flattened, the negative swings of the cross-terms are greatly reduced, as shown in the RID result. Notice that cross-terms always appear between auto-terms. It may appear that some of the cross-term components violate this rule. However, there are low-level auto-term components that are below the lowest contour in this illustration, and cross-terms, being even larger than the auto-terms, can pop up above the lowest contour, as seen here. Cross-terms also appear between different outlying parts of an auto-term and that can be observed as well.

The RID approach has revealed that there are chirp-like structures—short bursts at high and low frequencies—tonal components, and sharp transitions in ECoGs during seizures [27,28]. These structures are very difficult to identify using other techniques.

The σ parameter in the exponential distribution (ED) allows one to adjust the trade-off between resolution and cross-term suppression. The consequences of chan-

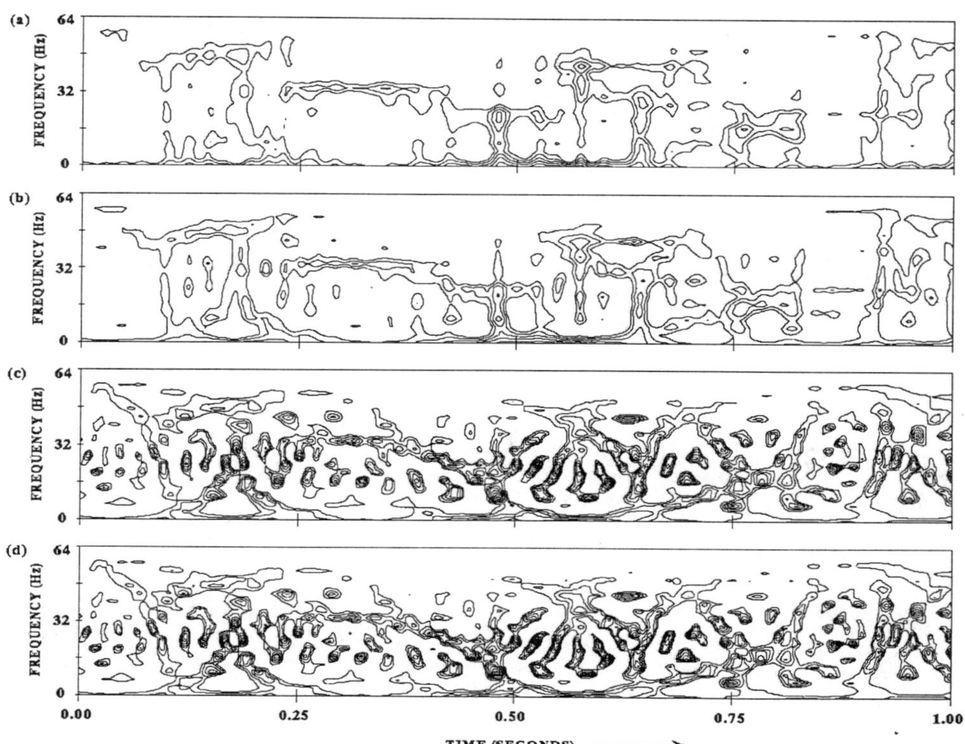

Figure 2-2 A demonstration of the control of cross-terms in the ED by the change of σ. For small σ, the energy in the interference terms is low, and increasing σ increases the energy in the interference terms. (a) $\sigma = 0.1$; (b) $\sigma = 10.0$; (c) $\sigma = 1000.0$ (here $\phi_{ED} \approx \phi_{WD}$); (d) WD of the same signal, showing that the ED is nearly identical for large σ [27,28].

ging σ are shown in Figure 2-2. Very low values of σ allow cross-terms to be suppressed very effectively. The kernel in the ambiguity plane is concentrated close to the axes. However, in the local autocorrelation domain, the kernel is spread out more along the t dimension. This tends to violate the time support constraint. Also, resolution is worse, since the kernel actually attenuates a portion of the auto-terms as well as the cross-terms. Large values of σ essentially "open-up" the ambiguity form of the kernel so that it encompasses a large part of the ambiguity plane, approaching the WD kernel. One can see from Figure 2-2 that a value of $\sigma = 1$ does a nice job of revealing the t-f structures without undue cross-term activity. A value of $\sigma = 1000$ allows cross-term activity to appear to such an extent that it is difficult to interpret what is going on and the ED has essentially been transformed into a WD.

Just as is the case with time-domain formulations such as cross correlation, and frequency-domain formulations such as cross spectral density, cross-RIDs [29] are useful in revealing consistent relationships between signal components. Cross-RIDs and cross-WDs can be formed in a manner very similar to other RIDs and WDs by replacing one of the signal terms, say $f(t)$, in the formulation by another signal $g(t)$, which may be related to the signal. Results from this approach are shown in Figure 2-3 for the cross-WD and Figure 2-4 for the cross-RID (ED form). Again, for the cross-WD, cross-terms obscure the result to a point where interpretation is very difficult. The cross-RID, on the other hand, provides a clear indication of the t-f similarity of components in the two signals. These techniques have not yet been thoroughly investigated, but they seem to be quite promising. The idea of a time-frequency coherence measure is also appealing, but some conceptual barriers to such a formulation have not yet been overcome.

EEGs from scalp electrodes are also used in the evaluation of seizure data. The "spike and slow wave" is a very interesting example of some of these results. The

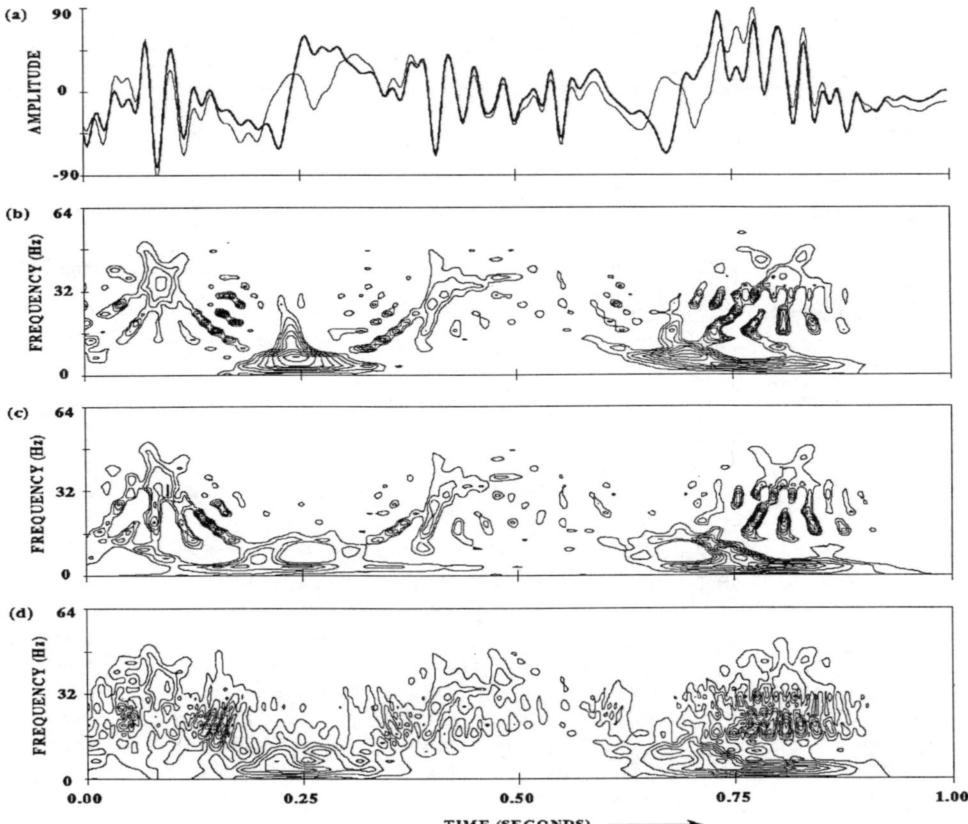

Figure 2-3   The auto-WDs and cross-WD of the analytic forms of seizure ECoG signal segments (a), recorded at two adjacent electrodes, are shown as contour representations in (b)–(d). The dashed line is $f(t)$ and the solid line is $g(t)$. The positive values of the auto-WD representations $C_f(t, \omega; \phi_{\mathrm{WD}})$ and $C_g(t, \omega; \phi_{\mathrm{WD}})$ are shown in (b) and (c), respectively. The magnitude of the cross-WD representation $C_{fg}(t, \omega; \phi_{\mathrm{WD}})$ is shown in (d). Interference terms exist between signal components in the auto-WDs and the cross-WD. The interference between signal components renders the cross-WD impossible to interpret [29].

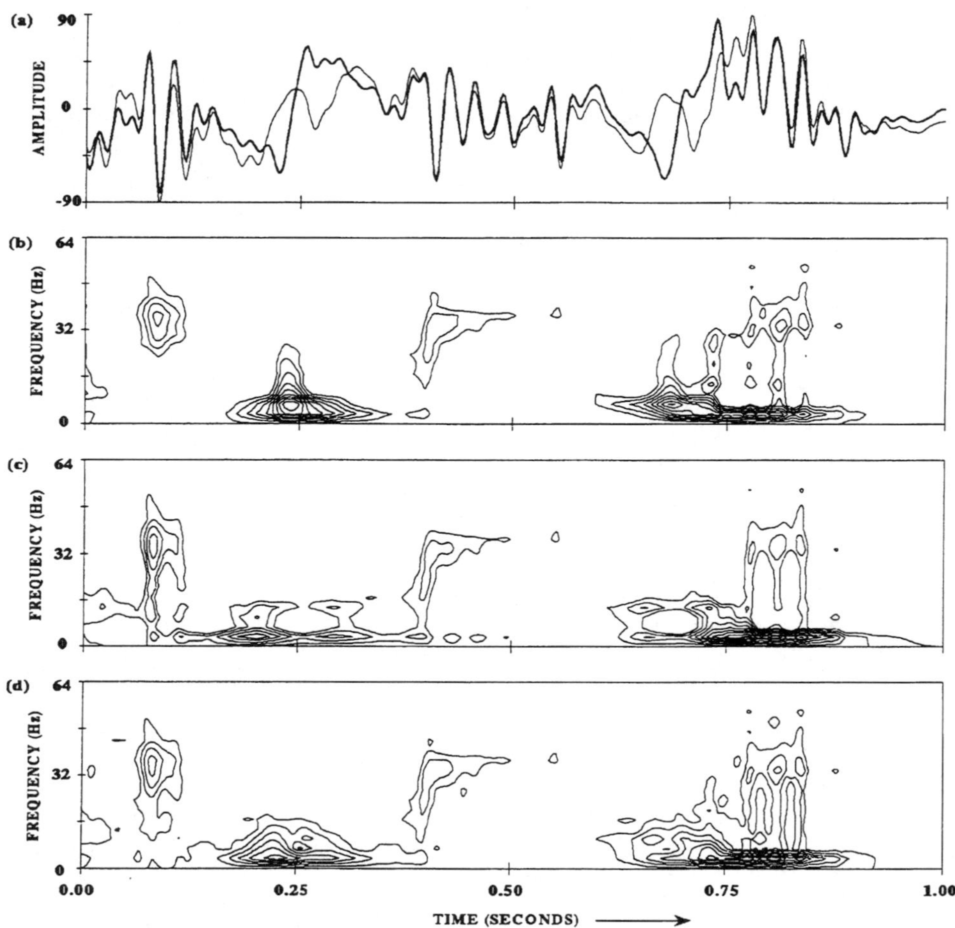

Figure 2-4   The auto-RIDs and cross-RID of the analytic forms of the signals in Fig. 2-3(a) are shown as contour representations in (b)–(d). The signals $f(t)$ and $g(t)$ are replicated in (a) to aid in viewing this figure. The positive values of the auto-RID representations $C_f(t, \omega; \phi_{ED})$ and $C_g(t, \omega; \phi_{ED})$ are shown in (b) and (c), respectively. There is less troublesome interference-term activity in the auto-RID representations than in the corresponding auto-WD representations. The magnitude of the cross-RID representation $C_{fg}(t, \omega; \phi_{ED})$ is shown in (d). The cross-RID of these signals is remarkably clearer than their cross-WD; t-f components common to $f(t)$ and $g(t)$ can be readily identified [29].

spike and slow wave consists of alternating, sharply peaking components (the spike) followed by a slowly varying component. The descriptions in neurology textbooks imply that these two components arise from different physiological mechanisms in different physical locations. An attempt to use the spectrogram for two different window sizes is shown in Figure 2-5. One can see that the short window spectrogram resolves what indeed does look like a "spike" and a "wave" component. The long

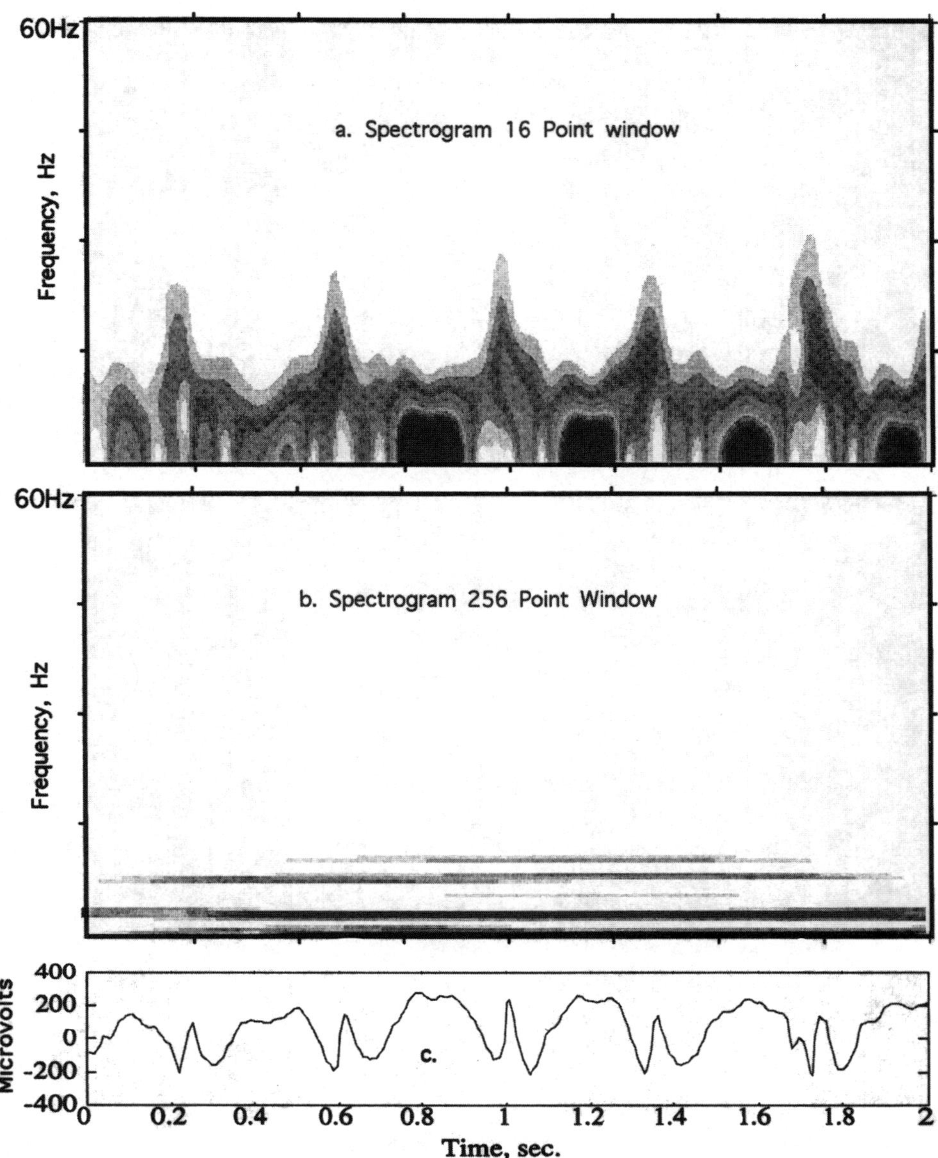

Figure 2-5   (a) Spectrogram result for spike and slow wave using a short (32-point) window; (b) spectrogram result for using a long (256-point) window, (c) the spike-and-slow-wave signal itself.

window spectrogram, on the other hand, simply captures the 3/s fundamental of the spike-wave signal and some of its harmonics. An RID example [28] is shown in Figure 2-6.

Spectrogram results obtained using long windows (256 points) simply show a tonal structure indicative of the harmonic content of this signal. Short window spectrograms (32 points) seem to verify the distinct spike and wave structure. It is

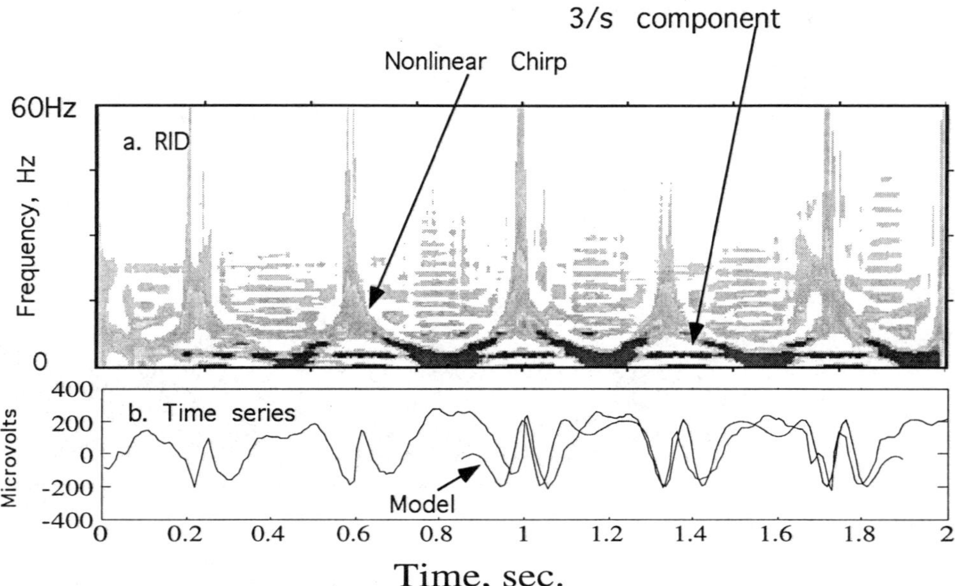

Figure 2-6 (a) RID result (BD realization) for spike and slow wave, showing nonlinear chirps, interference terms, and tonal structure; (b) the signal and the fit for a simple time-varying model are shown [28].

clear that the signal is rapidly and periodically changing frequency, however. The instantaneous frequency dwells at a low frequency for a short time and then rises to a high frequency very rapidly. Then, the instantaneous frequency again falls to a low frequency. This process is repeated almost periodically. Figure 2-6 illustrates several points concerning RID representation. First, the rapidly changing frequency (and consequently phase) produces a sharp, broadband structure, particularly evident and clear at 1 s. Such "peaky" behavior in signals is often indicative of rapid frequency changes in the signal and is not an artifact or interference term as has been suggested by some authors. Next, there are also several horizontally oriented or tonal structures present. These might be considered cross-terms between the prominent nonlinear chirp components. However, the strong 3-Hz component that is noted is strongly related to the character of the signal. These biopotentials are often called "three per second spike and slow waves." So, the 3-Hz component is indicative of this observation and results from the periodic structure of the signal. One might be tempted to insist upon only the nonlinear chirp structure. However, if a Fourier series were performed on this signal, one might expect a fundamental of about 3 Hz, a second harmonic of about 6 Hz, and so on. There is a hint of this in the RID result. This illustrates the fact that cross-terms are important keepers of a portion of the signal energy and sometimes organize themselves to reveal important attributes of the TFD not evident in the auto-terms.

A simple second-order differential equation with periodically time-varying coefficients is able to account for the observed behavior very well. Many of the analysis techniques applied to biological signals assume that the systems being analyzed are time-invariant, or at least time-invariant within the analysis window. Consequently, it is assumed that the eigenfunctions of the system are damped exponentials. There are many powerful techniques available which will characterize linear time-invariant systems very well. The signal components revealed by RID analysis cannot be the responses of linear time-invariant systems. Figure 2-6 shows the RID result for a typical spike and slow wave. One can see that the energy of the signal follows a highly repetitive series of chirps which increase in frequency and subsequently decrease in frequency. A linear time-invariant differential equation cannot exhibit such responses. However, a time-varying differential equation has been found to provide a very good fit to these results. A second-order time-varying differential equation of the following form was investigated.

$$x'' + a(t)x' + b(t)x = 0 \tag{2-2}$$

Mathieu–Hill systems are represented by linear differential equations with time-varying coefficients [30]. This is an example of the single Mathieu–Hill equation. In general, the solution is

$$x(t) = q(t)e^{-\int_0^t r(\theta)d\theta} \tag{2-3}$$

It has been found that a very simple time-varying system model suggested by the RID results can effectively yield results close to the actual result. A response of the form

$$x(t) = A + Be^{-\int_0^t a\cos(\omega_o\theta)d\theta} \tag{2-4}$$

yielding a second-order differential equation of the form

$$x''(t) + a\omega_o\sin(\omega_o t)x'(t) + a\omega_o^2\cos(\omega_o t)x(t) = 0 \tag{2-5}$$

does a nice job of modeling several cycles of the spike-and-slow-wave signal, as shown in Figure 2-6. The model is able to track the signal for some time before drifting off. Time-invariant models based on autoregressive moving average (ARMA) models were not nearly as effective in modeling these phenomena with so few terms.

The rapidly peaking and declining structure revealed in Figure 2-6 seems to be generally typical of much epileptic signal behavior. This pattern is seen in signals that are superficially not much alike when the time series is examined. We have had some success in using this pattern to detect seizure events. Improved TFD analysis has high potential for application in this area.

Recent application of the ED to EEG signals [31] has shown that the TFD approach may be superior to previous power spectra methods. The TFD was more localized in frequency, corresponding better to the patient's awareness under anesthesia than was the case with the power spectrum, which remained unchanged.

2.2.2 The Importance of Invariance in EEG Representation

The RID equipped with a product kernel exhibits the interesting and valuable property of scale invariance. It is well known that the FT of $z(at)$ is $Z(\omega/a)/a$, where a is a scale factor. One would like for t-f distributions to exhibit this property as well. The RID does exhibit this property, but the spectrogram does not [32]. In addition, the RID exhibits the important property of *information invariance* [33]. Members of Cohen's class whose kernel does not vary with time or frequency automatically exhibit time-shift and frequency-shift invariance. The continuous wavelet transform exhibits scale invariance and time-shift invariance. The spectrogram equivalent for the wavelet transform is called the scalogram [34]. Figure 2-7 compares the RID with the spectrogram. In Figure 2-7(a), a spectrogram with a 256-point window is used (von Hamm windows in each case). In Figure 2-7(b), a spectrogram with a 32-point window is used. Finally, in Figure 2-7(c), an RID (binomial distribution) is used. Three signal segments are concatenated. The first is the original EEG segment. The second is the original EEG segment time-scaled by a factor of 2 and amplitude-scaled by $\sqrt{2}$, the latter to preserve energy. The third segment is a frequency-shifted version of the original EEG segment. One can see that both the long window and short window significantly smear the t-f result. The long window spectrogram resolves frequency components well and the short window spectrogram resolves time components. The RID provides a high-resolution representation of both. At least four important components are present. Furthermore, the RID preserves this structure well even when the original signal is scaled or shifted in frequency. The spectrograms preserve the t-f structure under time-shift and frequency-shift, but fail to do so under scaling. Similarly, the wavelet transform and associated scalogram would preserve the structure under time-shift and scaling, but fail under frequency-shift. The scalogram suffers much the same problems as the spectrogram, but the trade-off is now between time and scale rather than time and frequency [34].

2.2.3 Event Related Potentials

Event-related potentials (ERPs) are biopotential events recorded using scalp electrodes. These biopotentials are evoked by stimuli and are thus time-locked to the occurrence of the stimuli. Conventionally, brain biopotentials are characterized in terms of frequencies, (e.g., $\alpha, \beta, \gamma, \delta$), particularly for the ongoing EEG, or in terms of "components" (e.g., $N_{200}, P_{300}, N_{400}$, etc.), which delineate certain positive or negative peaks in the signal. Component analysis is preferred in ERP analyses. It is clear that brain biopotentials are more complex than that. Although such concepts have proved to be useful, TFD analysis reveals a much richer signal complexity. We have applied the RID approach to ERPs evoked in response to briefly flashed words and pictures [35–37]. Previous classification of signals using RIDs [35,36] has depended on picking effective time-frequency bins from the RID image using forward sequential feature selection and various types of conventional pattern recognition schemes, including quadratic classifiers. Significant results have been obtained. These results reveal a difference between the classification success of different classes of words related to the patient's problems under supraliminal and subliminal

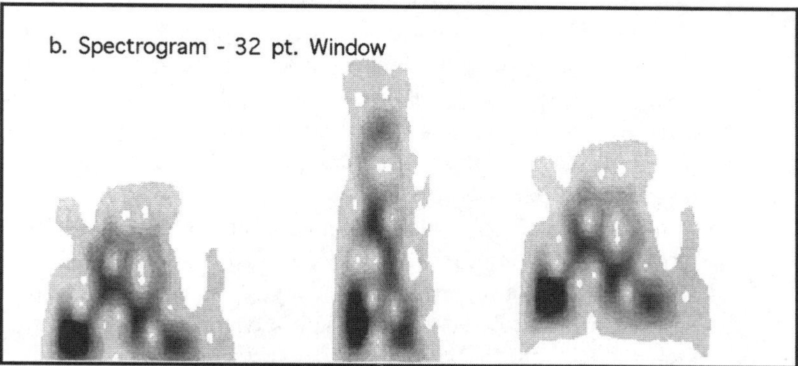

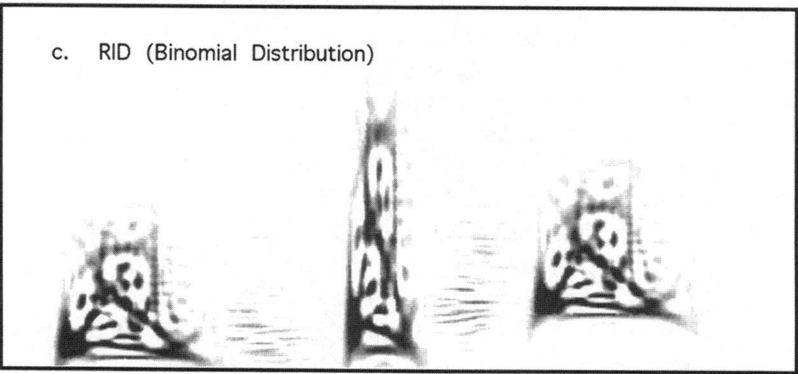

Figure 2-7   The effect of time scaling, time shifting, and frequency scaling on spectrograms and the RID. Original, scaled and time-shifted, and frequency-shifted and time-shifted EEG segments are shown: (a) a long window spectrogram is employed; (b) a short window spectrogram is used; (c) the RID is used. Clearly defined t-f structures are evident [28]. Kernel width is 256 points.

stimulus conditions. However, even though classification improves using the Gabor logons [38] as features, we believe that additional improvements may be made using more advanced adaptive techniques related to neural network concepts [37,39].

The Gabor logon (aka Gabor wavelet) may be mathematically expressed as follows:

$$f(t) = A \cos\left(2\pi\left\{\omega - \left[\omega_o + \frac{\beta}{2}(t - t_o)\right]\right\}(t - t_o) + \phi\right) \exp\left(-\frac{(t - t_o)^2}{2\sigma^2}\right) \tag{2-6}$$

The parameters describing the logon may be used as features in a pattern classification setting.

Although RID analysis was partially motivated by this problem area due to frustration with conventional approaches, it is difficult to discuss the research results in this forum due to the complexities of the experimental designs and the psychological/psychoanalytic hypotheses being explored. Suffice it to say that RID analysis has revealed important TFD components in ERPs which are critically related to patient and subject categorization of different classes of stimuli. These "direct" measures complement and extend the indirect measures of brain states obtained by psychoanalysis and pure psychological experiments which seek to address the responses of patients to stimuli.

The stimuli used in these experiments commonly consist of words and short phrases briefly presented visually for 30 ms (supraliminal) or 1 ms (subliminal). The words are pleasant, unpleasant, overtly related to a patient's phobias or anxieties (conscious) or related to a patient's deep conflicts (unconscious) which have, in fact, induced the problems. These words, carefully chosen by the clinical research team, provide four classes of stimuli. Significant t-f responses correlated with the word classes are obtained for both supraliminal and subliminal stimuli. The consciously troublesome words generally act more strongly supraliminally than the unconsciously troublesome words for supraliminal stimulation, and the unconsciously troublesome words act more strongly for subliminal stimuli. Important and revealing interactions with the patient's problem and personality are also revealed therein.

The general t-f pattern is depicted in Fig. 2-8. Here, the regions of support of Gabor logons as reflected by the contour obtained at half the peak amplitude of the Gabor logon are shown. The real TFD data are difficult to interpret by the uninitiated, due to noise and variability. However, statistically significant results are obtained. Generally, five Gabor logons are sufficient to fit the ERP data very well, as shown in Fig. 2-8. Earlier studies have shown that five features provide the best classification results in a development set/test set examination.

These methods may prove to be useful in providing an objective, direct method of measuring physiological responses as well as psychological responses in psychiatric research. This is an area that, unlike many other areas, has a paucity of objective measures for determining normative values and deviations from them in order to aid diagnosis.

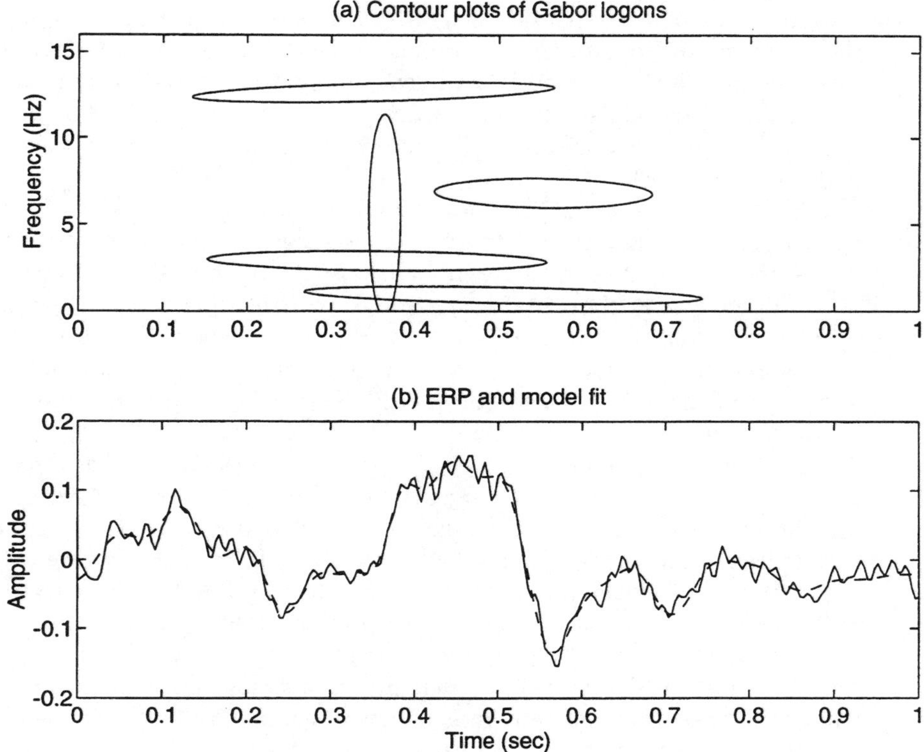

Figure 2-8   Gabor logon fit of an ERP: (a) contour plots of the five Gabor lagons at half of the peak value contour; (b) ERP (solid) and model fit (dashed). This is a supraliminal ERP. There is a 130-ms prestimulus interval in this case [38, 40].

2.2.4 Other Electrophysiological Results

Time-frequency analysis has recently been utilized in several additional areas. The electrophysiology of the heart is an obvious target for time-frequency analysis. The CKD (ZAM) [12] has been used in a comparison with the smoothed pseudo-Wigner–Ville distribution (SPWVD) and the spectrogram in an investigation of ventricular fibrillation detection. It was found that the CKD and the SPWVD had better t-f resolution than the spectrogram, and concluded that t-f approaches have future promise in detecting and classifying arrhythmias.

Muscle biopotential electromyographs (EMGs) have been investigated using modern TFDs. Some preliminary results were present fairly early on by our group [41]. The RID has also been of use in assessing EMG artifacts in EEG signals [42].

EMGs have been investigated using a comprehensive set of studies involving EMGs from uterine contractions [43]. Both real EMG signals and synthetic EMG signals were used. The SPWVD performed very well, in general. The ED also performed well. The CKD was felt to be disappointing in this application, however.

This group carried out another study [44], comparing a number of t-f approaches; the STFT, autoregressive (AR), Baraniuk–Jones signal dependent (SIG-DEP), the SPWVD, the CKD, and the ED. Synthetic uterine EMG signals were used. The AR approach and the SIG-DEP performed best over the six criteria chosen. Since the signals were generated basically using an AR model, it is not too surprising that AR performed very well.

Electrogastrogram (EGG) studies have been performed using the ED [45]. The electrogastrogram is recorded by placing electrodes on the abdominal skin. Slow waves and spikes are commonly observed. Both synthetic EGGs and real EGGs were subjected to analysis. AR and ARMA approaches have been used in studies of EGGs as well. It was concluded that the ED is reliable and accurate in this application, and provides accurate information about the frequency and amplitude variations of the EEG. In addition, the ED is free from the assumption of stationarity.

It is clear that no one approach clearly stands out in the studies reported here. In general, if the frequency of the signal is changing rather slowly and does not suffer discontinuities in frequency, then the SPWVD and the AR/ARMA and even the spectrogram approaches should be expected to perform well. However, if there is rapid change in frequency, then the CKD and the ED would be expected to perform well. The CKD is adept at tracking discontinuous tonelike activity and the ED is superior when the signal changes very rapidly in frequency or has a mixture of impulsive and tonal components. Baraniuk and Jones' data adaptive distribution will generally perform well due to its adaptive nature. However, computation time may be an issue here.

Testing techniques on synthetically generated signals is useful, but cannot be taken completely seriously, since there is a built-in bias toward the character of the true nature of the signal. One is usually trying to discover this character, so signal synthesis may be good or far from the mark. It will tell one whether the technique works well on synthetic signals. Of course, a modeling approach will work very well on its assumed model.

2.3. BIOACOUSTICS EXAMPLES

There are a number of acoustical signals in biology and medicine that are of considerable interest. TFD analysis is particularly effective in these applications.

2.3.1 Temporomandibular Joint Sounds

Many people have experienced the popping or clicking of a joint. Sometimes the temporomandibular joint (TMJ) clicks or pops when the jaw is moved. In many cases, this is a rare and harmless event. However, it may be indicative of a potentially serious condition. Such activity may be associated with a great deal of pain, perhaps even triggering headaches or other distressing symptoms. Even if one does not experience pain, these joint sounds may herald the development of more serious conditions. There is almost always someone with personal knowledge of pain and suffering from these problems, either in terms of themselves, friends, or family members when t-f results are presented on these TMJ sounds. Previous work aimed at analyzing these signals has been spotty and mixed in results. Usually, the clinician simply listens to the sound and tries to report the results in some objective way. However, this method is highly subjective and often is couched in terms such as likening the sound to "a dry stick breaking" or "the crunching of dry snow." Arthritic joints, in particular, often produce creaking or crunching sounds. This is often called *crepitation*.

TFD analysis is very useful in the analysis of these sounds [46,47]. Five distinct types of TMJ sounds have been identified during extensive research studies using RID techniques in the analysis. Two examples will be presented here. A particular TMJ click observed in these studies reveals something about the phenomenon as well as indicating the potential usefulness of RID in these studies. Spectrogram analysis has been tried in the past with little success. Figure 2-9(a) shows the typical spectrogram result for a TMJ click. The sampling rate was 7200 Hz and the window was 128 points long. One can see a large low-frequency component here. A short window spectrogram of 16 points reveals more in Fig. 2-9(b). The signal is appended in both cases in order to show how various aspects of the signal manifest themselves in the TFDs. Both the TFDs and the signal are normalized to allow comparison.

The pad that cushions the action of the condyle as it moves within the joint capsule is thought to be damaged in this example. Instead of moving smoothly over the pad, the condyle may push the damaged pad tissue ahead of it. Finally, the condyle slips over the bunched pad tissue, producing a click. The t-f results seem to reflect these events well. The long window spectrogram shows mostly prolonged low-frequency activity [Fig. 2-9(a)]. The short window spectrogram captures a second packet of energy at a higher frequency [Fig. 2-9(b)]. This higher frequency component most likely results from the release of the pad tissue as the condyle finally jumps over it.

The results of the TMJ click analysis using higher resolution techniques are shown in Fig. 2-10. The WD and the RID show similar results when used to analyze the click. The RID shows a very broad band component that is of very short time duration, followed by a rather prolonged low-frequency component and a shorter duration component at a higher frequency. The initial broadband component is produced by the discontinuity at the start of the click. The signal (plotted along the front margin) rises rapidly to a very sharp peak. The TFD of an impulse, $\delta(t - t_o)$, should produce a knife-edge aligned along the frequency. The Fourier transform of $\delta(t - t_o)$ is a constant with frequency, but exists only at the time of the impulse.* The Fourier transform of a step function also contains an impulse. Any rapid change in the signal will produce such an effect. This effect has been wrongly identified as a cross-term by several authors. It is not an artifact, but well reflects the short duration broadband character of such phenomena as just described. One can also see that the initial broadband component is followed by a few additional components of similar appearance. These may indeed be cross-terms.

---

*It is easy to work out the WD of $\delta(t - t_0)$ to be $\delta(t - t_0)$ for all (t, ω). This means it is precisely located at t_o and forms a "knife edge" along ω, which is infinitely high and infinitesimal in time width.

(a) Spectrogram -- 128 point window

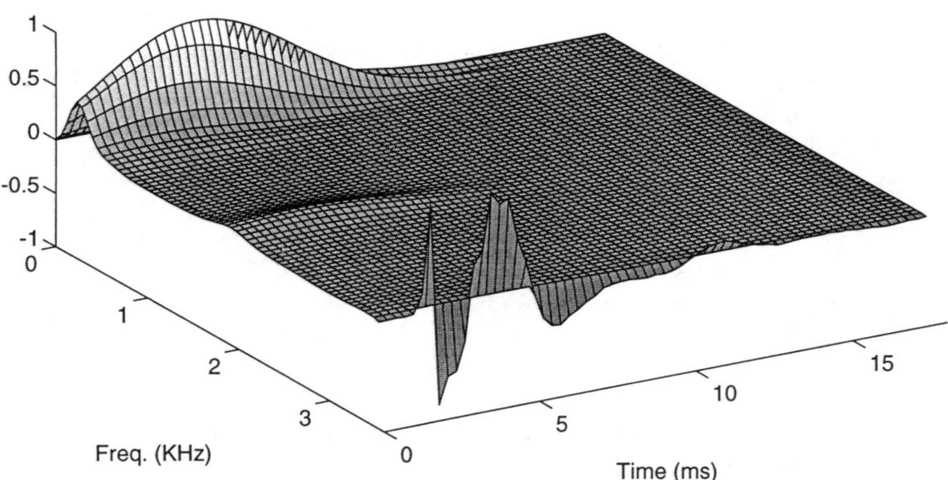

(b) Spectrogram -- 16 point window

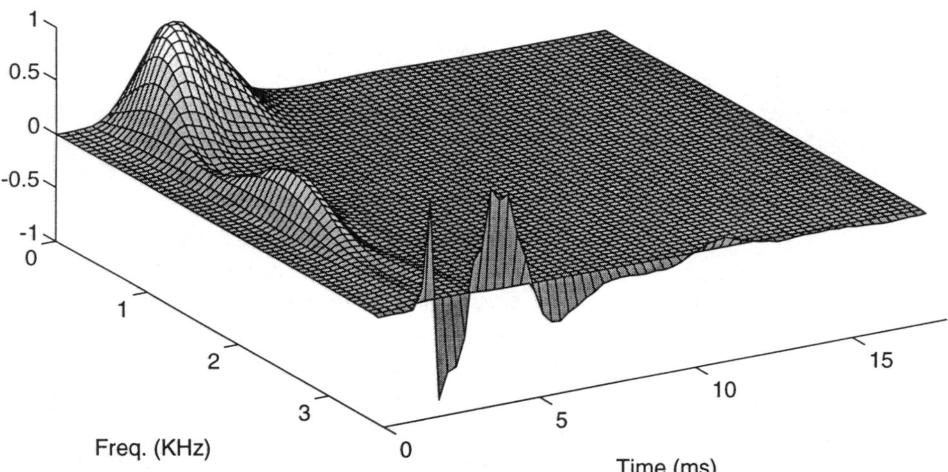

Figure 2-9　TMJ click analysis using (a) a long window and (b) short window spectrogram. The signal is appended at the front of the mesh plots as a reference.

Alternatively, the low- and high-frequency components might actually be composed of a succession of transient events, each producing a broadband component at its onset.

The WD is similar to the RID, but exhibits a lot of cross-term activity that tends to obscure the higher frequency component. The short window spectrogram reveals

(a) Wigner Distribution

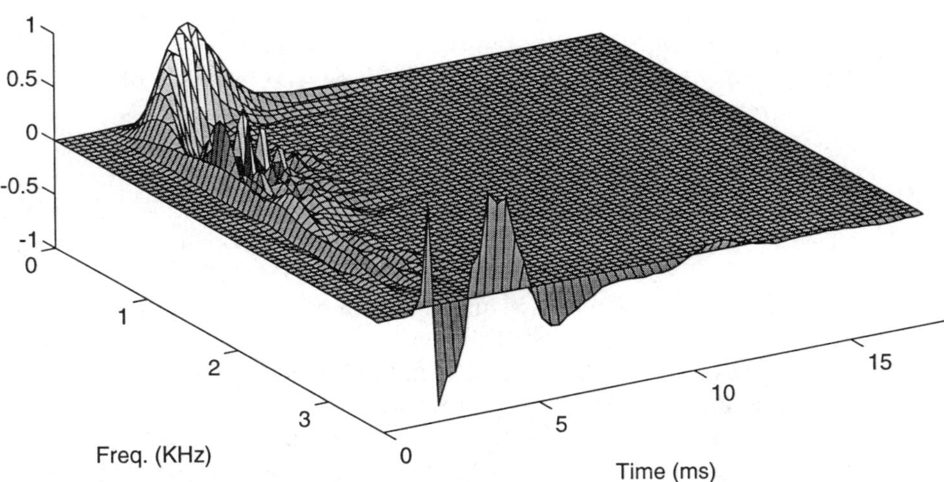

(b) Reduced Interference Distribution

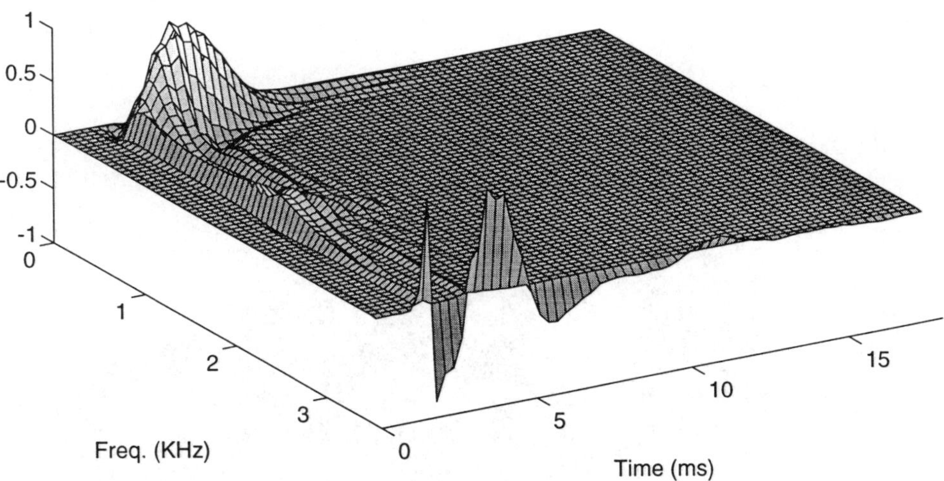

Figure 2-10   TMJ click analysis using (a) WD and (b) RID. The signal is appended
at the front of the mesh plots as a reference.

most of the essential structure of the TFD. However, it smears the frequency of the
low- and high-frequency components considerably. The RID components seem to
track the signal quite well in time, wherein notable changes in the signal are reflected
by accompanying changes in RID components. In some past studies, only the large
component in the spectrogram at low frequencies was considered. In fact, low

sampling rates and low-pass filtering probably served to eliminate the higher frequency components.*

Crepitation produces a very different TFD result. Figure 2-11 shows RID results for an arthritic joint. There are many peaks in this result. This is due, no doubt, to the roughness of the joint surfaces. At least five types of RID-based classes of joint responses can be distinguished at this writing [47]. Types 1–3 are characterized by a single peak in the t-f plane. The peak frequency of Type 1 is greater than 600 Hz. Type 2 has a peak within a range of 600–1200 Hz. Type 3 has a peak frequency greater than 1200 Hz. Types 4 and 5 exhibit multiple energy peaks characteristic of crepitation. The energy for Type 4 is predominantly below 600 Hz and the energy for Type 5 is predominantly above 600 Hz. Obviously, these type designations are a working definition and serve as a point of reference that may change as the research progresses. Types 1–3 may be easily characterized by a Gabor logon.

This has proved to be a useful form to fit many of the t-f components observed in applications of RID analysis. With the confines of the TFD representation in mind, this Gabor logon can be shifted to any point in time (t_o), any frequency (ω_o), given any desired spread (σ), and tilted in time-frequency to provide a chirp effect (β). The phase shift can be provided (ϕ) and the amplitude set (A). One logon is sufficient to fit Types 1–3 joint clicks. Generally, the β term is not needed with joint clicks and is set to zero.

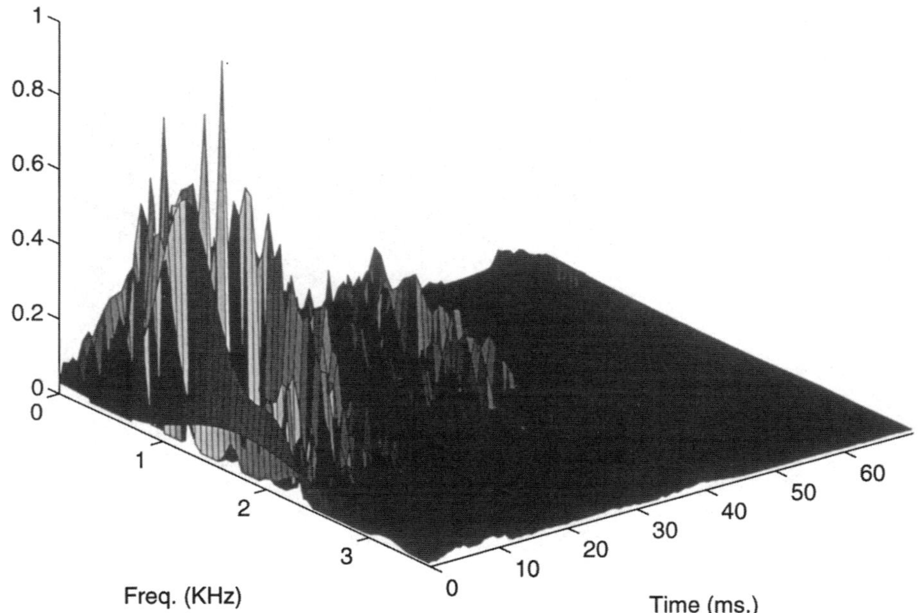

Figure 2-11   RID result for crepitation.

*Components up to several kilohertz can be detected with proper measurement, filtering, and sampling.

The Gabor logon does not provide an appropriate fit for Types 4 and 5. There, an information measure based on Rényi's generalized information is quite useful [3]. Rényi information is defined as

$$R_\alpha = \frac{1}{1-\alpha} \log_2 \left(\int \int \mathrm{RID}(t, \omega)^\alpha dtd\omega \right) \tag{2-7}$$

with the RID appropriately normalized such that

$$\int \int \mathrm{RID}(t, \omega)dtd\omega = 1 \tag{2-8}$$

Summations replace integrations for discrete application. The α parameter sets the order of the Rényi information. Shannon information is obtained for $\alpha = 1$. Higher orders of Rényi information are useful in evaluating TFDs. Shannon information, due to its form,* is inappropriate for this purpose, since many interesting TFDs exhibit negative values. Since the log of the distribution is taken before integration in Shannon information, the result is undefined. Rényi information provides a way of sidestepping this problem, since the integration is carried out before the log is taken. One would like an information measure which yields a value of 0 bit when one signal component is present, 1 when two are present, and so on for any n components wherein a value of $\log_2(n)$ bits is obtained. The Gabor logon is the signal with the minimum uncertainty or information. Accordingly, it also provides the smallest time-bandwidth product of $\frac{1}{2}$. If one sets this to be the reference point of 0 bits of information (one signal component), then all other signals can be referenced to this. The *information gain*, or amount of information above the base value of the Gabor logon, correctly reflects the information in a signal from this viewpoint. The Rényi measure provides a very handy means of separating Types 1–3 from Types 4 and 5. The Rényi measure is close to 0 for Types 1–3 and considerably above 1 for crepitation (Types 4 and 5). A value of 1 is taken to indicate that there may be two Type 1–3 clicks present.

Using the concepts outlined herein, it has been possible to classify TMJ sounds very well. In conjunction with a neural net classifier, a correct classification of 85-90 percent is typically achieved at this writing. These sounds are often very difficult to distinguish by ear. It is clear that these techniques may lead to an inexpensive diagnostic methodology which can be easily applied in a dentist's office. Other diagnostic methods presently available are much more expensive and time consuming. A simple screening test for incipient TMJ problems is much needed.

2.3.2 Animal Sounds

Animal sounds have long been of interest in biosignal analysis. Human speech is certainly an example. Speech processing has reached a certain maturity and the spectrogram is a valued tool in that application. The analysis of other types of sounds such as marine mammal sounds, bat sounds, bird sounds, and sounds from other animals is of considerable interest to biologists. These researchers have

---

*Shannon information would be $-\int \int \mathrm{RID}\,(t, \omega) \log_2[\mathrm{RID}\,(t, \omega)]dtd\omega$ for $\alpha = 1$.

sometimes recognized the limitations of conventional tools such as the spectrogram. William Watkins [3] recognized these problems a number of years ago. His article on the use and limitations of the spectrogram in bioacoustics must be regarded as a classic. He provided analyses of some very cleverly chosen synthesized and real sounds to prove his points. His years of experience with marine mammals have taught him to be skeptical about analysis tools such as the spectrogram which do not reveal what comes naturally to the trained ear. It is in this context that some of our joint research with William Watkins and his group at Woods Hole Oceanographic Institution is presented in this article. Marine mammal sounds are quite complex. These animals produce a variety of clicks, squeaks, moans and chirps. Space does not permit a full presentation of analyses which have been performed on these sounds. However, an example will be given which illustrates the complexity of one type of sound and also serves as a teaching tool concerning t-f analysis in general.

Dolphins, in this example the common bottlenose dolphin, produce a variety of sounds, including whistles and clicks. Figure 2-12 shows a typical click example. Both spectrogram and RID (binomial) analyses are provided. In order to simultaneously explore some other important ideas, some modified forms of the click are also included. One modification involves shifting the click in time, and compressing it in time by a 2:1 ratio as well. The amplitude is also multiplied by $\sqrt{2}$, in this case to preserve the energy of the signal. Another modification is to shift the original click in time and also shift it in frequency to twice the original center frequency. These modified clicks are shown in Fig. 2-12(b). Figure 2-12(a) shows the spectrogram obtained using a long window and Figure 2-12(c) shows the RID result using the binomial form.

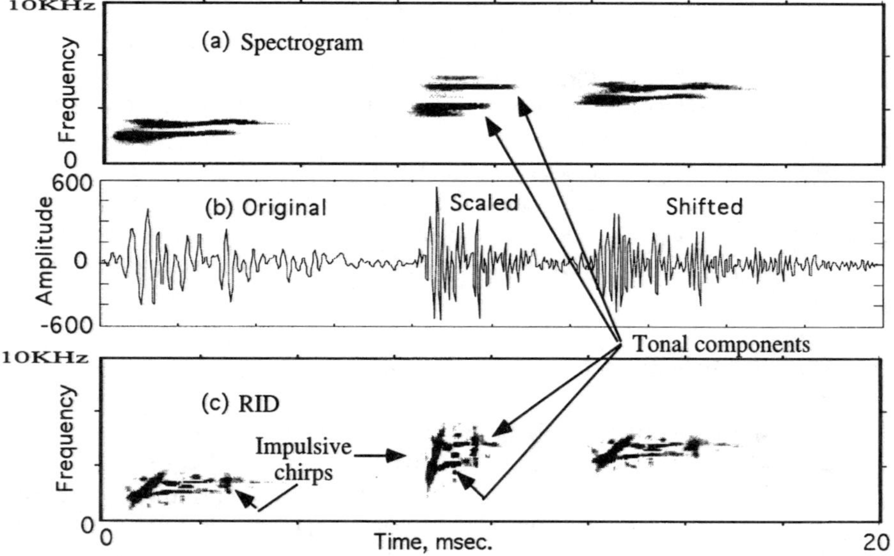

Figure 2-12   Dolphin clicks: (a) spectrogram result; (b) original, scaled, and frequency-shifted versions of a click; and (c) RID result.

One can see that the original, scaled, and frequency-shifted versions of the clicks look somewhat alike, but one might not recognize them as being the same basic signal entity, especially buried in other signals and noise. The spectrogram result extracts the tonal structure in the clicks. The RID extracts both the tonal and impulsive structures as well. The RID reveals two tones and two impulsive chirps as the main components of this signal. These findings have been essentially confirmed in another study by Wood and Barry [48], wherein a Radon-based modification to TFD analysis was employed. The frequency-shifted click produces very similar results when compared with the original in both the spectrogram and the RID example. The scaled version of the click produces a spectrogram that is somewhat different in appearance. The RID result is appropriately scaled and the spectrogram result is not. This is because the RID is scale-invariant as well as being time-shift and frequency-shift invariant, as has been previously mentioned.

One can fit these click structures very well with about five of the Gabor logons previously discussed in this chapter [38]. The RID result from the synthesized click is quite similar to the original, giving credence to the identification of the basic components of the signal.

Returning to biological considerations, one might ask what is the possible purpose of the click structure? The two impulsive components might form a good reference for unequivocal determination of range and the two tonal components could form a good reference for Doppler. This would be helpful in tracking objects using such clicks. Individual marine mammals also emit sounds that may be distinctive enough in structure so as to allow identification of that animal. This would be quite useful in tracking free-ranging animals and identifying them for census purposes. More research will be required to verify the effectiveness of this idea.

One can often identify other unique structures in marine mammal sounds using RID. Dolphins also produce rather prolonged signature "whistles" that vary in frequency over time. Individuals seem to have unique whistles. There is a great deal of fine structure associated with these whistles which is revealed by RID [49], but is lost by the spectrogram. There appear to be subtle FM and AM modulations of the whistle. In addition, it is possible to simultaneously explore both the tonal structure of these whistles and clicks which are also emitted at the same time [49,50].

Some work has been done on other types of animal sounds as well. Baraniuk and Jones [2] have applied their adaptive distribution to bat sounds with excellent results. Animal sound analysis is a wide-open arena for the application of TFD approaches.

2.3.3 Heart and Muscle Sounds

Heart sounds have long been utilized in diagnosis, using the simple stethoscope. Cole and Barry first used TFDs to study muscle sounds produced during contraction [51]. Wood and Barry and their colleagues [48, 52–54] have utilized time-frequency techniques extensively in the study of heart sounds. The binomial form of the RID has been found to be quite useful in these studies. Heart valve action produces sharp impulses and contracting heart muscle produces chirps. These features can be observed in t-f results.

Guo, Durand, and Lee [55,56] have devised a new kernel, the Bessel kernel, that generally falls into the RID class. It has some advantages in the study of heart sounds, particularly valve sounds, and ultrasonic signals related to blood flow measurements. They have found that a considerable improvement in the ability to identify important factors concerning this issue is possible using the newly emerging t-f techniques. Their distribution, using the Bessel kernel, performed better than other RID-type distributions, particularly if the time support constraint was relaxed. The Bessel kernel requires more computation, however, to gain this advantage. Apart from the value of their work in extending the theory and application to another area, their work verifies some important impressions concerning TFDs. First, in order to gain the benefits of improved TFDs, one must often increase the computational burden. Second, relaxing some of the constraints on the distribution may improve the results in a specific application. Relaxing the time support constraint means that the kernel is integrating (or summing) over a larger range of signal values. This tends to suppress the noise. This is something we have also observed with the original exponential kernel. The time and frequency support constraints can be imposed, but sometimes at the expense of good performance in a noisy environment.

2.4. CONCLUSIONS

The area of time-frequency analysis has reached a certain maturity at this point. It is gratifying to find that interest in these approaches is high. There is also a willingness to examine new ideas in the area and try them out in specific applications, in contrast to a few years past, when these ideas were met with doubt and even hostility. The proof of any concept or method is in its use. If there is a sustained interest and the techniques have merit, then they will be fully exercised and both their merits and their failings will be reported. This is all that one can ask. There are certainly competing methods around, some old, some new. This includes wavelet transform approaches, parametric modeling (AR, ARMA), and high-resolution techniques, among others. All of these methods have their merits and their failings. One needs to know something of this in order to choose and apply a technique carefully. One needs to understand the mathematics and theory (to some degree) and the art of applying such techniques. It is to be hoped that future research will draw some of the theories together and provide a more unified view of time-frequency analysis in all its manifestations and variations. There are sure to be some surprises still beyond the horizon.

ACKNOWLEDGMENTS

Many have contributed to the ideas and results in this paper, including colleagues who have been involved in founding the area, graduate students, and colleagues in biology, dentistry, and medicine. All of their contributions are gratefully acknowledged and appreciated.

REFERENCES

[1] L. Cohen, "Time-frequency distributions—A review," *Proc. IEEE,* vol. 77, pp. 941–981, 1989.

[2] L. Cohen, *Time-Frequency Signal Analysis.* Englewood Cliffs, NJ: Prentice Hall, 1995.

[3] W. A. Watkins, "The harmonic interval: Fact or artefact in spectral analysis of pulse trains," *Marine Bio-acoust.,* vol. 2, pp. 15–43, 1966.

[4] E. Wigner, " On the quantum correction for thermodynamic equilibrium," *Phys. Rev.,* vol. 40, pp. 749–759, 1932.

[5] J. Ville, "Theorie et applications de la notion de signal analytique," *Cables et Transmissions,* vol. 20A, pp. 61–74, 1948.

[6] T. C. M. Claasen and W. F. G. Mecklenbräuker, "The Wigner distribution—A tool for time-frequency signal analysis—Part I: Continuous-time signals," *Philips J. Res.,* vol. 35, pp. 217–250, 1980.

[7] T. C. M. Claasen and W. F. G. Mecklenbräuker, "The Wigner distribution—A tool for time-frequency signal analysis—Part II: Discrete time signals," *Philips J. Res.,* vol. 35, pp. 276–300, 1980.

[8] T. C. M. Claasen and W. F. G. Mecklenbräuker, "The Wigner distribution—A tool for time-frequency signal analysis—Part III: Relations with other time-frequency signal transformations," *Philips J. Res.,* vol. 35, pp. 372–389, 1980.

[9] F. Hlawatsch, "Interference terms in the Wigner distribution." In *Digital Signal Processing—84.* V. Cappellini and A. Constantinides (eds.), pp. 363–367, Amsterdam: Elsevier, 1984.

[10] P. Flandrin, "Some features of time-frequency representations of multicomponent signals," *IEEE Int. Conf. Acoust., Speech, Signal Proc.,* pp. 4.1–4.4, 1984.

[11] W. J. Williams and J. Jeong, "Reduced interference time-frequency distributions." In *Time-Frequency Signal Analysis.* B. Boashash (ed)., New York: Longman and Cheshire, Wiley, 1992.

[12] Y. Zhao, L. Atlas, and R. Marks, "The use of cone-shaped kernels for generalized time-frequency representations of nonstationary signals," *IEEE Trans. Acoust., Speech, Signal Proc.,* vol. 38, pp. 1084–1091, 1990.

[13] R. G. Baraniuk and D. J. Jones, "Signal-dependent time-frequency representation: Optimal kernel design," *IEEE Trans. Signal Proc.,* vol. 41, pp. 1589–1602, 1993.

[14] W. J. Williams, "Reduced interference distributions: Biological applications and interpretations," *Proc. IEEE,* vol. 84, pp. 1264–1280, 1996.

[15] L. Cohen, "Generalized phase-space distribution functions," *J. Math. Phys.,* vol. 7, pp. 781-786, 1966.

[16] F. Hlawatsch and G. F. Boudreax-Bartels, "Linear and quadratic time-frequency signal representations," *IEEE Signal Proc. Mag.,* vol. 9, pp. 21–67, 1992.

[17] P. Kellaway and I. Petersen (eds.), *Quantitative Analytic Studies in Epilepsy*, New York: Raven Press, 1976.

[18] J. S. Barlow, "Computerized clinical electroencephalography in perspective," *IEEE Trans. Biomed. Eng.*, vol. 26, pp. 377–391, 1979.

[19] A. S. Gevins, "Analysis of the electromagnetic signals of the human brain: Milestones, obstacles and goals," *IEEE Trans. Biomed. Eng.*, vol. 33, pp. 833–850, 1984.

[20] C. H. Page, "Instantaneous power spectra," *J. Appl. Phys.*, vol. 23, pp. 103–106, 1952.

[21] N. Kawabata, "A nonstationary analysis of the electroencephalogram," *IEEE Trans. Biomed. Eng.*, vol. 20, pp. 444–452, 1973.

[22] N. Kawabata, "Dynamics of the electroencephalogram during performance of a mental task," *Kybernetik*, vol. 15, pp. 237–242, 1974.

[23] J. P. C. de Weerd and J. I. Kap, "Spectro-temporal representations and time-varying spectra of evoked potentials: A methodological investigation," *Biol. Cybern.*, vol. 41, pp. 101–117, 1981.

[24] A. W. Rihaczek, "Signal energy distribution in time and frequency," *IEEE Trans. Informat. Theory*, vol. 14, pp. 369–374, 1968.

[25] N. Morgan and A. S. Gevins, "Wigner distributions of human event-related brain potentials," *IEEE Trans. Biomed. Eng.*, vol. 33, pp. 66–70, 1986.

[26] M. S. Scher, M. Sun, G. M. Hatzilabrou, N. L. Greenberg, G. Cebulka, D. Krieger, R. D. Guthrie, and R. J. Sclabbasi, "Computer analyses of EEG-sleep in the neonate: Methodological considerations," *J. Clinical Neurophys.*, vol. 7, pp. 417–441, 1989.

[27] H. P. Zaveri, W. J. Williams, L. D. Iasemidis, and J. C. Sackellares, "Time-frequency representations of electrocorticograms in temporal lobe epilepsy," *IEEE Trans. Biomed. Eng.*, vol. 39, pp. 502–509, 1992.

[28] W. J. Williams, H. P. Zaveri, and J. C. Sackellares, "Time-frequency analysis of electrophysiology signals in epilepsy," *IEEE Eng. Med. Biol. Mag.*, vol. 14, pp. 133–143, March–April, 1995.

[29] H. P. Zaveri, W. J. Williams, and J. C. Sackellares, "Cross time-frequency representation of electrocorticograms in temporal lobe epilepsy," *Proc. IEEE Int. Conf. Eng. Med. Biol.* vol. 13, pt. 1, pp. 437–438, 1991.

[30] R. E. Kronauer, "Oscillations," in *Handbook of Applied Mathematics: Selected Results and Methods*, 2nd Ed. Carl E. Pearson (ed.) New York: Van Nostrand Reinhold, 1990.

[31] A. Nayak, R. J. Roy, and A. Sharma, "Time-frequency spectral representation of the EEG as an aid in the detection of depth of anesthesia," *Ann. Biomed. Eng.*, vol. 22, pp. 501–513, 1994.

[32] W. J. Williams, "Time-frequency analysis of biological signals," *IEEE Int. Conf. Acoust. Speech, Signal Proc.*, vol. I, pp. 83-86, 1993.

[33] W. J. Williams, M. L. Brown, and A. O. Hero III, "Uncertainty, information and time-frequency distributions," *Advanced Signal Processing Algorithms, Architectures and Implementations II: SPIE Proc.*, vol. 1566, pp. 144–156, 1991.

[34] O. Rioul and P. Flandrin, "Time-scale energy distributions: A general class extending wavelet transforms," *IEEE Trans. Signal. Proc.*, vol. 40, pp. 1746–1757, 1992.

[35] H. I. Choi, W. J. Williams, and H. P. Zaveri, "Analysis of event related potentials: time-frequency energy distributions," *Proc. Rocky Mountain Bioeng. Symp.*, pp. 251–258, 1987.

[36] H. Shevrin, W. J. Williams, R. E. Marshall, R. K. Hertel, J. A. Bond, and L. A. Brakel, "Event-related potential indicators of the dynamic unconscious," *Consciousness and Cognition*, vol. 1, pp. 340–366, 1992.

[37] W. J. Williams, M. L. Brown, H. P. Zaveri, and H. Shevrin, "Feature extraction from time-frequency distributions," In *Intelligent Engineering Systems Through Artificial Neural Networks*, vol. 4. pp. 823–829, C. H. Dagli et al, (eds). New York: ASME Press, 1994.

[38] M. L. Brown, W. J. Williams and A. O. Hero III,"Non-orthogonal Gabor representations of biological signals," *Proc. Int. Conf. Acoustics, Speech, Signal Proc.*, vol. 4, pp. 305–308, 1994.

[39] W. J. Williams and T. H. Sang, "Adaptive RID kernels which minimize time-frequency uncertainty," *IEEE Int. Symp. Time-Frequency and Time-Scale Analysis*, pp. 96–99, 1994.

[40] M. L. Brown, W. J. Williams, and S.-E. Widmalm, "Automatic classification of temporomandibular joint sounds,"*Intelligent Engineering Systems Through Artificial Neural Networks*, vol. 4, pp. 725–730, C.H. Dagli et al (eds.). New York: ASME Press, 1994

[41] W. J. Williams and J. Jeong, "New time-frequency distributions : theory and applications," *Proc. IEEE Int. Symp. Circuits Sys.*, vol. 2, pp. 1243–1247, 1989.

[42] C. S. Zheng, W. J. Williams, and J. C. Sackellares, "RID time-frequency analysis of median filters and lowpass filters in reducing EMG artifacts in EEG recording," *Proc. IEEE Int. Conf. Eng. Med. Biol.*, vol. 15, pt. 1, pp. 350–351, 1993.

[43] J. Duchêne, D. Devedeux, S. Mansour and C. Marque," Analyzing uterine EMG: tracking instantaneous burst frequency," *IEEE Eng. Med. Biol. Mag.*, vol. 14, no. 2, pp. 125–132, 1995.

[44] D. Devedeux and J. Duchêne, "Comparison of various time/frequency distributions (classical and signal-dependent) applied to synthetic uterine EMG signals," *IEEE Int. Symp. Time-Frequency and Time-Scale Analysis*, pp. 572–575, 1994.

[45] Z.-Y. Lin and J. De Z. Chen, "Time-frequency representation of the electrogastrogram—application of the exponential distribution," *IEEE Trans. Eng. Med. Biol.*, vol. 41, pp. 267–275, 1994.

[46] S. E. Widmalm, W. J. Williams, and C. S. Zheng, "Time frequency distributions of TMJ sounds," *J. Oral Rehabilitation,* vol. 18, pp. 403–412, 1991.

[47] S. E. Widmalm, W. J. Williams, R. L. Christiansen, S. M. Gunn, and D. K. Park, "Classification of temporomandibular joint sounds based on their reduced interference distribution," *J. Oral Rehabilitation,* vol. 23, pp. 35–43, 1996.

[48] J. C. Wood and D. T. Barry, "Radon transformation of time-frequency distributions for analysis of multicomponent signals," *IEEE Trans. Signal Proc.,* vol. 42, pp. 3166–3177, 1994.

[49] P. L. Tyack, W. J. Williams, and G. S. Cunningham, "Time-frequency fine structure of dolphin whistles," *2nd Int. Conf. Time-frequency and Wavelet Techniques,* pp. 18–20, Victoria, B. C., Oct. 1992.

[50] J. Jeong, G. S. Cunningham, and W. J. Williams, "The discrete-time phase derivative as a definition of discrete instantaneous frequency and its relation to discrete time-frequency distributions," *IEEE Trans. Signal Proc.,* vol. 43, pp. 341–344, 1995.

[51] N. M. Cole and D. T. Barry, "Muscle sounds occur at the resonant frequency of skeletal muscle," *IEEE Trans. Biomed. Eng.,* vol. 37, pp. 525–531, 1990.

[52] J. C. Wood and D. T. Barry, "Time-frequency analysis of the first heart sound," *IEEE Eng. Med. Biol. Mag.,* vol. 14, pp. 133–143, March–April, 1995; vol. 14, pp. 144–151, 1995.

[53] J. C. Wood, A. J. Buda, and D. T. Barry,"Time-frequency transforms: A new approach to first heart sound frequency analysis," *IEEE Trans. Eng. Med. Biol.,* vol. 38, pp. 728–739, 1992.

[54] J. C. Wood, M. P. Festen, M. J. Lim, A. J. Buda, and D. T. Barry, "Differential effects of myocardial ischemia on regional first heart sound frequency," *J. Appl. Physiology,* vol. 36, pp. 291–302, 1994.

[55] Z. Guo, L.-G. Durand, and H. C. Lee, "Comparison of time-frequency distribution techniques for analysis of simulated Doppler ultrasound signals of the femoral artery," *IEEE Trans. Biomed. Eng.,* vol. 41, pp. 332–342, 1994.

[56] Z. Guo, L.-G. Durand, and H. C. Lee, "The time-frequency distributions of nonstationary signals based on a Bessel kernel," *IEEE Trans. Signal Proc.,* vol. 42, pp. 1700–1707, 1994.

Chapter 3

The Application of Advanced Time-Frequency Analysis Techniques to Doppler Ultrasound

S. Lawrence Marple, Jr., Tom Brotherton, Doug Jones

3.1. INTRODUCTION

Doppler ultrasound echoes from cardiac structures are rich in detail and highly nonstationary. The goal of time-frequency and time-scale analysis is to extract features from these echo signals for high-confidence visual and machine classification of a patient's condition. There is a wide variety of techniques available to researchers. Presented in this chapter are side-by-side comparisons of a selection of time-frequency representations (TFRs) applied to Doppler ultrasound data. The techniques illustrated include the short-time Fourier transform, the Wigner–Ville transform using the Choi–Williams kernel, the adaptive optimal kernel TFR, the adaptive cone kernel, a wavelet transform, and a model-based TFR. A discussion of how parameters are selected in order to achieve the "best" representation that can be achieved for each of these techniques is also given. The trade-offs among the various analysis representations are also discussed.

The classical method for analyzing nonstationary signals has been the short-time Fourier transform (STFT). As described in Chapter 1, the STFT has a time resolution fundamentally determined by the duration T of the analysis window selected for the STFT. The STFT has a frequency resolution fundamentally determined by the bandwidth F of the analysis window, which is nominally lower-bounded by the reciprocal of the window duration, $F \geq 1/T$. Thus the time-bandwidth product, $T \times F \geq 1$, is a measure of the joint time-frequency resolution of the STFT technique.

An example STFT TFR is shown in Fig. 3-1. The figure presents a complex in-phase/quadrature Doppler ultrasound data record taken from an Acuson 128XP ultrasound scanner system. The data are sampled at 6250 samples per second. Figure 3-1 is a 5-s sample of these data. The figure shows the STFT TFR (the large portion in the middle of the figure) as well as both the in-phase and quadrature components of the time series (the plots on the left and right of the STFT TFR). The STFT TFR was produced by taking 128-sample epochs spaced 64 samples apart and using a Hamming window. The figure shows slightly over six cardiac cycles. At the bottom of the figure is the long-term Fourier transform for comparison. As seen in Fig. 3-1, the data are highly nonstationary and lack significant coherent structure.

Some nonstationary signals have proven difficult to analyze with an STFT due to the STFT resolution limitation. Alternative TFRs have been introduced in an attempt to improve the time-bandwidth product performance bound, many of which are summarized by Hlawatsch and Boudreaux-Bartels [1] and Boashash [2]. Prominent among these alternative TFRs is the Wigner transform, sometimes called the Wigner–Ville function (WVF), described in Chapter 1. The WVF can be shown by both analytical and graphical means [3] to achieve *at most* a factor of 2 improvement in the time-bandwidth product over the STFT, i.e., the WVF can achieve $T \times F \geq \frac{1}{2}$. However, this improvement comes at a price, that price being the cross-term artifacts introduced by the WVF that make distinguishing actual signals from artifacts difficult in the presence of multiple frequency components and/or noise. These cross-term artifacts are often greater in magnitude than weaker signal components and can overlap or mask actual signal components. Although linear TFRs, such as the STFT, have somewhat less time-frequency resolution than quadratic TFRs like the WVF, their linear operations on the signal lead to more understandable effects in the presence of multiple signal components and noise. In addition, many signal processing chips have architectures optimized for performing fast Fourier transform (FFT) operations.

Alternative model-based techniques have been developed by Marple [4] and used for high-resolution spectral estimates with good success. These techniques give estimates by "extending" the data or the data's autocorrelation function by use of a model. The STFT technique implicitly assumes that the data outside of the window of interest are zero—which is not a very realistic assumption. The model is fit to the data to approximate the time series under consideration. The model fit presumes that the data outside the segment being modeled are nonzero. The spectrum of the data is then found based on the model rather than the data used to estimate the model.

Nonstationary signals may be categorized into two types: (1) momentary transient and (2) persistent. The *momentary transient* signal has a brief, finite duration which is characterized by a time-frequency gram with finite-duration time axis and finite-bandwidth frequency axis. Almost the entire technical literature on TFRs deals with this type of signal, typically no longer than a few hundred data samples at most. The *persistent* nonstationary signal, on the other hand, has continuous time-varying behavior over an indefinite time period, which is best characterized by a waterfall presentation of the time-frequency gram that scrolls the continuously updated TFR estimate through a finite-duration time window on the display. Most nonstationary

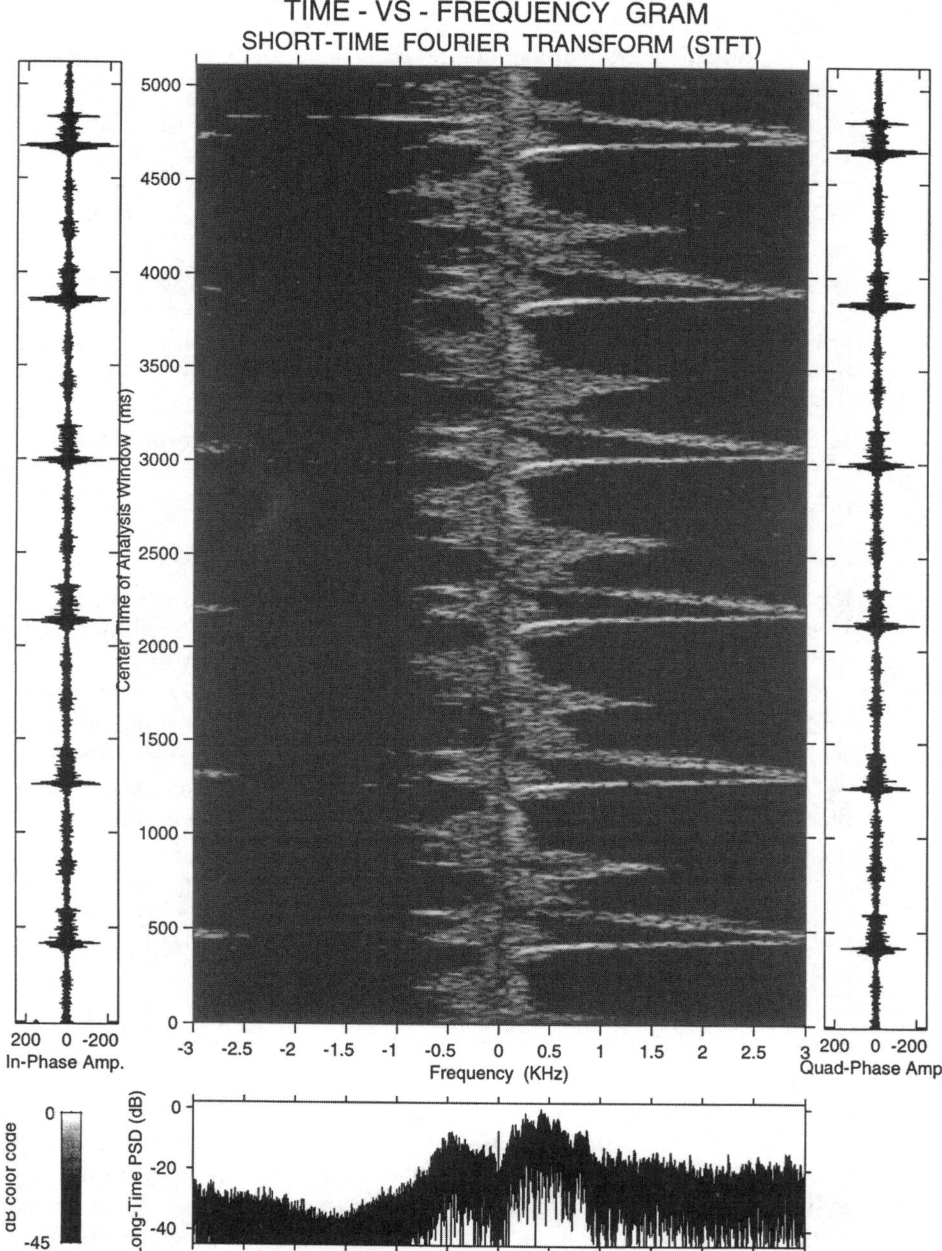

Figure 3-1   A traditional STFT TFR of 5 s of Doppler ultrasound data. The data were sampled at 6250 samples/s. The number of samples input to the FFT was 128. A Hamming window was applied and the data were zero padded to 512 points prior to taking a 512-point FFT. For each scan in the figure, 64 samples were skipped between FFTs.

signals of practical interest are the persistent type. It is not always obvious how to extend to the persistent case those methods created to handle the momentary transient case. The TFR techniques presented here are able to handle the persistent nonstationary signal case.

Figure 3-2 illustrates the steps needed to generate a generic time-frequency gram for analysis of persistent nonstationary signals. These steps involve five fundamental components: (1) computation of the base TFR from the available data; (2) performing any optional post-TFR operations to enhance visual presentation; (3) taking the squared-magnitude of complex TFR values, if necessary, to convert to energy values that the gram will display; (4) assigning grayscale or colorscale values to a subset of the available dynamic range of values; and (5) displaying the time-vs.-frequency gram on a monitor or outputting to a grayscale or color high-resolution printer.

For the Doppler ultrasound data considered here, we are interested in discovering more detail within a single cardiac cycle. A roughly 380-ms interval was analyzed to discover fine grain detail, if any, within the signal's nonstationary behavior.

3.1.1 Adaptive Quadratic Time-Frequency Representations

The Wigner transform, or Wigner–Ville function (WVF), is the best known of the quadratic TFRs. A detailed discussion of the WVF is contained in Chapter 1. Described here are improvements to the WVF TFR developed by Jones, Baraniuk, and Czerwinski [5–9]. These techniques apply a data-adaptive kernel for deriving quadratic TFRs. The WVF is not obtained directly or linearly from the signal under analysis, but as the transform of either a correlated function of the signal, $R(t, \mathbf{t}) = x(t + \mathbf{t}/2)x^*(t - \mathbf{t}/2)$, or a correlated function of the transform of the signal, $S(f, \mathbf{f}) = X(\mathbf{f} + f/2)X^*(\mathbf{f} - f/2)$. Here $x(t)$ is the signal of interest, $x^*(t)$ is the complex conjugate, and $X(\mathbf{f})$ is the Fourier transform. It is the correlation that provides the WVF with its factor of 2 improvement in time-frequency resolution over the STFT [3]. It is also the correlation that is the source of the cross-term artifacts.

In order to mitigate the deleterious effects of cross-terms, a variety of fixed kernels have been introduced to smooth the higher frequencies in the WVF, where it is anticipated that most cross-terms are present. However, suppressing the cross-terms will also degrade the time-frequency resolution so that the net improvement factor over the STFT will be less than the ideal factor of 2. This means that the smoothing operations will involve a trade-off between minimizing cross-term artifacts and maximizing time-frequency resolution.

The WVF and complex ambiguity function (CAF) can be generalized by two-dimensional time-frequency kernels that weight by multiplication the CAF (and smooth by convolution the WVF) in order to suppress interfering cross-term artifacts, as described in Chapter 1. Selection of the kernels involves a trade-off between auto-term resolution/concentration and cross-term suppression. Most kernels in the literature are data-independent functions that typically have one or two manually adjustable parameters and are usually easier to apply in the CAF domain rather than the WVF domain.

For comparison with the other techniques to be described, Fig. 3-3 shows the steps for computing the generalized quadratic TFR with a fixed kernel. The steps in

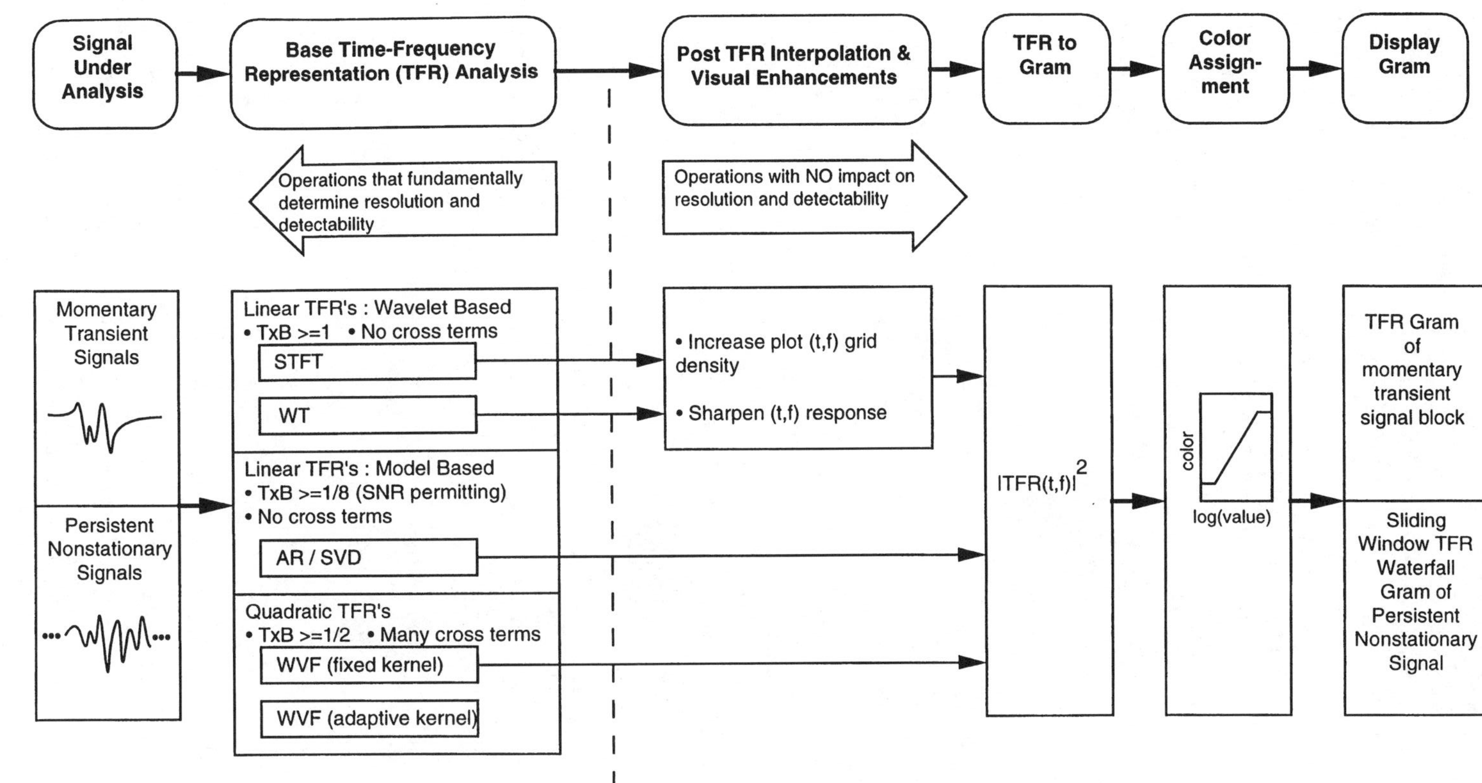

Figure 3-2 Generic flow diagram for computation and visualization of a variety of TFRs. AR, autoregressive; WT, wavelet transform.

Fig. 3-3(a) for generating the WVF or a smoothed WVF assume the case of a momentary transient signal so that the processing can be performed as a block of data. For persistent nonstationary signals, a sliding-window WVF analysis approach can be devised, as shown in Fig. 3-3(b), which uses a rectangular region within the diamond-shaped temporal correlation region in order to form a constant time- and frequency-resolution gram analysis. The sliding-window WVF was introduced by Marple and Friedlander [3] and is one of the first in the literature to devise a method of applying quadratic TFRs to persistent nonstationary signals.

The Adaptive Optimal Kernel (AOK) TFR.   The WVF described uses fixed, data-independent kernel functions of fixed shape to control cross-component sup-

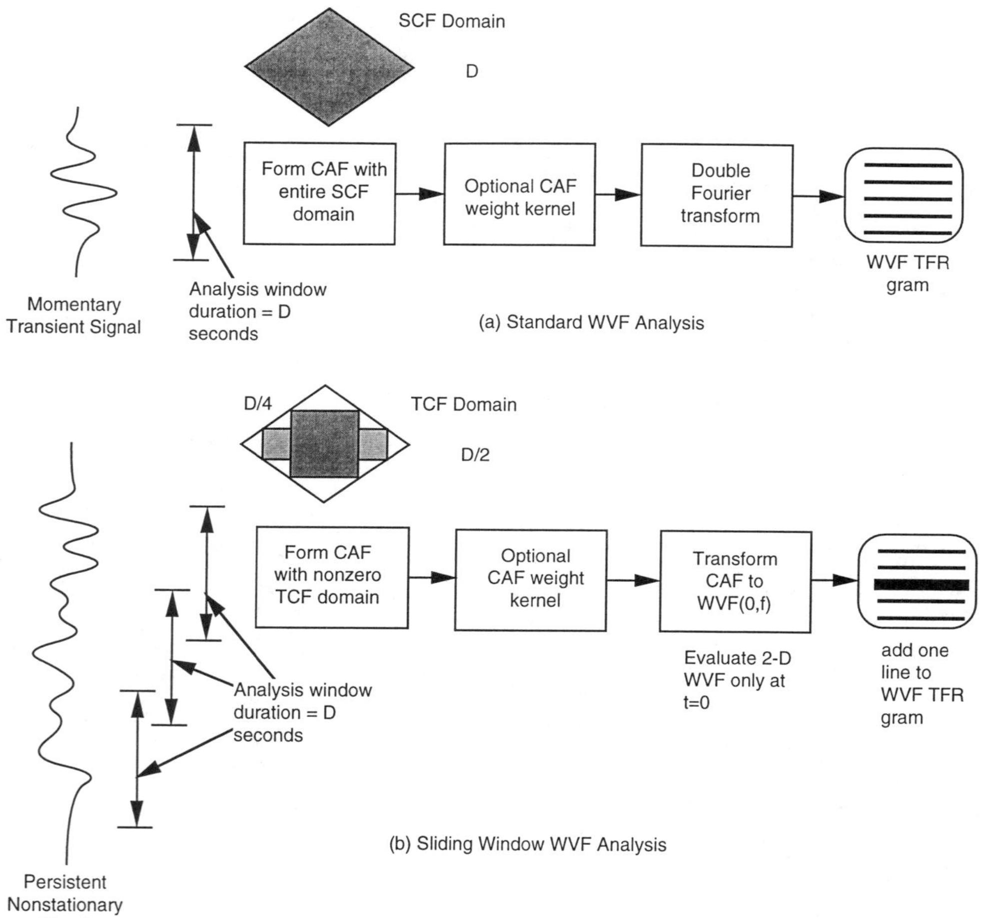

Figure 3-3   Detailed flow diagram for computation of the Wigner–Ville function (WVF) TFR: (a) the standard block WVF analysis for momentary transient signals; (b) sliding analysis window WVF for persistent nonstationary signals. SCF, spectral correlation function; TCF, temporal correlation function.

pression properties. Most of the TFRs described in the literature use such kernels, with fixed passband and stopband regions to remove cross-term interference. However, specification of a fixed kernel limits the class of signals for which the time-frequency representation reflects the true nature of the signal. That is, given any fixed kernel, it is always possible to find signals with either significant cross-component energy in the passband of the kernel, or significant auto-component energy in the stopband of the kernel.

Since the locations of the auto- and cross-components in the ambiguity function plane depend on the signal, we expect to obtain good performance for a broad class of signals only by using a *signal-dependent* kernel. Signal-dependent kernels have been considered by several authors; a number of references are given in [1].

Baraniuk and Jones [5, 6] have developed a procedure for selecting a signal-dependent kernel that, given a signal, automatically designs a kernel that is optimal with respect to a set of performance criteria. The TFR uses a radially Gaussian signal-dependent kernel that changes shape to optimally smooth the distribution. The procedure is called the adaptive optimal kernel (AOK) TFR.

The signal-dependent kernel design procedure can be formulated as an optimization problem. In general, the problem formulation involves the signal, and includes a set of constraints that defines a class of kernels from which the optimal kernel is chosen, and a performance index that measures the quality of the time-frequency representation for each kernel. The kernel that maximizes the value of the performance measure is selected as the optimal kernel for the signal. Possible constraints could include, for example, constraints that force the kernel to suppress cross-components, satisfy the marginal distributions, or satisfy Moyal's formula. Clearly, the choice of constraints and performance measure is crucial to the success of the method. However, once a satisfactory set of constraints and measure is found, kernel design for a wide range of signals reduces to solving an optimization problem. The optimal kernel, Φ, for a signal is defined as the solution to the following optimization problem:

$$\max_{\Phi} \int_0^{2\pi} \int_0^{\infty} |A(r, \Psi)\Phi(r, \Psi)|^2 \, r \, dr \, d\Psi \tag{3-1}$$

subject to

$$\Phi(r, \Psi) = e^{-\frac{r^2}{2\sigma^2(\Psi)}} \tag{3-2}$$

$$\frac{1}{2\pi} \int_0^{2\pi} \int_0^{\infty} |\Phi(r, \Psi)|^2 \, r \, dr \, d\Psi \leq \alpha, \qquad \alpha \geq 0 \tag{3-3}$$

Here $A(r, \Psi)$ is the complex ambiguity function (CAF) of the signal in polar coordinates. Once the optimal kernel is computed, the TFR is given by

$$P_{\text{AOK}}(t, \omega) = \frac{1}{2\pi} \int_{-\infty}^{\infty} \int_{-\infty}^{\infty} A(\theta, \tau)\Phi(\theta, \tau)e^{-j\theta t - j\tau\omega}\, d\theta\, d\tau \qquad (3\text{-}4)$$

The representation has worked well for characterizing short-duration and nonstationary events. However, the AOK TFR is computationally expensive. The flow diagram for computing the sliding-window AOK TFR is shown in Fig. 3-4.

The Adaptive Cone Kernel Distribution.   The cone kernel distribution (CKD) [10] is a time-frequency representation which has gained popularity because of its ability to resolve transient signal components or abrupt changes in signal characteristics. This capability stems from the more general CKD property of preserving the outer hull of a signal's time support, a property shared by any TFR defined by a kernel that is zero-valued outside a cone-shaped region of the time-lag plane.

The cone kernel is often parameterized by a single value, the *cone length*, which strongly controls the behavior of the resulting TFR. For example, a short cone allows the TFR to display quickly changing transient signal features, while a long cone permits high resolution of sinusoidal or other long-term signal components. Recent work has resulted in a technique to adaptively select the cone length at each point in time [7]. The technique is called the adaptive cone kernel (ACK) distribution.

The approach involves an optimization criterion similar to that used in [8, 9]. At each point in time, the cone length is selected which jointly maximizes the energy in the characteristic function, the pointwise product of the kernel, and the complex ambiguity function. The ACK distribution is able to select a distinct cone length at each point in time by using a fast algorithm which recursively computes a short-time ambiguity function, and approximates the energy in characteristic functions corresponding to every possible discrete cone length. The total computation required for the adaptation is $O(N)$, which is less than that of the $O(N \log N)$ fast Fourier transform used in producing each time slice. The flow diagram for computing the sliding-window AOK TFR is shown in Fig. 3-4.

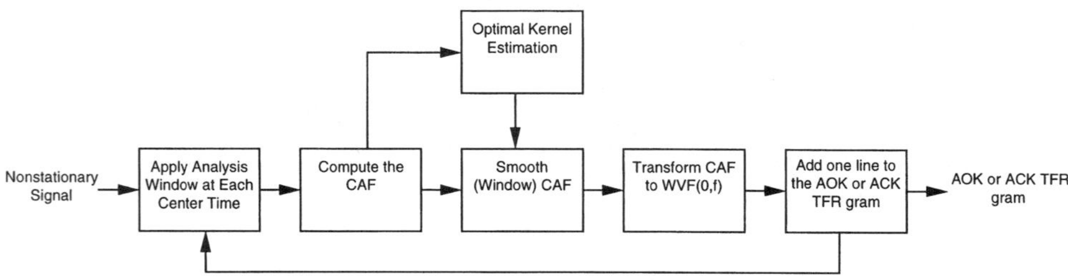

Figure 3-4 Detailed flow diagram for computation of the adaptive optimal kernel and adaptive cone kernel TFRs.

3.1.2 The Wavelet Transform Time-Frequency Representation

The primary focus in the development of wavelet techniques has been on the *synthesis* problem, which seeks to find minimal time-scale representations (TSRs) for one-dimensional (1-D) temporal and two-dimensional (2-D) spatial (image) signals [11, 12]. This usually involves orthogonal wavelets with octave (factor of 2) scalings that can reconstruct (synthesize) the original 1-D or 2-D signal from a grid of TFR samples by some optimality and/or uniqueness criteria. Algorithms that can evaluate these TSRs efficiently have also been extensively researched and implemented. This research has led to some useful compression techniques, especially for images (2-D signals) and certain classes of temporal (1-D) signals.

While a considerable body of knowledge is now available on multiresolution techniques, much of it is not relevant for signal *analysis* and identification purposes, because the desirable features required for wavelet/multiresolution *synthesis* are unnecessary for *analysis* problems. The wavelet analysis TSRs useful for signal detection and feature extraction are often quite different from the wavelet synthesis TSRs. Whereas the wavelet synthesis TSR will have a "grid" of evaluation points in the TSR plane at some minimal time- and scale-sampling intervals, the analysis TSR evaluation "grid" will be oversampled, perhaps greatly oversampled, in order to get the best detection or feature extraction performance. While oversampling of the TSR is desirable from an analysis viewpoint, this oversampling introduces redundancy in the time-scale representation, which is undesirable from a synthesis viewpoint (e.g., for compression applications).

A second limitation of wavelet transform techniques as presented in the technical literature is the selection of a scaling profile such that frequency resolution (temporal resolution) decreases (increases) monotonically as frequency increases— the so-called constant-Q property. While this choice of scaling leads to very nice mathematical structures and algorithms, there is typically no physical reason to assume that a constant-Q wavelet transform is the best way to analyze signals. It makes more sense, in our opinion, to adapt the temporal-resolution vs. frequency-resolution trade-off to the characteristics of the signals being analyzed, rather than to use the fixed constant-Q temporal-resolution vs. frequency-resolution properties of the wavelet transform, which are dictated by mathematical considerations rather than signal considerations.

Recall that the short-time Fourier transform is defined by

$$X_h(\mathbf{t}, f) = F\{x_h(\mathbf{t}, t)\} = \int_{-\infty}^{\infty} x_h(\mathbf{t}, t)\exp(-j2\pi ft)dt \qquad (3\text{-}5)$$

where $h(t)$ is the analysis window. The product $h(t)\exp(-j2\pi fnt)$ can be considered to be the analyzing "wavelet" corresponding to this linear transform. Thus the STFT may alternatively be interpreted as *analysis* by wavelets composed of modulated versions of the window weighting function, in which the frequency resolution remains constant across the entire frequency range. Overlap is a practical necessity when doing STFT *analysis*, in order to assure a greater chance of matching the time-frequency behavior of the signal. Overlap is not desirable when *synthesis* is the goal of the processing, because it introduces redundancy in the information provided by the samples of the time-frequency grid.

The classical formulation of the wavelet transform (WT) is

$$\mathrm{WT}(t, s) = \int_{-\infty}^{\infty} x(\tau)\left(\frac{1}{s}\right)^{1/2} h^*\!\left(\frac{\tau - t}{s}\right) d\tau \tag{3-6}$$

in which s is the scaling factor and $h(t)$ is the analyzing wavelet. The time-frequency version is obtained by making the substitution $s = f_0/f$

$$P_{\mathrm{WT}}(t,f) = \mathrm{WT}(t,f) = \int_{-\infty}^{\infty} x(\tau)\left\|\frac{f}{f_0}\right\|^{1/2} h^*\!\left(\frac{f}{f_0}(\tau - t)\right) d\tau \tag{3-7}$$

in which the analyzing wavelet $h(t)$ becomes essentially a prototype bandpass signal with center time $t = 0$ and center frequency f_0.

The effective time duration of the analysis window corresponding to the wavelet transform is a nonlinear function of frequency. As the frequency increases, the frequency resolution increases proportionately (the constant-Q) property, which means that the time duration of the analysis window decreases. The dependence of the time duration of the analysis window on frequency is depicted by the plot on Fig. 3-5. Henceforth we will call this type of plot the "profile" of the analysis window. Due to the frequency (or time) scaling, the wavelet transform is often plotted as a function of the logarithm of the frequency, although this will distort the patterns displayed in the time-frequency plane.

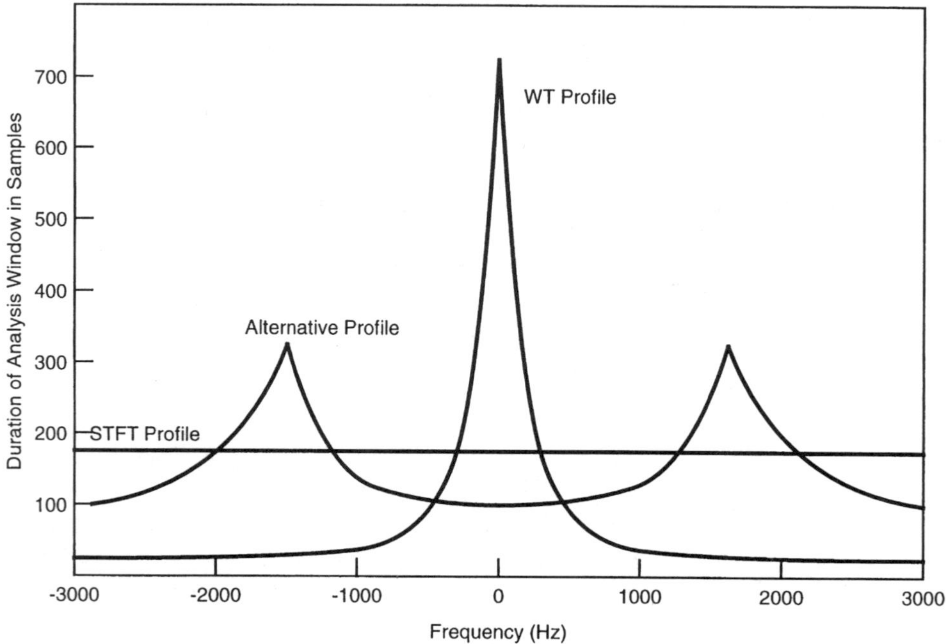

Figure 3-5 Example "profiles" for STFT TFR, the wavelet transform, and a user-selected "alternative" frequency-vs.-duration profile.

Note that the definition of the wavelet transform *always* assumes that the frequency with highest frequency resolution and least time resolution is near DC (0 Hz), whereas the frequency of least frequency resolution and highest time resolution is at the high-frequency end of the spectrum. This assumption is made regardless of whether or not this assignment of relative resolutions makes sense for the signal under analysis.

3.1.3 Model-Based Approaches

Short-Time Fourier Transform with Data Extrapolation.   Because high-resolution spectra like the autoregressive (AR) methods have peaks nonlinearly related to the strength of the signal component producing the peak, and many applications require that the spectral peaks have linear relationships to the strength, means have been devised for extrapolating from original data to create an enlarged data set of original + extrapolated data for which classical techniques like the STFT may be employed to yield spectra with peaks linearly related to signal strength. Such extrapolated data sets would increase the resolution over that achievable with only the original data set without the inherent nonlinearities of techniques like the AR methods. The concept of using the linear prediction parameters, once these are estimated from the original block of data, to extend (or extrapolate) the data off each end of the original block of data can be traced to Bowling and Lai [13, 14], who promoted the idea for processing short-duration radar data using one-dimensional linear prediction, and Frost and Sullivan [15], who promoted the idea for processing synthetic aperture radar data using two-dimensional linear prediction. These authors used both forward and backward linear prediction parameters to extend off each end of the original data. All of these extrapolation ideas were motivated by the work of Nuttall [16], who was the first to show how linear prediction approaches could be used to interpolate through regions with bad or missing data values. Our personal experience and demonstrations provided by the references suggest that (1) extrapolation to create a data set that is twice the size of the original data set duration is reasonably robust for many classes of signals, (2) extrapolation to create a data set three times the original set is sometimes possible if the signal-to-noise ratio (SNR) is high, and (3) extrapolation beyond a factor of three times the length of the actual data is often unreliable and should be used very cautiously.

Data Extrapolation with Covariance Method of Linear Prediction.   The simplest data extrapolation algorithm involves the use of the covariance method of linear prediction to estimate both the forward and backward linear prediction parameters based on the data within each analysis window. Once the parameters have been estimated given the values $x[1]$ to $x[N]$, extrapolation in the forward (in time) direction is obtained simply by forming

$$\hat{x}[n] = -\sum_{k=1}^{p} a_p^f[k]x[n-k] \tag{3-8}$$

for $n = N + 1$ to $n = N + N/2$, and extrapolation in the backward direction is obtained simply by forming

$$\hat{x}[n] = -\sum_{k=1}^{p} a_P^b[k]x[n + k] \tag{3-9}$$

for $n = 0$ to $n = -N/2 + 1$. This yields a total of $2N$ data values from the original N data values that can then be submitted for the usual STFT TFR analysis.

Data Extrapolation with Signal Subspace Enhancement. Signal subspace enhancements are also possible. In this method, the data extrapolation based on the forward and backward linear prediction parameters of the covariance method of linear prediction are replaced with their signal subspace estimated parameters. This is an eigenanalysis approach that attempts to remove the influence of noise so that the extrapolation accuracy can be improved in the presence of such noise. It turns out to be one of the most robust algorithms that has been investigated. We shall show in this section that there are three techniques that may be used to enhance the estimation of exponential signals in the data. These methods involve (1) the use of both forward and backward linear prediction polynomial zeros, (2) the use of high prediction orders, and (3) the application of singular value decomposition (SVD) for purposes of selecting the signal subspace. The algorithmic approach for use in TFR analysis is shown in Fig. 3-6.

In the absence of additive noise, it can be shown (chapter 11 of [4]) that m exponential signals may be generated by the forward linear prediction

$$\sum_{l=0}^{m} a^f[l]x[n - l] = 0 \tag{3-10}$$

in which $a^f[0] = 1$ and the characteristic polynomial

$$\mathbf{A}^f(z) = \sum_{l=0}^{m} a^f[l]z^{m-l} \tag{3-11}$$

has roots at $z_k = \exp(s_k)$ for $k = 1$ to $k = m$, where $s_k = (\alpha_k + j2\pi f_k)T$ contains the damping factor and frequency of the kth exponential. This is just another way of stating that the solution to a homogeneous difference equation involves the sum of exponential terms. The same m exponentials may also be generated in reverse time by the backward linear prediction

$$\sum_{l=0}^{m} a^b[l]x[n - m + l] = 0 \tag{3-12}$$

in which $a^b[0] = 1$. The characteristic polynomial

$$\mathbf{A}^b(z) = \sum_{l=0}^{m} a^{b*}[l]z^{m-l} \tag{3-13}$$

formed from the conjugated backward linear prediction coefficients, has roots at $z_k = \exp(-s_k^*) = \exp([-\alpha_k + j2\pi f_k]T)$ for $k = 1$ to $k = m$. For a decaying damping factor ($\alpha_k < 0$), the roots of the forward linear prediction characteristic polynomial $\mathbf{A}^f(z)$ fall *inside* the unit z-plane circle, whereas the roots of the backward linear prediction characteristic polynomial $\mathbf{A}^b(z)$ fall *outside* the unit circle due to the reciprocal damping factor $\exp(-\alpha_k T)$, which is a growing exponential. These characteristics of the root locations of polynomials $\mathbf{A}^f(z)$ and $\mathbf{A}^b(z)$ are properties of deterministic exponentials.

The application of signal subspace techniques via the SVD can provide further improvement. The forward and backward linear prediction errors may be expressed succinctly as

$$\mathbf{X}_p^f \mathbf{a}_p^f = -\mathbf{x}_p^f + \mathbf{e}_p^f \qquad \mathbf{X}_p^b \mathbf{a}_p^b = -\mathbf{x}_p^b + \mathbf{e}_p^b \tag{3-14}$$

in which the Toeplitz data matrices $\mathbf{X}_p^f$, $\mathbf{X}_p^b$ and data vectors $\mathbf{x}_p^f$, $\mathbf{x}_p^b$ are defined as

$$
\mathbf{X}_p^f = \begin{pmatrix} x[p] & \cdots & x[1] \\ \vdots & & \vdots \\ x[N-1] & \cdots & x[N-p] \end{pmatrix}, \qquad \mathbf{x}_p^f = \begin{pmatrix} x[p+1] \\ \vdots \\ x[N] \end{pmatrix}
$$

$$
\mathbf{X}_p^b = \begin{pmatrix} x[p+1] & \cdots & x[2] \\ \vdots & & \vdots \\ x[N] & \cdots & x[N-p+1] \end{pmatrix}, \qquad \mathbf{x}_p^b = \begin{pmatrix} x[1] \\ \vdots \\ x[N-p] \end{pmatrix} \tag{3-15}
$$

and the forward linear prediction coefficient vector $\mathbf{a}_p^f$, forward linear prediction error vector $\mathbf{e}_p^f$, backward linear prediction coefficient vector $\mathbf{a}_p^b$, and backward linear prediction error vector $\mathbf{e}_p^b$ are defined as

$$
\mathbf{a}_p^f = \begin{pmatrix} a^f[1] \\ \vdots \\ a^f[p] \end{pmatrix}, \quad \mathbf{e}_p^f = \begin{pmatrix} e^f[p+1] \\ \vdots \\ e^f[N] \end{pmatrix}, \quad \mathbf{a}_p^b = \begin{pmatrix} a^b[p] \\ \vdots \\ a^b[1] \end{pmatrix}, \quad \mathbf{e}_p^b = \begin{pmatrix} e^b[p+1] \\ \vdots \\ e^b[N] \end{pmatrix}
$$

$$\tag{3-16}$$

The data matrices have the following singular value decompositions:

$$\mathbf{X}_p^f = \sum_{n=1}^{p} \sigma_n^f \mathbf{u}_n^f (\mathbf{v}_n^f)^H \qquad \mathbf{X}_p^b = \sum_{n=1}^{p} \sigma_n^b \mathbf{u}_n^b (\mathbf{v}_n^b)^H \tag{3-17}$$

in which the σ_n^f are positive singular values of $\mathbf{X}_p^f$ and the σ_n^b are positive singular values of $\mathbf{X}_p^b$. The $\mathbf{u}_n$ and $\mathbf{v}_n$ are given eigenvectors of the respective data matrix. If a signal consists of m exponentials in additive noise, then the m eigenvectors associated with the m largest singular values primarily span the m exponential components. The p-m eigenvectors of the remaining smaller singular values primarily span the noise components. Assuming the singular values have been ordered by decreasing value, e.g., $\sigma_1^f > \sigma_2^f > \cdots > \sigma_p^f$, then a *reduced rank approximation* to each data matrix may be formed by truncating the SVD relationships of Eq. (3.17) to the m principal singular values (the so-called signal subspace):

$$\hat{\mathbf{X}}_p^f = \sum_{n=1}^{m} \sigma_n^f \mathbf{u}_n^f (\mathbf{v}_n^f)^H \qquad \hat{\mathbf{X}}_p^b = \sum_{n=1}^{m} \sigma_n^b \mathbf{u}_n^b (\mathbf{v}_n^b)^H \tag{3-18}$$

This will reduce the noise contribution to the data matrix, effectively enhancing the SNR. Minimizing the norms $\|\mathbf{a}_p^f\|_2$ and $\|\mathbf{a}_p^b\|_2$ with respect to the reduced rank data matrices will yield the solutions

$$\mathbf{a}_p^f = -\left(\hat{\mathbf{X}}_p^f\right)^{\#} \mathbf{x}_p^f \qquad \mathbf{a}_p^b = -\left(\hat{\mathbf{X}}_p^b\right)^{\#} \mathbf{x}_p^b \tag{3-19}$$

in which the pseudoinverse data matrices are defined as

$$\left(\hat{\mathbf{X}}_p^f\right)^{\#} = \sum_{n=1}^{m} \left(\sigma_n^f\right)^{-1} \mathbf{v}_n^f \left(\mathbf{u}_n^f\right)^H \qquad \left(\hat{\mathbf{X}}_p^b\right)^{\#} = \sum_{n=1}^{m} \left(\sigma_n^b\right)^{-1} \mathbf{v}_n^b \left(\mathbf{u}_n^b\right)^H \tag{3-20}$$

The order p must lie in the range m to $N\text{-}m$ in order to keep the rank of $\mathbf{X}_p^f$ and $\mathbf{X}_p^b$ greater than or equal to m, the assumed number of exponentials. If the number of exponentials is not known, it may be estimated by comparing the relative magnitudes of the singular values. The signal-related singular values generally will be larger than the noise-related singular values. Once the signal subspace forward and backward linear prediction parameters have been computed from Eqs. (3-15), the polynomial zeros may be found from these eigenanalysis-enhanced linear prediction parameters. Note that the data vectors $\mathbf{x}_p^f$ and $\mathbf{x}_p^b$ are not considered for noise effects themselves, even though they are themselves noisy. Figure 3-6 shows a flow chart for the extended data/model-based TFR.

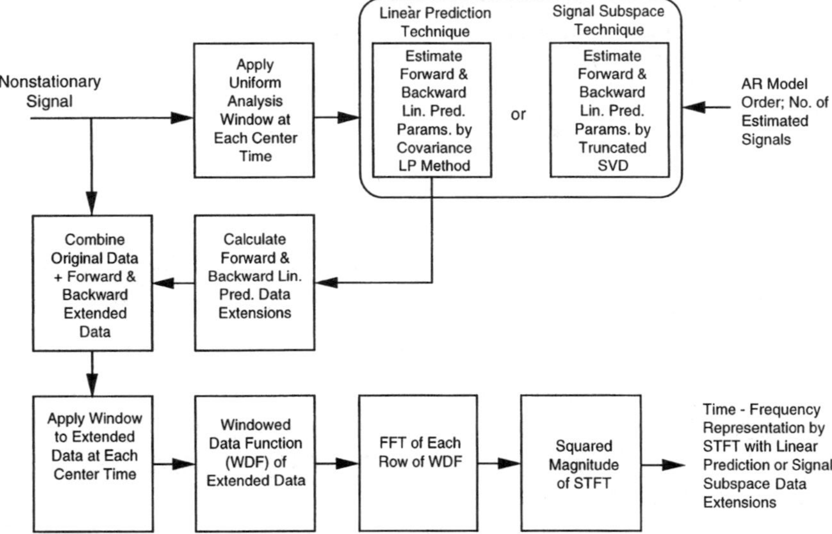

Figure 3-6   Detailed flowchart for the model-based TFR. The TFR is found as the STFT of the original plus extended data obtained using both the forward and backward linear model for prediction. The model was found using the least squares covariance method of linear prediction or the truncated singular value decomposition of the covariance data matrix.

3.2. DOPPLER ULTRASOUND DATA PROCESSING RESULTS

For the Doppler ultrasound data considered here, we are interested in discovering more detail within a single cardiac cycle. A roughly 380-ms interval was analyzed to discover fine grain detail, if any, within the signal's nonstationary behavior. In all of the plots that follow, the spectra are presented in logarithmic units of decibels (dB). Each TFR gram obtained is referenced to the maximum value in the TFR so that the maximum scale value is 0 dB. The data were then clipped at −45 dB so that the dynamic range that appears in the plots is between −45 and 0 dB. In all but the wavelet transform case, 512 frequency points per scan line are calculated and plotted for each of the spectra.

3.2.1 The Short-Time Fourier Transform (STFT)

Figures 3-7 through 3-9 show the results of STFT TFR processing for several different parameter settings. Figure 3-7 shows the best STFT result. To obtain those results, the analysis window contained 32 samples. The input vector is windowed with a Hamming window. The windowed vector input to the FFT was then zero-padded out to 512 points. There were four samples skipped between scans in the output TFR (i.e., the time resolution in the figure is four samples, or 0.64 ms). Figure 3-8 has all the same selections except the analysis interval is 128 samples rather than 32 as in Fig. 3-7. There is not much coherent signal except near 100 ms into the start of the processing record, after which there is a split for approximately 200 ms into low Doppler and high Doppler components; note the aliasing occurring about 150 ms into the processing. Using longer analysis intervals up to 512 samples and greater step sizes between gram lines, such as 16 samples rather than four, does not improve the resulting fine grain analysis (see Fig. 3-9). This suggests there is very little coherent signal in the nonstationary waveform.

3.2.2 Generalized Wigner–Ville and Complex Ambiguity
Functions

Figure 3-10 shows the results of processing the Doppler ultrasound data with the Choi–Williams kernel-smoothed WVF. The analysis window for each scan contains 128 points. Four samples between analysis window centers were used. The Choi–Williams weighting parameter is set to 0.3. Figure 3-11 is an alternative WVF TFR, using all the same parameters as in Fig. 3-11, except that the analysis window used is 32 points (the same analysis window size that gave rise to the best STFT TFR).

3.2.3 The Adaptive Optimal Kernel (AOK) TFR

Figure 3-12 shows the results for the AOK TFR. The analysis window for each line of the TFR contains 128 points to compute the ambiguity function and a kernel volume parameter of 1.25. Four samples between analysis window centers were used; 512 frequency samples were found and displayed.

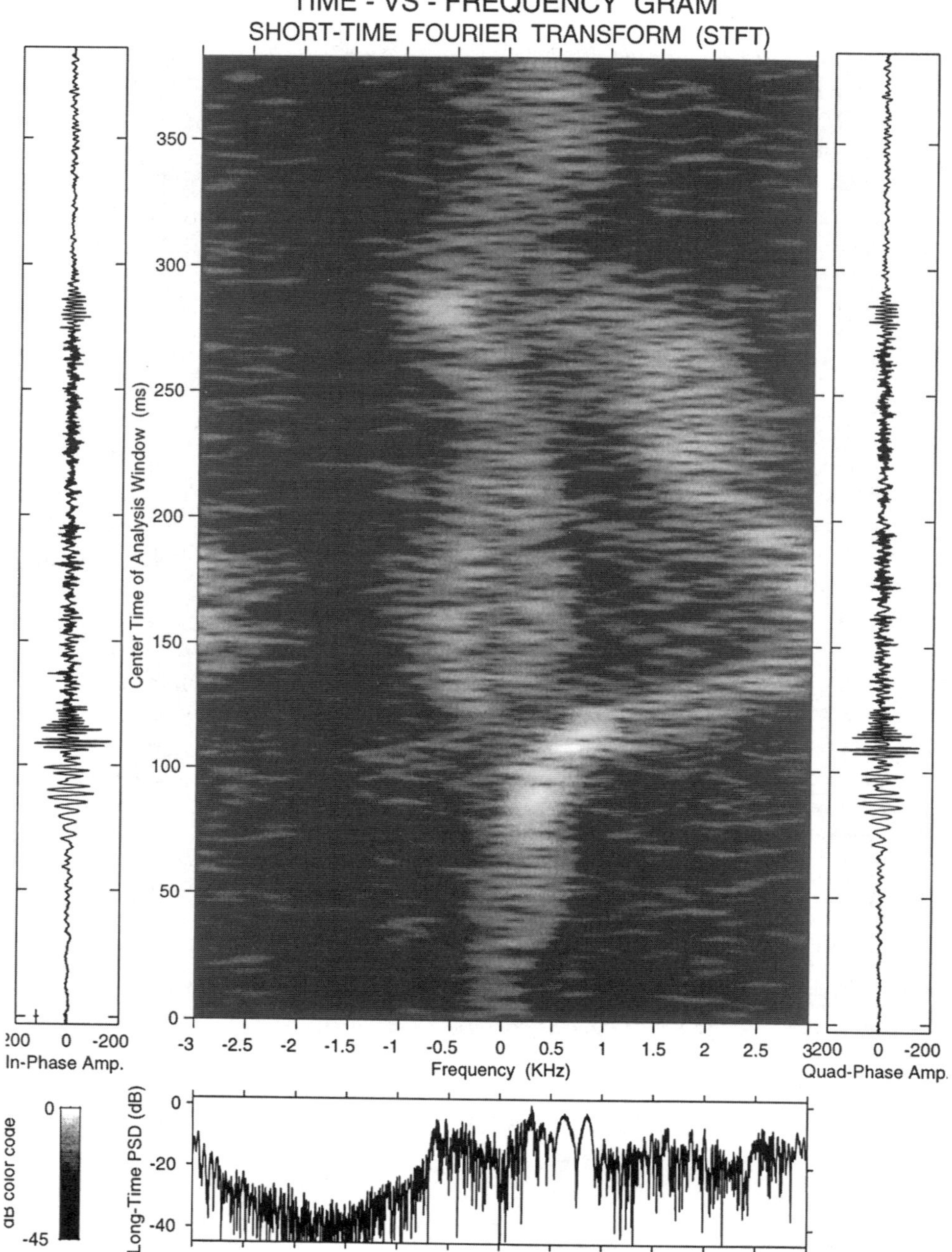

Figure 3-7   The "best" STFT TFR of selected ultrasound data. Thirty two points were input to the FFT. The input data were Hamming windowed and padded to 512 points prior to taking the FFT. Four samples are skipped in the data between successive scans of the TFR. Notice that there is fairly good time resolution, however, there is "streaking" in frequency.

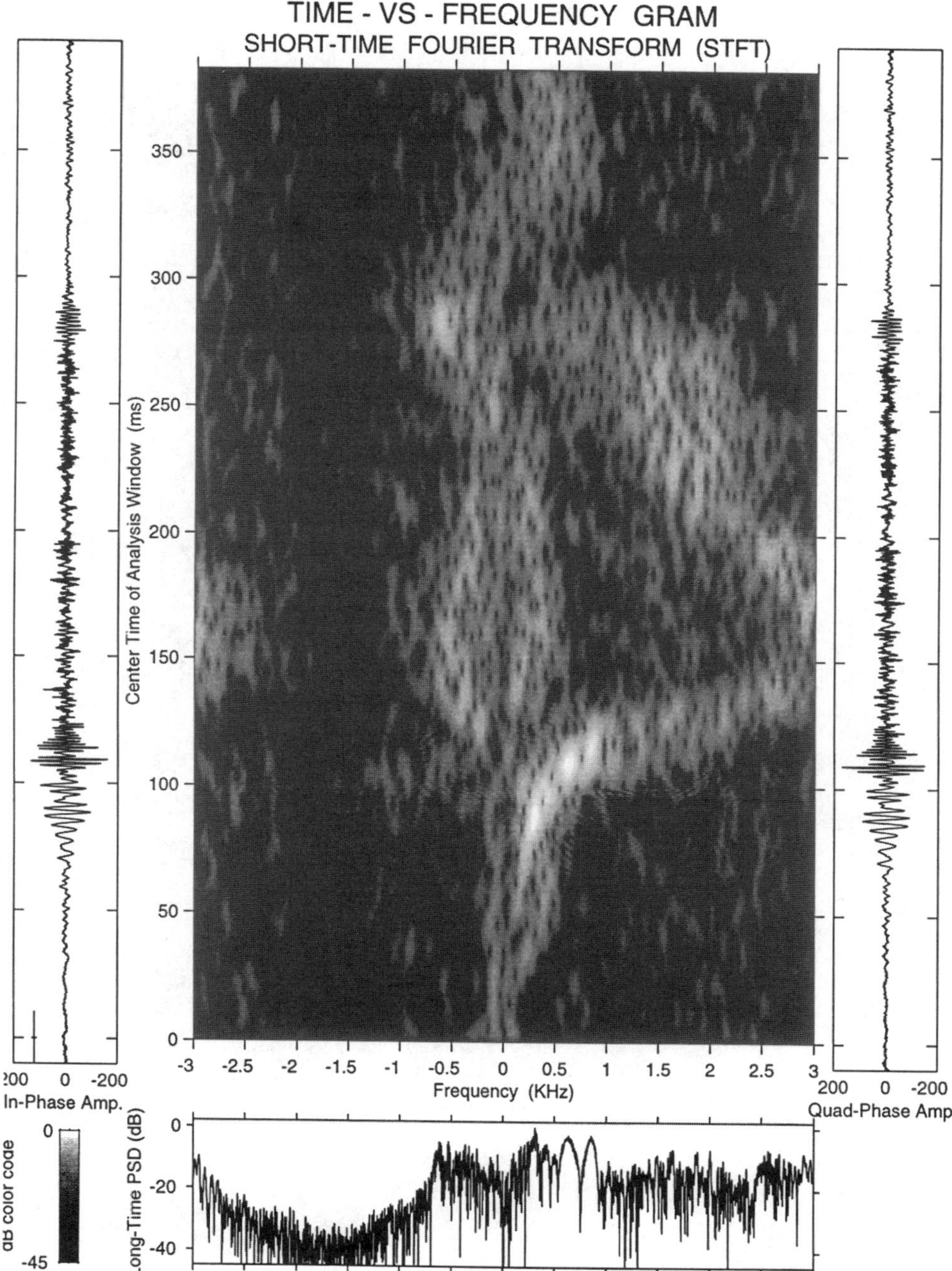

Figure 3-8 An "alternative" STFT TFR. Processing here is the same as that used for Fig. 3-7 except that 128 points (vs. 32) were used as inputs to the FFT. Notice that frequency resolution has improved, but at the expense of time resolution.

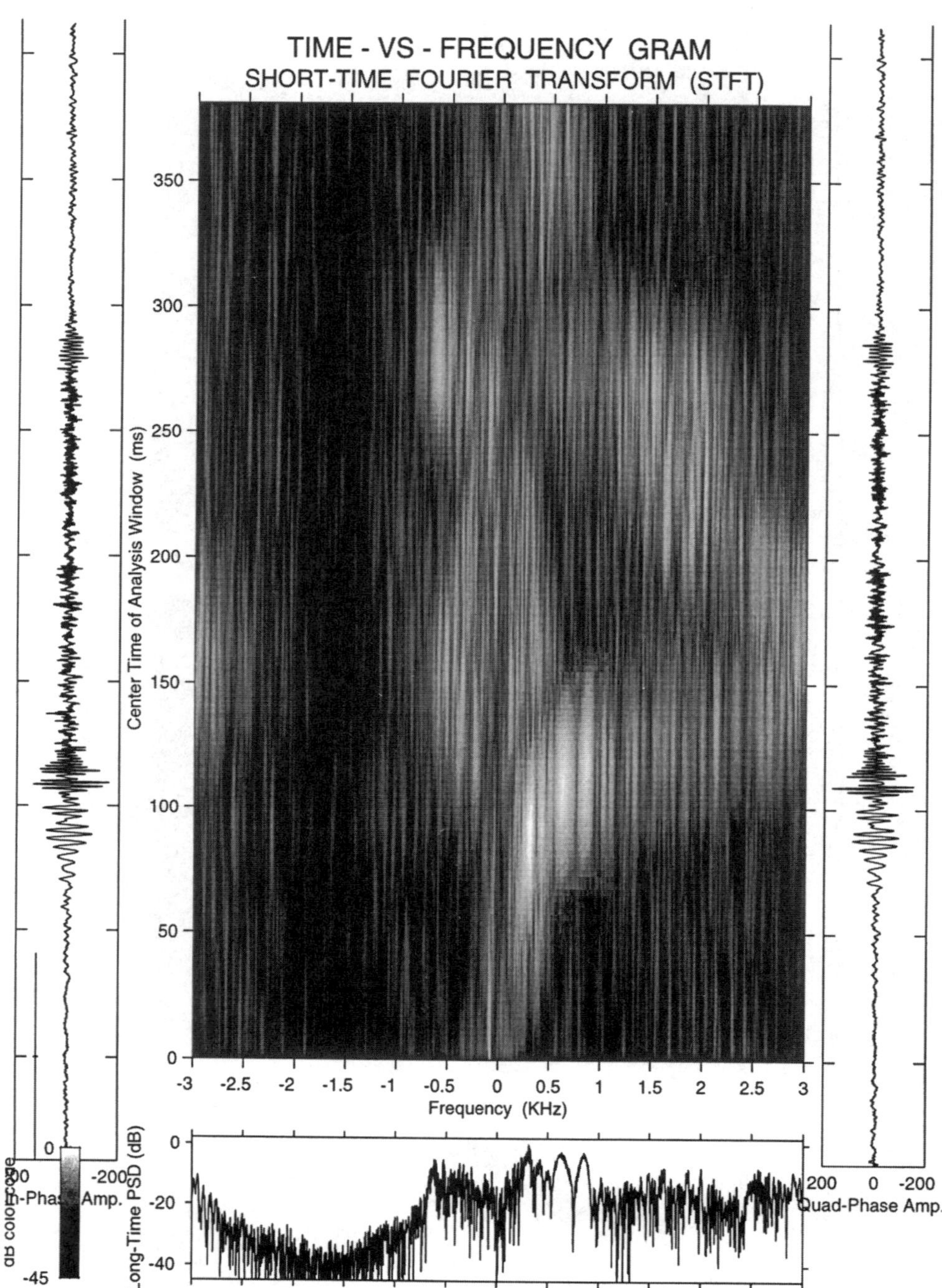

Figure 3-9   An "alternative" STFT. Processing is similar to that shown in Figs 3-7 and 3-8, however, now 512 points are used as input to the FFT and 16 samples are skipped between successive scans in the figure.

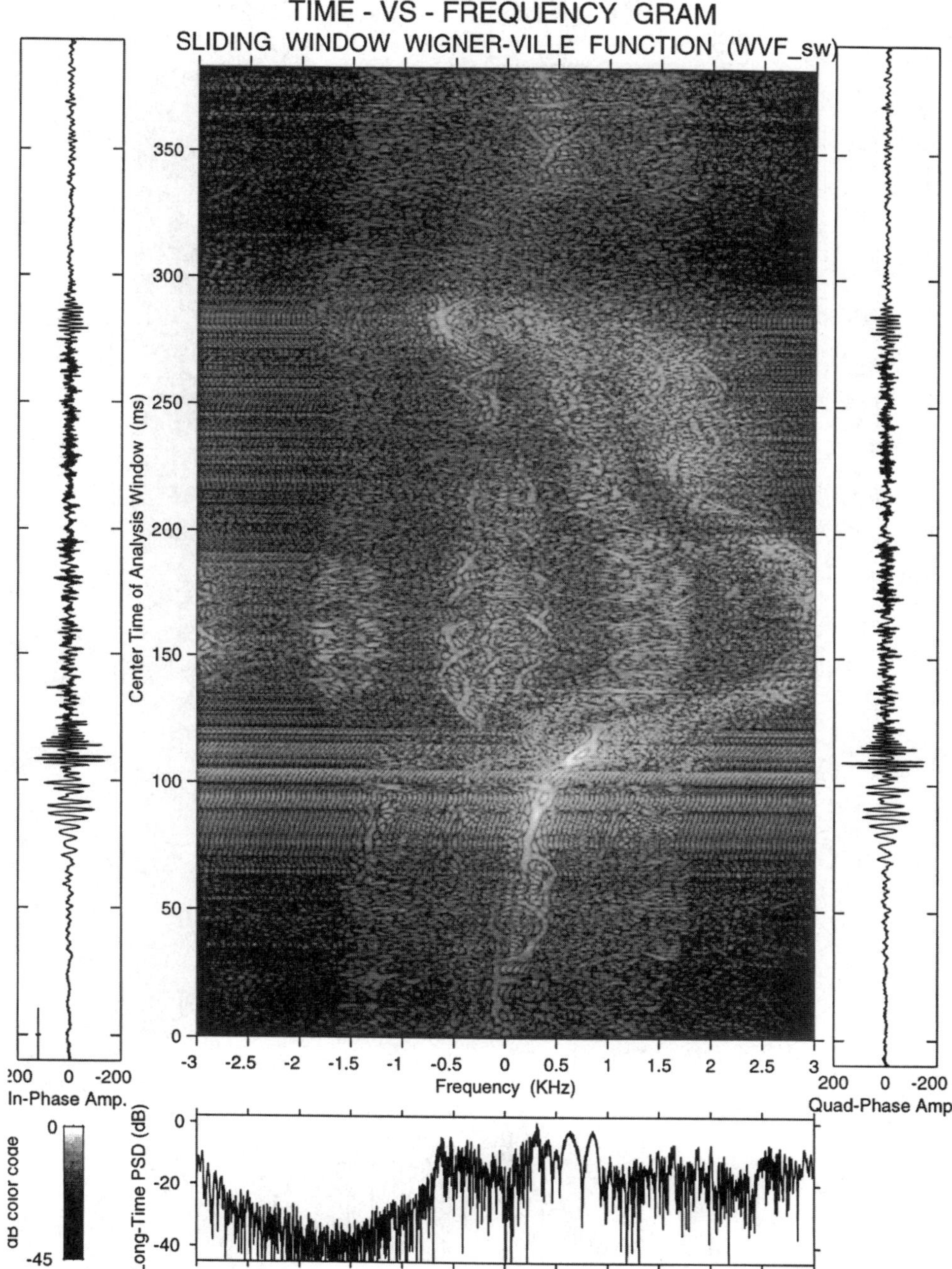

Figure 3-10   The smoothed WVF found using the Choi–Williams smoothing kernel. A 128-point analysis window was used. Four samples were skipped between successive scans. The Choi–Williams parameter was set to 0.3.

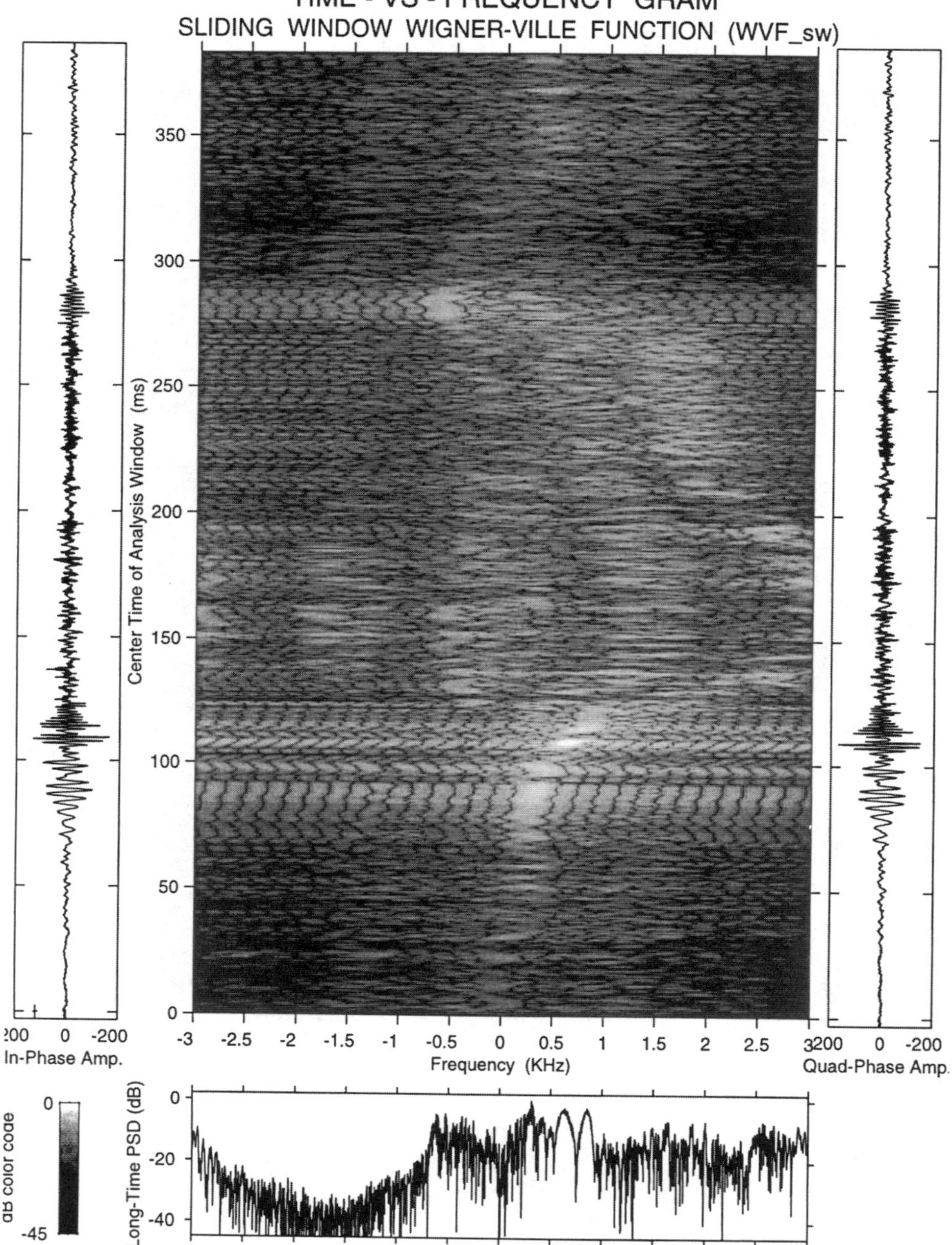

Figure 3-11   The smoothed WVF found using the Choi–Williams smoothing kernel. Here, a 32-point analysis window (vs. 128-point analysis window used for the results shown in Fig. 3-10) was used. Four samples were skipped between successive scans. The Choi–Williams parameter was set to 0.3.

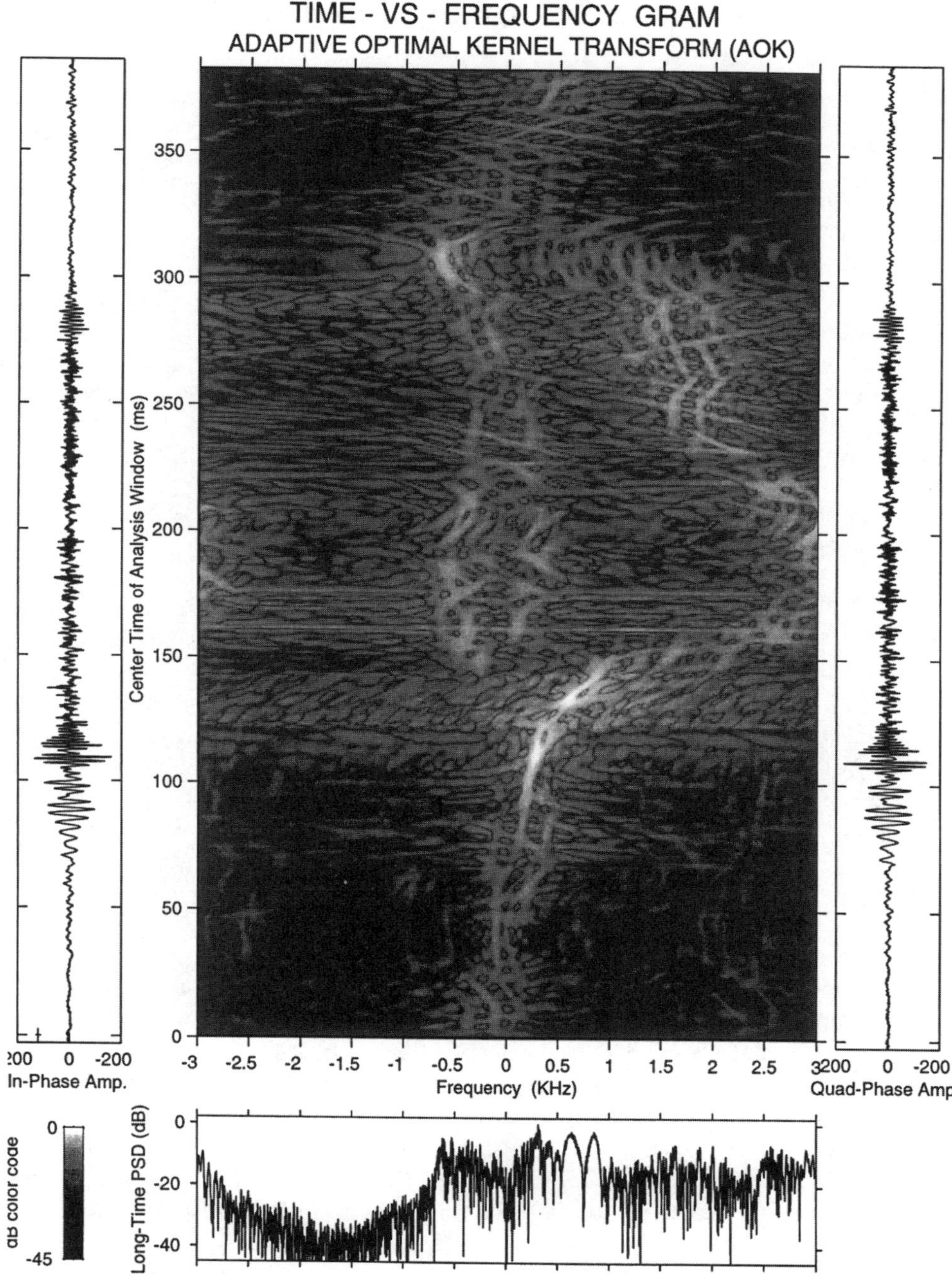

Figure 3-12   The AOK TFR: 128 points were used in the analysis window and four samples were skipped between successive output scans. The AOK kernel shape parameter was set as 1.25.

3.2.4 The Adaptive Cone Kernel (ACK) Distribution

Figure 3-13 shows the results for the ACK TFR. The adaptation is based on a 512×512 point short-time ambiguity function centered around each time; each time slice is separated by four samples, and contains 512 frequency points. For the results shown in Fig. 3-13, the maximum cone length was set to 64.

3.2.5 The Wavelet Transform Time-Frequency Representation

Figure 3-14 shows the results of applying the wavelet transform to the data. In this figure, 332 logarithmically spaced frequency bins are evaluated. A version of the wavelet transform using oversampling, a 1/32-octave spacing on frequencies (rather than the traditional octave spacing), and a Hamming wavelet produced the result shown in Fig. 3-14.

3.2.6 Model-Based Approaches: Signal Subspace Enhancement/Linear Prediction for Extended Data STFT

Figure 3-15 shows the results of using linear prediction parameters for extending the data. A 32-point analysis window was used. The model order considered for the data extrapolation was 8. The expected number of signals in the data (i.e., the dimension of the retained signal subspace) was set to 4. As with the previous techniques, a Hamming window was applied and a 512-point FFT taken of the original plus extrapolated data. Four samples between analysis window centers were used.

3.3. CONCLUSIONS

The varying abilities of these methods to resolve fine signal structure is apparent in the representations shown in Figs. 3-7 through 3-15. As seen in the figures, the gross appearance of each distribution is similar. Both the adaptive optimal kernel (ACK) (Fig. 3-13) and the model-based extended data technique (Fig. 3-15) appear to more sharply resolve the fine detail when compared with the other TFRs. A comparison of Figs. 3-13 and 3-15 show significant differences in the spectral shapes. The most interesting difference is at the start of the cardiac contraction (between 50 and 100 ms into the data). In the ACK TFR, we see three distinct spectral components. The three components can just be made out in the AOK TFR (Fig. 3-12), but are not seen in any of the other TFRs.

Figure 3-7 shows the best STFT result. Figure 3-8 has all the same selections except the analysis interval is 128 samples rather than 32 as in Fig. 3-7. There is not much coherent signal except near 100 ms into the start of the processing record, after which there is a split for approximately 200 ms into a lower Doppler and high Doppler components; note the aliasing centered about 150 ms into the processing. Using longer analysis intervals up to 512 samples and greater step sizes between

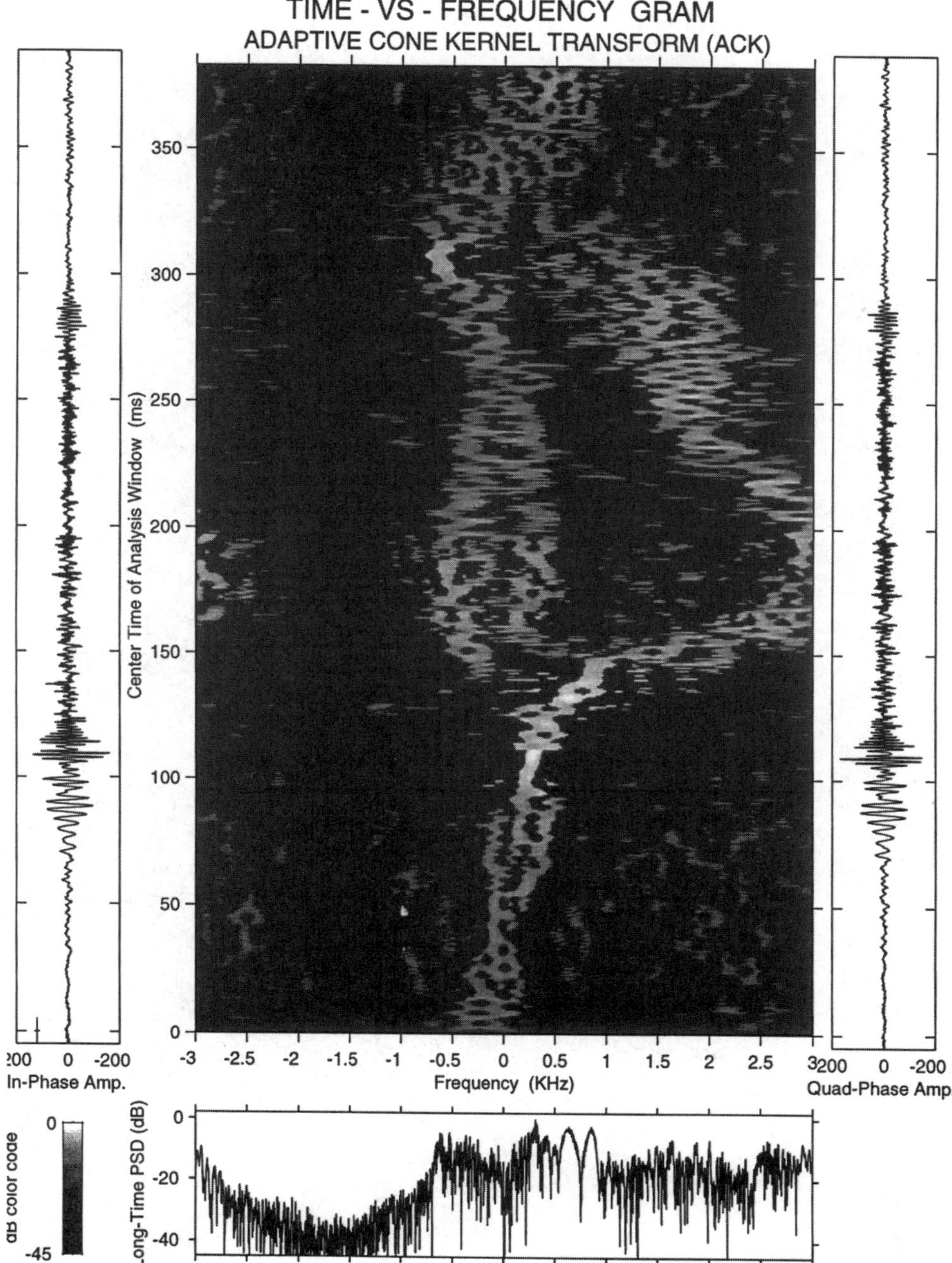

Figure 3-13   The ACK TFR: 128 points were used in the analysis window and four samples were skipped between successive output scans. The ACK maximum cone length was set to 64.

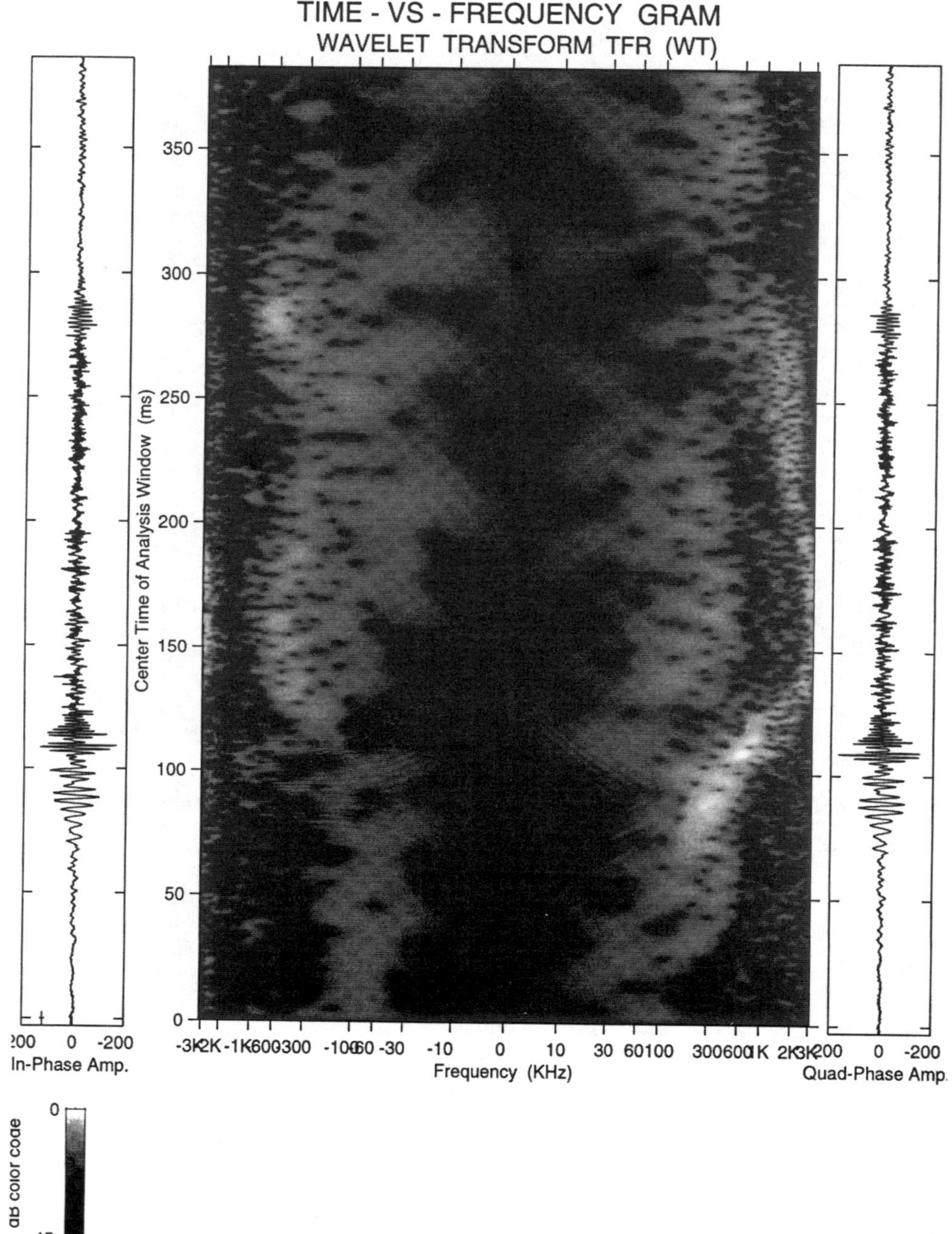

Figure 3-14   The WT TFR: note the logarithmic frequency scale.

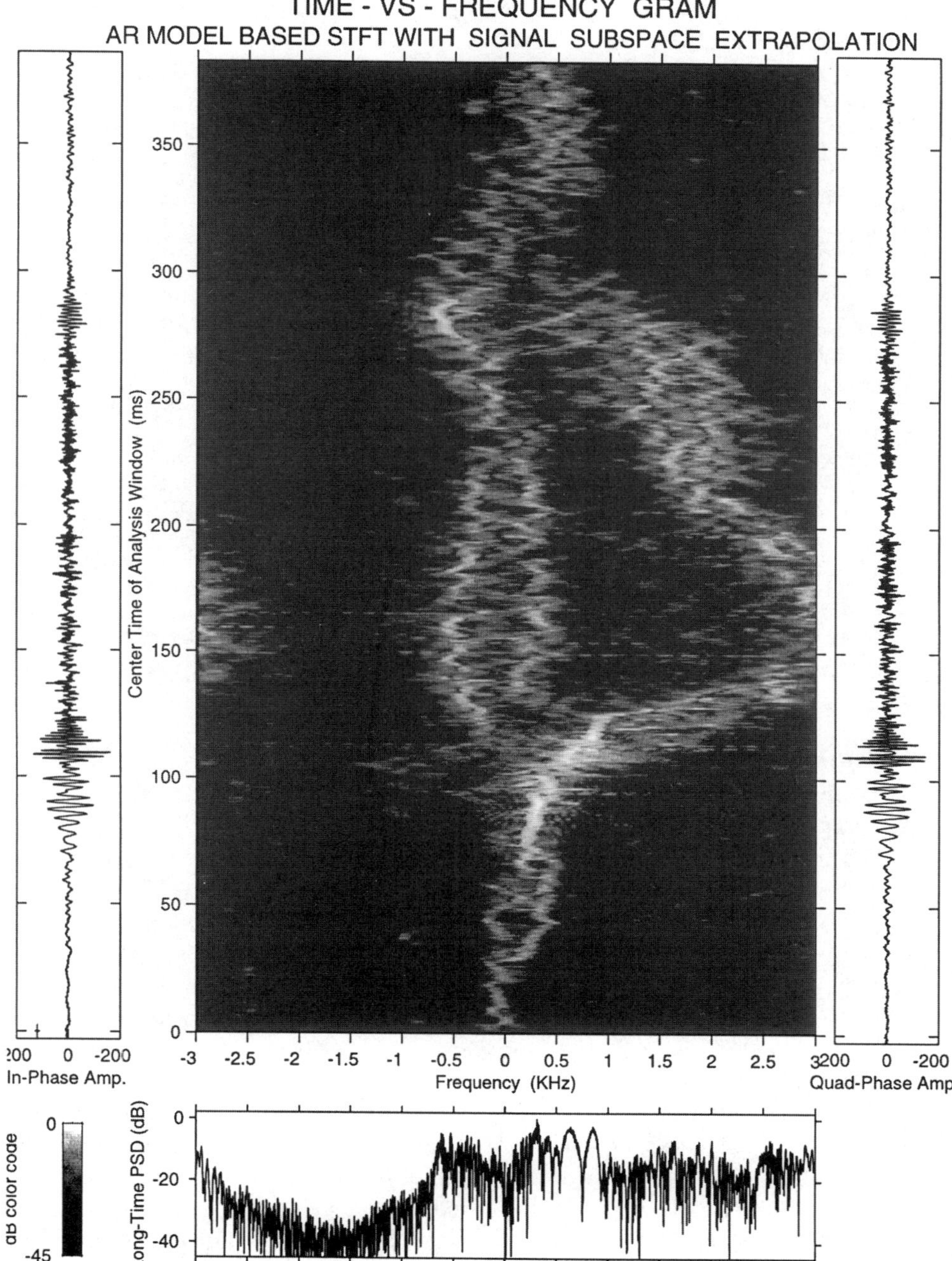

Figure 3-15   The model-based data extrapolation/signal subspace TFR. The analysis window used 32 samples, and four samples were skipped between successive output scans. The model order used for data extrapolation was 8 and the dimension of the signal subspace was set to 4.

gram lines, such as 16 samples rather than 4, does not enhance distinguishing fine structure (see Fig. 3-9).

Using a sliding-window version of the Wigner–Ville function produced the result shown in Fig. 3-10 for a 128-sample analysis window, which was the best setting. Despite the use of a Choi–Williams cross-terms suppression filtering, there are still significant cross-terms left that make it difficult to detect any detail in the gram. It may be that the Choi–Williams smoothing kernel is not well matched to the placement of the interference cross-terms for these data. Note the extreme graininess of the WVF, further reinforcing the observation that the signal is mostly incoherent and very noisy. Coherent features in the WVF will tend to show up as thin lines, and none are apparent. Figure 3-11 is another WVF gram in which the only change in setting is from 128 to 32 samples in the analysis window as used by the STFT TFR. The result it produces is not as good as that of Fig. 3-10.

Figure 3-12 shows the results of using the data adaptive AOK technique to develop a smoothing kernel for the quadratic TFR. As seen, many of the cross-interference terms are eliminated using the AOK technique, but some streaking occurs. The streaking is caused by the smoothing action on the "noisy" cross-terms. If the cross-terms were truly eliminated, there would be no streaking. The two terms near 0 Hz are easily distinguished in the AOK.

Similar results are obtained using the adaptive cone kernel transformation, as seen in Fig. 3-13. The streaking seen in the AOK is further reduced and the spectral peaks appear to be sharp. There does appear to be more detail by the presence of three different peaks in the ACK in the region between 50 and 100 ms. These three peaks are not distinguished in any of the other TFRs.

For the wavelet transform, shown in Fig. 3-14, no apparent advantage of the logarithmic analysis intervals is seen. In fact, the noisy signal conditions near 0 Hz are emphasized. Details in the high frequency are lost.

The best (highest resolution) results appeared to be obained using the ACK technique, as seen in Fig. 3-13. The signal subspace approach in conjunction with linear prediction extrapolation, followed by the STFT of the original plus extrapolated data record as seen in Fig. 3-15, is a close contender. As previously noted, there is a split in the spectral components between 50 and 100 ms that is clearly seen only on the ACK representation. It appears that there may be a split in the Doppler between 100 and 250 ms around the frequencies ± 500 Hz, which are not well separated by the other TFR techniques.

In many situations, and in this example in particular, the data adaptive algorithms (the model-based approach, AOK, and ACK TFRs) display better resolution of fine signal structure. As a rule of thumb, fixed-kernel/window constraints yield results most sensitive to details involving abrupt variations in the signal in time, whereas STFT TFRs tend to be most sensitive to relatively long-duration sinusoidal components. The adaptive methods will tend to break down at lower SNRs than the fixed-kernel methods. The STFT TFR and wavelet transforms are relatively insensitive to noise. The pertinent question in a real application is whether the potentially better resolution of the adaptive techniques could be of value. In addition, the adaptive TFRs and several of the model-based techniques are computationally intensive when compared to the STFT and wavelet transforms. If only general features of a time-frequency structure are of interest, an STFT TFR or a wavelet transform should be used. If details of the time-structure are important, a data-adaptive algorithm is likely to be best. Often the best results will include components from several different TFR analysis tools.

REFERENCES

[1] F. Hlawatsch and G. F. Boudreaux-Bartels, "Linear and quadratic time-frequency signal representations," *IEEE Signal Proc. Mag.*, vol. 9, pp. 21–67, April 1992.

[2] B. Boashash, "Time-frequency signal analysis." In *Advances in Spectrum Analysis and Array Processing—Volume I*. S. Haykin (ed.) Englewood Cliffs, NJ: Prentice Hall, 1991.

[3] S. L. Marple, Jr. and B. Friedlander, "Application to Gabor bases to extracting information from the Wigner–Ville transform," ORINCON Corp. Tech. Rep. OCR 92-U-0441, 30 September 1992, under contract DAAH01-92-C-R077 issued by U.S. Army Missile Command under ARPA Order no. 5916.

[4] S. L. Marple, *Digital Spectral Analysis with Applications*. Englewood Cliffs, NJ: Prentice-Hall, 1987.

[5] R. G. Baraniuk and D. L. Jones, "A radially Gaussian, signal-dependent time-frequency representation," *Signal Proc.*, vol. 32, no. 2, pp. 263–284, 1993.

[6] R. G. Baraniuk and D. L. Jones, "A signal-dependent time-frequency representation: Optimal kernel design," *IEEE Trans. Signal Proc.*, vol. 41, pp. 1589–1602, April 1993.

[7] R. N. Czerwinski and D. L. Jones, "Adaptive cone-kernel time-frequency analysis," *IEEE Trans. Signal Proc.*, vol. 43, no. 7, pp. 1715–1718, July 1995.

[8] R. G. Baraniuk and D. L. Jones, "Signal-dependent time-frequency analysis using a radially Gaussian kernel," *Signal Proc.*, vol. 32, no. 3, pp. 263–284, June 1993.

[9] R. G. Baraniuk and D. L. Jones, "A signal-dependent time-frequency representation: Optimal kernel design," *IEEE Trans. Signal Proc.*, vol. 41, no. 4, pp. 1589–1602, April 1993.

[10] Y. Zhao, L. E. Atlas, and R. J. Marks, "The use of cone-shaped kernels for generalized time-frequency representations of nonstationary signals," *IEEE Trans. Acoust, Speech, Signal Proc.*, vol. 38, no. 7, pp. 1084–1091, July 1990.

[11] O. Rioul and M. Vetterli, "Wavelets and Signal Processing," *IEEE Signal Proc. Mag.*, pp. 14–38, October 1991.

[12] I. Daubechies, "The wavelet transform: A method for time-frequency localization." In *Advances in Spectrum Analysis and Array Processing*. S. Haykin (ed) Englewood Cliffs, NJ: Prentice Hall, 1991.

[13] S. B. Bowling, "Linear prediction and maximum entropy spectral analysis for radar applications," Project Report RMP-122, M.I.T. Lincoln Laboratory, 24 May 1977; available as DTIC report DDC-AD-A042817/7.

[14] S. B. Bowling and S. T. Lai, "The use of linear prediction for the interpolation and extrapolation of missing data and data gaps prior to spectral analysis," Tech. Note TN-1979-46, M.I.T. Lincoln Laboratory, 1979; also presented as a paper in *Proc. Rome Air Development Center Spectrum Estimation Workshop*, pp. 39–49, 3–5 October 1979.

[15] O. L. Frost and T. M. Sullivan, "High-resolution two-dimensional spectral analysis," in *Proc. Int. Conf. Acoust., Speech and Signal Proc.*, pp. 673–676, 1979.

[16] A. H. Nuttall, "Spectral analysis of a univariate process with bad data points, via maximum entropy and linear prediction techniques," Tech. Rep. 5303, Naval Underwater Systems Center, 26 March 1976.

Chapter 4

Analysis of ECG Late Potentials Using Time-Frequency Methods

Hartmut Dickhaus, Hartmut Heinrich

4.1. INTRODUCTION

In recent years, the demand for noninvasive diagnostic procedures stimulated a remarkable increase of advanced signal analysis techniques and sophisticated computer algorithms. This chapter stresses some aspects of the latest time-frequency methods for pattern recognition in the context of a relevant clinical problem in cardiology: the identification of high-risk patients by electrocardiogram (ECG) signal analysis. Concerning this problem, much interest has been focused by cardiologists on ventricular late potentials (LPs) of the ECG [1–3].

In surface recordings, LPs appear as small signals of about 5-40 μV with a frequency content mainly above 40 Hz. Several groups have demonstrated the existence of such potentials in patients with different types of cardiac diseases and arrhythmias [4, 5]. The latter group of patients is of particular diagnostic interest, because sustained ventricular tachycardia (VT) is regarded as one of the reasons for sudden cardiac death. The hypothesis that LPs predict the extent to which sustained VT can be induced explains their significance as markers for life-threatening arrythmias [1].

There are serious problems in detecting and quantifying LPs in body surface recordings, however, signal-averaging and high-pass filtering, as proposed by many authors, are only standard preprocessing methods [5]. For a quantitative description, many groups prefer signal analysis of LPs in the time domain, for which guidelines are recommended [6]. Nevertheless, there are obvious problems in determining time intervals and amplitudes under varying noise levels. Therefore, some authors suggest approaches in the frequency domain, for example, fast Fourier transform (FFT) processing and windowing of specific segments of the QRS complex [7]. However, simple spectral analysis is based on the assumption of stationarity of the observed signal. Therefore, their results failed in the case of fast transient signal episodes and showed substantial disadvantages, mainly due to the fixed duration of the window applied. Precise temporal detection of a sudden frequency shift within the signals is impossible. The high-frequency content of the recorded signals, including LPs, obviously results in time-varying characteristics that render simple power spectral analysis of limited value. Nevertheless, these nonstationary signal characteristics are typical and carry the important information which has to be detected, decoded, and understood in the context of the patient's particular situation.

Signal representations in the time-frequency plane provide the chance to regard a phenomenon simultaneously under two different points of view [8–10]. As pointed out in [11], (Chapter 1 of this book), the frequency characteristics as well as the temporal behavior of the signal can be studied, depending on the uncertainty principle with adequate resolutions.

To overcome most of these problems, we have tried to study the characteristics of our LP signals using the time-frequency transformations already mentioned. Several approaches such as spectrotemporal mapping (or spectrogram), the wavelet transform, and the Wigner distribution are compared. In particular, we report our experiences involving two independent clinical studies with ECG recordings from patients with ventricular tachycardia and myocardial infarction at different hospitals. The results show that the ECG records can be successfully evaluated by time-frequency methods for the identification of high-risk cardiac patients.

4.2. METHODS

4.2.1 Data Acquisition and Preprocessing

Two groups of patients from different hospitals were considered. Group I, from the Heidelberg university hospital, contained 21 patients with documented ventricular tachycardia. The patients were studied during normal sinus rhythm without receiving antiarrhythmic medication. Group II consisted of 25 patients after myocardial infarction from the university hospital of Münster. For both groups, representative recordings of 29 healthy subjects from Heidelberg and 30 subjects from Münster were used as control standards.

Signal averaging of three orthogonal leads x, y, z was simultaneously performed during two minutes. The signals were sampled using 1000 Hz. There were 300 sample points, including the complete QRS complex, that were used for further signal processing. The block diagram of Fig. 4-1(a) demonstrates schematically the different preprocessing steps. In order to reduce the low-frequency components with high voltage level, the signals of each lead were passed through a digital 1-pole Butterworth filter with 100 Hz cut-off frequency. The leads x, y, z were then combined into a vector magnitude:

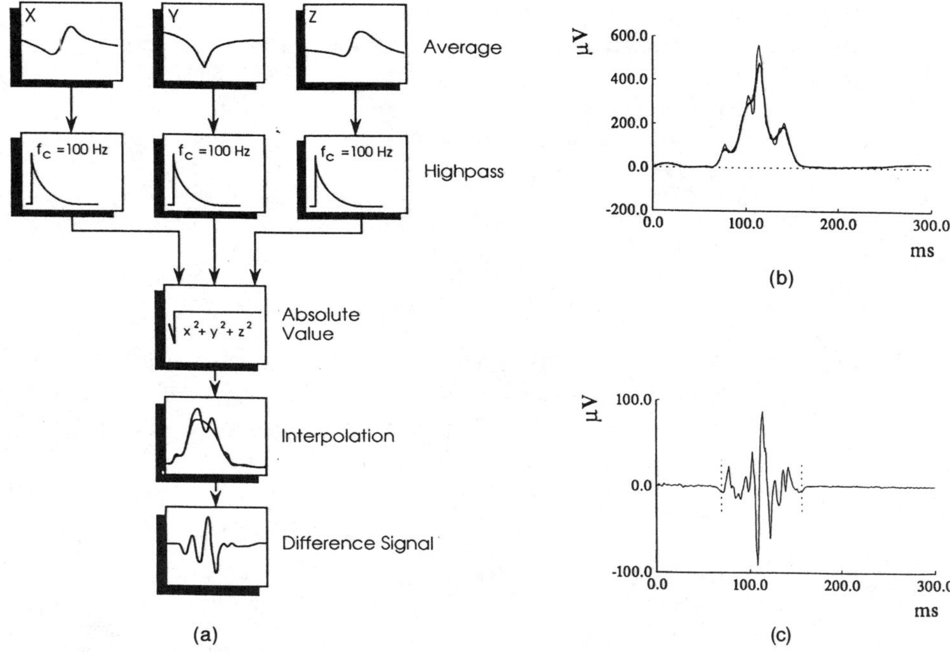

Figure 4-1 (a)   Block diagram of preprocessing three ECG leads x, y, and z; (b) shows the vector magnitude of 200 averaged ECG segments x, y, and z. The thick line indicates its smoothed version. Part (c) represents the difference signal of the vector magnitude and its smoothed version as shown in (b).

$$V = \sqrt{x^2 + y^2 + z^2} \qquad (4\text{-}1)$$

By this nonlinear processing step, low frequencies were enhanced. Therefore, further high-pass filtering was necessary. To prevent the filtering of data from creating artifacts such as ringing, we smoothed the vector magnitude by Bezier–Spline interpolation [12]. Figure 4-1(b) demonstrates an example of the vector magnitude and its smoothed version. Subtraction of the smoothed signal from the unsmoothed version yielded the so-called difference signal [see Fig. 4-1(c)]. The difference signals which emphasize the LPs were used for further nonstationary signal processing in the time-frequency plane [13].

4.2.2 Comparison of Time-Frequency Representations by Simulated ECG Test Signals

The different linear and quadratic approaches of time-frequency representations, such as the spectrogram, the wavelet transform, and the Wigner distribution, transform a one-dimensional signal $x(t)$ into a two-dimensional function $X(t, \omega)$ of time and frequency. This fundamental property permits the important interpretation of how spectral components of the signal vary with time. First, the results obtained with simulated test signals for ECGs with and without late potentials will be discussed.

In order to enable a realistic and comparable behavior of the three methods, the test signal $x(t)$ was adapted to the difference signal obtained from real data after preprocessing. $x(t)$ is defined as:

$$x(t) = a_1 \sin(2\pi f_1 t) + a_2 \sin(2\pi f_2 t) + a_3 \sin(2\pi f_3 t) \qquad (4\text{-}2)$$

$$f_1 = 30\,\text{Hz} \qquad f_2 = 35\,\text{Hz} \qquad f_3 = 40\,\text{Hz}$$

$$a_1 = 50\,\mu\text{V} \qquad a_2 = 50\,\mu\text{V} \qquad a_3 = 50\,\mu\text{V}$$

Figure 4-2(a) shows the time course of the test signal comprising the three different sinewaves. An additional transient sinewave signal with a frequency of 80 Hz and a much lower amplitude than the other sinewaves is added in the region

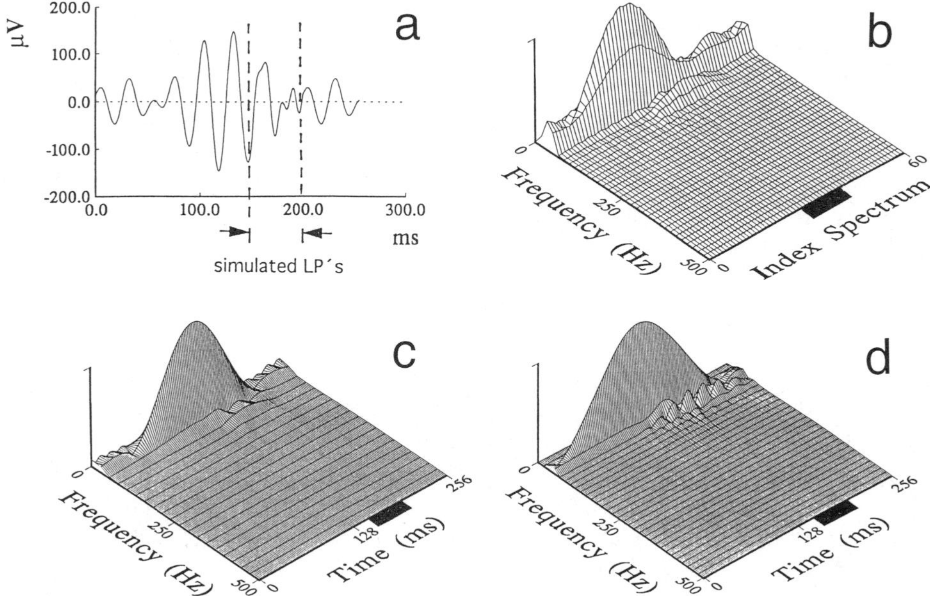

Figure 4-2 Time-frequency representations of the test signal are shown in (a). The test signal comprises three different sinewaves of 30 Hz, 35 Hz, and 40 Hz and a very low amplitude sinewave of 80 Hz between 150 ms and 200 ms. It simulates the difference curve with additional LPs. (b) The spectrogram is calculated for a segment length of 128 points, and a shift interval of 8 points. (c) The results from the wavelet transformation; (d) the results from the Wigner distribution. Details concerning the three different methodical approaches are explained in the text. The black bars on the time axis indicate the time intervals where the simulated LPs are present.

150 ms $\leq t \leq$ 200 ms to simulate LP activity during a fixed time interval. The results of the spectrogram, the scalogram, and the Wigner distribution computed for the test signal are demonstrated as three-dimensional (3-D) plots in the time-frequency plane. The short-time Fourier transform (STFT), defined as the Fourier transform of the windowed signal $x(\tau)g^*(\tau - t_i)$, was calculated for 60 different segments x_i up to a maximal frequency of 500 Hz:

$$\text{STFT}(t_i, \omega) = \int x(\tau)g^*(\tau - t_i)\exp(-j\omega\tau)d\tau \tag{4-3}$$

Because it is desirable to interpret the time-frequency representation in physical terms of energy distribution, a quadratic structure of the magnitude of the linear $\text{STFT}(t_i, \omega)$ is chosen as the spectrogram. In Fig. 4-2(b) the distribution of the spectrogram method is displayed. The time axis of the time-frequency plane is denoted by the "index spectrum" to indicate the consecutive number of overlapping segments x_i. The spectrogram was computed with a segment length of 128 points and a shift interval of 8 points. To reduce the effect of spectral leakage, we applied a Blackman–Harris window to each segment before computing the FFT. The simulated LP activity is located approximately in the frequency band 50 Hz $< f <$ 110 Hz and between spectrum index 24 and 48. As expected, the exact localization of the simulated late potentials activity, indicated by the bar on the index spectrum axis, was not correctly determined. Although an analysis of this type produces useful qualitative results regarding the simulated activity, careful attention must be paid to interpreting the quantitative features of the plot. To improve the time localization, a shorter window is required. However, the frequency resolution will then become worse. This is the consequence of the uncertainty principle, which states that the time resolution can only be improved at the expense of frequency resolution. Further details of this method are given in [11].

The highest time-frequency resolution is obtained by the quadratic Wigner distribution, which allows an energetic signal interpretation similar to the spectrogram:

$$W(t, \omega) = \int x\left(t + \frac{\tau}{2}\right)x^*\left(t - \frac{\tau}{2}\right)\exp(-j\omega\tau)d\tau \tag{4-4}$$

However, as mentioned in [11] and [14], due to the quadratic kernel, the Wigner distribution has the disadvantage of exhibiting interference, or so-called cross-terms, which carry redundant information, and show oscillatory behavior. Figure 4-2(d) demonstrates the Wigner distribution result. In contrast to the spectrogram, the time axis is scaled in milliseconds. It is obvious that the time-frequency resolution has been improved compared with the spectrogram. However, the simulated LP activity has not been precisely located and detected. There is still an inaccuracy in the on- and offset of the fluctuation. Furthermore, the detailed shape of the plot is due to the cross-terms of the Wigner distribution, which leads to redundant or irrelevant information.

To overcome the disadvantages of the cross-terms of the Wigner distribution and the resolution limitations of the STFT, the multiresolution analysis by wavelet transformation is an alternative. The mathematical background of this method is given in [15, 16]. The continuous wavelet transform of a signal $x(t)$ is defined as

$$\text{WT}(t, a) = \frac{1}{\sqrt{a}} \int x(\tau) h^* \left(\frac{\tau - t}{a} \right) d\tau = \sqrt{a} \int X(\omega) H^*(a\omega) \exp(j\omega t) d\omega \qquad (4\text{-}5)$$

where $h(t)$ is the so-called analyzing wavelet. $X(\omega)$ and $H(\omega)$ are the Fourier transforms of $x(t)$ and $h(t)$, respectively. The analysis can be viewed as a filter bank comprising bandpass filters with bandwidths proportional to frequency.

A frequently used analyzing wavelet, which we also applied, is the modulated Gaussian function, or Morlet wavelet:

$$h(t) = \exp \left(-\frac{t^2}{2} + j\omega_0 t \right), \qquad \omega_0 = 5.33 \qquad (4\text{-}6)$$

To optimize the computation time, the digital implementation of the $\text{WT}(t, a)$ is performed in the frequency domain.

The squared magnitude of the $\text{WT}(t, a)$ is called a *scalogram* [15]. The scalogram can be considered a modified version of the spectrogram characterized by the equivalent bandpass filter with a constant ratio of the center frequency/bandwidth. Remember that the filter bank, in the case of the spectrogram procedure, consists of bandpass filters with equal bandwidths.

The wavelet transform has been applied in Fig. 4-2(c). The scalogram illustrates the advantages of this type of energy distribution compared with the other methods. As indicated by the bar on the time axis, the simulated LP activity has been detected with sufficient accuracy. Although the frequency resolution has been slightly decreased, the wavelet transformation demonstrates an optimal compromise for signal separation regarding the late potentials.

4.3. APPLICATION OF TIME-FREQUENCY TRANSFORMATIONS TO CLINICAL ECG DATA

4.3.1 Evaluation of Time-Frequency Representations

The simulation results suggest the superiority of the wavelet transform concerning time and frequency resolution of our ECG data records. In particular, time localization of the simulated activity was most accurate using this method. This interesting property should be further studied with real clinical data.

The ECG recordings of the control and the patient groups were preprocessed according to the methods outlined in section 4.2.2. Further investigations were oriented to the processing procedures of the test signal, i.e., spectrograms, scalograms, and Wigner distributions were calculated for the control and patient groups of both hospitals. Figures 4-3, through 4-5 demonstrate the representative results of each method for a healthy subject and a patient with sustained ventricular tachycardia. All three approaches show significant differences in their energy distributions between patients and healthy subjects. Although the typical time-frequency structure

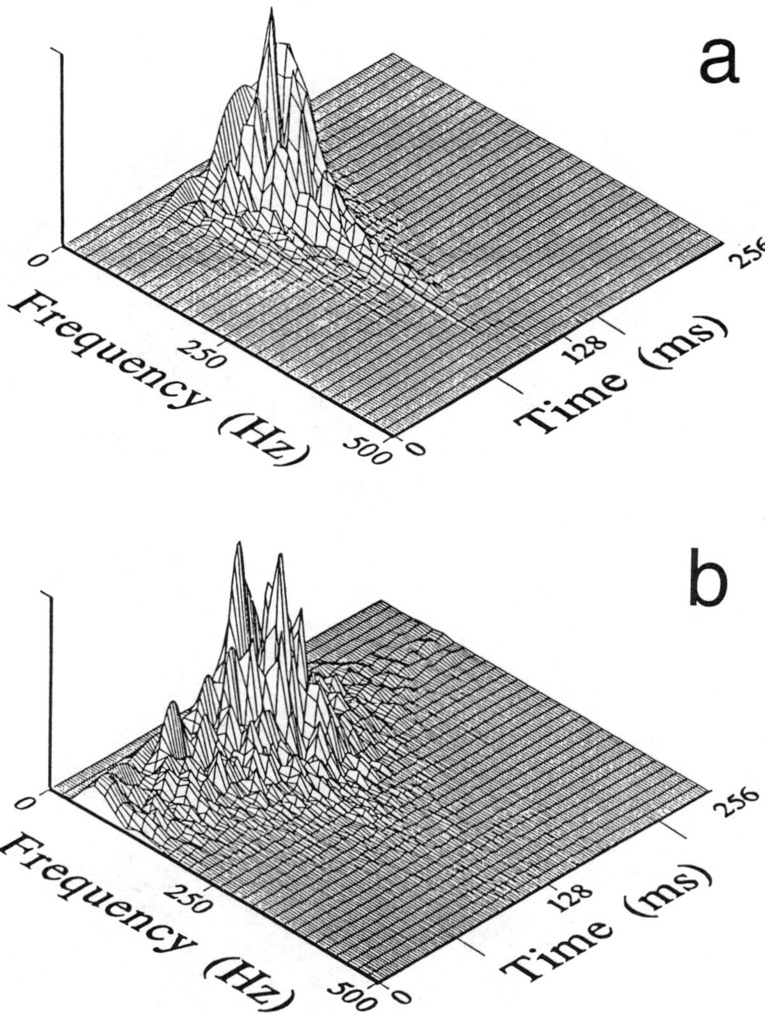

Figure 4-3 Wigner distribution of (a) a healthy control subject and (b) a patient with ventricular tachycardia.

of the QRS complex is basically similar, it is evident that for the patient with VT, the spread of energy in the time-frequency plane has been enhanced. Typically, frequency activity started earlier and, in most cases, it was persistent during the late phase of the QRS complex.

As demonstrated in Fig. 4-3, the refined signal structure of the control subject is best displayed by the Wigner distribution. However, due to the existence of cross-terms, the Wigner distribution reveals too much information and is rather confusing. Furthermore, comparisons of the three methods reveal that the best time resolution is achieved by the wavelet transform (see Fig. 4-4). The late potential activity is located mainly in the time interval from 120 ms to 190 ms with a frequency band-

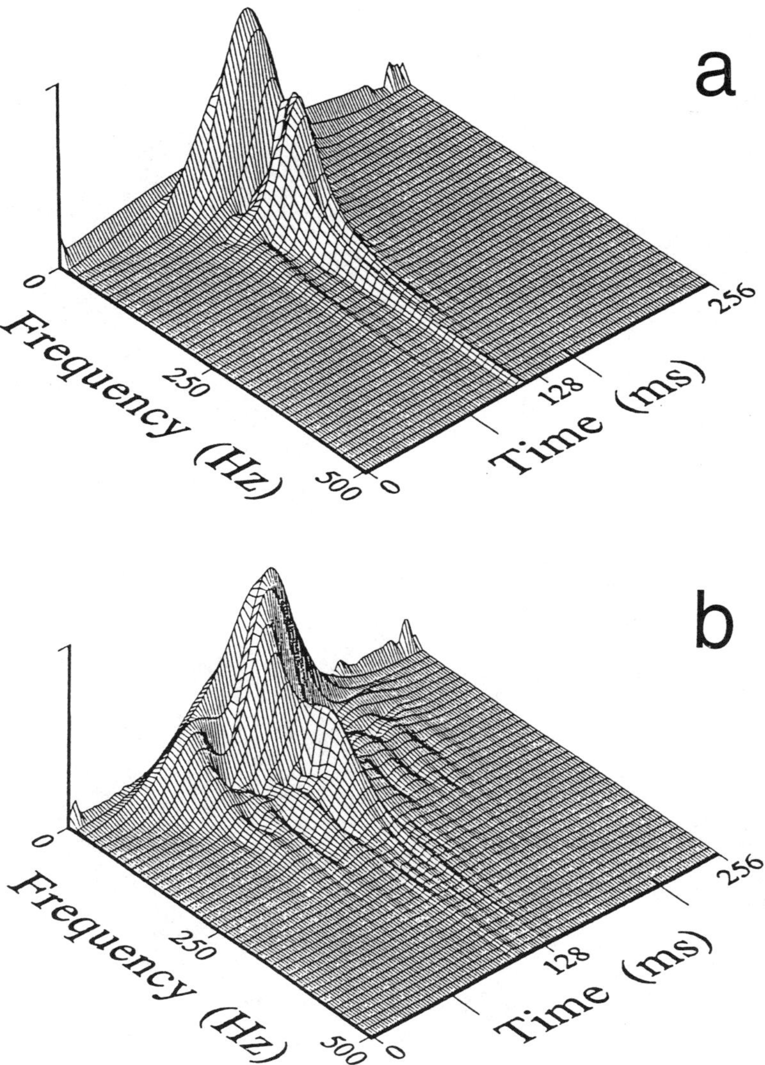

Figure 4-4   Scalograms calculated for (a) a healthy control subject and (b) a patient with ventricular tachycardia.

width of 80 Hz to 200 Hz. The results of the spectrogram method shown in Fig. 4-5 clearly demonstrate some disadvantages, particularly in the blurred temporal resolution compared with the distributions of the two other time-frequency representations.

The documented and examined results of all patients and control subjects of both hospitals more or less always show the same behavior. Thus, the conclusion suggested by the test signal study is confirmed by this real clinical data evaluation. Indeed, the wavelet transform seems to be an adequate method for quantitative

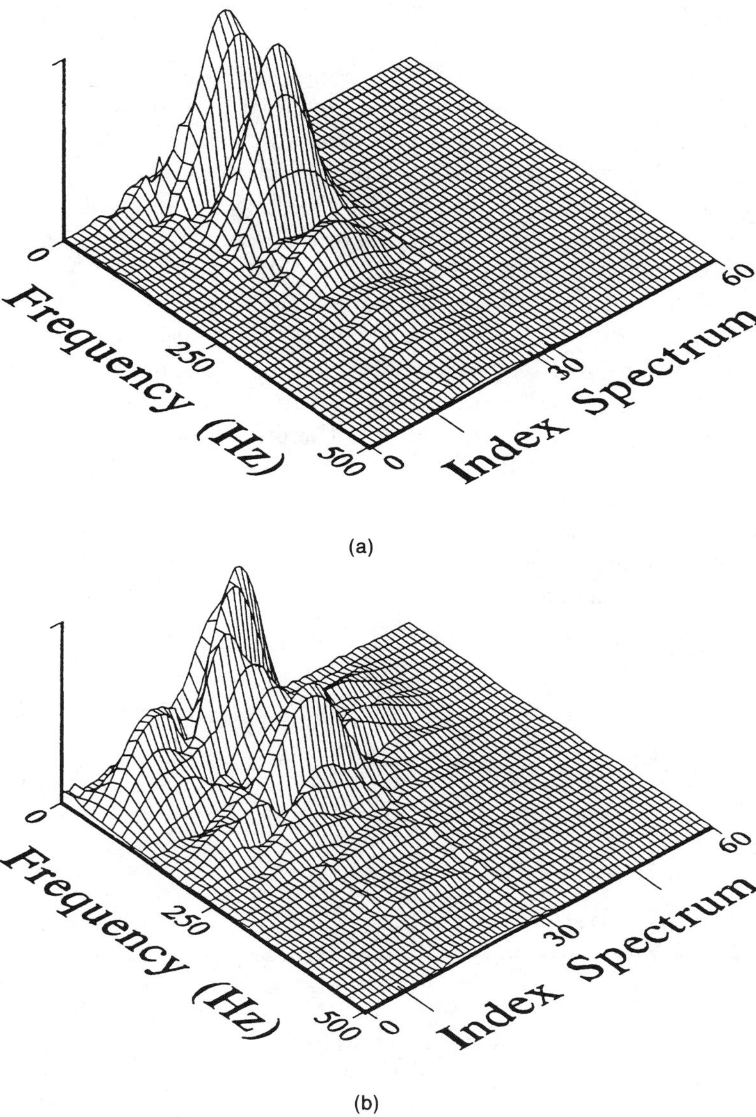

Figure 4-5 Spectrograms calculated for (a) a healthy control subject and (b) a patient with ventricular tachycardia.

energy representation of LPs in the time-frequency plane. However, if the outlined scalogram method is to be proposed as a diagnostic tool in cardiology, easy-to-calculate predictive parameters have to be derived for discrimination or classification purposes. Therefore, the next section introduces more quantitative investigations to illustrate the performance of the wavelet method using parameters extracted from the scalogram involving the different data samples of all investigated patients and healthy subjects.

4.3.2 Parameter Optimization for Classification Purposes

So far, the scalograms have been subjectively evaluated in their shape and extension relative to frequency and time. However, presumably there exists a particular region in the time-frequency plane for which the difference in energy between the control and patient group is most evident. To investigate this question, the time-frequency plane was divided into four regions, as illustrated in Fig. 4-6. Region R1 was defined in the low-frequency range (approximately up to 100 Hz) and for the first 100 ms from the onset of the QRS complex. The second area, R2, extended this region up to the end of the QRS complex. The third region, R3, enclosed the higher frequencies of the QRS complex and in R4, the high-frequency content of the late phase of the QRS complex should be localized. For every subject of the patient and control groups, the energy was calculated as a discriminating parameter for these different regions of the scalograms. The discrimination performance of the energy parameters was estimated by the statistical values of sensitivity and specificity for the classification between healthy subjects and patients. The mean value of sensitivity and specificity SS/2 was subsequently used as a criterion for the quality of the discrimination procedure.

First, examinations with the group I data from Heidelberg confirmed our hypothesis that the normalized energy of the R4 region (100 Hz–300 Hz and the

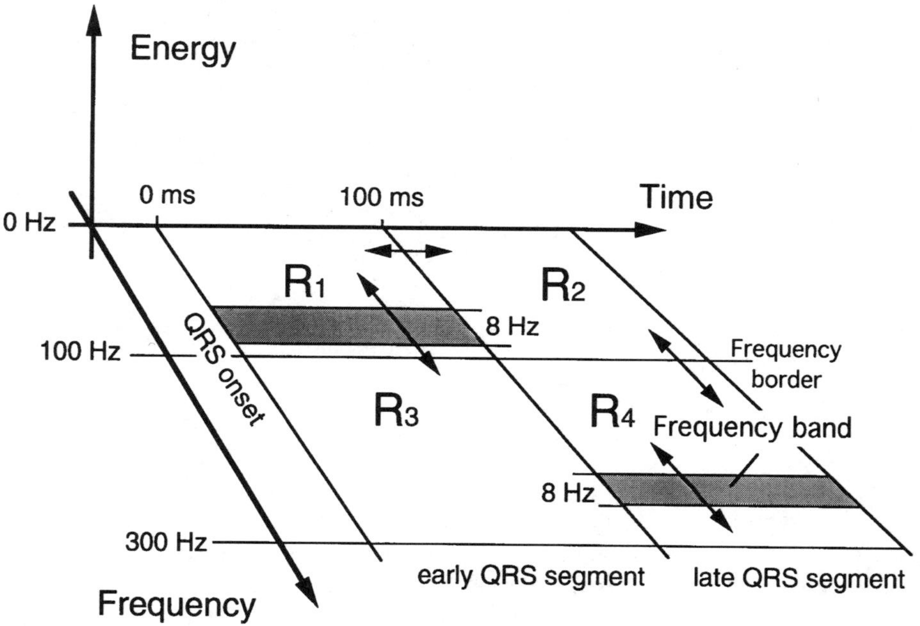

Figure 4-6   The time-frequency plane divided into four regions (R1 to R4) considering the typical time-frequency structure of ECG signals of patients with late potential activity. Further details are explained in the text.

last segment of the QRS complex) was particularly appropriate to discriminate between patients with VT and healthy subjects, with about 90% of correct classifications [17].

A second detailed study was carried out to look more systematically for the time-frequency interval with the best discrimination power. We varied the frequency border between the upper and lower bands. Whereas the upper frequency of areas R3 and R4 did not affect the results appreciably, the frequency border for values lower than 80 Hz and higher than 150 Hz showed significant shifts in discrimination power. However, for the energy of the R4 region, the SS/2 value was always higher than for R3, and yielded nearly constant high discrimination power of above 90% when the lower frequency border ranged between 80 Hz and 150 Hz.

We tried to further discriminate the different groups using the energy of a small frequency band. The position of this band was varied over the whole frequency range, as demonstrated in Fig. 4-6. The SS/2 value already mentioned was used as the criterion by which the energy calculated for the different frequency band positions was evaluated. As an example for these examinations, Fig. 4-7 shows a comparison between the data of the Heidelberg hospital group I [left side, Fig. 4-7(a), 4-7(c)] and the Münster hospital group II [right side, Fig. 4-7(b), 4-7(d)]. The tuning curves are calculated for a frequency bandwidth of 8 Hz. The plots of the upper row are estimated for the late QRS segment, starting at the time axis when the magnitude of the QRS vector is less than 60 μV. In comparison to this definition of the time segments, the plots of the lower row in Fig. 4-7 are achieved with a late QRS segment, which starts at a constant interval of 86 ms after QRS onset. The tuning plots show slightly different effects for the two patient groups of the Heidelberg and Münster study. If one takes into account that the patients of the Heidelberg study were strongly selected for VT and tested with programmed electrical stimulation, the higher discrimination power of at most 10% is easy to understand. On the other hand, even maximal numbers of 90% correct classification using a constant time border for the post-myocardial patients of the Münster study are encouraging.

A fixed amplitude threshold of 60 μV [upper row, Fig. 4-7(a), 4-7(b)] seems to yield a slightly lower maximal discrimination power for the late segment than a fixed length of the early QRS complex and a corresponding late segment with variable duration [lower row, Fig. 4-7(c), 4-7(d)]. This behavior was found in both data sets from the Heidelberg hospital as well as from Münster. Although the ECG signals of both patient groups recorded at different hospitals were independently acquired, the evaluations show a considerable similarity. This fact confirms the plausibility of the conclusions.

All demonstrated results of the scalogram method are calculated by the Morlet wavelet. However, many other wavelet functions with different characteristic features are proposed. In order to get optimal discrimination values, we tested other different functions with identical clinical data sets. As an example, Fig. 4-8 compares the calculated contour plots of a patient with VT [lower row, Fig. 4-8(c), 4-8(d)] and a healthy subject [upper row, Fig. 4-8(a), 4-8(b)] for the Morlet or Gaussian wavelet (left side) and the Hamming wavelet (right side). As shown in the plots, the results for the different wavelets are in principle quite similar. Without dealing with more

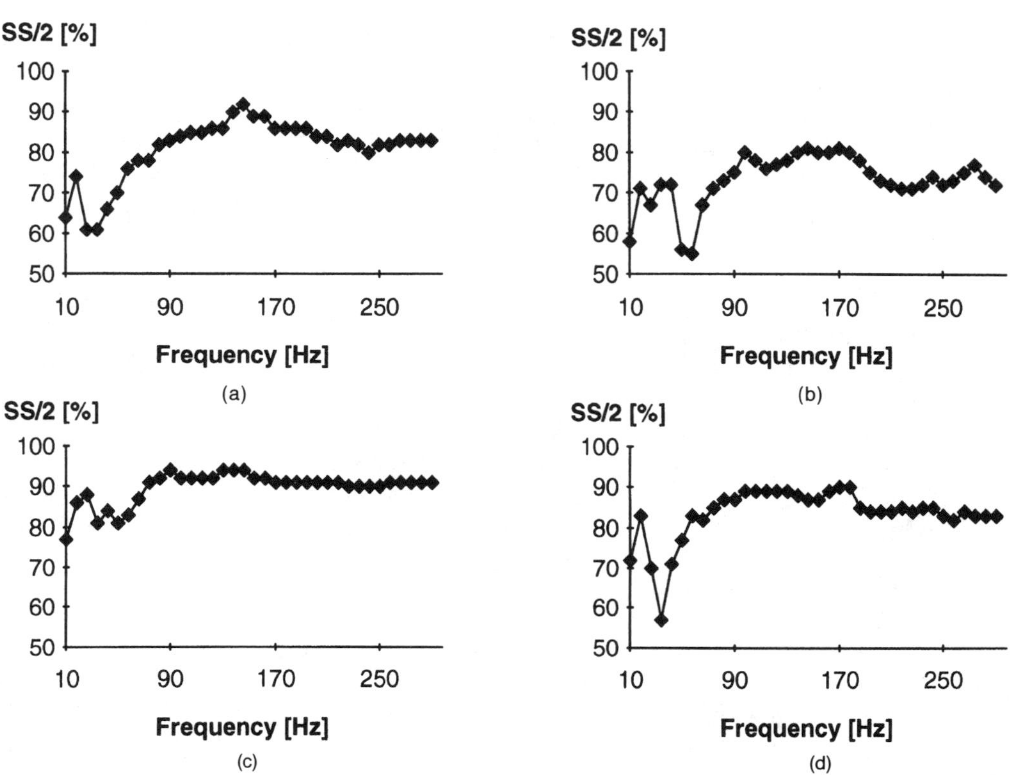

Figure 4-7 Mean value of sensitivity and specificity SS/2 calculated using the energy of the late QRS segment within a frequency band of 8 Hz at different positions along the frequency axis for the two patient and control groups. (a, b): $60\,\mu$V threshold defines the late segment; (c, d): fixed length of the early QRS segment causes the variable duration of the late segment.

details at this point of the investigation, the quantitative evaluation confirmed the superiority of the Morlet wavelet.

4.4. CONCLUSION

In summary, one can say that nonstationary signal analysis performed by time-frequency transformations is well suited to describe the complex behavior of cardiac late potentials in ECGs. The kind of transformation that is most appropriate is difficult to decide a priori, and requires explorative examinations of the problem. From our experimental simulation results as well as from the real data evaluation, it is

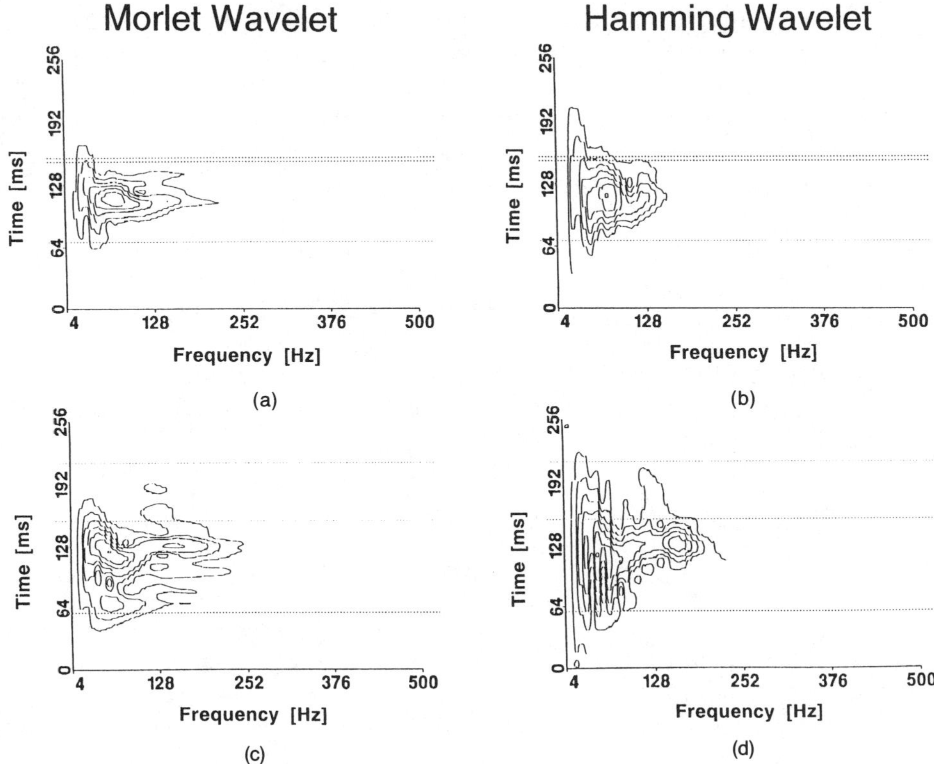

Figure 4-8   Contour plots of scalograms calculated for different wavelet functions. (a, c): Morlet wavelet; (b, d): Hamming wavelet. (a, b): healthy control subject; (c, d): patient with ventricular tachycardia.

evident that the wavelet transform method achieves a useful representation of the preprocessed ECG signal in the time-frequency plane. The plots of the scalograms enable an extended understanding of the electrophysiological basis of these phenomena. Detection accuracy of the wavelet transform, tested with low-amplitude sinusoidal waves, is higher than was achieved with the other two methods. The frequency resolution is much better than the common spectrogram technique allows, and only slightly worse compared with the Wigner distribution. In our application, the cross-terms of the Wigner distribution are due to the limited value of this approach.

Discrimination between patients with ventricular tachycardia and myocardial infarction as well as the control subjects was possible with remarkably good results. It could be demonstrated that the signal energy of a particular time-frequency region is an interesting feature by which more than 90% correct classifications for all patients and control subjects was achieved. The most sensitive region of the time-frequency plane for discrimination is characterized by a frequency band between 80 Hz and 150 Hz and a time segment 90 ms after QRS onset.

As the wavelet function, we used the modulated Gaussian function or Morlet wavelet. Other types of wavelets were tested, but with minor success. Unpublished studies with the same data sets using either time or frequency parameters alone, did not attain comparably good results. Therefore, our experiences suggest that ECG late potential analysis in the time-frequency plane provides an encouraging tool for the identification of patients with high cardiac risk. This method should be regarded as a powerful complementary noninvasive tool in addition to the well-established invasive method of electrical heart stimulation.

ACKNOWLEDGMENT

The authors are grateful to Dr. J. Brachmann from the University Hospital of Heidelberg and Dr. T. Fetsch from the University Hospital of Münster for putting the ECG data sets at our disposal.

REFERENCES

[1] E. J. Berbari and R. Lazzara, "The significance of electrocardiographic late potentials: Predictors of ventricular tachycardia," *Annu. Rev. Med.*, vol. 43, pp. 157–169, 1992.

[2] M. Malik, O. Odemuyiwa, J. Poloniecki, P. Kulakowski, T. Farrell, A. Stauton, and A. J. Camm, "Late potentials after acute myocardial infarction. Performance of different criteria for the prediction of arrythmic complications," *Eur. Heart. J.*, vol. 13, pp. 599–607, 1992.

[3] J. Schramm, C. Frumento, H. Dickhaus, C. Schmitt, and J. Brachmann, "Ventricular late potentials in patients with and without inducible sustained ventricular tachycardia: A combined analysis in the time and frequency domain," *Eur. Heart. J.*, vol. 12, pp. 563–572, 1991.

[4] E. J. Berbari, B. J. Scherlag, R. R. Hope, and R. Lazzara, "Recording from the body surface of arrhythmogenic ventricular activity during the ST-segment," *Am. J. Cardiol.*, vol. 41, pp. 697–702, 1978.

[5] M. B. Simson, "Use of signals in the terminal QRS complex to identify patients with ventricular tachycardia after myocardial infarction," *Circulation*, vol. 64, pp. 235–242, 1981.

[6] G. Breithardt, M. E. Cain, N. El-Sherif, N. C. Flowers, V. Hombach, M. Janse, M. B. Simson, and G. Steinbeck, "Standards for analysis of ventricular late potentials using high resolution or signal-averaged electrocardiography," *Eur. Heart J.*, vol. 12, pp. 473–480, 1991.

[7] M. E. Cain, H. D. Ambos, F. X. Witkowski, and B. E. Sobel, "Fast Fourier transform analysis of signal-averaged electrocardiograms for identification of patients prone to sustained ventricular tachycardia," *Circulation*, vol. 69, pp. 711–720, 1984.

[8] L. Khadra, H. Dickhaus, and A. Lipp, "Representations of ECG late potentials in the time frequency plane," *J. Med. Eng. Tech*, vol. 17, no. 6, pp. 228–231, 1993.

[9] N. V. Thakor, G. Xin-Rong, S. Yi-Chun, and D. F. Handley, "Multiresolution wavelet analysis of evoked potentials," *IEEE Trans. Biomed. Eng.*, vol. 40, no. 11, pp. 1085–1094, 1993.

[10] M. Akay, Y. Akay, W. Welkowitz, and S. Lewkowicz, "Investigating the effects of vasodilator drugs on the turbulent sound caused by femoral artery stenosis using short-term Fourier and wavelet transform methods," *IEEE Trans. Biomed. Eng.*, vol. 41, no. 10, pp. 921–928, 1994.

[11] W. Williams, "Recent advances in time-frequency representations: Some theoretical foundations," Chapter 1, this volume, pp. 3–43.

[12] C. Frumento, H. Dickhaus, and J. Schramm, "Are the so-called late potentials restricted to the terminal portion of the QRS-complex?" In *Proc. Medical Informatics Europe.* R. Hansen et al. (eds.), Berlin: Springer Verlag, pp. 415–419, 1988.

[13] H. Dickhaus, L. Khadra, A. Lipp, and M. Schweizer, "Ventricular late potentials studied by nonstationary signal analysis," in *Proc. Int. Conf. IEEE EMBS*, J. P. Morucci et al. (eds.), vol. 14, pp. 490–491, 1992.

[14] W. Martin and P. Flandrin, "Wigner–Ville spectral analysis of nonstationary processes," *IEEE Trans. Acoust., Speech, Signal Proc.*, vol. 33, pp. 1461–1470, 1985.

[15] O. Rioul and M. Vetterli, "Wavelets and signal processing," *IEEE Signal Proc. Mag.*, pp. 14–38, 1991.

[16] O. Rioul and P. Duhamel, "Fast algorithms for wavelet transform computation," Chapter 8, this volume, pp. 211–242.

[17] H. Dickhaus, L. Khadra, and J. Brachmann, "Time-frequency analysis of ventricular late potentials," *Meth. Inform. Med.*, vol. 33, no. 2, pp. 187–195, 1994.

Chapter 5

Time-Frequency Distributions Applied to Uterine EMG: Characterization and Assessment

Jacques Duchêne, Dominique Devedeux

5.1. INTRODUCTION

Despite the great medical strides that have been observed over the last 20 years, the frequency of preterm birth has not significantly decreased, and remains the first cause of perinatal morbidity and mortality. In addition to the medical impact of such perinatal problems, the economic importance of preterm delivery has been shown [1]. The longer stay of preterm infants in a neonatal intensive care environment notably increases health expenses. Some of them present neurological or physical troubles requiring long-term care and specialized education systems. Every gain in terms of *in utero* preservation allows better fetal maturation, and thus higher viability. The detection of preterm delivery risk seems to be one deciding factor for the prolongation of the *in utero* fetal stay, and this detection is strongly related to uterine contractility. However, current detection tools are still too invasive, too subjective, or too inaccurate to assess this muscle contractility.

The idea is to assess this factor using temporal and spectral parameters of the abdominal electromyography (EMG). As a matter of fact, several studies have recently shown the relevance of indices contained within the abdominal EMG. Externally recorded using classical surface electrodes on the abdomen, this signal can be shown as an addition of internal fiber electrical activities contaminated by uncorrelated noise, mechanical activities, and electrical signals coming from surrounding muscles. Now these internal fiber activities are directly generating the mechanical process. The analysis of such a signal is thus of great interest for uterine contractility characterization purposes.

The basic component of cellular electrical activity is the action potential. During a contraction, each cell generates a burst of electrical activity made of a series of action potentials. It is now well known that the firing frequency of the action potentials within a burst varies with the contraction phase. Increasing at the beginning of the contraction, it stabilizes and then decreases as soon as the mechanical effect reaches its maximum [2]. In terms of contractility indices, the contraction frequency is related to the cell excitability when the firing frequency of the action potentials within a burst has a direct effect on the contraction strength, classically measured by means of intrauterine pressure (IUP). The uterus presents some activity all during pregnancy, but this basic activity differs from labor in its lower appearing frequency and especially in its much lower amplitude and firing frequency. Therefore, it is crucial to obtain as much information as possible on the burst firing rate if the purpose is to predict possible premature labor.

Concerning the signal characteristics, as mentioned earlier, the EMG can be described as a nonstationary signal made of successive bursts. Within each burst, the carrier is made of pulses corresponding to the successive action potentials and the position modulation of these pulses is directly related to the contraction efficiency. Due to its nonsinusoidal carrier, this kind of frequency modulation consequently produces harmonic components which are easily detectable on classical spectral representations (Figure 5-1), whereas the frequencies of the modulating signal are

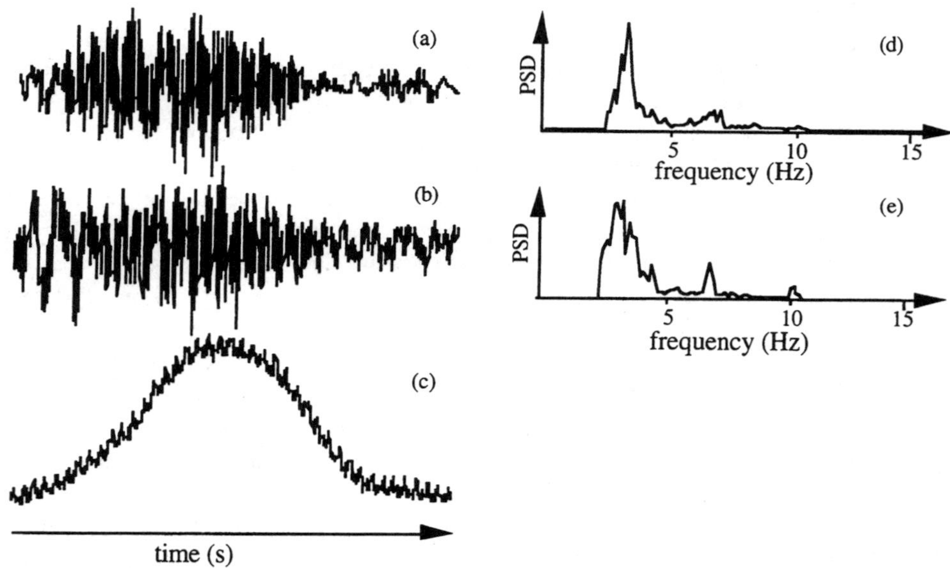

Figure 5-1   Signals simultaneously recorded on a cynomolgus monkey during the last third of gestation: (a) uterine EMG recorded internally by monopolar wire electrodes; (b) uterine EMG recorded externally by abdominal bipolar electrodes; (c) corresponding IUP recorded by means of a catheter inserted in the amniotic cavity; (d) PSD of (a); (e) PSD of (b).

located around the position of the carrier frequency as usual. The processing methods to be applied to this kind of signal are thus derived from nonstationary tools, allowing first the analysis of the instantaneous frequency variation within each burst, then the detection of harmonic components, attesting to the presence of the genuine internal nonsinusoidal carrier.

When speaking of instantaneous frequency, the primary idea is to refer to direct extraction using the classical methods related to the definition of the FM waveforms. Referring to sinewave carrier, an expression for the general FM waveform is

$$\text{FM}(t) = A \cos\left[\omega_0 t + k \int_0^t f(\tau)d\tau \right] \tag{5-1}$$

where A is the amplitude, $f(t)$ is the modulating signal, ω_0 is the carrier basic angular frequency, and k is a constant defining the modulation range. If the carrier is not a sinusoidal signal, it can, however, be written as a sum of sinewaves whose frequencies are equal to multiple integers of the fundamental, and whose amplitudes are given by the Fourier series coefficients. Since the Fourier transform is a linear operator, the spectrum of such a nonsinusoidal FM carrier consists of a combination of classical FM sinusoidal components. From this point, many methods are available to return to the modulating signal. The first can be derived from the classical demodulation technique: assuming that most of the signal power is concentrated around the fundamental component of the carrier (owing to the low-pass filtering effect of the tissues between signal generation and recording location) the demodulation could be achieved by a differentiation changing the FM modulation to AM modulation. Then, extraction of the modulating signal can be made by means of an envelope detector as long as $k \cdot f(t)$ is negligible compared to ω_0. An equivalent result can be obtained in an empirical manner by computing either the interpulse interval width or the number of zero crossings within a stationary time window. Specific algorithms are available for the measurement of the instantaneous frequency within this scope of FM assumption [3]. However, it can seem more justified to refer to the instantaneous frequency as the derivative of the phase of the analytic signal $z(t)$, composed of the input signal $x(t)$ as its real part and the Hilbert transform of $x(t)$ as its imaginary part:

$$z(t) = x(t) + j\,\mathbf{H}[x(t)]$$
$$\Omega(t) = \frac{d}{dt}\arg[z(t)] \tag{5-2}$$

As a matter of fact, this definition does not imply any assumption on any specific modulation form. In addition, efficient computational algorithms are available [4], and it is possible to prove a relationship between instantaneous frequency as previously described and classical time-frequency distributions [5]. However, the expression of the phase of $z(t)$ has no reason to be equivalent to the phase of the original modulated signal [6], and the definition of the instantaneous frequency derived from the analytic signal can lead to different results with respect to the definition derived from the FM expression. Nevertheless, in a previous study, we tried to extract the

EMG modulation by this class of methods [7,8]. Unfortunately, the results presented an important scatter around the true modulation, associated with a shift between true and computed modulation. Therefore, owing to this high level of variance, this method, relying on direct computation of the instantaneous frequency, obviously produced results which were difficult to use in terms of the quantitative relationship between the computed frequency and the behavior of the mechanical event represented by the IUP. In addition, it did not account for the presence of harmonic patterns which are specific to this kind of signal. These are the main reasons why we decided to apply time-frequency distributions on our signals in order to extract properly the main frequency component whose variation with respect to time seems to be directly related to contraction efficiency. Simultaneously, these methods are able to show the harmonic components, which are very useful in confirming the genuineness of the recorded signals in terms of physiological significance.

In order to validate the methods with respect to their objective of good frequency modulation detection, the signals used throughout this work are mainly divided into two parts. The first class is made of artificial signals where all parameters can be controlled to assess the main features of the various processing methods. As each EMG corresponds to the addition of many filtered cellular activities, an artificial signal contains the following features:

- Each action potential is simulated by a Dirac function.
- Each burst (simulating an individual cell) is made of successive Dirac functions whose position is modulated by a predefined shape (here, the inverse of a Blackman window), contaminated by a Gaussian white noise whose signal-to-noise-ratio (SNR) is perfectly known.
- The filtering effect of the tissues is simulated by a second-order low-pass Butterworth filter.
- The global signal is constituted by the addition of several elementary bursts whose onset time is taken as a Gaussian random parameter.

The final form of the artificial signal can be written as

$$s(t) = \sum_{k=1}^{K} \sum_{i=1}^{M} b\left(t - t_0^{(k)}(i) - t_k \right) \tag{5-3}$$

where $b(t)$ is the impulse response of the low-pass filter, M is the number of action potentials generated by the kth cell, K is the number of simulated cells, $t_0^{(k)}$ is the delay in position of each action potential related to each cell, and t_k is the onset delay of each cell. The parameters of interest for generation of various reference signals are the number of cells, the onset delay between cells, the SNR of the modulating signal, and the SNR of the whole signal.

The second group of signals is made of real recordings, which will be taken into account in a second step in order to confirm the results obtained on artificial signals. They are made of signals recorded internally on monkeys [9]. A preliminary report has been published in which external recordings from women are used, extracted either during pregnancy or parturition [10].

From that point, the problem is to define which time-frequency method is the best adapted for this specific problem of frequency modulation extraction, where two main features are specifically required: robustness (owing to high contamination level) and selectivity (right discrimination of spectral components). This latter feature is particularly important when the point is to classify the modulation shapes in terms of contraction efficiency. After a short description of each analyzed time-frequency distribution (TFD), we will define a set of criteria which are supposed to characterize properly these TFD with respect to our purpose. Some results on artificial signals will then confirm the efficiency of each method, and complementary results on real signals will confirm the interest of such processing techniques for pregnancy monitoring.

5.2. TIME-FREQUENCY DISTRIBUTIONS

5.2.1 The Parametric Approach: AR Modeling

Among various interesting properties, autoregressive (AR) modeling allows a direct estimation of the power spectral density (PSD) from the coefficient estimates, yielding a smoothed PSD estimate from the expression [11]:

$$S(\omega) = \frac{\sigma^2}{\left|1 + \sum_{k=1}^{p} a_k e^{-j\omega k}\right|^2} \tag{5-4}$$

where σ^2 is the variance of the white noise driving the process, p is the model order and a_k is the estimate of the kth coefficient. These estimates can be computed in many ways when referring to stationary processes (see, for instance, first Makhoul's tutorial review [12], then the reference work of Kay and Marple [13]). Otherwise, in the nonstationary case, there are several methods to describe the evolutive nature of an AR spectrum. Grenier [14] builds AR or autoregressive moving average (ARMA) models, where the coefficients are constructed from linear combinations of a set of functions defining the base. Martin [15] proposes an adaptive estimation of the spectral density, adapting the length of the memory to the stationarity of the signal. However, the most commonly used method is still the computation of the model on successive fixed-length windows after evaluation of the minimal stationarity duration [16]. The model order is classically determined from the final prediction error (FPE) or the Akaike information criterion (AIC) [17]. It can be defined once for the whole record, or specifically recomputed for each window [18]. A development of these order selection techniques can be found in Kaluzynski's work related to Doppler blood flow analysis [19].

In the present study, where no problem of optimization had to be considered at this stage, we decided to define AR models in successive stationary windows. A test for stationarity has been applied to all available signals in a preliminary study, in order to define the optimal common window length [20]. In the same way, a common optimal model order has been determined using the AIC criterion. As a result of these preliminary tests, the model order has been fixed to 15 and the window width to 6 s. Model coefficients have been computed for each window by the Burg lattice method.

5.2.2 Cohen's Class Distributions

The main developments corresponding to this TFD class are given in a previous chapter. For an exhaustive review, refer also to [21] or to Cohen's excellent synthesis [22]. Let us simply recall the general expression of a member of Cohen's bilinear class, which can be seen as the Fourier transform in the lag variable τ of a weighted nonstationary autocorrelation:

$$C_x(t, v; \phi) = \int\int\int \Phi(\eta, \tau) x\left(t' + \frac{\tau}{2}\right) x^*\left(t' - \frac{\tau}{2}\right) e^{-2j\pi\eta t} e^{-2j\pi\tau v} e^{2j\pi\eta t'} dt' d\tau \, d\eta \qquad (5\text{-}5)$$

The kernel $\Phi(\eta, \tau)$ defines the weighting and time averaging of the lag products [23]. Therefore, it determines the main properties of the resulting TFD: the choice of a specific distribution within the Cohen's class reduces to the choice of an adapted kernel.

From this point, the well-known Wigner–Ville distribution is derived from the general Cohen's class expression when taking the kernel as $\Phi(\eta, \tau) = 1$. Many authors start from this specific distribution to define kernels whose main property is to reduce the interference patterns induced by the distribution itself. Another approach to kernel design is to define it optimally in the sense of best resolution or best concentration [24,25], since the kernel effect on the auto-terms can be compared to the one-dimensional (1-D) window effect for the classical PSD evaluation. Recent developments now define signal-dependent kernels to offer improved time-frequency representations for specific classes of signals [26,27].

Among these various possibilities, we chose to test for the kernels which were as different as possible in terms of resulting effects on the distributions: the spectrogram, which is the most widely used yet; the smoothed pseudo-Wigner–Ville distribution, where filtering is defined separately in η and τ [28]; the Zhao–Atlas–Marks kernel, which emphasizes spectral peaks while preserving finite time support property [29,30]; and the Choi–Williams representation, where the optimization is thought of in terms of best cross-term reduction [31].

Spectrogram.   The classical definition of the spectrogram corresponds to the application of a specific short time window on the signal around each time sample before computing the squared-module of the Fourier transform. However, it is possible to show that the spectrogram is a special case of the Cohen's class distributions, where the kernel is related to the symmetrical ambiguity function of the short time window [21]:

$$\Phi(\eta, \tau) = \int h\left(t + \frac{\tau}{2}\right) h^*\left(t - \frac{\tau}{2}\right) e^{-2j\pi\eta t} dt \qquad (5\text{-}6)$$

Among all the possibilities, the Gaussian window is often used (whatever the TFD) because of its ability to produce good resolution. As a matter of fact, according to several authors, the Gaussian signal,

$$x(t) = Ae^{-t^2/\sigma^2} \tag{5-7}$$

is the most concentrated signal in time and frequency [25,32,33]. For this reason, we will make frequent use of this window in addition to others (especially Blackman), whatever the considered TFD.

Smoothed Pseudo-Wigner–Ville distribution.   Except for comparison purposes with respect to other methods [34] or preliminary studies [35], the direct use of the Wigner–Ville distribution as

$$\mathrm{WD}(t, v) = \int x\left(t + \frac{\tau}{2}\right) x^*\left(t - \frac{\tau}{2}\right) e^{-2j\pi v\tau} d\tau \tag{5-8}$$

is rarely encountered for biomedical applications, where the interference terms have classically no meaning in terms of physiological or clinical interpretation. It is usually associated with smoothing windows having separate effects on the time and frequency domains. In that form, it is called the smoothed pseudo-Wigner–Ville distribution, or in a shorter form, smoothed Wigner distribution (SW). Though it belongs to the Cohen's class, we prefer to use the following expression, where the separate effects of the filtering in time and frequency appear more clearly:

$$C_x(t, v) = \int \left| h\left(\frac{\tau}{2}\right) \right|^2 \left[\int g(t' - t) x\left(t' + \frac{\tau}{2}\right) x^*\left(t' - \frac{\tau}{2}\right) dt' \right] e^{-2j\pi v\tau} d\tau \tag{5-9}$$

Again, the Gaussian window is often encountered, especially for spectral smoothing, when a rectangular time window is sometimes taken [36]. The window choice is still determined by a good representation of the auto-terms as well as a strong attenuation of the cross-terms, which explains why one of the tested windows has been chosen as Gaussian for both time-averaging and frequency-smoothing in the present study.

Exponential kernel.   As indicated before, kernels based on Gaussian smoothing functions are widely used, essentially for their property of compactness [37]. The kernel (exponential distribution, ED) proposed by Choi and Williams [31] has the following form:

$$\Phi(\eta, \tau) = e^{-\eta^2 \tau^2/\sigma} \tag{5-10}$$

where σ is a scaling factor. It has been found by these authors that this distribution is very effective in decreasing the effects of the cross-terms while retaining most of the useful properties for a time-frequency distribution. The drawback of this distribution is the fact that it does not preserve time and spectral supports. This kind of representation is well used in the biomedical field. Some applications are very close to our problem since Zheng et al. [38] made use of ED for EMG (motor unit potentials) description, and Sahiner and Yagle [39] had the same kind of frequency tracking problem for blood flow speed determination in the domain of magnetic resonance imaging.

Cone-shaped kernel. This method seemed very attractive to us, because it was supposed to simultaneously preserve the property of finite time support, emphasize the spectral peaks, and smooth the cross-terms [29]. Though we did not find any application within the biomedical field, we tested this method on our signals as well, in order to have an idea of the influence of the finite time support on peak selectivity. The kernel can be defined as

$$\Phi(\eta, \tau) = \left| h\left(\frac{\tau}{2}\right) \right|^2 \frac{2|\tau|}{a} \operatorname{sinc}\left(\frac{2\pi\eta\tau}{a}\right) \tag{5-11}$$

for the equation of the distribution given as

$$C_x(\tau, v) = \int \left| h\left(\frac{\tau}{2}\right) \right|^2 \left[\int_{t-\frac{|\tau|}{a}}^{t+\frac{|\tau|}{a}} x\left(t' + \frac{\tau}{2}\right) x^*\left(t' - \frac{\tau}{2}\right) dt' \right] e^{-2j\pi v\tau} d\tau \tag{5-12}$$

We only considered the case where $a = 1$, applying a Gaussian window as $h(t)$.

5.2.3 Signal-Dependent Optimal Kernel

The choice of a specific distribution within the Cohen's class reduces to a simple kernel selection without any reference to signal features. We decided to test TFDs defined with an optimal kernel, in the sense of an optimal adaptation to the main signal features. This class of signal-dependent transformations has been widely used in different ways by many authors. Kadambe et al. [40] combined AR modeling and adaptive filtering to design an optimal kernel. Jones and Parks [24,25] made use of an index of local signal concentration to adapt the shape of a Gaussian kernel in time and frequency. This signal-dependence is also characteristic of the time-frequency representation proposed by Baraniuk and Jones [26,27]. They propose the automatic computation of a radially Gaussian kernel designed as adapted to the signal under study, and recomputed for each successive time window. The quality of the final time-frequency distribution is measured by a performance index based on a criterion of energy concentration:

$$\max_{\Psi} \int_0^{2\pi} \int_0^\infty |A(r, \Psi)\Phi(r, \Psi)|^2 r \, dr \, d\Psi \tag{5-13}$$

with

$$\Phi(r, \Psi) = e^{-r^2/2\sigma^2(\Psi)} \tag{5-14}$$

subject to a constraint of maximum kernel volume:

$$\frac{1}{2\pi} \int_0^{2\pi} \int_0^\infty |\Phi(r, \Psi)|^2 r \, dr \, d\Psi \leq \alpha \qquad \alpha \geq 0 \tag{5-15}$$

The specific configuration of the Gaussian kernel allows us to reduce the latter constraint to:

$$\frac{1}{2\pi}\int_0^\pi \sigma^2(\Psi)\,d\Psi \le \alpha \qquad (5\text{-}16)$$

As the computation has to be achieved in the radial directions, Baraniuk and Jones propose a fast algorithm based on a minimum-junction tree describing approximately a sampling of the radial directions [26].

The kernel can be computed globally on the whole time record, leading to a single time-frequency representation, or recomputed locally at each time position, yielding a different kernel, and then a frequency vector for each of these positions. This kind of signal-dependent TFD seems well adapted to our problem where the frequency modulation to be detected increases in the first phase of the contraction and decreases afterwards. We have to decide if the kernel has to be defined locally or globally and, in the latter case, whether it is necessary to compute the kernel for each individual signal or if a standard model can be defined for a whole class of signals. However, the locally defined kernel has been chosen for the current evaluation.

5.2.4 Reassignment Procedure

The previously described methods directly act on the definition of the kernel in order to optimize the resulting transformation, using various criteria and features such as filtering capabilities, optimal local concentration, or good adaptation to the signal ambiguity function. Another way to optimize the final time-frequency representation could be to apply any TFD first, then to transform the result in order to limit, for instance, the spreading of the autocomponents by an operation of reassignment. This method was introduced first by Kodéra et al. [41], but its use has been limited by implementation problems and an absence of theoretical proofs of its efficiency. It was recently revisited by Auger and Flandrin [42], who demonstrated the validity and the interest of such a method for TFD optimization.

The basic principle of the method can be expressed from the general expression of the Cohen's class TFD, using the Wigner–Ville distribution as a starting point:

$$C_x(t,\omega) = \frac{1}{2\pi}\int\int \Phi(u,\Omega)\left[\int x\!\left(t+\frac{\tau}{2}\right)x^*\!\left(t-\frac{\tau}{2}\right)e^{-j\omega\tau}d\tau\right]du\,d\Omega \qquad (5\text{-}17)$$

where $\Phi(u,\Omega)$ represents the kernel of a specific Cohen's class element. In this expression, the role of the kernel appears as a weighting of the Wigner–Ville distribution values, so that each point of the final representation is the result of an operation of averaging. When the kernel corresponds to a two-dimensional (2-D) low-pass filter, this process obviously leads to a reduction of the amplitude and the number of cross-components. However, it also spreads the autocomponents onto the time-frequency plane, which is much less desirable. The reassignment process consists of moving the value computed at position (t,ω) toward the mean position of the weighted values of the corresponding Wigner–Ville distribution:

$$\hat{t}(t, \omega) = t - \frac{\int\int u\Phi(u, \Omega)\left[\int x\left(t + \frac{\tau}{2}\right)x^*\left(t - \frac{\tau}{2}\right)e^{-j\omega\tau}d\tau\right]du\,d\Omega}{\int\int \Phi(u, \Omega)\left[\int x\left(t + \frac{\tau}{2}\right)x^*\left(t - \frac{\tau}{2}\right)e^{-j\omega\tau}d\tau\right]du\,d\Omega} \qquad (5\text{-}18)$$

$$\hat{\omega}(t, \omega) = \omega - \frac{\int\int \Omega\Phi(u, \Omega)\left[\int x\left(t + \frac{\tau}{2}\right)x^*\left(t - \frac{\tau}{2}\right)e^{-j\omega\tau}d\tau\right]du\,d\Omega}{\int\int \Phi(u, \Omega)\left[\int x\left(t + \frac{\tau}{2}\right)x^*\left(t - \frac{\tau}{2}\right)e^{-j\omega\tau}d\tau\right]du\,d\Omega} \qquad (5\text{-}19)$$

Thus the modified representation is the result of the reassignment at position (t', ω') of all values of the initial representation moved toward this new position:

$$C_x(t', \omega') = \frac{1}{2\pi}\int\int C_x(t, \omega)\,\delta(t' - \hat{t}(t, \omega))\delta(\omega' - \hat{\omega}(t, \omega))dt\,d\omega \qquad (5\text{-}20)$$

This reassignment process presents the drawback that the transformation is no longer bilinear, but still preserves the support in time and frequency as well as the signal energy, so that it can still be considered as an energy distribution. The authors underscore the fact that the reassignment method yields a perfectly localized representation of a linearly modulated signal whatever the kernel, whereas this property is only verified by the Wigner–Ville distribution within the Cohen's class.

Though the reassignment principle can be applied on most of the TFDs, we chose to analyze its effect on the spectrogram, since the main problem of this representation corresponds to the necessary compromise between spreading in time and frequency. The former expressions can be modified according to the specific form of the spectrogram definition as related to the square module of the Fourier transform:

$$S_{h,x}(t, \omega) = \left|\int x(u)h^*(t - u)e^{-j\omega u}du\right|^2 \qquad (5\text{-}21)$$

which can be expressed in the same way as before, using the Rihaczek distribution:

$$S_{h,x}(t, \omega) = \frac{1}{2\pi}\int\int \text{Ri}_h^*(u, \Omega)\text{Ri}_x(t - u, \omega - \Omega)\,du\,d\Omega \qquad (5\text{-}22)$$

$$\text{Ri}_x(t, \omega) = x(t)X^*(\omega)e^{-j\omega t} \qquad (5\text{-}23)$$

Finally, it is interesting to note that the effect of the window width differs before and after reassignment: on spectrograms, the limit when this width approaches zero corresponds to the instantaneous power without any reference to spectral localization. On the other hand, after reassignment, all frequency points will be moved toward the instantaneous frequency position. The same observation could be made for the reassignment on the group delay position when $h(u)$ is taken as a constant.

5.3. CRITERIA FOR DETERMINING THE REPRESENTATION QUALITY

5.3.1 Back to the Initial Problem: Modulation Extraction

In this part of the work, the aim is not to define criteria approaching the overall representation quality, but to relate the quality of the results to the problem of optimal extraction of the modulating signal within a burst of uterine EMG. Thus the first point is to implement an algorithm for the extraction of this specific component. Then the quality assessment will be achieved using, simultaneously, criteria expressing the information concentration around the true modulation and the ability of a given method to discriminate between near components.

Two methods of modulation extraction have been implemented. The first one (PEAK) is derived from a classical peak detection applied to any time position in the time-frequency representation. However, the genuine component to be detected in a real signal does not necessarily exhibit the highest peak of the computed representation. Then the automatically detected maximum does not necessarily correspond to the right one when using a simple peak detector. The algorithm has been improved by inserting restrictions in the interval where the peak has to be detected. In addition, some conditions are also specified on the first derivative in order to detect local low peaks as well. The second method (MEDIAN) is related to classical image processing techniques. As a matter of fact, the former method does not take into account the 2-D nature of the representations. After TFD computation, the region of interest is manually defined. Then the region is transformed into a binary picture by applying a threshold manually adjusted as well. Finally, applying the tools of mathematical morphology, a skeleton is deduced from the binary image, and the peak position is detected as the median point of the skeleton with respect to each time position. Both methods will be tested on reference images first. Only one will be retained afterwards, according to the comparison between the genuine modulation position and the result obtained by both algorithms.

5.3.2 Criteria Definition

In this section, various criteria are presented for TFD characterization. The choice is not so easy to achieve because only a few references to such criteria are addressed in the literature. Some of the retained indices can be referred to as overall quality criteria, some others are specifically designed in order to quantify the ability of the TFD to track the peak evolution efficiently. A global score for a given signal and a given transformation will be computed by simple addition of the rank of the configuration for every criterion.

The first criterion (CRITI) computes an estimation of the overall representation concentration as the fourth power of the L_4 norm divided by the squared L_2 norm of the TFD magnitude:

$$\text{CRITI} = \frac{\int\int |\text{TFD}(t, \omega)|^4 dt\, d\omega}{\left(\int\int |\text{TFD}(t, \omega)|^2 dt\, d\omega\right)^2} \qquad (5\text{-}24)$$

This definition has been locally used by Jones and Parks [24] as a criterion to optimize the parameters of their data-adaptive time-frequency representation.

The second criterion (CRIT2) reflects the mean width of the peaks detected by the methods described in the previous paragraph. After peak detection, the width is computed for each time position at 50% of the peak level. Eventually, the average width is calculated from all time positions.

Combined with CRIT2, the mean peak magnitude is a complementary index of the sharpness of the peak. It is computed as follows:

$$\text{CRIT3} = \sqrt{\frac{\sum\limits_{i=1}^{N} \frac{(\text{mag}(f_i) - m)}{s^2}}{N}} \tag{5-25}$$

where $\text{mag}(f_i)$ represents the peak magnitude at the ith time position, and m and s are the estimates of the mean and the standard deviation of the representation, respectively. The two other criteria are quantifying the relationship between the true modulation (well known thanks to the simulation parameters) and the extracted one.

CRIT4 is a measure of the bias of the estimate, whereas CRIT5 estimates the correlation between true and extracted modulations:

$$\text{CRIT4} = \frac{\sum\limits_{i=1}^{N}(\text{mod}_i - f_i)}{N} \tag{5-26}$$

$$\text{CRIT5} = 1 - \frac{\sum\limits_{i=1}^{N}(\text{mod}_i - f_i)^2}{\sum\limits_{i=1}^{N}(f_i - \bar{f})^2} \tag{5-27}$$

mod_i corresponds to the true (and f_i to the estimated) position of the modulation at the ith time position. These latter criteria do not take advantage of the bidimensional configuration of the time-frequency representations, owing to the extraction process. However, this limitation probably does not influence the comparisons between methodologies and between signals of various features.

5.4. RESULTS

5.4.1 Method Validation

Choice of the Best Modulation Extraction Method.   Except for CRIT1, all criteria are directly related to the extracted peak position. Before deducing any conclusion about the quality of any time-frequency representation, it is crucial to define the best estimate of the true modulation position. Two estimates were retained as a priori candidates: median and peak. An estimate of the mean was not considered because its value could be widely influenced by the associated definition of the peak limits. The opportunity of defining such an estimate will be discussed after the preliminary results. Thus in a first approach, the implemented methods were only related to the peak position (PEAK), or approach the median position (MEDIAN). The test procedure implies the generation of N realizations of the same random process referred to as a simulated signal. Recalling its mathematical expression:

$$s(t) = \sum_{k=1}^{K} \sum_{i=1}^{M} b(t - t_0^{(k)}(i) - t_k) \qquad (5\text{-}28)$$

the random process selected for that purpose (SS1) has been taken as

$$K = 50 \qquad P(-0.5 < t_k < +0.5) = 0.95 \qquad \text{SNR}_{\text{MOD}} = 30\,\text{dB}$$
$$\text{Modulation Range} = [2.5\text{-}3.75]\,\text{Hz}$$

It was not necessary to apply this set of realizations SS1 to every TFD. Therefore, four TFDs were selected to participate in the test: AR, spectrogram (SP), smoothed Wigner–Ville (SW) associated with Gaussian windows, and the method of optimal kernel developed by Baraniuk and Jones (BJ). This choice was made taking into account the various approaches of these time-frequency representations. In addition, only CRIT4 (bias) and CRIT5 (correlation) were retained to quantify the efficiency of each extraction method, since they are taking into account the true and the estimated modulations simultaneously. Table 5-1 shows the mean results for $N = 8$, and Fig. 5-2 shows an extraction example for SW.

The MEDIAN method systematically overestimates the modulation value and presents weaker levels of correlation with respect to PEAK. In addition, in some cases it was nearly impossible to extract any skeleton from the images, which explains some surprising negative results in the correlation values for SW. The bias is easily explained by the asymmetrical shape of the area around the peak. This remark a posteriori justifies the elimination of the mean as a complementary modulation position estimate, since it probably leads to a larger bias than the median. In the following, the only extraction method will be the PEAK algorithm.

Criteria Validation.   The criteria have to be used in a comparison of several representations. For a given TFD, they have to be as robust as possible so that they do not generate a large scatter, masking the differences between representations.

Table 5-1: Comparison of Both Extraction Methods: Results on CRIT4 and CRIT5

	CRIT4		CRIT5	
	PEAK	MEDIAN	PEAK	MEDIAN
AR	−0.01	−3.79	0.96	0.86
SP	−6.34	−12.22	0.78	0.29
SW	−3.43	−13.48	0.87	< 0
BJ	1.97	−5.07	0.9	0.54

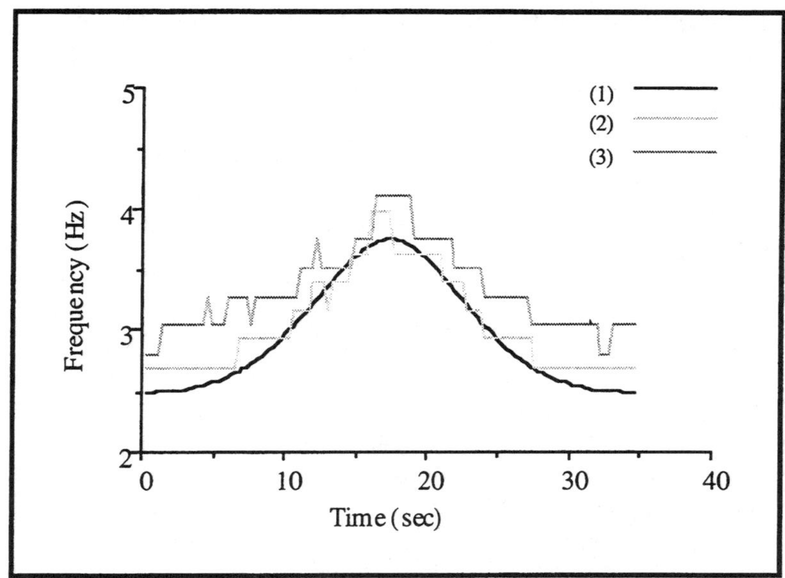

Figure 5-2   Modulation extraction from SW: (1) true modulation, (2) Peak extraction, (3) Median extraction.

This validation was achieved using the same process SS1 as before. The same set of TFDs (AR, SP, SW, BJ) was selected. As our purpose was to measure the scatter induced by the criteria themselves due to the random character of SS1, the comparison was achieved using the coefficient of variation $C_v = \sigma/\mu$ as a robustness index, where σ and μ are the standard deviation and the mean estimate, respectively. Figure 5-3 shows the results on the selected DFTs. The criteria Crit2 and Crit5 (indicating peak width and true vs. extracted correlation, respectively) produce the lowest values of C_v. The scatter on the peak value was expected, due to the estimation process of taking into account a single peak value for each time position, whereas concentration as well as correlation are estimated as sample means. The result on Crit4 (expressing the bias between true and extracted modulations) was expected as well. It shows how our measure of scatter is unadapted in this case, since a good representation is supposed to produce a low bias. Therefore, this criterion has to be kept without regarding its bad results with respect to the coefficient of variation.

As a conclusion concerning this analysis, most of the criteria are classified approximately identically whatever the TFD, except for the BJ representation for which Crit2 and Crit5 present a much higher scatter. This fact will be discussed when the purpose is to compare the representations among themselves.

Time-Frequency Representation Optimization.   As it was necessary to choose the optimal extraction method or to assess the robustness of the selected criteria, a comparison between various representations implies that the representative of each TFD is first optimized. As a matter of fact, several parameters have to be defined for

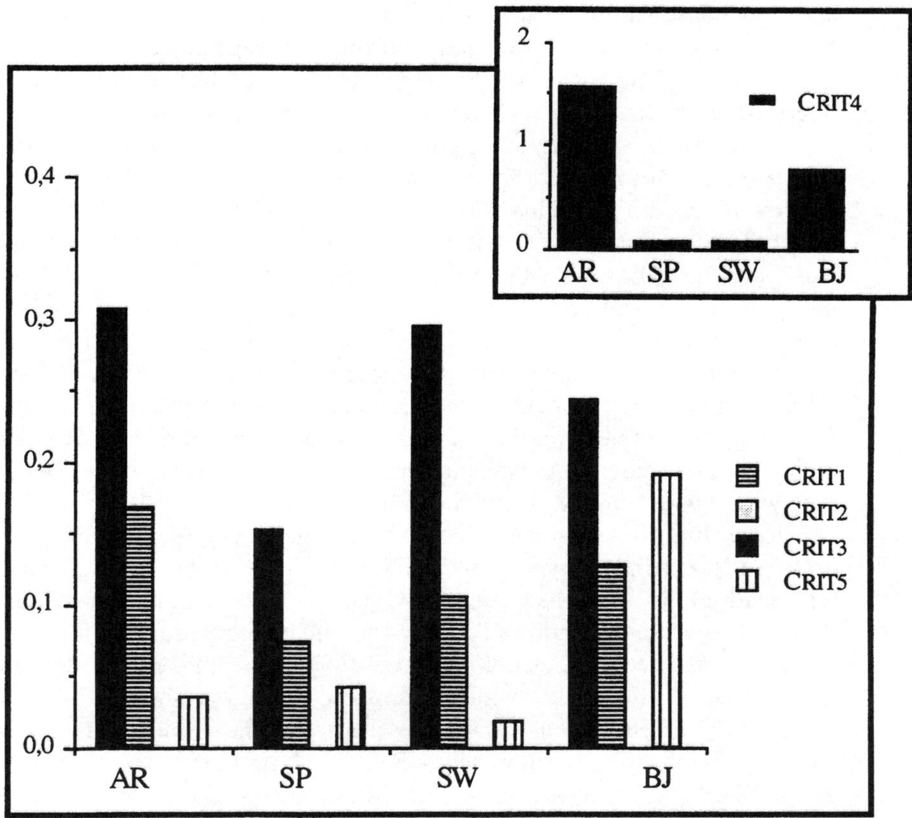

Figure 5-3   Analysis of the criteria scatter. CRIT4 has been represented separately
because of its much larger scale.

each representation, such as the AR order, the window width, etc. The procedure of
the test is based on the same SS1 process, and the optimization criterion is the score
of the various versions of each type of TFD. The type of AR representation
described in this work was chosen as the simplest to implement, i.e., computation
of AR coefficients in successive stationary windows, each modeling being defined
with the same model order. As basic as this model is, it has been proved very efficient
and produces encouraging results. In addition, we know that it can be improved in
many ways (adapted order, iterative procedure, etc.) after obtaining the first results.
At the moment, two parameters have to be optimized: the model order and the
analysis window width. As previously indicated, Mansour et al. [7] have determined
the optimal order value using real signals extracted from experiments on monkeys,
where the AIC criterion was selected as the optimization index. For all tested signal
samples, a 15th-order model has been retained in that work. In the same study, the
stationarity of the uterine EMG has been analyzed simultaneously using a RUN test
and a test on the autocorrelation function [20] on 95 real signals. The result corresponds to a mean stationary window of 17 s and 90% of stationary windows beyond

a 10-s window width. This 10-s limit has been taken as an upper bound for the following test. From SS1, we computed four AR representations depending on the window width. The results are given in Table 5-2. The best choice corresponds to an intermediate width of 3.3 s, corresponding to 100 signal samples.

As far as the spectrogram is concerned, the only degrees of freedom are related to type and width of the short-term window. It is classically shown that the type of window (Gaussian, Blackman, Hanning, etc.) does not influence the representation quality to a large extent. Our study only considers two types of window: Blackman and Gaussian. Three different window widths have been tested for Blackman windows. As a verification, a Gaussian window equivalent to one of the Blackman examples has been tested as well. The results are given in Table 5-3 where it can be verified that the window shape does not influence the final score deduced from the selected criteria. A reasonable choice of window width seems to be around 4 s. In the following, for implementation simplicity reasons (keeping in mind the possible design of a real-time monitoring system), the selected window will be of Blackman type with a width of 128 points, i.e., about 4.2 s.

Regarding these results, it did not seem necessary to carry on further studies using various window types again. Thus the analysis of the best parameters for representing the smoothed pseudo-Wigner–Ville distribution has been achieved only with Blackman windows for time- as well as for frequency-filtering. The same reasons as before (implementation simplicity) justify this choice. For this kind of transformation, the window width is supposed to widely influence the resulting shape of the TFD, especially as far as the peak sharpness and width are concerned. However, within our window specifications, CRIT2 (computed as a mean number of samples) does not exhibit any modification, the sharpness of the peak being sufficient to make this criterion insignificant. All results are indicated in Table 5-4. However, we must underscore again the fact that most of the criteria are computed

Table 5-2 Results of the Comparison of Various AR Modeling Choices

Width	CRIT1		CRIT2		CRIT3		CRIT4		CRIT5		Score
1.6 s	0.42	2	2.75	2	5.38	4	−1.37	3	0.63	2	13
3.3 s	0.69	4	2.29	4	4.6	2	−0.01	4	0.96	4	18
6.6 s	0.527	3	2.63	3	4.79	3	−2.45	2	0.86	3	14
9.9 s	0.339	1	3.52	1	4.43	1	−4.62	1	0.46	1	5

Table 5-3 Results of the Comparison of Various Spectrogram Specifications

Type	CRIT1		CRIT2		CRIT3		CRIT4		CRIT5		Score
Blackman 8.4 s	0.1	1	9.31	1	3.38	2	−11.9	1	0.12	1	6
Blackman 4.2 s	0.147	3	4.5	3	3.47	3	−6.34	3	0.781	4	16
Blackman 2.1 s	0.17	4	3.48	4	2.6	1	−8.02	2	0.67	2	13
Gaussian 4.2 s	0.14	2	4.5	3	3.61	4	−6.15	4	0.78	3	16

Table 5-4   Results of the Comparison of Various Smoothed Wigner–Ville Specifications

h	g	CRIT1		CRIT2		CRIT3		CRIT4		CRIT5		Score
1	2.1	0.193	7	2	7	1.88	4	−3.92	4	0.86	6	28
2.1	2.1	0.188	4	2	7	1.94	6	−4.57	3	0.82	5	25
0.5	2.1	0.193	7	2	7	1.81	3	−3.43	5	0.87	7	29
1	1	0.19	5	2	7	1.33	1	−2.34	7	0.82	5	25
1	4.2	0.18	3	2	7	2.5	7	−4.98	2	0.82	5	24
0.25	2.1	0.18	3	2	7	1.63	2	−2.43	6	0.6	1	19
4.2	2.1	0.175	1	2	7	1.85	5	−5.84	1	0.61	2	16

directionally, i.e., at each time position all along the frequency axis. Thus the filtering effect is obviously masked in the time direction.

Except for a possible additional filtering whose effect has been analyzed already, the Choi–Williams TFD is associated with a 1-parameter kernel, defined similarly in any time-frequency direction by means of its unique parameter σ. We tested only a few values of σ because we never succeeded in detecting the modulation properly. The results are reported in Table 5-5 as illustrating the effect of σ, but the discussion is obviously limited, regarding the poor results in terms of proper detection.

The cone-shaped kernel (Zhao–Atlas–Marks) is designed in a unique manner. Therefore, it does not require any specific optimization. As far as the signal-dependent kernel proposed by Baraniuk and Jones is concerned, the only parameter to adjust is related to the global kernel volume since the kernel deformation according to the time-frequency directions is defined automatically by the signal features. In their original paper [26], Baraniuk and Jones recommended an α value within the range

$$0.69 \leq \alpha \leq 3.0$$

The results exhibited an increasing score with respect to α, even for higher values than the upper bound first proposed by the authors (oversmoothing). A value $\alpha = 4$ has been retained for the remaining analyses.

The last method is a little different in the sense that it uses results of previous TFDs to optimize the corresponding representation by a reassignment operation. As

Table 5.5   Effect of the Kernel Size of the Choi–Williams Distribution

Size σ	CRIT1		CRIT2		CRIT3		CRIT4		CRIT5		Score
0.5	0.15	2	2	5	1.42	2	4.22	2	< 0	1	12
1	0.161	4	2	5	1.5	4	4.03	3	< 0	1	17
10	0.164	5	2	5	1.54	5	1.19	4	< 0	1	20
20	0.159	3	2	5	1.46	3	−0.31	5	< 0	1	17
50	0.143	1	2.14	1	1.11	1	−6.2	1	< 0	1	5

we decided to apply this method on the spectrogram only, the optimization process only relates to the spectrogram window, and this point has been analyzed already.

From this point, we dispose of the optimal configuration of every TFD. Figure 5-4 presents the results obtained for the same realization of the SS1 process. The

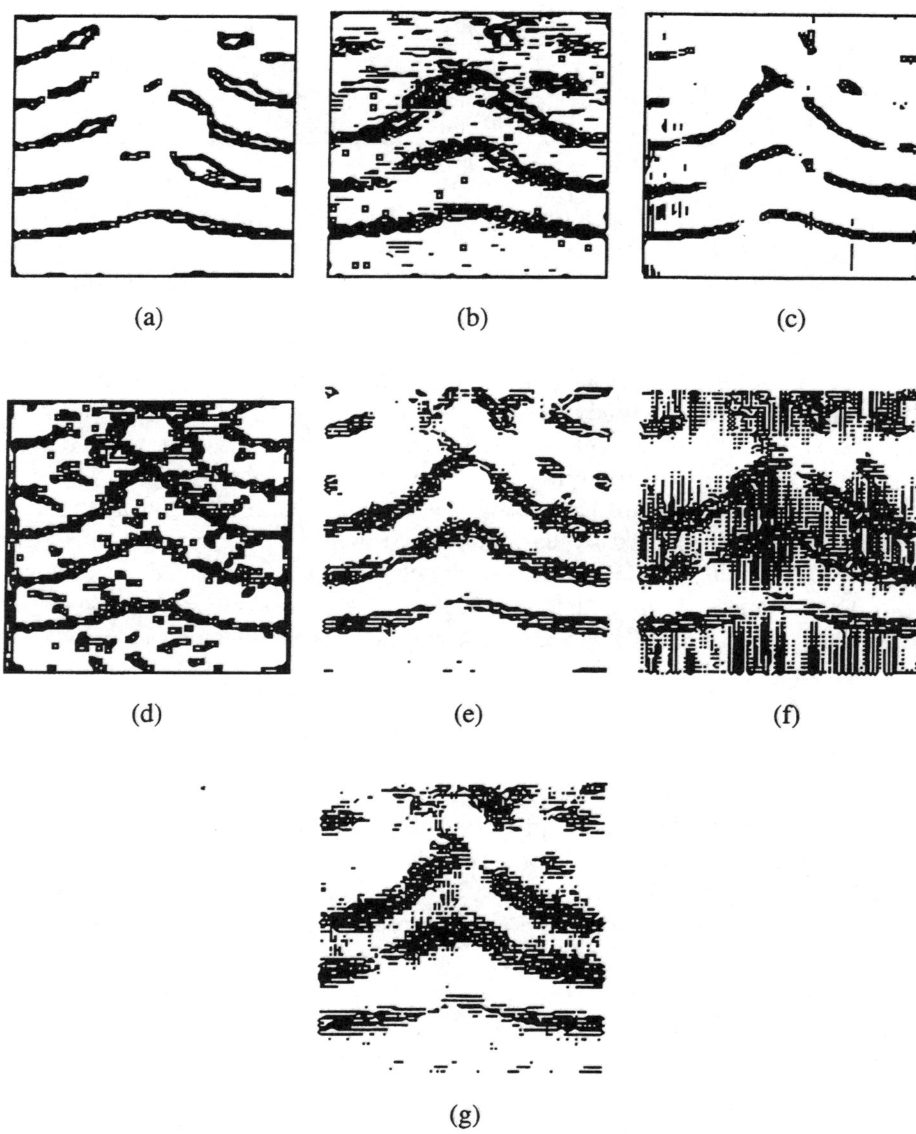

Figure 5-4   Representation of one realization of SS1 produced by a selection of "optimal" TFDs: (a) AR; (b) spectrogram; (c) Baraniuk-Jones; (d) reassigned spectrogram; (e) smoothed pseudo-Wigner–Ville; (f) Choi–Williams; (g) Zhao–Atlas–Marks. Time duration, 30 s, frequency range, 0–15 Hz.

drawings are represented here after edge detection, in order to allow a qualitative assessment of the peak sharpness and neatness in a black-and-white manner. Now, the next step consists of the comparison of all these representations according to the quality criteria previously defined.

5.4.2 Results on the Comparison Between Representations

After optimization of the various representations retained for our work, it is now necessary to compare them with respect to each other. The procedure is to compute the criteria again and to deduce a score referring to each TFD, the reference process still being SS1. The results are summarized in Table 5-6, together with the drawings of Fig. 5-4. As preliminary comments, regarding the criteria related to the representation itself, without considering the reference to the time modulation, it can be seen first that all representations directly related to the Wigner–Ville distribution (SW: smoothed pseudo-Wigner–Ville, CW: Choi–Williams, ZAM: Zhao–Atlas–Marks) behave nearly identically. Therefore, they are grouped with similar score values. In this first evaluation, the AR modeling appears to score much higher than any other representation, even though it was computed in an elementary manner in successive windows which were supposed to be stationary. The highest score in concentration (CRIT1) is surprising, taking into account the severe filtering effect of this sort of modeling in the frequency domain (a 15th order was retained in the present work). Nevertheless, this high score can be explained by the smoothing effect itself because it suppresses any possibility of scatter along the frequency axis, modeling the main frequency components only. This remark applies to CRIT3 as well, this criterion representing the relative amplitude of the peak. The less the model order, the wider the frequency peaks, then the more scattered is the energy. The peak width is quantified by CRIT2, for which the AR modeling effectively scores a little lower. However, in CRIT3 the peak height is weighted by the overall variance of the representation, which widely explains the good score exhibited by AR modeling again.

As far as the comparison between true and extracted modulations is concerned, AR modeling scores much higher again. Its computation method was a priori not in

Table 5-6   Score of the Various TFDs under Study with Respect to the
Retained Criteria

TFD	CRIT1		CRIT2		CRIT3		CRIT4		CRIT5		Score
AR	0.69	7	2.29	4	4.6	7	−0.01	7	0.96	7	32
SP	0.14	1	4.54	1	3.47	5	−6.34	1	0.78	4	12
SW	0.193	4	2	7	1.81	2	−3.43	3	0.87	5	21
CW	0.164	2	2	7	1.54	1	−0.31	6	< 0	1	17
ZAM	0.189	3	2	7	2.05	3	−4.26	2	0.62	2	17
ASS	0.31	6	2.29	4	4.11	6	−2.35	4	0.76	3	23
BJ	0.237	5	3.31	2	3.13	4	1.97	5	0.9	6	22

favor of a good continuity with respect to time, the frequency estimate being computed for each time position separately, while classical TFDs are computed globally in time and frequency simultaneously. However, the frequency smoothing associated with a large time window overlapping leads to this very good result.

The bias in modulation position estimate (CRIT4) is surprisingly high for the spectrogram representation. The extraction method can probably be in question in order to explain the result, but it is noted that the reassignment process highly improves the corresponding score. As expected, it acts in the same way for representation concentration (CRIT1) and peak sharpness (CRIT2 and CRIT3). Then the reassignment process seems very efficient for improving the criterion values when related to the overall representation.

Concerning the correlation values (CRIT5), the automatic peak extraction is the reason for the disappointing results obtained on some TFDs, especially the Choi–Williams distribution. As a matter of fact, the result obtained with this kernel can be very accurate in some modulation direction, but presents discontinuities in wide parts of the representations. Now the automatic peak detection does not take into account these parts where the peak is not present. This remark has to be related to the very good score of the group of "Wigner–Ville based" distributions in terms of peak width: the drawback of a narrow peak representation is the fact that this peak can remain undetected in large parts of the TFD.

As an intermediate conclusion, from these results it can be deduced that the AR modeling presents much higher qualities regarding the selected criteria. This has to be confirmed by an analysis of its selectivity capabilities, for which it is a priori less adapted. Another group made up of SW, BJ, and the reassigned spectrogram (ASS) exhibits acceptable results as well. However, for the next analyses, only the BJ distribution will be considered in addition to AR modeling. Concerning SW, after optimization, it still presents many discontinuities in peak detection. This fact is not crucial when using simulated signals, but when the point is to analyze real ones, it seems that the efficiency of the extraction process is rather limited for this kind of distribution. Regarding ASS (reassignment), our knowledge on the reassignment process is still too limited to apply it to real signals. As a matter of fact, we first have to ensure that this process does not move low peaks originally in their right location toward near peaks of much higher energy. In addition, as underscored in a previous section, this method can be applied in a further step as an improvement on any TFD. This improvement has been noticed already on the spectrogram. It can be assessed on other distributions later.

5.4.3 Robustness and Selectivity

This analysis has been achieved on the two remaining distributions, i.e., AR and BJ. For robustness assessment, the varying parameters of the simulation process are the signal-to-noise ratios of the modulating and of the whole signal, respectively. In the present work, the results have only been obtained on one realization of the process in each SNR condition. Therefore, the results presented have to be taken as preliminary and to be interpreted very carefully. As far as the selectivity is concerned, a modified SS1 process has been considered, where the modulation is now made of the addition of two identical shapes shifted in frequency. Many shift values have been tested in preliminary analyses before selection of the final ranges, chosen here as illustrating differences between TFDs as well as possible.

The results on robustness are summarized in Table 5-7. As indicated before, these criterion values do not come from a mean of many realizations of the random processes, but represent the behavior of only one event randomly selected. The comments can therefore only be proposed in terms of a general trend. It seems that the BJ representation does not vary in concentration with respect to the additive noise, whereas AR modeling is significantly affected by poorer signal-to-noise ratios. The same comment can be applied on the peak width where the BJ distribution remains remarkably stable. On the other hand, concerning the criteria related to a correct modulation extraction, the latter distribution seems more sensitive to noise, especially regarding the correlation criterion. The variations on the bias values seem due to the normal effect of process randomness rather than a genuine influence of the presence of noise. Regarding these first results, it thus seems difficult to select a specific distribution using robustness criteria, and a more precise analysis on this basis does not appear as useful to learn more about the respective qualities of each representation in that way.

The last point which could discriminate between both representations concerns their capability to properly separate near-frequency components. For this purpose, two signals have been derived from SS1, shifting the frequency range of the modulation from [2.5–3.75 Hz] to [1.9–3.15 Hz] and [2.2–3.45 Hz], respectively. The results (Fig. 5-5) are not discussed any longer in terms of score related to the previous criteria, but qualitatively observing the presence or absence of well-separated peaks all along the time axis. Each time-frequency image has been represented after the best processing to enhance the two components of the reference signal as much as possible. It was not relevant to make use of edge detection any longer, since this method automatically mixed both peaks. Therefore, we chose to adapt the grayscale of each resulting image separately.

Though the results were more or less expected, the difference in behavior between both representations is highly remarkable. For a shift which was supposed

Table 5-7 Results of the Robustness Analysis. When No Value is Indicated in the SNR Columns, No Noise was Applied to the Simulated Process or the Corresponding Modulation

SNR (dB) Sig.	SNR (dB) Mod.	CRIT1		CRIT2		CRIT3		CRIT4		CRIT5	
		AR	BJ	AR	BJ	AR	BJ	AR	BJ	AR	BJ
—	—	0.41	0.21	2.2	3.05	9.12	6.83	0.05	1.6	0.97	0.94
—	30	0.69	0.23	2.29	3.31	4.6	3.13	−0.01	1.97	0.96	0.9
—	20	0.42	0.22	2.32	3.53	8.58	4.92	0.17	0.41	0.94	0.64
—	10	0.39	0.25	4.32	3.95	7.91	5.48	−1.82	0.3	0.69	0.64
15	30	0.67	0.25	3.00	3.46	10.6	7.53	−1.32	0.75	0.73	0.68
10	30	0.38	0.25	4.29	4.24	8.43	4.37	−2.35	−4.79	0.8	0.12

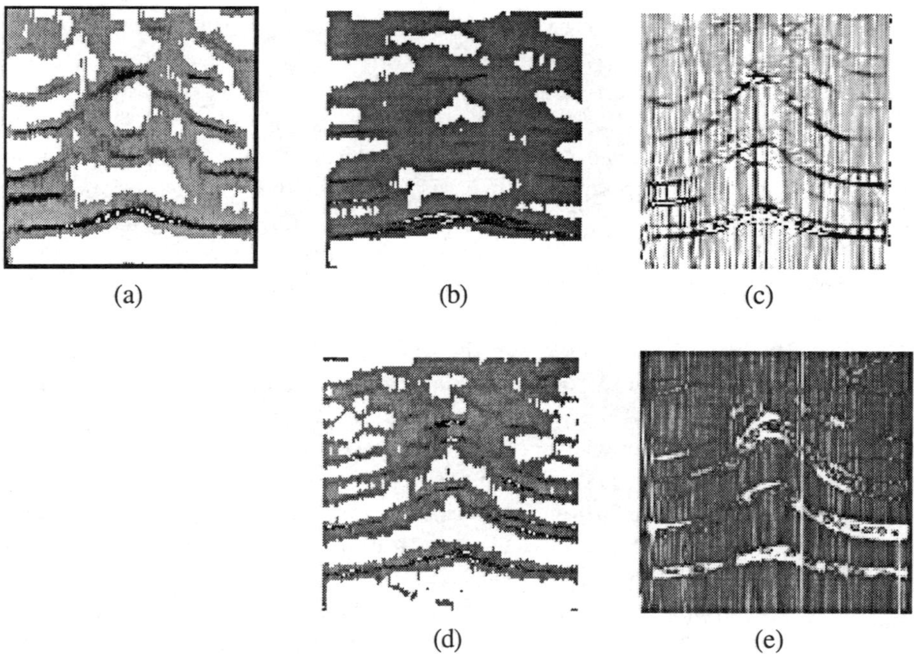

Figure 5-5   Results on AR and BJ in terms of selectivity: (a), (b), and (c) are computed from the larger shift, with (a) AR 15th order; (b) AR 25th order; (c) BJ; (d) and (e) are computed from the narrower shift, with (d) AR 25th order, (e) BJ.

to be relatively large, the signal-dependent representation provides very satisfactory results, since the AR modeling in its first version (Fig. 5-5(a): 15th order) does not detect any difference except to a certain extent right in the center of the representation. The model order obviously contributes to a large extent to this poor discrimination. Therefore, we tried to increase this order, leading to the result of Fig. 5-5(b). The representation discriminates both frequency modulations a little better, but remains far from the quality obtained by the BJ distribution. Going a little further into the discussion concerning this latter representation, we can observe the interference phenomenon in two situations. The first one was obviously expected and corresponds to components linking the two autocomponents representing both modulations. The second creates high-level interferences between the low part of the second harmonic component and the high part of the first, inducing possible misunderstanding. In further attempts to improve these results, it could be desirable to modify the kernel volume to some extent in order to test if a better result could be obtained in terms of interference suppression without losing anything in selectivity.

The second part of Fig. 5-5 relates to the simulated process where the modulation components are closer together. In view of the previous results, it was useless to compute the representation corresponding to the 15th-order AR modeling. Figure 5-5(d) indicates the result with a 25th-order modeling. The discrimination between both frequency modulations are only detectable on the harmonic components. On the other hand, the results on the signal-dependent representation are satisfactory enough, though the mixing of both peaks is locally important due to the role of the cross-terms. The kernel volume is probably a little too large here to allow an optimal discrimination.

5.4.4 Toward a Possible Final Choice

From all results previously presented, it appears that two types of time-frequency representations globally produce satisfactory results when the problem is to properly detect a modulation whose shape is close to a genuine frequency variation within a uterine contraction. They are both perfectly complementary in the sense that one ensures a good continuity thanks to wide frequency-smoothing and large time-overlapping, while the other is constructed such that it adapts as well as possible to the specific modulation direction in the time-frequency plane. In terms of possible methodological improvements, AR modeling has been chosen as very rudimentary in the present work. Therefore, many improvements can be introduced to make the modeling more efficient, especially by using adaptive methods. The Baraniuk–Jones optimal kernel is really optimally designed to take the signal specificities into account. In that way, no improvements can be viewed. On the other hand, a more detailed analysis could be achieved to optimize the kernel volume α, including selectivity criteria. In addition, the first results on robustness are favorable for designing a unique kernel for a specific class of modulation shapes. This point has to be considered for studies where the main modulation directions are always the same whatever the signal, which is the case for uterine EMGs.

For simplicity, in order to illustrate the modulation significance on real signals in terms of contraction efficiency, we have selected AR modeling as a representation method in the next section.

5.5. SOME EXAMPLES ON REAL SIGNALS

The aim of this last section is to show some results obtained from real signals recorded on monkeys in various experimental situations. As a matter of fact, if each specific physiological situation corresponds to a specific modulation shape, then the modulation extraction can be interpreted and used for monitoring purposes. These results are illustrative, again, but give meaning to the methodological developments described and a posteriori justify the choice of the criteria which determined the selection of the best set of time-frequency representations. Figure 5-6 produces an example of real time-frequency representation using AR modeling. The modulation is well recognizable, and it is possible to see the harmonic components as well. This observation is quite interesting because it confirms our assumption of nonsinusoidal carrier. Unfortunately, this typical phenomenon can be observed only by internal recordings, i.e., by putting wire electrodes directly on the uterine muscle. That explains why the results presented here do not come from recordings on women, but only relate to recordings on monkeys.

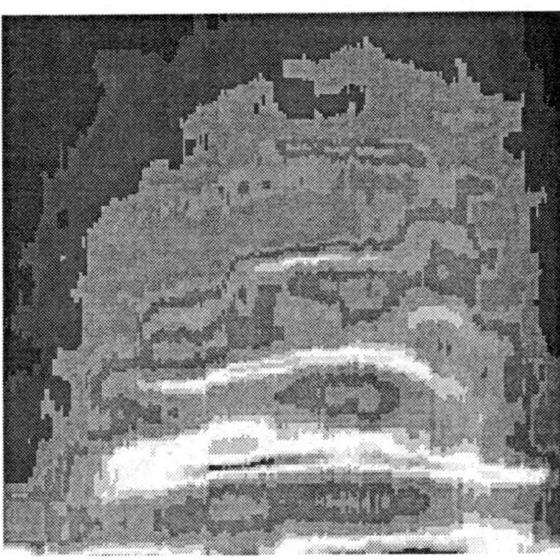

Figure 5-6 Example of time-frequency representation of a real signal. The record comes from internal electrodes on a monkey just before parturition. Ranges are 0–3 s in time (horizontal axis) and 0–15 Hz in frequency (vertical axis).

The shape of the representation is supposed to vary with respect to many factors. The moment of the recording with respect to full-term is one of the main factors in normal pregnancies because the electrical generation and conduction structures dramatically change just before parturition [43], involving deep modifications in the shape of the recorded signal, especially in terms of frequency shift toward higher values. As indicated in the introduction, the firing frequencies inside a burst increase notably. Consequently, the mechanical contraction becomes more and more efficient as attested by the evolution of intrauterine pressure. Another factor which exhibits an important effect on the uterine contractility is the presence of specific hormones such as oxytocin. This hormone is known as a powerful uterine stimulating factor [44] and is frequently administered to trigger labor, or to speed up the normal labor process. Experiments have been achieved in these various conditions on monkeys: normal pregnancy, contractions at full-term, oxytocin injection during normal pregnancy. It is not the purpose of this chapter to describe the experimental protocol and the corresponding results in detail, but we extracted some illustrating results to show how the modulation shape effectively produces useful information in these various cases. Figure 5-7 gathers these results together. The modulation frequency has been computed automatically on the whole record, but only the parts between the vertical dashed lines have to be taken into account. Again, they are presented as selected illustrations and have to be discussed very carefully, all the more so since they come from different animals. However, notice first the difference in frequency level between pregnancy and parturition. Also observe the difference in shape for these two situations. Concerning the signals recorded during pregnancy at two-thirds of full-term, note the large difference in modulation behavior in experiments (c) with or (b) without oxytocin injection, and how (c) resembles (a), the normal parturition condition.

These few examples illustrate the possible role of the modulation detection and analysis in discriminating between normal and abnormal situations. For a long time [45], it has been proved that a frequency analysis on EMGs gives useful information on contraction efficiency, but this was demonstrated using all components of the recorded signals. In the present work, the only component selected for

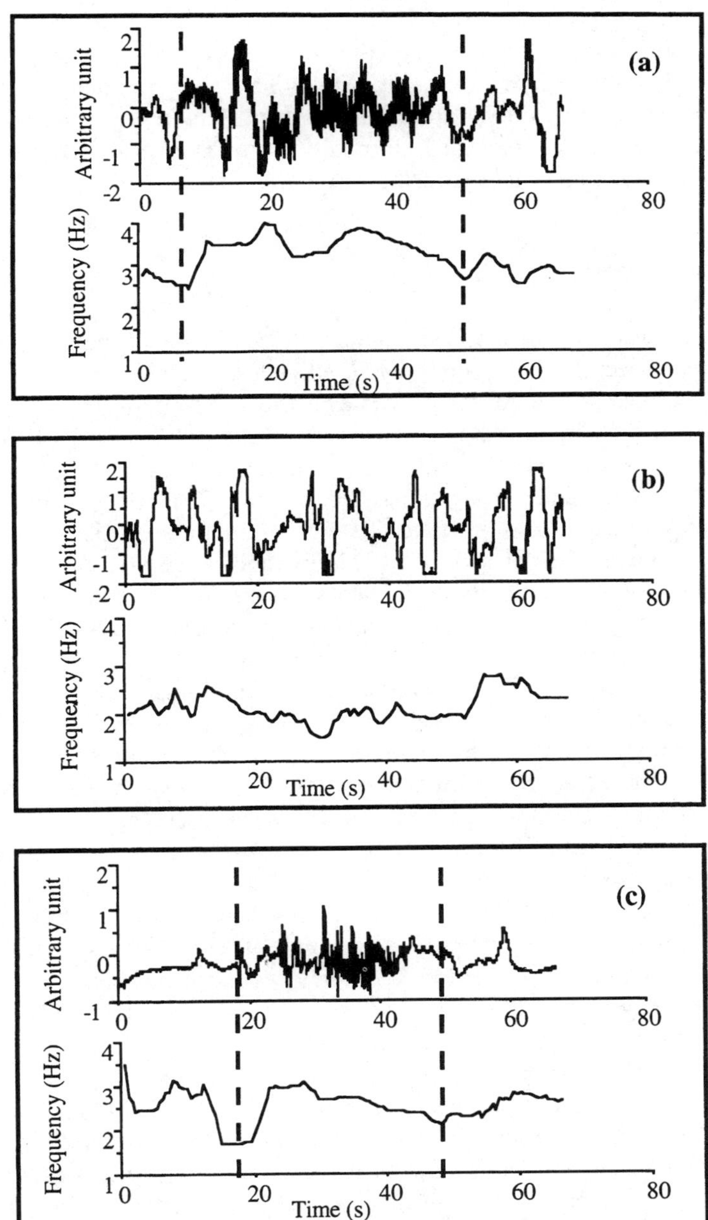

Figure 5-7 Some examples of modulation extraction: (a) normal labor; (b) normal pregnancy; (c) induced labor by oxytocin administration. For each case, upper curve: uterine EMG; lower curve: extracted modulation.

the analysis is related to the firing rate of the action potentials inside the bursts of electrical activity, therefore describing a specific internal phenomenon instead of a global observation.

Physiologically speaking, the results have to be discussed in depth using all available records. We have only shown here the potential capability of the modulation shape to describe various situations differently. In addition, we take into consideration signals recorded externally on women to extend the results on animals toward humans for monitoring purposes.

5.6. CONCLUSION

This study was an attempt to characterize various time-frequency representations with respect to their capability to solve a given problem. Many other approaches could have been considered, whatever the application. From the user point of view, we can draw many lessons from the results of this work. First, the best method cannot be defined in an absolute manner, but with respect to precise objectives related to specific applications. Second, the signal-dependent kernels seem a very good approach to solve the problem of modulation extraction when this modulation is nonlinear. The drawback of this method, if any, is the definition of a suitable kernel volume as a compromise between filtering and selectivity.

Concerning the relative failure of the classical TFDs derived from the Wigner–Ville distribution, the reason for the bad results has to be found essentially in the criteria definition. These criteria have been designed in a precise way and for a precise aim, therefore, they do not reflect the overall qualities of the various representations or their capabilities in reproducing the time-frequency characteristics of the analyzed process as accurately as possible. To go to the limits of this reasoning, a method which could preserve only the main peak while suppressing or deeply modifying the other components can be as suitable as the previously analyzed ones. Our approach was related to measures rather than to theoretical overall properties. We must admit that our approach is limited, but well adapted to our real processes. To reinforce this point of view, let us return to the selectivity results. Only one realization randomly constructed was sufficient to confirm the superiority of the BJ representation, whereas considerable efforts would have been necessary to prove it theoretically.

Another conclusion is related to the relevance of simple AR modeling for this kind of process. It has been shown that elementary AR models can detect the modulation very efficiently. The subsequent interest of such simple methods becomes evident when one has to design a real-time monitoring system. Now, our results show that a model assuming stationarity in windows of fixed length and being defined with a fixed order can act perfectly as a good modulation extractor. Many years ago, we had designed a real-time system for parturition monitoring purposes. It was based on an overall spectral analysis, the spectral features being selected by analyzing reference women classes. Here, the result has been refined since we know that the tracked frequency is straightforwardly related to the internal burst firing rate, then to the contraction strength. The definition of a new monitoring system thus necessarily goes through the use of a real-time algorithm. Again, AR modeling is well adapted to this kind of instrumentation, either using fixed windows or computed in an adaptive manner. That is not the case for most of the time-frequency distributions, which need the knowledge of the whole process before distribution computation. A further step in terms of methodological developments could be to search for adaptive time-frequency transformations. In parallel, another step could be to analyze if the reassignment process, which showed promising results on the spectrogram, could be successfully transferred to other distributions such as signal-dependent ones.

REFERENCES

[1] C. Y. Kao, "Electrophysiological properties of the uterine smooth muscle," In *Biology of the Uterus.* Wynn, R. M. (ed.) New York: Plenum Press, pp. 423–496, 1977.

[2] C. Marque and J. Duchêne, "Human abdominal EHG processing for uterine contraction monitoring," In *Applied Biosensors.* Wise, D. L. (ed.) Stoneham: Butterworth, pp. 187–226, 1989

[3] L. J. Griffiths, "Rapid measurement of digital instantaneous frequency," *IEEE Trans. Acoust., Speech, Signal Proc.*, 23, pp. 207–222, 1975.

[4] M. Sun and R. J. Sclabassi, "Discrete-time instantaneous frequency and its computation," *IEEE Trans. Signal Proc.* 41, pp. 1867–1880, 1993.

[5] L. B. White and B. Boashash, "On estimating the instantaneous frequency of a Gaussian random signal by use of the Wigner–Ville distribution," *IEEE Trans. Acoust., Speech, Signal Proc.*, 36, pp. 417–420, 1988.

[6] P. Flandrin, *Temps-Fréquence.*, Paris: Hermès, pp. 29–31, 1993.

[7] S. Mansour, D. Devedeux, J. Duchêne, G. Germain, and C. Marque, "Uterine EMG spectral characteristics and instantaneous frequency measurement," *Proc. 14th Annual Int. Conf. IEEE EMBS*, 14, pp. 2602–2603, 1992.

[8] J. Duchêne, D. Devedeux, S. Mansour, and C. Marque, "Analyzing uterine EMG: Tracking instantaneous burst frequency", *IEEE EMB*, vol. 14, pp. 125--132, 1995.

[9] S. Mansour, D. Devedeux, J. Duchêne, G. Germain, and C. Marque, "Relevance of external uterine electromyography to contractile process characterization," *Proc. 2nd European Conf. Eng. Medicine*, 2, pp. 370–371, 1993.

[10] J. Gondry, C. Marque, J. Duchêne, and D. Cabrol, "Uterine EMG processing during pregnancy: preliminary report," *Biomed. Instrum. Technol.*, vol. 27, pp. 318–324, 1993.

[11] A. Papoulis. *Probability, Random Variables and Stochastic Processes*, 3rd ed. New-York: McGraw-Hill, 1991.

[12] J. Makhoul, "Linear prediction: A tutorial review," *Proc. IEEE*, 63, p. 561, 1975.

[13] S. M. Kay and S. L. Marple, "Spectrum analysis—A modern perspective," *Proc. IEEE*, 69, p. 1380, 1981.

[14] Y. Grenier, "Modèles ARMA à coefficients dépendant du temps: estimateurs et applications," *Traitement du Signal*, 3, pp. 219–233, 1986.

[15] N. Martin, "An AR spectral analysis of non-stationary signals," *Signal Processing*, 10, pp. 61–74, 1986.

[16] Z. Guo, L. G. Durand, L. Allard, G. Cloutier, H. C. Lee et al., "Cardiac doppler blood-flow signal analysis. Part 1: Evaluation of the normality and stationarity of the temporal signal," *Med. Biol. Eng. Comput.*, 31, pp. 237–241, 1993.

[17] H. Akaike, "A new look at the statistical model identification," *IEEE Trans. Automat. Control*, 19, p. 716, 1974.

[18] Z. Guo, L. G. Durand, L. Allard, G. Cloutier, H. C. Lee et al., "Cardiac doppler blood-flow signal analysis. Part 2: Time/frequency representation based on autoregressive modelling," *Med. Biol. Eng. Comput.*, 31, pp. 242–248, 1993.

[19] K. Kaluzynski, "Order selection in doppler blood flow signal spectral analysis using autoregressive modelling," *Med. Biol. Eng. Comput.*, 27, pp. 89–92, 1989.

[20] J. S. Bendat and A. G. Piersol, *Random Data. Analysis and Measurement Procedures*, 2nd Ed., New York: Wiley, 1986.

[21] R. K. Dunn and L. Y. Lacy, "The sound spectrograph," *J. Acoust. Soc. Am.*, 18, pp. 19–49, 1946.

[22] L. Cohen, "Time-frequency distributions—a review," *Proc. IEEE*, 77, pp. 941–981, 1989.

[23] T. A. C. M. Claasen and W. F. G. Mecklenbräuker, "The Wigner distribution—a tool for time-frequency signal analysis. Part III: Relations with other time-frequency signal transformations," *Philips J. Res.*, 35, pp. 372–389, 1980.

[24] D. L. Jones and T. W. Parks, "A high resolution data-adaptive time-frequency representation," *IEEE Trans. Acoust., Speech, Signal Proc.*, 38, pp. 2127–2135, 1990.

[25] D. L. Jones, and T. W. Parks, "A resolution comparison of several time-frequency representations," *IEEE Trans. Signal Proc.*, 40, pp. 413–420, 1992.

[26] R. G. Baraniuk and D. L. Jones, "A signal-dependent time-frequency representation: optimal kernel design," *IEEE Trans. Signal Proc.*, 41, pp. 1589–1602, 1993.

[27] R. G. Baraniuk and D. L. Jones, "Signal-dependent time-frequency analysis using a radially Gaussian kernel," *Signal Proc.*, 32, pp. 263–284, 1993.

[28] W. Martin and P. Flandrin, "Wigner–Ville spectral analysis of nonstationary processes," *IEEE Trans. Acoust., Speech, Signal Proc.*, 33, pp. 1461–1470, 1985.

[29] Y. Zhao, L. E. Atlas, and R. J. Marks, "The use of cone-shaped kernels for generalized time-frequency representations of nonstationary signals," *IEEE Trans. Acoust., Speech, Signal Proc.*, 38, pp. 1084–1091, 1990.

[30] S. Oh, and R. J. Marks, "Some properties of the generalized time frequency representation with cone-shaped kernel," *IEEE Trans. Signal Proc.*, 40, pp. 1735–1745, 1992.

[31] H. Choi and W. J. Williams, "Improved time-frequency representation of multicomponent signals using exponential kernels," *IEEE Trans. Acoust., Speech, Signal Proc.*, 37, pp. 862–871, 1989.

[32] A. J. E. M. Janssen, "On the locus and spread of pseudodensity function in the time-frequency plane," *Philips J. Res.*, 37, pp. 79–110, 1982.

[33] P. Flandrin, "Maximum signal energy concentration in a time-frequency domain," *Proc. IEEE ICASSP*, pp. 2176–2179, 1988.

[34] H. P. Zaveri, W. J. Williams, and J. C. Sackellares, "Cross time-frequency representation of electrocorticograms in temporal lobe epilepsy," *Proc. 13th Annual Int. Conf. IEEE EMBS*, 13, pp. 437–438, 1991.

[35] R. M. S. S. Abeysekera, R. J. Bolton, L. C. Westphal, and B. Boashash, "Patterns in Hilbert transforms and Wigner–Ville distributions of electrocardiogram data," *Proc. IEEE ICASSP*, pp. 1793–1796, 1986.

[36] P. Novak and V. Novak, "Time/frequency mapping of the heart rate, blood pressure and respiratory signals," *Med. Biol. Eng. Comput.*, 31, pp. 103–110, 1993.

[37] J. C. Andrieux, M. R. Feix, G. Mourgues, P. Bertrand, B. Izrar, et al., "Optimum smoothing of the Wigner–Ville distribution," *IEEE Trans. Acoust., Speech, Signal Proc.*, 35, pp. 764–769, 1987.

[38] C. Zheng, S. E. Widmalm, and W. J. Williams, "New time-frequency analyses of EMG and TMJ sound signals," *Proc. 11th Annual Int. Conf. IEEE EMBS*, 11, pp. 741–742, 1989.

[39] B. Sahiner and A. E. Yagle, "Application of time-frequency distributions to magnetic imaging of non-constant flow," *Proc. IEEE ICASSP*, pp. 1865–1868, 1990.

[40] S. Kadambe, G. F. Boudreaux-Bartels, and P. Duvaut, "Window length selection for smoothing the Wigner distribution by applying an adaptive filter technique," *Proc. IEEE Int. Conf. Acoust., Speech, Signal Proc., ICASSP'89*, pp. 2226–2229, 1989.

[41] K. Kodéra, R. Gendrin, and C. de Villedary, "Analysis of time-varying signals with small BT values," *IEEE Trans. Acoust., Speech, Signal Proc.*, vol. 34, no. 1, pp. 64–76, 1978.

[42] F. Auger and P. Flandrin, "Generalization of the reassignment method to all bilinear time-frequency and time-scale representations," *Proc. IEEE Int. Conf. Acoust., Speech, Signal Proc., ICASSP'94*, 1994.

[43] R. E. Garfield and R. H. Hayashi, "Appearance of gap junctions in the myometrium of women during labor," *Am. J. Obstet. Gynecol.*, vol. 140, pp. 254–260, 1981.

[44] G. M. J. A. Wolfs and M. Van Leeuwen, "Electromyographic observations on the human uterus during labor," *Acta Obstet. Gynecol. Scand.*, Suppl vol. 90, pp. 1–62, 1979.

[45] C. Marque, J. Duchêne, S. Leclercq, G. Panzcer, and J. Chaumont, "Uterine EHG processing for obstetrical monitoring," *IEEE Trans. Biomed. Eng.*, vol. 33, pp. 1182–1187, 1986.

Chapter 6

Time-Frequency Analyses
of the Electrogastrogram

Zhiyue Lin, Jiande Z. Chen

6.1. INTRODUCTION

Electrogastrography is usually referred to as the noninvasive technique of recording gastric myoelectrical activity by placing electrodes on the abdomen. The surface recording obtained using the electrogastrographic technique is called the electrogastrogram (EGG).

Although the first measurement of the EGG was reported 70 years ago [1], progress has been relatively slow and the application of this noninvasive method has been limited, in particular when compared to the progress made in electrocardiography. One of the main problems is the difficulty in data analysis and extraction of useful and relevant information from the EGG. Unlike other surface electrophysiological recordings such as the ECG, the EGG has a low signal-to-noise ratio. It contains noise, such as respiratory, motion artifacts, and possible myoelectrical activity from other organs [2, 3]. As a result, direct visual interpretation of the EGG time signal is almost impossible. Since the EGG signal is more or less sinusoidal, with a periodicity determined by that of the gastric slow wave, it is not surprising that most attention has been paid to the periodic nature of the signal [4]. Most quantitative analyses of the EGG have relied on spectral methods, including fast Fourier transform (FFT) [5], phase-lock filtering [6], autoregressive modeling [7], and smoothed power spectral analysis [8–10]. Unfortunately, all of these methods process EGG data in a batch manner, leaving temporal analysis concerning rhythmic variations obscured.

To simultaneously reveal both the frequency components in an EGG signal and the temporal features of variations in frequency components, time-frequency analysis methods have been introduced or developed for EGG analyses. These include the short-time Fourier transform (STFT), or spectrogram [11, 12], the adaptive spectral analysis method [13], and the exponential distribution (ED) [14]. The first method introduced into the area was the short-time Fourier analysis by van der Schee and Grashuis [15]. It was called *running spectral analysis* by some investigators. The adaptive spectral analysis was developed by Chen [13] for EGG data analysis. The exponential distribution was introduced for EGG analysis only very recently [16].

Time-frequency analyses have at least two advantages in the application of the EGG. The first is their ability to separate the gastric signal from interferences or noise. In time-domain (tracing), the interferences are superimposed on the gastric signal, hindering accurate analysis of the EGG. In frequency-domain (spectrum), the signal and interferences are separated due to their different frequencies. For example, an interference of 12 cycles per minute (cpm) is superimposed on the gastric slow wave in an EGG recording, as shown in Fig. 6-1 (top panel), and makes it very difficult to accurately measure the amplitude and frequency of the gastric signal.

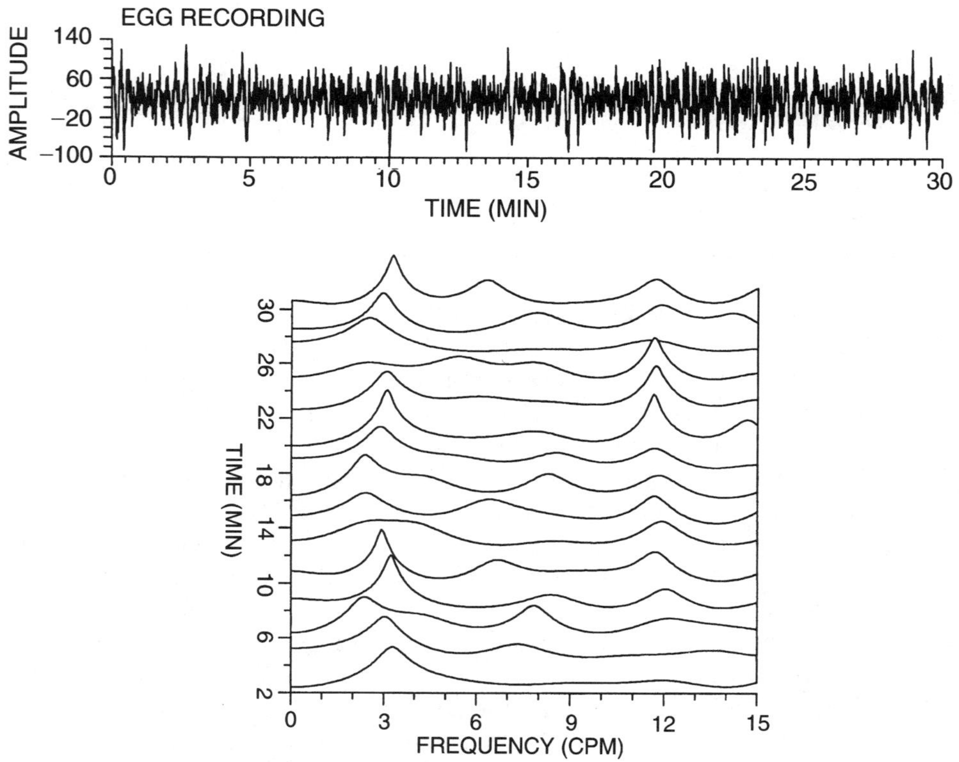

Figure 6-1 An EGG recording with an interference of 12 cpm (top) and its time-frequency representation calculated by the adaptive method (bottom). Each spectrum (from bottom to top) stands for 2 min of EGG data.

However, the gastric signal and the interference are separated in its time-frequency representation calculated by the adaptive method (see Fig. 6-1, bottom panel). It is seen that the gastric signal has a frequency of 3 cpm, whereas the interference has a frequency of 12 cpm and does not affect the power and frequency of the gastric signal. Thus quantitative parameters such as the frequency and power of the gastric signal can be accurately assessed from the time-frequency representation. The second is that the time-frequency representation provides not only information about the frequency of the EGG, but also information about time variations of the frequency. The periodogram method [17] provides the power spectrum of a whole data set, from which one obtains information only about frequency. Time information is completely missing. It has been shown that the frequency of the gastric electrical signal may be time-varying and the EGG often contains dysrhythmias which are usually of brief duration, such as tachygastria and bradygastria [18]. Figure 6-2 presents a typical example of a 15-min EGG recording. It is seen from the tracing and its running power spectra computed by the adaptive method (Fig. 6-2, top and right bottom panel) that the frequency of the gastric signal is about 4.6 cpm within the first 8 min, then it shifts to 2 cpm, and gradually it shifts to 3.0 cpm. From the power spectra

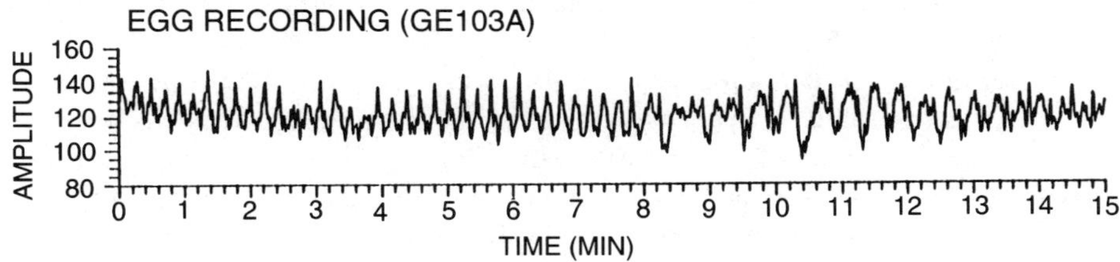

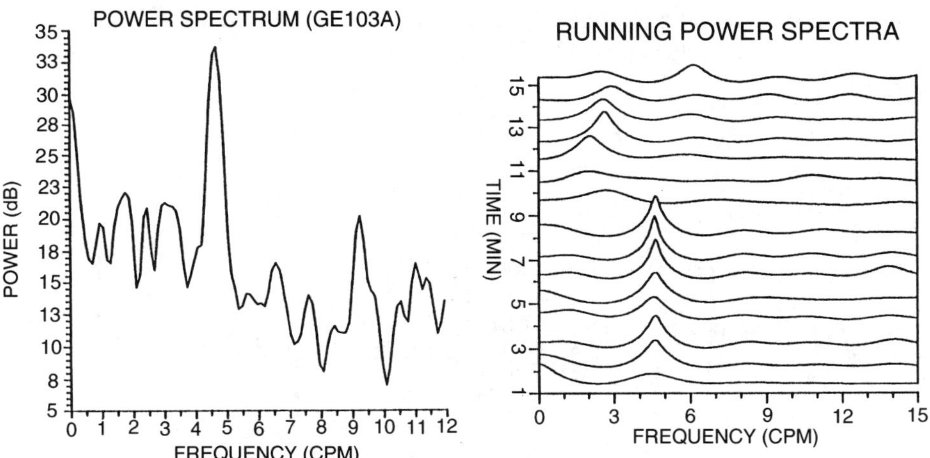

Figure 6-2   A typical EGG recording (top), its power spectrum (left bottom), and its time-frequency representation (right bottom) calculated by the adaptive method. Each spectrum (from bottom to top) stands for 1 min of EGG data.

computed by the periodogram method, however, only the dominant frequency of the recorded signal can be appreciated (Fig. 6-2, left bottom panel). It is the time-frequency analysis methods that simultaneously reveal both the frequency components of an EGG signal and the temporal features of variations in the frequency components.

This chapter provides an overview of the applications of the three time-frequency analysis methods (the spectrogram, the adaptive spectral analysis, and the exponential distribution) to the study of the EGG. The primary objective is to compare their performances and to provide some guidance for the selective use of these methods.

6.2. ELECTROGASTROGRAPHY

6.2.1 Myoelectrical Activities in the Stomach

As in the heart, electrical activities exist in the human stomach. Two kinds of gastric electrical activities can be measured from internal electrodes: the slow wave (or electrical control activity, basic electrical rhythm, or pacesetter potentials) and spikes (or electrical response activity). The slow wave is present all the time and originates in a region near the junction of the proximal one-third and distal two-thirds of the gastric corpus along the great curvature. It is characterized by regular recurring changes in potentials, propagating circumferentially and distally toward the pylorus with increasing velocity and amplitude. The frequency of the normal slow wave is about 3 cpm (or 0.05 Hz) in humans. Gastric contractions occur when spikes are superimposed on gastric slow waves. The spikes have a frequency of about 60 cpm and random phases [19–21]. Figure 6-3 presents gastric myoelectrical activities measured with a pair of serosal electrodes in a human subject. Both gastric slow waves and superimposed spikes can be appreciated in this recording.

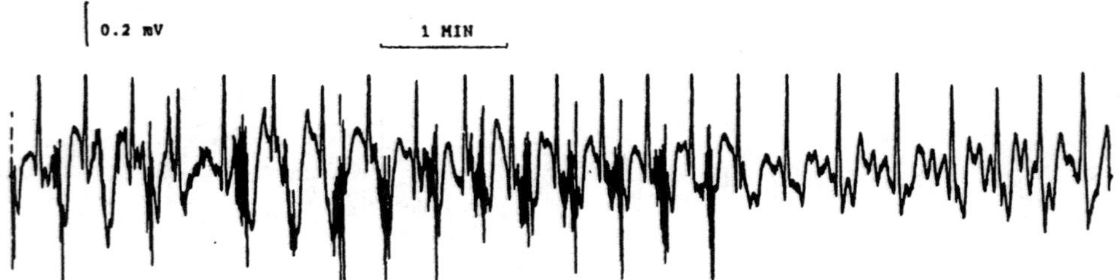

Figure 6-3 Electrical activity in a human subject measured from the serosal surface of the stomach. The trace shows slow waves of about 3 cpm with and without superimposed spikes. (from J.D.Z. Chen et al., *Am. J. Physiol.*, vol. 266, p. G95, 1994. With permission.)

6.2.2 Electrogastrogram (EGG)

Myoelectrical activity in the stomach can be measured serosally, intraluminally, or cutaneously. The serosal recording is obtained by sewing needle electrodes on the serosal surface of the stomach while performing an abdominal surgery. The intraluminal recording is acquired by asking the subject to swallow a tube with electrodes into the stomach. Serosal and intraluminal electrodes can record both slow waves and spikes. However, these methods are invasive and their applications are limited. The cutaneous recording obtained by placing surface electrodes on the abdomen is called the *electrogastrogram* [1]. The placement of the surface electrodes is shown in Figure 6-4. The EGG is very attractive because it is noninvasive and does not disturb ongoing activity of the stomach. Simultaneous recordings of serosal and cutaneous [22, 23] or mucosal and cutaneous [24] myoelectrical signals have shown that cutaneous electrodes are able to pick up the rhythm of the gastric slow wave, and the dominant frequency of the EGG is of gastric origin and is the same as that of the gastric slow wave. Spikes are reflected in the EGG as an increase in amplitude. That is, a relative amplitude change of the EGG reflects spikes, or contractions of the stomach. When correctly recorded, the noninvasive EGG provides reliable information about gastric myoelectrical activity.

Like other surface electrophysiological recordings, the EGG is a combination of the gastric signal and noise. Compared with other surface recordings, such as the ECG, the quality of the EGG is usually poor. The gastric signal in the EGG is disturbed or may even be completely obscured by noise [2]. The noise in the EGG is composed of respiratory and motion artifacts, the ECG, and electrical interference

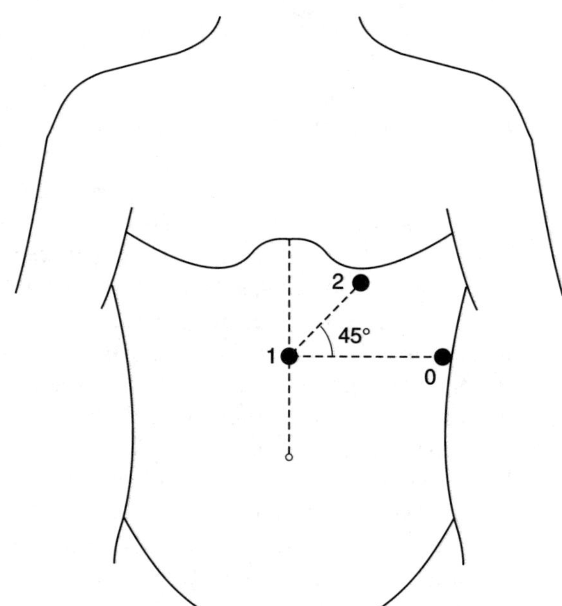

Figure 6-4 Position of the abdominal electrodes. Electrode 1: the midway between the xiphoid process and the umbilicus; electrode 2: 5 cm away from electrode 1; electrode 0: reference electrode. Bipolar EGG signal is derived from electrodes 1 and 2.

of the small intestine. The respiratory artifact is particularly noteworthy because it is usually severe and its frequency is close to that of the gastric slow wave. It can be canceled using an adaptive filtering technique [25]. Motion artifacts should be minimized during recording. The ECG has no effect on spectral analysis of the gastric slow wave. The small intestinal slow wave has a frequency gradient of 12 cpm to 8 cpm from the duodenum to the ileum. To avoid recording small intestinal slow waves, the electrodes should be placed accurately over the gastric outline. In addition, the gastric electrical signal may be time-varying. That is, it may consist of not only normal slow waves (regular frequency of 2–4 cpm), but also tachygastria (regular frequency of 4–9 cpm), bradygastria (regular frequency of 0.5–2 cpm), and arrhythmia (irregular rhythmic activities). Arrhythmia may be divided further into bradyarrhythmia and tachyarrhythmia. It is the time variation of the frequency that may provide useful information in the assessment of patients with motility disorders. Therefore, not only information about the frequency of the gastric signal, but also information about the time variation of the frequency should be extracted from the EGG.

Despite many attempts made over the decades, visual inspection of the EGG signal has not led to the identification of waveform characteristics that would help the clinician to diagnose functional or organic diseases of the stomach. All EGG data are usually subjected to computerized data analysis. Among numerous data analysis methods, the time-frequency analysis technique is widely accepted.

6.3. SHORT-TIME FOURIER TRANSFORM AND SPECTROGRAM

The short-time Fourier transform (STFT) and spectrogram have been extensively discussed in Chapter 1 of this book. In this section, we will introduce the advantages, limitation, and applications of the STFT and spectrogram in the analysis of the EGG.

6.3.1 Advantages and Limitations

Applying the spectrogram to the EGG is simple. For a given data set of the EGG, first a time window with a length of L samples is applied, then an FFT with a length of N (assuming $N = L$) is calculated and a power spectrum is obtained. To calculate the power spectrum at the next time step, the window is shifted some samples ahead and the same process is repeated [15]. The Hamming window is the most commonly used window in the EGG application [9, 15, 26] mainly due to its simple form and relatively good performance. The main advantage of this method is its ease of implementation. It is the most efficient method in computations.

The major drawback inherent in the STFT, or spectrogram, is that a trade-off is inevitable between temporal and spectral resolution. To increase the frequency resolution, one must use a longer observation duration (longer window), which means that variations occurring during this interval will be smeared, lowering temporal resolution. In EGG applications, several minutes of data are required to accurately compute the spectrum of the EGG. Each spectrum provides ensemble information of the signal during these minutes. Therefore, any rhythmic variations within these several minutes may not be detected and the exact time information of the rhythmic variation is not available. The type and length of the window are the main issues in the practical use of this method. Figure 6-5 shows the effects of different windows on the time-frequency representation of a simulated EGG signal. The EGG signal was

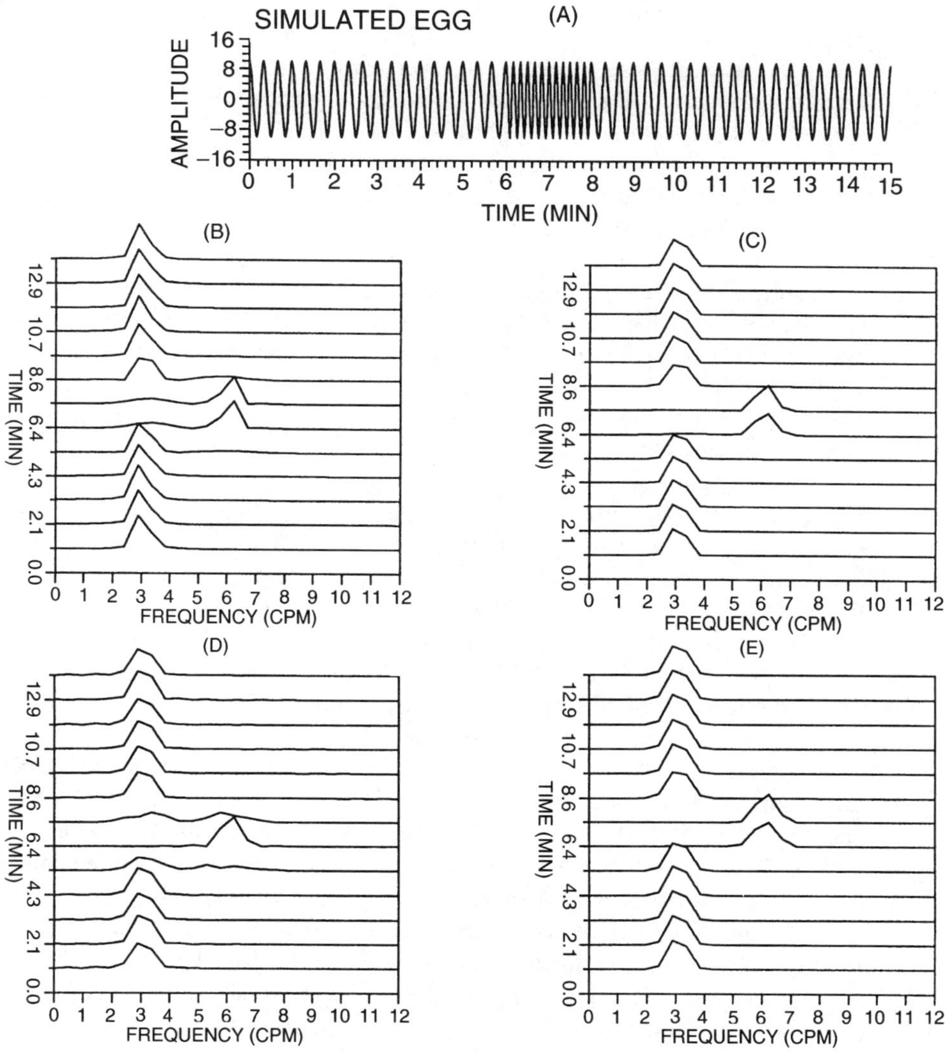

Figure 6-5   Effects of different windows on the time-frequency representation of a simulated EGG signal calculated by the spectrogram method ($N = 128$): (a) simulated signal with an interposed 2-min episode of 6 cpm between segments of 3 cpm sinusoid; (b) using rectangular window; (c) using Hamming window; (d) using Gaussian window; (e) Using Kaiser window.

simulated by interposing a 2-min episode of 6 cpm tachygastria between segments of 3 cpm sinusoid [Fig. 6-5(a)]. The tachygastria occurs between minutes 6 and 8. The time-frequency representations of this simulated signal with different windows are shown in Fig. 6-5(b)–(e) (the length of window: 128 samples). It is seen that the rectangular window [Fig. 6-3(b)] generated the narrowest spectral peaks (highest frequency resolution) at 3 cpm. Its temporal resolution is, however, very poor. Three cpm waves are falsely detected during minutes 6 and 8 due to the apparent "leakage" of a 3-cpm component. The Kaiser window [Fig. 6-5(e)], on the other hand, provides accurate detection of tachygastria in time (high temporal resolution). The spectral peaks are broader, however (low frequency resolution). The use of a Hamming window [Fig. 6-5(c)] or a Gaussian window [Fig. 6-5(d)] is considered a reasonable compromise between leakage reduction and main-lobe broadening [27].

Another drawback of the spectrogram method is the cross-term, which does not represent the signal components. The difference between the spectrogram and STFT is that the STFT is a linear signal decomposition and there are no cross-terms between signal components. However, the spectrogram is a bilinear signal energy distribution due to the magnitude-squaring operation. Thus the spectrogram has cross-terms. In other words, the cross-terms that do not exist in the STFT may appear visible in the spectrogram through the magnitude-squaring process [28, 29]. Examples will be given later.

6.3.2 Applications

Gastric dysrhythmias have been found to be associated with several clinical symptoms, such as nausea, vomiting, motion sickness, and early pregnancy [9, 30-33]. The detection and assessment of gastric dysrhythmias from the EGG are, therefore, of great clinical significance.

To detect gastric dysrhythmia, van der Schee and Grashuis [15] introduced a running spectral analysis (RSA) method (based on the STFT), and showed that a dysrhythmic event of 64 s can be identified from the EGG using this method. Since then, a number of investigators have used this method to detect the dysrhythmia in the EGG [9, 26, 30–34]. To study gastric myoelectrical activity in patients with unexplained nausea and vomiting, Geldof et al. [32] analyzed EGG data in both fasting and postprandial states in 48 patients with unexplained nausea and vomiting, and in 52 control subjects using the spectrogram method. Based on this method, they defined and calculated a mean gastric frequency with standard deviation and its power content, an instability factor, and a ratio of the power in the postprandial state to the power in the fasting state for each subject. They reported that in 48% of the patients, abnormal myoelectrical activity was found, which was characterized by: (1) instability of the gastric frequency; (2) tachygastria in both fasting and postprandial states; (3) the absence of the normal amplitude increase in the postprandial EGG.

To investigate the relationship between onset of tachygastria and onset of the symptoms of motion sickness, Stern et al. [9] used the spectrogram method to analyze EGG data obtained from 15 healthy human subjects who were seated inside a circular vection drum. The EGG data were made for three 15-min periods: baseline, during drum rotation, and after drum rotation. They found that five subjects showed continuation of normal 3-cpm activity during drum rotation and reported no symptoms of motion sickness. Ten subjects showed a shift from 3 cpm to 4–9 cpm during drum rotation and reported symptoms of motion sickness. For example, Fig. 6-6 shows the time-frequency representation of the EGG of one subject who reported that during rotation he was sweating, dizzy, and had a queasy stomach. Whereas 3- and 1-cpm components are the dominant frequency before drum rotation, 6 min after the onset of rotation, the dominant frequencies are in the tachygastria range (5–9 cpm). The results showed that the spectrogram method revealed a close correspondence over time between tachygastria and reports of symptoms of motion sickness.

Koch et al. [33] applied the spectrogram method to measure gastric myoelectrical activity in women with and without nausea during the first trimester of pregnancy. They made EGG recordings in 32 pregnant women for 30–45 min. Gastric dysrhythmias were found in 26 pregnant subjects: seventeen had tachygastrias, five had bradygastria, and four had arrhythmia. They believed that the spectrogram method was able to detect dysrhythmias in the EGG.

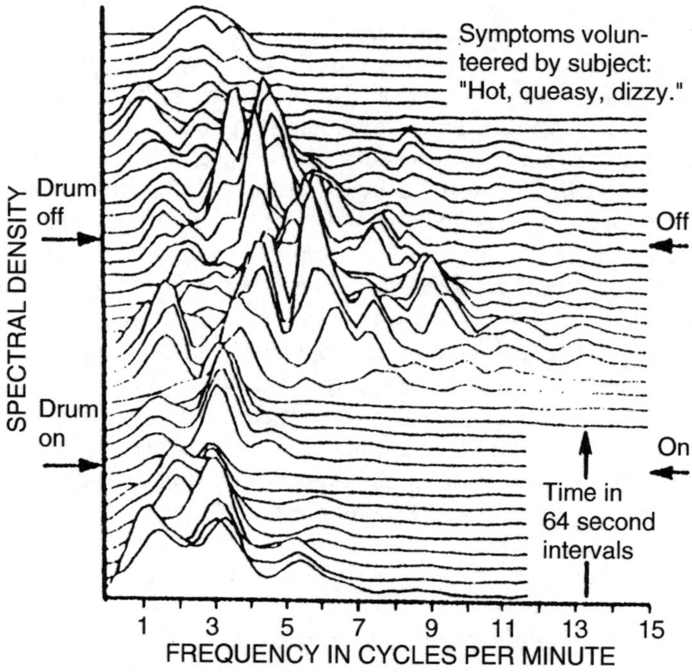

Figure 6-6   Time-frequency representations of the EGG of a subject who reported his symptoms of motion sickness during drum rotation. Whereas 3- and 1-cpm activity dominate the spectral analysis before drum rotation, 6 min after the onset of rotation spectral density showed a peak at 6 cpm, with additional activity in the tachygastria range (5–9 cpm). (From Stern et al., *Gastroenterology*, vol. 92, p. 94, 1987. With permission.)

Pfister et al. [26] applied the spectrogram method to EGG recordings from normal and diabetic subjects to detect frequency differences. It was shown that there were significant differences between the EGGs of diabetic gastroparetic patients and normal subjects. These differences were reflected in the higher mean peak frequency of the diabetics over normals in the postfed state and in the increase of occurrence of high-range peak frequencies (tachygastria) from pre- to postfed states in diabetics. These differences were not readily observable from manual scoring of the data, but became apparent using the spectrogram technique.

6.4. EXPONENTIAL DISTRIBUTION

Exponential distribution (ED) [14] is a new Cohen's class of distributions [35] (refer to Chapter 1 of this book). The advantages, limitations, and applications of this method in analysis of the EGG will be discussed in this section.

6.4.1 Advantages and Limitations

Theoretically, the ED is free from the assumption of stationarity and is well suited for the time-frequency representation of the time-varying signal [36–39]. It provides high resolution in time and frequency while suppressing cross-terms by way of controlling a single parameter. By definition, the centroid frequency of the ED at each time is equal to the instantaneous frequency of the signal [14]. The ED is periodic in π. This will cause undesirable spectral overlap or aliasing if the signal is sampled at the Nyquist frequency. To avoid aliasing in this representation, it is necessary to either sample the signal at a frequency which is at least twice the Nyquist rate or use the analytic form of the signal. In this chapter, the analytic form of the signal is used. Since the analytic form has energy only at positive frequencies, interference between negative and positive frequency components is avoided [40, 41]. Although cross-terms are being substantially suppressed, they are still the main drawback of the ED method, which suffers from an inherent trade-off between simultaneous cross-term reduction and auto-term preservation [42]. Figure 6-7 shows the effects of the parameter M in the ED ($N = 512$, $\sigma = 0.5$) on the cross-term and the autoterms of the simulated signal with two sinusoidal components. The cross-term is reduced as M becomes larger.

6.4.2 Applications

Understanding the correlation between the EGG and gastric contractions is very important for clinical applications of the EGG in gastroenterology. Because of the relatively poor signal-to-noise ratio of the EGG, direct visual inspection of the EGG cannot be reliably used to identify gastric contractions [32, 43].

The ED method was used to characterize the EGG in different motility states of the stomach [16]. EGG recordings were made in five volunteers for 2 hours in the fasting state. Gastric contractions were simultaneously measured by placing an intraluminal probe via the nose to the distal stomach. The probe had three solid-

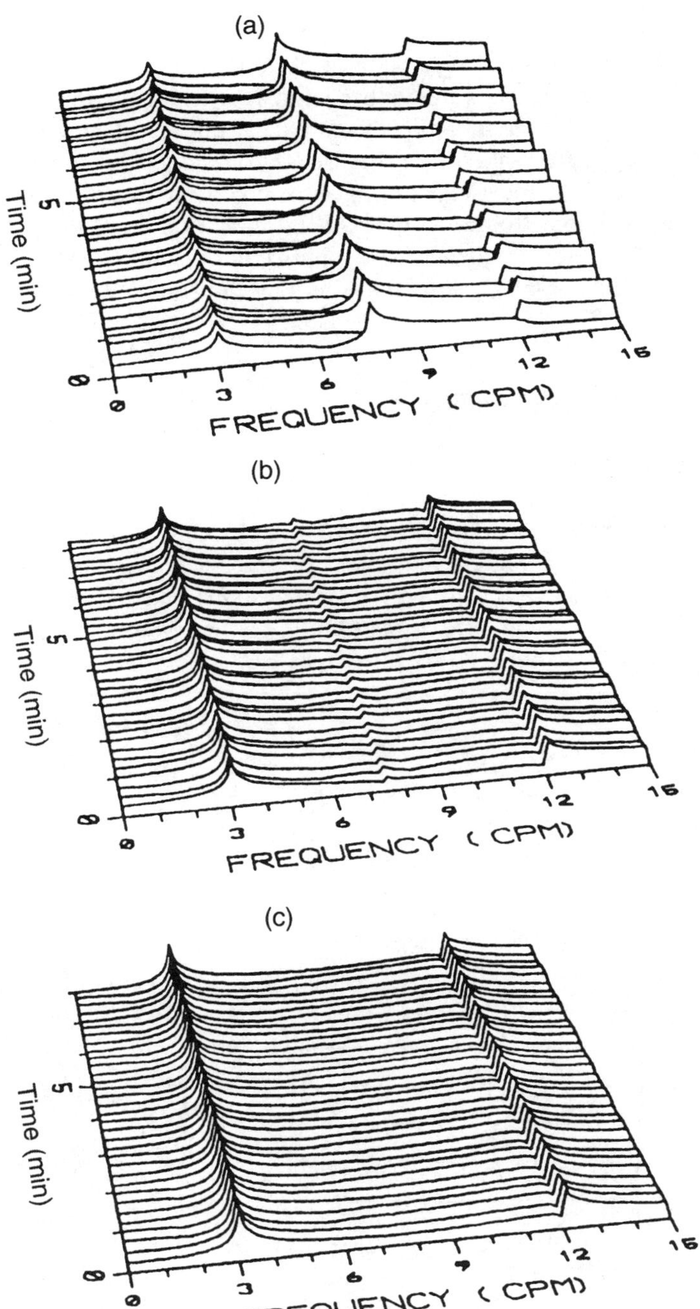

Figure 6-7   Effects of the parameter M in the ED ($N = 512$, $\sigma = 0.5$) on the cross-term and the auto-terms of the simulated signal with two sinusoidal components ($f_1 = 0.05\,\text{Hz}$, $f_2 = 0.2\,\text{Hz}$): (a) $M = 2$, (b) $M = 20$, (c) $M = 40$. The cross-term is reduced as M becomes larger.

state pressure sensors recording gastric contractions. A period of 20-min EGG during gastric contractions and a period of 20-min EGG during gastric quiescence in each of the five subjects were digitized and analyzed using the ED method ($N = 512$, $M = 40$, $\sigma = 0.5$). Figure 6-8 presents typical time-frequency representations of the EGG during gastric quiescence [Figure 6-8(a)] and gastric contractions [Fig. 6-8(b)]. It was found that the EGG during gastric contractions had a significantly higher power at 2–4 cpm (total energy: 49.8 ± 1.8 vs. 39.8 ± 3.3 dB, $p < 0.01$, t-test) and significantly lower frequency (0–2 cpm) components (total energy: 48.2 ± 3.3 vs. 37.4 ± 3.1 dB, $p < 0.01$, t-test) than the EGG during gastric quiescence.

6.5. ADAPTIVE ARMA MODELING

6.5.1 Definition and Implementation

The adaptive spectral analysis proposed by Chen [13] is a parametric method. It is based on the autoregressive moving average (ARMA) model [44]. In this method, it is assumed that a signal s_j (j = time instant) can be generated by exciting an ARMA process using a random time series, n_j. Mathematically, a time series s_j can be modeled as an ARMA process as follows:

$$s_j = -\sum_{k=1}^{p} a_k s_{j-k} + \sum_{k=1}^{q} c_k n_{j-k} + n_j \tag{6-1}$$

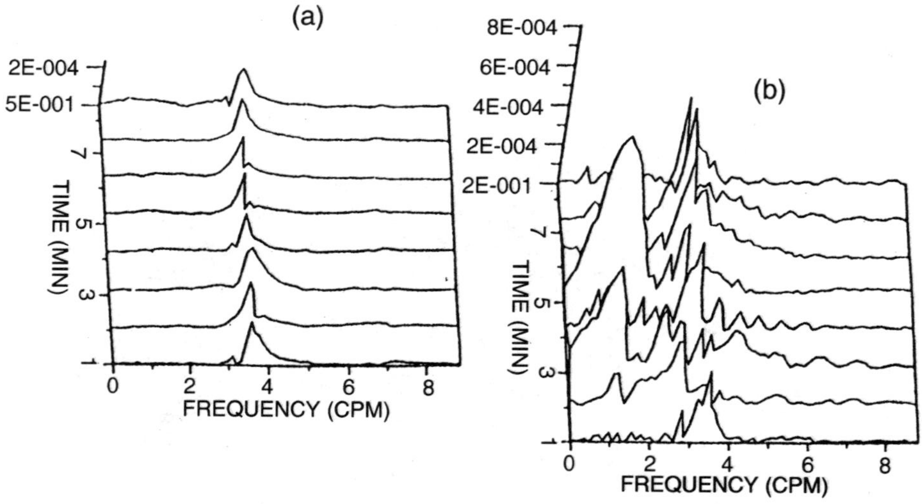

Figure 6-8   Time-frequency representations of the EGG during gastric quiescence (left) and gastric contractions (right) calculated by the ED method ($N = 512$, $M = 40$, $\sigma = 0.5$). The EGG during quiescence has a relatively single frequency, whereas the EGG during contractions shows more low frequencies and an increased power.

where a_k $(k = 1,2,\ldots,p)$ and c_k, $(k = 1,2,\ldots q)$ are called the ARMA parameters. n_j is a white noise process.

Once the parameters of the ARMA model are identified, the power spectrum of the signal, s_j, can be calculated from these ARMA parameters:

$$P_s(\omega) = \frac{\sigma^2 |1 + \sum_{k=1}^{q} c_k \exp(-i\omega k)|^2}{|1 + \sum_{k=1}^{p} a_k \exp(-i\omega k)|^2} \tag{6-2}$$

where σ^2 is the variance of the white noise, $i = \sqrt{-1}$, and ω is the angular frequency.

To model an EGG signal x_j, one simply proceeds in the opposite direction and constructs a so-called adaptive ARMA filter, as shown in Fig. 6-9. z^{-1} in the figure stands for one sample delay. x_j is the input signal at time j that is to be analyzed. The sets $a_{k,j}$ and $c_{k,j}$ are, respectively, the feedforward and feedback weights of the adaptive ARMA filter. y_j is the estimate of the input signal by the ARMA filter and is expressed as

$$y_j = \sum_{k=1}^{p} a_{k,j} x_{j-k} + \sum_{k=1}^{q} c_{k,j} e_{j-k} \tag{6-3}$$

where e_j is the estimation error expressed as

$$e_j = x_j - y_j \tag{6-4}$$

These equations imply that y_j is an ARMA estimate of the signal x_j if the error signal e_j is a white noise. After the convergence of the ARMA filter weights, the power spectrum of the signal x_j can be estimated.

A least mean square (LMS) algorithm for the adaption of the weights of the adaptive ARMA filter was derived in [13] and is described as follows.

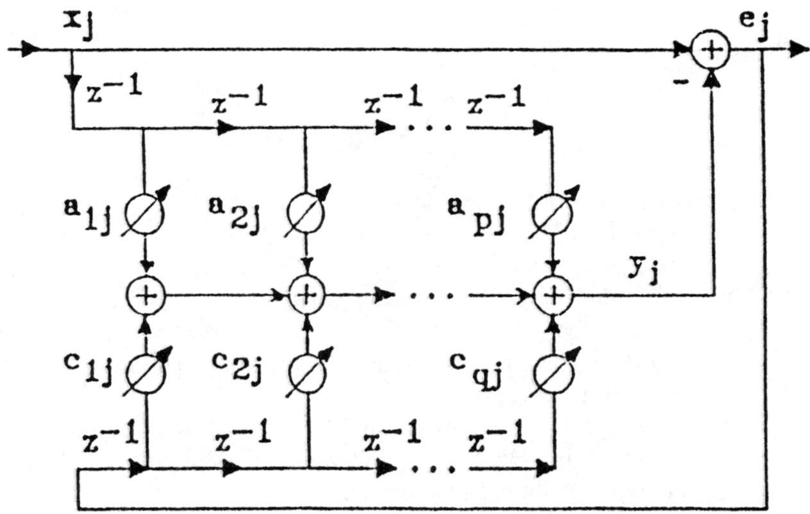

Figure 6-9 The detailed structure of the adaptive ARMA filter.

Defining feedforward and feedback weight vectors A_j and C_j and feedforward and feedback input vectors X_j and E_j as

$$A_j^T = [a_{1,j} \ldots, a_{p,j}], \qquad C_j^T = [c_{1,j}, \ldots, c_{q,j}] \tag{6-5}$$

$$X_j^T = [x_{j-1}, \ldots, x_{j-p}], \qquad E_j^T = [e_{j-1}, \ldots, e_{j-q}] \tag{6-6}$$

the output of adaptive ARMA filter y_j can be written as

$$y_j = A_j^T X_j + C_j^T E_j \tag{6-7}$$

and the error signal as

$$e_j = x_j - A_j^T X_j - C_j^T E_j \tag{6-8}$$

Applying the steepest descent method and approximating the gradient of the mean square error with respect to the filter weight vectors by the gradient of the squared error, we have

$$A_{j+1} = A_j - \mu_a \frac{\partial E_j^2}{\partial A_j} \tag{6-9}$$

$$C_{j+1} = C_j - \mu_c \frac{\partial E_j^2}{\partial C_j} \tag{6-10}$$

To have a simple adaption algorithm, we assume that the feedback input vector E_j is not a function of the filter parameters. Under this assumption, the LMS algorithm can be easily derived and is expressed as follows:

$$A_{j+1} = A_j + 2\mu_a e_j X_j \tag{6-11}$$

$$C_{j+1} = C_j + 2\mu_c e_j E_j \tag{6-12}$$

or

$$a_{k,j+1} = a_{k,j} + 2\mu_a e_j x_{j-k}, \qquad k = 1, 2, \ldots p \tag{6-13}$$

$$c_{k,j+1} = c_{k,j} + 2\mu_c e_j e_{j-k}, \qquad k = 1, 2, \ldots, q \tag{6-14}$$

where step-sizes μ_a and μ_c are small constants controlling the adaptation speed of the LMS algorithm. The algorithm states that the filter parameters at each successive time step, $a_{k,j+1}$ and $c_{k,j+1}$, are equal to their current values, $a_{k,j}$ and $c_{k,j}$, plus a modification term. The number of filter parameters is equal to $q + p$. The best value for q was found to be in the range of 2–10 for spectral analysis of the EGG [45]. The value of p must be greater than or equal to the number of digitized points that span the longest rhythmic cycle of interest in a signal. For example, if the period of the rhythmic component of interest in an EGG is 20 s (0.05 Hz, or 3.0 cpm), and the sampling frequency is 1 Hz, the smallest value of p should be 20. The requirement of this large value is attributed to the nature of the LMS algorithm [46].

Once the adaptive filter converges, the power spectrum of the input signal x_j can be calculated from the filter parameters. At any point in a time series, a power spectrum can be calculated instantaneously from the updated parameters of the model. Similarly, the power spectrum of the signal for any particular time interval can be calculated by averaging the filter parameters over that time interval. Since the parameters are initially set to zero, the adaptive filter needs an initial period of time to converge, thus the power spectrum for the initial time period is unavailable. This problem is circumvented by processing the initial section of a time series twice. The values of the parameters reached at the end of the first run can be utilized to initialize the model during the second run.

6.5.2 Advantages and Limitations

The main advantage of this method is the increased spectral and temporal resolution. In addition, it has no problems of cross-terms. Numerous experiments have shown that the adaptive spectral analysis method provides narrow frequency peaks, permitting more precise frequency identification, enhancing the ability to determine frequency changes at any time point [47]. This method is especially powerful in detecting dysrhythmic events of brief duration and rhythmic variations of any biomedical signals.

The drawback of the adaptive spectral analysis method is that it does not preserve the relative amplitudes of multicomponent signals due to the nonlinearity of the ARMA modeling [48, 49]. The amplitude of spectral peaks is largely dependent on the accuracy of the modeling rather than the energy in the signal. When a signal containing multiple frequency components is analyzed, the relative amplitude of the individual component may not be preserved.

6.5.3 Applications

To determine whether the EGG could be used to differentiate patients with gastroparesis, or delayed emptying of the stomach, from asymptomatic healthy controls, the adaptive spectral analysis method was used for quantitative analysis of EGG data obtained from 24 asymptomatic healthy controls and 27 patients with gastroparesis [31]. The procedure of data analysis was as follows. Each EGG recording was divided into 2-min blocks. The power spectrum of each 2-min piece of EGG data was calculated by the adaptive spectral analysis method, yielding 15 consecutive power spectra for each 30-min preprandial recording and 60 consecutive power spectra for each 120-min postprandial recording. If there was no peak at 2–4 cpm, the 2-min EGG was said to be abnormal. If more than 30% of the power spectra showed no peaks at 2–4 cpm or there was a decrease in power at the dominant frequency after the meal, the EGG recording was defined as abnormal. Based on this method, it was found that all 24 controls showed slow wave normality in both fasting and fed states. Approximately 75% of the patients had abnormal pre- and/or postprandial EGG. It was shown that the cutaneous EGG may be used to differentiate gastroparetic patients from asymptomatic normals using the adaptive spectral analysis method. A typical example of a normal EGG obtained in a normal control is shown in Fig. 6-10. The normal slow wave activity (3 cpm) is clearly visible, and an increase in the amplitude of the EGG is evident. More convincingly,

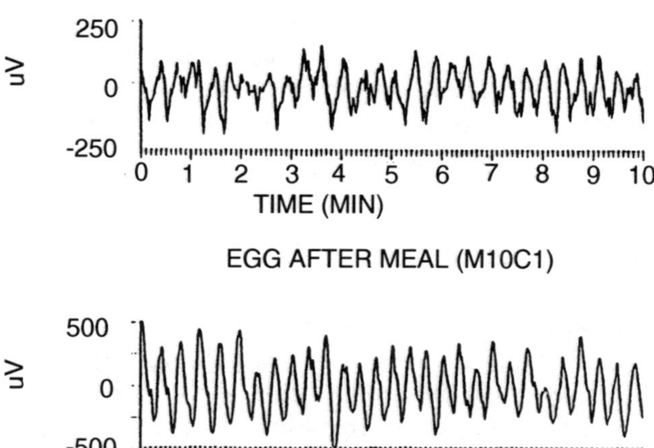

Figure 6-10   Normal cutaneous EGGs obtained in a normal control: 10-min pre-prandial EGG right before meal (top) and 10-min postprandial EGG right after a testing meal (bottom). Amplitude increase after the meal is very evident in these recordings.

as shown in Fig. 6-11, 100% of the power spectra has predominant peaks at the normal frequency (2–4 cpm). Figures 6-12 and 6-13 show an example of typical EGG recordings obtained in a gastroparetic patient and its time-frequency representation, respectively. It can be seen from Fig. 6-13 that sharp peaks at 8–9 cpm in this figure indicate tachygastria and regular 3-cpm slow waves are almost completely missing.

Recently, Lin et al. [50] have extracted some parameters from the EGG using the adaptive spectral analysis method to investigate whether gastric emptying could be predicted from the EGGs. The EGG recordings were obtained in 97 patients, each for 30 min in the fasting state and for 2 hours simultaneously with gastric emptying being monitored after an isotope-labeled beef stew meal. Gastric emptying was defined as abnormal if the gastric retention at 2 hours was more than 70% or the $T_{1/2}$ was longer than 150 min. The following EGG parameters were extracted: (1) δP, the difference of the EGG power at the dominant frequency after and before the meal; (2) the percentage of 2–4 cpm waves in the fasting and fed states. The results showed that 16 patients had a decrease in EGG peak power after the meal and 75% of them had abnormal gastric emptying. Twenty-seven patients had a dysrhythmic postprandial EGG (2–4 cpm waves < 70%) and 80% of them had abnormal gastric emptying. Seven patients had both a decrease in postprandial EGG power and a dysrhythmic postprandial EGG, and all had abnormal gastric emptying. Thirty patients had a dysrhythmic preprandial EGG and 57% of them had abnormal gastric emptying. It was concluded that a decrease in postprandial power or a dysrhythmic postprandial EGG analyzed by the adaptive method is suggestive of delayed gastric emptying.

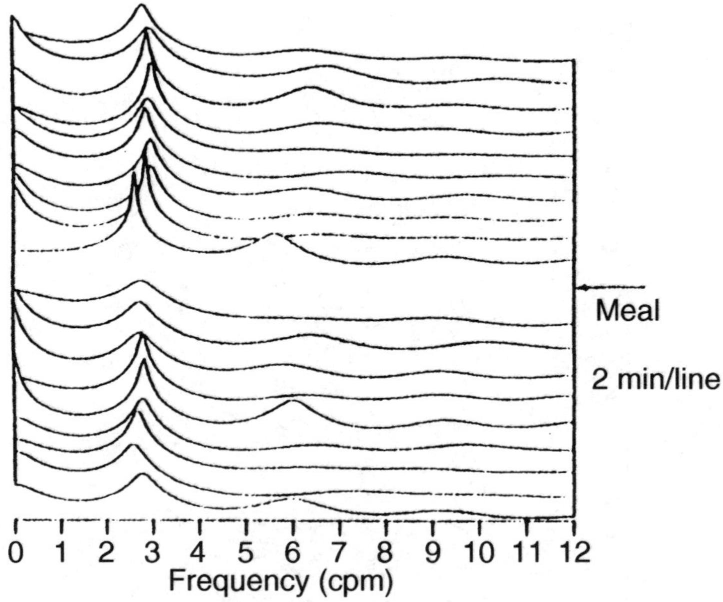

Figure 6-11   Running power spectra of normal EGGs in Fig. 6-10 calculated by the adaptive method. Each spectrum represents analysis of serial 2-min periods of EGG data. Sharp peaks at 3 cpm indicate regular slow wave activity.

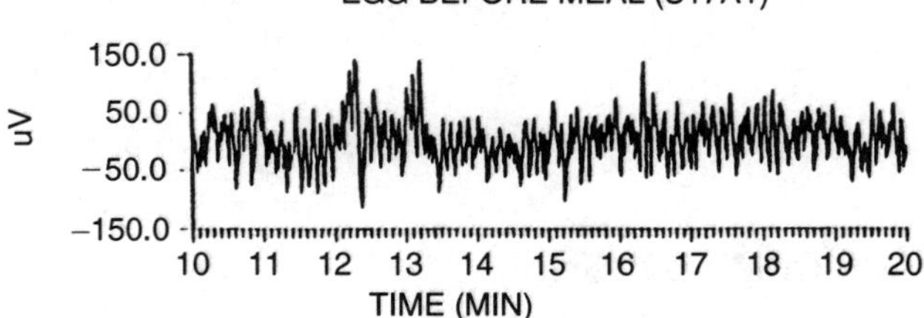

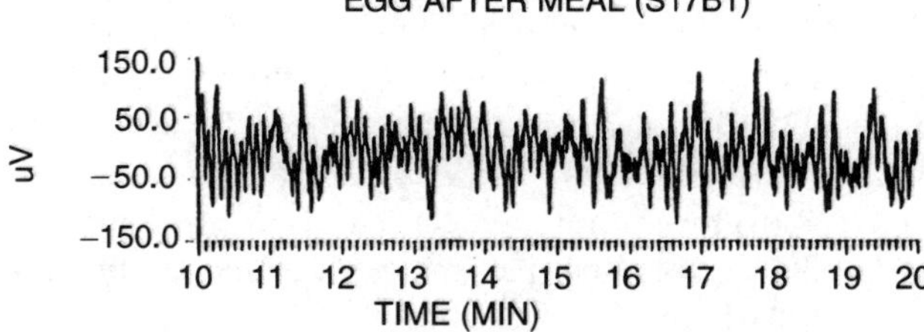

Figure 6-12   Abnormal cutaneous EGGs obtained in a patient with idiopathic gastroparesis: 10-min preprandial EGG (top) and 10-min postprandial EGG (bottom). No regular 3-cpm slow waves are seen in these tracings.

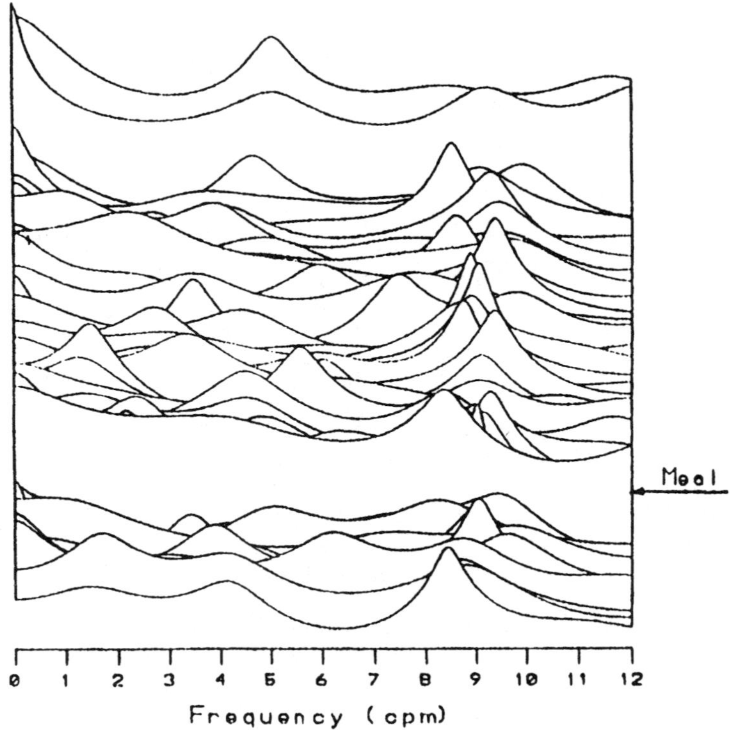

Figure 6-13 Time-frequency representation of the abnormal postprandial EGG in Fig. 6-12. Each spectrum represents analysis of serial 2-min periods of EGG data. Sharp peaks at 8–9 cpm indicate abnormal slow wave activity, tachygastria. Regular 3-cpm slow waves are almost completely missing.

6.6. PERFORMANCE COMPARISON

6.6.1 Simulation Results

A series of computer simulations has been conducted to investigate and compare the performances of the aforementioned three methods in the time-frequency analysis of the EGG. If not particularly specified, the window for the spectrogram is the Hamming window and its length is 256 samples; the parameters for the ED method were as follows: both $W_N(\tau)$ and $W_M(\mu)$ are rectangular windows, $N = 256$, $M = 30$, $\sigma = 1$; the parameters of the adaptive method are $p = 40$, $q = 2$, $\mu_a = \mu_c = 0.01/$(total input energy).

Cross-Terms in the Spectrogram and the ED. To investigate the effects of cross-terms on the performance of the spectrogram and the ED, we consider a chirp signal

$x(t)$ [Fig. 6-14(c)], which has two frequency components, $x_1(t)$ and $x_2(t)$, at each time instant. The frequency of one component, $x_1(t)$, increases over time [see Fig. 6-14(a)] while the frequency of the other decreases over time (Fig. 6-14(b)). Figure 6-15 shows the effect of the cross-terms on the spectrogram. The spectra in Fig. 6-15(a) illustrate the "ideal" spectra of the signal $x(t)$, without cross-terms, i.e., $\mathrm{SPEC}_{x1} + \mathrm{SPEC}_{x2}$. The spectra in Fig. 6-15(b) are the actual running power spectra computed by the STFT method with the Hamming window. The contribution of the cross-terms can be clearly observed by comparing the "ideal" spectra (a) and the actual spectra (b). It can be observed that the cross-terms appear when the frequencies of the two individual components are close or overlap. Figure 6-15 (c) and (d) presents the time-frequency representations of the signal using the Kaiser window and the rectangular

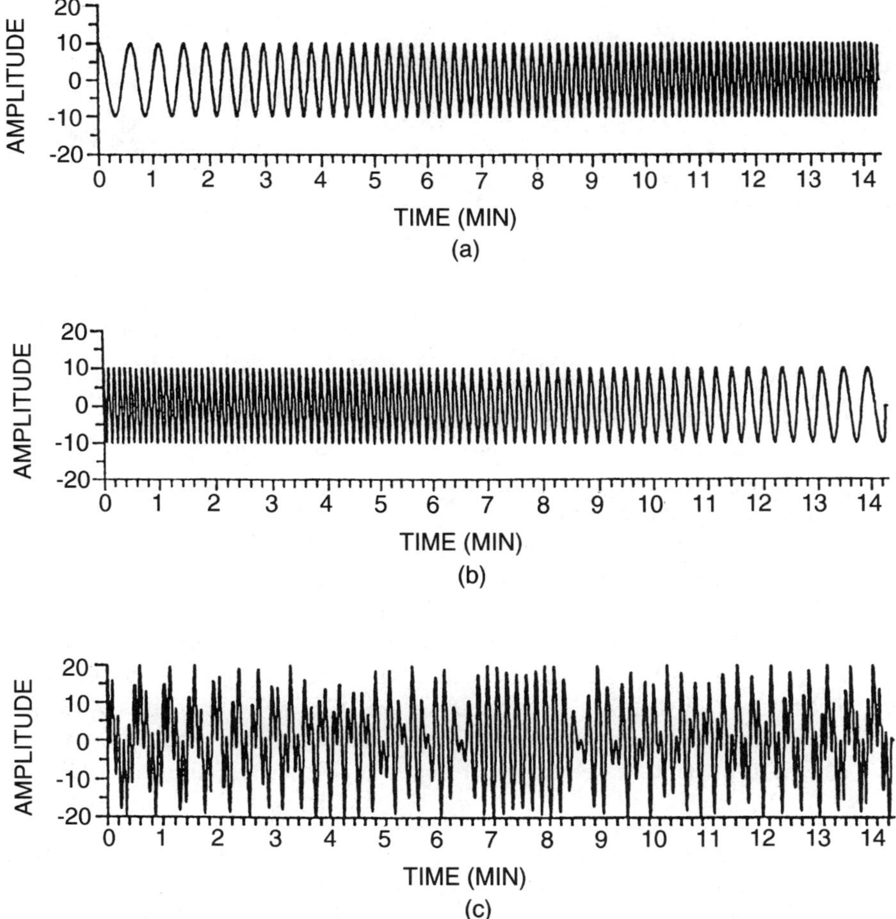

Figure 6-14 Simulated signals: (a) a chirp signal whose frequency increases with time; (b) a chirp signal whose frequency decreases with time; (c) the sum of these two chirp signals. (From Zhiyue Lin et al., *Med. Biol. Eng. Comput.*, vol. 33, p. 598, 1995. With permission.)

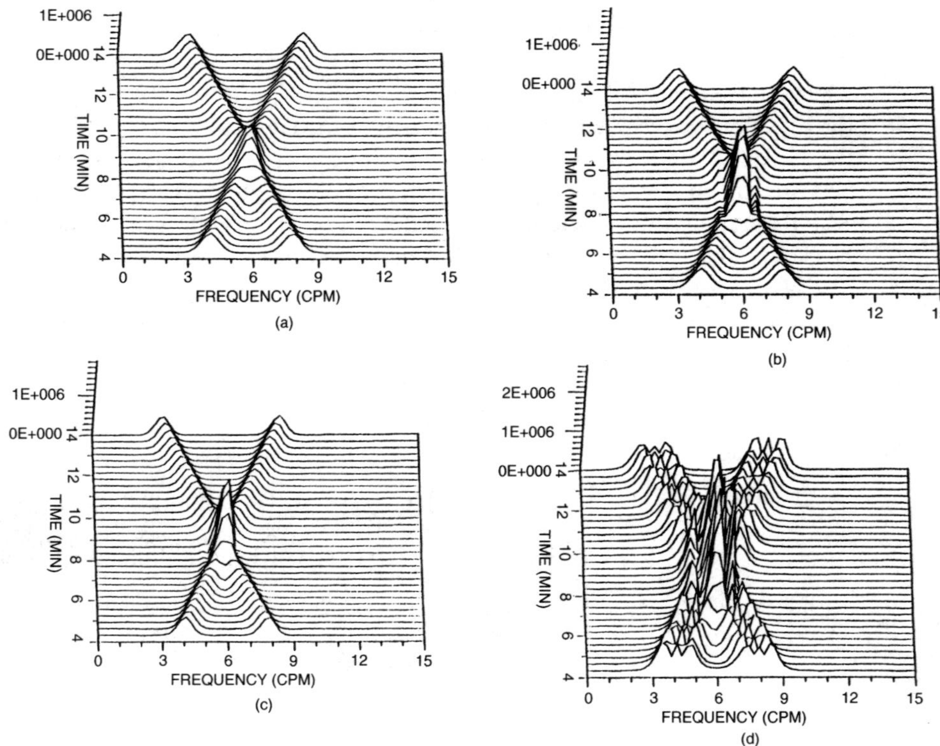

Figure 6-15 Spectrograms of the simulated signals in Fig. 6-14: (a) the sum of two spectrograms in Fig. 6-14(a) and (b); (b)–(d) spectrograms of the sum of two chirp signals in Fig. 6-14(c): using (b) Hamming window, (c) Kaiser window, and (d) rectangular window.

window, respectively. It is seen that the Kaiser window results in fewer cross-terms than the other two windows.

The effects of the cross-terms on the performance of the ED method was similarly investigated. The test signal was the same as the one previously described, $x(t) = x_1(t) + x_2(t)$. Figure 6-16(a) illustrates the "ideal" time-frequency representation of the signal $x(t)$ with no cross-terms. It was computed by adding the ED of $x_1(t)$ and the ED of $x_2(t)$. The actual time-frequency representation of the signal $x(t)$ computed by the ED method is shown in Fig. 6-16(b). False spectral peaks are attributed to the cross-terms. Compared with the spectrogram method, the ED method provides narrower spectral peaks (higher resolution) but more cross-terms.

Tracking Amplitude Changes of the Signal. Accurate estimation of the EGG amplitude or the power at the dominant frequency is very important. This is because relative amplitude changes (or 3-cpm power changes) in the EGG are related to gastric contractile activity and thus provide useful information for the clinical diagnosis of patients with motor disorders of the stomach. To investigate the ability of the three time-frequency analysis methods in tracking the amplitude changes of the

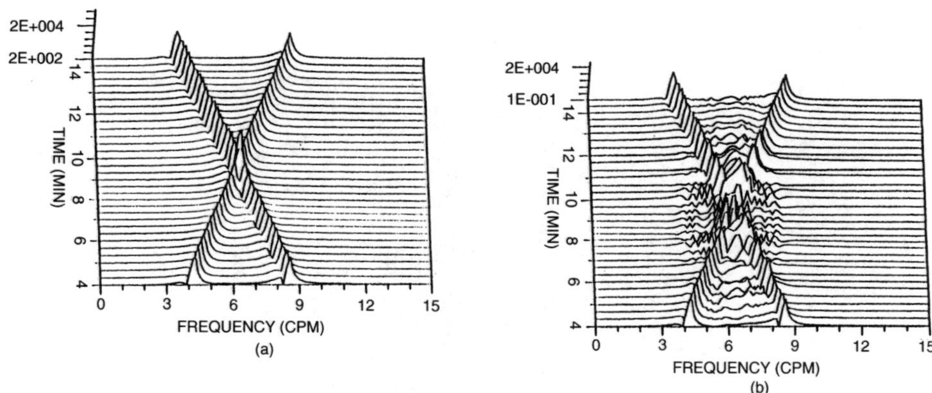

Figure 6-16   The EDs of the simulated signals in Fig. 6-14: (a) the sum of two EDs
in Fig.6-14(a) and (b); (b) the ED of the sum of two chirp signals in Fig.
6-14(c).

signal, an amplitude-modulated signal was simulated [see Fig. 6-17(a)]. This artificially generated signal has a persistent frequency of 3 cpm. Its amplitude change
follows a sinusoidal wave. The time-frequency representations shown in Fig.
6-17(b)–(d) were computed by the spectrogram, ED, and adaptive method, respectively. It can be seen that (1) all three methods provided an accurate estimation of the
signal frequency at 3 cpm and (2) the 3-cpm power computed by the spectrogram
and ED methods was linearly proportional to that of the signal [see Fig. 6-17(e)]. The
3-cpm power calculated by the adaptive method, however, was not linearly proportional to that of the signal.

Tracking Frequency Changes of the Signal.   The normal frequency of the gastric slow wave is about 3 cpm. After eating or drinking, the frequency of the gastric
slow wave changes. Postprandial transitional frequency changes have been documented in several previous studies [51–53]. To investigate the ability of the three
spectral methods in tracking frequency changes of the EGG, a 20-min computer-
generated signal was used, representing a simulated rhythmic shift from 1.5 to 4.5
cpm in the gastric slow wave. Fig. 6-18(a) shows the simulated signal. The time-
frequency representations shown in Fig. 6-18(b)–(d) were computed by the spectrogram, the ED, and adaptive method, respectively. The linear unit was used for the
spectrogram and ED method, and the decibel (dB) unit for the adaptive method. It
can be observed from this figure that (1) the adaptive method provides a more
accurate estimation of the instantaneous signal frequency and highest frequency
resolution (narrowest frequency peaks) and (2) the performance of the ED method
was better than the spectrogram.

Detection of Dysrhythmia with Brief Duration.   A simulated signal was used to
investigate the performance of the three time-frequency analysis methods in the
detection of dysrhythmia of brief duration. Figure 6-19(a) shows a computer-gener-
ated EGG signal with typical dysrhythmias, which were composed of 6 min normal

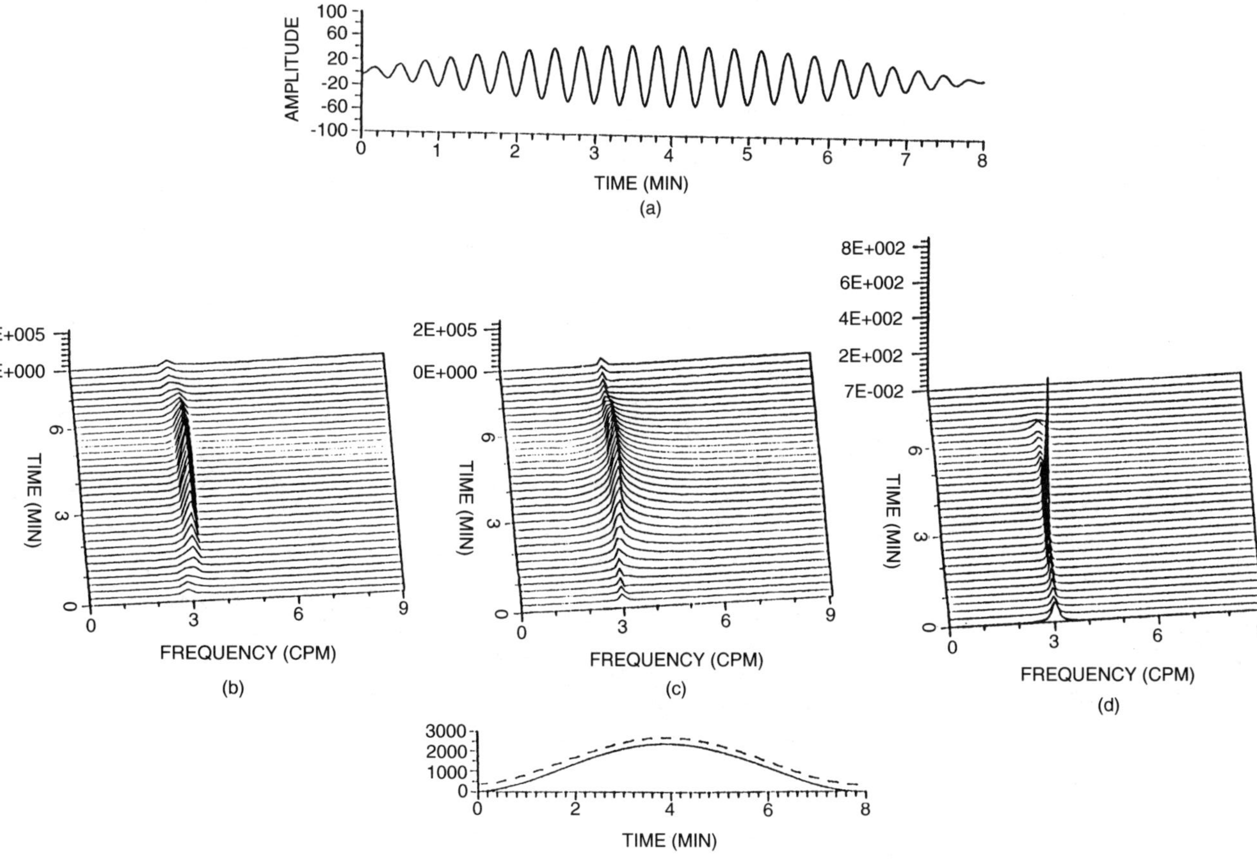

Figure 6-17 (a) A simulated amplitude-modulated (AM) signal and its time-frequency representation using (b) the spectrogram; (c) the ED; and (d) the adaptive method. (e) Real power of the AM signal (solid curve) and the estimated power (broken curve) using the ED method.

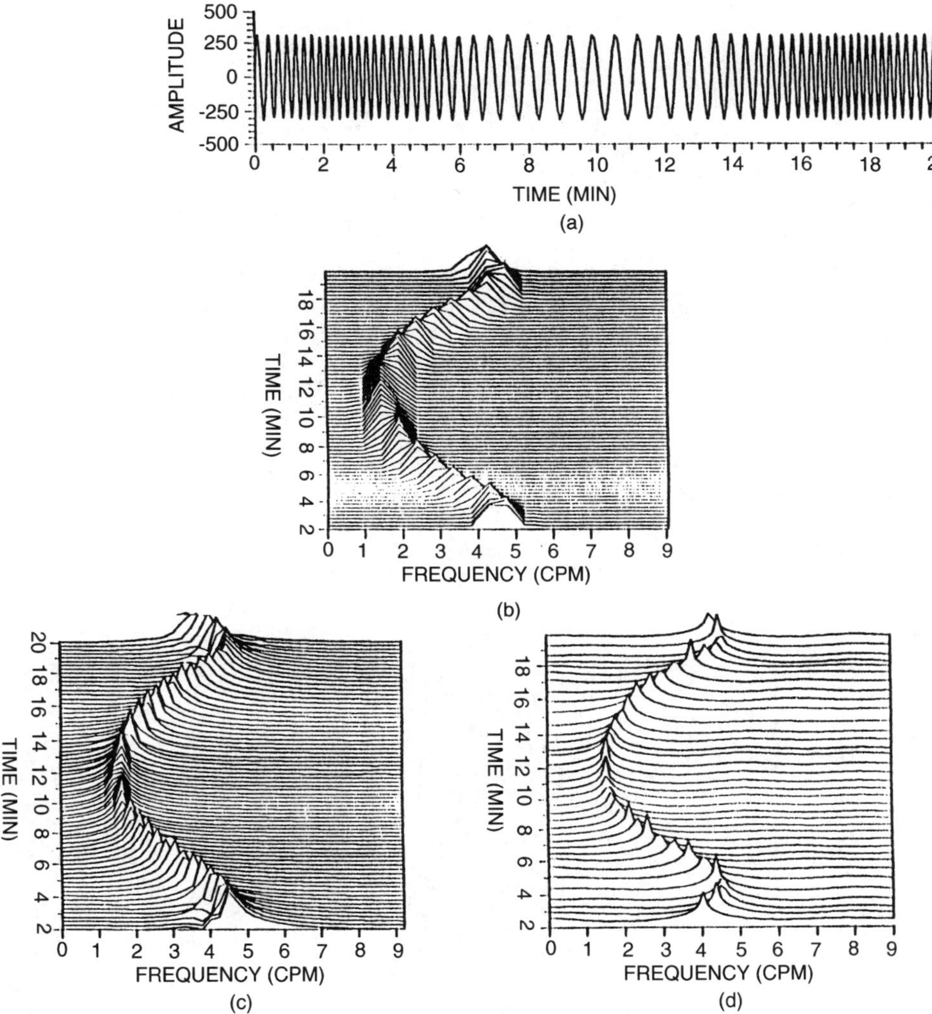

Figure 6-18   (a) A simulated frequency-modulated signal and its time-frequency representation using (b) the spectrogram; (c) the ED; and (d) the adaptive method.

slow waves (3 cpm), 2 min tachygastria (7 cpm), 2 min bradygastria (1 cpm), and 2 min pause (0 cpm). The time-frequency representations of this signal computed by the spectrogram, ED, and adaptive methods are shown in Fig. 6-19(b)–(d), respectively. It is seen in this figure that the adaptive method provides the best performance. It provides the highest temporal and frequency resolution. The STFT method provides broadest spectral peaks (lowest frequency resolution). The performance of the ED method is between the STFT method and the adaptive method for both time and frequency resolution.

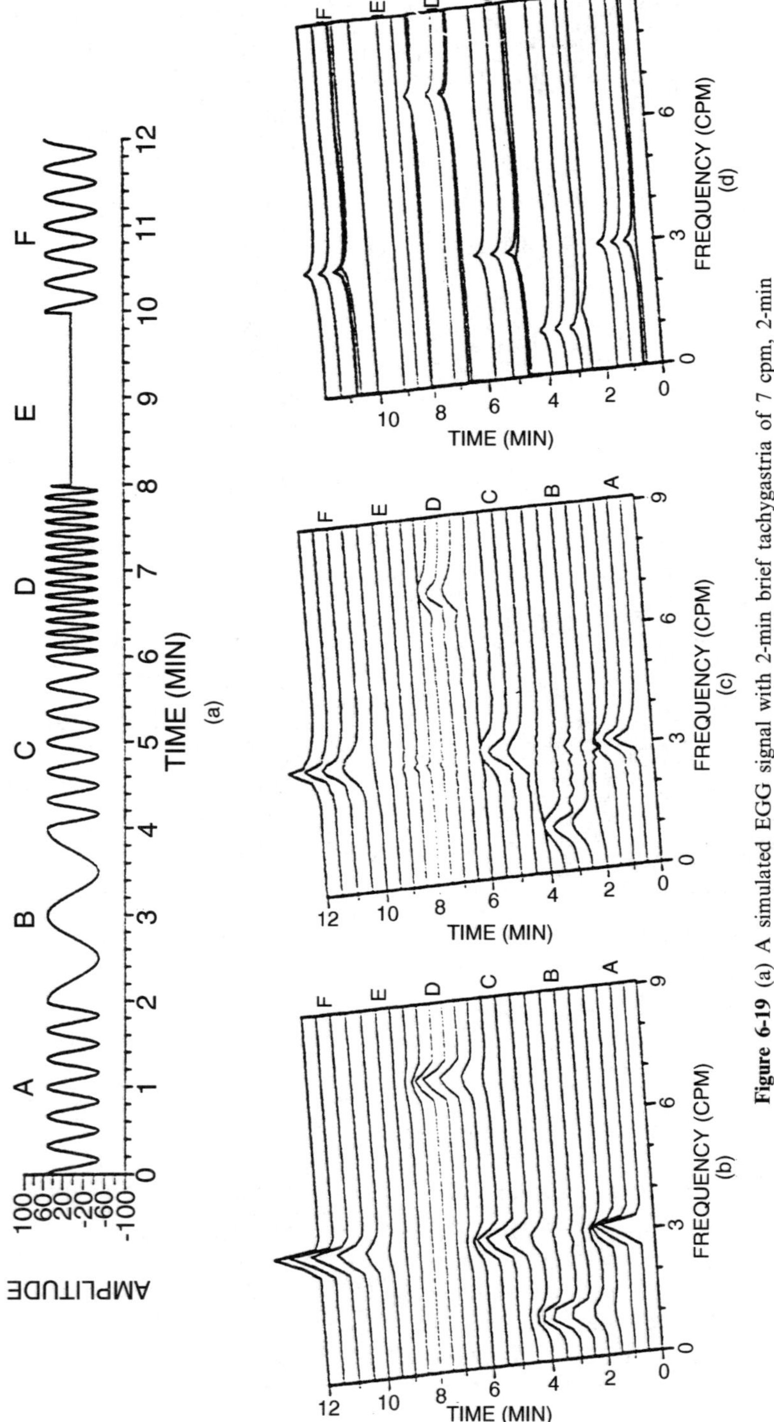

Figure 6-19 (a) A simulated EGG signal with 2-min brief tachygastria of 7 cpm, 2-min bradygastria of 1 cpm, and 2-min pause of 0 cpm interposed between the normal 3-cpm slow waves and its time-frequency representation using (b) the spectrogram; (c) the ED; and (d) the adaptive method. (From Zhiyue Lin et al., *Med. Biol. Eng. Comput.*, vol 33, p. 601, 1995. With permission.)

6.6.2 Clinical Applications

The three time-frequency analysis methods already mentioned were applied on real EGG recordings obtained from selected patients. The EGG recordings were obtained as follows: Prior to the attachment of electrodes, the abdominal surface where electrodes were to be positioned were shaved when necessary and cleaned with sandy skin-prep paste to reduce the impedance between the pair of electrodes to below 10 kΩ. Three silver/silver chloride electrodes were placed on the abdomen, as shown in Fig. 6-4. Two active electrodes were placed over the stomach with an interval of 4.0 cm. A referencing electrode was placed on the left-most side of the belly. The recordings were made using an ambulatory EGG recorder with low- and high-cutoff frequencies of 1 cpm (0.016 Hz) to 18 cpm (0.3 Hz), respectively. The ambulatory unit has a built-in analog/digital converter and is capable of storing 24-hour EGG recordings with a sampling frequency of 1 Hz [54]. The patients lay in a supine position and were asked not to talk or move during recording to avoid motion artifacts. Unless otherwise specified, the parameters of the three methods were the same as used for simulated signals.

Assessment of the regularity of an EGG Recording.   The assessment of the normality/abnormality of an EGG recording is an important part of a clinical EGG study. This is because the percentage of normal slow waves is a quantity of the regularity of gastric slow waves measured from the EGG [55]. It can be computed based on time-frequency representations of an EGG recording using one of the three methods already described. A spectrum from a certain period of EGG data is defined as normal if it has a clear peak in the 2–4 cpm range, bradygastria if it has a clear peak in the 0.5–2 cpm range, tachygastria if it has a clear peak in the 4–9 cpm range, and arrhythmia if it does not have a clear peak in the 0.5–9 cpm range. Figure 6-20 shows time-frequency representations of a 15-min EGG (top) derived from the spectrogram (left bottom), ED (middle), and adaptive method (right bottom), respectively. Table 6-1 presents results of the assessment of this 15-min EGG recording based on these three time-frequency representations and visual examination of the EGG tracing. We can see that the adaptive analysis method is more precise than the other two methods.

Detection of Brief Tachygastria.   Figure 6-21 presents the performance of the three methods in the time-frequency representation of the real EGG signal with brief tachygastria. The EGG [Fig. 6-21(a)] was obtained in a normal subject immediately after a drink of 140 ml of milk (2% fat). Gastric dysrhythmias were noted during the first 9 min of the recording. The time-frequency representations of this signal are shown in Fig. 6-21(b)–(d), computed by the spectrogram, the ED, and adaptive method, respectively. Subtle changes are detected by the adaptive method, but not by the other two methods.

Detection of Bradygastria.   Figure 6-22 shows time-frequency representations of a 14-min EGG (top) derived from the spectrogram (left bottom), ED (middle), and adaptive method (right bottom), respectively. In this example, the linear scales

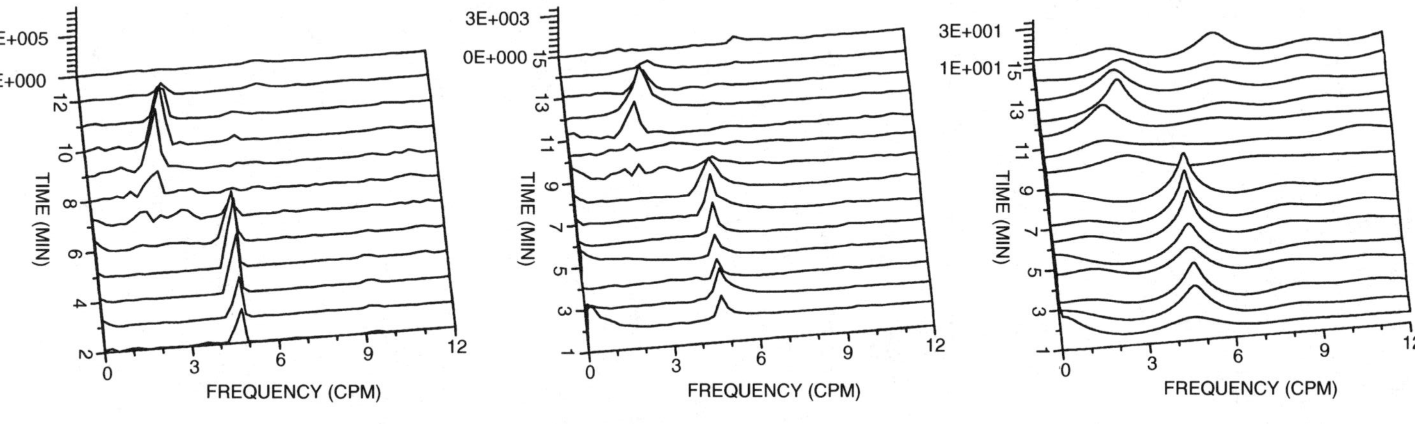

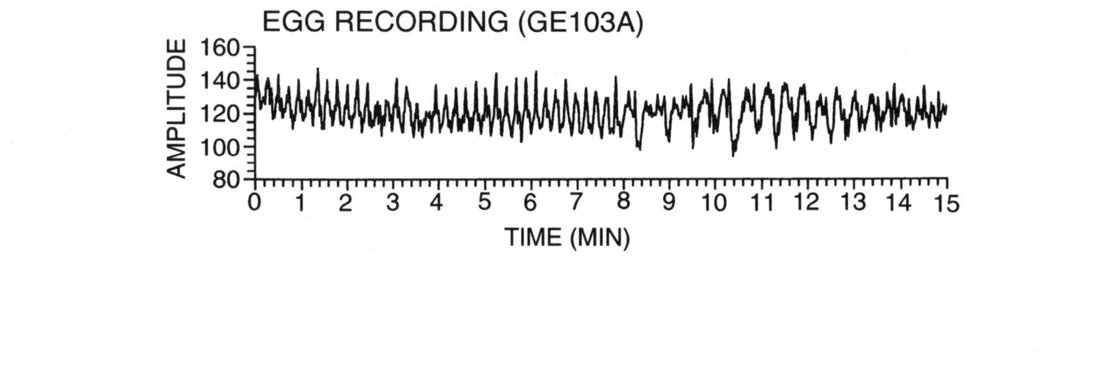

Figure 6-20 A typical EGG signal with dysrhythmia (top) and its time-frequency representation using the spectrogram (left bottom), the ED (middle), and the adaptive method (right bottom).

TABLE 6-1   Results of Assessment of a 15-min EGG Recording Based on Three Time-Frequency Analysis Methods and Visual Analysis (Gold Standard)

Time (min)	Visual Exam.	Adaptive	STFT	ED
1–2	Tachygastria	+	+	+
2–3	Tachygastria	+	+	+
3–4	Tachygastria	+	+	+
4–5	Tachygastria	+	+	+
5–6	Tachygastria	+	+	+
6–7	Tachygastria	+	+	+
7–8	Tachygastria	+	+	+
8–9	Arrhythmia	+	?	+
9–10	Bradygastria	+	−	−
10–11	Bradygastria	+	−	+
11–12	Normal	+	+	+
12–13	Normal	+	+	+
13–14	Normal	+	+	+

Note: + = the same as visual; − = not the same as visual; ? = not clear.

are used in the spectrogram method and the ED method, and the decibel (dB) unit is used in the adaptive method to present the power spectra of the EGG. It can be seen that for a relatively long period of bradygastria recording, all three of these methods were able to detect them correctly.

Pre- and Postprandial EGG Analysis.   Typical recordings of pre- and postprandial EGG of a normal subject are shown in Fig. 6-23. Both the preprandial EGG [Fig. 6-23(a)] and the postprandial EGG Fig. 6-23(b) showed regular rhythmic activities. The amplitude of the EGG is higher after eating a solid meal (500 kcal) than in the fasting state. The time-frequency representations of the EGG computed by the spectrogram, ED, and adaptive methods are shown in Fig. 6-23(c)–(e), respectively. It can be seen that all three of these methods can represent 3-cpm activities in both pre- and postprandial EGGs.

6.7. CONCLUSIONS

In this chapter, we have summarized three time-frequency analysis methods, including the STFT (or spectrogram), the ED, and the adaptive method, and their applications in analyzing EGG signals. Their advantages and limitations were discussed and their performances were compared using a series of simulations and real EGG signals. The following conclusions can be made.

- The adaptive method provides the highest frequency resolution and the most accurate temporal information. It is well suited for the detection of gastric

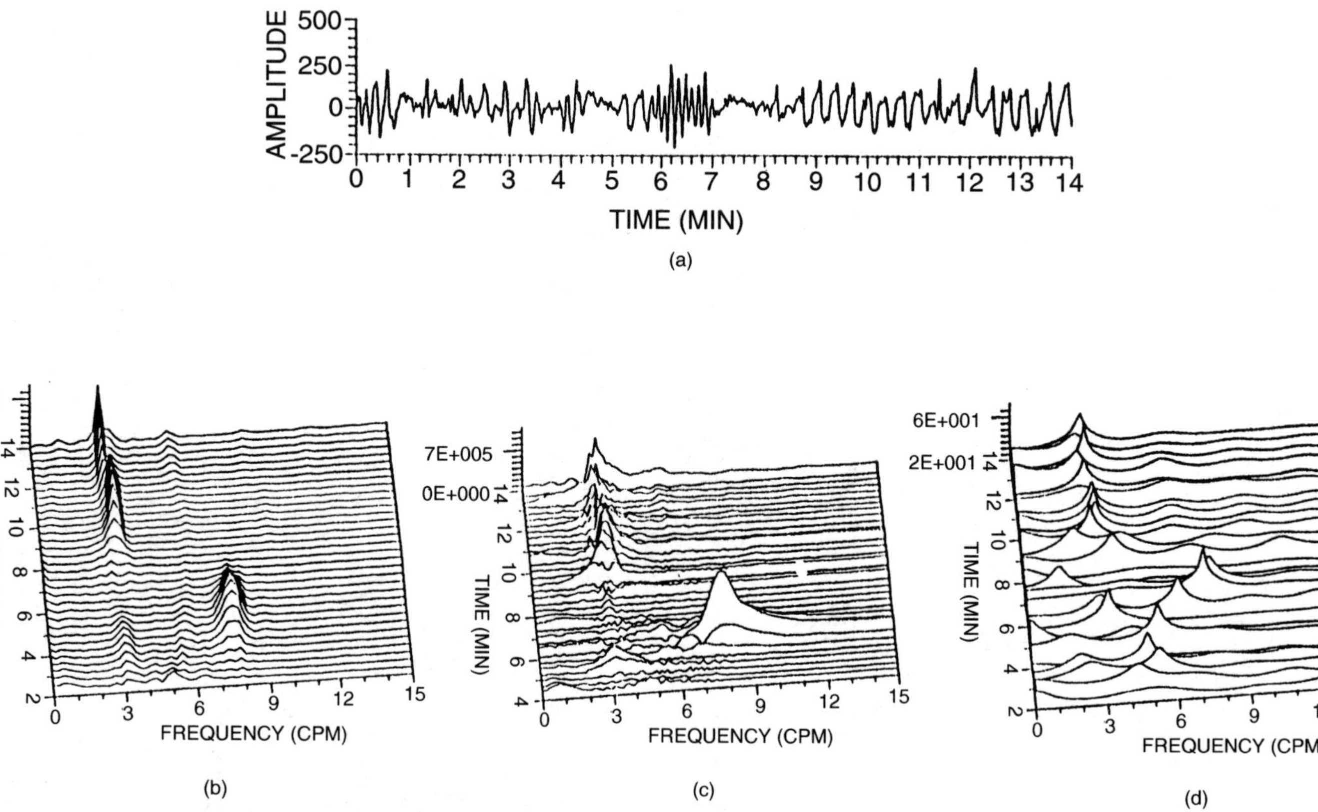

Figure 6-21 (a) A real EGG signal with a brief duration of tachygastria and its time-frequency representation using (b) the spectrogram; (c) the ED; and (d) the adaptive method. (From Zhiyue Lin et al., *Med. Biol. Eng. Comput.*, vol 33, p. 602, 1995. With permission.)

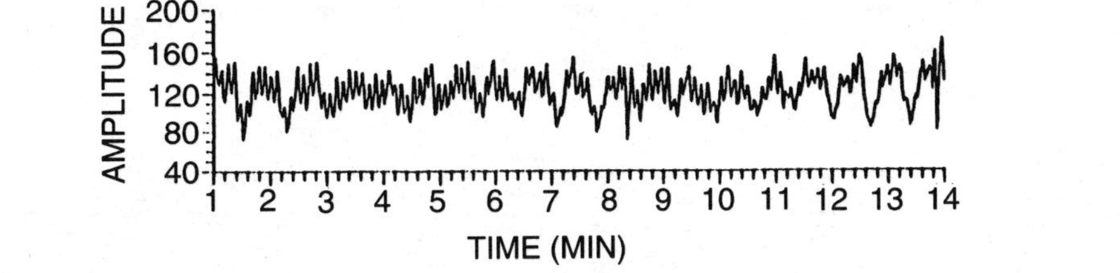
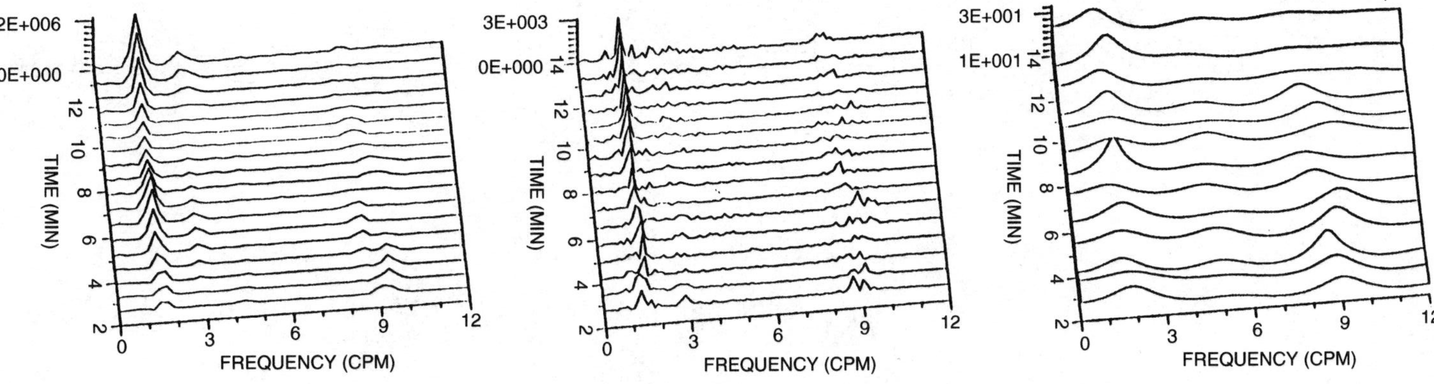

Figure 6-22 A typical EGG signal with bradygastria (top) and its time-frequency representation using the spectrogram (left bottom), the ED (middle), and the adaptive method (right bottom).

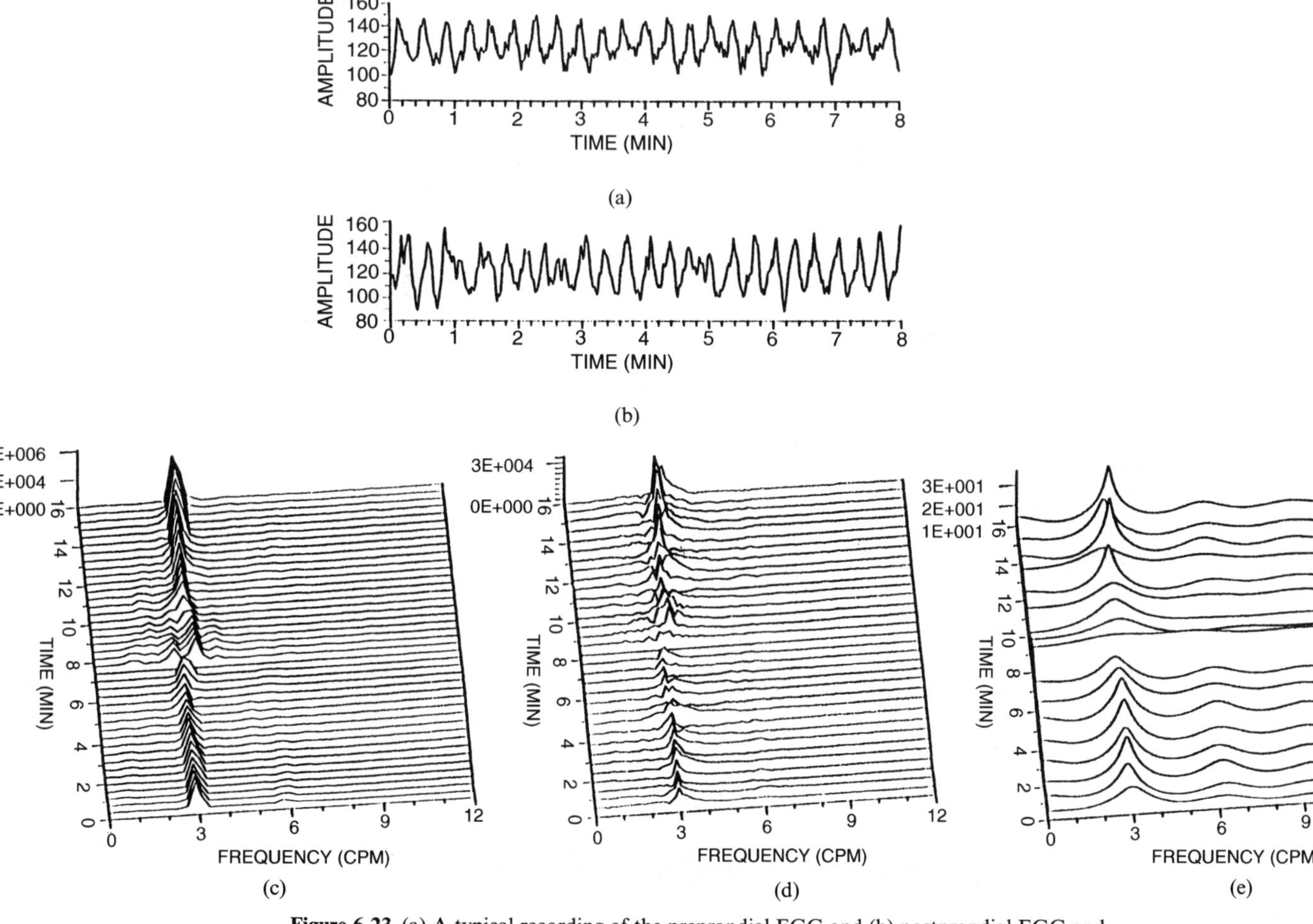

Figure 6-23 (a) A typical recording of the preprandial EGG and (b) postprandial EGG and their time-frequency representations using (c) the spectrogram; (d) the ED; and (e) the adaptive method. (From Zhiyue Lin et al., *Med. Biol. Eng. Comput.*, vol. 33, p. 603, 1995. With permission.)

dysrhythmias of brief duration, but may not be a good choice for the estimation of the EGG power.

- The spectrogram generates the lowest frequency resolution among the three methods. However, its estimation in EGG power is more accurate than the adaptive method. It provides reliable results when the signal is relatively stationary (no swift changes in signal frequency). It may not be adequate for the detection of gastric dysrhythmias of brief duration. Windows may be selected based on the purposes of analysis. The Kaiser window provides more accurate temporal information at the expense of frequency resolution (broader in spectral peaks). Rectangular windows may not be adequate for the detection of gastric dysrhythmias.

- The performance of the ED method falls between the spectrogram method and the adaptive method. The ED method may perform better than the spectrogram method when the EGG signal has a high signal-to-noise ratio (a clean signal) and contains a single main frequency component. The cross-terms may deteriorate the performance of the ED method if the EGG signal contains several different frequency components.

ACKNOWLEDGMENTS

The authors would like to thank the Whitaker Foundation for the support of this work.

REFERENCES

[1] W. C. Alvarez, "The electrogastrogram and what it shows", *J. Am. Med. Assoc.*, vol. 78, pp. 1116–1118, 1922.

[2] J. Chen and R. W. McCallum, "Electrogastrography: Measurement, analysis and prospective applications", *Med. Biol. Eng. Comput.*, vol. 29, pp. 339–350, 1991.

[3] J. Z. Chen and R. W. McCallum, *Electrogastrography: Principles and Applications*, New York: Raven Press, 1994.

[4] A. J. P. M. Smout, H. J. A. Jebbink, and M. Samsom, "Acquisition and analysis of electrogastrographic data". In *Electrogastrography: Principles and Applications*. J. Z. Chen and R. W. McCallum (eds.), New York: Raven Press, 1994, pp. 3–30.

[5] B. H. Brown, R. H. Smallwood, H. L. Duthie, and C. J. Stoddard, "Internal smooth muscle electrical potentials recorded from surface electrodes", *Med. Biol. Eng. Comput.*, vol. 13, pp. 97–102, 1975.

[6] R. H. Smallwood, "Analysis of gastric electrical signals from surface electrodes using phaselock techniques", *Med. Biol. Eng. Comput.*, vol. 16, pp. 507–518, 1978.

[7] D. A. Linkens and S. P. Datardina, "Estimation of frequencies of gastrointestinal electrical rhythms using AR modeling", *Med. Biol. Eng. Comput.*, vol. 16, pp. 262–268, 1978.

[8] M. A. Kentie, E. J. van der Schee, J. L. Grashuis, and A. J. P. M. Smout, "Adaptive filtering of canine electrogastrographic signals", *Med. Biol. Eng. Comput.*, vol. 19, pp. 759–769, 1981.

[9] R. M. Stern, K. L. Koch, W. R. Stewart, and I. M. Lindblad, "Spectral analysis of tachygastria recorded during motion sickness", *Gastroenterology*, vol. 92, pp. 92–97, 1987.

[10] J. Chen, "A computerized data analysis system for electrogastrogram", *Computers in Biology and Medicine*, vol. 22, pp. 45–58, 1992.

[11] J. B. Allen and L. R. Rabiner, "A unified theory of short-time spectrum analysis and synthesis", *Proc. IEEE*, vol. 65, pp. 1558–1564, 1977.

[12] S. N. Nawab and T. F. Quatieri, "Short-time Fourier transform", In *Advanced Topics in Signal Processing*. J. S. Lim and A. V. Oppenheim (eds.) Englewood Cliffs, NJ: Prentice-Hall, 1988.

[13] J. Chen, "Adaptive filtering and its applications in echo cancellation and biomedical signal processing", Ph.D. thesis, Katholieke Universiteit Leuven, Leuven, Belgium, 1989.

[14] H.-I. Choi and W. J. Williams, "Improved time-frequency representation of multicomponent signals using exponential kernel", *IEEE Trans. Acoust., Speech, Signal, Proc.*, vol. 37, pp. 862–871, 1989.

[15] E. J. van der Schee and J. L. Grashuis, "Running spectral analysis as an aid in the representation and interpretation of electrogastrographic signals", *Med. Biol. Eng. Comput.*, vol. 25, pp. 57–62, 1987.

[16] Z. Y. Lin and J. D. Z. Chen, "Time-frequency representation of the electrogastrogram—application of the exponential distribution", *IEEE Trans. Biomed. Eng.*, vol. 41, pp. 267–275, 1994.

[17] A. V. Oppenheim and R. W. Schafer, *Digital Signal Processing*, Englewood Cliffs, NJ: Prentice-Hall, 1975.

[18] C. M. Kim, R. B. Hanson, T. L. Abell, and J-R Malagelada. "Effect of inhibition of prostaglandin synthesis on epinephrine-induced gastroduodenal electromechanical changes in humans", *Mayo Clinic Proc.*, vol. 64, pp. 149–157, 1989.

[19] E. E. Daniel, "Electrical activity of gastric musculature", In *Handbook of Physiology*, section 6, Alimentary canal, Vol IV. Washington, DC: American Physiological Society, 1968, pp. 1969–1984.

[20] S. K. Sarna, "Gastrointestinal electrical activity: Terminology", *Gastroenterology*, vol. 68, pp. 1631–1635, 1975.

[21] R. A. Hinder and K. A. Kelly, "Human gastric pacesetter potential. Site of origin, spread and response to gastric transection and proximal gastric vagotomy", *Am. J. Surg.*, vol. 133, pp. 29–33, 1978.

[22] A. J. P. M. Smout, E. J. van der Schee and J. L. Grashuis, "What is measured in electrogastrography?" *Dig. Dis. Sci.*, vol. 25, pp. 179–187, 1980.

[23] B. O. Familoni, K. L. Bowes, Y. J. Kingma, et al., "Can transcutaneous recordings detect gastric electrical abnormalities?" *Gut*, vol. 32, pp. 141–146, 1991.

[24] J. W. Hamilton, B. Bellahsene, M. Reichelderfer, et al., "Human electrogastrograms: Comparison of surface and mucosal recordings", *Dig. Dis. Sci.*, vol. 31, pp. 33–39, 1986.

[25] J. Chen, J. Vandewalle, W. Sansen, et al., "Adaptive method for cancellation of respiratory artifact in electrogastric measurement", *Med. Biol. Eng. Comput.*, vol. 27, pp. 57–63, 1989.

[26] C. J. Pfister, J. W. Hamilton, N. Nangel, P. Bass, J. G. Webster and W. J. Tompking, "Use of spectral analysis in the detection of frequency differences in the EGGs," *IEEE Trans. Biomed. Eng.*, vol. 35, pp. 935–941, 1988.

[27] R. B. Blackman and J. W. Tukey, *The Measurement of Power Spectra*, New York: Dover Publications, 1959.

[28] F. Hlawatsch and G. F. Boudreaux-Bartels, "Linear and quadratic time-frequency signal representations", *IEEE Trans. Signal Proc. Mag.*, pp. 21-67, April 1992.

[29] S. Kadambe and G. F. Boudreaux-Bartels, "A comparison of the existence of 'cross terms' in the Wigner distribution and the squared magnitude of the wavelet transform and the short-time Fourier transform", *IEEE Trans. Signal Proc.*, vol. 40, pp. 2498–2517, 1992.

[30] C. H. You, K. Y. Lee, W. Y. Chey, and R. Menguy, "Electrogastrographic study of patients with unexplained nausea, bloating and vomiting", *Gastroenterology*, vol.79, pp. 311–314, 1980.

[31] J. Chen and R. W. McCallum, "Gastric slow wave abnormalities in patients with gastroparesis," *Am. J. Gastroenterology*, vol. 97, pp. 477–482, 1992.

[32] H. Geldof, E. J. van der Schee, M. Van Blankenstein, and J. L. Grashuis, "Electrogastrographic study of gastric myoelectrical activity in patients with unexplained nausea and vomiting," *Gut*, vol.27, pp. 799–808, 1986.

[33] K. L. Koch, R. M. Stern, M. Vasey, J. J. Botti, G. W. Creasy, and A. Dwyer, "Gastric dysrhythmias and nausea of pregnancy," *Dig. Dis. Sci.*, vol. 35, pp. 961–968, 1990.

[34] R. J. Taylor and L. E. A. Troncon, "The use of Fourier transform and spectral analysis in the detection of distension-induced gastric arrhythmias in dysoeotic patients," *Physiol. Meas.*, vol. 14, pp. 137–144, 1993.

[35] L. Cohen, "Time-frequency distributions—review," *Proc. IEEE*, vol. 77, pp. 941–981, 1989.

[36] E. P. Wigner, "On the quantum correction for thermodynamic equilibrium," *Phys. Rev.*, vol. 40, pp. 749–759, 1932.

[37] T. A. C. M. Claasen and W. F. G. Mecklenbräuker, "The Wigner distribution—A tool for time-frequency signal analysis—Part I: Continuous-time signals," *Philips J. Res.*, vol. 35, pp. 217–250, 1980.

[38] T. A. C. M. Claasen and W. F. G. Mecklenbräuker, "The Wigner distribution—A tool for time-frequency signal analysis—Part II: Discrete-time signals," *Philips J. Res.*, vol. 35, pp. 276–300, 1980.

[39] T. A. C. M. Claasen and W. F. G. Mecklenbräuker, "The Wigner distribution—A tool for time-frequency signal analysis—Part III: Relations with other time-frequency signal transforms, *Philips J. Res.*, vol. 35, pp. 372–389, 1980.

[40] B. Boashash, "Note on the use of the Wigner distribution for time-frequency signal analysis," *IEEE Trans. Acoust., Speech, Signal Proc.*, vol. 36, pp. 1518–1521, 1988.

[41] B. Boashash and P. J. Black, "An efficient computation of the Wigner–Ville distribution," *IEEE Trans. Acoust., Speech, Signal Proc.*, vol. 35, pp. 1611–1618, 1987.

[42] A. Papand and G. F. Boudreaux-Bartels, "Generalization of the Choi–Williams distribution and the Butterworth distribution for time-frequency analysis," *IEEE Trans. Signal Proc.*, vol. 41, pp. 463–472, 1993.

[43] E. J. van der Schee and J. L. Grashuis, "Contraction-related, low-frequency components in canine electrogastrographic signals," *Am. J. Physiol.*, vol. 245, pp. G470–G475, 1983.

[44] S. M. Kay and S. L. Marple, Jr., "Spectrum analysis: A modern perspective," *Proc. IEEE*, vol. 69, pp. 1380–1419, 1981.

[45] J. Chen, J. Vandewalle, W. Sansen, G. Vantrappen, and J. Janssens, "Adaptive spectral analysis of cutaneous electrical signals using autoregressive moving average modelling," *Med. Biol. Eng. Comput.*, vol. 28, pp. 531–536, 1990.

[46] B. Widrow, J. R. Glover, Jr., J. M. McCool, J. Kaunits, C. S. Williams, R. H. Nearn, J. R. Zeidler, E. J. Dong, Jr., and R. C. Goodlin, "Adaptive noise canceling: Principles and applications," *Proc. IEEE*, vol. 63, pp. 1692–1716, 1975.

[47] J. Chen, W. R. Stewart, and R. W. McCallum, "Adaptive spectral analysis of episodic rhythmic variations in gastric myoelectric potentials," *IEEE Trans. Biomed. Eng.*, vol. 40, pp. 128–135, 1993.

[48] B. Boashash, B. Lovell, and L. White, "Time-frequency analysis and pattern recognition using singular value decomposition of the Wigner–Ville distribution," *Advanced Algorithms and Architectures for Signal Processing II*, vol. 826, pp. 104–114, 1987.

[49] S. L. Marple, *Digital Spectral Analysis with Applications.* Englewood Cliffs, NJ: Prentice-Hall, 1987.

[50] Z. Y. Lin, J. Pan, R. W. McCallum, and J. D. Z. Chen, "Do gastric myoelectrical abnormalities predict delayed gastric emptying?" *Gastroenterology*, vol. 108, p. A639, 1995.

[51] H. Geldof, E. J. van der Schee, A. J. P. M. Smout, J. P. van de Merwe, M. van Blankenstein, and J. L. Grashuis, "Myoelectrical activity of the stomach in gastric ulcer patients: An electrogastrographic study," *J. Gastrointest. Mot.*, vol. 1, pp. 122-130, 1989.

[52] K. L. Koch, W. R. Stewart, and R. M. Stern, "Effect of barium meals on gastric electromechanical activity in man," *Dig. Dis. Sci.*, vol. 32, pp. 1217–1222, 1987.

[53] J. Chen and R. W. McCallum, "The response of electrical activity in normal human stomach to water and solid meals," *Med. Biol. Eng. Comput.*, vol. 29, pp. 351–357, 1991.

[54] J. D. Z. Chen, Z. Y. Lin, and R. W. McCallum, "Toward ambulatory recording of electrogastrogram." In *Electrogastrography: Principles and Applications.* J. Z. Chen and R. W. McCallum (eds.), New York: Raven Press, pp. 127–153, 1994.

[55] J. D. Z. Chen and R. W. McCallum, "Electrogastrographic parameters and their clinical significance." In *Electrogastrography: Principles and Applications.* J. Z. Chen and R. W. McCallum (eds.), New York: Raven Press, pp. 45–73, 1994.

Chapter 7

Recent Advances in Time-Frequency and Time-Scale Methods

Claudia Mello, Metin Akay

7.1. INTRODUCTION

Mappings between the time and the frequency domains have been widely used in signal analysis and processing. Since Fourier's work in the nineteenth century, we have learned a great deal about the natural world using his powerful tool. However, because Fourier methods may not be appropriate to nonstationary signals, or signals with short-lived components, alternative approaches have been sought.

Among the early works in this area, one can cite Gabor's development of the short-time Fourier transform, a procedure in which a window function is passed through a signal, with the assumption that inside the window the signal is stationary.

Another approach was taken by Wigner [18], and was later adapted to signal processing by Ville [19]. In this case, a quadratic distribution of the time and frequency characteristics of the signal was derived. The major drawback of this representation was in its interpretation. Namely, the representation not only contained the signal components but also interference terms generated by the interaction of these signal components with each other.

Many suggestions were made to improve the Wigner–Ville distribution. All of them primarily used some kind of filtering process to enhance the signal components and to attenuate the interference terms.

Cohen [21] unified the quadratic time-frequency representations. He showed that most of them belonged to a general class, in which each member was generated by the choice of an appropriate kernel function.

In the early 1980s, a theory that unified a set of ideas about analyzing a signal at different resolutions was proposed, and was called the wavelet representation. An interesting characteristic of this method relied on its ability to behave like a "mathematical microscope," that is, it could "zoom in" on short-lived signal components. A derivation similar to Cohen's was obtained for the wavelet representation, originating the affine class.

More recently, a unification of Cohen's and the affine classes was proposed, and a class of representations that are invariant under time or frequency shifts and under time-frequency scalings was derived. Furthermore, unitary transformations can be performed in any member of these defined classes, generating representations that may suit the signal at hand even better.

In this chapter these methods will be reviewed. Section 7.2 deals with Fourier representations; section 7.3 discusses Cohen's class operators. It also deals with the geometric location of the interference terms, as well as with an interesting method for designing the most appropriate kernel for the application being considered. The Wigner–Ville distribution and the Cohen's class are also expressed in terms of a higher-order moment spectrum.

Section 7.4 discusses the general idea of the wavelet representation, in its continuous and discrete versions, as well as in terms of a multiresolution approximation. In addition, the general expression for the affine class, and the relationship between the affine and Cohen's classes are presented. Also, the shift-scale invariant class is defined. This class basically combines the properties of both classes.

Finally, section 7.5 briefly discusses a recent development, namely, the use of unitary transformations in both Cohen's and the affine classes, with the consequent generation of even more specific tools for signal analysis. Afterwards the conclusions are presented.

7.1.1. Notation

Throughout this paper the following notation will be used:
- $L^p(I)$ is the Hilbert space of p-integrable functions defined on the interval I
- if x and $g \in L^p(I)$, then $\langle x, g \rangle$ is the inner product of x and g, that is, $\langle x, g \rangle = C \int_I x(\tau)\bar{g}(\tau)d\tau$, where C is a constant.
- if x and $g \in L^p(I)$, then x and g are orthogonal to each other if and only if $\langle x, g \rangle = C$ if $x = g$; 0 otherwise.

7.2. FOURIER REPRESENTATION

Consider $[O_i(t)]_{0 \leq i \leq \infty}$ to be an orthogonal system on the interval I. If the system is complete, then any periodic function $x(t) \in L^2(I)$ can be written as

$$x(t) = \sum_{i=0}^{\infty} c_i O_i(t) \tag{7-1}$$

This is called the Fourier series expansion of $x(t)$ [1], and the constants c_i, expressed by

$$c_i = \frac{\langle x(t), O_i(t) \rangle}{\langle O_i(t), O_i(t) \rangle} \tag{7-2}$$

are the Fourier series coefficients.

Energy conservation is guaranteed due to the completeness of the system. This is written as Parseval's equation,

$$\int_I x^2(t)\,dt = \sum_{i=0}^{\infty} c_i^2 \|O_i(t)\|^2 \tag{7-3}$$

Note that, by conjecture, the series on the right-hand side of (7-1) requires an infinite number of terms for an accurate representation of $x(t)$. In practice, this is not feasible; a finite number of terms must be used for the expansion of $x(t)$. Let us call $x'(t)$ the representation of $x(t)$ using only N terms, that is,

$$x'(t) = \sum_{j=0}^{N-1} c_j O_j(t)$$

The Fourier series converges to $x(t)$ in the mean, that is,

$$\lim_{N \to \infty} \int_I [x(t) - x'(t)]^2\,dt = 0 \tag{7-4}$$

It is possible to show [1] that the Fourier series of a given signal may converge to the signal itself [$x(t)$] and to $g(t)$, provided that $x(t) = g(t)$ almost everywhere on I (for example, $x(t)$ and $g(t)$ may differ in a finite number of points, which is a set of measure zero).

To extend the Fourier representation to any function $x(t) \in L^2(R)$ (without the periodicity requirement), one must use the Fourier transform. It is defined as

$$X(w) = \int_R x(t)e^{-iwt}\,dt \tag{7-5}$$

where $-\infty < w < \infty$ and i stands for the imaginary unit.

The signal can be recovered by taking the inverse Fourier transform of (7-5),

$$x(t) = \frac{1}{2\pi} \int_R X(w)e^{iwt}\,dw \tag{7-6}$$

for $t \in R$.

Since the integral in (7-5) is performed in the Lebesque sense, changes in $x(t)$ on an arbitrary finite set of points will have no effect on its Fourier transform [1].

Fourier methods are important to signal analysis because they provide a tool to relate temporal with frequency contents of a given signal. They also have a number of drawbacks. Among them, one can cite the necessity of knowing all temporal information in order to perform the spectral analysis—thus real-time applications are out of the question, because future information is needed [2].

Also, due to Gibbs' phenomenon, discontinuities or abrupt transitions in the signal (which may be due to particular characteristics of the signal or to the addition of noise) in the time domain are spread over the entire frequency range. This makes the analysis of transient signals, which often have short-lived time components, even more difficult. Therefore, an underlying assumption behind Fourier analysis methods is stationarity [3,4]. For long-term signals, this is a hard condition to satisfy. To overcome this difficulty, Gabor [5] suggested a windowing process over the Fourier transform. This is known as the short-time Fourier transform (STFT).

In the STFT, a window function with compact support, $h(t) \in L^2(R)$, is chosen, and then by translations this window is slid throughout the whole signal. In this case, it is assumed that inside the window the signal is stationary.

It is possible to show [6] that the Fourier coefficients, which are defined by the following in the case of the STFT ,

$$c_{m,n}(x) = \int_{-\infty}^{\infty} e^{imtw_0} h(t - n\tau_0)x(t)dt \tag{7-7}$$

for $m, n \in Z$, correspond to the inner product of the signal $x(t)$ with a discrete family of square integrable functions, $h_{m,n}(t)$, generated by a single function $h(t)$ by phase-space translations, that is,

$$h_{m,n}(t) = e^{imtw_0} h(t - n\tau_0) \tag{7-8}$$

In other words,

$$c_{m,n} = \langle h_{m,n}, x \rangle \tag{7-9}$$

Originally, Gabor chose Gaussians as window functions [5]. Although these window functions do not have compact support, they are maximally concentrated in the time and frequency domains, which assures good localization properties. Gaussians also achieve the lowest bound of the joint entropy (effective spatial extent times bandwidth), that is, they provide the best spectral information for every point along the signal variation.

However, as pointed out in [7,8], the use of Gaussians as window functions has some drawbacks. First, it depends on the choice of the "time-frequency sampling grid" used:

- If $w_0\tau_0 = 2\pi$, as originally used, then the reconstruction algorithm is numerically unstable [6].
- If $w_0\tau_0 > 2\pi$, it is possible to find well-behaved signals $x(t) \in L^2(R)$ that cannot be represented using this method.
- Finally, if $w_0\tau_0 < 2\pi$, then the discrete family $h_{m,n}(t)$ is nonorthogonal, and the Fourier coefficients are not uniquely determined [9].

Furthermore, in order to achieve a good time resolution, it is necessary to use a short time window $h(t)$. In this case, however, one will get a poor frequency resolution, since the number of samples inside the window is small. On the other hand, good frequency resolution requires a narrow-band window, but in this case it will have a long time response.

The STFT possesses a property called the "resolution of identity" [2,6], which guarantees the reconstruction of $x(t)$ using its phase-space projections. Mathematically [6]

$$\int dp \int dq |\langle h_{p,q}, x \rangle|^2 = 2\pi ||h||^2 ||x||^2 \qquad (7\text{-}10)$$

This can be rewritten as [6]

$$\frac{1}{2\pi} \int dp \int dq \langle h_{p,q}, x \rangle h_{p,q} = x \qquad (7\text{-}11)$$

Fourier analysis methods have been used in several applications [9–12], but, due to the stationarity hypothesis, they may not be suitable to the analysis of biological signals [13,14], since these are often nonstationary or contain short-lived components that carry important information [15,16].

An attempt to extend the use of Fourier methods to nonstationary signals is Priestley's theory of evolutionary spectra [4]. In this approach, the nonstationary signal is assumed to be composed of various stationary parts, each of which has different statistical characteristics. Thus the spectrum of the signal corresponds to a local time-frequency distribution, that is, its value at each instant is an averaged spectrum which considers all the processes in the neighborhood of that instant. However, the statistical characteristics of the processes are supposed to vary smoothly over time, which may not be the case in biomedical applications.

Another approach is the multiresolution Fourier transform [17]. This can be viewed as an STFT in which the window has variable width. It was used to analyze music and images [17], providing interesting results. Notwithstanding, it does not take into account transient information [17].

7.3. COHEN'S CLASS OPERATORS

In addition to the methods based on traditional Fourier analysis, a time-frequency energy distribution has been used to analyze complex signals, such as nonstationary or multicomponent signals.

A joint time-frequency energy distribution was proposed by Wigner [18], and later adapted to signal processing by Ville [19]. The Wigner–Ville Distribution (WVD) is defined in the time domain as

$$W_x(t, w) = \frac{1}{2\pi} \int x^* \left(t - \frac{\tau}{2} \right) x \left(t + \frac{\tau}{2} \right) e^{-iw\tau} d\tau \qquad (7\text{-}12)$$

and in the frequency domain as

$$W_X(t, w) = \frac{1}{2\pi} \int X \left(w - \frac{\zeta}{2} \right) X^* \left(w + \frac{\zeta}{2} \right) e^{it\zeta} d\zeta \qquad (7\text{-}13)$$

This quadratic distribution is such that it provides information about the energy of the signal per unit time per unit frequency. This is attained by the satisfaction of the "marginal properties"* [20], which concern the instantaneous power of the signal ($|x(t)|^2$) and the energy density spectrum ($|X(w)|^2$),

$$\int W_x(t, w)dw = |x(t)|^2 \tag{7-14}$$

$$\int W_X(t, w)dt = |X(w)|^2 \tag{7-15}$$

To further understand how the WVD works, let us consider a signal $x(t)$, composed by N terms [20], $x(t) = \sum_{i=0}^{N-1} c_i x_i(t)$. Calculating the WVD of this signal, two types of terms will be obtained: one, called *autocomponents* (or signal terms), which are due to the individual characteristics of each subcomponent $x_i(t)$ in $x(t)$, and another, due to the interferences between each pair of subcomponents in the original signal. These are called *interference terms*, or cross-components. These interference terms are responsible for the nonzero values of the WVD at the points in the time-frequency domain where the signal is zero.

To calculate the range of the Wigner–Ville distribution, one folds the part of the signal to the left of a given point over the part to the right of that point [21]. If there is any superposition, then the WVD will not be zero—not even if the signal is zero in that interval. Furthermore, even if the signal exhibits low-intensity components at some interval, they may be obscured by the distribution [22]. In the case of finite-time signals, the WVD will be zero before the signal starts and after it ends [21].

Another critical point of the distribution refers to noise. As pointed out in [22], if in a particular interval the signal is not mixed with noise, but it was in the past or it will be in the future, then the WVD may exhibit noise characteristics in that interval [21,22]. When the signal is of infinite length, it does not matter where the noise occurs, the WVD will spread the noise throughout the whole signal [21].

Several methods have been proposed to deal with the cross-terms. They mostly involve either the design of other quadratic time-frequency distributions with kernel functions with desirable characteristics or the smoothing of the WVD by filtering. For a very thorough review, see chapter 1.

As an example, we will review one of these methods, namely, the Radon–Wigner transform [23], a process in which the auto-terms are enhanced by operations performed in the Radon-Wigner domain.

The Radon transform is defined as [24]

$$R[f(x, y)] = P_\theta(t) = \int_{-\infty}^{\infty} \int_{-\infty}^{\infty} f(x, y)\delta(x \cos \theta + y \sin \theta - t)dxdy$$

---

*It is important to mention that not all quadratic distributions that satisfy the marginal properties can have an energetic interpretation [20] and, in fact, there are some energetic distributions which do not satisfy the marginals [21].

where

- $f(x, y)$ is the function being transformed;
- θ is the rotation angle;
- t is the horizontal distance, relative to the origin of the rotated axis;
- $P_\theta(t)$ is the projection formed by combination of all line integrals at a constant angle θ.

Moreover, this representation satisfies the remarkable Fourier slice theorem, which states that the Fourier transform of $P_\theta(t)$ gives the value of $F(w_1, w_2)$ along a line on the $w_1 \times w_2$ plane. This line makes an angle θ with the w_1 axis.

The Radon–Wigner transform is defined as the Radon transform of the WVD [23], that is,

$$R[W_x(t, w)] = \int W_x(t, w_0 + mt)dt$$

where $r = x\cos(\theta) + y\sin(\theta)$, $m = \cot(\theta)$, and $w_0 = r/\sin(\theta)$.

The line integration of the WVD is known as "dechirping," that is, a frequency modulation that varies linearly with time, followed by a backprojection onto the frequency axis. Therefore, in accordance with the Fourier slice theorem, filtering of the WVD can be performed by simple operations in the Rado–Wigner domain. A generalized version of the WVD (GWVD) was proposed by Claasen and Mecklenbräuker [25,26] as

$$\text{GWVD}_x(t,f) = \int_{-\infty}^{\infty} x(t + (0.5 + \rho)\tau)x^*(t - (0.5 - \rho)\tau)e^{-i2\pi f\tau}d\tau \qquad (7\text{-}16)$$

where ρ is a real-valued parameter that controls the distance of the interference terms to the middle point—in time and frequency—between each two signal components [27].

For example, consider a multicomponent signal $x(t) = x_1(t) + x_2(t)$. As its previous counterpart, the GWVD will produce signal components centered at (t_1, f_1) and (t_2, f_2), due to $x_1(t)$ and $x_2(t)$, respectively, and cross-terms around $\{(t_1 + t_2)/2 - \rho(t_1 - t_2), \ (f_1 + f_2)/2 + \rho(f_1 - f_2)\}$ and $\{(t_1 + t_2)/2 + \rho(t_1 - t_2), \ (f_1 + f_2)/2 - \rho(f_1 - f_2)\}$. It is easy to see that as $|\rho|$ increases, the farther from the middle position the interference terms will be.

After the Wigner–Ville distribution, other time-frequency energy distributions were proposed [28–32], each with interesting properties. Cohen [21,33] unified them in a general functional, given by [21]

$$C_x(t, w, \phi)$$
$$= \frac{1}{4\pi^2} \int_{-\infty}^{\infty} \int_{-\infty}^{\infty} \int_{-\infty}^{\infty} e^{-it\theta - iw\tau + i\theta\mu}\phi(\theta, \tau)x^*(\mu - 0.5\tau)x(\mu + 0.5\tau)d\mu d\tau d\theta \qquad (7\text{-}17)$$

In (7-17), $\phi(., ., .)$ is a kernel function [26] whose type defines the distribution. In this way, desirable properties of the final distribution can be obtained as constraints on the kernel.

The general expression given by (7-17) can be rewritten in terms of the WVD as [26]

$$C_x(t,f,\phi) = \frac{1}{2\pi} \int_{-\infty}^{\infty} \int_{-\infty}^{\infty} \Phi(t - t', f - f') W_x(t',f') dt' df' \tag{7-18}$$

where

$$\Phi(t,f) = \frac{1}{2\pi} \int_{-\infty}^{\infty} \int_{-\infty}^{\infty} e^{i(t\chi - w\tau)} \phi(\chi, \tau) d\chi d\tau \tag{7-19}$$

Therefore the members of Cohen's class can be viewed as smoothed versions of WVD, where the smoothing function corresponds to the two-dimensional (2-D) Fourier transform of the original kernel [26].

Because the constraints will determine the kernel, it is clear that the "best distribution" is a totally application-dependent idea. Therefore, some techniques to design the desired kernel have been proposed.

One approach is the optimal kernel distribution (OKD), a procedure that takes into account the signal being analyzed [34]. Given a signal and its ambiguity function, $A(\theta, \tau)$* the optimal kernel will be the real, nonnegative function $\phi'(\theta, \tau)$, which solves [34]

$$\text{maximize}_\phi \int_{-\infty}^{\infty} \int_{-\infty}^{\infty} |A(\theta, \tau)\phi(\theta, \tau)|^2 d\theta d\tau \tag{7-20}$$

subject to

- $\phi(0, 0) = 1$
- $\phi(\theta_1, \tau) \geq \phi(\theta_2, \tau) \qquad \forall\, \theta_1 < \theta_2 \text{ and } \forall\tau$
- $\frac{1}{2\pi} \int_{-\infty}^{\infty} \int_{-\infty}^{\infty} |\phi(\theta, \tau)|^2 d\theta d\tau \leq \alpha$

where $\alpha \geq 0$ and α is the scaling parameter.

The last condition controls the relative importance between auto-term enhancement and interference term suppression, by adjusting the volume under the optimal kernel.

As described elsewhere [34], these constraints guarantee that the optimal kernel is a low-pass function, which enhances autocomponents because they are centered in the origin of the ambiguity plane.

Another approach to find the best kernel was proposed in [36]. In this signal-independent approach, the problem of finding the optimal kernel is posed as an optimization problem, where the constraints are the properties that one wishes the desirable kernel will have. Thus if there is a nonempty intersection among the constraint sets—which, for almost all desired properties, form convex sets—by

---

*The ambiguity function can be seen as the dual of the WVD [35], given by

$$A(\theta, \tau) = \int_{-\infty}^{\infty} \int_{-\infty}^{\infty} W_x(t, w) e^{-i(t\tau - w\theta)} dt dw$$

alternating projections onto these sets, the desired kernel will be found. It is possible to show [36] that this process is independent of the initialization point, and that, if two constraint sets do not intersect, the kernel will converge to the minimum mean-squared solution. However, if three or more sets do not intersect, then limit cycles among these sets may exist. It is also possible to verify that a kernel that satisfies all possible desirable properties is not feasible—that is, the perfect distribution does not exist. Optimal kernels satisfying sets of interesting properties were designed in [36].

Due to their bilinear nature, the distributions belonging to Cohen's class suffer from the presence of cross-terms, which, as in the WVD case, are a result of the interaction among the signal components. Cohen's finite support properties guarantee that the distributions will be zero before the signal begins and after it ends, but they do not say anything for finite periods in which the signal vanishes.

This situation was deeply discussed in [37], where the structure of the interference terms was studied. It was proved that, for a two-tone signal with frequencies f_1 and f_2, the interference will be the 2-D Fourier transform of the kernel of the distribution, $\Phi(\theta, \tau)$, evaluated along the axis $(f_2 - f_1)$ and shifted in frequency by $(f_1 + f_2)/2$, that is,

$$\Phi\left(f_2 - f_1, f - \frac{f_1 + f_2}{2}\right)e^{i2\pi(f_2 - f_1)t}$$
$$= \int_{-\infty}^{\infty}\int_{-\infty}^{\infty} \phi(\theta - \theta', \tau)e^{i\pi f_1(\theta' + \theta)}e^{-i\pi f_2(\theta' - \theta)}e^{-i2\pi f\tau}\,d\theta'\,d\tau \qquad (7\text{-}21)$$

It is possible, then, by analysis of expression (7-21), to derive restrictions on the kernel in order to avoid the interference terms. However, under these restrictions, the generated distributions will have poor marginal properties [37]. Thus a trade-off exists, and the worthiness of imposing these new constraints is application dependent.

Distributions belonging to Cohen's class have been used in a variety of applications. Some examples are as follows: to present time-frequency mapping of heart rate, blood pressure, and respiration signals [38], to analyze cardiac late potentials [39], to analyze bioacoustical signals [40], brain potentials [41], to design ultrasound transducers [42], to analyze muscle sounds [43] or electrogastrogram [44], to represent event-related potentials [45], to characterize acoustic transients [46], and so on.

Interesting comparative papers among different members of Cohen's class are available [32,47]; comparisons between WVD and Priestley's running spectra theory were discussed in [48].

Members of Cohen's class are based on second-order moments of the signal. Another class stands for distributions based on higher-order moment spectra [49].

As in Cohen's formulation, there is a basic representation, in which the kernel is the unity, and a general extension. As in Cohen's class, the basic "building block" is based on the WVD, but in this case it is the Wigner-Ville higher-order moment spectra (WVHOMS).

Given a signal $x(t)$, the WVHOMS of order k is defined as [49]

$$\text{WVH}_{k,x}(t, f_1, \ldots, f_k)$$

$$= \int_{\tau_1} \cdots \int_{\tau_K} x^*\left(t - \frac{1}{k+1}\sum_{m=1}^{k}\tau_m\right)\prod_{l=1}^{k} x\left(t + \frac{k}{k+1}\tau_l - \frac{1}{k+1}\sum_{j=1, j\neq l}^{k}\tau_j\right) e^{-i2\pi f_l \tau_l}\, d\tau_l$$

$$(7\text{-}22)$$

It is possible to show [49] that this distribution preserves most of the properties of the WVD. Foremost among the properties are the following.

1. Time support of $x(t)$: it assures that if $x(t) = 0$, for all $t \in [T_1, T_2]$ then $\text{WVH}_{k,x}(t, f_1, \ldots, f_k) = 0$, $\forall t \in [T_1, T_2]$. Notice that it does not say anything about finite intervals in which the signal assumes zero values.

2. Frequency support of $X(f)$: if $X(f) = 0$ for all $\in [F_1, F_2]$, then the WVHOMS will have nonzero values for the range $-(F_2 - F_1) \geq f_i - f_j \geq (F_2 - F_1)$.

Consider $\phi_{\text{HOS}}(\Omega, \tau_1, \ldots, \tau_k)$ a multidimensional kernel; the characteristic function in the higher-order moment spectra domain can be defined as

$$C_{\text{HOS}}(\Omega, \tau_1, \ldots, \tau_k)$$

$$= \phi_{\text{HOS}}(\Omega, \tau_1, \ldots, \tau_k)\int_{f_1}\cdots\int_{f_k}\int_u \text{WVH}_{k,x}(u, f_1, \ldots, f_k)e^{i2\pi u\Omega}\prod_{l=1}^{k}e^{-i2\pi f_l \tau_l}\, df_l\, du \qquad (7\text{-}23)$$

The expression (7-23) allows the definition of a general higher-order class of time-frequency distributions—$\text{HOC}(t, f_1, \ldots, f_k)$—as

$$\text{HOC}(t, f_1, \ldots, f_k)$$

$$= \int_\Omega \int_{\tau_1}\cdots\int_{\tau_k} C_{\text{HOS}}(\Omega, \tau_1, \ldots, \tau_k)e^{i2\pi t\Omega}\prod_{l=1}^{k}e^{-i2\pi f_l \tau_l}\, d\tau_l\, d\Omega \qquad (7\text{-}24)$$

The properties of the members of $\text{HOC}(t, f_1, \ldots, f_k)$ can be obtained using the properties of $\text{WVH}_{k,x}(t, f_1, \ldots, f_k)$ and the particular kernel $\phi_{\text{HOS}}(\Omega, \tau_1, \ldots, \tau_k)$. The influence of interference terms as well as smoothing processes over them are the subject of current research [49].

7.4. WAVELETS: FRAMES, MULTIRESOLUTION APPROXIMATION, AND BEYOND

The wavelet transform provides a representation of the signal in a lattice of "building blocks" which have good frequency localization (that is, they are part of a trigonometric system, as Fourier methods), but also have good time localization, as the functions of the Haar system [50].

Even though the idea of analyzing a signal at different scales or resolutions has existed since the beginning of the century—with the works of Haar [51], Franklin [52] and Calderon [53]—wavelet theory was tied together only in the early 1980s.

Coifman and Weiss [54] created the "building blocks" of various functional spaces, with well-determined construction rules; orthonormal wavelet bases were constructed [50]; the first wavelets were used [55,56].

The wavelet transform (WT) is a signal decomposition on a set of basis functions, obtained by dilations, contractions, and shifts of a unique function, the wavelet prototype. As pointed out in [8], a very basic distinction between wavelet transforms and Fourier methods, like STFT, is that while the basis functions of the latter consist of a function of constant width translated in time and "filled in" with high-frequency oscillations, the former has a frequency-dependent width. In other words, it is narrow at high frequencies and broad at low frequencies. This gives to the WT the ability to "zoom-in" on transitory phenomena, which usually are short-lived components of a signal.

Let $x(t) \in L^2(R)$ be a given signal. Its continuous wavelet transform (CWT) is defined by [8]

$$\mathrm{CWT}_x(a, b) = \int x(\gamma)\psi_{a,b}^*(\gamma)d\gamma \qquad (7\text{-}25)$$

where * denotes complex conjugation, a represents the scaling factor, and b represents the time.

$\psi_{a,b}(\cdot)$ is obtained by scaling the prototype wavelet $\psi(t)$ at time b and scale a:

$$\psi_{a,b}(t) = \frac{1}{\sqrt{|a|}} \psi\left(\frac{t-b}{a}\right) \qquad (7\text{-}26)$$

$\forall a, b \in R^+$.

In the expression (7-26), the factor $1/\sqrt{|a|}$ was introduced to guarantee energy preservation [57].

A brief analysis of (7-26) shows that when the scale factor a becomes large, the basis function $\psi_{a,b}(t)$ becomes a stretched version of the prototype, which is useful for the analysis of low-frequency components of the signal. On the other hand, when the scale factor is small, the basis function will be contracted, which is interesting for the analysis of high-frequency components.

The signal $x(t)$ can be reconstructed as [57]

$$x(t) = c^{-1} \int_{a>0} \int \mathrm{CWT}_x(a', b')\psi_{a',b'}(t)\frac{da'\,db'}{a'^2} \qquad (7\text{-}27)$$

where the admissibility condition is given by

$$c = 2\pi \int_0^\infty |\Psi_{a,b}(\gamma)|^2|\gamma|^{-1}d\gamma = 2\pi \int_{-\infty}^0 |\Psi_{a,b}(\gamma)|^2|\gamma|^{-1}d\gamma < \infty \qquad (7\text{-}28)$$

A wavelet is a function that behaves like a wave, but it is also localized in time-frequency. To be a wavelet, a function must satisfy

$$\int_0^\infty \psi(t)dt = 0 \qquad (7\text{-}29)$$

that is, the function corresponds to a bandpass filter. Also, the function has to decay, to guarantee localization; thus

$$\lim_{t \to \infty} |\psi(t)| = 0 \tag{7-30}$$

An interesting result of WT comes in the face of the Heisenberg inequality (or uncertainty principle). Consider $\bar{t}$ and $\bar{\Omega}$ as the center of mass of $\psi(t)$ and $\Psi(\Omega)$, respectively, where [58]

$$\bar{t} = \frac{\int_{-\infty}^{\infty} t|\psi(t)|^2 dt}{\int_{-\infty}^{\infty} |\psi(t)|^2 dt}$$

and

$$\bar{\Omega} = \frac{1}{\pi} \frac{\int_{0}^{\infty} \Omega|\Psi(\Omega)|^2 d\Omega}{\int_{0}^{\infty} |\Psi(\Omega)|^2 d\Omega}$$

The rms spreads are given by

$$\sigma_t^2 = \frac{\int_{-\infty}^{\infty} (t - \bar{t})|\psi(t)|^2 dt}{\int_{-\infty}^{\infty} |\psi(t)|^2 dt}$$

and

$$\sigma_{\Omega}^2 = \frac{1}{\pi} \frac{\int_{0}^{\infty} (\Omega - \bar{\Omega})^2|\Psi(\Omega)|^2 d\Omega}{\int_{0}^{\infty} |\Psi(\Omega)|^2 d\Omega}$$

Because the wavelet function is centered at $\bar{t}$ in the time domain and at $\bar{\Omega}$ in the frequency domain, the adoption of a resolution step equal to 2 makes $\psi_{j,k}(t)$ centered at $2^j \bar{t} + k$ in time and at $\bar{\Omega}/2^j$ in frequency, with spread

$$\sigma_{t(j,k)}^2 = (2^j)^2 \sigma_t^2$$

$$\sigma_{\Omega(j,k)}^2 = (2^{-j})^2 \sigma_{\Omega}^2$$

and the resolution cell is $\sigma_{t(j,k)}\sigma_{\Omega(j,k)} \geq 1/2$. This implies that the resolution in the time-frequency domain depends on the scale; although the area of the cell is constant, its shape depends on j. This is called a tiling of the plane. Therefore, the method will have good time resolution at high frequencies and good frequency resolution at low frequencies.

To computationally perform the method, a discretization of the parameters is required. This can be done by first discretizing the scale parameter

$$a = a_0^m$$

for $a_0 > 1$ and then requiring that the other parameter guarantees the appropriate lengths of the steps, that is, it should be narrow for high frequencies (large m) and broader for low frequencies (small m) [8]. A natural choice is

$$b = nb_0 a_0^m$$

for $b_0 > 0$ and $n \in Z$.

The discrete-parameters wavelet can now be written as [8]

$$\psi_{m,n}(t) = a_0^{-m/2}\psi(a_0^{-m}t - nb_0) \tag{7-31}$$

In this case, the signal can be reconstructed as

$$x(t) = \sum_m \sum_n \langle x, \psi_{m,n}\rangle \tilde{\psi}_{m,n} \tag{7-32}$$

where $<,>$ denotes the inner product.

Unlike its continuous counterpart, in the discrete case, the problem is not so simple. It is possible to show [8] that a stability condition to justify (7-32) is that there should exist constants $A > 0$ and $B < \infty$ such that

$$A||x||^2 \le \sum_m \sum_n |\langle x, \psi_{m,n}\rangle|^2 \le B||x||^2 \tag{7-33}$$

The condition expressed by (7-33) is satisfied if $\psi_{m,n}(t)$ constitute a frame [8]. The constants A and B are called frame bounds, and $\tilde{\psi}_{m,n}$ is the dual frame.

It is noteworthy to mention that a frame is not a basis. In general, a frame is a crowded set of vectors, that is, although they span the interval in which they are defined, they are not linearly independent. The information contained in one vector may differ just slightly from the information in another. Even tight frames (that is, frames where $A = B$) may not be a basis, except if the tight frame is normalized ($A = B = 1$), in which case they constitute an orthonormal basis [6].

The introduction of frames at this point serves a twofold purpose: in one way, it guarantees that it is possible to recover the signal $x(t) \in L^2(R)$ from its wavelet coefficients $\langle x, \psi_{m,n}\rangle$. On the other hand, it also assures stability to the reconstruction process, in the sense that if two sequences of coefficients $\langle x_1, \psi_{m,n}\rangle$ and $\langle x_2, \psi_{m,n}\rangle$ are numerically close, then the signals $x_1(t)$ and $x_2(t)$ are close as well [6].

The admissibility conditions for reconstruction are that the wavelet has to satisfy the conditions given by (7-29) and (7-30) and, if $\psi_{m,n}(t)$ constitutes a frame for $L^2(R)$ with frame bounds A and B, then [8]

$$\frac{b_0 ln(a_0)}{2\pi}A \le \int_0^\infty |\Psi(\eta)|^2 |\eta|^{-1} d\eta \le \frac{b_0 ln(a_0)}{2\pi}B \tag{7-34}$$

and

$$\frac{b_0 ln(a_0)}{2\pi}A \le \int_{-\infty}^0 |\Psi(\eta)|^2 |\eta|^{-1} d\eta \le \frac{b_0 ln(a_0)}{2\pi}B \tag{7-35}$$

Furthermore, if the frame is almost tight [that is, $(B/A) - 1 << 1$] [8], then the reconstruction expressed in (7-32) can be conveniently approximated by

$$x(t) = \sum_m \sum_n \langle x, \psi_{m,n}\rangle \psi_{m,n} \tag{7-36}$$

A similar method, in which the frame is not almost tight but the dual frame is generated by a single function, was proposed by Frazier and Jawerth [59,60], under the denomination ϕ-transform.

A particularly interesting subdivision of the frequency axis is in octaves, as in the musical score. Thus $a_0 = 2$, and the $\psi_{m,n}(t)$ associated with this parameter is called a dyadic wavelet. In this case, the admissibility condition is [2] that there exist two constants A and B, $0 < A \le B < \infty$, such that

$$A \le \sum_{j=-\infty}^{\infty} |\Psi(2^{-j}w)|^2 \le B \tag{7-37}$$

almost everywhere.

The signal $x(t)$ can be expressed as

$$x(t) = \sum_{j=-\infty}^{\infty} 2^{3j/2} \int_{-\infty}^{\infty} \mathrm{CWT}_x(2^{-j}, b)\tilde{\psi}[2^j(t-b)]db \tag{7-38}$$

where $\mathrm{CWT}_x(\cdot, b)$ is given by (7-25);
$\tilde{\psi}(\cdot)$ is called a dyadic dual wavelet and is such that it satisfies $\sum_{j=-\infty}^{\infty} \Psi^*(2^{-j}w)\tilde{\Psi}(2^{-j}w) = 1$ almost everywhere

If one chooses the sampling rate $b = 1$, then it is possible to show that, for any $x(t) \in L^2(R)$ [2],

$$x(t) = \sum_{m \in Z}\sum_{n \in Z} c_{m,n}\psi_{m,n}(t) = \sum_{m \in Z}\sum_{n \in Z} d_{m,n}\tilde{\psi}_{m,n}(t) \tag{7-39}$$

Expression (7-39) defines the wavelet series expansion.

It is interesting to mention that, under the restrictions $a_0 > 0$, $a_0 \neq 1$, and $b_0 > 0$ in (7-31), it is possible to find $\psi(t)$ with good localization properties such that an orthonormal basis for $L^2(R)$ is generated. This approach, described in the works of Meyer [50] and Mallat [61], constitutes a multiresolution approximation of $L^2(R)$.

A multiresolution analysis is an approach that allows different information to be expressed at different levels. This is interesting when the signal is composed of various features with different sizes or, accordingly, occurring in frequency bands with different widths. Thus by choosing a resolution beforehand, we may corrupt some features and even risk losing others.

Consider a signal $x(t) \in L^2(R)$. A multiresolution approximation of $L^2(R)$ is defined as an increasing sequence of nested and closed linear subspaces $V_j \in L^2(R)$, $\forall j \in Z$, which satisfies the following properties [50]:

1. These subspaces cover completely $L^2(R)$, that is, $\cup_{j=-\infty}^{\infty} V_j$ is dense in $L^2(R)$; also, the subspaces contain different information, that is, $\cap_{j=-\infty}^{\infty} = 0$.

2. The approximation does not favor any subspace, that is, $x(t) \in V_j \Leftrightarrow x(2t) \in V_{j+1}$, $\forall x \in L^2(R)$, $\forall j \in Z$.

3. The approximation is invariant to time shifts, $x(t) \in V_0 \Leftrightarrow x(t-k) \in V_0$, $\forall x \in L^2(R)$, $\forall k \in Z$.

4. It is possible to find a Riesz basis* (or unconditional basis) for the subspace V_0, that is, it is guaranteed that there exists $g(t) \in V_0$ such that the sequence $g(t - k), \forall k \in Z$, is a Riesz basis of V_0.

Another important characteristic is the regularity of the approximation. A multiresolution approximation is r-regular ($r \in N$) if the function $g(t)$, in property 4, is such that [50]

$$|\partial^{\alpha} g(t)| \leq C_m (1 - |t|)^{-m} \tag{7-40}$$

where $\forall m \in N$, $\forall \alpha \in R$ with $|\alpha| \leq r$, ∂^{α} is the derivative operator given by $\partial^{\alpha} = (\partial / \partial t)^{\alpha}$.

It is also possible to show [50] that the function $\phi(t)$, whose Fourier transform is $\hat{\phi}(\xi)$, is such that the sequence $2^{mj/2} \phi(2^j t - k), k, j \in Z$ forms an orthonormal basis of V_j:

$$\hat{\phi}(\xi) = \hat{g}(\xi) \left(\sum |\hat{g}(\xi + 2k\pi)|^2 \right)^{-1/2} \tag{7-41}$$

Notice that the basis for $L^2(R)$ is still unknown. In order to obtain this basis, consider E_j as the orthogonal projection of $L^2(R)$ onto V_j, given by

$$E_j[x(t)] = \sum_{k \in Z} \alpha(j, k) \phi(2^j t - k) \tag{7-42}$$

where

$$\alpha(j, k) = 2^{nj} \int x(\tau) \phi^*(2^j \tau - k) d\tau \tag{7-43}$$

As pointed out in [50], (7-43) corresponds to the sampled values of $x(t)$ in the lattice $\Gamma_j = 2^{-j} Z$, and (7-42) is the extrapolation from this sampling.

It is now possible to prove that there exists a function $\psi(t) \in L^{\infty}(R)$ such that

$$|\psi(t)| \leq C_m'(1 - |t|)^{-m} \tag{7-44}$$

$\forall m \in N$, and

$$\int_{-\infty}^{\infty} \psi(\tau) d\tau = 0 \tag{7-45}$$

Furthermore, considering that

$$\psi_{j,k}(t) = 2^{-j/2} \psi(2^{-j} t - k) \tag{7-46}$$

---

*If a sequence $[e_i]_{0 \leq i \leq k} \in H$ is a Riesz basis of H, then $\exists$ two constants, $C_2 \geq C_1 \geq 0$, such that

$$C_1 \left(\sum_{j=1}^{k} |\alpha_j|^2 \right)^{1/2} \leq \left\| \sum_{j=1}^{k} \alpha_j e_j \right\| \leq C_2 \left(\sum_{j=1}^{k} |\alpha_j|^2 \right)^{1/2}$$

and the vector space of finite sums $\sum_{j=1}^{k} \alpha_j e_j$ is dense in H.

then it is possible to show that the orthogonal projection of $L^2(R)$ onto W_j (where W_j is the orthogonal complement of V_j in $L^2(R)$) is given by

$$D_j = E_{j+1} - E_j = \sum_{k \in Z} \langle x, \psi_{j,k} \rangle \psi_{j,k} \qquad (7\text{-}47)$$

Moreover,

$$L^2(R) = \bigoplus_{j=-\infty}^{\infty} W_j \qquad (7\text{-}48)$$

that is, the sequence $\psi_{j,k}(t)$, $\forall j, k \in Z$ constitutes an orthonormal basis for $L^2(R)$ [8].

Also, let $x(t) \in H^s(R)$, where H is the Sobolev space* with $-r \leq s \leq r$, and let V_j be an r-regular multiresolution approximation of $L^2(R)$. Then $E_j(x(t))$ converges to $x(t)$ in the $H^s(R)$-norm. This states the effectiveness of the multiresolution approximation [50].

In the foregoing, the function $\phi(t)$ is called scaling function, or father wavelet, whereas $\psi(t)$ is the wavelet prototype, or mother wavelet. This approach has interesting properties. First of all, Eq. (7-45) says that $\Psi(0) = 0$, where Ψ is the Fourier transform of $\psi(t)$. This shows that $\psi(t)$ behaves as the impulse response of a band-pass filter.

Wavelets have been widely used in biomedical signal processing, and among the applications it is possible to cite: detection of heart disease [62,63], for monitoring fetus maturation [15], for magnetic resonance imaging [64], for detection of late potentials [65], for analysis of ECG of postinfarction patients [66]. Also, signal recovery [67–69], for analysis of evoked potentials [70], to texture extraction [71,72], for analysis of oscillatory patterns in the CNS [73], for detection of regions of interest in a scene [74], localization of Radon transform [75], and so on. A review of wavelet theory applied to signal processing is in [76]. Theoretical foundations are in [2,8,50]. Relations between wavelets and filter banks are in [77], and fast algorithms to compute wavelets in [78]. A very detailed analysis of the error in using the approximation provided by the WT is presented in [79].

A derivation similar to Cohen's of time-frequency-shift invariant distributions was proposed [80] for time-scale invariant representations, such as WT, under the name affine class. It was shown that, if $A_x(t, a)$ is any bilinear distribution such that

$$A_{x_{\theta,a'}}(t, a) = A_x\left(\frac{t - \theta}{a'}, \frac{a}{a'}\right) \qquad (7\text{-}49)$$

then $A_x(t, a)$ is necessarily of the form

$$A_x(t, a; \Pi) = \int_{-\infty}^{\infty} \int_{-\infty}^{\infty} W_x \tau, \eta) \Pi\left(\frac{\tau - t}{a}, a\eta\right) d\tau d\eta \qquad (7\text{-}50)$$

---

*A Sobolev space is such that if $s = 0$, $H^s(R) = L^2(R)$; if $s \in N$, then $x \in H^s(R)$ if x and all of its derivatives up to order s are in $L^2(R)$ [50].

where $W_x(.,.)$ is the Wigner–Ville distribution of x(t) given by (7-12);
$\Pi(t,f)$ is an arbitrary time-frequency function.

An interesting result can be observed when the following equality is set [80]:

$$\Pi(t,f) = \Pi_0(t, f - f_0) \Leftrightarrow \hat{\Pi}(\nu, \eta) = \hat{\Pi}_0(\nu, \eta)e^{-i2\pi f_0 \eta} \tag{7-51}$$

where the hat denotes the 2-D Fourier transform. In this particular case, the affine class and Cohen's class are the same, with parameters

$$A_x(t, a, \Pi) = C_x(t, f_0/a, \Pi_0) \tag{7-52}$$

The identification of the affine class with the CWT is given by the following relation:

$$|\mathrm{CWT}_x(a, b)|^2 = A_x(t, a, W_\psi) \tag{7-53}$$

where $\mathrm{CWT}_x(.,.)$ is given by (7-25) and W_ψ is the WVD of the mother wavelet. The value in the left-hand side of (7-53) is called the scalogram, and it measures how the energy of the signal $x(t)$ is distributed in the time-scale plane.

The procedure to choose a particular distribution is application-dependent, as in Cohen's class. It is necessary to choose what are the properties of interest (some are cited in [80]) and then use the restrictions associated with these properties to find the appropriate distribution.

Due to the use of WVD in the general expression of formation of the distributions in the affine class, the same problem experienced by members of Cohen's class, namely, interference terms, appears. Thus a smoothing process may be required. This is called the affine-smoothed Wigner–Ville [80]. In this case, the original kernel $\Pi(t,f)$ is substituted by a product of kernels

$$\Pi(t,f) = g(t)H(f - f_0) \tag{7-54}$$

and the resulting general expression is

$$A_x(t,f,\Pi) = \int_{-\infty}^{\infty} \int_{-\infty}^{\infty} W_x(\tau, \eta)g\left(\frac{\tau - t}{a}\right)H(a\eta - f_0)d\tau d\eta \tag{7-55}$$

The previous smoothing process can offer a good trade-off between suppression of cross-terms and time-frequency resolution.

A step further in this generalization process was taken with the definition of the shift-scale invariant class [27], which combines the properties of Cohen's and affine classes. Therefore, members of this class are invariant under time or frequency shifts and under time-frequency scalings. Furthermore, the kernel used to generate these distributions is one-dimensional.

The class D [27] of shift-scale invariant time-frequency representations can be written as

$$D_x(t,f) = \int_{-\infty}^{\infty} \int_{-\infty}^{\infty} \left[\int_{-\infty}^{\infty} g(\eta)\frac{1}{|\eta|} e^{-i2\pi(t-t')(f-f')/\eta} d\eta \right] W_x(t', f')dt'\, df' \tag{7-56}$$

Once again, desirable properties of the final distribution can be translated as restrictions on the kernel $g(\cdot)$. A list of possible desired properties and their corresponding kernel restrictions can be found in [27].

A more appropriate way to express the class D is by using the generalized WVD as follows:

$$D_x(t,f) = \int_\eta g(-\eta)W_x^{(\eta)}(t,f)d\eta \tag{7-57}$$

Due to the interference geometry displayed by the GWVD, it is possible to show [27] that for each pair of signal components $x_1(t)$ and $x_2(t)$, concentrated around (t_1,f_1) and (t_2,f_2), respectively, and a finite support kernel function $g(\eta) \neq 0$, $\eta \in [-\eta_0, \eta_0]$, two types of terms will be produced by (7-57):

- Autoterms, centered respectively at (t_1,f_1) and (t_2,f_2);
- An interference band, a set of points satisfying

$$\left[\frac{t_1 + t_2}{2} - \eta(t_1 - t_2), \frac{f_1 + f_2}{2} + \eta(f_1 - f_2)\right]$$

for $\eta \in [-\eta_0, \eta_0]$

A smoothing process can be defined, as before, to attenuate these interference terms.

7.5. MORE TRANSFORMATIONS

A very interesting recent development is the concept of unitary equivalence [81]. In this approach the chosen tool—either a time-frequency representation or a time-scale representation—is pre- and postprocessed by unitary transformations. In this way, the characteristics of interest can be enhanced by the appropriate choice of the unitary transform operators.

A chosen tool P is cradled between two unitary transformations, U and V, such that the equivalent system is given by

$$E = UPV$$

For example, consider the reconstruction of a signal $x(t)$ from its WT coefficients (expressed by (7-32)), where the wavelet $\psi_{m,n}(t)$ is given by (7-31), and repeated here for convenience:

$$\psi_{j,k}(t) = a_0^{-m/2}\psi(a_0^{-m}t - nb_0) = (D'_{a_0^m}T_{nb_0}\psi)(t) \tag{7-58}$$

where D = dilation operator: $(D_d x)(t) = e^{-d/2}x(e^{-d}t)$
   T = time shift: $(T_\tau x)(t) = x(t - \tau)$
   $D' = D_{\log d}$

Consider now that a unitary transformation $U^{-1}: L^2(R) \to L^2(R)$ is applied to this basis. Thus

$$U^{-1}\psi_{m,n} = U^{-1}(D'_{a_0^m}T_{nb_0}\psi) = (U^{-1}D'_{a_0^m}U)(U^{-1}T_{nb_0}U)(U^{-1}\psi) \qquad (7\text{-}59)$$

Note that because U^{-1} is a unitary transformation, $[U^{-1}\psi_{m,n}]$ is still a basis of the $L^2(R)$. Moreover, the time, frequency, and scale characteristics, as well as the wavelet $\psi(t)$ itself, were mapped into something else. This, obviously, is only good when U^{-1} is chosen such that it enhances the interesting features.

Following the notation of [81], one can rewrite (7-12) as

$$(W_x)(t,f) = 2\langle F_{-f}T_{-t}x, F_f T_t x* \rangle$$

where $(F_f x)(t) = e^{j2ft\pi}x(t)$ is the frequency shift.

Also, considering the Cohen's class to be composed of functions $C: L^2(R) \to L^2(R)$, which are covariant under translation of time and frequency, it is possible to write [81]

$$(CF_\gamma T_\tau x)(t,f) = (Cx)(t-\tau, f-\gamma)$$

Now, defining

$$\tilde{F} = U^{-1}FU$$

$$\tilde{T} = U^{-1}TU$$

The U-Wigner distribution is given by

$$(WUx)(a,b) = 2\langle \tilde{F}_{-b}\tilde{T}_{-a}x, \tilde{F}_b\tilde{T}_a x* \rangle \qquad (7\text{-}60)$$

The U-Cohen's class is composed of functions $CU: L^2(R) \to L^2(R)$ covariant under translation by the operators $\tilde{T}$ and $\tilde{F}$, defined previously. In other words, the U-Cohen class distribution CU is obtained by convolving the chosen kernel ϕ with the U-Wigner distribution:

$$(CUx)(a,b) = \int_{-\infty}^{\infty} \int_{-\infty}^{\infty} (WUx)(\tau,\gamma)\phi(a-\tau, b-\gamma)d\tau d\gamma \qquad (7\text{-}61)$$

Furthermore, it is possible to define a postprocessing transformation $V: L^2(R) \to L^2(R)$ such that it maps back from the (a,b) domain to the (t,f) domain. This defines the VU-Cohen's class.

Similarly, it is possible to define U-affine and VU-affine classes (for details see [81]).

Another interesting result is given in [82]. Their work aimed to improve the readability of the time-frequency or of the time-scale representations. In this case, a given representation has its values relocated, in time and/or scale, to positions where they can produce a better localization of the signal components. This was called the reassignment method.

7.6. CONCLUSIONS

In this chapter, traditional methods for signal analysis were discussed, as well as the hypotheses underlying each of them, and their limitations. General classes of representations were also presented and discussed, in an effort to show that no representation is a priori better for all types of signals, and that research in this field is still extremely fruitful. Interesting applications of these methods in bioengineering were also cited.

ACKNOWLEDGMENT

This work was partially supported by CAPES (Coordenacao de Aperfeicoamento de Pessoal de Nivel Superior, Brazil) under the number 618/93-2.

REFERENCES

[1] H. J. Weaver. *Theory of Discrete and Continuous Fourier Analysis.* John Williams and Sons, 1989.

[2] C. K. Chui. *An Introduction to Wavelets, Wavelet Analysis and its Applications* (1). New York: Academic Press, 1992.

[3] A. Papoulis. *Probability, Random Variables and Stochastic Processes.* New York: McGraw-Hill, 1984.

[4] M. B. Priestley. *Multivariate Series, Prediction and Control.* vol. 2 in: *Spectral Analysis and Time Series.* New York: Academic Press, 1981.

[5] D. Gabor, "Theory of communications," *J. IEE*, 93(III):429–457, 1946.

[6] I. Daubechies, "The wavelet transform, time-frequency localization and signal analysis," *IEEE Trans. Informat. Theory*, 36(5):961–1005, 1990.

[7] A. J. E. M. Janseen," Gabor representation of generalized functions," *J. Math. Anal. Appl.*, 83:377–394, 1981.

[8] I. Daubechies. *Ten Lectures on Wavelets.* Society for Industrial and Applied Mathematics, Pennsylvania, 1992.

[9] A. J. E. M. Janseen, "Gabor representation and Wigner distribution of signals," *Proc. IEEE Int. Conf. Acoust., Speech Signal Proc.*, 41B.2.1–41.B.2.4, 1984.

[10] S. Akselrod, D. Gordon, F. A. Ubel, D. C. Shannon, A. C. Barger, et al., "Power spectrum analysis of heart rate fluctuation: A quantitative probe of beat to beat cardiovascular control," *Science*, 213:220–222, 1981.

[11] W. Craelius, M. Akay, and M. Tangella, "Heart rate variability as an index of autonomic imbalance in patients with recent myocardial infarction," *Med. Biol. Comput.* 30:385–388, 1992.

[12] P. J. Vaitkus, R. S. C. Cobbald, and K. W. Johnston, "A comparative study and assessment of Doppler ultrasound spectral estimation techniques, Part 2: Methods and results," *Ultrasound Med. Biol.*, 14(8):673–688, 1988.

[13] K. W. Johnson, W. H. Baker, S. J. Burnham, A. C. Haynes, C. A. Kupper, et al., "Quantitative analysis of continuous-wave Doppler spectral broadening for the diagnosis of carotid disease: results of a multicenter study," *J. Vasc. Surg.*, 4:453–504, 1986.

[14] D. Schild and H. A. Scchultus, "The Fourier transform of a peristimulus time histogram can lead to erroneous results," *Brain Res.*, 369:353–355, 1986.

[15] M. Akay, Y. M. Akay, P. Chung, and H. H. Szeto, "Time-frequency analysis of the electrocortical activity during maturation using wavelet transform," *Biol. Cybern.*, 71:169–176, 1994.

[16] M. Akay. *Detection and Estimation of Biomedical Signals.* New York: Academic Press, 1996.

[17] R. Wilson, A. D. Calway, and E. R. S. Pearson, "A generalized wavelet transform for Fourier analysis: The multiresolution Fourier transform and its application to image and audio signal analysis," *IEEE Trans. Informat. Theory*, 38(2):674–690, 1992.

[18] E. P. Wigner, "On the quantum correction for thermodynamic equilibrium," *Phys. Rev.*, 40:749–759, 1932.

[19] J. Ville, "Theorie et applications de la notion de signal analytique," *Cables et transmission*, 2A:61–74, 1948.

[20] F. Hlawatsch and G. F. Boudreaux-Bartels, "Linear and quadratic time-frequency signal representations," *IEEE Signal Proc. Mag.*, 21–67, 1992.

[21] L. Cohen, "Time-frequency distributions—A review," *Proc. IEEE* 77(7):941–981, 1989.

[22] L. Cohen, "On a fundamental property of the Wigner distribution," *IEEE Trans. Acoust., Speech, Signal Proc.*, ASSP-35(4):559–561, 1987.

[23] J. C. Wood and D. T. Barry, "Radon transformation of time-frequency distributions for analysis of multicomponent signals," *IEEE Trans. Signal Proc.*, 42(11):3166–3177, 1994.

[24] A. C. Kak and M. Slaney. *Principles of Computerized Tomographic Imaging.* New York: IEEE Press, 1988.

[25] T. A. C. M. Claasen and W. F. G. Mecklenbräuker, "The Wigner distribution—A tool for time-frequency signal analysis, part I: Continuous time signals," *Philips J. Res.*, 35:217–250, 1980.

[26] T. A. C. M. Claasen and W. F. G. Mecklenbräuker," The Wigner distribution—A tool for time-frequency signal analysis, part III: Relations with other time-frequency signal transformations," *Philips J. Res.*, 35:372–389, 1980.

[27] F. Hlawatsch and R. L. Urbanke, "Bilinear time-frequency representations of signals: The shift-scale invariant class," *IEEE Trans. Signal Proc.*, 42(2):357–366, 1994.

[28] J. E. Moyal, "Quantum mechanics as a statistical theory," *Proc. Cambridge Phil. Soc.*, 45:99–124, 1949.

[29] J. G. Kirkwood, "Quantum statistics of almost classical ensembles," *Phys. Rev.*, 44:31–37, 1993.

[30] A. W. Rihaczek, "Signal energy distribution in time and frequency," *IEEE Trans. Informat. Theory*, 14(3):369–374, 1968.

[31] C. H. Page, "Instantaneous power spectra," *J. Appl. Phys.*, 23:103–106, 1952.

[32] Z. Guo, L.-G. Durand, and H. C. Lee, "The time-frequency distributions of nonstationary signals based on a Bessel kernel," *IEEE Trans. Signal Proc.*, 42(7):1700–1707, 1994.

[33] L. Cohen, "Generalized phase-space distribution functions," *J. Math. Phys.*, 7(5):781–786, 1966.

[34] R. G. Baraniuk and D. L. Jones, "A signal-dependent time-frequency representation: Optimal kernel design," *IEEE Trans. Signal Proc.*, 41(4):1589–1601, 1993.

[35] F. Hlawatsch and P. Flandrin, "The interference structure of the Wigner distribution and related time-frequency signal representation." In *The Wigner Distribution: Theory and Applications in Signal Processing.* W. Mecklenbräuker (ed). Amsterdam: North Holland Elsevier, 1992.

[36] S. Oh, R. J. Marks II, and L. E. Atlas, "Kernel synthesis for generalized time-frequency distributions using the method of alternating projections onto convex sets," *IEEE Trans. Signal Proc.*, 42(7):1653–1661, 1994.

[37] P. J. Loughlin, J. W. Pitton, and L. E. Atlas, "Bilinear time-frequency representation: New insights and properties," *IEEE Trans. Signal Proc.*, 41(2):750–767, 1993.

[38] P. Novak and V. Novak, "Time/frequency mapping of the heart rate, blood pressure and respiratory signals," *Med. Biol. Eng. Comput.*, 31:103–110, 1993.

[39] D. J. Waldo, P. R. Chitrapu, B. R. J. Reddy, K. Jepsen, G. A. Kidwell, et al., "Use of Wigner–Ville distribution to analyze cardiac late potentials," *Annual Int., Conf. IEEE Eng. Med. Biol. Soc.*, 12(2):6825–6826, 1990.

[40] W. Martin and K. Kruger-Aly, "Application of Wigner–Ville spectrum to the spectral analysis of a class of bioacoustical signals blurred by noise," *Acoustica*, 61(3):176–183, 1986.

[41] N. H. Morgan and A. S. Gevins, "Wigner distribution of human event-related brain potentials," *IEEE Trans. Biomed. Eng.*, 33(1):66–70, 1986.

[42] N. M. Marinovic and W. A. Smith, "Application of joint time-frequency distribution to ultrasonic transducers," *Proc. IEEE Int. Symp. Circuits Syst.*, 50–54, 1986.

[43] D. T. Barry and N. M. Cole, "Muscle sounds are emitted at the resonant frequency of skeletal muscle." *IEEE Trans. Biomed. Eng.*, 37(5):525–531, 1990.

[44] Z.-Y. Lin and J. D. Z. Chen, "Time-frequency representation of the electro-gastrogram—Application of the exponential distribution," *IEEE Trans. Biomed. Eng.*, 41(3):267–275, 1994.

[45] H. I. Choi, W. J. Williams, and H. Zaveri, "Analysis of event-related potentials: Time-frequency energy distribution," *Biomed. Sci. Instrument.*, 23:251–258, 1987.

[46] W. J. Williams and J. Jeong, "New time-frequency distribution: Theory and applications," *Proc. IEEE Int. Symp. Circuit. Syst.*, 2:1243–1247, 1989.

[47] D. L. Jones and T. W. Parks, "A resolution comparison of several time-frequency representations," *IEEE Trans. Signal Proc.*, 40(2):413–420, 1992.

[48] J. K. Hammond and R. F. Harrison, "Wigner–Ville and evolutionary spectra for covariance equivalent nonstationary random process," *Proc. IEEE Int. Conf. Acoust., Speech, Signal Proc.*, 1025–1027, 1985.

[49] R. J. R. Fonolossa and C. L. Nikias, "Wigner higher order moment spectra: Definition, properties, computation and applications to transient signal analysis," *IEEE Trans. Signal Proc.*, 41(1):245–265, 1993.

[50] Y. Meyer. *Wavelets and Operators.* Cambridge Studies in Advanced Mathematics, Cambridge: Cambridge University Press, 1992.

[51] A. Haar, "Zur Theorie der Orthogonalen Funktionen-system" [in German], *Math. Analysis*, 69:331–371, 1910.

[52] P. Franklin, "A set of continuous orthogonal functions," *Math. Analysis*, 100:522–529, 1928.

[53] A. Calderon, "Intermediate spaces and interpolation, the complex method," *Studia Math.*, 24:113–190, 1964.

[54] R. R. Coifman and G. Weiss, "Extensions of Hardy spaces and their use in analysis," *Bull. Am. Math. Soc.*, 83:569–645, 1977.

[55] X. Rodet, "Time-domain formant-wave-function synthesis," *Comput. Music J.*, 8, Part 3, 1985.

[56] A. Grossman and J. Morlet, "Decomposition of Hardy functions into square integrable wavelets of constant shape," *SIAM J. Math. Anal.*, 15:723–736, 1984.

[57] R.-K. Martinet, J. Morlet, and A. Grossman," Analysis of sound patterns through wavelet transforms," *Int. J. Patt. Recog. Art. Int.*, 1(2):273–302, 1987.

[58] R. A. Haddad, A. N. Akansu, and A. Benyassine, "Time-frequency localization in transforms, subbands, and wavelets: A critical review," *Opt. Eng.*, 32(7):1411–1428, 1993.

[59] M. Frazier and B. Jawerth, "The ɸ-transform and applications to distribution spaces." In *Function Spaces and Applications*, M. Cwikel et al. (eds), Lecture Notes in Mathematics, 1302:233–246, New York: Springer Verlag, 1988.

[60] A. Kumar, D. R. Fuhrmann, M. Frazier, and B. D. Jawerth, "A new transform for time-frequency analysis," *IEEE Trans. Signal Proc.*, 40(7): 1697–1707, 1992.

[61] S. G. Mallat, "A theory for multiresolution signal decomposition: The wavelet representation," *IEEE Trans. Patt. Anal. Mach. Intell.*, 11(7):674–693, 1989.

[62] M. Akay, W. Welkowitz, J. L. Semmlow, and J. Kostis, "Application of the ARMA method to acoustic detection of coronary artery disease," *Med. Biol. Eng. Comput.* 29:365–372, 1991.

[63] M. Akay, Y. M. Akay, G. Landrsberg, W. Welkowitz, and D. Sapoznikov, "Time-frequency analysis of heart rate fluctuations during carotid surgery using wavelet transform." In *Comparative Approaches in Medical Reasoning.* New York: Springer Verlag, 1993.

[64] D. M. Healy, Jr. and J. B. Weaver, "Two applications of wavelet transform in magnetic resonance imaging," *IEEE Trans. Informat. Theory,* 38:840–860, 1992.

[65] O. Meste, H. Rix, R. Jau, and P. Cardinal, "Detection of late potentials by means of wavelet transform," *Proc. IEEE Eng. Med. Biol. Soc.,* 28–29, 1989.

[66] D. Morlet, F. Peyrin, P. Desseigne, and P. Rubel, "Wavelet analysis of high-resolution signal averaged ECGs in postinfarction patients," *J. Electrocardiology* 26(4):311–320, 1993.

[67] J. Lu, D. M. Healy, Jr., and J. B. Weaver, "Signal recovery and wavelet reproducing kernels," *IEEE Trans. Signal Proc.,* 42(7):1845–1848, 1994.

[68] A. E. Cetin and R. Ansari, "Signal recovery from wavelet transform maxima," *IEEE Trans. Signal Proc.,* 42(1):194–196, 1994.

[69] S. Mallat and S. Zhong, "Wavelet maxima representations," In *Wavelets and Applications.* Y. Meyer (ed), pp. 207–285, New York: Springer Verlag, 1992.

[70] O. Bertrand, J. Bohorquez, and J. Pernier, "Time-frequency digital filtering based on an invertible wavelet transform: An application to evoked potentials," *IEEE Trans. Biomed. Eng.,* 41(1):77–88, 1994.

[71] G. W. Rogers, J. L. Solka, C. E. Priebe, and H. H. Szu, "Optoelectronic computation of waveletlike-based features," *Opt. Eng.* 31(9):1886–1892, 1992.

[72] S. Pluvan, T. K. Oh, N. Caviris, Y. Li, and H. H. Szu, "Texture analysis by space-filling curves and one-dimensional Haar wavelets," *Opt. Eng.* 31(9):1899–1906, 1992.

[73] A. W. Przybyszewski, "An analysis of the oscillatory patterns in the central nervous system with the wavelet method," *J. Neurosci. Meth.,* 38:247–257, 1991.

[74] D. P. Casasent, J.-S. Smokelin, and A. Ye," Wavelet and Gabor transform for detection," *Optical Eng.,* 31(9):1893–1898, 1992.

[75] T. Olson and J. DeStefano, "Wavelet localization of the Radon transform," *IEEE Trans. Signal Proc.,* 42(8):2055–2067, 1994.

[76] O. Rioul and M. Vetterli, "Wavelets and signal processing," *IEEE Signal Proc. Mag.,* pp. 14–38, 1991.

[77] M. Vetterli and C. Henley, "Wavelets and filter banks: Theory and design," *IEEE Trans. Signal Proc.,* 40(9):2207–2232, 1992.

[78] O. Rioul and P. Duhamel, "Fast algorithms from discrete and continuous wavelet transforms," *IEEE Trans. Informat. Theory,* 38(2):569–586, 1992.

[79] S. Cambanis and E. Masry, "Wavelet approximation of deterministic and random signals: Convergence properties and rates," *IEEE Trans. Informat. Theory,* 40(4):1013–1029, 1994.

[80] O. Rioul and P. Flandrin, "Time-scale energy distributions: A general class extending wavelet transforms," *IEEE Trans. Signal Proc.*, 40(7):1746–1757, 1992.

[81] R. G. Baraniuk and D. L. Jones, "Unitary equivalence: A new twist on signal processing," *IEEE Trans. Signal Proc.*, 43(10):2269–2282, 1995.

[82] F. Auger and P. Flandrin, "Improving the readability of time-frequency and time-scale representations by the reassignment method," *IEEE Trans., Signal Proc.*, 43(5):1068–1089, 1995.

Wavelets, Wavelet Packets, and Matching Pursuits with Biomedical Applications

In this part, we will focus on the fundamentals of wavelets, wavelet packets, and matching pursuit methods and their biomedical applications.

Chapter 8 by Rioul and Duhamel summarizes the various wavelet transform methods with their fast implementations.

Chapter 9 by Teich et al. also reviews the continuous wavelet transform and short-time Fourier transform methods and their applications to the analysis of cellular vibrations in the living cochlea.

Chapter 10 by Matalgah et al. discusses a new iterative approach based on Gabor wavelets for the analysis of phonocardiogram signals.

Chapter 11 by Sun and Sclabassi presents the feature extraction from neurophysiological signals using the wavelet transform and denoising process.

Chapter 12 by Coifman and Wickerhauser summarizes the wavelets, wavelet packets, and local sines and cosines which are more suitable to represent the biological signals. Then, the denoising process based on the adapted wavelets is presented for speech and simulated signals with different SNRs.

Chapter 13 by Rutledge describes a novel speech-enhancement approach based on denoising and amplitude compression using the wavelet transform method for hearing aids.

Chapter 14 by Karrakchou and Kunt summarizes wavelet packets and proposes a new mutual wavelet packet scheme for subband adaptive filtering and estimation of pulmonary capillary pressure signals.

Chapter 15 by Durka and Blinowska discusses the fundamentals of the matching pursuit method and its application to EEG signal transients.

Finally, in this part, Chapter 16 by Senhadji et al. presents the applications of several methods, including the wavelet transform-based estimator for detecting transients embedded in stationary background EEG activity.

Chapter 8

Fast Algorithms for Wavelet Transform Computation

Olivier Rioul, Pierre Duhamel

8.1. INTRODUCTION

Wavelet transforms have a wide range of applications, from signal analysis to image or data compression. Compared to the classical Fourier-based transformations, it can play either the role of the short time Fourier transform—or the Gabor transform—or that of a discrete Fourier transform, or even that of a discrete cosine transform. Therefore, it is not astonishing that the tool referred to as "wavelet transform" can take very different forms, depending on the application.

The continuous wavelet transform is best suited to signal analysis [1–3, 5–7]. Its semi-discrete version (wavelet series) and its fully discrete one (discrete wavelet transform) have been used for signal coding applications, including image compression [4–6] and various tasks in computer vision [8,9]. Wavelet transforms also find applications in many other fields, too numerous to be listed here (see e.g., [5]).

8.1.1 Classification of Wavelet Transforms

In a general sense, a wavelet transformation of a time-varying signal $x(t)$ consists of computing coefficients that are inner products of $x(t)$ against a family of "wavelets." These wavelets $\psi_{a,b}(t)$ are labeled by scale and time location parameters a and b. In a *continuous* wavelet transform, the wavelet corresponding to scale a and time location b is

$$\psi_{a,b}(t) = \frac{1}{\sqrt{|a|}} \psi\left(\frac{t-b}{a}\right) \tag{8-1}$$

where $\psi(t)$ is the wavelet "prototype," which can be thought of as a bandpass

211

function (the factor $|a|^{-1/2}$ is there to ensure energy preservation [2,5]). There are various ways of discretizing time-scale parameters, each one yielding a different type of wavelet transform. We adopt the following terminology, which parallels the one commonly used for Fourier transforms.

The continuous wavelet transform (CWT) was originally introduced by Goupillaud, Grossmann, and Morlet [2], and is given by

$$\text{CWT}\{x(t); a, b\} = \int x(t)\psi_{a,b}^{*}(t)dt \tag{8-2}$$

where the asterisk stands for complex conjugation. Time t and the time-scale parameters (b, a) vary continuously.

Wavelet series (WS) coefficients are sampled CWT coefficients. Time remains continuous but time-scale parameters are sampled on a "dyadic" grid in the time-scale plane (b, a) [4,5,8–12]. A usual definition is

$$C_{j,k} = \text{CWT}\{x(t);\ \ a = 2^{j},\ \ b = k2^{j}\} \quad \text{for}\ \ j, k \in Z \tag{8-3}$$

The wavelets are, in this case,

$$\psi_{j,k}(t) = 2^{-j/2}\psi(2^{-j}t - k) \tag{8-4}$$

and the original signal can be recovered through the following formula:

$$x(t) = \sum_{j \in Z}\sum_{k \in Z} C_{j,k}\tilde{\psi}_{j,k}(t) \tag{8-5}$$

where wavelets $\tilde{\psi}_{j,k}(t)$ are also of the form (8-4).

Wavelet series have been popularized under the form of a signal decomposition onto "orthogonal wavelets" by Meyer, Mallat, Daubechies, and other authors [5,8–11,13]. In the orthogonal case, the functions $\psi_{j,k}(t)$ and $\tilde{\psi}_{j,k}(t)$ are equal, and form an orthogonal basis. If, more generally, (8-3) and (8-5) hold exactly for $\psi_{j,k}(t)$ and $\tilde{\psi}_{j,k}(t)$ not necessarily equal, we are in the so-called "biorthogonal" case; the two sets of wavelet functions form two "mutually orthogonal" bases [4,12,14].

The discrete-time wavelet transform (DTWT) corresponds to the (continuous) Wavelet Transform of a sampled sequence $x_n = x(nT)$. Assuming sampling period T to be unity leads us to consider only integer time shifts in the analysis, resulting in

$$\text{DTWT}\{x_n; a, m\} = \sum_{n} x_n\psi_{a,m}^{*}(n) \tag{8-6}$$

The discrete wavelet transform (DWT) (see e.g., [12,13,15]) applies to discrete-time signals—both time and time-scale parameters are discrete. A DWT output on J "octaves" consists of "wavelet coefficients" c_{jk} computed for $j = 1, \ldots, J$:

$$\text{DWT}\{x(n); 2^{j}, k2^{j}\} = c_{j,k} = \sum_{n} x_n h_j^{*}(n - 2^{j}k) \tag{8-7}$$

and "residual coefficients" at octave J given by

$$r_{J,k} = \sum_{n} x_n g_J^{*}(n - 2^{J}k) \tag{8-8}$$

The $g_J(n - 2^J k)$ are the analysis *scaling sequences*: They are used to bring the input signal from the initial scale to scale 2^J. The $h_j(n - 2^j k)$ are the analysis wavelets, the discrete equivalent to the $2^{-j/2}\psi[2^{-j}(t - 2^j k)]$. The connection between both versions (discrete and continuous) is clarified later.

The reconstruction formula by which the inverse DWT reconstructs the signal from its coefficients is given by

$$x_n = \sum_{j=1}^{\infty} \sum_{k \in Z} c_{j,k} \tilde{h}_j(n - 2^j k) + \sum_{k \in Z} r_{J,k} \tilde{g}_J(n - 2^J k) \tag{8-9}$$

This formula is to be compared with (8-5). The main difference, apart from discretization, is the additional (low-pass) term: it is there to ensure perfect reconstruction, due to the finite iteration on the scale ($j = 1, \cdots, J$ in place of $j \in Z$). "Scaling functions" similar to the $g_J(n - 2^J k)$ can be defined for wavelet series [5,8–12,14–16] as shown in section 8.2.1.

8.1.2 Note on the Choice of the Wavelet

Orthogonality and biorthogonality properties, as defined earlier in the WS case, hold also for the DTWT and DWT, using appropriate (continuous or discrete) definitions of the inner product. The choice of particular orthogonal or biorthogonal wavelets is sometimes of importance in particular applications.

Here we focus on implementation issues, not on wavelet design. Therefore, even though design constraints on the shape of wavelets can sometimes be used to reduce the computational load, we do not take advantage of them so as to be as general as possible. Note, however, that orthogonality can bring slight computational gains, at the cost of a more involved implementation [17], and linear phase wavelets (possible only in the biorthogonal case) can be used to cut the number of multiplications by 2 in the straightforward implementation of the DWT described in section 8.4, by a simple use of symmetry in impulse responses.

We shall also restrict our focus to the (most frequent) case of wavelets with finite support. The issue of designing a wavelet with finite support is somewhat similar to a situation found in classical spectral analysis: when analyzing time-varying signals with Fourier-based tools, one cannot use the continuous Fourier transform directly, since it involves the whole signal of infinite support. Hence, the signal is restricted to a short segment around the instant of analysis by applying some window, and this windowed segment is then analyzed by a Fourier transform. Here, the design of a wavelet with finite support includes that of the window. This explains why one often chooses the wavelet in some library, just like the window for the short-time Fourier transform. The problem of designing some wavelet transform with specific properties is not addressed here. Note, however, that the spline implementation of wavelet transform offers much flexibility for this purpose [27, 28].

8.2. MULTIRESOLUTION AND TWO-SCALE EQUATIONS

If we stay with our previous definitions of wavelet transforms, the problem of choosing a wavelet is almost totally unconstrained, and full flexibility is possible—particularly in the case of the CWT. However, one of the main concepts of wavelet theory is the interpretation of wavelet transforms in terms of *multiresolution decomposition*. Of course, wavelets can exist without the multiresolution interpretation. However, this concept is so enlightening that we shall briefly outline its underlying concepts. As shown in the following sections, this is especially useful for fast wavelet algorithms and for the initial approximations that usually have to be performed on the signal and wavelets.

8.2.1 Multiresolution Spaces

A *multiresolution analysis* of $L^2(\mathbf{R})$ is a sequence $\{V_j\}$ ($j \in \mathbf{Z}$) of subspaces of $L^2(\mathbf{R})$, having the properties listed here (see [11] for mathematical details). The V_j's model spaces of signals having resolution at most 2^{-j}.

- Every signal lies in some V_j, and no signal—except the null signal—belongs to all V_j.
- V_j contains V_{j+1}.
- V_j is closed under time shifts $t \to t - k2^{-j}$, and $x(t) \in V_0$ is equivalent to $x(2^{-j}t) \in V_j$.
- There exists a function $\phi(t) \in V_0$ such that the set $\{\phi(t - k), \ k \in \mathbf{Z}\}$ forms a basis of V_0.

The function $\phi(t)$ is the *scaling function*. It is easily seen that the set of functions, defined in a "dyadic wavelet style" as

$$\phi_{j,k}(t) = 2^{-j/2}\phi(2^{-j}t - k) \tag{8-10}$$

forms a basis of V_j. Hence, all elements of V_j can be defined as linear combinations of $\phi_{j,k}(t)$.

Wavelet spaces W_j are orthogonal complements to V_j in V_{j-1}. They contain the necessary information to go from resolution 2^{-j} to $2^{-(j-1)}$. By construction, the subspaces $\{W_j\}$ are mutually orthogonal, and their direct sum spans the whole signal space $L^2(\mathbf{R})$.

One of the main results of the multiresolution theory is the existence of a function $\psi(t)$—the "mother wavelet"—constructed from the scaling function, and such that the set $\psi(t - k)$ is an orthonormal basis of W_0. Hence it follows from the definition of the W_j's that

$$W_j = \left\{ x(t) = \sum_{k \in \mathbf{Z}} c_{j,k} 2^{-j/2} \psi(2^{-j}t - k), \quad c_{j,k} \in l_2(\mathbf{Z}^2) \right\} \tag{8-11}$$

and the set $2^{-j/2}\psi(2^{-j}t - k)$ forms an orthonormal basis of $L^2(\mathbf{R})$. This corresponds exactly to the definition of an orthonormal wavelet series: the coefficients of the WS at scale j are the coordinates of the signal in space W_j.

The biorthogonal case is slightly more complicated. It involves two sequences of multiresolution spaces, one (V_j) for the analysis, and the other $(\tilde{V}_j)$ for the synthesis.

8.2.2 Examples

Classical examples of multiresolution related with the topic of fast algorithms are

- Haar wavelet: the scaling function $\phi(t)$ is a rectangle of value 1 on the interval $[0, 1)$, and V_0 is the space of the functions of $L^2(\mathbf{R})$, which are constant by parts.

- The dual situation in frequency: scaling functions $\phi(t)$ have a compact support spectrum e.g., on $[-.5, .5]$. A natural candidate for $\phi(t)$ is the sinc function, and a corresponding wavelet $\psi(t)$ can be obtained as a linear combination of sinc functions.

  This leads to an interpretation of Shannon's theorem in terms of multiresolution: Sampling corresponds to the projection into a multiresolution space, while the signal with the next coarser resolution has a spectrum twice as small. This will lead to an interesting interpretation of the initial approximation of a fast DWT algorithm as being similar to the half-band prefiltering made prior to sampling.

- Spaces of spline functions, built by parts, using polynomials with degree lower or equal to d. V_0 can be obtained through the use of the B-spline function $\beta^d(t)$ of order d, the dth iterative convolution of the unit rectangular pulse. These functions naturally lead to multiresolution spaces, since they are imbricated:

$$\beta^d(t/2) = \sum_{k \in Z} c_k^d \beta^d(t - k) \tag{8-12}$$

These functions play an important role in many respects. The obtained spline wavelets have many useful properties, among them the possibility of convergence toward Gabor functions, which have an optimal mapping of the time-frequency plane. Second, they can easily be used to approximate a wavelet of any (time domain) shape, while building a multiresolution analysis, hence allowing the use of fast algorithms.

8.2.3 Two-Scale Equations

By construction, $\phi(t/2) \in V_1 \subset V_0$, and $\psi(t/2) \in V_1 \subset V_0$. These functions can therefore be expressed as linear combinations of $\{\phi(t - k)\}$, the basis functions of V_0. We obtain *two-scale difference equations*:

$$\frac{1}{\sqrt{2}}\phi(t/2) = \sum_{k \in Z} g_k \phi(t - k) = g * \phi \qquad (8\text{-}13)$$

$$\frac{1}{\sqrt{2}}\psi(t/2) = \sum_{k \in Z} h_k \phi(t - k) = h * \phi \qquad (8\text{-}14)$$

It is known in multiresolution theory that the scaling function and the wavelet are fully characterized by the set of coefficients g_k and h_k. These coefficients are, in fact, impulse responses of filters used in the implementation of a DWT. They correspond, in our notation, to scaling sequences $g_1(n - k)$ and wavelets $h_1(n - k)$. It is therefore natural to consider the DWT as a natural implementation of WS, as explained in the next section.

8.3. THE INITIAL SIGNAL APPROXIMATION

Assume that an approximation of a CWT [defined as in (8-2)] has to be computed, and consider the following analogy with Fourier transforms. When implementing the short time Fourier transform of some continuous signal, one first *samples* the continuous signal. Information is not lost under the assumption that the signal has a finite spectrum, by Shannon's sampling theorem. This finite spectrum property is ensured by some prefiltering to avoid spectrum aliasing. It is well known that this corresponds to a projection of the initial signal onto the space of finite spectrum signals, which minimizes the mean square error of the frequency estimates.

This section is concerned with the same problem in the wavelet case [18]: Given some wavelet, which continuous signals can be represented by wavelet series without loss of information? Intuitively, this class of signals will be the only ones for which there will be no possibility of misinterpretation when exploiting the wavelet coefficients (think of the Fourier analogy: spectrum aliasing). Also, which procedure has to be applied in order to minimize the reconstruction error (formally equivalent to prefiltering in the Fourier case)?

Note that this problem can be stated in the context of a generalized sampling theory, in which the sampler no longer takes its ideal form $x_n = x(nT)$ (see [19]). But a direct use of such a generalized sampling theorem would require the knowledge (or worse, the design) of some precise sampling device. However, one usually knows only the samples of the signal, which are assumed to be sampled according to Shannon's theorem. Therefore, we follow here the approach of Abry and Flandrin [18], which is closely related to practical applications.

Assuming the continuous time is normalized such that $T = 1$, the continuous signal is related to its samples x_n by

$$x(t) = \sum_n x_n \text{sinc}(t - n) \qquad (8\text{-}15)$$

$$x_n = \int x(t)\text{sinc}(t - n)dt \qquad (8\text{-}16)$$

However, the initial signal, in a discrete implementation of a multiresolution procedure such as DWT or WS, is assumed to belong to V_0. Hence, the initialization should consist of projecting the signal $x(t)$ into V_0, as follows.

$$\hat{x}_k = \int x(t)\phi(t-k)dt \qquad (8\text{-}17)$$

$$= \sum_n x_n \int \text{sinc}(t-n)\phi(t-k)dt$$

$$= \sum_n x_n \int \text{sinc}(u)\phi(u-k+n)du$$

$$= \sum_n x_n f_{k-n}, \qquad \text{with } f_k = \langle \text{sinc}, \phi(\cdot-k)\rangle \qquad (8\text{-}18)$$

This initialization takes the form of a digital prefiltering, which has to be applied before any computation involving multiresolution. When this computation is possible, it will ensure the estimation of the wavelet coefficients with the least distortion. However, these coefficients f_k are obtained through an integral involving the (continuous) wavelet, which may be computationally intensive if several wavelets are to be used on the signal. In this case, a cheap approximation has been proposed in [18], which we now summarize.

Quite often, no approximation is made prior to the wavelet computation, i.e., one uses the implicit choice $\hat{x}_n = x_n$. It is then possible to show that the errors made on the approximations at the various scales and on the additional "details" come from the distance of the scale function ϕ to an ideal low-pass filter. This makes sense, since if the initial projection were an ideal sampler, the initial projection would be this ideal low-pass filter. The idea, explained in [18], is to make use of the fundamental low-pass character of ϕ. Since most of its energy lies in the frequency range $[-0.5, 0.5]$, the result of its convolution by the sinc function will not change much of its spectrum. Hence, a reasonable approximation is

$$f_k \approx \phi(-k) \qquad (8\text{-}19)$$

$$\hat{x}_n = \sum_n x_n \phi(n-k) \qquad (8\text{-}20)$$

Abry and Flandrin [18] provide convincing examples showing the necessity of the initialization. Note, however, that the initialization is not compulsory in the special case where the scaling function has $[-1, 1]$ as time support (e.g., Haar wavelet, splines of order 0 or 1), or when the input data are largely oversampled.

8.3.1 Remarks on Initialization and Sampling

Sampling a continuous signal consists of representing the whole information carried by this signal by means of a discrete sequence of numbers $\hat{x}_n$. In the case of Shannon sampling, there is the additional requirement that these numbers $\hat{x}_n = x(nT)$. It is well known that this operation is feasible if the input signal is ensured to have a finite spectrum by some prefiltering. This prefiltering is a projection of the original signal into the subspace of $L^2(\mathbf{R})$ of the finite spectrum functions.

This space can be generated by a linear combination of translated sinc functions (the interpolation formula: from the samples to the continuous function), thus forming a multiresolution, whose scale function generates an orthogonal basis.

If, however, one does not constrain the "samples" $\hat{x}_n$ to be values taken by the signal at regularly spaced instants, the projection of $x(t)$ into any multiresolution space V_0 takes the general form (8-18), which is also some kind of sampling procedure. In what follows, the prefiltering will be assumed to have been applied prior to the fast algorithm computation

8.4. THE DISCRETE WAVELET TRANSFORM (DWT)

Most fast algorithms for WT computation use the DWT as a basic building block [5,8–14,16]), hence its importance. As a transform of its own, the DWT mainly finds application in image compression [4–6,8,9] (in a two-dimensional form), but is also another description of octave-band filter banks that were used for some time in one-dimensional coding schemes [17,20].

The DWT is very much like a WS but applies to discrete-time signals x_n, $n \in Z$. More than a simple discretization of the DTWT to the dyadic grid, we assume that it achieves a multiresolution decomposition of x_n on J octaves labeled by $j = 1, \cdots, J$. It is precisely this requirement for a multiresolution—hence hierarchical—structure that makes fast computation possible. The requirement for a multiresolution computation can be stated as follows: Given some signal, at scale j, one decomposes it in a sum of details, at scale $j + 1$ (the true wavelet coefficients), plus some residual, representing the signal at resolution $j + 1$ (twice as coarse). A further analysis at coarser scales involves only the residual (think of the imbrication of subspaces in section 8.2). This requirement relies on the wavelet and on the signal: whether such a computation corresponds exactly to a sampling of the DTWT or not depends on properties of the wavelet (two-scale difference equation) and of the signal (initialization).

The efficient DWT computational structure can be obtained by observing that, due to the multiresolution requirement, wavelets and scaling sequences can be deduced from one octave to the next by some two-scale difference equation. Consider the analysis part (the treatment of synthesis "basis functions" is similar), and proceed by analogy with the multiresolution defined on WS in section 8.2. Consider two filter impulse responses $g(n)$ (corresponding to some low-pass interpolating filter—the scaling function) and $h(n)$ (corresponding to a high-pass filter—the discrete wavelet). The wavelets and scaling sequences are obtained iteratively as

$$g_1(n) = g(n) \qquad h_1(n) = h(n)$$

$$h_{j+1}(n) = \sum_k h_j(k)g(n - 2k) \tag{8-21}$$

$$g_{j+1}(n) = \sum_k g_j(k)g(n - 2k) \tag{8-22}$$

i.e., one goes from one octave j to the next $(j+1)$ by applying the interpolation operator

$$f(n) \to \sum_k f(k)g(n-2k) \tag{8-23}$$

which should be thought of as the discrete equivalent to the dilation $f(t) \to 2^{-1/2}f(t/2)$.

Consider, for example, the computation of $c_{j,k}$ as given by (8-7). For fixed j, $c_{j,k}$ is the result of filtering the input signal by $h_j(n)$ and then decimating the output by discarding one every 2^jth sample. Now the z-transform of filter $h_j(n)$ can be easily deduced from (8-21), which reads $H_{j+1}(z) = H_j(z^2)G(z)$ in z-transform notation. We obtain

$$H_{j+1}(z) = G(z)G(z^2)\cdots G(z^{2^{j-1}})H(z^{2^j}) \tag{8-24}$$

and, similarly for $g_j(n)$,

$$G_{j+1}(z) = G(z)G(z^2)\cdots G(z^{2^j}) \tag{8-25}$$

The computations of a DWT are now easily reorganized in the form of a binary tree, as shown in Fig. 8-1.

It is thus easily recognized that the structure of computations in a DWT is exactly an octave-band filter bank [8,12,13,15,17,20] as depicted in Fig. 8-1. The DWT corresponds to the analysis filter bank with filters $g(n)$ and $h(n)$, whereas the inverse DWT (IDWT) corresponds to the synthesis filter bank with filters $\tilde{g}(n)$ and $\tilde{h}(n)$.

Note that this filter bank is critically sampled: given N input samples, the DWT computes about $N/2 + N/4 + \cdots + N2^{-J} + N2^{-J} = N$ coefficients. In keeping with the critical sampling, the octave parameter j is restricted to $j \geq 1$ so that the sampling rate of wavelet coefficients is always less than that of the signal. Whenever the inverse DWT is used in the following, we assume that the filters $g(n), h(n), \tilde{g}(n)$, and $\tilde{h}(n)$ have been suitably designed so that (8-7) and (8-9) hold exactly. That is, the filter bank of Fig. 8-1 allows perfect reconstruction (this corresponds to the biorthogonal case). The reader is referred to [10,12,14,17,20] for more details on the design.

8.5. THE DWT FOR WS COMPUTATION

8.5.1 WS Computation: Mallat and Shensa Algorithm

It is well known since Mallat [8,9] that orthogonal wavelet series can be implemented using an orthogonal DWT, provided the discrete input is related to the original signal $x(t)$ by (8-17). The resulting algorithm, using filter banks, has been popularized as the Mallat algorithm. It was first derived using particular orthonormal wavelets.

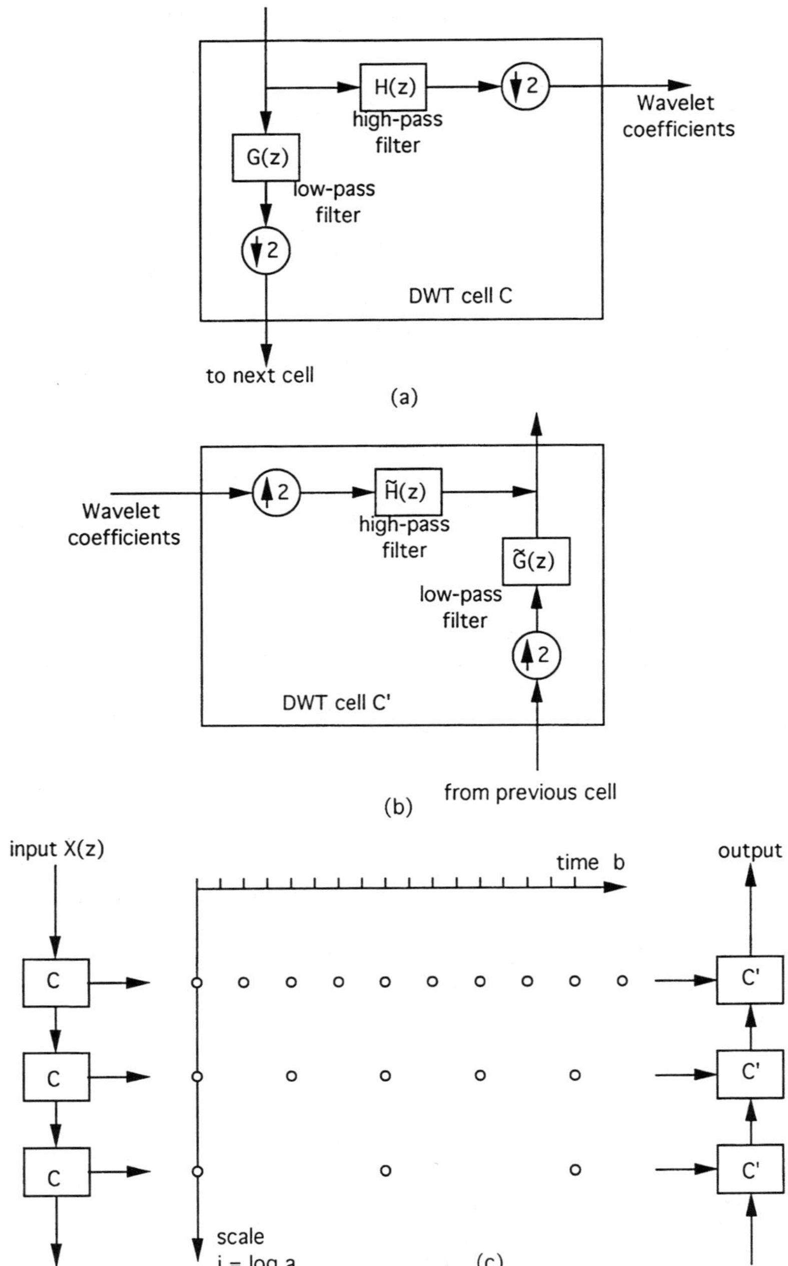

Figure 8-1   An octave-band filter bank. Basic computational cell of (a) the DWT and (b) the inverse DWT. (c) Overall organization takes the form of an octave-band filter bank. The analysis part gives wavelet coefficients that correspond to a dyadic grid in the time-scale plane. Signal is reconstructed using the transposed scheme (b) (synthesis filter bank).

The general algorithm of interest to us now, derived by Shensa [15], can be described as follows. Given the continuous-time wavelet $\psi(t)$, one first approximates it by $\hat{\psi}(t)$ in such a way that the following equation holds:

$$2^{-j/2}\hat{\psi}(2^{-j}t) = \sum_n h_j(n)\phi(t-n), \qquad j = 1, \cdots, J \tag{8-26}$$

where $h_j(n)$ are discrete wavelets present in a DWT, and $\phi(t)$ is some interpolating function (the scale function). The precise way these J simultaneous approximations can be accomplished is outlined in section 8.5.2.

The derivation of the algorithm is now straightforward. Substituting (8-26) into the equation defining the WS coefficients (8-3), and assuming that the initialization (8-17) has been done gives:

$$\hat{C}_{j,k} = \int \hat{x}(t)2^{-j/2}\hat{\psi}^*(2^{-j}t - k)dt \tag{8-27}$$

$$= \sum_n \hat{x}_n h_j^*(n - 2^j k) \tag{8-28}$$

$$= DWT\{\hat{x}_n, 2^j, k2^j\} \tag{8-29}$$

This ends the derivation of the Shensa algorithm: the WS coefficients with respect to the approximated wavelet $\hat{\psi}(t)$ are computed exactly for all signals using a DWT, provided that the input is appropriately prefiltered. The accuracy of this algorithm is balanced by the approximations made for the input (8-17) and for the wavelets (8-26); the algorithm is exact only when the input and the wavelets have been replaced by their approximations.

Note that we have three different types of inputs at work: the original analog signal, its approximation introduced by the original sampling, with discrete-time samples x_n, and the filtered version $\hat{x}_n$ defined by (8-18). They involve two successive approximations: the first one is made regardless of the parameters in the algorithm (initial sampling). The second one is the prefiltering, which depends on the parameters of the algorithm, and amounts to a nonorthogonal projection of $x(t)$.

8.5.2 The Wavelet Approximation

One may wonder how (8-26) can be computed. These approximations are important because their accuracy determines that of the whole algorithm. First, note that this whole set of equations is equivalent to assuming that Eqs. (8-14) [which is (8-26) rewritten for $j = 1$)], plus (8-13) hold. These two-scale difference equations were studied in detail by Daubechies and Lagarias in [21].

There are two steps involved. First, determine a low-pass filter $g(n)$ and an interpolating function $\phi(t)$ satisfying (8-13). Second, approximate $\psi(t)$ by linear combinations of integer translates of $\phi(t)$ (8-14). This step determines the high-pass filter $h(n)$. Of course, it is crucial to choose a good interpolating function $\phi(t)$ so that $\psi(t)$ can be accurately approximated. Note, however, that once $\psi(t)$ is accurately approximated by $\hat{\psi}(t)$ for which (8-13) and (8-14) hold, the J approximations at all scales (8-26) are satisfied automatically; for example, minimizing the error's energy $\int |\psi(t) - \hat{\psi}(t)|^2 dt$ minimizes the maximum error $|C_{j,k} - \hat{C}_{j,k}|$ of the wavelet coefficients at all scales. Several "standard" choices for $\phi(t)$ were cited in section 8.2.2.

8.5.3 Using the Inverse DWT to Compute the Inverse WS (IWS)

We have seen that wavelet series coefficients (8-3) can be computed using a DWT (8-29). Similarly, its inverse transform (8-5) can be computed using an inverse DWT (8-9), under a condition similar to (8-26), but written for *synthesis* wavelets $\tilde{\psi}_{j,k}(t)$

$$2^{-j/2}\tilde{\psi}(2^{-j}t) = \sum_n \tilde{h}_j(n)\tilde{\phi}(t - n), \qquad j = 1, \cdots, J \tag{8-30}$$

Of course, this condition is, in practice, replaced by more tractable conditions as explained earlier. Substituting (8-30) for $2^{-j/2}\tilde{\psi}(2^{-j}t)$ in the formula defining the inverse WS (8-5) results in

$$\text{IWST}\{C_{j,k}\} = \sum_n y_n \tilde{\phi}(t - n) \tag{8-31}$$

where the $C_{j,k}$ are the WS coefficients (8-3), and y_n is defined by

$$y_n = \text{IDWT}\{C_{j,k}\} \tag{8-32}$$

Thus, the inverse DWT, followed by a D/A converter with characteristic $\tilde{\phi}(t)$, computes the IWS exactly.

The accuracy of the algorithm again depends on that of the signal and wavelet approximation. The resulting analysis/synthesis WS scheme is depicted in Fig. 8-2. First, the analog signal $x(t)$ is discretized according to (8-15). The discrete-time signal x_n is then prefiltered (8-18) and fed into the DWT algorithm. During synthesis, the signal is reconstructed using an inverse DWT, followed by the interpolation (or D/A conversion) (8-31).

Note that in this WS/IWS Shensa algorithm, the analysis and synthesis discrete wavelets do not necessarily form a perfect reconstruction filter bank pair. However, we now restrict the focus to the perfect reconstruction case to derive conditions under which the original signal $x(t)$ is recovered exactly.

When the DWT allows perfect reconstruction, one has $y_n = \hat{x}_n$. It can be shown that we are in fact in the "biorthogonal" case [12,14], and that one has

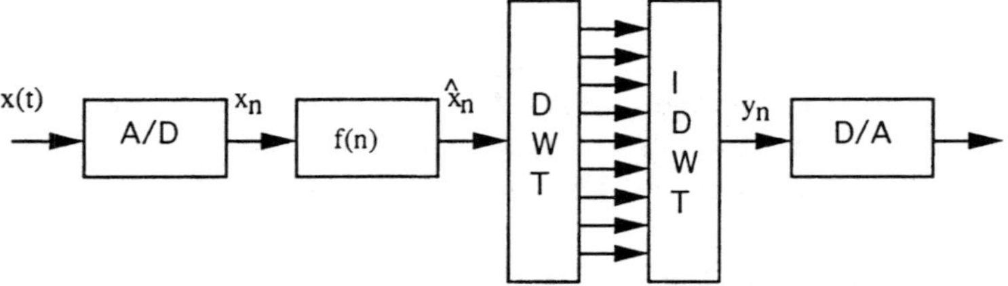

Figure 8-2   Full analysis/synthesis WS scheme. Exact reconstruction holds under certain conditions on $x(t)$ (see text).

$$\int \phi(t-n)\tilde{\phi}^*(t-m)dt = \delta_{m,n} \tag{8-33}$$

Since $y_n = \hat{x}_n$, we also have

$$\mathrm{IWS}\{\mathrm{WS}\{x(t)\}\} = \sum_n \left(\int \hat{x}(u)\phi^*(u-m)du \right) \tilde{\phi}(t-n) \tag{8-34}$$

The right-hand side of (8-34) is easily recognized to be a projection of $x(t)$ onto the subspace $\tilde{V}_0$ spanned by linear combinations of the $\tilde{\phi}(t-n)$: if $x(t)$ belongs to $\tilde{V}_0$, i.e., if $\hat{x}(u) = \sum_n c_k \tilde{\phi}(u-k)$, then using (8-33), Eq. (8-34) simplifies to $x(t)$. Therefore, only the projected approximation of $x(t)$ onto $\tilde{V}_0$ is recovered. However, since we recover $\hat{x}_n$, we may attempt to reconstruct $x(t)$ (or its projection onto V_0) directly from $\hat{x}_n$.

8.6. THE DWT FOR CWT COMPUTATION

The DWT, as well as WS, are nonredundant transforms. However, it may be useful to obtain samples of the CWT at denser places of the time-scale plane than the dyadic grid. It is, therefore, sometimes appropriate to generalize (8-29) in order to obtain more samples in the time-scale plane. This is especially useful for signal analysis, where one usually "oversamples" the discretization (8-3), in two ways: First, one may want to evaluate the scale output at any time sample, whatever the scale (see section 8.6.2), instead of a coarser sampling when increasing the scale. Then it is often useful to have a finer sampling in scale, in order to obtain, e.g., "M voices per octave" [5] (see section 8.6.1). Finally, one could wish a time-scale parameter sampling as follows:

$$a = a_0^j \tag{8-35}$$
$$b = k \tag{8-36}$$

where $1 < a_0 \leq 2$. Note that a is restricted to positive values. This implicitly assumes that the signal and wavelets are either both real-valued or both complex analytic (i.e., their Fourier transforms vanish for negative frequencies). One interest of (8-35) is the possibility to approximate a nearly continuous CWT representation in the time-scale plane for analysis purposes.

The full discretization previously defined is addressed separately, by working with one parameter sampled as in the WS transform, while the other one takes denser, regular sampling values. The general case is obtained by combining both techniques.

8.6.1 Finer Sampling in Scale

Here, we stay with $b = k2^j$, while the scale parameters are sampled according to

$$a = 2^{j+m/M}, \quad m = 0, \cdots, M-1 \tag{8-37}$$

where m is called the "voice." In other words, a_0 in (8-35) is chosen as an Mth root of 2.

The following simple method [3] allows one to compute WS coefficients on M voices per octave, using the standard "octave-by-octave" algorithm (8-29) as a building block. For each m, replace $\psi(t)$ by the slightly stretched wavelet $2^{-m/2M}\psi(2^{-j+m/M}t)$ in the expression of $\psi_{j,k}(t) = 2^{-j/2}\psi(2^{-j}t - k)$. The wavelet basis functions become

$$2^{-(j+m/M)/2}\psi(2^{-(j+m/M)}(t - k2^j)), \qquad j, \mathrm{k} \in Z, \qquad m = 0, \cdots, M - 1 \qquad (8\text{-}38)$$

The grid obtained in the time-scale plane (b, a) is shown in Fig. 8-3. Now, a computation on M voices per octave is done by applying the octave-by-octave algorithm M times, with M different prototypes.

Of course, the parameters of each octave-by-octave algorithm must be recomputed for each m using the procedure previously described. Clearly, the whole algorithm requires about M times the computational load of one octave-by-octave algorithm.

This method is certainly not the best one for an "M voices per octave" computation if M is large, because it does not take advantage of the fact that the various prototypes (8-38) are related in a simple manner. It would be more appropriate to devise a method that takes advantage of both time redundancy and scale redundancy (with more scales than in the octave-by-octave case). The algorithm devised by Bertrand et al. in [1] is based on scale redundancy but is suited for another type of computation (see section 8.9.3).

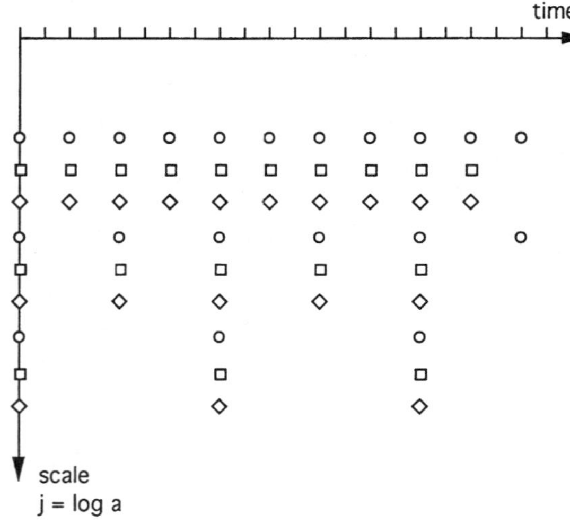

Figure 8-3 Sampling of the time-scale plane corresponding to three voices per octave in a WS. The imbrication of the computation is shown using points labeled by circles, squares, and crosses, which can be computed separately using octave-by-octave DWT algorithms.

8.6.2 Finer Sampling in Time: Modified Shensa and "à trous" Algorithms

Here, we restrict our study to an octave-by-octave computation, i.e., $a = 2^j$, while considering all possible values for the time parameter $b = k$. First, note that the computation of the WS coefficients treated in section 8.5.1 is nothing but part of the computation required here, since

$$C_{j,k} = \text{CWT}\{x(t);\, 2^j, k2^j\} \tag{8-39}$$

Now, the Shensa algorithm for the WS coefficients can be readily extended to the required computation of $\text{CWT}\{x(t);\, 2^j, k\}$ [15]. We have a result similar to (8-29), namely,

$$\text{CWT}\{x(t);\, 2^j, k\} = \text{DWT}\{\hat{x}_n;\, 2^j, k\} \tag{8-40}$$

where $\hat{x}_n$ is a prefiltered discrete input defined by (8-17), more easily computed using (8-18). The only difference is, of course, that the DWT is computed for all integer values of b, instead of $b = k2^j$, as in the standard description of the DWT. Equation (8-40) indicates that CWT coefficients sampled on an arbitrary grid in the time-scale plane can be computed using a filter bank structure derived from the initial DWT. This fact was mentioned by Gopinath and Burrus in [16] and subsequently discussed in detail by Shensa in [15]: The resulting CWT algorithm was recognized to be identical with the "à trous" algorithm of Holschneider et al. [3,5].

This "à trous" structure is pictured in Fig. 8-4(a). It can be easily derived as follows. For fixed j, the result of (8-40) is simply the discrete input filtered by $h_j(n)$, whose transfer function is given by (8-24). The difference with section 8.4 is the absence of decimation. Now, reorganize the computation in a hierarchical way as follows. The input is iteratively filtered by $G(z)$, $G(z^2)$, and so on. At the jth step, it is enough to filter by $H(z^{2^{j-1}})$ in order to obtain the expected coefficients (8-40), as shown in Fig. 8-4(a). The term "à trous"—with holes—was coined by Holschneider et al. in reference to the fact that only one every 2^{j-1} coefficients is nonzero in the filter impulse responses at the jth octave.

8.6.3 A Slightly Different Building Block

We now consider another variation of filter bank implementation of the CWT—which was also derived by Shensa in [15]—because it is more suited to further reduction of complexity using fast filtering techniques than the one using DWT. Consider the filter bank structure of Fig. 8-4(c), where the elementary cell is depicted in Fig. 8-4(b). This filter bank structure is easily deduced from the one of Fig. 8-4(a) [15].

The advantage of this slightly different structure is easily understood as follows: Consider the computation performed at the first octave ($j = 1$) of Fig. 8-4 and compare it to Fig. 8-1(a). In the latter structure, half the wavelet coefficients required for the CWT at this octave are computed: the missing ones are the outputs of $H(z)$ that are discarded by the decimation process. It is sufficient to remove the subsampling on $H(z)$ to obtain the required wavelet coefficients of the first octave, as shown

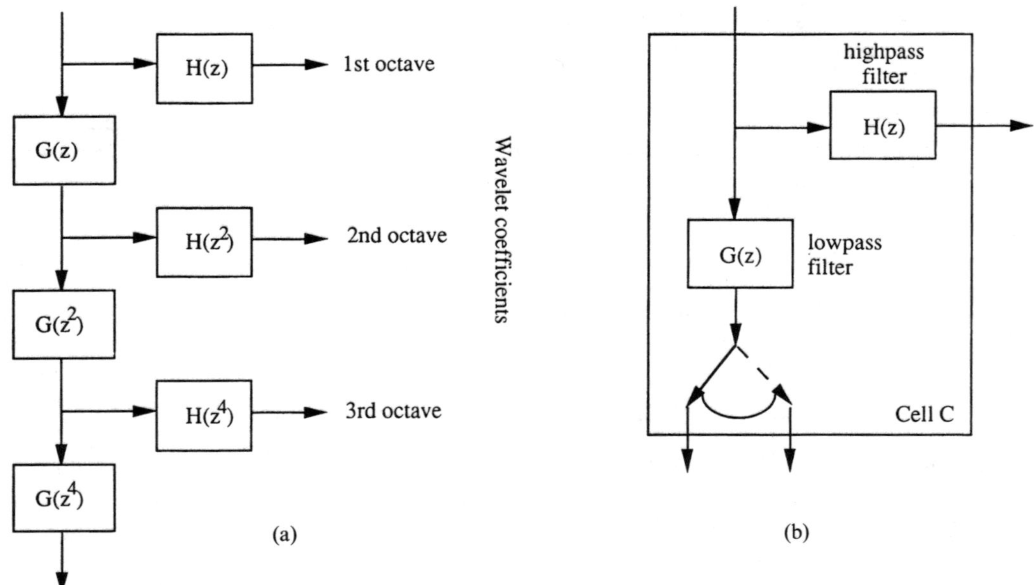

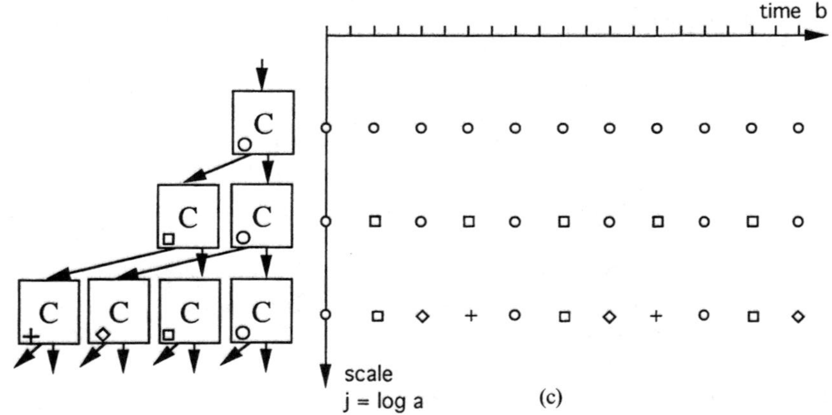

Figure 8-4 (a) "A trous" structure as derived by Holschneider et al; (b) basic computational cell used for computing CWT coefficients octave by octave; (c) connection of the cells used in this paper and corresponding location of the wavelet coefficients in the time-scale plane.

in Fig. 8-4(a). Also, in Fig. 8-1(a), the output of the filter $G(z)$ is used to compute the wavelet coefficients for the next stage ($j = 2$) for even values of the time-shift parameter b. The missing sequence, which allows one to obtain the coefficients with odd values of b is nothing but the discarded subsampled sequence; it is recovered in Fig. 8-4(a).

At the next octave, $j = 2$, both inputs are processed separately using identical cells. One provides the same coefficients as in the WS computation [round dots in

Fig. 8-4(c)], while the other allows one to start a new computation of the same type, shifted in time, and beginning at the next scale [squared dots in Fig. 8.4(c)]. The whole process is iterated as shown in Fig. 8-4(c).

In the overall organization, all outputs of both filters have to be computed, those of $G(z)$ being used to build two interleaved sequences, while those of $H(z)$ are simply the desired samples of the CWT at the given scale. This is in contrast to the basic computational cells of the fast DWT algorithms. Hence the reorganization of the computations described in section 8.7.2 should not be used in this case.

8.6.4 Inner Product Implementation of the CWT

Consider the filter bank implementation of Fig. 8-4(c), and assume that both filters $g(n)$ and $h(n)$ are finite impulse response (FIR) filters and have same length L. When the filters are directly implemented as inner products, the octave-by-octave CWT algorithm requires

$$2L \text{ mults/input point/cell} \qquad 2(L-1) \text{ adds/input point/cell} \qquad (8\text{-}41)$$

Note that there are 2^{j-1} elementary cells at the jth octave in Fig. 8-4(c), which are identical but "work" at a different rate: a cell at the jth octave is fed by an input which is subsampled by 2^{j-1} compared to the original input $x(t)$. Therefore, the total complexity required by an octave-by-octave CWT algorithm on J octaves, is exactly J times the complexity of one cell. Thus the complexity of any filter bank implementation of a CWT grows linearly with the number of octaves. This results then, for a CWT on J octaves, in

$$2LJ \text{ mults/input point} \qquad 2(L-1)J \text{ adds/input point} \qquad (8\text{-}42)$$

As mentioned in [3], this is a significant improvement compared to the naive method that would consist of directly implementing the CWT and would not take advantage of the fact that wavelets are easily related by dilation (this direct implementation would require a complexity exponentially increasing with J). Since the whole CWT algorithm requires J times the complexity of one cell, the latter is the total complexity of the CWT per input point and per octave. Hence the complexity of one cell is also the total complexity of the CWT per output point, i.e., per computed wavelet coefficient.

Since the elementary cell contains filters, its arithmetic complexity can be reduced using any fast filtering technique. This is explained in the following section.

8.7. EFFICIENT IMPLEMENTATIONS OF THE DWT

In the following, we derive efficient implementations of the DWT, which can be used to compute WS coefficients using the Shensa algorithm. Hence most of the content of this section also applies to the implementation of tree-structured two-band filter banks iterated on the low-pass filter.

8.7.1 Preliminaries

It is important to note that the standard DWT algorithm, implemented directly as a filter bank, is already "fast." This fact was mentioned by Ramstad and Saramaki in the context of octave-band filter banks [22]. What makes the DWT "fast" is the decomposition of the computation into elementary cells and the sub-sampling operations (decimations), which occur at each stage. More precisely, the operations required by one elementary cell at the jth octave [Fig. 8-1(a)] are counted as follows. There are two filters of equal length L involved. The "wavelet filtering" by $h(n)$ directly provides the wavelet coefficients at the considered octave, while filtering by $g(n)$ and decimating is used to enter the next cell. A direct implementation of the filters $g(n)$ and $h(n)$ followed by decimation requires $2L$ multiplications and $2(L-1)$ additions for every set of two inputs. That is, the complexity per input point for each elementary cell is

$$L \text{ mults/point/cell} \quad \text{and} \quad L-1 \text{ adds/point/cell} \tag{8-43}$$

Since the cell at the jth octave has input subsampled by 2^{j-1}, the total complexity required by a filter bank implementation of the DWT on J octaves is $(1 + \frac{1}{2} + \frac{1}{4} + \cdots + \frac{1}{2^{J-1}}) = 2(1 - 2^{-J})$ times the complexity (8-43). That is,

$$2L(l - 2^{-J})\text{mults/point} \quad \text{and} \quad 2(L-1)(1 - 2^{-J})\text{adds/point} \tag{8-44}$$

The DWT is therefore roughly equivalent, in terms of complexity, to one filter of length $2L$. Note that the complexity remains bounded as the number of octaves, J, increases [22].

In contrast, a naive computation of the DWT, which would implement (8-7) exactly as written, with precomputed discrete wavelets $h_j(n)$, would be very costly. This lack of efficiency is due to the fact that (8-7) does not take advantage of the dilation property of wavelets, summarized by the two-scale difference equation: Since the length of $h_j(n)$ is $(L-1)(2^J - 1) + 1$, one would have, at the jth octave, $(L-1)(2^j - 1) + 1$ real multiplications and $(L-1)(2^j - 1)$ real additions for each set of 2^j inputs. For a computation on J octaves ($j = 1, \cdots, J$), this gives

$$J(L-1) + 1 \text{ mults/point} \quad \text{and} \quad J(L-1) \text{ adds/point} \tag{8-45}$$

This complexity increases linearly with J, while that of the "filter bank" DWT algorithm is bounded as J increases. The use of the filter bank structure in the DWT computation thus reduces the complexity from JL to L. This is a huge gain; the DWT already deserves the term "fast."

8.7.2 Reorganization of the Computations

The derivation of faster algorithms described in section 8.8 is primarily based on the reduction of computational complexity. Here, "complexity" means the number of real multiplications and real additions required by the algorithm, per input point. In the DWT case, this is also the complexity per output point since the DWT is critically sampled. Of course, complexity is not the only relevant criterion. For example, regular computational structures (i.e., repeated application of identical computational cells) are also important for implementation issues. However, since most algorithms considered in this paper have regular structures, a criterion based on complexity is fairly instructive for comparing the various DWT algorithms. We have chosen the total number of operations (multiplications + additions) as the criterion. With today's technology, this criterion is generally more useful than the sole number of multiplications [23], at least for general-purpose computers (another choice would have been to count the number of multiplication-accumulations, for implementation on digital signal processors).

From the operation counts given earlier (8-44), it is clear that if all elementary cells require the same complexity, then a filter bank implementation of the DWT requires $2(1 - 2^J)$ times the complexity of one cell. Therefore, any fast convolution technique applied to the elementary cell will further reduce the computational load of the DWT. Section 8.8 proposes two classes of fast algorithms: one based on the fast Fourier transform (FFT) [24] and the other on short-length FIR filtering algorithms [23].

The basic DWT elementary cell, depicted in Fig. 8-1(a), contains two filters. However, they are always followed by subsampling (or decimation), which discards every other output. It is well known that reducing the arithmetic complexity of an FIR filter implementation is obtained by gathering the computations of several successive outputs [24]. Since the filter outputs are decimated in Fig. 8-1(a), it is necessary to reorganize the computations in such a way that "true" filters appear. To do this, we apply a biphase decomposition, [17] to all signals involved, which consists of separating them into even- and odd-indexed sequences. The biphase decomposition expresses the z-transform of the input sequence x_n as:

$$X(z) = \sum_n x_n z^{-n} \tag{8-46}$$

$$= X_0(z^2) + z^{-1} X_1(z^2) \tag{8-47}$$

where $X_0(z) = \sum_n x_{2n} z^{-n}$ and $X_1(z) = \sum_n x_{2n+1} z^{-n}$.

Similarly, apply the biphase decomposition to the L-tap filters $G(z)$ and $H(z)$ involved in the computation. The cell output $Y(z)$ that enters the next stage is obtained by first filtering by $G(z)$, then subsampling. Picking out the even part of $G(z)X(z)$ results in

$$Y(z) = G_0(z)X_0(z) + z^{-1} G_1(z)X_1(z) \tag{8-48}$$

Now that this rearrangement has been made, the output $Y(z)$ is obtained differently: First the even- and odd-indexed input samples $X_0(z)$ and $z^{-1}X_1(z)$ are extracted as they flow by (hence, the delay factor z^{-1} for odd-indexed samples). Then, $L/2$-tap filters $G_0(z)$ and $G_1(z)$ are applied to the even and odd sequences, respectively. Finally, the results are added together. The other output of the elementary cell (the one corresponding to the filter $H(z)$) is obtained similarly using $H_0(z)$ and $H_1(z)$.

The resulting flow graph is depicted in Fig. 8-5 (the corresponding IDWT cell is simply obtained by flow graph transposition). Compare with Fig. 8-1(a): there are now four "true" filters of length $L/2$, whose impulse responses are the decimated

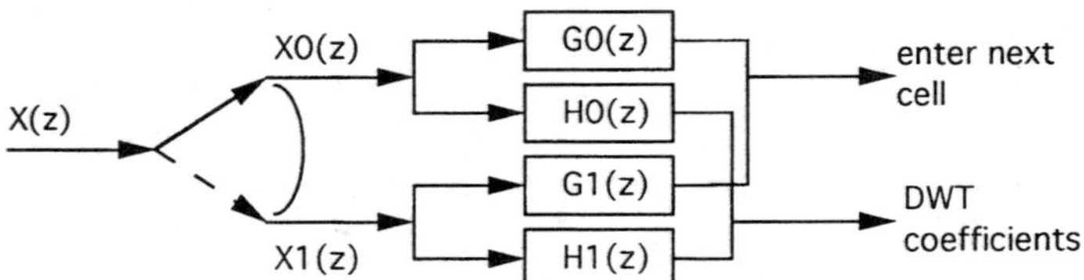

Figure 8-5   Rearrangement of the DWT cell of Fig. 8-1(a) that avoids subsampling, hence allows the application of fast filtering techniques.

initial filters $G(z)$ and $H(z)$. The complexity has not changed, but the resulting structure is easily improved by the use of classical fast filtering algorithms, as shown in the next section.

8.8. FASTER DWT ALGORITHMS

The aim of this section is to further reduce the computational load of the DWT. We briefly motivate this with a brief analogy to fast filtering. FFTs are used for implementing long filters (typically $L \geq 64$) because they greatly reduce the complexity: Compared to a direct implementation of the filter, the number of operations per input point is reduced from L to $\log_2 L$, hence the term "fast." For short filters, however, the FFT is no longer efficient and other fast filtering techniques are used [23,24]; the resulting gain is fairly modest, but still interesting when heavy computation of short filters is required, provided that the accelerated algorithm does not require a much more involved computation compared to the initial one. The situation of the DWT is identical: using FFTs, the complexity of the DWT can be reduced from $2L$ to $4 \log_2 L$, when the filter length L is large. However, DWTs have been mostly used with short filters so far (although nothing ensures that this will last forever). For them, using different techniques, smaller gains are obtained, typically a 30% saving in the number of computations, which can still be useful.

We assume real data and filters (of finite length), but the results extend easily (if necessary) to the complex-valued case. A quick evaluation of the corresponding number of operations can be obtained from the results provided in the following real-valued case: the FFT-based algorithms described next require about twice as many multiplications in the complex case as in the real case, a property shared by FFT algorithms [24]. However, a straightforward filter bank implementation of the DWT (Fig. 8-1), or the "short-length" algorithms described in section 8.8.3, require about three times as many multiplications in the complex case, assuming that a complex multiplication is carried out with three real multiplications and additions [24].

We shall not derive algorithms explicitly for the inverse DWT. However, an inverse DWT algorithm is easily obtained from a DWT algorithm as follows: If the wavelets form an orthogonal basis, the exact inverse algorithm is obtained by taking the Hermitian transpose of the DWT flowgraph. Otherwise, only the structure of the inverse algorithm is found that way, the filter coefficients $g(n), h(n)$ have to be replaced by $\tilde{g}(n)$, $\tilde{h}(n)$, respectively. In both cases, any DWT algorithm, once transposed, can be used to implement an inverse DWT. It can be shown that this implies that the DWT and inverse DWT require exactly the same number of operations (multiplications and additions) per point.

The filters involved in the computation of the DWT (cf. Fig. 8-1) usually have equal length L. This is true in the orthogonal case, while in the biorthogonal case the filter lengths may differ by a few samples only. Although an implementation of "Morlet-type" wavelets given in [3,5] uses a short low-pass filter $g(n)$ and a long high-pass filter $h(n)$, we restrict our focus in this section to the case of equal filter lengths for simplicity. If lengths differ, one can pad the filter coefficients with zeros.

8.8.1 An FFT-Based DWT Algorithm

This method consists of computing the four $L/2$-tap filters of Fig. 8-5 using the overlap-add or overlap-save FFT. Operation counts are done using the "split radix" FFT algorithm which, among all practical FFT algorithms, has the best known complexity for lengths that are powers of 2: $N = 2^n$ ($n = \log_2 N$ should not be confused here with the sample index n). For real data, the split radix FFT (or inverse FFT) requires exactly

$$2^{n-1}(n - 3) + 2 \quad \text{(real) mults} \tag{8-49}$$

$$2^{n-1}(3n - 5) + 4 \quad \text{(real) adds} \tag{8-50}$$

We now briefly recall the standard method for computing filters using the FFT. The input of the DWT cell is blocked B samples by B samples (the decimated sequences input to the filters therefore flow as blocks of length B/2). Each discrete filter is performed by computing the inverse FFT (IFFT) of the product of the FFTs of the input and filter. Since the latter FFT can be precomputed once and for all, only one IFFT and one FFT are required per block for one filter. However, this results in a cyclic convolution [24], and the overlap-add and overlap-save methods [24] can be used in order to avoid wraparound effects. One is the transposed form of the other and both require exactly the same complexity. For one filter of length $L/2$, with input block length $B/2$, wraparound effects are avoided if the FFT length N satisfies $N \geq L/2 + B/2 - 1$. Here, we assume $B = 2N - (L - 2)$.

Assume that each elementary cell has the same structure, pictured in Fig. 8-6. The input is first split into even- and odd-indexed sequences. Then, a length-N FFT is performed on each decimated input, and four frequency-domain convolutions are performed by multiplying the (Hermitian symmetric) FFT of the input by the (Hermitian symmetric) FFT of the filter. This requires $4N/2$ complex multiplications for the four filters. Finally, two blocks are added ($2N/2$ additions) and two IFFTs are applied. Assuming that a complex multiplication is done with three real multiplications and three real additions [24], this gives a total of

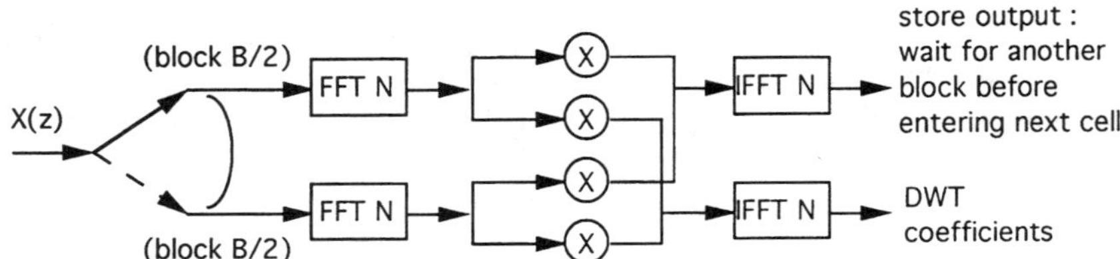

Figure 8-6   FFT-based implementation of the DWT cell of Fig. 8-4. Overlap-add (or overlap-save) procedure is not explicitly shown.

$$\frac{n2^{n+1} + 8}{2^{n+1} - (L - 2)} \qquad \text{mults/point/cell} \qquad (8\text{-}51)$$

$$\frac{(3n - 1)2^{n+1} + 16}{2^{n+1} - (L - 2)} \qquad \text{adds/point/cell} \qquad (8\text{-}52)$$

Note that for a given length L, there is an optimal value of N that minimizes the complexity. Tables 8-1 and 8-2 show the resulting minimized complexities for different lengths L in comparison with the inner product implementation of the filter bank. The comparison is clearly in favor of the FFT version of the DWT algorithm for medium to large filter lengths ($L \geq 16$). The asymptotic gain brought by the FFT-based DWT algorithm is about $L/(2 \log_2 L)$. However, as seen in Table 8-1, the FFT implementation of the DWT is not effective for short filters.

There is a subtlety to keep in mind when wraparound effects at the cell output are eliminated in the time-domain. One could immediately take the output blocks

TABLE 8-1:   FFT-Based DWT Algorithms: Arithmetic Complexity Per Point and Per Octave

Filter Length	Inner Product Filter Bank	FFT-Based Algorithm	Vetterli, 2 Octaves Merged	Vetterli, 3 Octaves Merged	Vetterli, 4 Octaves Merged
2	2 + 1	3 + 6 (2)	3.17 + 5.83 (2)	3.07 + 6.07 (4)	3.17 + 6.17 (4)
4	4 + 3	4 + 9.33 (4)	4.56 + 10.97 (16)	5.17 + 12.43 (32)	5.58 + 14.00 (128)
8	8 + 7	5.23 + 14.15 (16)	5.68 + 14.67 (64)	6.10 + 15.33 (128)	6.61 + 16.90 (256)
16	16 + 15	6.56 + 18.24 (32)	6.61 + 17.41 (128)	6.88 + 18.10 (512)	7.25 + 19.06 (1024)
32	32 + 31	7.92 + 22.37 (64)	7.50 + 20.05 (256)	7.56 + 20.14 (1024)	7.90 + 21.01 (2048)
64	64 + 63	9.12 + 26.20 (256)	8.25 + 22.55 (1024)	8.23 + 22.13 (2048)	8.54 + 22.90 (4096)
128	128 + 127	10.27 + 29.67 (512)	9 + 24.79 (2048)	8.89 + 24.10 (4096)	9.16 + 24.76 (8192)

Each entry gives the number of operations per input or output point in the form *mults* + *adds*, and the initial FFT length. Complexities should be multiplied by $2(1 - 2^{-J})$ for a computation of the DWT on J octaves.

TABLE 8-2:   Arithmetic Complexity Per Point and Per Cell: DWT
Algorithms

Filter Length L	Straightforward Filter Bank	FFT-Based Algorithm	Short Length Algorithm
4	$4 + 3$	$4 + 9.33$ (4)	$3 + 4$ (2)
6	$6 + 5$	$4.67 + 12$ (8)	$4 + 6.3$ (3)
8	$8 + 7$	$5.23 + 14.15$ (16)	$4.5 + 8.5$ (2×2)
10	$10 + 9$	$5.67 + 15.33$ (16)	$4.8 + 14.2$ (5)
12	$12 + 11$	$6.18 + 16.73$ (16)	$6 + 12$ (2×3)
16	$16 + 15$	$6.56 + 18.24$ (32)	$9 + 13$ (2×2)
18	$18 + 17$	$6.83 + 19$ (32)	$8 + 17$ (3×3)
20	$20 + 19$	$7.13 + 19.83$ (32)	$7.2 + 21.4$ (5×2)
24	$24 + 23$	$7.32 + 20.68$ (64)	$12 + 18$ (2×3)
30	$30 + 29$	$7.76 + 21.92$ (64)	$9.6 + 27$ (5×3)
32	$32 + 31$	$7.92 + 22.37$ (64)	$18 + 22$ (2×2)

Each entry gives the number of operations per input or output point in the form *mults* + *adds*, and either the FFT length or the type of fast-running FIR algorithm used. Complexities should be multiplied by $2(1 - 2^{-J})$ for a computation of the DWT on J octaves.

(now of length $B/2$ instead of B) as inputs to the next cell, but this would halve the block length at each stage. This method is not effective eventually because the FFT is most efficient for an optimized value of the block length B (at fixed filter length L). It is therefore advisable to work with the same optimized degree of efficiency at each cell, by waiting for another block before entering the next cell, so that each cell has the same input block length B and FFT length N. This method involves strictly identical cells: they not only have the same computational structure, but they also process blocks of equal length. As usual, the resulting total complexity of the DWT is $2(1 - 2^J)$ times the complexity of one cell, as shown in section 8.7.1.

8.8.2 A Generalization: The Vetterli Algorithm

The FFT-based DWT algorithm just described can be improved by gathering J_0 consecutive stages, using a method due to Vetterli (originally in the filter bank context [25], and then applied to the computation of the DWT [12]). The idea is to avoid subsequent IFFT's and FFT's by performing the subsampling operation in the frequency domain. This is done by inverting the last stage of a decimation-in-time radix-2 FFT algorithm. The FFT length is then necessarily halved at each DWT stage, whereas the filter lengths remain constant, equal to $L/2$.

Unfortunately, this class of algorithms has two major limitations. First, the structure of computations is less regular than for the simple FFT algorithm of the preceding section because FFTs have different lengths. Second, the relative efficiency of an FFT scheme per computed point decreases at each stage.

Table 8-1 lists the resulting complexities for $J_0 = 2, 3$, and 4, minimized against $N = 2^n$. Vetterli algorithms are more efficient than the initial FFT-based computation of the DWT ($J_0 = 1$) only for long filters ($L \geq 32$) and small J_0. Efficiency is lost in any case when J_0 is greater than 3.

8.8.3 DWT Algorithms for Short Filters

We have seen that for small filter lengths ($L < 16$), FFT-based algorithms do not constitute an improvement compared to the initial filter bank computation. Therefore, it is appropriate to design a specific class of fast algorithms for short filters. Here, "fast running FIR" algorithms [23] are applied to the DWT computation. The class of "fast running FIR algorithms" is interesting because the multiply/ accumulate structure of computations is partially retained, hence these algorithms are very efficiently implemented [23].

A detailed description of fast running FIR algorithms can be found in [23]. Basically, a filter of length L is implemented as follows. The involved sequences (input, output, and filters) are separated into subsequences, decimated with some integer ratio R. Assuming L is a multiple of R, filtering is done in three steps:

1. The input is decimated and the resulting R sequences are suitably combined, requiring A_i additions per point, to provide M subsampled sequences.
2. The resulting sequences serve as inputs to M decimated subfilters of length L/R.
3. The outputs are recombined, with A_o additions per point, to provide the exact decimated filter outputs.

Fig. 8-7 provides an example for $R = 2$, $A_i = 2$, $M = 3$, and $A_o = 2$. Other algorithms derived in [23] were also applied, corresponding to $R = 3$ and $R = 5$.

This computation can be repeated: the subfilters of length L/R are still amenable to further decomposition. For example, in order to implement a 15-tap filter, one can either use a fast running FIR algorithm for $R = 3$ or $R = 5$, or decompose this filter by a "3 × 5 algorithm," which first applies the procedure with $R = 3$, then again decomposes the subfilters using the procedure associated with $R = 5$. Alternatively, a "5 × 3 algorithm" can be used. Each of these algorithms yields different complexities, which are discussed in detail in [23]. The short-length DWT algorithm is derived as follows. One applies fast running FIR algorithms to the four filters of length $L/2$ in the elementary cell of the DWT (Fig. 8-5). Here, since two

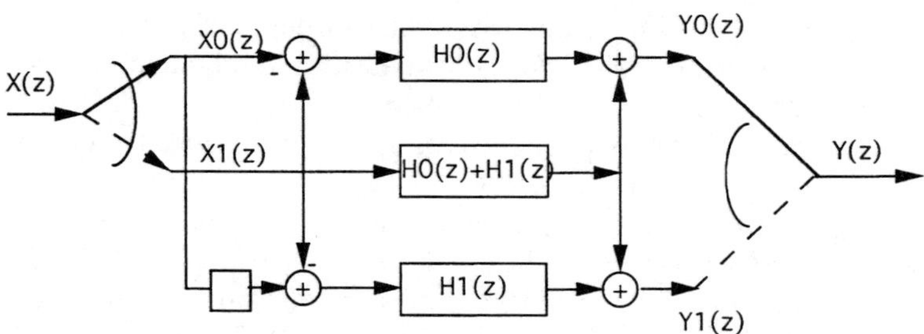

Figure 8-7   Simple example of fast-running FIR filtering algorithm with decimation ratio $R = 2$ [22]. Subscripts 0 and 1 indicate biphase decomposition.

pairs of filters share the same input, all preadditions can be combined together on a single input.

Table 8-2 lists the resulting complexities, using the fast running FIR algorithm that minimizes the criterion (multiplications + additions). When two different decompositions yield the same total number of operations, we have chosen the one that minimizes the number of multiplications. Table 8-2 shows that short-length DWT algorithms are more efficient than the FFT-based DWT algorithms for lengths up to $L = 18$.

Since, in practice, DWTs are generally computed using short filters [8,9], the short-length algorithms probably give the best practical alternative when heavy DWT computation is required. As an example, for $L = 18$, the short-length algorithm requires a total of 25 operations per point instead of 35 for the direct method.

8.8.4 Other Considerations

- *The Orthogonal Case:* In our derivations, we did not take advantage of orthogonality constraints [5,8–11,13] so as to be as general as possible. However, orthogonality is worthy of consideration because of its simplicity: the analysis and synthesis filters coincide (within time reversal and complex conjugation). Furthermore, it allows one to further reduce the complexity of the DWT: Using a lattice implementation of the DWT filter bank cell of Fig. 8-1(a), Vaidyanathan has shown [17] that the complexity can be reduced by a factor of 50% in the orthogonal case.

  Whether or not this reduction can be attained while preserving the inner products (unlike the lattice structure implementation) is an open problem. In any case, Tables 8-1 and 8-2 do not provide a fair and detailed comparison between various algorithms in the orthogonal case.

- *Unequal Filter Lengths:* In the previous derivations, we have restricted ourselves to filters of equal lengths for simplicity. However, it may happen that one uses a low-pass interpolation filter $g(n)$ of small length ($L_g \ll 16$) and a very long high-pass filter $h(n)$ of length $L_h \gg 16$. This is the case in [3,5], where one typically uses a first-order interpolation filter $g(n)$ ($L_g = 3$) to approximate the "Morlet wavelet," a modulated Gaussian.

Obviously, for a direct implementation of the DWT filter bank, it is in this case absurd to assume equal filter lengths since the complexity then becomes $(L_g + L_h)/2$ mults and $(L_g + L_h - 1)/2$ adds.

However, FFT-based DWT algorithms are still efficient when one of the filters is very long. In this case, some efficiency of FFT-based algorithms is lost, but they still yield a substantial gain over a standard, straightforward filter bank implementation of the DWT. As an example, for a wavelet of length $L_h = 64$ and interpolation filters of length $L_g = 3, 7$, and 11, the FFT-based algorithms give respective gains over a standard DWT of 46.9%, 49.9% and 52.6%.

- *Linear phase:* In the previous discussion, we did not take other properties of filters into account, such as the linear phase property, which holds for the Morlet wavelet. In this case, rather than using involved fast algorithms, we recommend straightforward use of the symmetry in the inner product implementation of the algorithms, which cuts by 2 the number of multiplications.

8.8.5 Faster CWT Algorithms

The same fast convolution tools can be applied on the CWT, slightly modified building block described in section 8.6.2. The main difference is that the filters involved are comparatively twice as long as in the WS case, due to the absence of decimation. This increases the efficiency of the "faster" algorithms. Being applications of the same techniques, they are not described, but the arithmetic complexities are given in Table 8-3, in order to allow the reader to evaluate their potential compared to straightforward inner product implementation.

8.9. OTHER ALGORITHMS FOR CWT COMPUTATION

Several algorithms for computing CWT coefficients, which differ notably from those already described, have been proposed recently (see e.g., [1,16,26–28]). Several of them are outlined in this section.

8.9.1 Reproducing Kernels

Gopinath and Burrus [16] proposed a method that also uses DWTs. The signal is assumed to be completely determined from its WS coefficients. Therefore, these alone can be used to compute all CWT coefficients by some reproducing kernel equation. The introduction of an auxiliary wavelet moreover allows one to precompute the kernel and to obtain a method particularly suited to the computation of CWT coefficients with respect to several wavelets. However, the kernel expansion in [16] seems to be computationally expensive.

TABLE 8-3:   Arithmetic Complexity Per Computed Point for Various CWT Algorithms

Filter Length L	Straightforward Filter bank	FFT-Based Algorithm	FFT-Based (2 Octaves Merged)	Short Length Algorithm
2	4 + 2	4 + 10 (4)	4.8 + 12 (16)	3 + 3 (2)
3	6 + 4	5 + 14 (8)	5.8 + 15.2 (32)	4 + 5.3 (3)
4	8 + 6	6 + 16.8 (8)	6.5 + 17.2 (32)	4.5 + 7.5 (2 × 2)
5	10 + 8	6.5 + 19 (16)	6.9 + 18.7 (64)	4.8 + 13.2 (5)
6	12 + 10	7.1 + 20.7 (16)	7.3 + 19.8 (64)	6 + 11 (2 × 3)
8	16 + 14	7.9 + 23.5 (32)	7.8 + 21.6 (128)	9 + 12 (2 × 2)
9	18 + 16	8.2 + 24.5 (32)	8.1 + 22.3 (128)	8 + 16 (3 × 3)
10	20 + 18	8.6 + 25.6 (32)	8.3 + 22.9 (128)	7.2 + 20.4 (5 × 2)
12	24 + 22	9.2 + 27.4 (64)	8.6 + 24.2 (256)	12 + 17 (2 × 3)
15	30 + 28	9.7 + 29 (64)	9 + 25.2 (256)	9.6 + 26 (5 × 3)
16	32 + 30	9.9 + 29.6 (64)	9.1 + 25.5 (256)	18 + 21 (2 × 2)
18	36 + 34	10.3 + 30.9 (64)	9.4 + 26.3 (256)	16 + 24 (3 × 3)
20	40 + 38	10.6 + 31.8 (128)	9.6 + 27 (512)	14.4 + 27.6 (5 × 2)
24	48 + 46	11 + 33 (128)	9.8 + 27.8 (512)	24 + 29 (2 × 3)
25	50 + 48	11.1 + 33.3 (128)	9.9 + 27.9 (512)	11.5 + 44.9 (5 × 5)
27	54 + 52	11.3 + 34 (128)	10 + 28.3 (512)	24 + 32 (3 × 3)
30	60 + 58	11.7 + 35 (128)	10.2 + 28.9 (512)	19.2 + 35.6 (5 × 3)
32	64 + 62	11.9 + 35.7 (128)	10.4 + 29.4 (512)	36 + 39 (2 × 2)
64	128 + 126	13.7 + 41.1 (512)	11.6 + 33.1 (2048)	72 + 75 (2 × 2)
128	256 + 254	15.4 + 46.2 (1024)	12.7 + 36.4 (4096)	144 + 147 (2 × 2)

Each entry gives the number of operations per computed coefficient (i.e., per input point per octave) in the form *mults + adds*, and either the FFT length or the type of fast-running FIR algorithm used.

8.9.2 Algorithms Using Splines

We have already emphasized the importance of splines for WT computation in section 8.2.2. In fact, there is another remarkable property which makes them useful for CWT computation: B-splines of degree d follow a generalized two-scale difference equation (8-12) (generalized to an m-scale equation), valid for any $m > 0$ if d is odd, and m odd only if d is even:

$$\beta^d(t/m) = \sum_{k \in Z} c_k^{m,d} \beta^d(t - k). \tag{8-53}$$

with $c_k^{m,d}$ defined (by identification) as

$$\sum c_k^{m,d} z^{-k} = \frac{z^{(d+1)(m+1)/2}}{m^d} \left(\sum_{k=0}^{m-1} z^{-k} \right)^{d+1} \tag{8-54}$$

This has led Unser *et al.* [28], following a generalization of Shensa's algorithm, to use B-splines in order to compute a DTWT in which the scale parameter a can take any integer value. While this property is the key to an increased flexibility, the fast algorithm is obtained following the same steps as in the Shensa algorithm:

- First, approximate the input signal as its spline approximations of degree d_1:

$$x(t) = \sum_{k \in \mathbf{Z}} x_k \beta^{d_1}(t - k) \tag{8-55}$$

- Then, specify the wavelet by its B-spline expansion of degree d_2:

$$\psi(t) = \sum_{k \in \mathbf{Z}} p_k \beta^{d_2}(t - k) \tag{8-56}$$

Thus due to the generalized two-scale difference equation, the wavelet, when expanded by a factor m, can be expressed as

$$\psi(t/m) = \sum_{k \in \mathbf{Z}} (\{[p]_{\uparrow m}\} * \{c^{m,d_2}\})(k) \beta^{d_2}(t - k) \tag{8-57}$$

where $(\{[p]_{\uparrow m}\} * \{c^{m,d_2}\})(k)$ denotes the kth term of the convolution of sequences p_l, as defined in (8-56), upsampled by a factor m, and of sequence c_l^{m,d_2}, as defined in (8-54).

- Finally, the CWT of $x(t)$ at scale m is given by

$$\text{CWT}\{x(t), m, b\} = \sum_{k \in \mathbf{Z}} (\{[p]_{\uparrow m}\} * \{c^{m,d_2}\} * \{x_k\})(k) \beta^{d_1+d_2+1}(b - k) \tag{8-58}$$

which, when evaluated at integer time samples, simplifies to:

$$\text{CWT}\{x(t), m, k\} = (\{[p]_{\uparrow m}\} * \{c^{m,d_2}\} * \{b^{d_1+d_2+1}\} * \{x_k\})(k) \tag{8-59}$$

where $\{b^{d_1+d_2+1}\}$ is the discrete B-spline of order $d_1 + d_2 + 1$.

The filter bank at work in the algorithm has very simple low-pass filters owing to the special structure of B-splines. As seen from (8-54), they are iterated discrete convolutions of moving sums, and therefore can be computed without any multiplication. This remarkable feature thus results in very efficient algorithms.

8.9.3 Mellin-Transform–Based Algorithms

Another beautiful CWT algorithm, which uses the scaling property of wavelets $\psi(t) \to a^{-1/2}\psi(t/a)$ rather than the convolutional form of (8-1), (8-2) has been proposed by Bertrand et al. [1]. This algorithm makes use of some redundancy between the computations of the various scales of a signal around some time location, while the previously described algorithms make use of redundancy between the computations of several successive outputs of the same scale.

This algorithm is briefly outlined here. Write (8-2) in the frequency domain, assuming that the signal $x(t)$ and wavelet $\psi(t)$ are complex analytic. This gives

$$\text{CWT}\{x(t); a, b\} = \int_0^\infty X(f)e^{2i\pi f b}\sqrt{a}\psi^*(af)df \tag{8-60}$$

where $X(f) = \int x(t)e^{-2i\pi f t}dt$ and $\psi(f)$ are the Fourier transforms of $x(t)$ and $\psi(t)$, respectively. Then perform the changes of variable $\phi = \ln f$. A correlation form in $\alpha = \ln a$ appears in the integral.

$$\text{CWT}\{x(t); a, b\} = \int_R X(e^\phi)e^{\phi/2}e^{2i\pi e^\alpha b}\psi(e^{\alpha+\phi})e^{\frac{\alpha+\phi}{2}}d\phi \tag{8-61}$$

After suitable discretization, this correlation can be performed using an FFT algorithm. As stated in [1], the Mellin transform, $M_x(\beta)$ of $x(t)$, plays a central role, since it turns out to be exactly the inverse Fourier transform of $\sqrt{f}X(f)$ in the variable $\phi = \ln f$:

$$M_x(\beta) = \int_{f>0} X(f)f^{-1/2+2i\pi\beta}df \tag{8-62}$$

$$= \int e^{\phi/2}X(e^\phi)e^{2i\pi\beta\phi}d\phi \tag{8-63}$$

As a result, the FFTs involved in the computation of (8-61) are "discrete Mellin transforms," as defined in [1].

This algorithm requires the precomputation of the whole Fourier transform of $x(t)$, which makes a running implementation (in case of infinite duration signals) cumbersome. To overcome this difficulty, we propose a variation on the Bertrands–Ovarlez algorithm, based on the time domain rather than on the frequency domain. Assume that the signal and wavelets are causal (i.e., supported by $t \geq 0$), and make the change of variable $\tau = \ln t$ in (8-2). One obtains a convolution in $\alpha = \ln a$:

$$\text{CWT}\{x(t); a, b\} = \int e^{\tau/2}x(e^\tau + b)e^{(\tau-\alpha)/2}\psi^*(e^{\tau-\alpha})d\tau \tag{8-64}$$

The CWT coefficients are obtained, for a given b, by discretizing the convolution (8-64), resulting in a discrete filtering operation that can be implemented for running data.

Both algorithms (8-61), (8-64) have common characteristics. Some of them can be considered as drawbacks: First, they involve a geometric sampling of either $X(f)$ or $x(t)$. Second, the approximation error made by discretizing (8-61) or (8-64) is difficult to estimate. Finally, in contrast to the octave-by-octave CWT implementation previously described, the time shift structure of b has completely disappeared, and the input has to be recomputed for each value of b. As a result, the complexity of such algorithms (about two FFTs of length $2JM$ per input point, where J is the number of octaves and M is the number of voices per octave) is found higher than the one obtained for the more classical algorithms described earlier.

However, a nice property of the Mellin-based algorithms is that the CWT coefficients are computed for all desired values of $\ln a$ at the same time (for given value of b), while the efficiency of the classical algorithms requires the computation of long signals. It makes the Bertrands-Ovarlez algorithms very useful when a "zoom," or a refinement, of the wavelet analysis in a short extent around some time location b is desired.

8.10. CONCLUSION

This chapter has reviewed several methods for efficiently implementing various kinds of wavelet transforms, from the fully discrete version to the fully continuous one, and for any type of wavelet.

Emphasis has been put on the various approximations required for the algorithms to be efficient, and on their link with multiresolution analysis. As a result, prefiltering the signal allows one to use the DWT as an intermediate computation for any type of wavelet transform. Guidelines were given for the design of the appropriate prefilter.

Fast DWT algorithms were derived for computing WS coefficients and were modified to compute wavelet coefficients with oversampling in the time-scale plane ("CWT algorithms").

While the inner product implementation of these transforms is already efficient, a further improvement has been obtained by using fast convolution algorithms, adapted to the situation. The availability of both FFT-based and fast-running-FIR-based algorithms allows one to reduce the complexity of the existing algorithms in any case of interest. Tables are provided for the reader to evaluate whether the decrease in computation is worth the complexity of the implementation.

Other fast algorithms were also outlined, either using splines, or using discrete Mellin transforms, each one offering specific advantages: The splines-based algorithms can easily approximate some given wavelet, while still allowing a fast implementation. Mellin-based transforms are more suited to the situation where one is able to sample the signal in a geometric manner (either in the time or in the frequency domain), in which case the redundancy between all scales can efficiently be exploited.

REFERENCES

[1] J. Bertrand, P. Bertrand, and J. P. Ovarlez, "Discrete Mellin transform for signal analysis," in *Proc. 1990 IEEE Int. Conf. Acoust., Speech, Signal Processing*, Albuquerque, NM, April 3–6, 1990, pp. 1603–1606.

[2] P. Goupillaud, A. Grossmann, and J. Morlet, "Cycle-octave and related transforms in seismic signal analysis," *Geoexploration*, vol. 23, pp. 85–102, 1984/85.

[3] M. Holschneider, R. Kronland-Martinet, J. Morlet, and Ph. Tchamitchian, "A real-time algorithm for signal analysis with the help of the wavelet transform," in [5], pp. 286–297.

[4] M. Vetterli and J. Kovačević, *Wavelets and Subband Coding*, Englewood Cliffs, NJ: Prentice Hall, 1995.

[5] J. M. Combes, A. Grossmann, and Ph. Tchamitchian, Eds., *Wavelets, Time-Frequency Methods and Phase Space*, Berlin: Springer, IPTI, 1989.

[6] Y. Meyer Ed., *Wavelets and Applications*, Paris: Masson/Berlin: Springer Verlag, 1992.

[7] O. Rioul and P. Duhamel, "Fast algorithms for discrete and continuous wavelet transforms," *IEEE Trans. Inform. Thoery*, vol. 38, pp. 569–586, March 1992.

[8] S. Mallat, "A theory for multiresolution signal decomposition: The wavelet representation," *IEEE Trans. Pattern Anal. Machine Intell.*, vol. 11, pp. 674–693, July 1989.

[9] S. Mallat, "Multifrequency channel decompositions of images and wavelet models," *IEEE Trans. Acoust., Speech, Signal Process*, vol. 37, pp. 2091–2110, December 1989.

[10] I. Daubechies, "Orthonormal bases of compactly supported wavelets," *Comm. Pure Applied Math.*, vol. 41, no. 7, pp. 909–996, 1988.

[11] Y. Meyer, *Ondelettes et Operateurs*, Tome 1. Paris: Herrmann, 1990.

[12] M. Vetterli and C. Herley, "Wavelets and filter banks: Theory and design," *IEEE Trans. Acoust., Speech, Signal Process*, vol. SP-40, pp. 2207–2232, 1992.

[13] G. Evangelista, "Orthogonal wavelet transforms and filter banks," presented at Proc. 23rd Asilomar Conf., IEEE, November 1989.

[14] A. Cohen, I. Daubechies, and J. C. Feauveau, "Biorthogonal bases of compactly supported wavelets," *Comm. Pure Applied Math.*, vol 45, pp. 485–560, 1992.

[15] M. J. Shensa, "Affine wavelets: Wedding the Atrous and Mallat algorithms," *IEEE Trans. Signal Proc.*, vol. 40, pp. 2464–2482, October 1992.

[16] R. A. Gopinath and C. S. Burrus, "Efficient computation of the wavelet transforms," in *Proc. IEEE Int. Conf. Acoust., Speech, Signal Processing*, Albuquerque, NM, April 3–6, 1990, pp. 1599–1601 .

[17] P. P. Vaidyanathan, *Multirate Systems and Filter Banks*, Englewood Cliffs, NJ: Prentice Hall, 1993.

[18] P. Abry and P. Flandrin, "On the initialization of the discrete wavelet transform algorithm," *IEEE Sig. Proc. Letters*, vol. 1, pp. 32–34, February 1994.

[19] M. Unser and A. Aldroubi, "A general sampling theory for non ideal acquisition devices," *IEEE Trans. Signal Proc.*, vol. 42, pp. 2915–2925, November 1994.

[20] M. J. T. Smith and T. P. Barnwell, "Exact reconstruction for tree-structured subband coders," *IEEE Trans. Acoust., Speech, Signal Process*, vol. ASSP-34, pp. 434–441, June 1986.

[21] I. Daubechies and J. C. Lagarias, "Two-scale difference equations 1. Existence and global regularity of solutions," *SIAM J. Math. Anal.*, vol. 22, no. 5, pp. 1388–1410, September 1991.

[22] T. A. Ramstad and T. Saramah, "Efficient multirate realization for narrow transition-band FIR filters," in *IEEE 1988 Int. Symp. Circ. Syst.*, 1988, pp. 2019–2022.

[23] Z. J. Mou and P. Duhamel, "Short length FIR filters and their use in fast nonrecursive filtering," *IEEE Trans. Signal Proc.*, vol. 39, pp, 1322–1332, June 1991.

[24] H. J. Nussbaumer, *Fast Fourier Transform and Convolution Algorithms*. Berlin: Springer, 1981.

[25] M. Vetterli, "Analyse, Synthese et Complexité de Calcul de Bancs de Filtres Numériques," Ph.D. thesis, Ecole Polytechique Federale de Lausanne, 1986.

[26] D. L. Jones and R. G. Baraniuk, "Efficient computation of densely sampled wavelet transforms," in *Advanced Signal-Processing Algorithms, Architectures, and Implementations II*, F. T. Luk (ed.), *Proc. SPIE 1566*, San Diego, CA, July 1991.

[27] M. Unser, "Fast Gabor-like windowed Fourier and continuous wavelet transforms," *IEEE Signal Proc. Letters*, vol. 1, pp.76–79, May 1994.

[28] M. Unser, A. Aldroubi, and S.J. Schiff, "Fast implementation of the continuous wavelet transform with integer scales," *IEEE Trans. Signal Proc.*, vol. 42, pp. 3519–3523, December 1994.

Chapter 9

Analysis of Cellular Vibrations in the Living Cochlea Using the Continuous Wavelet Transform and the Short-Time Fourier Transform

M. C. Teich, C. Heneghan, S. M. Khanna

9.1. INTRODUCTION

In the process of hearing, sound waves travel to the eardrum (tympanic membrane) through the external ear and ear canal. The sound pressure acting on the tympanic membrane produces mechanical vibrations that are transmitted, via the ossicular chain in the middle ear, to the inner ear (cochlea). The cochlea, which is encased in a bony shell, consists of three fluid-filled canals: scala vestibuli, scala media, and scala tympani. A thin membrane (Reissner's membrane), running the length of the cochlea, separates the scala vestibuli from the scala media (middle canal). The basilar membrane forms the base of the middle canal, separating it from the scala tympani. The cochlea is coiled; it has a diameter that is widest at the base and narrowest at the apex. There are two openings in the bony shell near the base: (1) the oval window, through which the stapes drives the fluid in the scala vestibuli, and (2) the round window, which is covered by a thin membrane that accommodates the movement of fluid in the cochlea. The sensory organ of hearing (the organ of Corti) is located on the scala media side of the basilar membrane. It consists of several types of specialized cells that are organized in precise transverse and longitudinal arrangements. The transverse morphological arrangement is the same from base to apex, though the width and stiffness of the basilar membrane and the dimensions of most of the cells change over this region [1,2].

The velocity of vibration of individual cells, selected as desired, can be measured with a specially designed confocal heterodyne interferometer in response to sound applied to the ear canal. The details of the stimulus-generation and measuring techniques are described elsewhere [2]. We have devoted a good deal of attention to describing the character of the dynamical vibrations of individual sensory cells in the third and fourth turns of the guinea-pig cochlea using a temporal-bone preparation excised from a freshly sacrificed animal and maintained in oxygenated tissue-culture medium [3–13]. In this chapter, we turn to a similar analysis of the recordings from the basal cochlear turn of the living cat [14].

A suitable method for studying level-dependent changes in the dynamical response of these cells is to use sinusoidal-carrier amplitude-modulated (AM) acoustic waves with low-frequency modulation. This provides an opportunity for studying the change in nonlinear dynamical response over a broad range of carrier levels, as the envelope slowly increases and decreases. The AM format is also useful because the heterodyne interferometer can measure the velocity of an object but not its absolute position.

This chapter describes the application of time–scale and time–frequency representation techniques in the analysis of cellular velocity data. The relative advantages of the two techniques are compared for several data sets. The modulation depth of the AM acoustic signal was unity and the modulation frequency was 6.1 Hz. The carrier frequency ranged from 1000 Hz to 40000 Hz, and the total duration of each data set collected was fixed at 0.16384 s (representing 16384 samples at 10-μs intervals).

In general, we also record the cellular vibration in the absence of acoustic stimuli. Using this technique we have already established the presence of spontaneous vibrations in the third turn of the guinea-pig cochlea [8,9]. We have seen similar spontaneous vibrations in the basal turn of the cat cochlea [15]. However, since the amplitude of the spontaneous components is approximately time-invariant over the period of recording, there is no particular advantage to using time-frequency or time-scale methods as compared to a purely spectral technique such as the periodogram. Accordingly, we will not discuss these results further in this chapter.

We examined both the continuous wavelet transform (CWT) and the short-time Fourier transform (STFT) of the velocity responses elicited by the AM stimuli just described. Both analysis techniques were useful in discriminating the frequency components present in the responses, though the wavelet basis for the CWT had to be carefully chosen to provide the desired frequency resolution. CWTs using a high-Q Morlet wavelet basis were found to be particularly useful for discriminating among the various response components. Octave-band–based CWTs (using low-Q Morlet, Meyer, and Daubechies 4-tap wavelets) were largely ineffective in analyzing these signals, inasmuch as their frequency resolution was too poor to distinguish among the frequency components present in the velocity responses.

9.2. METHODS

Optical access to the basal turn of the cochlea was obtained by carefully opening the cat's bulla. The measurements shown here were made after the removal of the round-window membrane. A Nikon SLWD 20X objective lens with a 19.9-mm working distance was used to view the cochlea with an optical sectioning microscope. A laser interferometer was coupled with this optical sectioning microscope, in such a way that their planes of focus coincided. This allowed identification of the structure lying at the focus of the microscope/interferometer. The 2-μm-diameter laser spot on the object, seen through the microscope eyepiece, was placed on a selected cell by moving the animal's head with an x-y-z micropositioning system [2]. The sectioning depth was 10 μm so that vibrations were recorded from small regions of single cells. The interferometer measures the vibration of the object on which it is focused [2].

The measurements were carried out in a soundproof chamber. Sound generated by a high-fidelity acoustic driver was applied to the ear canal. The electrical signal applied to the acoustic driver was generated using a 486-microprocessor–based computer system coupled to a 16-bit D/A converter. The signal from the interferometer was passed through an antialiasing filter, a 16-bit A/D converter, and then stored in the hard disk of the computer system. The same system can be used to measure cellular vibration in the absence of a stimulus by simply disconnecting the sound stimulus system.

9.3. THEORY

9.3.1 The Continuous-Time Fourier Transform

Signal analysis addresses the problem of extracting information from a given signal $x(t)$ and converting it into a recognizable form. One approach to this problem is to transform $x(t)$, using an information-preserving mapping, to a different domain (viz., a dual domain), where it is easier to interpret the signal. The best-known of these approaches, perhaps, is the continuous-time Fourier transform (CFT). The relations between $x(t)$ (where t is chosen to represent time) and its dual representation $X(f)$ are

$$X(f) = \int_{-\infty}^{\infty} x(t)\exp(-j2\pi ft)\, dt \tag{9-1}$$

$$x(t) = \int_{-\infty}^{\infty} X(f)\exp(j2\pi ft)\, df \tag{9-2}$$

$X(f)$ is referred to as a spectral representation of $x(t)$, with the dual variable f defined as (global) frequency.

The CFT reveals how the energy in the signal $x(t)$ is distributed in frequency. One limitation of the CFT is that the value of $X(f)$ is affected by all values of $x(t)$ from $t = -\infty$ to $+\infty$. As a result, any particular feature in $X(f)$ cannot be linked with a specific time region of $x(t)$. The CFT provides a totally global perspective on how a signal's energy is distributed as a function of frequency; in other words, $X(f)$ is a completely nonlocal spectral representation of $x(t)$.

9.3.2 The Short-Time Fourier Transform

In many cases, the CFT is a most useful representation, particularly if $x(t)$ is stationary, or "steady," in time (for a discussion of the notion of stationarity, see

[16]). However, for many signals, the nature of $x(t)$ changes with time. For example, if $x(t)$ represents the vertical motion of a point on the wheel of an automobile over the course of a journey, its frequency of motion will vary depending on whether the car is moving at constant speed, accelerating, decelerating, or at rest. The CFT of this signal tells us about the range of frequencies that the motion achieved over the course of the entire journey, but it fails to provide information about the time order in which they occurred. A representation of $x(t)$ that tells us *when* a particular frequency was present provides a more useful account of the journey. To achieve this, a spectral representation that includes some explicit dependence on time is needed; a function of the form $X(f, \tau)$, where f again represents frequency and τ represents time, would be useful.

The first attempt to construct such a function was carried out by Gabor in 1946 [17]. His approach retained the frequency variable f defined by the CFT, but ensured that only values of $x(t)$ in the near vicinity of $t = \tau$ would be able to influence $X(f, \tau)$. He achieved this by multiplying the original signal $x(t)$ by window functions that are localized in time at $t = \tau$. In this way, he constructed a local spectral representation of the signal in the vicinity of time τ.

Gabor's approach was later shown to be a special case of the STFT [18,19], which is expressed as

$$\text{STFT}_x^g(f, \tau) = \int_{-\infty}^{\infty} x(t)g^*(t - \tau)\exp(-j2\pi ft)\, dt \tag{9-3}$$

Here, t and τ are time variables, $x(t)$ represents the time waveform being analyzed, $g(t)$ represents a window function, f is the frequency variable, and the superscript star denotes complex conjugation. The Gaussian $[g(t) = \exp(-\beta t^2/2)]$ is a typical choice for the window function since it falls smoothly and symmetrically to zero around the time $t = 0$. Accordingly, the function $g^*(t - \tau)$, which in this case is equal to $g(t - \tau)$, is centered about the time $t = \tau$, and falls away quickly to zero for times away from τ. It is apparent that in the absence of a window function $[g(t) = 1]$, the STFT in Eq. (9-3) reduces to the CFT given in Eq. (9-1).

There are several alternative ways of expressing the STFT that are useful in different circumstances. For example, it can be written as an integral in the frequency domain, viz.,

$$\text{STFT}_x^g(f, \tau) = \exp(-j2\pi f\tau)\int_{-\infty}^{\infty} X(u)G^*(u - f)\exp(j2\pi u\tau)\, du \tag{9-4}$$

where $X(u)$ and $G(u)$ represent the CFTs of $x(t)$ and $g(t)$, respectively, and u is a dummy frequency variable. Here, $\text{STFT}_x^g(f, \tau)$ is seen to be a frequency-shifted version of the inverse CFT of $[X(u)G^*(u - f)]$ [compare with Eq. (9-2)]. If the function $G(u)$ is taken to represent a low-pass filter in frequency, then $X(u)G^*(u - f)$ is the CFT of $x(t)$ after filtering by a bandpass filter whose shape is simply that of $G(u)$, translated in frequency so that it is centered about f instead of 0. The factor $\exp(-j2\pi f\tau)$ in Eq. (9-4) simply frequency shifts the filtered output back down to zero frequency. The STFT can thus be viewed as the frequency-shifted output from a bank of filters $G^*(u - f)$, each with constant bandwidth but different center frequency. This filter-bank interpretation of the STFT [20–22] is illustrated in Fig. 9-1(a).

The STFT can also be written as a convolution in either the time domain:

$$\text{STFT}_x^g(f, \tau) = \exp(-j2\pi f\tau)[x(\tau) * g^*(-\tau)\exp(j2\pi f\tau)] \tag{9-5}$$

or in the frequency domain:

$$\text{STFT}_x^g(f, \tau) = X(f) * [G^*(-f)\exp(-j2\pi f\tau)] \tag{9-6}$$

where $*$ denotes the convolution integral operator [i.e., $u(s) = \int_{-\infty}^{+\infty} v(r)w(s - r)dr \equiv v(s) * w(s)$]. The convolution formalism represented in Eq. (9-5) identifies a limitation of the STFT. In calculating $STFT_x^g(f, \tau_0)$, the value of $x(t)$ at time $t = \tau_0$ is smeared over time by the convolution integral. Therefore, any sharp change in the value of $x(t)$ at time $t = \tau_0$ will not appear in the STFT solely at τ_0, but rather will be spread over a region of time in the vicinity of $\tau = \tau_0$. The range of time over which information is spread depends on the width (time duration) of $g^*(-t)\exp(j2\pi ft)$. (The width of a function can be defined in many ways [23]; in this chapter, we define it as the full-width at $1/e$-maximum, whether it be a time duration or a bandwidth. Alternative definitions of bandwidth would, of course, be acceptable for the transform properties we discuss provided they are dealt with consistently.)

According to Eq. (9-6), a sharp spectral feature at frequency $f = f_0$ will similarly be blurred by convolution with $G^*(-f)\exp(-j2\pi ft)$.

The ability to resolve fine features in either the time or frequency domains is referred to as the time or frequency resolution of the transform operation, respectively. From the discussion in the preceding paragraph, it is apparent that the time and frequency resolutions of the STFT are dependent on the widths of the functions $g(t)$ and $G(f)$. These widths are denoted by Δt and Δf, respectively, and it would be ideal if both of these quantities could go to zero.

This is not possible. To illustrate this, consider choosing $g(t)$ as short in time as possible. In the limit, we obtain $g(t) = \delta(t)$, where $\delta(t)$ is the Dirac delta function, which transforms to $G(f) = 1$. Substitution into Eq. (9-3) then shows that $\text{STFT}_x^g(f, \tau) = x(\tau)\exp(-j2\pi f\tau)$, which is simply the original signal $x(t)$ translated down in frequency. This STFT has therefore exactly preserved the time information in the signal $x(t)$, but it provides no frequency information whatsoever. This is because the width of the function $G(f)$ is infinite.

The inability to simultaneously access information at arbitrarily small values of Δt and Δf is an inherent property of the transform. In fact, by using the Schwarz inequality for any function $g(t)$, it can be shown that an uncertainty principle ensues [23], i.e., that $\Delta t \Delta f = C$, where C is a nonzero constant whose precise value depends on the definition of width that is selected. Thus once $g(t)$ is chosen, the time and frequency resolutions of the STFT are fixed for all values of t and f. This is shown schematically in Fig. 9-1(b) by drawing regions in the τ-f plane where a set of functions $g^*(\tau - \tau_i)\exp(-j2\pi f_j\tau)$ are concentrated, since it is functions of this form that set the time and frequency resolutions of the STFT. These regions are illustrated as rectangles of fixed area and dimensions for all values of τ_i and f_j, and are said to tile the time-frequency plane [20–22].

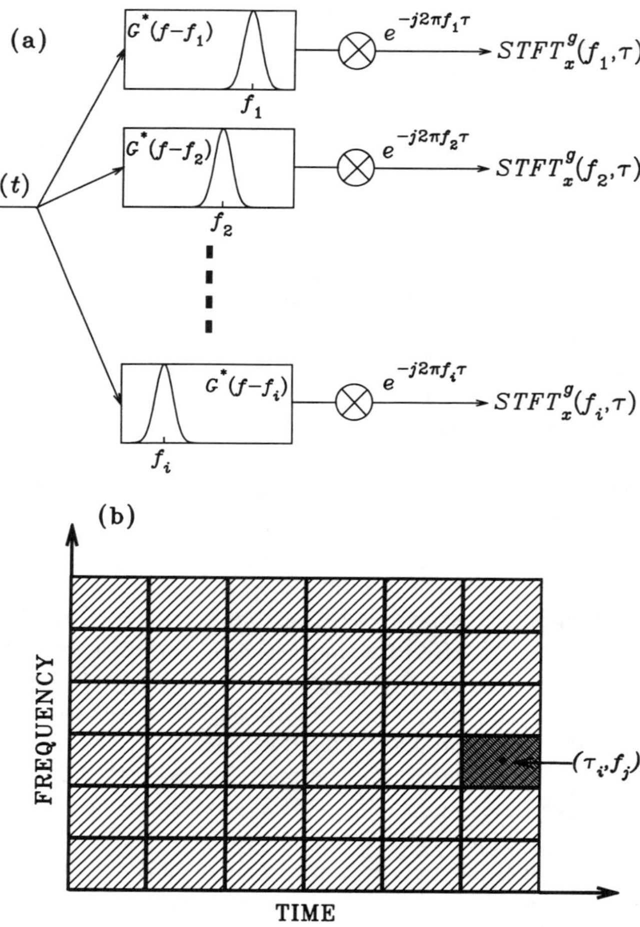

Figure 9-1   (a) Representation of the STFT in terms of filter-bank operations. The signal $x(t)$ to be analyzed is passed through a bank of filters, each with the same shape and bandwidth, but slightly different center frequencies f_i. The output from each filter is then multiplied by the factor $\exp(-j2\pi f_i \tau)$, which has the effect of shifting the output down to zero frequency and thereby providing the envelope. (b) Tiling of the time–frequency plane by the STFT. The rectangles centered at (τ_i, f_j) represent regions of the time–frequency plane where the functions $g^*(\tau - \tau_i)\exp(-j2\pi f_j \tau)$ are concentrated. These rectangles therefore also indicate the time and frequency resolution of the STFT.

9.3.3 The Continuous Wavelet Transform

A characteristic of the STFT is that both the time and frequency resolutions of the transform are fixed over the entire time–frequency plane. The time resolution Δt is fixed for the function $g(t)\exp(-j2\pi ft)$, whatever the value of f. As a result,

$\Delta f = C/\Delta t$ is also fixed over the entire time–frequency plane. In certain circumstances it is desirable to relax this restriction. Consider, for example, a signal with a mixture of short-lived high-frequency events that are closely spaced in time together with long-duration low-frequency components that are closely spaced in frequency. A suitable transform for this signal would have sufficient time resolution to distinguish the brief high-frequency events, and at the same time, enough frequency resolution to separate the closely spaced low-frequency components. These two aims are incompatible with the STFT since the time and frequency resolutions are both fixed.

One possible approach is to calculate two STFTs with different choices of $g(t)$: a short-lived $g(t)$ with a small value of Δt for good time resolution, and a long-lived $g(t)$ for good frequency resolution. An alternative solution is to use a representation that has variable time–frequency resolution over the (τ, f) plane, chosen in such a way that it provides good time resolution at high frequencies and good frequency resolution at low frequencies. One such representation is the continuous wavelet transform [19–22,24,25]. The CWT is expressed as

$$\mathrm{CWT}^h_x(r, \tau) = \frac{1}{\sqrt{|r|}} \int_{-\infty}^{\infty} x(t)h^*\left(\frac{t - \tau}{r}\right)dt \tag{9-7}$$

where t and τ are time variables, $x(t)$ is the time waveform being analyzed, $h(t)$ is the wavelet basis function, and r is a variable known as scale. As with the STFT, the CWT can also be expressed as an integral in the frequency domain [compare with Eq. (9-4)]:

$$\mathrm{CWT}^h_x(r, \tau) = \sqrt{|r|} \int_{-\infty}^{\infty} X(u)H^*(ur) \exp(j2\pi u\tau)du \tag{9-8}$$

or as a convolution in the time domain [compare with Eq. (9-5)]:

$$\mathrm{CWT}^h_x(r, \tau) = \frac{1}{\sqrt{|r|}} \left[x(\tau) * h^*\left(\frac{-\tau}{r}\right)\right] \tag{9-9}$$

where, as before, $X(u)$ and $H(u)$ denote the CFTs of $x(t)$ and $h(t)$, respectively.

The prefactor of $\sqrt{|r|}$ that appears in Eq. (9-8) illustrates that the standard CWT does not map equal-amplitude sinusoids of different frequencies to CWTs of the same magnitude; rather, it suppresses low-scale (high-frequency) components relative to those at high-scale (low-frequency). To facilitate comparison of the CWT results with those obtained with the STFT, it is useful to eliminate this difference. We therefore generally plot $|r|^{-1/2}|\mathrm{CWT}|$, which we refer to as the modified CWT. This has no effect on the time–frequency resolution characteristics of our analysis.

To understand how the CWT differs from the STFT, consider Eq. (9-9), which is the formulation of the CWT as a convolution in the time domain. As with the STFT, the value of $x(t)$ at $t = \tau_0$ is smeared over a time equal to the width of the function $h(\tau/r)$. In this case, however, the width of $h(\tau/r)$ is not fixed, but rather depends on the value of r. As an example, $h(2t)$ has half the width of $h(t)$, while $h(t/2)$ has twice the width of $h(t)$. The larger the value of r, the wider the function $h(\tau/r)$. Since time resolution depends on the width of this function, the following situation obtains: as r decreases, $h(\tau/r)$ becomes narrower in time so that the time resolution improves. Conversely, as r increases, the time resolution is degraded, but the frequency resolution is simultaneously enhanced because the quantity $\Delta t \Delta f$ must be maintained constant. This is also apparent from Eq. (9-8), where $H(ur)$ becomes narrower as r increases, thus improving the frequency resolution. The reason the variable r is called scale is that it stretches and contracts the function $h(\tau/r)$. The net result is that such a transform is in fact useful for analyzing the kind of mixed signal discussed earlier.

The CWT is strictly defined as a time–scale representation; however, it often proves easier to interpret CWTs in terms of time and frequency rather than time and scale. A short-lived function (r small) inherently contains high frequencies, so that r is inversely related to frequency. For a given wavelet transform, the mapping $f = K/r$ can be used, allowing the CWT of a signal to be interpreted in terms of frequency rather than scale. This mapping is discussed further under the section that deals with the details of implementation (section 9.3.5).

As with the STFT, a filter-bank interpretation [20–22] can be invoked for the CWT, as illustrated in Fig. 9-2(a). In this case, the CWT is obtained by filtering the original signal by a bank of filters with fixed *relative* bandwidth rather than fixed absolute bandwidth, as for the STFT. The relative bandwidth (BW_{rel}) of a filter (or function) is defined as the absolute bandwidth (Δf) of the bandpass region surrounding the filter's center frequency divided by the center frequency (f_h) itself. It is the inverse of the Q-factor:

$$\text{BW}_{\text{rel}} = \frac{\Delta f}{f_h} = \frac{1}{Q} \tag{9-10}$$

For the CWT filter bank illustrated in Fig. 9-2(a), the relative bandwidth of the filters remains fixed, since both the absolute bandwidth and the center frequency of the functions $H(ur)$ vary in inverse proportion to r. The tiling of the time–frequency plane associated with the CWT is shown in Fig. 9-2(b); it consists of rectangles of fixed area but variable shape. At low frequencies, the rectangles are broad in time but narrow in frequency, since for large r, the time resolution is poor and the frequency resolution is good. The converse is true at high frequencies.

9.3.4 Wavelet Bases

Many different functions $h(t)$ can be used as prototypes in forming a CWT. Much work in recent years has focused on the issue of defining and using different wavelet bases for a variety of purposes. In this chapter we examine three particular wavelet bases—the Morlet, Meyer, and Daubechies 4-tap. It is useful to calculate CWTs using these different bases, with the aim of determining which are most useful for analysis purposes.

All wavelet basis sets should satisfy an "admissibility condition" [20, 22, 24], which states that if $h(t)$ is a wavelet basis for $\mathcal{L}^2$ (the set of square-integrable functions) then

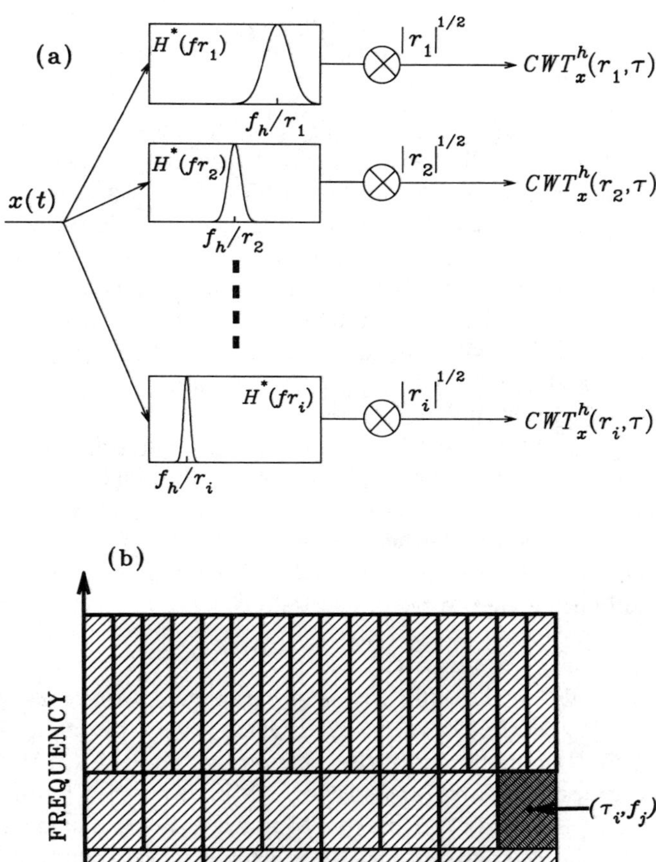

Figure 9-2   (a) Representation of the CWT in terms of filter-bank operations. The signal $x(t)$ to be analyzed is passed through a bank of filters, each of which is simply a scaled version of some prototype filter. Each filter has a fixed shape and relative bandwidth, but in absolute terms the bandwidths of the filters increase as the center frequency increases. The center frequency of the analysis filters is inversely proportional to scale r_i. The output from each filter is then multiplied by the gain factor $\sqrt{|r_i|}$. Since this gain factor increases with scale r, low-frequency components are accentuated with respect to high-frequency components. (b) Tiling of the time–frequency plane by the CWT. The rectangles centered at (τ_i,f_j) represent regions of the time–frequency plane where the functions $h^*([\tau - \tau_i]/r_j)$ are concentrated, with $f_j = K/r_j$. These rectangles also indicate the time and frequency resolution of the CWT. The CWT cannot resolve components at DC, since the basis functions do not extend to zero frequency. The figure shown here represents a dyadic grid, i.e., the rectangles double in length as we move along the frequency axis, but of course their areas are all the same. Our choice of a factor of 2 is illustrative; other ratios could equally well be used.

$$\int_{-\infty}^{+\infty} \frac{|H(f)|^2}{f}\, df < \infty \tag{9-11}$$

A consequence of the admissibility condition is that $H(0) = 0$ (otherwise $|H(f)|^2/f$ diverges at $f = 0$). In general, it is also true that $H(f) \to 0$ as $|f| \to \infty$. Combined, these two conditions show that the CFTs of wavelet basis functions represent bandpass filters since they remove components at both low and high frequencies.

In Fig. 9-3 we show four different wavelet bases in the time domain, along with the magnitudes of their CFTs in the frequency domain. For ease of comparison, we have normalized the CFTs so that their maxima always have unity magnitude and lie at $f = 1$. The Morlet and Meyer wavelet bases [Figs. 9-3(a), (c), and (e)] are complex and only the real part of the wavelet is plotted; the Daubechies 4-tap wavelet [Fig. 9-3(g)] is purely real.

Figure 9-3(a) shows the real part of the single-sided Morlet wavelet [24] given by $h(t) = \exp(jct)\exp(-\alpha t^2/2)$ [with $c = 2\pi$ and $\alpha = 0.0151$], and the magnitude of its Fourier transform [Fig. 9-3(b)], which is given by $\sqrt{2\pi/\alpha}\,\exp[-(2\pi f - c)^2/2\alpha]$. For the values of c and α that we have chosen here, $h(t)$ is a windowed sinusoid and $H(f)$ is a narrow bandpass filter centered at $f = 1$. The relative bandwidth of this wavelet is readily calculated to be

$$\mathrm{BW}_{\mathrm{rel}} = \frac{2\sqrt{2\alpha}}{c} \tag{9-12}$$

which equals 0.055 for this choice of parameters. Consequently, the Q-factor for this wavelet is $Q = 1/0.055 = 18.2 \gg 1$, which is why it is referred to as a high-Q Morlet wavelet in the remainder of this chapter.

Figures 9-3(c) and (d) also show a Morlet wavelet, but this time with $c = 2\pi$ and $\alpha = 0.151$. As in Fig. 9-3(b), this Morlet wavelet is also a bandpass function; however, its relative bandwidth ($= 0.55$) is much larger than that for the previous choice of parameters, and the number of oscillations in the time domain is lower than in Fig. 9-3(a). This wavelet is referred to as a low-Q Morlet wavelet in the text and figures that follow since its Q-factor is only 1.82.

Figures 9-3(e) and (f) show the Meyer wavelet similarly normalized (for details concerning the construction of the Meyer wavelet see [22]). Its structure is similar in both the frequency and time domains to the Morlet wavelet shown in Figs 9-3(c) and (d). However, unlike the Morlet wavelet, there are no free parameters that can be used to alter the Meyer wavelet's relative bandwidth. This is a consequence of the constraints under which the Meyer wavelet is constructed. The relative bandwidth of the Meyer wavelet is 1.18.

Finally, in Figs 9-3(g) and (h) we show a wavelet generated by an infinite iteration of a 4-tap finite impulse response (FIR) filter proposed by Daubechies (for details pertaining to the construction of this wavelet see [21,22]). Unlike the previous three wavelets, this basis is not symmetric, it has multiple peaks in the frequency domain, and there are no closed-form expressions for $h(t)$ or $H(f)$. The effect of these properties is discussed in section 9-4, which includes a CWT calculated using the Daubechies 4-tap wavelet. Like the Meyer wavelet, the relative bandwidth of this wavelet is also fixed, as a result of the manner in which it is calculated. The relative bandwidth of the Daubechies 4-tap wavelet (where the center frequency of the filter is defined as the frequency at which the maximum of the first bandpass region occurs) is 1.5.

The principle feature of the Morlet wavelet of interest to us is that its relative bandwidth is easily adjusted by choice of the parameters c and α. We later show that this allows us complete flexibility in setting the CWT to have a desired frequency resolution at any particular frequency. However, the Morlet wavelet, unlike the Meyer and Daubechies wavelets, has two theoretical limitations. Though these are worthy of mention, they have no bearing on the usefulness of this basis set for our purposes. First, the Morlet wavelet does not strictly satisfy the admissibility condition since $H(0) \neq 0$. However for $\mathrm{BW}_{\mathrm{rel}} \leq 0.8$, the value of $H(0)$ is close to zero, and the Morlet wavelet is deemed to be practically admissible. For the Morlet wavelets shown in Figs 9-3(a) and (c), $\mathrm{BW}_{\mathrm{rel}} \approx 0.055$ and 0.55, respectively. Second, the Morlet wavelet cannot be used as the prototype wavelet $h(t)$ to create an orthonormal basis for $\mathcal{L}^2$ of the form $\{h_{ij}(t) = 2^{i/2}h(2^i t - j)\}$, $i, j \in \mathcal{Z}$ (the set of natural numbers) [20–22, 25]. The ability to form such a basis is central to the design of wavelet bases for use in perfect reconstruction filter banks.

For the functions shown in Figs 9-3(c)–(h) (low-Q Morlet, Meyer, and Daubechies 4-tap wavelets), the relative bandwidth of the wavelet's CFT is of the order of unity. Such wavelets are loosely termed octave-band, since the functions $H(f)$ and $H(2f)$, whose center-frequencies are separated by an octave, are just about far enough apart in frequency to be resolved. This octave-band property arises naturally in wavelets that are designed to satisfy two-scale equations [21,22]. Both the Meyer and Daubechies wavelets satisfy such equations, and can be rigorously used as prototype wavelets to form an orthonormal basis for $\mathcal{L}^2$ of the form indicated above. However, while orthonormal-basis-generating wavelets do provide a useful set of functions for constructing efficient wavelet series expansions of $x(t)$ [22], they are not always suited for use as a CWT basis, as our examples will show.

9.3.5 STFT and CWT Implementation

A sampled version of the STFT, often referred to as the discrete STFT, was calculated using a summation approximation of Eq. (9-3):

$$\mathrm{STFT}_x^g[k, n] = \sum_{m=0}^{L-1} x[n + m]g^*[m]\exp\left(\frac{-j2\pi mk}{N}\right), \qquad 0 \leq k \leq N - 1 \qquad (9\text{-}13)$$

where k is the discrete frequency index, n is the discrete time index, L is the window length in samples, and $g[m]$ is chosen to be samples of a Gaussian window $g(t) = \exp(-\beta t^2/2)$, with $g(t)$ falling to e^{-4} at the sampled endpoints:

$$g[m] = \exp\left[-\left(-2 + \frac{4m}{L-1}\right)^2\right], \qquad 0 \leq m \leq L - 1 \qquad (9\text{-}14)$$

The formulation presented in Eq. (9-13) reminds us that the discrete STFT is simply a sequence of discrete Fourier transforms (DFTs) of the windowed signal segments.

HIGH-Q MORLET WAVELET

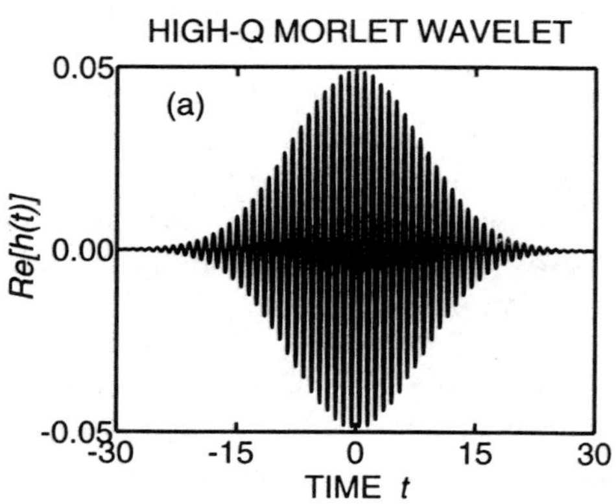

HIGH-Q MORLET WAVELET

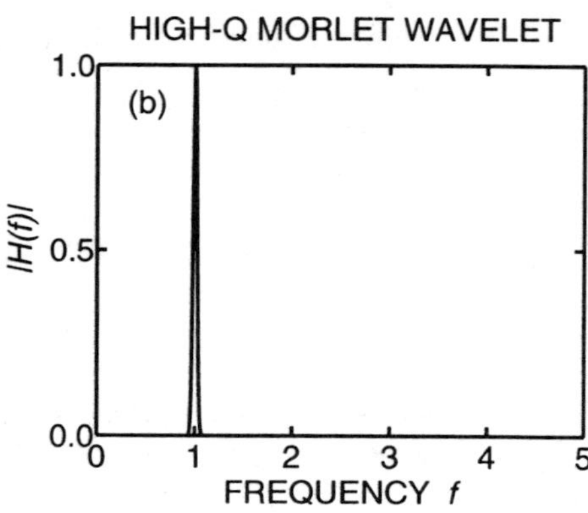

LOW-Q MORLET WAVELET

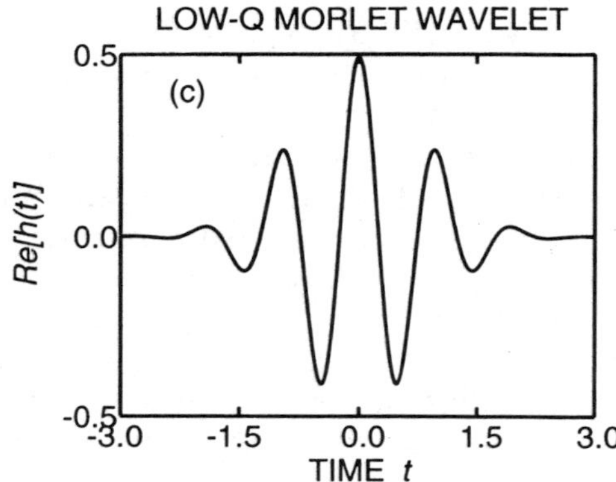

LOW-Q MORLET WAVELET

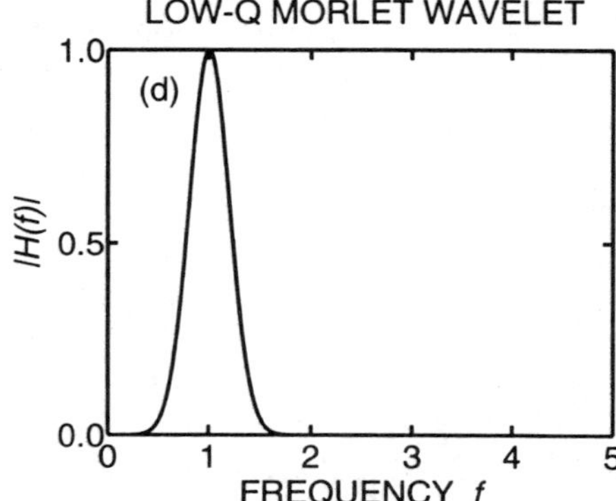

MEYER WAVELET

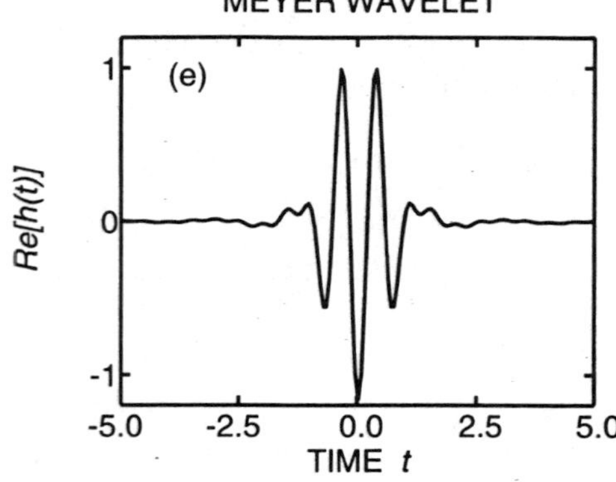

MEYER WAVELET

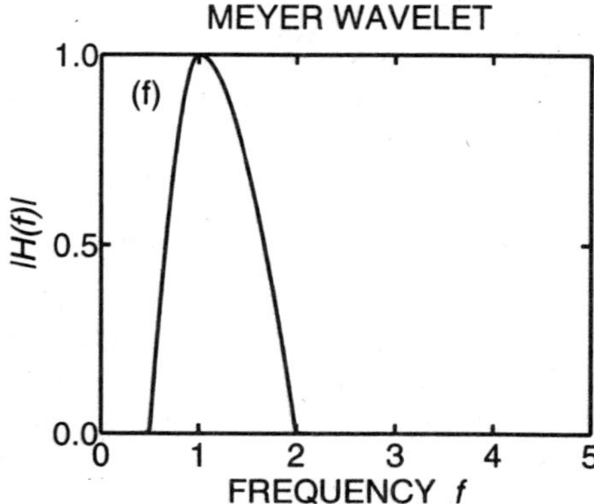

Figure 9-3 Four wavelet bases in the time domain and their respective CFTs. (a) Real part of a high-Q Morlet wavelet, $h(t) = \exp(jct)\exp(-\alpha t^2/2)$, with $c = 2\pi$ and $\alpha = 0.0151$. There are many (> 50) oscillations within the envelope of the wavelet, and its overall duration is approximately 60. (b) Magnitude of the CFT of the wavelet shown in (a). The relative bandwidth of this function is narrow ($= 0.055$). It is symmetric about the frequency $f = 1$. (c) Real part of a low-Q Morlet wavelet, $h(t) = \exp(jct)\exp(-\alpha t^2/2)$, with $c = 2\pi$ and $\alpha = 0.151$. There are relatively few oscillations within the envelope of the wavelet, and its overall duration is approximately 6. (d) Magnitude of the CFT of the wavelet shown in (c). The relative bandwidth of this function is 0.55. It is symmetric about the frequency $f = 1$. (e) Real part of a Meyer wavelet calculated as outlined in [22] (there is no closed-form expression for this wavelet in the time domain). As for the Morlet wavelet pictured in (c), there are relatively few oscillations within the envelope. The overall duration of the wavelet is approximately 5. (f) Magnitude of the CFT of the wavelet shown in (e). The relative bandwidth of this function is approximately 1.18. The function is asymmetric about the frequency $f = 1$. (g) Time-waveform of a wavelet calculated using an iterative procedure based on a 4-tap filter with coefficients as proposed by Daubechies (the Daubechies 4-tap filter) [22]. This wavelet has no closed-form expression, is purely real, and is not symmetric in time. (h) Magnitude of the CFT of the wavelet shown in (g). The relative bandwidth of the first bandpass region is 1.5. This function has multiple lobes in the frequency domain.

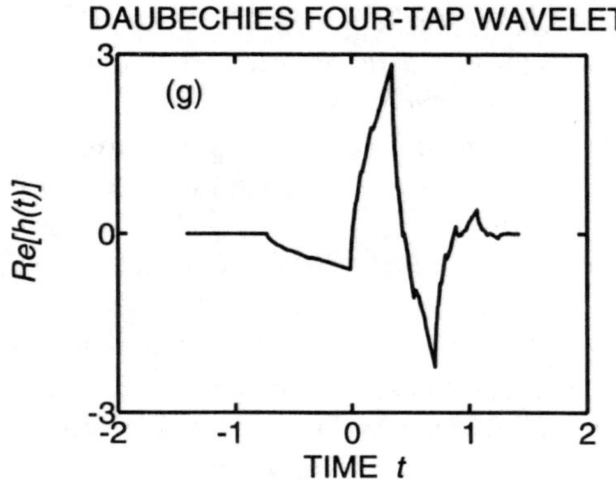

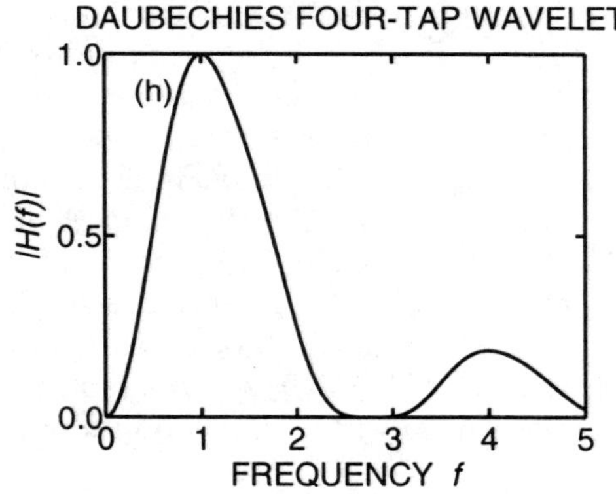

Once a window length L is chosen, the time–frequency uncertainty product is fixed—a good rule of thumb is to choose the window length so that the signal appears "relatively stationary" within it. For example, in the analysis of AM responses, as discussed subsequently, a window length $L = 128$ samples was appropriate. This corresponds to 1.28 ms at the 100 kHz sampling rate used in recording our data. The value of N (which sets the number of discrete frequencies at which the STFT is sampled in the frequency domain) was chosen equal to L. The STFT was not evaluated for all values of n; moving the time window through 256 time samples for successive evaluations of the STFT provided a sufficiently detailed picture for our purposes. For our particular choice of window length and sampling rate, the time resolution of the STFT was $\Delta t = 0.64$ ms, and the frequency resolution was $\Delta f = 1989.4$ Hz. The uncertainty product $\Delta t \Delta f = C = 4/\pi \approx 1.27$.

We present the STFT magnitude in two visual formats. The first is a three-dimensional representation, often referred to as a 3-D spectral plot. In this format, time and frequency form the bottom plane, and the STFT magnitude is represented on a linear axis in the third dimension. The second format provides 2-D contour plots, on which contours of equal STFT magnitude are traced on a time–frequency plane.

To calculate the CWT, we implemented the fast-CWT algorithm proposed by Jones and Baraniuk [26]. This technique avoids carrying out a time-consuming direct time-convolution of the data with the scaled wavelet time waveform; instead we express the time-convolution as multiplication in the frequency domain. Efficient algorithms are then exploited to carry out the calculation.

To see explicitly how this is done, reconsider Eq. (9-8), in which the CWT is written as an integral in the frequency domain:

$$\text{CWT}_x^h(r, \tau) = \sqrt{|r|}\left[\int_{-\infty}^{\infty} X(u)H^*(ur)\exp(j2\pi u\tau)du\right] \tag{9-15}$$

where r is scale, τ is time, and u represents frequency. The term inside the square brackets is the inverse CFT of $X(u)H^*(ur)$ [compare with Eq. (9-2)], which suggests that an inverse fast Fourier transform (FFT) can be used to evaluate the CWT. Specifically, consider $x[n]$ as a well-sampled version of the continuous-time function $x(t)$ with sampling time equal to T_x, and $h[n]$ as a well-sampled version of $h(t)$, normalized as shown in Fig. 9-3, with a sampling time equal to T_h. There are N_x samples of $x[n]$ and N_h samples of $h[n]$. The discrete-time Fourier transforms (DTFTs) of $x[n]$ and $h[n]$ are denoted by $X(e^{j\omega})$ and $H(e^{j\omega})$, respectively, where ω represents digital angular frequency [18]. The sampling theorem tells us that for well-sampled signals, $X(e^{j\omega}) = X(f)/T_x$ at $f = \omega/2\pi T_x$ over the range $\omega = [-\pi, \pi]$. Since we assume that $x(t)$ is well-sampled, $X(f)$ is essentially zero for $f > 1/2T_x$, which allows us to write Eq. (9-15) in terms of the DTFTs of $x[n]$ and $h[n]$:

$$\text{CWT}_x^h(r, \tau) = \sqrt{|r|}\left[\frac{T_x}{2\pi}\int_{-\pi}^{+\pi} X(e^{j\omega})H^*(e^{j\omega r})\exp(j\omega\tau/T_x)\,d\omega\right] \tag{9-16}$$

Restricting ourselves to evaluating the CWT at a discrete set of $r = [r_i]$ and $\tau = nT_x$ brings us to a sampled version of the CWT:

$$\text{CWT}_x^h[r_i, n] = \sqrt{|r_i|}\,\frac{T_x}{2\pi}\int_{-\pi}^{+\pi} X(e^{j\omega})H^*(e^{j\omega r_i})\exp(j\omega n)\,d\omega \tag{9-17}$$

which is recognizable as a multiple of the inverse DTFT of the function $Y(e^{j\omega}, r_i) = X(e^{j\omega})H^*(e^{j\omega r_i})$. This inverse DTFT is efficiently implemented by the inverse FFT

$$\text{CWT}_x^h[r_i, n] = \sqrt{|r_i|}\,\frac{T_x}{M}\sum_{k=-M/2}^{M/2-1} Y[k, r_i]\exp(j2\pi kn/M) \tag{9-18}$$

where $Y[k, r_i] = Y(e^{j\omega}, r_i)$ evaluated at $\omega = 2\pi k/M$, with k ranging from $-M/2$ to $M/2 - 1$. Since efficient computation algorithms exist for the inverse FFT, only two questions remain: how to calculate $Y[k, r_i]$, and what is an appropriate value of M?

We can write $Y[k, r_i]$ as $X[k]H^*[k, r_i]$, where $X[k] = X(e^{j\omega})$ evaluated at $\omega = 2\pi k/M$, and $H^*[k, r_i] = H^*(e^{j\omega r_i})$, also evaluated at $\omega = 2\pi k/M$. $X[k]$ is now simply the FFT of the sequence $x[n]$, and $H[k, r_i]$ is the chirp z-transform of the sequence $h[n]$. Efficient algorithms exist for both the FFT and the chirp z-transform, allowing us to evaluate both $X[k]$ and $H[k, r_i]$, which in turn leads us to $Y[k, r_i]$ and ultimately to $\mathrm{CWT}^h_x[r_i, n]$ via the inverse FFT. (For a complete discussion of the chirp z-transform, see pp. 623–628 of [18].)

To choose an appropriate value of M, we must ensure that the frequency multiplication of $H(e^{j\omega r})$ and $X(e^{j\omega r})$ really gives us the desired linear convolution in time from Eq. (9-15). This is ensured by selecting M greater than the combined lengths of the sequences $x[n]$ and the longest wavelet basis we use (which is N_h multiplied by $\max(r_i)$, since the largest value of r will produce the most stretched wavelet function). In practice, the next highest power of 2 greater than $N_x + \max(r_i)N_h - 1$ is chosen, so that power-of-2 FFTs can be used.

There is a remaining subtlety in the fast-CWT algorithm. At the outset, we assumed that $H(f) = 0$ for all values of $|f| > 1/2T_x$. Therefore, to correctly carry out the multiplication of Eq. (9-15), we must ensure that aliased versions of $H(e^{j\omega r_i})$ are not brought into the range $[-\pi, \pi]$. Therefore, we set $H(e^{j\omega r_i}) = 0$ for $|\omega r_i| > \pi$. Since we are using $\omega = 2\pi k/M$, this implies $Y[k, r_i] = 0$ for $k > M/2r_i$. As k ranges from $-M/2$ to $M/2 - 1$, this only occurs for $r_i < 1$.

As stated earlier, scale r is inversely related to frequency. For convenience in interpreting CWTs, we have mapped scale to frequency using the mapping $f = K/r$ as indicated earlier. We choose this mapping to assign a given scale r_i to a frequency f_i equal to the center frequency of the filter $H(fr_i)$. The proportionality constant K is evaluated by obtaining the center frequency of the function $H(f)$, since this corresponds to $H(fr_i)$ at $r_i = 1$. Therefore, we must find what $H(f)$ the samples $h[n]$ represent. The set $h[n]$ is constructed by taking N_h samples of the continuous function $h(t)$, normalized as shown in Fig. 9-3. To provide a complete representation, but without undue oversampling, $h[n]$ is constructed by sampling $h(t)$ at a different rate $(1/T_h)$ as compared to the sampling rate $1/T_x$ for $x[n]$. However, in evaluating the CWT as described previously, the $h[n]$ are taken as a set of samples at rate $1/T_x$. This means that $h[n]$ represents samples of a continuous function $h(tT_h/T_x)$. The center frequency of this function's CFT, which is $|T_x/T_h| H(fT_x/T_h)$, occurs at $f = T_h/T_x$ rather than at $f = 1$. This center frequency T_h/T_x corresponds to scale $r = 1$. Substituting these values into $f = K/r$, we find that $K = T_h/T_x$. This allows us to map the CWTs on a time–frequency plane, but we remind the reader that this is only an interpretational convenience; strictly speaking, the CWT is a time–scale representation.

A word is in order describing the link between the CWT and its discrete counterpart, the discrete wavelet transform (DWT) [21,22]. The DWT provides a multiresolution approximation of the sequence $x[n]$. This approximation involves repeated high-pass and low-pass filtering of the original signal $x[n]$, with downsampling by a factor of 2 after each filtering operation. The various high-pass sequences are retained as a useful approximation of the signal. The number of samples contained in the complete set of high-pass sequences plus the residual low-pass sequence is equal to the number in the original signal, and the high-pass sequences contain nearly all the information of the original signal (however the DC component is lost as it is contained in the low-pass residual). This process decomposes the sequence $x[n]$ into various frequency bands, as represented by the set of high-pass sequences. The center frequency of each band differs by a factor of 2 as a result of the down-sampling factor used in calculating the DWT.

In a similar manner, the CWT decomposes the original signal $x(t)$ into an infinite set of time sequences $\mathrm{CWT}_x^h(r, \tau)$, also distributed in various frequency bands across the time–frequency plane. However, unlike the DWT, the various frequency bands are not constrained to differ by a factor of 2; rather, they can be evaluated at arbitrary values of r. Essentially, the CWT is an interpolated version of the DWT in which a decomposition at arbitrary scales can be examined. This is analogous to considering the CFT as an interpolated version of the FFT. The DWT and FFT may alternatively be viewed as sampled versions of the CWT and CFT, respectively.

As in the case of the STFT, we use two visual formats for the CWT. The first is a three-dimensional (3-D) representation, in which time and frequency form the bottom plane, and CWT magnitude (multiplied by the factor of $|r|^{-1/2}$ for ease of comparison with the STFT) is represented on a linear axis in the third dimension. The second format provides 2-D contour plots, on which contours of equal modified CWT magnitude are traced on a time–frequency plane.

9.4. RESULTS

We used the modified CWT and STFT to examine the behavior of the velocity response of auditory sensory cells in response to the AM acoustic signals described in section 9.1. These AM signals (with unity modulation depth and a fixed modulation frequency of 6.1 Hz) were applied to the ear canal, using different carrier frequencies and levels. This allowed us to examine the nonlinear velocity response of a given cell as the carrier frequency was altered from well below to above the CF, and as the stimulus intensity was varied. The sampling frequency was fixed at 100 kilosamples/s in recording the responses. The data reported here were collected at the level of the basilar membrane, just below the first row of outer hair cells, at the edge of the tunnel region of the cochlea. The CF was 36 kHz, a typical value for cells in the basal turn of the cat cochlea.

Figure 9-4 shows the response of the cell when an AM tone with a carrier frequency $f_c = 13$ kHz (well below CF), and a peak sound pressure level of ≈ 108 dB:re .0002 dyne/cm^2, is applied to the ear. The velocity response of the cell during one cycle of the modulation envelope is displayed in Fig. 9-4(a). It roughly reflects the underlying shape of the modulation envelope.

The modified CWT magnitude of the velocity response, shown in 3-D and 2-D formats in Figs 9-4(b) and (c), respectively, reveals the time behavior of the signal at different scales, which are represented as different frequencies as discussed earlier. The mapping from scale to frequency was accomplished using $f = K/r$ with $K = 1.27 \times 10^4$. This CWT is based on the high-Q Morlet wavelet shown in Figs 9-3(a) and (b). The modified CWT shows three spectral components at multiples of

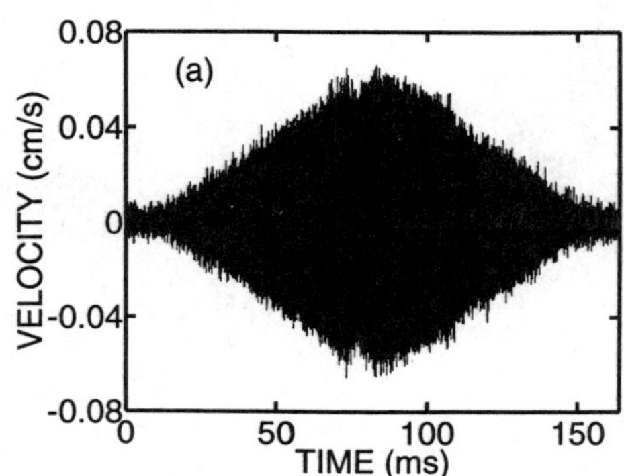

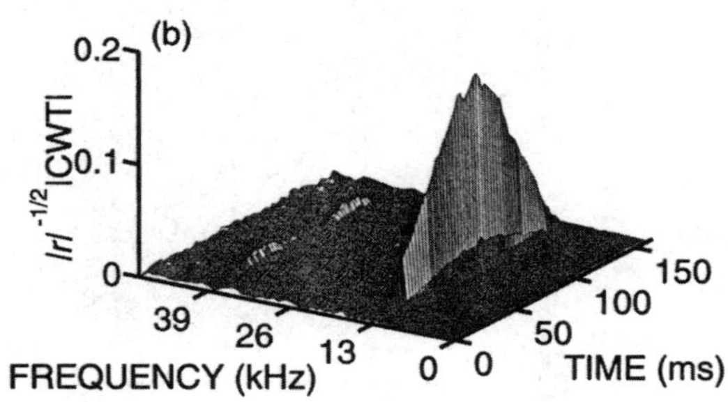

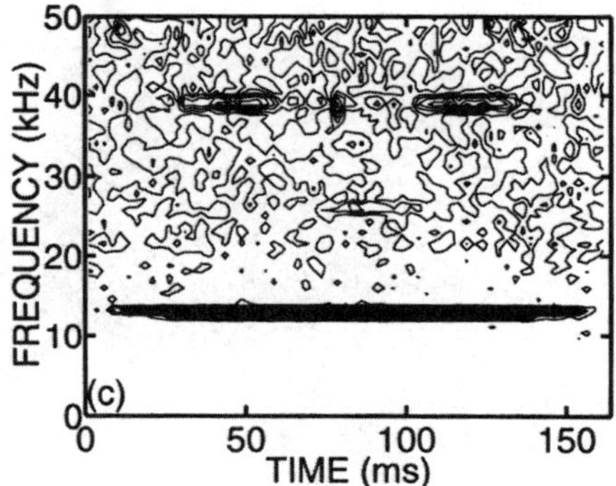

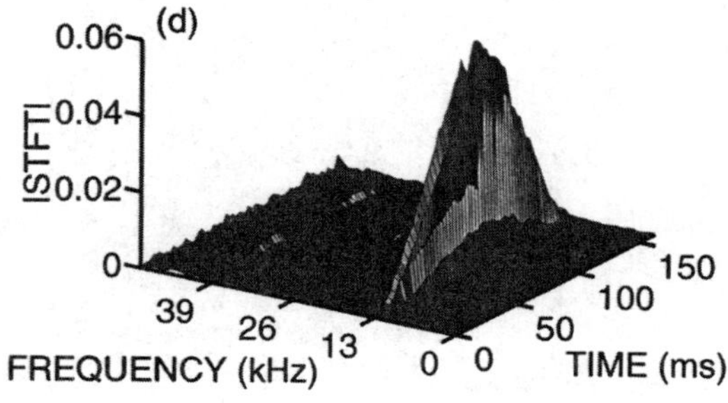

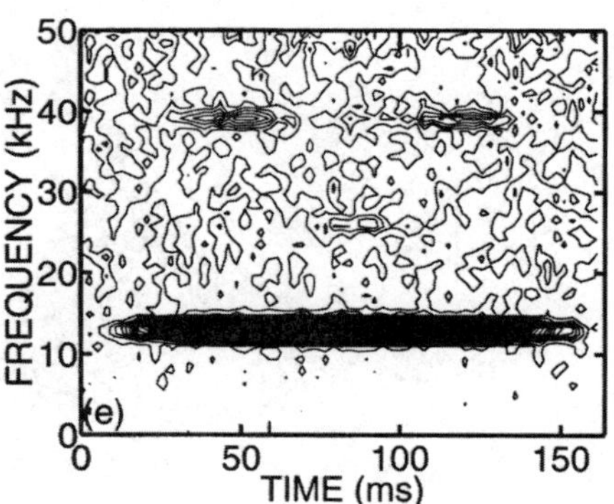

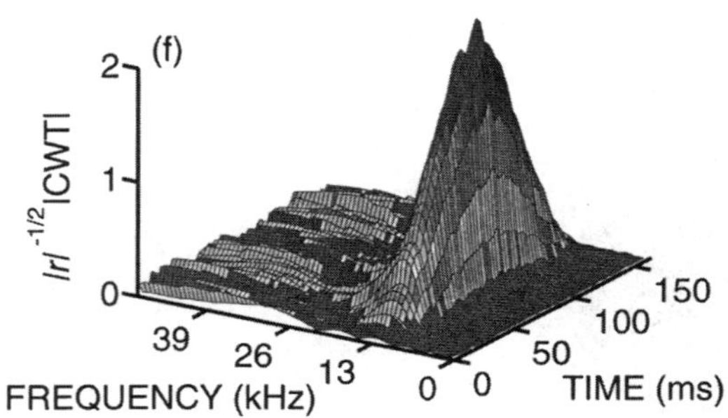

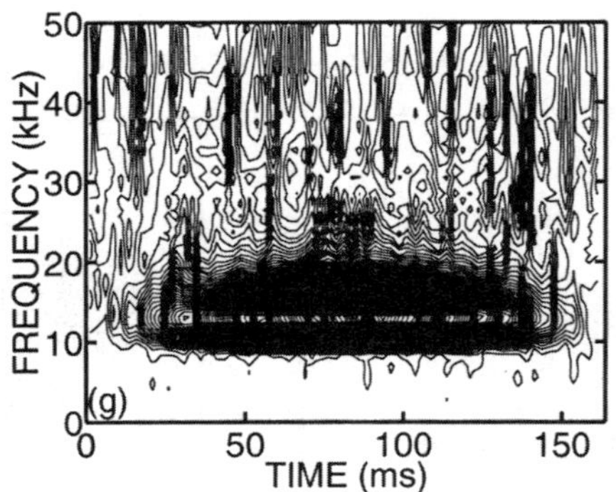

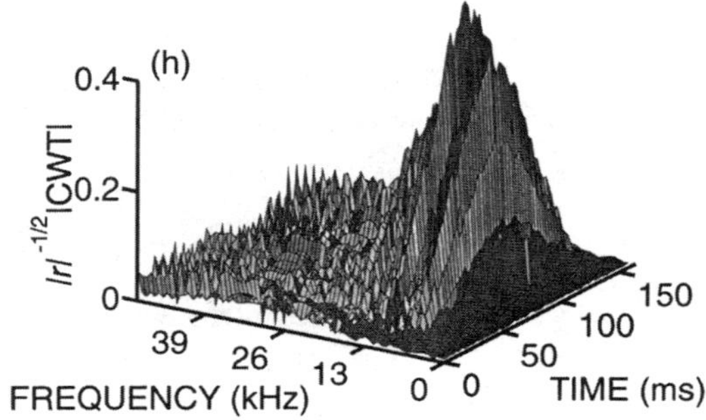

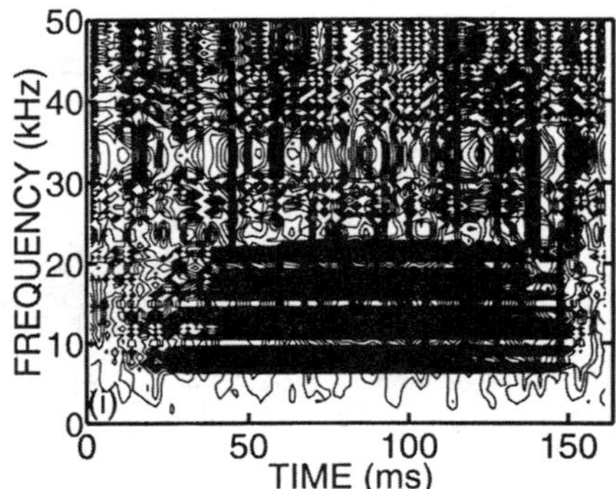

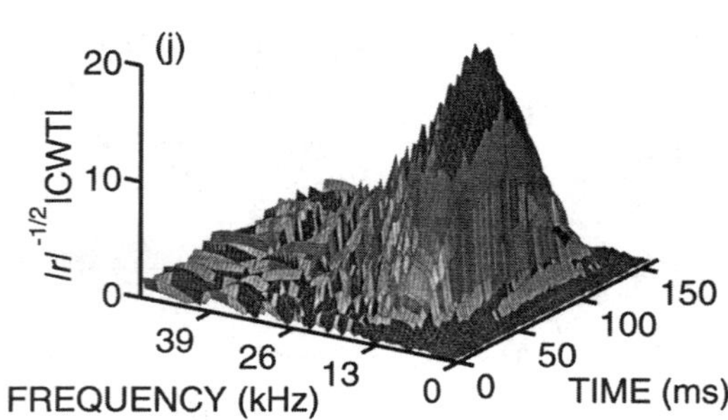

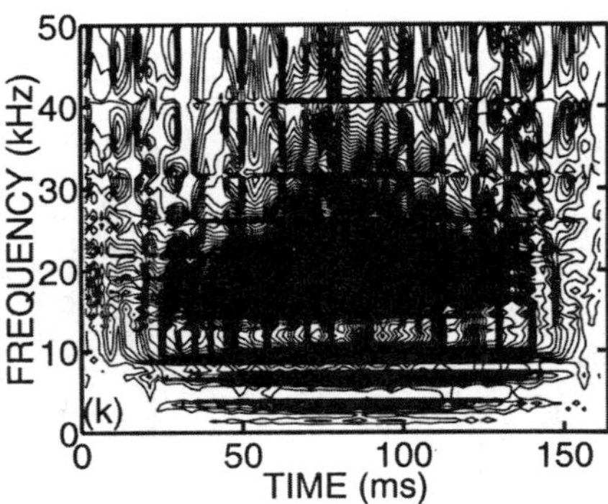

Figure 9-4   Velocity response at the level of the basilar membrane, under the first row of outer hair cells in the basal turn of the living cat cochlea (file number 38260710.mat). The CF of the cell was 36 kHz, and this response is to an AM stimulus with carrier frequency $f_c = 13$ kHz (well below CF) and modulation frequency 6.1 Hz. The highest sound pressure level, occurring at the center of the input envelope was ≈ 108 dB:re .0002 dyne/cm^2. (a) Time waveform of the velocity response (in cm/s) of the cell. (b) 3-D plot of the modified CWT magnitude (viz., the CWT multiplied by the factor $|r|^{-1/2}$) of the velocity response shown in (a). The high-Q Morlet wavelet basis shown in Figs 9-3(a) and (b) was used. The x and y axes represent time (ms) and frequency (kHz), respectively, and modified CWT magnitude is plotted on a linear scale on the z axis. This plot shows spectral components at the carrier frequency f_c and at two higher harmonic frequencies, $2f_c$ and $3f_c$. (c) Same modified CWT magnitude as shown in (b), but now plotted in 2-D format with 80 equally spaced (in modified CWT magnitude) contour lines joining points of constant magnitude. (d) 3-D plot of the STFT magnitude of the velocity response shown in (a). Spectral components are present at the carrier frequency f_c and at two higher harmonics, $2f_c$ and $3f_c$. (e) Same STFT magnitude as shown in (d), but now plotted in 2-D format with 80 equally spaced (in STFT magnitude) contour lines joining points of constant magnitude. (f) 3-D plot of the modified CWT magnitude of the velocity response shown in (a). The low-Q Morlet wavelet basis shown in Figs 9-3(c) and (d) was used. This plot shows a large-bandwidth spectral component centered at the carrier frequency f_c. Energy is present at higher frequencies but the resolution is very poor. (g) Same modified CWT magnitude as shown in (f), but now plotted in 2-D format with 80 equally spaced (in modified CWT magnitude) contour lines joining points of constant magnitude. (h) 3-D plot of the modified CWT magnitude of the velocity response shown in (a). The Meyer wavelet basis shown in Figs 9-3(e) and (f) was used. As in (f), this plot shows a large-bandwidth spectral component centered at the carrier frequency f_c, and energy at higher frequencies that cannot be clearly resolved. (i) Same modified CWT magnitude as shown in (h), but now plotted in 2-D format with 80 equally spaced (in modified CWT magnitude) contour lines joining points of constant magnitude. (j) 3-D plot of the modified CWT magnitude of the velocity response shown in (a). The Daubechies 4-tap wavelet basis shown in Figs 9-3(g) and (h) was used. As in (f) and (h), this plot shows a large-bandwidth spectral component at the carrier frequency f_c and energy at higher frequencies that cannot be clearly resolved. (k) Same modified CWT magnitude as shown in (j), but now plotted in 2-D format with 80 equally spaced (in modified CWT magnitude) contour lines joining points of constant magnitude.

the carrier frequency. At the beginning and end of the modulation cycle, when the magnitude of the envelope is low, the response closely follows the input and is exclusively at the carrier frequency. As the magnitude of the envelope increases, a spectral component at $3f_c$ appears, and right at the center of the modulation envelope, where the sound pressure is greatest, a component at $2f_c$ is just discernible above the noise floor. These components subsequently disappear as the envelope decreases. The spectral component at f_c is more clearly resolved in frequency as compared to the component at $3f_c$, as expected for the CWT, where frequency resolution improves at lower frequencies. Of course, this improvement of frequency resolution is accompanied by a degradation in time resolution. The presence of harmonic generation clearly indicates nonlinearity in the cellular response. The generation of multiple harmonic components for carrier frequencies below CF is also typical of sensory cells in the guinea-pig cochlea [12,13]. In particular, we have often seen a tendency for odd-harmonic components to be more significant than their neighboring even-harmonic components (e.g., see Fig. 8 of [13]).

For comparison, Figs 9-4(d) and (e) show the STFT magnitude of the velocity response, calculated as described earlier. The time and frequency resolution of the STFT were chosen to match those of the CWT shown in Figs 9-4(b) and (c) at a frequency of 36 kHz. Like the CWT, the STFT also shows spectral components at f_c, $2f_c$, and $3f_c$, with the component at $2f_c$ barely visible.

As expected, the spectral widths represented in the STFT are constant whatever the frequency of the component; unlike the CWT, the frequency resolution is fixed for all frequencies. Similarly, the time resolution remains constant at all frequencies.

To evaluate the relative merits of different wavelet bases, the CWT of the same velocity-response waveform shown in Fig. 9-4(a) was calculated using the three other wavelet bases shown in Figs 9-3(c)–(h).

Figures 9-4(f) and (g) display the modified CWT magnitude calculated using the low-Q Morlet wavelet basis displayed in Figs 9-3(c) and (d). The proportionality constant K, mapping scale to frequency, was 1.27×10^3. These plots show the presence of a component at approximately 13 kHz, but little else besides. The component at $3f_c$ is not apparent, since the frequency resolution of this CWT is so poor at that frequency. This is because the relative bandwidth of this wavelet basis is a factor of ten larger than that of the high-Q Morlet wavelet, so the frequency resolution at any given frequency is diminished by a factor of ten. For instance, at the frequency 39 kHz ($= 3f_c$ for this example), the frequency resolution of the high-Q Morlet wavelet is 2155 Hz; that of the low-Q Morlet wavelet is 21550 Hz, so that it is not surprising that no clearly resolved component can be distinguished at that high frequency. The improved time resolution accruing at higher frequencies is of little advantage for our purposes, as the poor frequency resolution makes it hard to ascribe events in time to a particular frequency component. Such information would be useful in determining the underlying dynamics of the response.

Figures 9-4(h) and (i) show the modified CWT magnitude calculated using the Meyer wavelet basis shown in Fig. 9-3(e) and (f). The proportionality constant K, mapping scale to frequency, was 6.26×10^3. The results are similar to those for the low-Q Morlet wavelet, as expected. However, there is more roughness in the CWT; this reflects the roughness of the absolute value of the Meyer wavelet, whose real part is displayed in Fig. 9-3(e).

Finally, Figs 9-4(j) and (k) show the modified CWT magnitude calculated using the Daubechies 4-tap wavelet basis shown in Figs 9-3(g) and (h). The proportionality constant K, mapping scale to frequency, was 281. Unlike the previous three wavelets, this wavelet is real and asymmetric. The CWT is therefore also real, and is capable of switching between positive and negative values. As a consequence, the absolute value of the CWT often goes to zero, leading to a highly scalloped structure for the CWT magnitude. A further difficulty in using the Daubechies 4-tap wavelet basis as an analysis tool is the presence of multiple bandpass regions in the CFT of the wavelet basis, as is evident in Fig. 9-3(h). This allows the energy from a single frequency component to enter the CWT at a variety of different frequencies. In the CWT shown here, for example, the energy from the component at 13 kHz appears in the CWT magnitude not only at 13 kHz, but also at $13/4 = 3.25$ kHz and $13/8 = 1.625$ kHz (the component at 13 kHz arises when the main lobe of the analysis wavelet is at 13 kHz, the component at 3.25 kHz appears when the second lobe of the analysis wavelet is at 13 kHz, and the component at 1.625 kHz appears when the third lobe of the analysis wavelet is at 13 kHz).

The most useful wavelet basis for analyzing the responses recorded in our experiments is the high-Q Morlet wavelet. This is because the frequency separation of the spectral components we need to resolve is relatively small compared to the absolute values of the frequencies themselves. The Morlet wavelet basis works well since we can control its relative bandwidth by adjusting the ratio $\sqrt{\alpha}/c$. The other wavelet bases we have investigated are unsuitable as analysis tools for the class of data examined here since they are octave-band in nature. In the further data analysis that follows, therefore, we restrict ourselves to use of the high-Q Morlet wavelet CWT and the STFT.

Figure 9-5 shows the response of the same cell when an AM tone with a carrier frequency $f_c = 22$ kHz (below CF), and a peak sound pressure level of ≈ 107 dB:re .0002 dynes/cm^2, is applied to the ear. The velocity response of the cell over one modulation cycle is displayed in Fig. 9-5(a). Figures 9-5(b) and (c) show the modified CWT magnitude of the velocity response. In this case a component at $2f_c$ can be distinguished; the presence of components at higher harmonics cannot be established, as the sampling frequency does not allow us to view components above 50 kHz.

Figures 9-5(d) and (e) show the STFT magnitude of the velocity response. Again, the STFT and modified CWT magnitudes display similar information.

Figure 9-6 shows the response of this cell when an AM tone with a carrier frequency $f_c = 36$ kHz (at CF), and a peak sound pressure level of ≈ 99 dB:re .0002 dynes/cm^2, is applied to the preparation. The velocity response of the cell over one cycle of the modulation envelope is displayed in Fig. 9-6(a). The response does not follow the shape of the input envelope, but rather shows substantial amplitude variations. We have often observed similarly uneven time waveform responses in the guinea-pig cochlea for AM signals with carrier frequencies near CF (e.g., see Fig. 9(a) of [13] and Fig. 3(a) of [12]; these responses often contain harmonic, half-harmonic, and quarter-harmonic components). Figures 9-6(b) and (c) show the modified CWT magnitude of the velocity response calculated using the high-Q Morlet wavelet. The modified CWT shows that the amplitude of the component at f_c varies significantly with time near the center of the modulation envelope. The STFT magnitude of the velocity response [Figs 9-6(d) and (e)] shows a similar effect. This variation in time of the response at f_c is likely accompanied by the presence of energy at higher harmonic frequencies which we did not measure.

Similar results are obtained for carrier frequencies above the CF, at least up to the highest frequency we examined, which was 40 kHz.

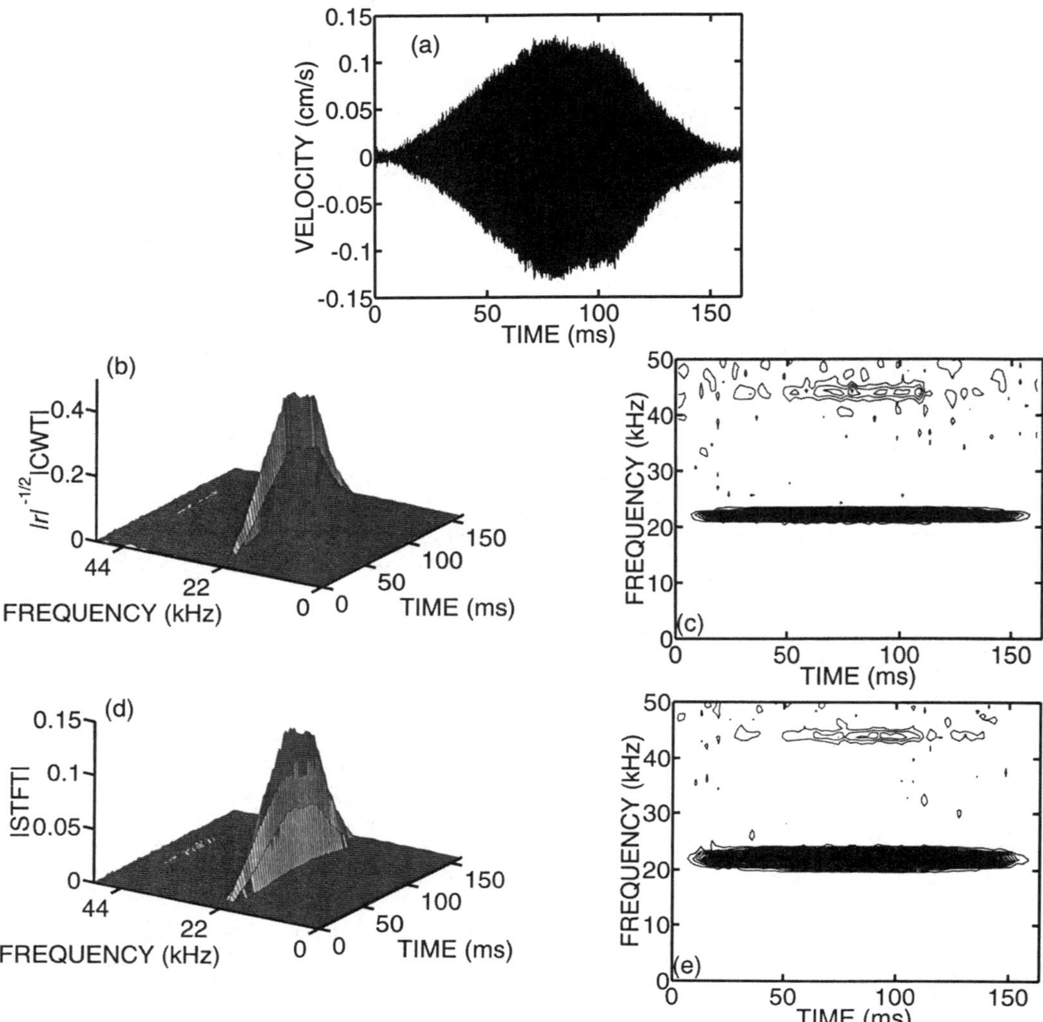

Figure 9-5   Velocity response of the same cell as shown in Fig. 9-4, but now to an AM stimulus with carrier frequency $f_c = 22$ kHz (below CF) and modulation frequency 6.1 Hz. The highest sound pressure level, occurring at the center of the input envelope was ≈ 107 dB:re .0002 dyne/cm^2. (a) Time waveform of the velocity response (in cm/s) of the cell. (b) 3-D plot of the modified CWT magnitude of the velocity response shown in (a). The high-Q Morlet wavelet basis shown in Figs 9-3(a) and (b) was used. This plot shows spectral components at the carrier frequency f_c and at the second harmonic, $2f_c$. (c) Same modified CWT magnitude as shown in (b), but now plotted in 2-D format with 80 equally spaced (in modified CWT magnitude) contour lines joining points of constant magnitude. (d) 3-D plot of the STFT magnitude of the velocity response shown in (a). Spectral components are present at the carrier frequency f_c, and at the second harmonic, $2f_c$. (e) Same STFT magnitude as shown in (d), but now plotted in 2-D format with 80 equally spaced (in STFT magnitude) contour lines joining points of constant magnitude.

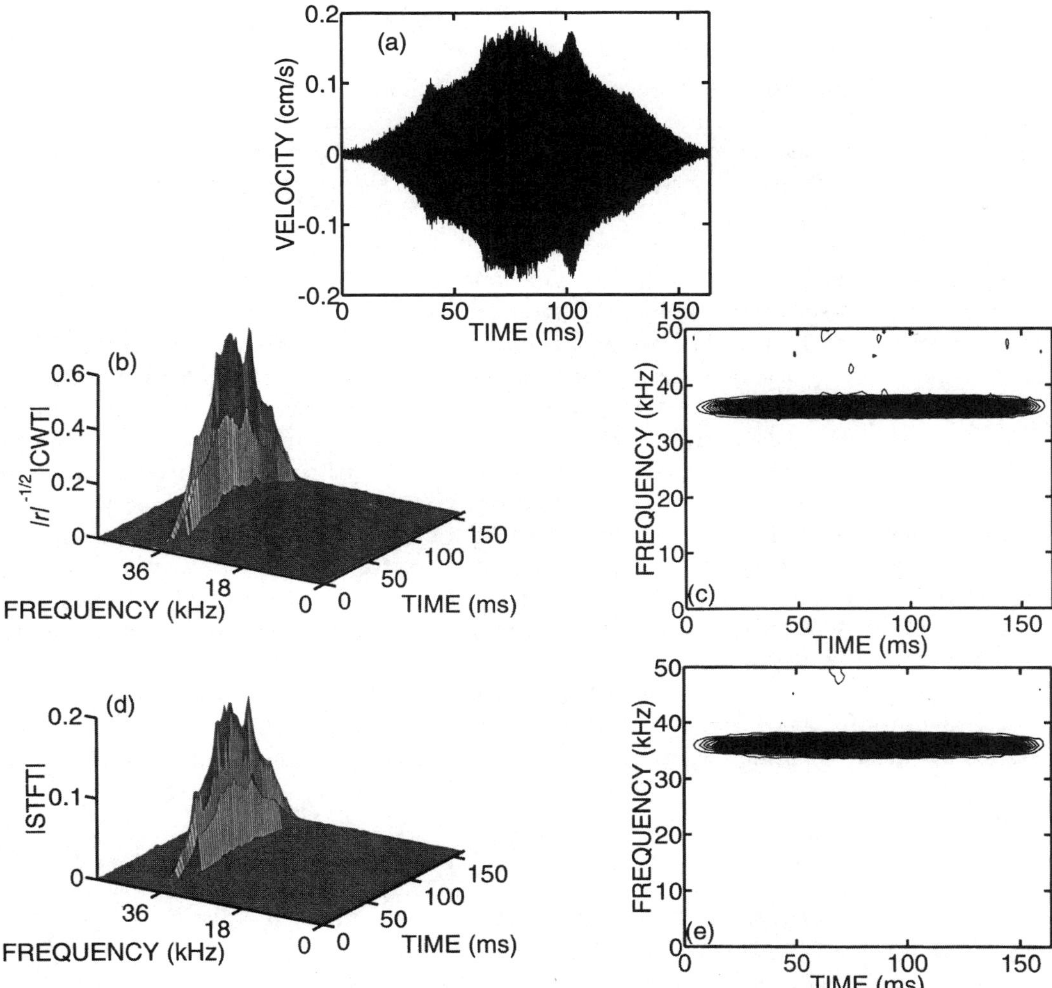

Figure 9-6   Velocity response of the same cell as shown in Figs 9-4 and 9-5, but now to an AM stimulus with carrier frequency $f_c = 36\,\text{kHz}$ (at CF) and modulation frequency 6.1 Hz. The highest sound pressure level, occurring at the center of the input envelope, was ≈ 99 dB:re .0002 dyne/cm^2. (a) Time waveform of the velocity response (in cm/s) of the cell. (b) 3-D plot of the modified CWT magnitude of the velocity response shown in (a). The high-Q Morlet wavelet basis shown in Figs 9-3(a) and (b) was used. This plot shows the irregular variation with time of the spectral component at f_c, particularly near the center of the modulation envelope. (c) Same modified CWT magnitude as shown in (b), but now plotted in 2-D format with 80 equally spaced (in modified CWT magnitude) contour lines joining points of constant magnitude. (d) 3-D plot of the STFT magnitude of the velocity response shown in (a). This plot also shows the time-variation of the spectral component at the carrier frequency f_c. (e) Same STFT magnitude as shown in (d), but now plotted in 2-D format with 80 equally spaced (in STFT magnitude) contour lines joining points of constant magnitude.

9.5. DISCUSSION

It is apparent from the foregoing results, and the data analyses presented in Figs 9-4–9-6, that the CWT and STFT have different frequency- and time-resolution properties. It is these properties, and their relationship to the characteristics of the signal itself, that determine the relative advantages of the two techniques for analyzing a given signal.

The time and frequency resolutions of the CWT vary with scale, though their product remains fixed. Whatever the choice of wavelet basis, the frequency resolution is worst at small scales, and improves (i.e., decreases in width) with increasing scale. Since frequency is proportional to inverse scale [20–22], this means that CWT frequency resolution is best for low frequencies, and worst for high frequencies.

Given this fundamental property of the CWT, one can inquire whether a desired frequency resolution Δf can be selected at any arbitrary frequency f. This is equivalent to asking whether the relative bandwidth $(\Delta f/f)$ of a wavelet basis function can be freely chosen. Many wavelet bases (including the Meyer and Daubechies 4-tap wavelets) were designed to provide a useful multiresolution framework onto which a signal could be efficiently decomposed [22], and therefore satisfy two-scale equations. As a consequence, their relative bandwidths are fixed (near unity) and cannot be freely chosen. These wavelets (and others like them) were not primarily meant for use in signal analysis.

The Morlet wavelet, however, was designed as an analysis tool, and its relative bandwidth can be freely set by choosing the value of $\sqrt{\alpha}/c$. Explicitly, given a desired frequency resolution Δf at the frequency f, set $\sqrt{\alpha}/c$ by using Eq. (9-12), i.e.,

$$\frac{\sqrt{\alpha}}{c} = \frac{\Delta f}{2\sqrt{2}f} \qquad (9\text{-}19)$$

The choice of c and α individually is unimportant; rather it is the ratio $\sqrt{\alpha}/c$ that determines the relative bandwidth.

STFT analysis, on the other hand, requires that the frequency resolution and time resolution remain constant at all frequencies. The frequency resolution is determined by the choice of the window function $g(t)$. In our case, the window function is a sampled version of $g(t) = \exp(-\beta t^2/2)$, so that the frequency resolution (at all frequencies) is given by $\sqrt{2\beta}/\pi$. In our discrete implementation of the STFT, β was a function of the sampled window length L and the sampling time T_x, in accordance with $\beta = 32/L^2 T_x^2$. For $L = 128$ and $T_x = 1/100\,000$, which were used in calculating the STFTs shown here, $\beta = 1.95 \times 10^7$ and the frequency resolution of the STFT was $\Delta f = 1989.4\,\text{Hz}$ at all frequencies.

With this kind of control over the time and frequency resolutions of both the CWT and the STFT, the final choice of which technique to use is signal-dependent. From a signal-analysis point of view, the CWT fares best when the required frequency resolution (or spacing between spectral components) varies as inverse frequency. On the other hand, the STFT is best suited to situations where the spectral components are linearly spaced in frequency. Thus, the CWT is the preferred tool when the analysis requires good frequency resolution at low frequencies together with good time resolution for impulsive (high-frequency) events. The STFT is appropriate when the required frequency resolution (and time resolution) remains fixed across the time–frequency plane.

Other time–frequency analysis techniques can be used aside from the CWT and the STFT. Indeed it can be shown that the absolute squares of the CWT and the STFT are members of more general classes of quadratic time–frequency representations (the affine class for the absolute square of the CWT, and the Cohen class for the absolute square of the STFT [19]). Indeed, other wavelet-based analysis techniques exist. Of particular interest are wavelet packets [25,27], which provide a more general tiling of the time-frequency plane than does the CWT. Rather than being restricted to good frequency resolution at low frequencies and poor frequency resolution at high frequencies, wavelet packets permit good frequency resolution to be achieved at arbitrary analysis frequencies. The product of the time and frequency resolutions, of course, remains fixed. Yet other signal-dependent wavelet techniques have also been developed [27].

9.6. CONCLUSION

Time–frequency analysis has been found to be useful for the analysis of cochlear cellular responses to AM acoustic stimuli. The time course of cellular vibration in the inner ear of the living cat has been successfully studied using both the STFT and an appropriately chosen CWT. Similar results, but with lower carrier-to-noise ratios, were obtained with the round-window membrane intact. Both analysis techniques reveal the time course of the nonlinear dynamics.

ACKNOWLEDGMENTS

This work was supported by the Office of Naval Research under Grant N00014-92-J-1251, by the National Institutes of Health through NIDCD Program Project Grant DC00316, and by the Emil Capita Foundation. We thank C. Herley, S. B. Lowen, and R. G. Turcott for helpful suggestions in the preparation of the manuscript.

REFERENCES

[1] G. von Békésy, *Experiments in Hearing*. New York, NY: McGraw-Hill, 1960; Huntington, NY: Krieger, 1980.

[2] International Team for Ear Research (ITER), "Cellular vibration and motility in the organ of Corti, *Acta Otolaryngologica* (Stockholm) Supplement 467, pp. 1–279, 1989.

[3] M. C. Teich, S. M. Khanna, and S. E. Keilson, "Nonlinear dynamics of cellular vibrations in the organ of Corti," *Acta Otolaryngologica* (Stockholm) Supplement 467, pp. 265–279, 1989.

[4] M. C. Teich, S. E. Keilson, S. M. Khanna, L. Brundin, M. Ulfendahl, and Å. Flock, "Chaos in the cochlea." In *Abstracts 14th Midwinter Res. Mtg. Assoc. Res. Otolaryngology,* February 3–7, p. 50, Abstract No. 155, 1991.

[5] M. C. Teich, S. E. Keilson, S. M. Khanna, L. Brundin, M. Ulfendahl, and Å. Flock, "Chaotic vibrations of outer hair cells and Hensen's cells in the cochlea." In *Abstracts 15th Midwinter Res. Mtg. Assoc. Res. Otolaryngology,* February 2–6, p. 17, Abstract No. 41, 1992.

[6] M. C. Teich, C. Heneghan, S. M. Khanna, Å. Flock, L. Brundin, and M. Ulfendahl, "Analysis of dynamical motion of sensory cells in the organ of Corti using the spectrogram." In *Biophysics of Hair Cell Sensory Systems.* H. Duifhuis, J. W. Horst P.van Dijk, and S. van Netten (eds.), Singapore: World Scientific, pp. 272–279, 1993.

[7] C. Heneghan, M. C. Teich, S. M. Khanna, and M. Ulfendahl, "Nonlinear dynamical motion of cellular structures in the cochlea." *Proc. SPIE,* vol. 2036, pp. 183–197, 1993.

[8] S. E. Keilson, S. M. Khanna, M. Ulfendahl, and M. C. Teich, "Spontaneous cellular vibrations in the guinea-pig cochlea," *Acta Otolaryngologica* (Stockholm), vol. 113, pp. 591–597, 1993.

[9] S. M. Khanna, S. E. Keilson, M. Ulfendahl, and M. C. Teich, "Spontaneous cellular vibrations in the guinea-pig temporal-bone preparation," *Br. J. of Audiol.,* vol. 27, pp. 79–83, 1993.

[10] C. Heneghan, M. C. Teich, S. M. Khanna, and M. Ulfendahl, "Analysis of nonlinear cellular dynamics in the cochlea using the continuous wavelet transform and the short-time Fourier transform." In *Proc. IEEE-Signal Processing Int. Symp. Time–Frequency and Time–Scale Analysis,* October 25–28, 1994.

[11] M. C. Teich, C. Heneghan, S. M. Khanna, and M. Ulfendahl, "Investigating cellular vibrations in the cochlea using the continuous wavelet transform and the short-time Fourier transform." In *Proc. 16th Ann. Int. Conf. IEEE Eng. Med. Biol. Soc.,* November 1–6, 1994.

[12] C. Heneghan, S. M. Khanna, Å. Flock, M. Ulfendahl, L. Brundin, and M. C. Teich, "Investigating the nonlinear dynamics of cellular motion in the inner ear using the short-time Fourier transform and the continuous wavelet transform." *IEEE Trans. Signal Proc.,* vol. 42, pp. 3335–3352, 1994.

[13] M. C. Teich, C. Heneghan, S. M. Khanna, Å. Flock, M. Ulfendahl, and L. Brundin, "Investigating routes to chaos in the guinea-pig cochlea using the continuous wavelet transform and the short-time Fourier transform," *Ann. Biomed. Eng.,* vol. 23, pp. 583–607, 1995.

[14] S. M. Khanna, C. Heneghan, and M. C. Teich, "Dynamical motion of cellular structures in the basal turn of the living cat cochlea." In *Abstracts 17th Midwinter Res. Mtg. Assoc. Res. Otolaryngology,* February 6–10, p. 88, Abstract No. 351, 1994.

[15] M. C. Teich, C. Heneghan, and S. M. Khanna, "Spontaneous cellular vibrations in the basal turn of the living cat cochlea." In *Abstracts 17th Midwinter Res. Mtg. Assoc. Res. Otolaryngology,* February 6–10, p. 88, Abstract No. 352, 1994.

[16] A. Papoulis, *Probability, Random Variables, and Stochastic Processes.* New York: McGraw-Hill, 1984.

[17] D. Gabor, "Theory of communication," *J. IEE,* vol. 93, pp. 429–457, 1946.

[18] A. V. Oppenheim and R. W. Schafer, *Discrete-Time Signal Processing.* Englewood Cliffs, NJ: Prentice-Hall, pp. 713–726, 1989.

[19] F. Hlawatsch and G. F. Boudreaux-Bartels, "Linear and quadratic time–frequency signal representations," *IEEE Signal Proc. Mag.,* vol. 9, no. 2, pp. 21–67, April 1992.

[20] O. Rioul and M. Vetterli, "Wavelets and signal processing," *IEEE Signal Proc. Mag.,* vol. 8, no. 4, pp. 14–38, October 1991.

[21] M. Vetterli and C. Herley, "Wavelets and filter banks: theory and design," *IEEE Trans. Signal Proc.,* vol. 40, pp. 2207–2232, 1992.

[22] M. Vetterli and J. Kovačević, *Wavelets and Subband Coding.* Englewood Cliffs, NJ: Prentice-Hall, 1995.

[23] B. E. A. Saleh and M. C. Teich, *Fundamentals of Photonics.* New York: Wiley, pp. 921–924, 1991.

[24] A. Grossmann, R. Kronland-Martinet, and J. Morlet, "Reading and understanding continuous wavelet transforms." In *Wavelets: Time–Frequency Methods and Phase Space.* J. M. Combes, A. Grossmann, and Ph. Tchamitchian (eds.), New York, NY: Springer-Verlag, pp. 2–20, 2nd edn., 1989/1990.

[25] C. K. Chui, *Wavelet Analysis and its Applications.* Boston, MA: Academic Press, 1992.

[26] D. L. Jones and R. G. Baraniuk, "Efficient approximation of the continuous wavelet transform," *Electronics Letts.,* vol. 27, pp. 748–750, 1991.

[27] C. Herley, J. Kovačević, K. Ramchandran, and M. Vetterli, "Tilings of the time–frequency plane: construction of arbitrary orthogonal bases and fast tiling algorithms," *IEEE Trans. Signal Proc.,* vol. 41, pp. 3341–3359, 1993.

Chapter 10

Iterative Processing Method Using Gabor Wavelets and the Wavelet Transform for the Analysis of Phonocardiogram Signals*

Mustafa Matalgah, Jerome Knopp, Salah Mawagdeh

10.1. INTRODUCTION

Heart sound analysis by auscultation is a qualitative method and insufficient to diagnose some heart diseases. It does not enable the analyst to obtain both qualitative and quantitative characteristics of a phonocardiogram (PCG) [1–3]. Abnormal heart sounds may contain, in addition to the first sound S1 and second sound S2, murmurs and aberrations caused by different pathologies of the cardiovascular system [3]. These aberrations confuse the human ear, obscuring the main sounds of the heart. Studying the physical characteristics of heart sounds and human hearing has shown that the human ear is poorly suited for cardiac auscultation [4]. Therefore, physician capabilities to diagnose heart sounds are limited. Other qualitative characteristics, such as muffled components of a sound, musical murmurs, rumble, or whiff, may be hard to measure or quantify. These characteristics and other features, such as timing of heart sounds and their components, frequency content, location in cardiac cycle and envelope shape of murmurs, can be quantified using digital signal processing techniques. A review of different digital signal processing applications in phonocardiogram analysis was accomplished by Rangayyan and Lehner [3].

The need to detect transient signals arises in various applications such as underwater acoustics, seismic surveillance, and biomedical signals. The processing of these

*Part of the work in this chapter (results in sections 10.4.1, 10.4.2, 10.4.3, and 10.4.4) was initiated at Jordan University of Science & Technology and completed at The University of Missouri (see references [8, 9]). The rest of the chapter is documented by references [12, 13].

highly nonstationary signals often starts by transforming the time-domain signals using a time-frequency (or time-scale) transform. Time-frequency representations, for instance spectrograms, scalograms, or bi-linear representations like the Wigner or smoothed Wigner distributions are now commonly used in signal analysis to investigate the time-frequency content of an analyzed signal. Most of the algorithms that have been developed in this context are essentially based on the study of the "time-frequency energy localization." The representation or its squared-modulus in the linear case is then interpreted as an energy density in the time-frequency plane. Among the linear representations, the Gabor representation [5] (or sliding window Fourier transform), based on time and frequency translations, has been the most popular for a long time. More recently, Grossmann and Morlet [6] proposed an alternative representation, the wavelet transform, which basically has the same structure [7]. The frequency translations are replaced by dilations (the dilation parameter being interpreted as the quotient of a reference frequency by a frequency shift). Both methods have been applied to signal analysis, with comparable performance levels, in a different context. The wavelet transform, in particular, had been applied to analyze normal PCGs [8, 9]. One objective of this chapter is to present a comparison between the wavelet transform performance and other time-frequency analysis methods and generalize its application to normal and abnormal PCG signal analysis.

The role of time-frequency (or time-scale) transforms in transient signal *detection* is somewhat less clear. While practical detection algorithms are often based on these transforms [10], their performance characteristics are not fully understood. In particular, it is not clear which transform offers the best performance for a given class of transient signals. Friedlander and Porat [11] developed an analytical framework within which transforms such as the STFT, the Gabor transform, and the wavelet transform can be compared in a systematic way, in terms of their detection performance. It was observed that the wavelet transform seems generally to be the better transient detector for signals with large time-bandwidth product (broadband signals). This means the wavelet transform gives better time resolution for signals with high-frequency components. On the other hand, the STFT gives better frequency resolution. Therefore, the other objective of this chapter is to develop an algorithm to efficiently analyze the broadband nonstationary signals. This algorithm is a combination of the wavelet and Fourier transforms. It makes use of the wavelet transform (WT) to display the time-domain signal components and the Fourier transform (FT) to display the frequency-domain signal components.

This chapter is organized as follows. First, we start with a theoretical review of the standard FT, short-time Fourier transform (STFT), Wigner distribution (WD), and WT. We compare the WT performance with other time-frequency transforms and generalize its application to normal and abnormal PCG signals. Then, a new algorithm is presented to efficiently analyze the nonstationary signals. It is based on combined WTs and FTs. This algorithm makes use of the WT to display the time-domain signal components and the FT to display its spectral components. A new iterative processing method is also suggested to eliminate the interference terms between the different scales in wavelet analysis [12, 13]. A theorem for this new method is stated and proved. This new algorithm is applied to computer simulation and real PCG examples. Results suggest that it provides comparable temporal and frequency resolution of transients in biomedical signals.

10.2. THEORETICAL BACKGROUND [14–22]

In this section we present the definitions and some theoretical background of the FT, the STFT, the WDs, and the WT.

10.2.1 The Fourier Transform and the STFT

The FT $X(\omega)$ of a signal $x(t)$ is defined as:

$$X(\omega) = \int x(t)\, e^{-j\omega t} dt \tag{10-1}$$

where t and ω are the time and frequency parameters, respectively. It defines the spectrum of $x(t)$ which consists of components at all frequencies over the range for which it is nonzero. The STFT is obtained from the usual FT by multiplying the time signal $x(t)$ by an appropriate sliding time window $w(t)$. The location of the sliding window adds a time dimension and one gets a time-varying frequency analysis. Thus, instead of the usual FT expression one gets a time-frequency expansion of the form:

$$X(t, \omega) = \int x(t)\, w(\tau - t)\, e^{-j\omega\tau} d\tau \tag{10-2}$$

where $w(t)$ is the time window applied to the signal, ω is the frequency, and t is the time.

In the STFT, the signal under study is subdivided into a number of small records where it is assumed that each sub-record is stationary. To reduce the effect of leakage (the effect of having finite data), each sub-record is then multiplied by an appropriate window and then the FT is applied to each sub-record. As long as each sub-record does not contain rapid changes, the spectrogram will give an excellent idea of how the spectral composition of the signal has changed during the whole time record. However, there exist signals in nature whose spectral content is changing so rapidly that finding an appropriate short-time window is problematic since there may not be any time interval for which the signal is stationary. To deal with these time changes properly, it is necessary to keep the length of the time window as short as possible. This, however, will reduce the frequency resolution in the time-frequency plane. Hence, there is a trade-off between time and frequency resolutions.

10.2.2 The Wigner Distribution

The WD of a signal $x(t)$ is defined as:

$$WD(t, \omega) = \int x\!\left(t + \frac{\tau}{2}\right) x^*\!\left(t - \frac{\tau}{2}\right) e^{-j\omega\tau} d\tau \tag{10-3}$$

where t and ω are the time and frequency variables, respectively.

The Wigner Distribution and the corresponding Wigner–Ville Distribution have been satisfactorily applied in the analysis of nonstationary signals [14–16]. This comes from the ability of the WD to separate signals in both time and frequency domains. One advantage of the WD over the STFT is that it does not suffer from the time-frequency trade-off problem. On the other hand, the WD has the disadvantage that it is limited by the appearance of cross-terms. These cross-terms are due to the nonlinearity property of the WD. One way to remove these cross-terms is by smoothing the time-frequency plane [15], but this will be at the expense of decreased resolution in both time and frequency.

10.2.3 The Wavelet Transform

The WT of a signal $x(t)$ with respect to an analyzing wavelet $g(t)$ is defined as:

$$WT(t, a) = \frac{1}{\sqrt{(a)}} \int x(\tau) g^* \left(\frac{t - \tau}{a} \right) d\tau$$

$$= \sqrt{(a)} \int X(\omega) G^*(a\omega) e^{j\omega t} d\omega \tag{10-4}$$

where * denotes a complex conjugate, and $g(t)$ is the so-called analyzing wavelet. $X(\omega)$ and $G(\omega)$ are the FTs of $x(t)$ and $g(t)$, respectively. The parameter 'a' is a scaling parameter which is inversely proportional to the frequency.

The analyzing wavelet function $g(t)$ must satisfy a certain number of properties. The most important are continuity, integrability, square integrability, progressivity and admissibility (zero DC components). Moreover, the wavelet $g(t)$ must be concentrated in both time and frequency as much as possible. It is well known that the smallest time-bandwidth product is achieved by the Gaussian function [17, 18]. Hence the most suitable analyzing wavelet for time-frequency analysis is the complex exponential modulated Gaussian function. If we choose the analyzing wavelet that has the FT of the form:

$$G(\omega) = Ae^{-(\omega-\omega_0)^2/2} + \epsilon \tag{10-5}$$

where ϵ is a small correction term, theoretically necessary to satisfy the admissibility condition of wavelets, and ω_0 is chosen large enough so that the correction term is negligible and can be ignored. Then, in the time-domain, with the correction term ignored and $G(\omega)$ normalized to 1, the analyzing wavelet is the modulated Gaussian function:

$$g(t) = \frac{1}{\sqrt{(2\pi)}} e^{-t^2/2} e^{j\omega_0 t} \tag{10-6}$$

This is known as the Gabor wavelet. It was shown that $\omega_0 = 5.33$, which is enough to make the correction term negligible and gives an optimal time-bandwidth product, and this value was used by others [17, 19, 21]. The main properties of the wavelet transform are found in references [17, 19–22].

10.3 COMBINED WAVELET–FOURIER TRANSFORM [12, 13]

We describe in this section an algorithm developed to improve the time and frequency resolutions in the *spectral analysis* of nonstationary signals [12, 13]. This analysis is limited to the particular case of broadband signals.

As indicated in the introduction, the WT gives better time-resolution than the STFT and Gabor transform in the analysis of broadband signals. On the other hand, the STFT (which depends on the standard FT) gives better frequency resolution. These two transforms can be combined together to produce better resolution in both time and frequency. In this algorithm, the WT role is to display the time-domain signal components of the nonstationary signal since it gives better time resolution; whereas, the frequency-domain signal components are displayed using the standard FT since it gives better frequency resolution. The algorithm can be summarized as follows. First, the nonstationary signal is decomposed into subband stationary signals using the WT with the appropriate scales. Second, the inverse WT is applied to each of these stationary subband signals to obtain sub-stationary signals in the time-domain. Finally, the classical fast Fourier transform (FFT) is applied to each sub-stationary signal to find its spectrum. This algorithm can be generalized for both discrete and continuous WTs. In this chapter, we are concerned with the continuous wavelet transform (CWT) since it is more appropriate than the discrete wavelet transform (DWT) for the spectral analysis of nonstationary signals [21–28]. Applications of the DWT are found in signal coding, including image compression and computer vision [29, 30]. This chapter is concerned with the spectral analysis of nonstationary signals, not with the signal-coding applications. Therefore, only CWTs will be used in the algorithm when applied to our examples throughout this chapter. In the following subsection, we introduce a theorem for multiscale analysis of nonstationary signals then we apply it in the analysis of PCGs.

10.3.1 Theorem and Proof

In this section, we first discuss the phenomena of the effect of interference terms on the WT then we introduce a theorem to eliminate those interference terms. The CWT was originally introduced by Grossmann and co-workers [6, 7, 31]. Time t and time-scale parameters vary continuously as defined in Eq. (10-4). Define $g_{a,b}$ (b,a), the wavelet corresponding to scale a and time location b, as $g_{a,b} = a^{-1/2} g((t\text{-}b)/a)$, and $g(t)$ is the wavelet "prototype," which can be thought of as a bandpass function. Then the WT can be defined as the inner product of the signal $x(t)$ and the wavelet function $g_{a,b}(b,a)$:

$$\text{WT}(b, a) = \langle x(t), \frac{1}{\sqrt{(a)}} g_{a,b}\left(\frac{(t - b)}{a}\right)\rangle \tag{10-7}$$

This is the convolution of $x(t)$ with $(1/\sqrt{(a)})g^*(t/a)$. If we use the convolution property of the FT which states that convolving two signals in the time-domain corresponds to their multiplication in the frequency domain, then in the frequency domain of $x(t)$ and $g(t)$, $\text{WT}(\omega, a)$ can be written as:

$$\text{WT}(\omega, a) = \sqrt{(a)}\, X(\omega)\, G^*(a\omega) \tag{10-8}$$

For the case of a single tone signal $x(t) = \cos(\omega_1 t)$, the $WT\,(\omega, a)$ is

$$\text{WT}(\omega, a) = \frac{\sqrt{(a)}}{2}\, G(a\omega_1) \tag{10-9}$$

Linear combinations of such signals show the expected interference effects [21]. To explain this effect, let $x(t) = \cos(\omega_1 t) + \cos(\omega_2 t) + \cdots + \cos(\omega_n t)$, then the wavelet transform of $x(t)$ is

$$\text{WT}(\omega, a) = \frac{\sqrt{(a)}}{2}[G(a\omega_1) + G(a\omega_2) + \cdots + G(a\omega_n)] \tag{10-10}$$

In multiscale analysis, the scale parameter a is chosen to adapt a specific tone component (e.g., of frequency ω_i). We choose the function that has an FT of the form of Eq. (10-5) and a time-domain of the form of Eq. (10-6), assuming $A = 1$ and $\epsilon = 0$, to be our analysis wavelet. This is known as the Gabor wavelet [5]. In multiscale (wavelet) analysis, to detect a certain signal of frequency ω_i , we choose scale parameter $a = \omega_0/\omega_i$. Then the modulus of the wavelet transform in Eq. (10-10) is

$$|\text{WT}(\omega, a_i)| = \frac{\sqrt{(a_i)}}{2}\left[G\left(\frac{\omega_0}{\omega_i}\omega_1\right) + G\left(\frac{\omega_0}{\omega_i}\omega_2\right) + \cdots + G(\omega_0) + \cdots + G\left(\frac{\omega_0}{\omega_i}\omega_{n-1}\right) \right.$$
$$\left. + G\left(\frac{\omega_0}{\omega_i}\omega_n\right) \right] \tag{10-11}$$

The spectral main peak associated with scale $a = \omega_0/\omega_i$ comes from the factor $G(\omega_0)$ in the above equation. All other terms are interference effects resulting from the other frequency components in the signal. This phenomenon was indicated by Kronland-Martinet et al. [21]. The main contribution of this chapter is to eliminate those interference terms. We introduce the following theorem.

Theorem.   Let $x(t)$ be a combination of N signals of different frequency components ω_i , $i = 1, 2, \ldots, N$ $(x(t) = \cos(\omega_1 t) + \cos(\omega_2 t) + \ldots + \cos(\omega_n t))$. The spectrum of any component of frequency ω_i can be detected from $x(t)$ with reduced interference terms by repeating the process of consecutive forward and backward wavelet transform M number of times with the same scale parameter $a_i = \omega_0/\omega_i$, provided that the analysis wavelet is normalized to 1 in the frequency domain, and as M approaches infinity the interference terms approach 0.

Proof.

1. Let $\text{WT}_{(1)}(\omega, a_i)$ be the forward wavelet transform of a signal $x(t)$ in the frequency domain as given in Eq. (10-8).
2. The inverse wavelet transform in the time-domain, $\text{IWT}_{(1)}(b, a_i)$, can be written as the convolution of $\text{WT}_{(1)}(b, a_i)$ with the wavelet function $(1/\sqrt{(a_i)})g^*(b/a_i)$ as follows:

$$\mathrm{IWT}_{(1)}(b, a_i) = \mathrm{WT}_{(1)}(b, a_i) * \frac{1}{\sqrt{(a_i)}} g^*\left(\frac{b}{a_i}\right) \qquad (10\text{-}12)$$

3. The second forward wavelet transform, $\mathrm{WT}_{(2)}(b, a_i)$, of the inverse wavelet transform function, $\mathrm{IWT}_{(1)}(b, a_i)$, can be written as follows:

$$
\begin{aligned}
\mathrm{WT}_{(2)}(b, a_i) &= \mathrm{IWT}_{(1)}(b, a_i) * \frac{1}{\sqrt{(a_i)}} g^*\left(\frac{b}{a_i}\right) \\
&= \mathrm{WT}_{(1)}(b, a_i) * \frac{1}{\sqrt{(a_i)}} g^*\left(\frac{b}{a_i}\right) * \frac{1}{\sqrt{(a_i)}} g^*\left(\frac{b}{a_i}\right)
\end{aligned}
\qquad (10\text{-}13)
$$

4. The second inverse wavelet transform $\mathrm{IWT}_{(2)}(b, a_i)$ of the second forward wavelet transform function, $\mathrm{WT}_{(2)}(b, a_i)$, is:

$$
\begin{aligned}
\mathrm{IWT}_{(2)}(b, a_i) &= \mathrm{WT}_{(2)}(b, a_i) * \frac{1}{\sqrt{(a_i)}} g^*\left(\frac{b}{a_i}\right) \\
&= \mathrm{WT}_{(1)}(b, a_i) * \frac{1}{\sqrt{(a_i)}} g^*\left(\frac{b}{a_i}\right) * \frac{1}{\sqrt{(a_i)}} g^*\left(\frac{b}{a_i}\right) \\
&\quad * \frac{1}{\sqrt{(a_i)}} g^*\left(\frac{b}{a_i}\right)
\end{aligned}
\qquad (10\text{-}14)
$$

5. Since convolution in the time-domain is equivalent to multiplication in the frequency domain, and taking the FT of both sides of Eq. (10-14) with respect to the time parameter b, we can end with the following form:

$$
\begin{aligned}
\mathrm{IWT}_{(2)}(\omega, a_i) &= \mathrm{WT}_{(1)}(\omega, a_i) \frac{1}{\sqrt{(a_i)}} a_i G^*(a_i\omega) \frac{1}{\sqrt{(a_i)}} a_i G^*(a_i\omega) \frac{1}{\sqrt{(a_i)}} a_i G^*(a_i\omega) \\
&= \mathrm{WT}_{(1)}(\omega, a_i)\left(\sqrt{(a_i)}\, G^*(a_i\omega)\right)^3
\end{aligned}
\qquad (10\text{-}15)
$$

6. If we continue this consecutive WT-IWT process M number of times, we get the following general form for IWT after M number of times of WT-IWT:

$$\mathrm{IWT}_{(M)}(\omega, a_i) = \mathrm{WT}_{(1)}(\omega, a_i) \cdot \left(\sqrt{(a_i)} \cdot G^*(a_i\omega)\right)^{M+1} \qquad (10\text{-}16)$$

7. Substituting for $\mathrm{WT}_{(1)}(\omega, a_i)$ from Eq. (10-8), we get

$$
\begin{aligned}
\mathrm{IWT}_{(M)}(\omega, a_i) &= \sqrt{(a_i)}X(\omega)\, G^*(a_i\omega)\left(\sqrt{(a_i)}\, G^*(a_i\omega)\right)^{M+1} \\
&= X(\omega)\left(\sqrt{(a_i)}\, G^*(a_i\omega)\right)^{M+2}
\end{aligned}
\qquad (10\text{-}17)
$$

The modulus $\left|\mathrm{IWT}_{(M)}(\omega, a_i)\right|$ is

$$\left|\mathrm{IWT}_{(M)}(\omega, a_i)\right| = \left|X(\omega)\left(\sqrt{(a_i)}\, G^*(a_i\omega)\right)^{M+2}\right| \qquad (10\text{-}18)$$

$G(\omega)$ is given by Eq. (10-5) with $A = 1$ and $\epsilon = 0$, and $X(\omega)$ is the FT of $x(t)$:

$$X(\omega) = \frac{1}{2}[\delta(\omega - \omega_1) + \delta(\omega + \omega_1)] + \frac{1}{2}[\delta(\omega - \omega_2) + \delta(\omega + \omega_2)]$$
$$+ \cdots + \frac{1}{2}[\delta(\omega - \omega_n) + \delta(\omega + \omega_n)] \tag{10-19}$$

The WT considers only positive scales ($a_i = \omega_0/\omega$), i.e., $X(\omega)$ will take only positive frequencies. Therefore, $X(\omega)$ will be rewritten as:

$$X(\omega) = \frac{1}{2}[\delta(\omega - \omega_1) + \delta(\omega - \omega_2) \cdots + \delta(\omega - \omega_n)] \tag{10-20}$$

8. By substituting for $X(\omega)$ in Eq. (10-18), we get:

$$\left|\text{IWT}_{(M)}(\omega, a_i)\right| = \frac{1}{2}\left(\sqrt{(a_i)}\right)^{M+2}(\delta(\omega - \omega_1) + \delta(\omega - \omega_2) + \cdots + \delta(\omega - \omega_i)$$
$$+ \cdots + \delta(\omega - \omega_n))(G^*(a_i\omega))^{M+2}$$
$$= \frac{1}{2}\left(\sqrt{(a_i)}\right)^{M+2}[(G^*(a_i\omega_1))^{M+2} + (G^*(a_i\omega_2))^{M+2}$$
$$+ \cdots + (G^*(a_i\omega_i))^{M+2} + \cdots + (G^*(a_i\omega_n))^{M+2}] \tag{10-21}$$

The spectrum of a component, $x_i(t)$, of frequency ω_i can be detected by choosing $a_i = \omega_0/\omega_i$. Then, $X_i(\omega)$ can be found from Eq. (10-21) as:

$$|X_i(\omega)| = \left|\text{IWT}_{(M)}(\omega, a_i)\right| = \frac{1}{2}\left(\sqrt{(a_i)}\right)^{M+2}\left[\left(G^*\left(\frac{\omega_0}{\omega_i}\omega_1\right)\right)^{M+2}\right.$$
$$+ \left(G^*\left(\frac{\omega_0}{\omega_i}\omega_2\right)\right)^{M+2} + \cdots + \left(G^*\left(\frac{\omega_0}{\omega_i}\omega_i\right)\right)^{M+2} \tag{10-22}$$
$$\left. + \cdots + \left(G^*\left(\frac{\omega_0}{\omega_i}\omega_n\right)\right)^{M+2}\right]$$

The analysis wavelet $G(\omega)$ is equal to 1 at $\omega = \omega_0$, and is less than 1 at all other values. Therefore, by taking the limit as M goes to infinity in Eq. (10-22), all terms $(G^*(\frac{\omega_0}{\omega_i}\omega_k))^{M+2}$ $\{k = 1, 2, \ldots, n; \omega \neq \omega_i\}$ will converge to zero, except when ω approaches ω_i then one of the terms will stay constant at $(G^*(\omega_0))^{M+2} = 1$, (Eq. (10-5), for any value of M, with $A = 1$ and $\epsilon = 0$). Therefore, all cross-terms will converge to zero except the main spectral peak associated with the scale $a = \omega_0/\omega_i$ which comes from the factor $G^*(\omega_0)$ in the above equation. This main peak will converge to 1. This completes the proof of the theorem. $\qquad\square$

10.4. COMPUTER SIMULATION AND REAL DATA

In this section we present the experimental results and discuss the applications of each of the FFT, STFT, WD, and WT to the analysis and diagnosis of PCG signals. Two cases, normal and abnormal PCG signals, are considered. The sampling rate used is 800 samples/s. Both the time and frequency axes are scaled linearly. The frequency scan is from 1 Hz to 400 Hz. Different methodologies regarding heart sound transmission, acquisition, and localization of their sources are discussed in detail in reference [3].

10.4.1 The Fourier Transform

Figure 10-1(a) shows one second of a normal heart sound signal which contains the two major sounds S1 and S2. The sampling rate used is 800 samples/s. The FFT is applied to the first half of this signal to analyze the frequency contents of S1 as shown in Fig. 10-1(b), and then to the second half to analyze the frequency contents of S2 as shown in Fig. 10-1(c). A 512-point FFT is applied to S1 for 160 samples (0.20 s), and a 512-point FFT is applied to S2 for 80 samples (0.10 s); (zero-padding is used). The basic frequency components are obviously detected by the FT but not the time delay between these components. The two components A2 and P2 of the second sound S2 are obvious in Fig. 10-1(c). On the other hand, FFT analysis of S2 cannot tell which of A2 and P2 precedes the other. For a normal heart usually A2 precedes P2, but due to some pathological conditions A2 and P2 may be reversed in time order [3]. Therefore, the usual FFT is unable to accurately diagnose heart diseases. It is essential to look for a transform which will describe a kind of "time-varying" spectrum. A common approach to estimating time-varying spectra is the short-time Fourier transform (STFT).

10.4.2 The Short-Time Fourier Transform

Figure 10-2(a), (b), and (c) show the STFT of the normal heart sound signal of Fig. 10-1(a) for three different window lengths for 32, 64, and 128 samples, respectively. A 128-sample FFT is applied to each windowed segment in each of the three cases (zero-padding is used to improve the resolution in the frequency domain). It is shown that the four components of the first sound S1 are not detected in all cases, and also the two components A2 and P2 of the second sound are not accurately represented. In order to analyze this signal more accurately using the STFT, one may think of increasing the sampling rate of the original signal; but we will see later that the WT gives better results under the same conditions and same sampling rate.

10.4.3 The Wigner Distribution

Figure 10-3(a) and (b) show the surface and contour plots of the pseudo-WD of the normal PCG signal of Fig.10-1(a). As shown in the figure, the four components of S1 are not accurately detected and S2 seems to have only 1 component rather than the 2 components A2 and P2. The pseudo-WD is being used since it does not suffer from the cross-terms problem.

10.4.4 The Wavelet Transform

Figure 10-4(a) and (b) (surface and contour plots) show the WT of the normal PCG signal of Fig. 10-1(a). The WT of S1 is displayed in Fig. 10-4(c) and (d) (surface

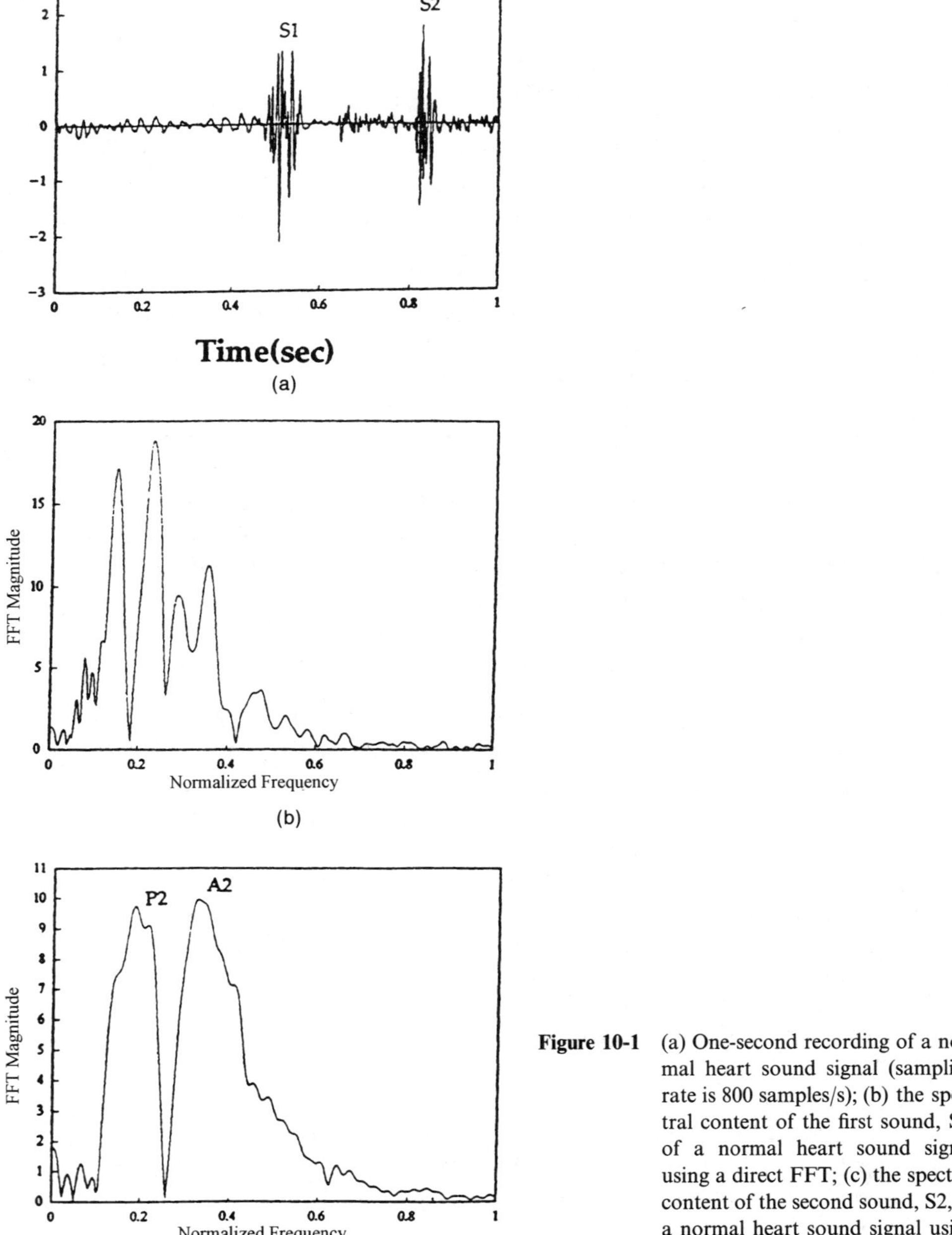

Figure 10-1   (a) One-second recording of a normal heart sound signal (sampling rate is 800 samples/s); (b) the spectral content of the first sound, S1, of a normal heart sound signal using a direct FFT; (c) the spectral content of the second sound, S2, of a normal heart sound signal using a direct FFT.

and contour plots). The Wavelet Transform of S2 is displayed in Figs. 4(e) and 4(f) (surface and contour). As is clear from Fig. 10-4(a) and (b), S2 has a higher frequency content than S1. This is expected since the amount of blood present in the

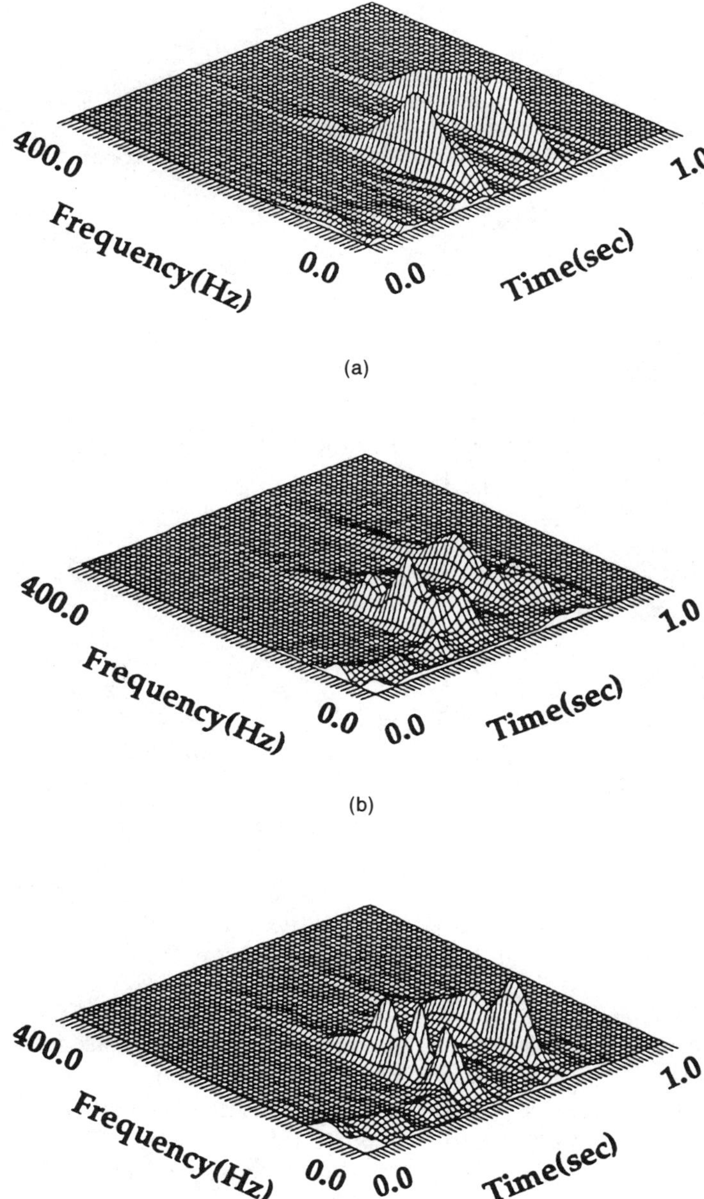

(a)

(b)

(c)

Figure 10-2   (a) The STFT of the normal heart sound signal in Fig.10-1(a) using a 32-point window length. (b) The STFT of the normal heart sound signal in Fig.10-1(a) using a 64-point window length. (c) The STFT of the normal heart sound signal in Fig.10-1(a) using a 128-point window length.

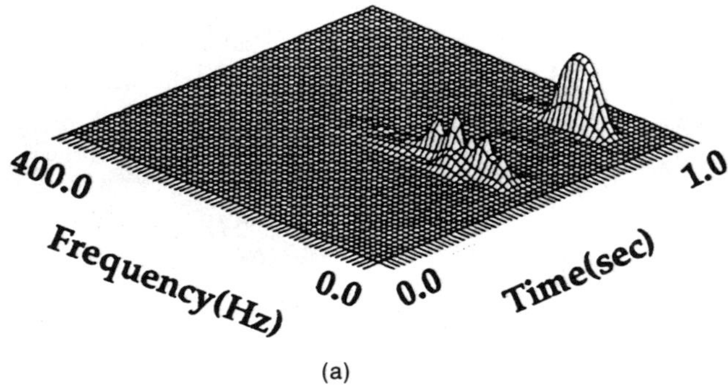

(a)

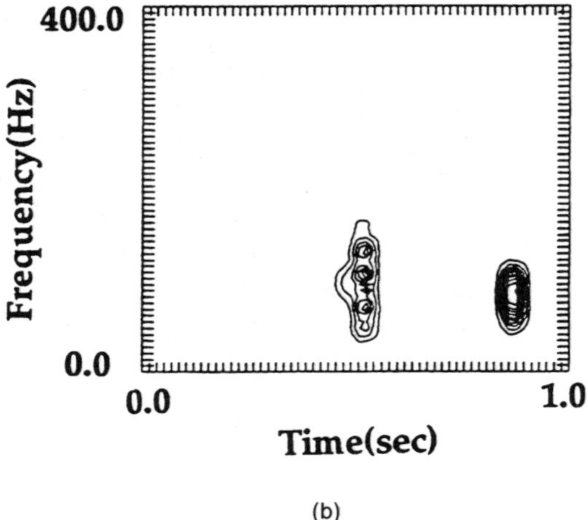

(b)

Figure 10-3   (a) The pseudo-WD of the normal heart sound signal in Fig.10-1(a).
(b) A contour plot of the surface of Fig.10-3(a).

cardiac chambers is less [4]. The spectrum of S1 has reasonable values in the range 10–300 Hz; and it is clearly resolved in time into four components. Most of the energy of S1 seems, however, to be concentrated in its second and third components. S2 is clearly resolved into two components, A2 and P2, which are due to closure of the aortic and pulmonary valves, respectively. As seen in Fig. 10-4(e) and (f), A2 has a higher frequency content than P2, and the time delay between A2 and P2 can be estimated as 20 ms. The aortic valve normally closes before the pulmonary valve and hence A2 should precede P2 [4]. The length of this delay is important in diagnosis since its length is directly related to different pathologies. There are even cases where the delay is negative, that is, the order of occurrence of A2 and P2 is reversed [3]. Therefore, the WT allows us to measure and determine this time difference and thus produce a diagnostic process regarding this parameter. To generalize the evaluation

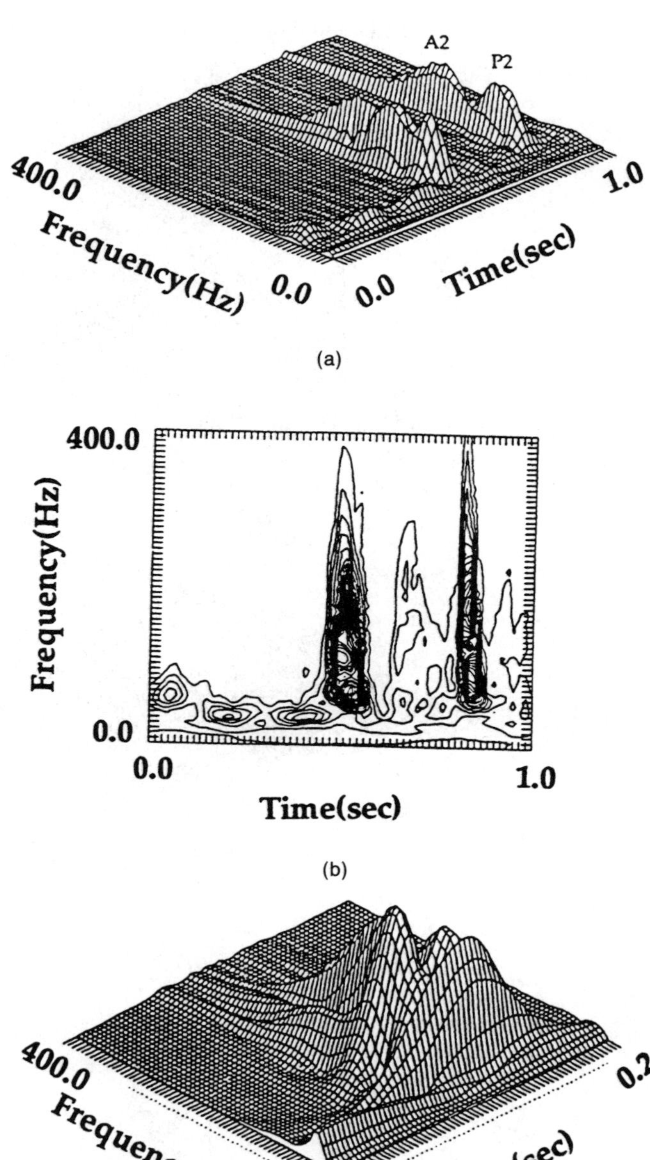

Figure 10-4   (a) The WT of the normal heart sound signal in Fig.10-1(a). (b) A contour plot of the surface of Fig.10-4(a). (c) The WT of the first sound, S1, of the normal heart sound signal of Fig.10-1(a). (d) A contour plot of the surface of Fig.10-4(c). (e) The WT of the second sound, S2, of the normal heart sound signal of Fig.10-1(a). (f) A contour plot of the surface of Fig. 10-4(e).

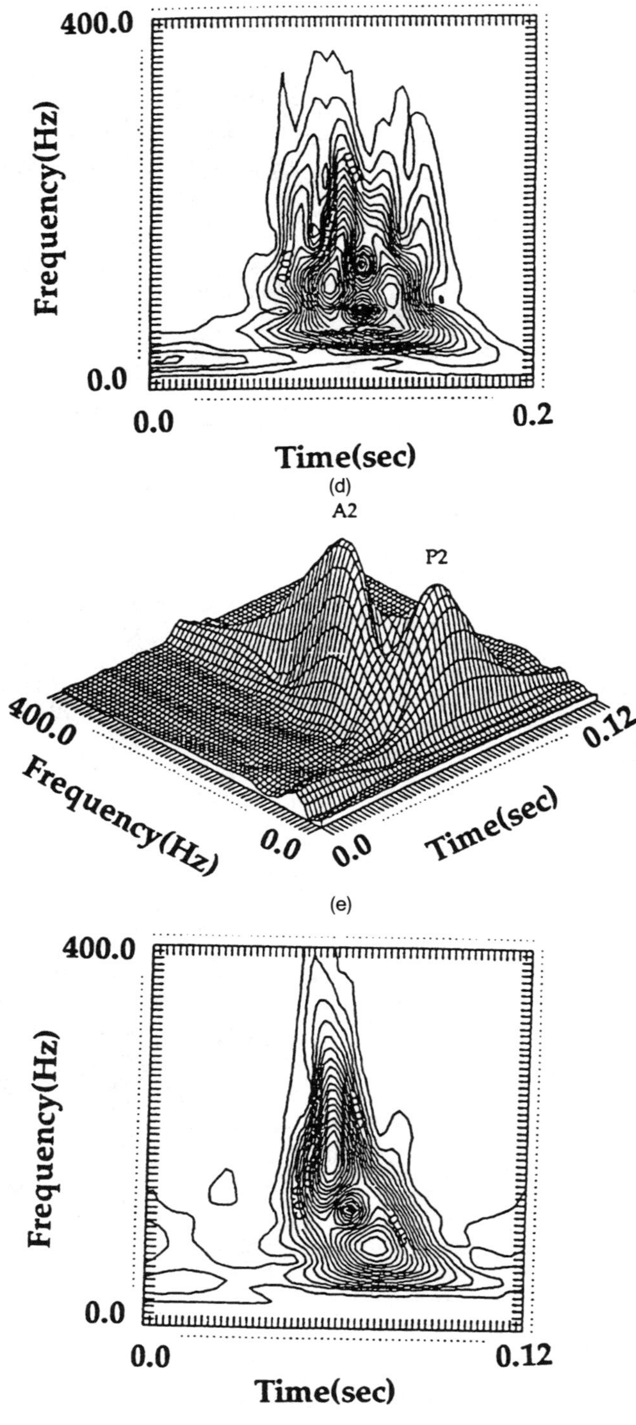
400.0
Frequency(Hz)
0.0
0.0
Time(sec)
0.2
(d)
A2
P2
400.0
Frequency(Hz)
0.0
0.0
Time(sec)
0.12
(e)
400.0
Frequency(Hz)
0.0
0.0
Time(sec)
0.12
(f)

of the WT, we analyzed two more normal PCG signals. These signals and their WTs are shown in Fig. 10-5(a), (b), and (c), and Fig. 10-6(a), (b), and (c). In these figures, it is shown that S2 is clearly resolved into two components, A2 and P2, and S1 is resolved into four components.

To generalize the utility of the WT in the analysis of PCGs, we applied it to an abnormal PCG signal. Figure 10-7(a) shows a 1 s recording of the PCG signal of a patient with "atrium septal defect (ASD)" sampled at a rate of 800 Hz. On investigation of the time-domain PCG signal, one may think it is a normal PCG, but actually, as shown in the figure, there is a small additional component in the period between S1 and S2. This abnormality in the PCG signal comes from an ASD. Moreover, as shown in Fig. 10-7(b) and (c) the two components A2 and P2 of the second sound are so close that they merge into one component. This indicates that this patient may have another kind of disease which cannot be detected from the time-domain of the PCG signal of Fig. 10-7(a). In addition, some components in the period between S1 and S2 are clearly obvious in Fig. 10-7(b) and (c).

10.4.5 Iterative Processing Method

In this section we apply our algorithm to analyze simulated and real data examples. It is shown how these signals can be resolved into their temporal and frequency components using this algorithm. Gabor wavelets [5, 32] are used in our examples for the WT analysis.

EXAMPLE 1 (COMPUTER SIMULATION SIGNAL):

In this example several weak transients embedded into a strong gaussian modulated signal are to be analyzed using the algorithm we have developed. The mathematical representation of this nonstationary signal is:

$$x(t) = [0.5\cos(2\pi f_1(t - t_1)) + 0.5\cos(2\pi f_2(t - t_1))]\exp(-(t - t_1)^2/1000)$$
$$+ 10\cos(2\pi f_3(t - t_2))\exp(t - t_2)^2/10\,000)$$
$$+ 0.5\cos(2\pi f_4(t - t_2))\exp(-(t - t_2)^2/100)$$
$$+ [0.5\cos(2\pi f_1(t - t_3)) + 0.5\cos(2\pi f_2(t - t_3))]\exp(-(t - t_3)^2/1000)$$

where $f_1, f_2, f_3,$ and f_4, are chosen to be 0.40, 0.45, 0.30, and 0.35, respectively, as fractions of the sampling rate. For a sampling rate of 1024 samples/s, which is used in our example, they correspond to $f_1 = 409.6$ Hz, $f_2 = 460.8$ Hz, $f_3 = 307.2$ Hz, and $f_4 = 358.4$ Hz. The occurrence times of the different transient signals are chosen to be $t_1 = 0.25$ s, $t_2 = 0.5$ s, $t_3 = 0.75$ s.

One second of this computer simulation transient signal is shown in Fig. 10-8. A sampling rate of 1024 samples/s was used. The weak transient signals can be enhanced by repeating the WT-IWT process several times. This suggested repetitive process also helps to suppress the interference terms in the wavelet transform, which will be illustrated in this example by hundred-fold repetitions of the WT. Figure 10-9(a) shows the WT of this signal for the scale $a = 0.002\,76$. The WT after 100 iterations for the same scale is shown in Fig. 10-9(b). Figure 10-10(a) and (b), show the WTs for the scale $a = 0.002\,37$ after 1 and 100 iterations, respectively. The WT for the scale $a = 0.002\,07$ is shown in Fig. 10-11(a) and (b), after 1 and 100 iterations, respectively. Figure 10-12(a) and (b), show the WTs for the scale $a = 0.001\,84$ after 1 and 100 iterations, respectively. Figure 10-13(a) shows the FT of the subband reconstructed

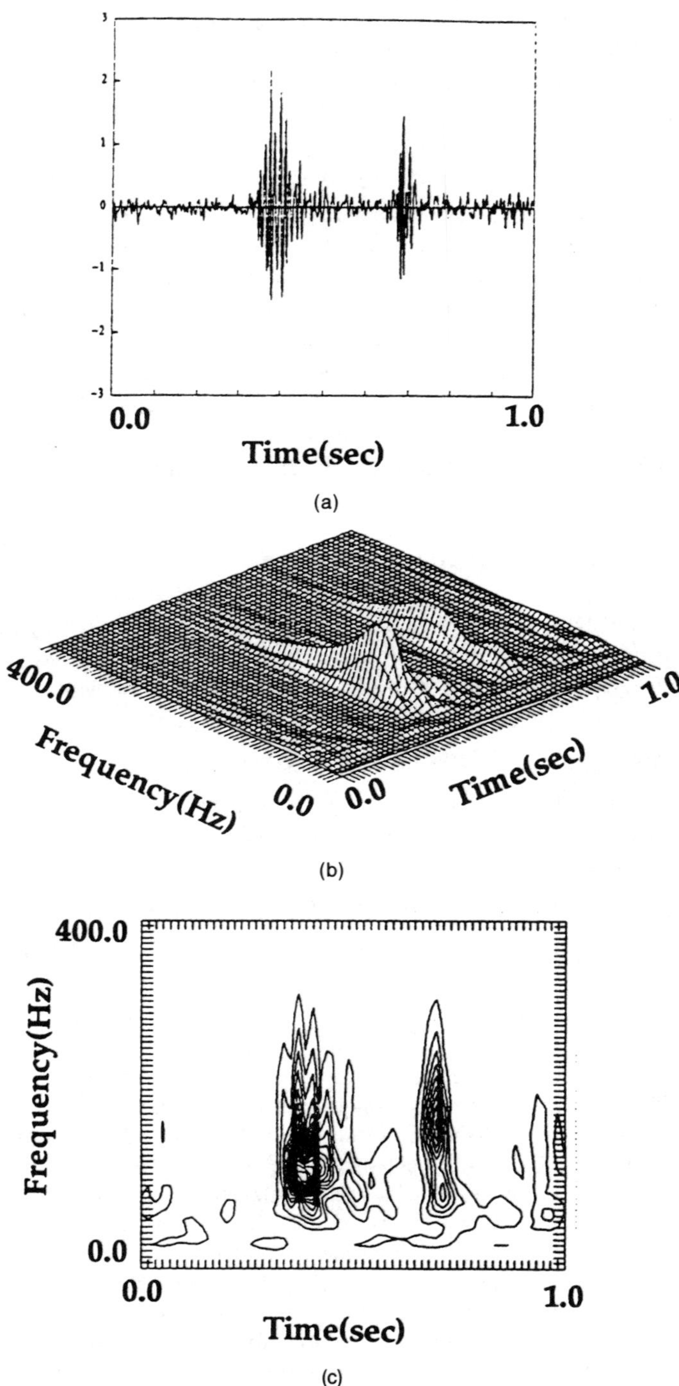

Figure 10-5   (a) One-second recording of a normal heart sound signal (sampling rate is 800 samples/s). (b) The WT of the normal heart sound signal in Fig.10-5(a). (c) A contour plot of the surface of Fig.10-5(b).

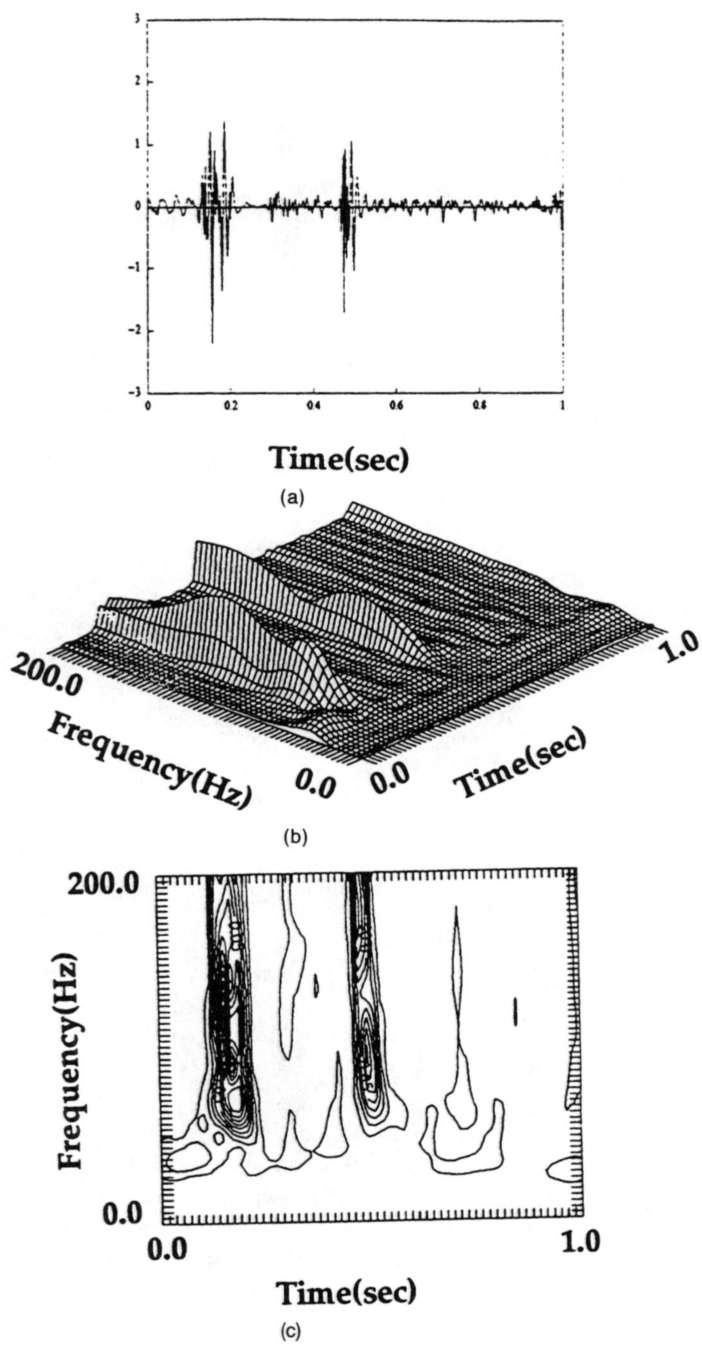

Figure 10-6   (a) One-second recording of a normal heart sound signal (sampling rate is 400 samples/s). (b) The WT of the normal heart sound signal in Fig. 10-6(a). (c) A contour plot of the surface of Fig. 10-6(b).

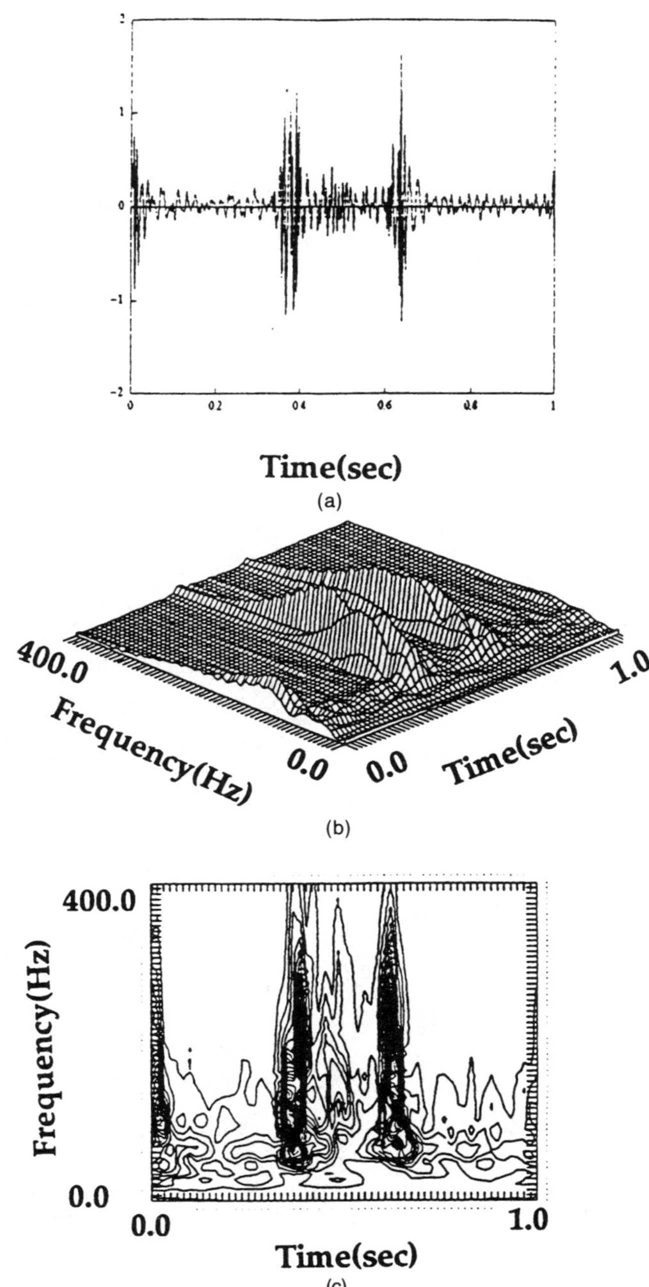

Figure 10-7   (a) One-second recording of the heart sound signal of a patient with ASD (sampling rate is 800 samples/s). (b) The WT of the abnormal heart sound signal in Fig.10-7(a). (c) A contour plot of the surface of Fig.10-7(b).

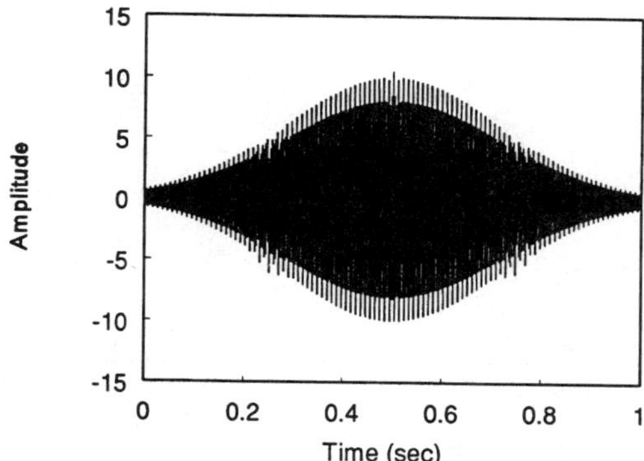

Figure 10-8 One second of the computer simulated transient signal used in example 1. The sampling rate is 1024 samples/s.

signal that corresponds to the scale $a = 0.00276$. Figure 10-13(b) shows the FT after 100 iterations for the same scale used for Fig. 10-13(a). Figure 10-14(a) and (b), shows the FTs of the reconstructed signal after 1 and 100 iterations, respectively, for the scale $a = 0.00237$. The FTs of the reconstructed signal that corresponds to the scale $a = 0.00207$ after 1 and 100 iterations are shown in Fig. 10-15(a) and (b), respectively. The FTs after 1 and 100 iterations using the scale $a = 0.00184$ are shown in Fig. 10-16(a) and (b), respectively.

EXAMPLE 2 (REAL SIGNAL):

Figure 10-17 shows a 1 s real PCG signal sampled at 800 samples/s. Its WT is shown in Figs. 10-18(a), 19(a), 20(a), 21(a), 22(a), and 23(a), for scales $a = 0.014727$, 0.009261, 0.007364, 0.005960, 0.011411, and 0.006590, respectively. The WTs after 100 iterations are shown in Figs. 10-18(b), 19(b), 20(b), 21(b), 22(b), and 23(b), using the same scales in the same order as for part (a) of the figures. The FTs of the subband signals that correspond to the same scales are shown in Figs. 10-24(a), 25(a), 26(a), 27(a), 28(a), and 29(a), for 1 iteration of the WT, and in Figs. 10-24(b), 25(b), 26(b), 27(b), 28(b), and 29(b), for 100 iterations of the WT.

10.5. DISCUSSION AND CONCLUSIONS

We have presented the applications of the STFT, WD, and WT to PCG signal analysis. These time-frequency analysis techniques can provide better diagnostic information than the normal FT. This is due to the fact that the PCG signals are nonstationary. A comparison of the three methods has shown the resolution differences between them. It is found that the STFT cannot detect the four components of the first sound, S1, of the PCG signal. Moreover, the two components of the second sound are inaccurately detected. The WD can provide time-frequency characteristics of the PCG signal, but with insufficient diagnostic information. The four components of the first sound S1 are not accurately detected and the two components of the second sound S2 seem to be one component. It is found that the WT is capable of detecting the two components, the aortic valve component A2 and the pulmonary valve component P2, of the second sound S2 of a normal PCG signal. These components are not detectable using the STFT or the WD. However, the standard FT can display the frequencies of these two components but cannot display the time delay between them. The WT provides more features and characteristics of the PCG

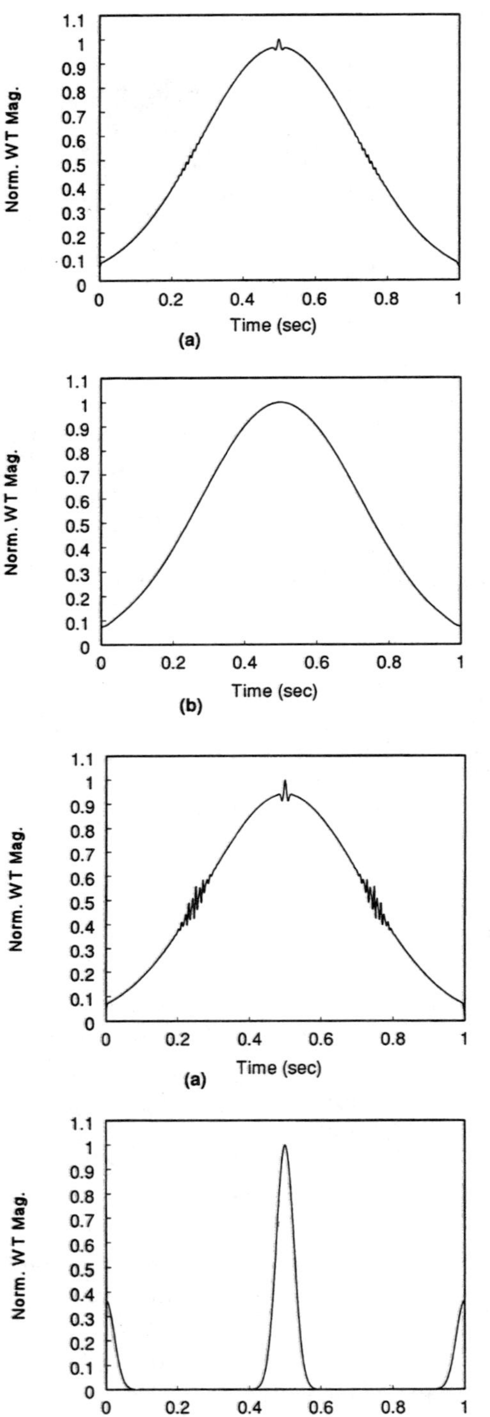

Figure 10-9 (a) WT of the signal in example 1 at scale 0.002 76. (b) WT after 100 repetitive WT-IWT processes of the signal in example 1 at scale 0.002 76.

Figure 10-10 (a) WT of the signal in example 1 at scale 0.002 37. (b) WT after 100 repetitive WT-IWT processes of the signal in example 1 at scale 0.002 37.

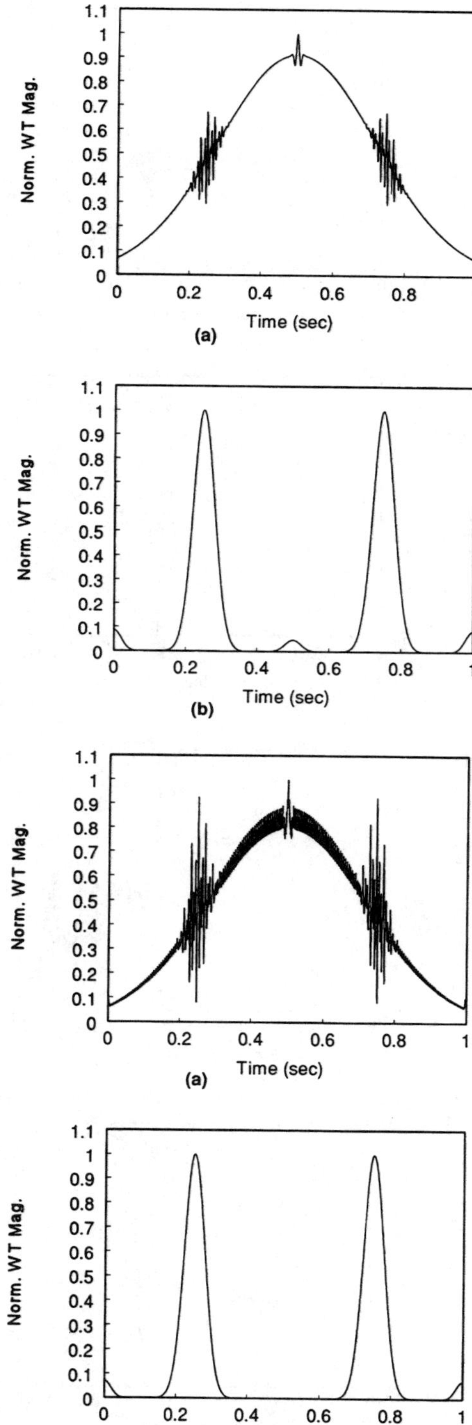

Figure 10-11   (a) WT of the signal in example 1 at scale 0.002 07. (b) WT after 100 repetitive WT-IWT processes of the signal in example 1 at scale 0.002 07.

Figure 10-12   (a) WT of the signal in example 1 at scale 0.001 84. (b) WT after 100 repetitive WT-IWT processes of the signal in example 1 at scale 0.001 84.

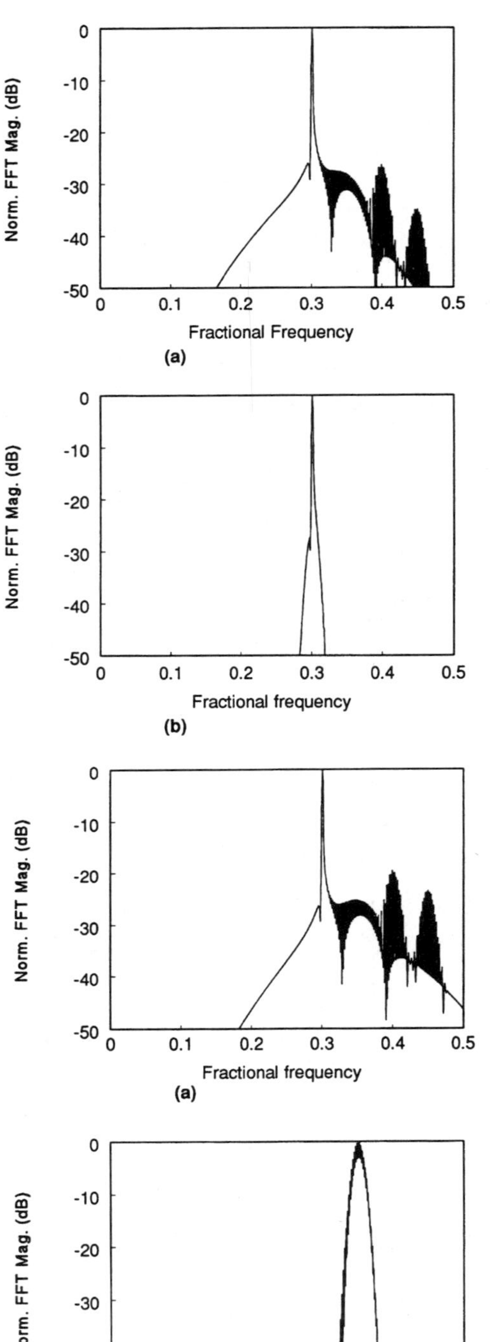

Figure 10-13   (a) FT of the reconstructed signal from the signal in example 1 after the WT-IWT process at scale 0.002 76. (b) FT of the reconstructed signal from the signal in example 1 after 100 repetitive WT-IWT processes at scale 0.002 76.

Figure 10-14   (a) FT of the reconstructed signal from the signal in example 1 after the WT-IWT process at scale 0.002 37. (b) FT of the reconstructed signal from the signal in example 1 after 100 repetitive WT-IWT processes at scale 0.002 37.

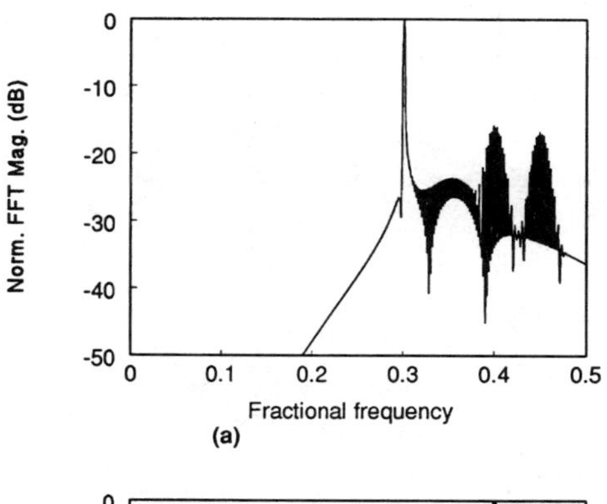

Figure 10-15   (a) FT of the reconstructed signal from the signal in example 1 after the WT-IWT process at scale 0.00207. (b) FT of the reconstructed signal from the signal in example 1 after 100 repetitive WT-IWT processes at scale 0.00207.

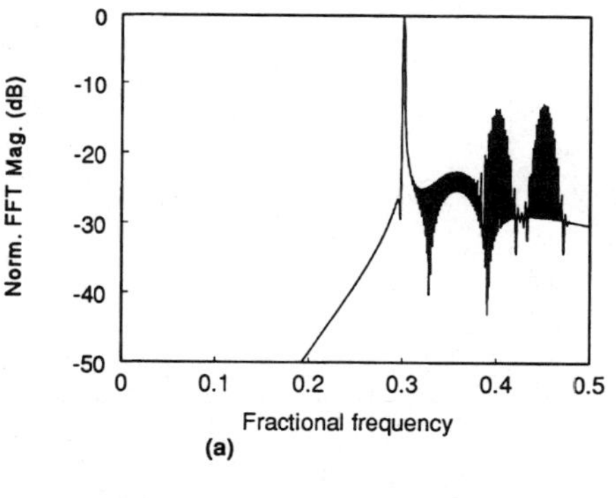

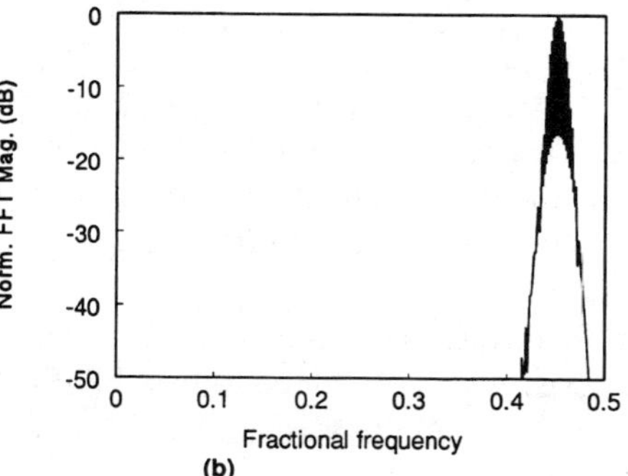

Figure 10-16   (a) FT of the reconstructed signal from the signal in example 1 after the WT-IWT process at scale 0.00184. (b) FT of the reconstructed signal from the signal in example 1 after 100 WT-IWT processes at scale 0.00184.

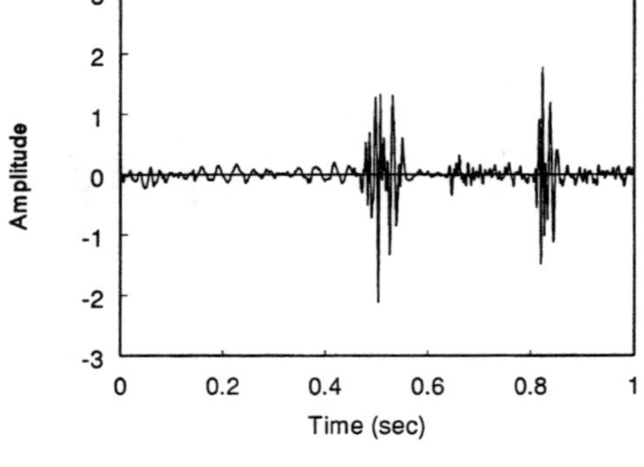

Figure 10-17 One second of the real data transient signal used in example 2. The sampling rate is 800 samples/s.

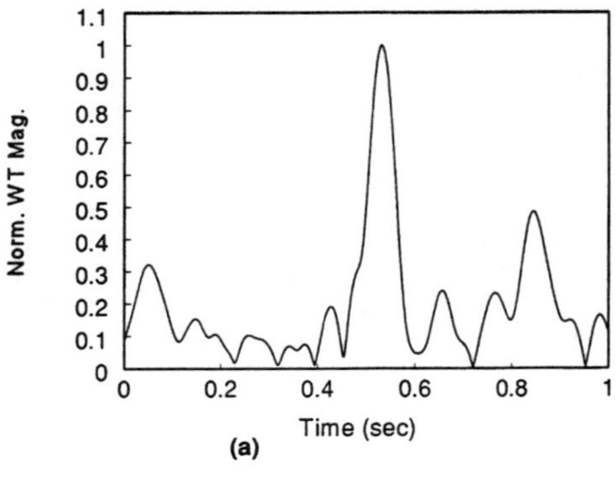

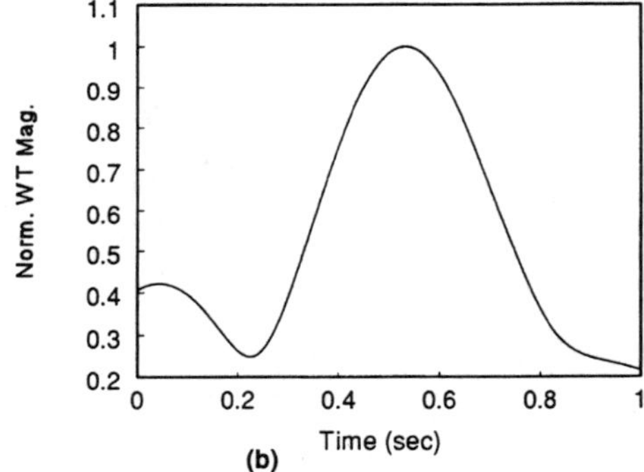

Figure 10-18 (a) WT of the signal in example 2 at scale 0.014 727.(b) WT after 100 repetitive WT-IWT processes of the signal in example 2 at scale 0.014 727.

signals. This will help physicians to obtain qualitative and quantitative measurements of the time-frequency characteristics of the PCG signals. Normal and abnormal signals have been considered to give some idea of the generality of the evaluation.

In this work, the standard FFT was first applied to a normal heart sound signal, which gives a basic understanding of the frequency content of the heart sounds. However, FFT analysis remains of limited value if the stationarity assumption is violated. Since heart sounds exhibit marked changes with time and frequency, they

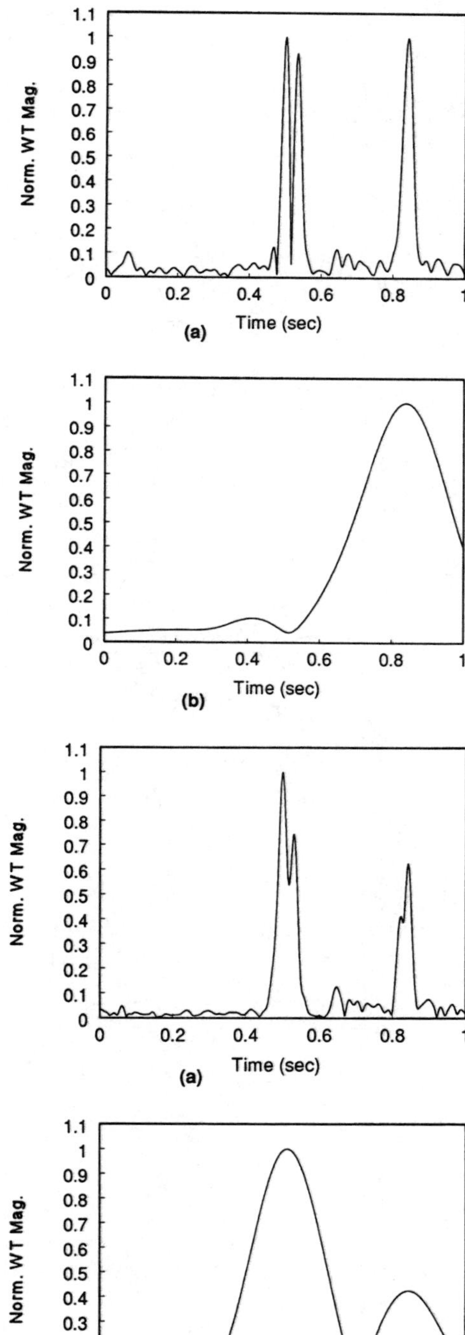

Figure 10-19   (a) WT of the signal in example 2 at scale 0.009 261. (b) WT after 100 repetitive WT-IWT processes of the signal in example 2 at scale 0.009 261.

Figure 10-20   (a) WT of the signal in example 2 at scale 0.007 364. (b) WT after 100 repetitive WT-IWT processes of the signal in example 2 at scale 0.007 364.

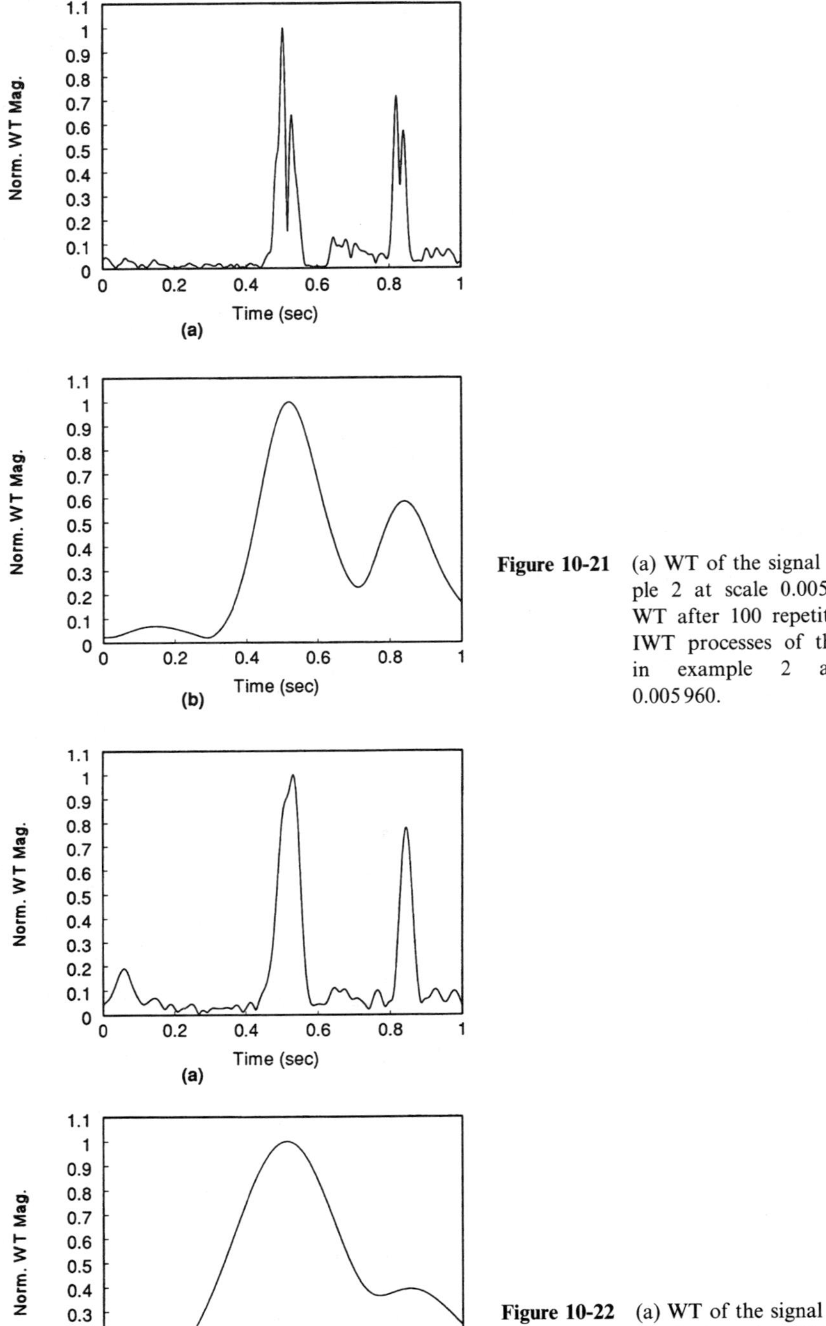

Figure 10-21   (a) WT of the signal in example 2 at scale 0.005 960. (b) WT after 100 repetitive WT-IWT processes of the signal in example 2 at scale 0.005 960.

Figure 10-22   (a) WT of the signal in example 2 at scale 0.011 411. (b) WT after 100 repetitive WT-IWT processes of the signal in example 2 at scale 0.011 411.

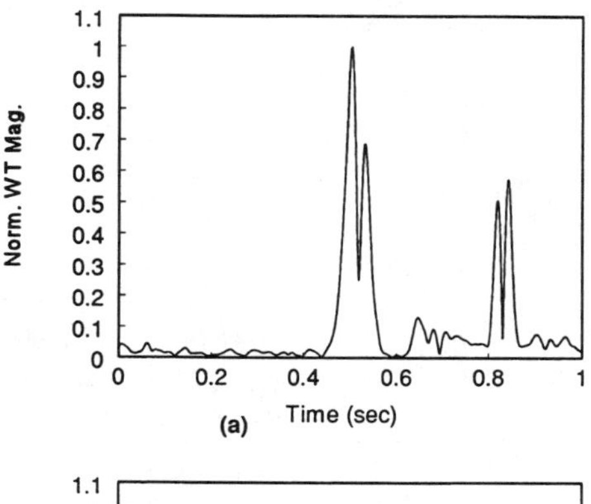

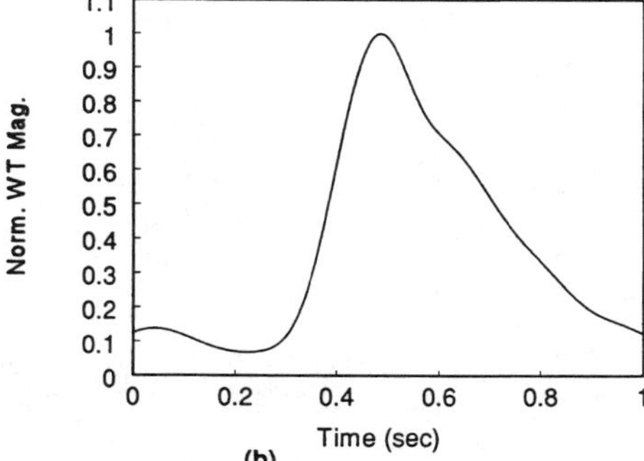

Figure 10-23 (a) WT of the signal in example 2 at scale 0.006 590. (b) WT after 100 repetitive WT-IWT processes of the signal in example 2 at scale 0.006 590.

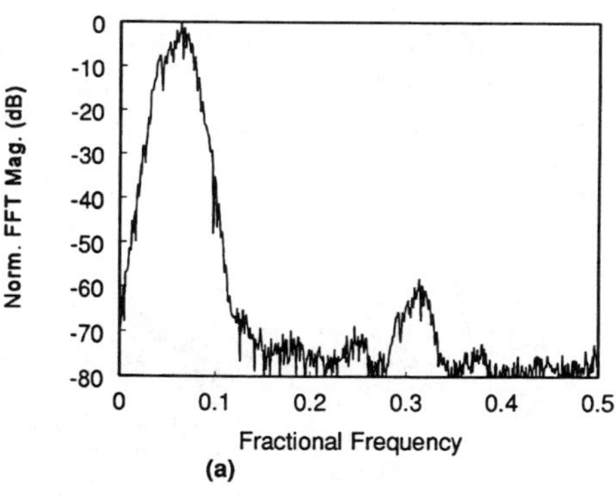

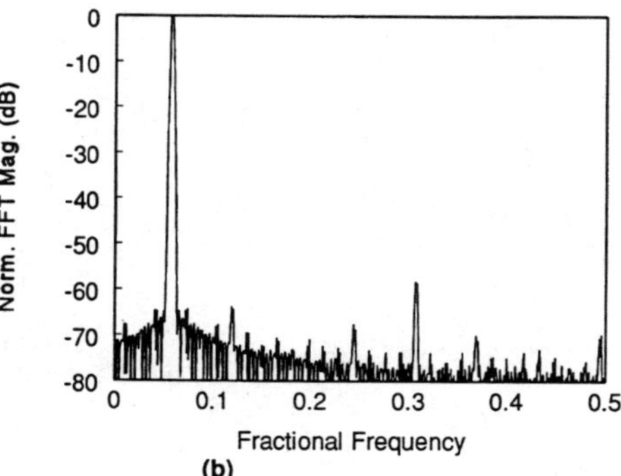

Figure 10-24 (a) FT of the reconstructed signal from the signal in example 2 after the WT-IWT process at scale 0.014 727. (b) FT of the reconstructed signal from the signal in example 2 after 100 repetitive WT-IWT processes at scale 0.014 727.

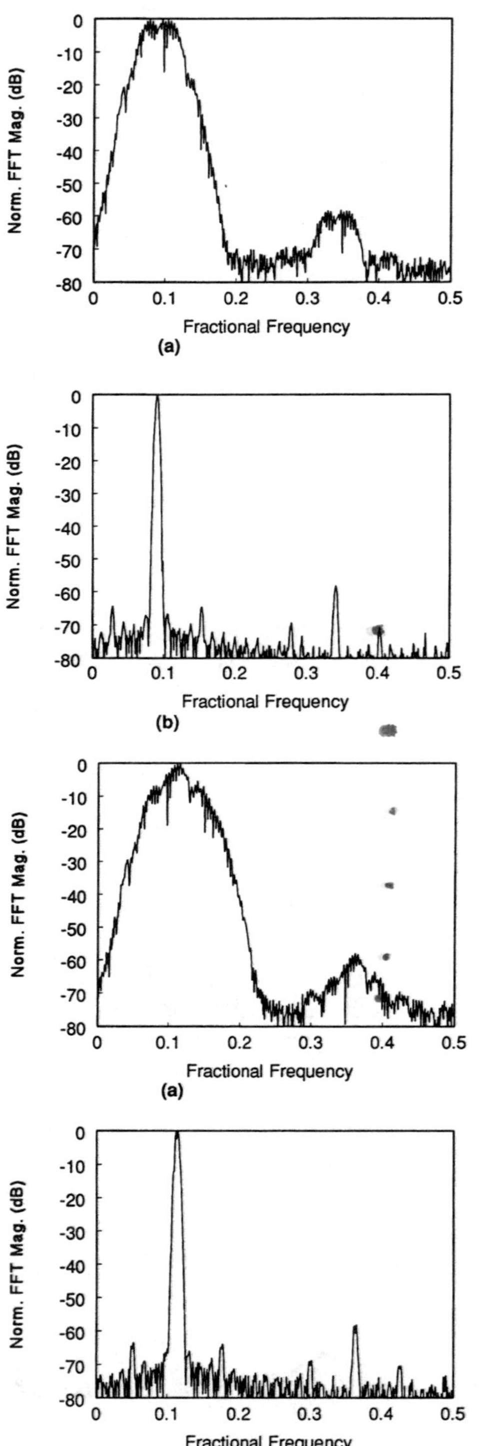

Figure 10-25 (a) FT of the reconstructed signal from the signal in example 2 after the WT-IWT process at scale 0.009 261. (b) FT of the reconstructed signal from the signal in example 2 after 100 repetitiveWT-IWT processes at scale 0.009 261.

Figure 10-26 (a) FT of the reconstructed signal from the signal in example 2 after the WT-IWT process at scale 0.007 364. (b) FT of the reconstructed signal from the signal in example 2 after 100 repetitive WT-IWT processes at scale 0.007 364.

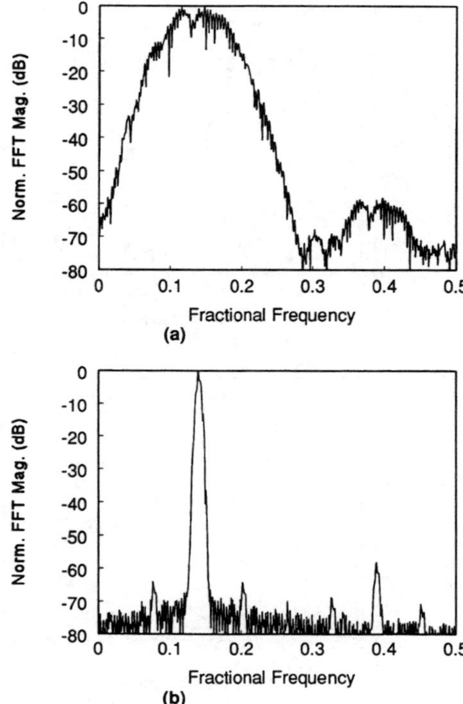

Figure 10-27 (a) FT of the reconstructed signal from the signal in example 2 after the WT-IWT process at scale 0.005960. (b) FT of the reconstructed signal from the signal in example 2 after 100 repetitive WT-IWT processes at scale 0.005960.

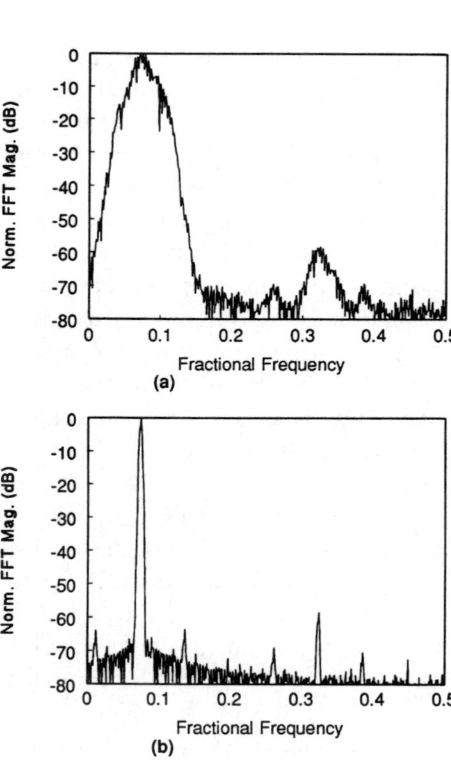

Figure 10-28 (a) FT of the reconstructed signal from the signal in example 2 after the WT-IWT process at scale 0.011411. (b) FT of the reconstructed signal from the signal in example 2 after 100 repetitive WT-IWT processes at scale 0.011411.

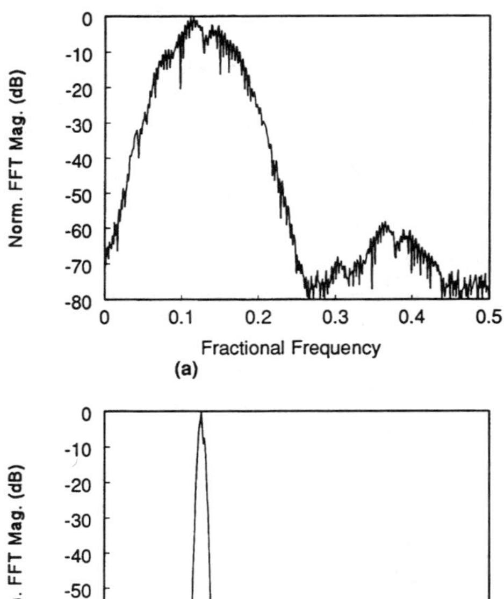

Figure 10-29 (a) FT of the reconstructed signal from the signal in example 2 after the WT-IWT process at scale 0.006 590. (b) FT of the reconstructed signal from the signal in example 2 after 100 repetitive WT-IWT processes at scale 0.006 590

are classified as nonstationary signals. To understand the exact feature of such signals, it is important to study their time-frequency characteristics. The short-time FT was applied to the same heart sound signal as a method to analyze nonstationary signals. The STFT cannot, however, track very sensitive sudden changes in the time-domain. Next, the pseudo-WD was applied to the heart sound signal and showed no success in displaying or separating the signal components in either the time or the frequency directions. Finally the WT was used to analyze the normal heart sound in the time and frequency domains. It was shown to have a very good time resolution for high-frequency components. The time resolution increases as the frequency increases and the frequency resolution increases as the frequency decreases [6, 17, 19, 21, 22]. The WT demonstrated the ability to analyze heart sound more accurately than the other transforms, and it was applied to an abnormal PCG signal to generalize its performance evaluation in the analysis of PCGs.

As a first conclusion, we have applied four different digital signal processing methods to the analysis of PCG signals. These are the standard FFT, STFT, WD, and WT. The WT is found to be the most suitable technique for the analysis of PCG signals due to the fact that the PCG signals are characterized by transients and fast changes in frequency as time progresses. Moreover, the WT has a very good time resolution for high-frequency components. It allows exact measurement of the time difference between the A2 and P2 components in the second sound, S2, of the PCG signal. Therefore, the WT provides a process which can diagnose abnormalities in the aortic and pulmonary valves. The exact frequency contents of A2 and P2, however, are better determined using the FFT; but the time difference between A2 and P2 is better determined using WTs. The STFT and the WD failed to display, correctly, the time-frequency characteristics of the PCG signals. The basic frequency contents of S1 and S2 of a PCG signal can be detected using the FFT. However, a time-frequency analysis obtained by the WT gives more details about the characteristics and features of the PCG signals.

A new algorithm for the analysis of nonstationary signals and transients based on combined wavelet-Fourier transforms has been presented. This approach suppresses the interference terms of the WT and enhances subband signals. The proposed algorithm is suitable for the analysis of signals that have a transient behavior such as speech, underwater acoustics, seismic surveillance, and biomedical signals. It has been applied to PCGs in this chapter. Compared with the usual WT, our algorithm improves the resolution in both time and frequency directions. In the examples in which we used our algorithm, the signals were completely resolved into their components in the time and frequency domain. Our claims are clear from the figures of the previous section. In Figs. 10-9(a), 10(a), 11(a), and 12(a), the transient signals were resolved into their subband components in the time-domain through the WT. In Figs. 10-13(a), 14(a), 15(a), and 16(a), the transient signals were resolved into their subband components in the frequency domain through the FT after one iteration of the WT-IWT. It is also clear that the WT suffers from a large amount of interference from the strong component when trying to detect the weak components using the appropriate scales. Those interference terms were suppressed using the repetitive WT-IWT process introduced in this chapter. The results are shown in Figs. 10-9(b), 10(b), 11(b), 12(b), and Figs. 10-13(b), 14(b), 15(b), 16(b). In *Example 2 (Real Data PCG signal)*, the signal is clearly resolved into six major components. The time and frequency locations of these components are shown in Figs. 10-18(b), 19(b), 20(b), 21(b), 22(b), 23(b), and Figs. 10-24(b), 25(b), 26(b), 27(b), 28(b), 29(b) after suppression of the interference terms associated with each subband component using the repetitive WT-IWT process with the appropriate scale.

In addition to the attractive features of the proposed algorithm, it can be used in many applications in filtering and signal processing especially in areas where it is important to distinguish a variety of time-frequency characteristics such as in PCGs or passive sonar applications. It can also be extended to the 2-D case and may find applications in image processing and computer vision. It can also be applied in coding schemes if it uses the DWT.

REFERENCES

[1] M. B. Rappaport and H. B. Sprague, "Physiologic and physical laws that govern auscultation, and their clinical application—the acoustic stethoscope and the electrical amplifying stethoscope and stethograph," *Am. Heart J.*, vol. 21, pp. 257, 1941.

[2] A. A. Luisada, *The Sounds of the Normal Heart.* St. Louis: W. H. Green, 1972.

[3] R. M. Rangayyan and R. J. Lehner, "Phonocardiogram signal analysis: A review," *CRC Critical Rev. Biomed. Eng.*, vol. 15, no. 3, pp. 211–236, 1988.

[4] L. P. Feigen, "Physical characteristics of sound and hearing," *Am. J. Cardiology*, vol. 28, no. 130, 1971.

[5] D. Gabor, "Theory of communication," *J. Inst. Elect. Eng.*, vol. 903, pp. 429–457, 1946.

[6] A. Grossmann and J. Morlet, "Decomposition of Hardy functions into square integrable wavelets of constant shape," *SIAM J. Math. Ann.*, vol. 15, pp. 723–736, 1984.

[7] A. Grossmann, J. Morlet, and T. Paul, "Transforms associated with square integrable group representations I," *J. Math. Phys.*, vol. 27, pp. 2473, 1985.

[8] L. Khadra, M. Matalgah, B. El-Asir, and S. Mawagdeh, "The wavelet transform and its applications to phonocardiogram signal analysis," *Medical Informatics*, vol. 16, no. 3, pp. 271–277, 1991. (Also appeared in the annual year book of Medical Informatics.)

[9] M. M. Matalgah, "The wavelet transforms in signal detection and applications." Master thesis submitted to the faculty of graduate school at Jordan University of Science and Technology as a partial fulfillment for the degree of M.S. in Electrical Engineering, April 1990.

[10] J. J. Wolcin, "Maximum likelihood detection of transient signals using sequenced short-time power spectra TM 831138, Naval Underwater Systems Center, August 1983.

[11] B. Friedlander and B. Porat, "Performance analysis of transient detectors based on a class of linear data transforms," *IEEE Trans. Inform. Theory*, vol. 38, pp. 665–673, 1992.

[12] M. M. Matalgah and J. Knopp, "Time varying spectral analysis of nonstationary signals based on combined wavelet and Fourier transforms," *Int. J. Electronics*, vol. 78, no. 3, pp. 463–476, March 1995.

[13] M.M. Matalgah and J. Knopp, "Hybrid wavelet-Fourier transform: a new algorithm for spectral analysis of nonstationary signals." In *Intelligent Engineering Systems Through Artificial Neural Networks* (Proc. *Artificial Neural Networks in Engineering (ANNIE'94) Conf.* St. Louis, MI,) November 13–16, 1994. C. H. Dagli et al. (eds.) vol. 4, pp. 529–534, Warrendale, PA: ASME Press.

[14] T. A. C. M. Claasen and W. F. G. Mecklenbräuker, "The Wigner Distribution—A tool for time-frequency signal analysis," *Philips J. Res.*, vol. 35, no. 3, pp. 217–250, 1980.

[15] L. Cohen, "Time-Frequency Distributions—A review," *Proc. IEEE*, vol. 77, no. 7, pp. 941–980, 1989.

[16] B. Boashash, *"Time-frequency signal analysis."* In *Advances in Spectrum Estimation.* S. Haykin (ed.). Englewood Cliffs, NJ: Prentice-Hall, 1990.

[17] F. B. Tuteur, "Wavelet transforms in signal detection," *IEEE Int. Con. Acoust., Speech, Signal Proc.*, vol. CH2561-9, pp. 1435–1438, 1988.

[18] A. Papoulis, *The Fourier Integral and its Applications*. New York: McGraw-Hill, 1962.

[19] A. Grossmann and R. Kronland-Martinet, "Time-and-scale representations obtained through continuous wavelet transforms." In *Signal Processing IV: Theories and Applications*. J. L. Lacoume et al. (eds). Amsterdam: Elsevier Science Publishers B. V. North-Holland, EURASIP, 1988.

[20] J. B. Allen, "Applications of the short-time Fourier transform to speech processing and spectral analysis." In *Proc. Int. Conf. Acoust., Speech, Signal Proc.*, Paris, pp. 1012–1015, 1982.

[21] R. Kronland-Martinet, J. Morlet, and A. Grossmann, "Analysis of sound patterns through wavelet transforms," *Int. J. Pattern Recognition Artificial Intel.*, vol. 1, no. 2, pp. 273–302, 1987.

[22] A. Grossmann, M. Holschneider, R. Kronland-Martinet, and J. Morlet, "Detection of abrupt changes in sound signals with the help of the wavelet transform." In *Inverse problems: An Interdisciplinary Study (Advances in Electronics and Electron Physics*, Supplement 19). New York: Academic Press, pp. 298–306, 1987.

[23 J. Bertrand, P. Bertrand, and J. P. Ovarlez, "Discrete Mellin transform for signal analysis." *Proc. 1990 IEEE Int. Conf. on Acoustics, Speech, and Signal Processing*, Albuquerque, NM, pp. 1603–1606, April 1990.

[24] P. Flandrin, "Some aspects of nonstationary signal processing with emphasis on time-frequency and time-scale methods." In *Wavelets, Time-Frequency Methods and Phase Space*. J. M. Combes, A. Grossmann, and Ph. Tchamitchian (eds.). Berlin: Springer, pp. 68–69, 1989.

[25] P. Goupillaud, A. Grossmann, and J. Morlet, "Cycle-octave and related transforms in seismic signal analysis," *Geoexploration*, vol. 23, pp. 85–102, 1984/85.

[26] A. Grossmann and R. Kronland-Martinet, "Time and scale representations obtained through continuous wavelet transforms." In *Proc. Int. Conf. EUSIPCO'88, Signal Processing IV: Theories and Applications*. J. L. Lacoume et al. (eds.). New York: Elsevier Science Publishers, pp. 475–482, 1988.

[27] M. Holschneider, R. Kronland-Martinet, J. Morlet, and Ph. Tchamitchian, "A real-time algorithm for signal analysis with the help of the wavelet transform." In *Wavelets, Time-Frequency Methods and Phase Space*. J. M. Combes, A. Grossmann, and Ph. Tchamitchian (eds.) Berlin: Springer, vol. IPTI, pp. 286–297, 1989.

[28] J. M. Combes, A. Grossmann and Ph. Tchamitchian, *Wavelets, Time-Frequency Methods and Phase Space*. Berlin: Springer & IPTI, 1989.

[29] M. Antonini, M. Barlaud, P. Mathieu, and I. Daubechies, "Imaging coding using vector quantization in the wavelet transform domain." *Proc. 1990 IEEE Int. Conf. Acoustics, Speech, and Signal Processing*, Albuquerque, NM, pp. 2297–2300, April 1990.

[30] S. Mallat and S. Zhong, "Signal characterization from multi scale edges." In *Proc. 10th Int. Conf. Pattern Recognition, Systems, and Applications*, Los Alamitos, CA, pp. 891–896, June 1990.

[31] P. Goupillaud, A. Grossmann, and J. Morlet, "Cycle-octave and related transforms in seismic signal analysis," *Geoexploration*, vol. 23, pp. 85–102, 1984/1985.

[32] A. C. Bovik, N. Gopal, T. Emmoth, and A. Restrepo, "Localized measurement of emergent image frequencies by Gabor wavelets," *IEEE Trans. Inform. Theory*, vol. 38, pp. 691–712, March 1992.

Chapter 11

Wavelet Feature Extraction
from Neurophysiological Signals

Mingui Sun, Robert J. Sclabassi

11.1. INTRODUCTION

The human central nervous system (CNS) is a biological information processing system dependent on the functioning of large numbers of neurons integrated into a complex network. The CNS receives information from afferent sensory data, forms memories, learns, and produces efferent signals which allow us to manipulate our external environment. Although much progress has been achieved in developing an understanding of the functioning of the CNS, many mechanisms of this system are still poorly understood. In recent years advances in computer technology and signal processing have provided powerful alternative tools for the study of the CNS. With the help of highly computerized medical equipment, the electroencenphalogram (EEG), magnetoencephalogram (MEG), and event-related potentials (ERPs), can be recorded simultaneously at a large number of locations, providing the basis for both assessment of function and source localization within the human brain [1–3]. In addition, intracranial blood flow and pressure can also be observed during neuro-surgery and neuro intensive care [4–7], which may be used to assess the clinical status of patients undergoing these treatments.

A significant step in the computational processing of data from the CNS is that of feature extraction; i.e., the identification and isolation of the attributes of the data which require study. A heavily utilized feature extraction tool has been the fast Fourier transform (FFT) which facilitates the quantification of the frequency-domain features of a signal, but provides no data compression. The transformed results, having the same number of samples as the input, tend to fluctuate significantly when signals are noisy. In addition, the computation of the spectral characteristics of the signals using the FFT approach may take considerable time when large numbers of data channels at high sampling frequencies are being investigated.

The Fourier transform builds on the concept of frequency; while the wavelet transform builds on the concept of scale. However, it is clear that the two concepts are related. Thus, we have been investigating an alternative to characterize the spectral information contained within these signals, obtained from the CNS, by using the scale information contained within the wavelet transform.

In our approach, features of signals are extracted from the wavelet coefficients. In particular, we first compute several scale levels of wavelet transforms on the neurophysiological data. Then the resulting wavelet coefficients are segmented into equally spaced intervals. The length of each interval is determined by considerations between the reliability in statistical estimation and the stationarity of the signal. For each segment, the wavelet coefficients are soft-thresholded toward zero [8], and then, at each scale, the statistical values of the variance of the wavelet coefficients, the average distance between adjacent zero-crossings (or extrema) in the wavelet coefficients, and the variance of the wavelet coefficients between the current and next scales are computed. These statistical values serve as the extracted features to form a set of pattern vectors.

Our analysis shows that the spectral information of the signal is well represented by the computed statistics, and that the variance values computed within and across scale levels correspond to an overlapped and interlaced frequency division on the frequency axis, providing a description of the entire spectrum of the signal. In addition, the mean values of the distances between zero-crossings and extrema represent the root mean-square frequency of the smoothed and differentiated signal. This feature extraction approach requires no spectral estimation procedures and the results are less affected by noise.

Our analysis also addresses the questions concerning how: (1) the wavelet is designed, (2) noise behaves in the wavelet domain, and (3) the noise removal process affects the spectral characteristics in the extracted features.

This chapter is organized as follows: in section 11.2, the particular wavelet transforms suitable for analyzing the CNS data are presented. Section 11.3 derives the signal-to-noise ratio in the wavelet domain, section 11.4 discusses the effects of the wavelet transform on the division of the frequency axis. The relationship between the wavelet features and the spectrum of the signal is presented in sections 11.5 and 11.6 and the Appendix, where we investigate the variances within and across wavelet scales, as well as the zero-crossings and extrema expressed in terms of the frequency-domain statistics. The implementation algorithms are presented in section 11.7. An example of the wavelet feature extraction from intracranial pressure measurements is presented in section 11.8, where the design of the wavelet, methods of suppressing noise, and prediction of the future values of the signal features are described.

11.2. WAVELET TRANSFORMS

We utilized a general form of the wavelet transform in which the wavelet coefficients are simply the samples of a set of continuous convolutions. The mathematical theory of this transform has been well-documented (see [9] and previous chapters for details). Let $x(t)$ be a square-integrable function of the continuous time t, the wavelet transform $wx_j(t)$ is given by

$$wx_j(t) = (x * \psi_j)(t) \tag{11-1}$$

where "$*$" denotes convolution and $\psi_j(t) = 2^{-j}\psi(2^{-j}t)$, with $\psi(t)$ being the wavelet. The wavelet scale is given by a positive integer j, $j = 1, 2, \cdots, J$, where $j = 1$ and J, respectively, correspond to the coarsest and finest scale levels. The Fourier transform of (11-1) is

$$WX_j(\omega) = X(\omega)\Psi_j(\omega) = X(\omega)\Psi(2^j\omega) \tag{11-2}$$

where the upper case variables denote the Fourier transforms of the corresponding lower case variables.

The choice of the wavelet involves many consideration [10–12]. In the case of processing neurophysiological signals, we require the wavelet to be highly "localized" in the joint time-frequency plane in order to effectively discriminate the temporal and spectral domain features present in the data. To meet this requirement, the wavelet must be sufficiently smooth in both domains. We also require the wavelet to be symmetric or antisymmetric with respect to an axis located at, or as close as possible to, the origin. This requirement minimizes the distortion and translation of the extracted features relative to the input signal.

11.3. SIGNAL-TO-NOISE RATIO

It is usually difficult to model the noise precisely because of its unknown origin and nonstationarity. However, the noise usually has a much wider bandwidth than the neurophysiological signal, allowing us to assume that the noise is white. Writing $x(t)$ as a sum of a signal, $s(t)$, and a noise, $n(t)$ with variance σ^2, we may express the signal-to-noise ratio (SNR), η_j, at wavelet scale level j by

$$\eta_j = \frac{\mathbf{E}[ws_j^2(t)]}{\mathbf{E}[wn_j^2(t)]} = \frac{\mathbf{R}_{ws_j}(0)}{\mathbf{R}_{wn_j}(0)} = \frac{\mathbf{E}[(s * \psi_j)^2(t)]}{\mathbf{E}[(n * \psi_j)^2(t)]} \tag{11-3}$$

where $\mathbf{E}$ and $\mathbf{R}$ denote, respectively, the expectation operator and the correlation function. The SNR may be alternatively expressed in the frequency domain:

$$\eta_j = \frac{\int_{-\infty}^{\infty} S_s(\omega)|\Psi(2^j\omega)|^2\, d\omega}{\int_{-\infty}^{\infty} S_n(\omega)|\Psi(2^j\omega)|^2\, d\omega} = \frac{1}{2\pi\sigma^2||\psi||^2} \int_{-\infty}^{\infty} S_s(2^{-j}\omega)|\Psi(\omega)|^2\, d\omega \tag{11-4}$$

where $S_s(\omega)$ and $S_n(\omega)$ are the Fourier transforms of $\mathbf{R}_s(\tau)$ and $\mathbf{R}_n(\tau)$, respectively, and $||\psi||^2$ is the squared norm $||\psi||^2 = \int_{-\infty}^{\infty} |\psi(t)|^2\, dt$. Although (11-4) cannot be computed since the pure signal, $s(t)$, is unknown, some knowledge about the spectral profile of $S_s(\omega)$ may be available. Equation (11-4) indicates that the SNR reaches a maximum when $\int_{-\infty}^{\infty} S_s(2^{-j}\omega)|\Psi(\omega)|^2 \, d\omega$, for any j, reaches a maximum. In this case, the signal is the least affected by the noise.

11.4. WAVELET SPECTRAL DIVISION

Equations (11-1) and (11-2) indicate that the wavelet transform produces a set of filtered signals indexed by the scale level, j. As implied by (11-2), an incremental dilation of the wavelet leads to an equivalent incremental reduction in the bandwidth. This effect is illustrated in Fig. 11-1 [based on Eq. (11.38)] where the solid and dashed lines, respectively, correspond to $|\Psi_j(\omega)|^2$ and $|\Psi_j(\omega)\Psi_{j+1}(\omega)|$ (to be discussed further in section 11.5). It is observed that the entire frequency axis is divided in a unique fashion having the following characteristics:

1. *The divisions are highly overlapped.* This overlap is useful in the case where the signal being analyzed is noisy and has a time-varying spectrum. In this case the signal's spectral peaks are very likely to be captured near the central frequencies of certain bandpass filters where the product $S_s(2^{-j}\omega)|\Psi(\omega)|^2$ [see Eq. (11-4)] is large, providing a high SNR. On the other hand, any spectral variation in time can be observed simultaneously by several bandpass filters—some produce increasing values, while others produce decreasing values. As a result, these values activate a series of changes in the feature space, providing an effective input to the pattern recognition process.

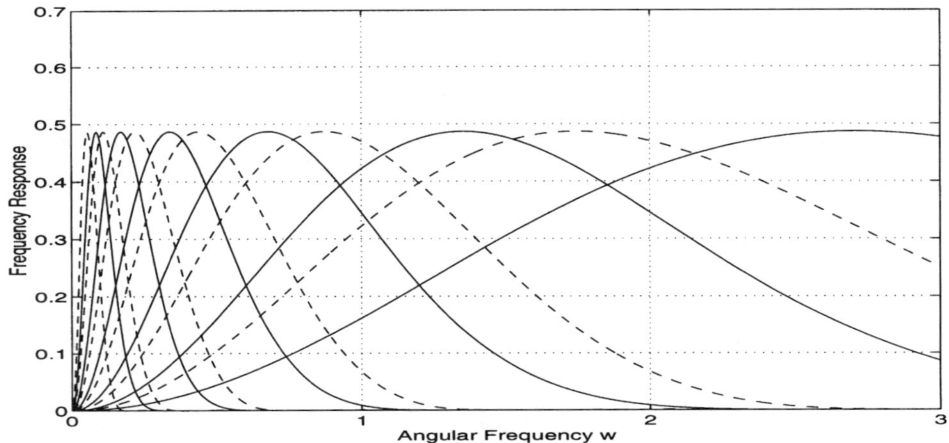

Figure 11-1   Frequency responses of the $|\Psi_j(\omega)|^2$ (solid lines) and $\eta|\Psi_j(\omega)\Psi_{j+1}(\omega)|$ (dashed lines) for scale levels one through six. The wavelet is given in Eq. (11-38) and the scale factor, $\eta \approx 1.231$, is determined by numerically solving $\max\{|\Psi_j(\omega)|^2\} = \eta \max |\Psi_j(\omega)\Psi_{j+1}(\omega)|$.

2. *The bandwidth increases as the frequency increases.* The wavelet transformation is a "constant-Q" bandpass filtering process [11], where the ratio between the bandwidth and center frequency is invariant among all bands. Therefore, the bandwidth is narrower for low frequencies, and broader for high frequencies.

3. *The frequency responses are smooth without distinct cut-offs.* This characteristic ensures that, when the signal's spectrum varies slowly, so do the extracted features. Hence, there are no undesired sharp changes in the filter's outputs which may trigger false alarms. The bell-shaped curves are also desirable for real-time implementation since the filters utilized to compute the wavelet transform are short for these types of curves [13].

11.5. VARIANCE

We assume that the signals are stationary within each short segment in time. Thus, within the segment, the variance of the wavelet transform $wx_j(t)$ can be considered as a value unrelated to t

$$\mathbf{E}[wx_j(t)]^2 = \mathbf{E}[(x * \psi_j)^2(t)] = \mathbf{R}_{wx_j}(0) = \sigma_{xj}^2 \qquad (11\text{-}5)$$

In the frequency domain

$$\sigma_{xj}^2 = \frac{1}{2\pi} \int_{-\infty}^{\infty} S_x(\omega) |\Psi_j(\omega)|^2 \, d\omega \qquad (11\text{-}6)$$

which may be written as

$$\sigma_{xj}^2 = 2^{-j} \|\psi\|^2 \frac{\int_{-\infty}^{\infty} S_x(\omega) |\Psi_j(\omega)|^2 \, d\omega}{\int_{-\infty}^{\infty} |\Psi_j(\omega)|^2 \, d\omega} \qquad (11\text{-}7)$$

From (11-7), it is clear that σ_{xj}^2 is proportional to the weighted average of the spectrum $S_x(\omega)$, where $|\Psi_j(\omega)|^2$ acts as a weighting function.

In practice, the raw signal $x(t)$ contains noise. As previously stated we assume that $x(t) = s(t) + n(t)$, where $s(t)$ and $n(t)$ are independent and $n(t)$ is white with a variance σ^2. Now $\sigma_{x_j}^2 = \sigma_{sj}^2 + \sigma_{n_j}^2$ for any j. If $\sigma^2 = \sigma_{n_0}^2$ is known, $\sigma_{n_j}^2$, for any $j \geq 1$, is also known because

$$\sigma_{n_j}^2 = \frac{\sigma^2}{2\pi} \int_{-\infty}^{\infty} |\Psi_j(\omega)|^2 \, d\omega \qquad (11\text{-}8)$$

This knowledge is useful for removing noise in the wavelet domain. Donoho and Johnstone [8] developed a soft-thresholding scheme using the operation

$$w\hat{s}_j = \mathrm{sgn}(wx_j)(|wx_j| - T_h)_+ \qquad (11\text{-}9)$$

where $w\hat{s}_j$ and T_h are, respectively, the estimate for ws_j and the threshold value, and $(a)_+$ equals a for $a > 0$ and zero otherwise. It has been shown [8] that, under some conditions and normalization, $T_h = \sigma \sqrt{2 \log N}$ provides the least estimation error for ws_j, where N is the length of the signal.

The relationship between the variances $\sigma_{\hat{s}j}^2$ and $\sigma_{x_j}^2$ can be derived by considering $w\hat{s}_j$ in (11-9) as a random variable, for which the probability density function is given by

$$p_{\hat{s}j}(s) = \begin{cases} p_{x_j}(s+T), & s \geq 0 \\ p_{x_j}(s-T) & s < 0 \end{cases} \tag{11-10}$$

Then, expressing $\sigma_{\hat{s}j}^2 = E[w\hat{s}_j(t)^2]$ in terms of $p_{\hat{s}j}$

$$\begin{aligned} \sigma_{\hat{s}j}^2 &= \int_{-\infty}^{\infty} s^2 p_{\hat{s}j}(s)\, ds \\ &= \int_{-\infty}^{-T} (s+T)^2 p_{x_j}(s)\, ds + \int_{T}^{\infty} (s-T)^2 p_{x_j}(s)\, ds \\ &= \sigma_{x_j}^2 - \int_{-T}^{T} s^2 p_{x_j}(s)\, ds + T^2 \text{Prob}_{x_j}(|s| > T) - 4T \int_{T}^{\infty} s p_{x_j}(s)\, ds \end{aligned} \tag{11-11}$$

For the Gaussian signals, (11-11) may be further simplified:

$$\sigma_{\hat{s}j}^2 = 2(\sigma_{x_j}^2 + T^2)\mathcal{N}(t/\sigma_{x_j}^2) - \sqrt{\frac{2}{\pi}} \sigma_{x_j} T \exp\left(-\frac{T^2}{2\sigma_{x_j}^2}\right) \tag{11-12}$$

where

$$\sigma_{x_j}^2 = \frac{1}{2\pi} \int_{-\infty}^{\infty} S_x(\omega) |\Psi_j(\omega)|^2\, d\omega$$

$\mathcal{N}(t)$ is the standard Gaussian one-side tail probability

$$\mathcal{N}(t) = \frac{1}{\sqrt{2\pi}} \int_{t}^{\infty} \exp(-x^2/2)\, dx$$

and

$$p_{x_j}(x) = \frac{1}{\sqrt{2\pi}\sigma_{x_j}} e^{-x^2/(2\sigma_{x_j}^2)}$$

Thus, the variance $\sigma_{\hat{s}j}^2$ can be estimated directly from (11-12).

The solid lines in Fig. 11-1 place the highest weight around the central frequencies of the scaled wavelets; however, the spectral component of interest may be located anywhere in the frequency axis, even in the neighborhood of the cross-point between two adjacent frequency bands. At this location, the spectral component is assigned with a small gain, signaling a low detection sensitivity. This problem can be approached by considering the cross-correlation between $wx_j(t)$ and $wx_{j+1}(t)$, i.e., $\mathbf{R}_{wx_j,wx_{j+1}}(\tau) = E[wx_j(t+\tau)wx_{j+1}(t)]$. In order to relate $\mathbf{R}_{wx_j,wx_{j+1}}(\tau)$ to the wavelet and the autocorrelation of the signal, we first calculate the cross-correlation, $\mathbf{R}_{x,wx_{j+1}}(\tau)$, between $x(t)$ and $wx_{j+1}(t)$:

$$\mathbf{R}_{x,wx_{j+1}}(\tau) = \mathbf{E}\left[\int_{-\infty}^{\infty} x(t+\tau)x(t-\alpha)\psi_{j+1}(\alpha)\,d\alpha\right]$$

$$= \int_{-\infty}^{\infty} \mathbf{R}_x(\tau+\alpha)\psi_{j+1}(\alpha)d\alpha$$

$$= \mathbf{R}_x(\tau) * \psi_{j+1}(-\tau) \tag{11-13}$$

Then, taking $wx_j(t)$ into account, we have

$$\mathbf{R}_{wx_j,wx_{j+1}}(\tau) = \mathbf{E}\left[\int_{-\infty}^{\infty} x(t-\alpha)\psi_j(\alpha)wx_{j+1}(t-\tau)\,d\alpha\right]$$

$$= \int_{-\infty}^{\infty} \mathbf{R}_{x,wx_{j+1}}(\tau-\alpha)\psi_j(\alpha)d\alpha$$

$$= \mathbf{R}_x(\tau) * \psi_{j+1}(-\tau) * \psi_j(\tau) \tag{11-14}$$

Expressing (11-14) in the frequency domain yields

$$\mathbf{R}_{wx_j,wx_{j+1}}(\tau) = \frac{1}{2\pi}\int_{-\infty}^{\infty} S_x(\omega)\bar{\Psi}_{j+1}(\omega)\Psi_j(\omega)e^{i\omega\tau}d\omega \tag{11-15}$$

where the over-bar denotes complex conjugation. As mentioned in section 11.2, we require the wavelet to be symmetric or antisymmetric with respect to a certain axis. This requirement implies that $\Psi_j(\omega)$ must take the following form:

$$\Psi_j(\omega) = A(\omega)e^{-i[\alpha\omega+r(\pi/2)]} \tag{11-16}$$

where $A(\omega)$ is either symmetric or antisymmetric with respect to zero, r may be even (for symmetric case) or odd (for antisymmetric case), and α determines the axis of symmetry for the wavelet. By (11-15) and (11-16)

$$\mathbf{R}_{wx_j,wx_{j+1}}(0) = \mathbf{E}[wx_j(t+\tau)wx_{j+1}(t)] = \frac{1}{2\pi}\int_{-\infty}^{\infty} S_x(\omega)A(\omega)A(2\omega)e^{i\alpha\omega}d\omega \tag{11-17}$$

which indicates that, when the modulus of $\Psi_j(\omega)$ has a single peak, $\mathbf{E}[wx_j(t+\tau)wx_{j+1}(t)]$ is most sensitive to the intermediate frequency component between the peaks of $|\Psi_j(\omega)|$ and $|\Psi_{j+1}(\omega)|$, as shown in Fig. 11-1 by the dashed lines. It is clear that these lines complement the frequency responses of the scaled wavelets (solid lines). In order to equalize the maximum gains between $|\Psi_j(\omega)|^2$ and $|\Psi_j(\omega)\Psi_{j+1}(\omega)|$, a scale factor η is included so that $\max\{|\Psi_j(\omega)|^2\} = \eta\max|\Psi_j(\omega)\Psi_{j+1}(\omega)|$. For the wavelet shown in Fig. 11-1, $\eta \approx 1.231$.

11.6. SPECTRAL FEATURES IN THE WAVELET EXTREMA AND ZERO-CROSSINGS

It has been shown [9] that the extrema (where the derivative values of the signal are zero) or zero-crossings (where the values of the signal are zero) in the wavelet transform contain a rich set of information about the signal. In many cases, the input signal can be reconstructed from these zero-crossings or extrema [9,14]. This result is very attractive for feature extraction because, in practice, the extrema (or zero-crossings) represent a much smaller data set than the raw signals, but consist of approximately the same amount of information. This observation leads us to consider the average distance between the adjacent extrema (or zero-crossings) as one of the signal features because: (1) the average distance is easy to evaluate; and (2) the information provided by this feature is quite different from that discussed in the previous section.

The mean distance between adjacent zero-crossings is given by the mean of $\tau_z = t_i - t_{i-1}$, with t_i and t_{i+1} being the zero-crossing points and $t_i, t_j \in T$, where T is a subset of the time interval where the input signal, $x(t)$, is defined. We will determine the mean of τ_z by solving the zero-crossing problem for a stochastic process [15]. Let $y(t) = wx_j(t)$ and let $m(T)$ be the number of zero-crossings of $y(t)$ in the time interval T. In the Appendix, we show that, when $y'(t)$, exists, the expected number of zero-crossings of $y(t)$ in the interval T, $\mathbf{E}[m(T)]$, is given by

$$\mathbf{E}[m(T)] = T p_y(0)\mathbf{E}[|y'|, \text{given } y = 0] \tag{11-18}$$

where $p_y(0)$ is the probability density function of $y(t)$ evaluated at $y = 0$. The linearity of the convolution and differentiation operators implies that, if, by assumption, the input signal $x(t)$ is a Gaussian and zero-mean random process, so are the signals $y(t)$ and $y'(t)$. Furthermore, since $y(t)$ is real-valued, the correlation function $\mathbf{R}_y(\tau)$ must be even. Using the equality

$$\mathbf{R}_{yy'}(\tau) = -\mathbf{R}'_y(\tau) \tag{11-19}$$

it is clear that $\mathbf{R}'_y(\tau)$ must be an odd function. Hence

$$\mathbf{R}'_y(0) = -\mathbf{E}[y(t)y'(t)] = 0 \tag{11-20}$$

Since both $y(t)$ and $y'(t)$ are Gaussian and zero-mean, the orthogonality between $y(t)$ and $y'(t)$ in (11-20) implies that $y(t)$ and $y'(t)$ must be statistically independent. Therefore, the expected value on the right side of (11-18) must be unconditional. Substituting $p_y(0) = 1/\sqrt{2\pi \mathbf{R}_y(0)}$ and $E(|y'(t)|) = \sqrt{-2\mathbf{R}''_y(0)/\pi}$ into (11-18), we have

$$\mathbf{E}[\tau_z] = \frac{T}{\mathbf{E}[m(T)]} = \pi\sqrt{-\frac{\mathbf{R}_y(0)}{\mathbf{R}''_y(0)}} \tag{11-21}$$

In order to connect the expected value $\mathbf{E}[\tau_z]$ to the spectral features of the input signal, we utilize the Fourier transform pairs

$$\frac{d^k}{dt^k}f(t) \longleftrightarrow (j\omega)^k F(\omega) \tag{11-22}$$

and define the the weighted root mean square (rms) frequency at wavelet scale level j as $\langle \omega_j \rangle_z = 2\pi/(2\mathbf{E}[\tau_z])$. By (11-21) and (11-22), the desired connection between the mean zero-crossing interval and the signal's spectrum can be established:

$$\langle \omega_j \rangle_z = \frac{\pi}{T}\mathbf{E}[m(T)] = \sqrt{\frac{\int_{-\infty}^{\infty} \omega^2 S_x(\omega)|\Psi_j(\omega)|^2\, d\omega}{\int_{-\infty}^{\infty} S_x(\omega)|\Psi_j(\omega)|^2\, d\omega}} \tag{11-23}$$

Eq. (11-23) reveals, in conjunction with (11-2), that $\langle \omega_j \rangle_z$ is equal to the square root of the moment of inertia for $|WX_j(\omega)|^2$ with respect to the origin. This quantity is always larger than or equal to the mean Fourier frequency (or, equivalently, the mean instantaneous frequency) [16], given by

$$[\omega_j]_z = \frac{\int_0^\infty \omega |WX_j(\omega)|^2 \, d\omega}{\int_0^\infty |WX_j(\omega)|^2 \, d\omega} \tag{11-24}$$

because the difference between $\langle \omega_j \rangle_z^2$ and $[\omega_j]_z^2$ can be written as

$$\langle \omega_j \rangle_z^2 - [\omega_j]_z^2 = \frac{\int_0^\infty (\omega - [\omega_j]_z)^2 |WX_j(\omega)|^2 \, d\omega}{\int_0^\infty |WX_j(\omega)|^2 \, d\omega} \tag{11-25}$$

which is always nonnegative.

The derivation for the extrema case is identical except that $x(t)$ is replaced by $x'(t)$ since, at the extrema of $x(t)$, $x'(t)$ must be equal to zero. Thus, the weighted rms frequency for the wavelet extrema, $\langle \omega_j \rangle_e$, is given by

$$\begin{aligned}
\langle \omega_j \rangle_e &= \sqrt{\frac{\int_{-\infty}^\infty \omega^2 |i\omega WX_j(\omega)|^2 \, d\omega}{\int_{-\infty}^\infty |i\omega WX_j(\omega)|^2 \, d\omega}} \\
&= \sqrt{\frac{\int_{-\infty}^\infty \omega^4 S_x(\omega) |\Psi_j(\omega)|^2 \, d\omega}{\int_{-\infty}^\infty \omega^2 S_x(\omega) |\Psi_j(\omega)|^2 \, d\omega}}
\end{aligned} \tag{11-26}$$

These derivations reveal an interesting result that the mean distance between adjacent zero-crossings (or extrema) actually estimates the weighted rms frequency, rather than the weighted mean frequency. Furthermore, the mean distance $\langle \omega_j \rangle_e$ is related to the wavelet transform of the differentiated signal $dx(t)/dt$. Since $S_x(\omega)$ is weighted by either ω^2 (in the case of zero-crossings) and ω^4 (in the case of extrema), an increased sensitivity is obtained for the high-frequency portion of the signal. The high-frequency components of the noise will also be amplified; however, $\psi_j(t)$ is a bandpass filter which curbs the noise effects. Figure 11-2 shows a spectrum (solid line) computed from the EEG data recorded from a male adult during sleep. The frequency components around 12 Hz are closely related to sleep spindles [17]. These components have lower amplitude values than the background EEG activities. In the processed spectrum (dashed line) computed by weighting the original spectrum with ω^2, the spindle components are significantly emphasized.

11.7. COMPUTATION

The process of feature extraction begins with a buffering of the raw signals into epochs. Three to six scale levels of the wavelet transform are then applied to these epochs. Next, we divide the resulting wavelet coefficients into segments and compute, for each segment and scale level, the statistical quantities described in the above sections. In this process, the computational efficiency of the wavelet transform must be optimized since this process takes the most computing time. The definition

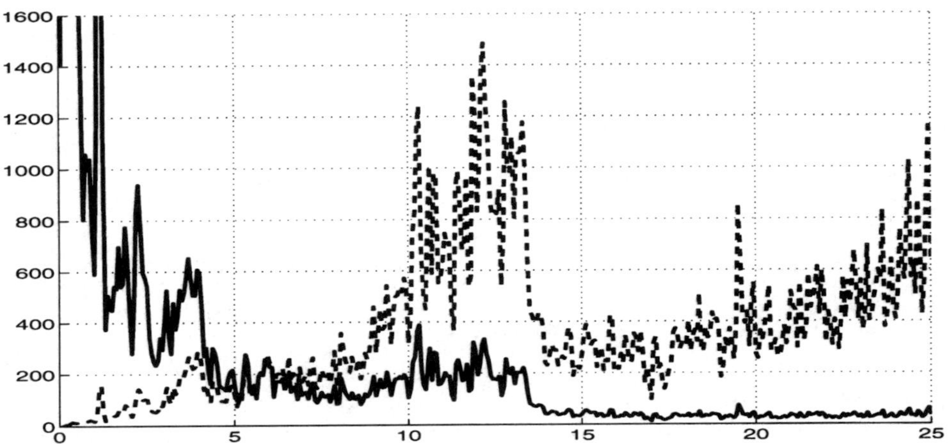

Figure 11-2 An EEG spectrum computed from the data recorded from a male adult during sleep. The frequency components around 12 Hz are closely related to sleep spindles. In the original spectrum (solid line), the spindle components have smaller amplitude values than the low-frequency components. In the processed spectrum (dashed line) computed by weighting the original spectrum with ω^2, the spindle components are emphasized.

of the wavelet transform in (11-1) is not a suitable form for computation due to the involvement of a time-consuming convolution. Instead, the Fourier transform of the wavelet, $\Psi(\omega)$, can be rewritten in terms of the Fourier transforms of two digital filters $h(n)$ and $g(n)$

$$\Psi(\omega) = G(\omega/2)\Phi(\omega/2) \tag{11-27}$$

where $G(\omega)$ is the Fourier transform of $g(n)$, $\Phi(\omega)$ is the Fourier transform of the *scaling function*

$$\Phi(\omega) = \prod_{p=1}^{+\infty} H(2^{-p}\omega) \tag{11-28}$$

and $H(\omega)$ is the Fourier transform of $h(n)$. It has been shown [9,12] that $H(n)$ and $G(n)$ must be constrained by

$$|H(\omega)|^2 + |H(\omega + \pi)|^2 \le 1, \qquad |H(\omega)|^2 + |G(\omega)|^2 = 1 \quad \text{and} \quad H(\pi) = G(0) = 0 \tag{11-29}$$

Equations (11-27) and (11-28) imply the following time-domain relationships:

$$W_{2^{j+1}}(n) = S_{2^j}(n) * g_{2^j}(n) \tag{11-30}$$

and

$$S_{2^{j+1}}(n) = S_{2^j}(n) * h_{2^j}(n) \tag{11-31}$$

where $g_{2^j}(n)$ and $h_{2^j}(n)$ are defined by inserting $2^j - 1$ zeros between any two successive samples of $g(n)$ and $h(n)$ [9]. This algorithm is initialized by setting $S_{2^0}(n) = x(n)$ and is then implemented recursively using (11-30) and (11-31) for $1 \leq j \leq J$, where J is the largest scale level desired.

The estimation of the variance values, $\mathbf{R}_{wx_j, wx_{j+1}}(0)$ and $\mathbf{R}_{wx_j}(0)$, and the mean distance, τ_e (or τ_z), is carried out by computing

$$C_{j,j+1} = \eta \sum_{m=1}^{N} wx_j(m)wx_{j+1}(m) \tag{11-32}$$

$$A_j = \frac{1}{N} \sum_{i=1}^{N} (wx_j(m_i))^2 \tag{11-33}$$

and

$$D_j = \frac{1}{M} \sum_{i=1}^{M} (m_{i+1} - m_i) \tag{11-34}$$

where η is the scale factor discussed in section 11.5, m is the discrete index corresponding to the integer samples of t, N is the number of samples in each segment, and M is the number of zero-crossings (or extrema) detected by examining sign changes in the signal (or in the differentiated signal). When soft-thresholding is used to remove noise, $wx_j(m)$ can be pre-processed by (11.9). The corresponding reduction in variance can be estimated by (11.12), if the Gaussian assumption is applicable.

The wavelet, $\psi(t)$, must satisfy the following admissible condition in order for the wavelet transform to be invertible:

$$\int_{-\infty}^{\infty} \frac{|\Psi(\omega)|^2}{|\omega|} \, d\omega < \infty \tag{11-35}$$

Hence, if $\psi(t)$ is admissible, $\Psi(\omega)$ must be zero at $\omega = 0$. This implies $\int_{-\infty}^{\infty} wx_j(t) \, dt = 0$ and that the sampled version $wx_j(m)$ for $j \geq 1$ does not contain offset values. Thus, subtracting the mean from the data is not necessary for computing Eqs. (11-32) and (11-33).

11.8. EXPERIMENTAL RESULTS

We provide an example using the intracranial pressure (ICP) data from a patient in the Neurotrauma ICU at the University of Pittsburgh Medical Center. The ICP provides important information about the state of the CNS for patients with severe head injuries [6,7]. Usually, only the moving average of the ICP is used for on-line evaluation; but the moving average reflects only the extremely low-frequency components in the signal. However, other frequency components are also important since they are related to the cerebrovascular dynamics [18]. Figure 11-3 (a) shows a 30-min segment of the ICP with a sharp increase at minute 5 and tapering at

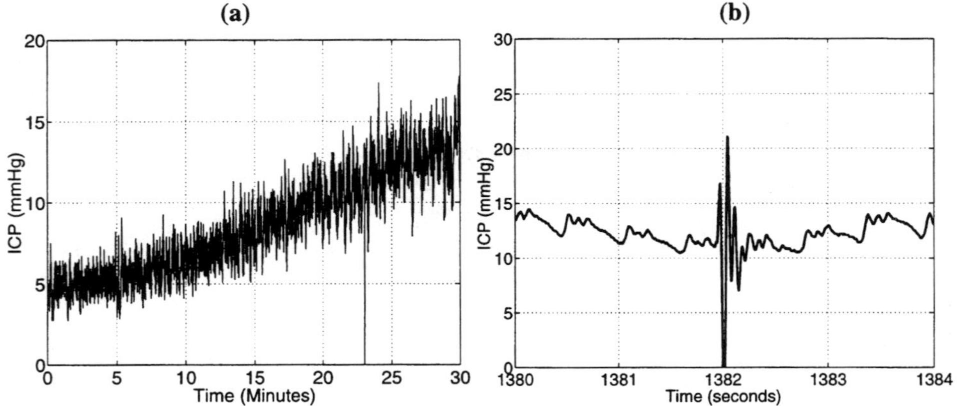

Figure 11-3   (a) A 30-min segment of the ICP data recorded from a brain trauma patient. An increase in the ICP can be observed. The rate of increase is slightly reduced in the later portion of the signal. This data set contains artifacts in several short periods. (b) An amplified view of the largest artifact occurring near minute 23 in (a). The details of the ICP waves are shown on both sides of the artifact.

minute 25, along with some variations in the amplitude values. This set of data contains several short periods of artifacts, with the largest one taking place near minute 23 (sharp positive and negative peaks) shown in Fig. 11-3(b). The first 20 min of the ICP signal were processed by the wavelet transform in which the wavelet was constructed from the low-pass filter $h(n)$, given by:

$$h(n) = (s * s * s)(n + 2) \qquad (11\text{-}36)$$

where "*" denotes convolution, and $s(n)$ is a square-like function:

$$s(n) = \begin{cases} 0.5, & n = 0, 1 \\ 0, & \text{otherwise} \end{cases} \qquad (11\text{-}37)$$

The high-pass filter, $g(n)$, and the scaling function may be determined using the frequency domain relationships given in (11-27) and (11-28); however, the Fourier transform of $g(n)$ cannot be obtained uniquely by (11-29). This problem can be solved by maximizing the attenuation rate in the coefficients of $g(n)$ with respect to their axis of symmetry [13]. This leads to the wavelet in the following closed form (specified in the frequency domain):

$$\Psi(\omega) = ie^{-i\omega/4}\text{sgn}(\omega)\left(\frac{\sin(\omega/4)}{\omega/4}\right)^3 \sqrt{1 - \cos^6(\omega/4)} \qquad (11\text{-}38)$$

In the process of computing the wavelet transform, we divided the acquired ICP data into 60 consecutive segments, each containing 4000 samples. We then computed, for each segment, statistics $C_{j,j+1}$, A_j, and D_j using the formulae given in (11-32), (11-33) and (11-34), respectively. Figure 11-4 shows a part of the extracted features $C_{j,j+1}$, A_j, and D_j (in dotted lines with circular markers). It can be observed that these features provide some identifiable trends, but fluctuate considerably,

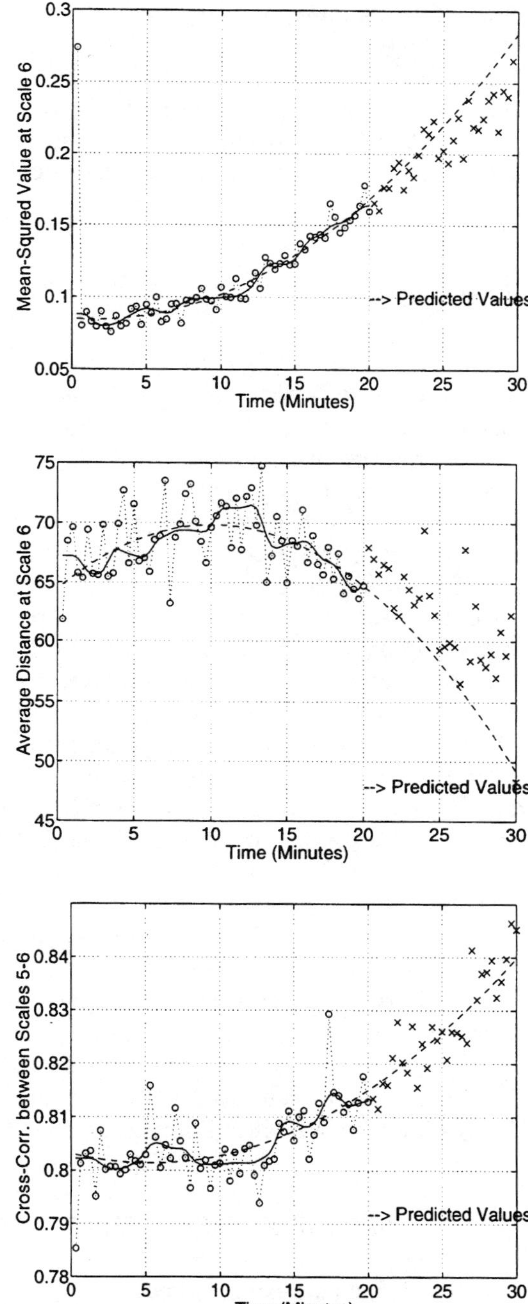

Figure 11-4 Three of the extracted features from the first 20 min of the data shown in Fig. 11-3(a), including variance estimate A_6 (top trace), average distance between adjacent extrema D_6 (middle trace), and cross-correlation estimate $C_{5,6}$ (bottom trace). In each trace, the circular patterns show the raw feature values; the solid lines show the results of nonlinear smoothing using the running medians; the dashed lines, which extend beyond minute 20, provide the results of the second-degree polynomial fitting; and the crosses from minutes 20 to 30 show the actually extracted feature values.

especially in the areas where artifacts occur. We then smoothed these features using a nonlinear smoothing algorithm based on the computation of running medians [19]. The results are shown by the solid lines. In order to predict future values, the baselines of the smoothed curves were computed by fitting the smoothed curves with polynomials of the second degree (the dashed lines extended beyond minute 20 for 10 min). For comparison, the actually measured values (separately computed from minute 20 to minute 30) are plotted as the crosses. Note that feature A_6 (in the middle of Fig. 11-4) deviates considerably from the predicted curve, possibly caused by the neurophysiological process that regulates the ICP.

11.9. DISCUSSION

A wavelet method of extracting frequency-related features from neurophysiological signals is investigated. This method, implemented by efficient algorithms, estimates the mean and variance of the wavelet coefficients. We have shown that, although the Fourier transform is not computed, the statistical estimates provide spectral information concerning the signal.

The form of frequency band division obtained from the wavelets is useful for analyzing neurophysiological signals which are generated by a large network of neurons. When the electrical pulses produced by neuronal circuits propagate from their generating sites to the recording electrodes, the high-frequency components are highly attenuated [20]. Consequently, these signals usually demonstrate a decaying spectral profile for which an increasing bandwidth proportional to the frequency helps to equalize the signal energy in each band. As a result, the extracted features have approximately equal average amplitude values, which are favored by artificial neural networks and other pattern classifiers. Our wavelet feature extraction approach is particularly useful for real-time analysis and the monitoring of data from the CNS in the operating room and the intensive care unit.

ACKNOWLEDGMENTS

The authors gratefully acknowledge Xiaopu Yan for his helpful discussions, and Neal D. Ryan, Ronald E. Dahl, and Fu-Chang Tsui for providing test data. This work was supported in part by the National Institute of Health Grants No. ND30318-01 and MH41712-06, and the Biomedical Engineering Grant Program of the Whitaker Foundation.

APPENDIX: EXPECTED NUMBER OF ZERO CROSSINGS

We first define an auxiliary function $g_\alpha(y)$

$$g_\alpha(y) = \begin{cases} \frac{1}{4\alpha}\cos(y/2\alpha), & |y| < \alpha\pi \\ 0, & \text{otherwise} \end{cases} \tag{11-39}$$

Let us show that the limit of $g_\alpha(y)$ is a δ-function when $\alpha \longrightarrow 0$.

For any continuous $f(y)$, the limit

$$\lim_{\alpha \to 0} \int_{-\infty}^{\infty} f(y)g_\alpha(y)\,dy = \lim_{\alpha \to 0} \int_{-\pi/2}^{\pi/2} f(2\alpha y)\frac{1}{2}\cos y\,dy \tag{11-40}$$

Since $f(y)$ is continuous, the limit can be taken inside the integral, yielding

$$\int_{-\pi/2}^{\pi/2} \lim_{\alpha \to 0} f(2\alpha y)\frac{1}{2}\cos y\,dy = f(0) \tag{11-41}$$

Thus, $\lim_{\alpha \to 0} g_\alpha(y) = \delta(y)$. In particular, setting $f(y) = 1$ implies $\int_{-\infty}^{\infty} \delta(y)\,dy = 1$. Let $y = y(t)$ be a differentiable function which crosses the t-axis m times at t_i, $i = 1, 2, \cdots, m$, in the time interval $[T_0, T_1]$ for fixed T_0 and T_1. We consider an arbitrary t_i at which $y'(t_i)$ is monotonic in its neighborhood, $[t_i - \epsilon, t_i + \epsilon]$, and define

$$\chi_\epsilon(t) = \begin{cases} 1, & t \in [t_i - \epsilon, t_i + \epsilon] \\ 0, & \text{otherwise} \end{cases} \tag{11-42}$$

For any crossing point t_i, we may write

$$1 = \int_{-\infty}^{\infty} \delta(y)\,dy = \int_{-\infty}^{\infty} \delta(y)|y'(t)|\chi_\epsilon(t)\,dt$$

$$= \int_{t_i-\epsilon}^{t_i+\epsilon} \delta[y(t)]|y'(t)|\,dt \tag{11-43}$$

where ϵ is an arbitrarily small positive number. Combining all zero-crossings and recognizing that $\delta(y) = 0$ for $y \neq 0$, we may write

$$m = \sum_{i=1}^{m} \int_{-\infty}^{\infty} \delta(y)\,dy = \int_{T_0}^{T_1} \delta[y(\tau)]|y'(\tau)|\,d\tau \tag{11-44}$$

Now, we consider the stochastic case where $y(t)$ is a stationary and differentiable random function. Then, m becomes a random variable whose mean is proportional to the length of the interval $T = T_1 - T_0$, i.e., $m = m(T)$ and

$$\begin{aligned}
\mathbf{E}[m(T)] &= \int_{T_0}^{T_1} \mathbf{E}[\delta[y(t)]|y'(t)|]\,dt \\
&= T\mathbf{E}[\delta[y(t)]|y'(t)|] \\
&= T\int_{-\infty}^{\infty}\int_{-\infty}^{\infty} |y'|\delta(y)p(y, y')\,dy\,dy' \\
&= T\int_{-\infty}^{\infty} |y'|\delta(y)p(0, y')\,dy' \\
&= T\int_{-\infty}^{\infty} |y'|p_y(0)p_{y'}(y'|y = 0)\,dy \\
&= Tp_y(0)\mathbf{E}[|y'||y = 0]
\end{aligned} \tag{11-45}$$

where $p(y, y')$, $p_y(y)$, $p_{y'}(y'|y = 0)$ are, respectively, the joint, marginal, and conditional probability density functions. The equality $p(0, y') = p_y(0)p_{y'}(y'|y = 0)$ follows Bayes's theorem.

REFERENCES

[1] A. Gevins, "Dynamic functional topography of cognitive tasks," *Brain Topography,* vol. 2, no. 12, pp. 37–56, 1989.

[2] T. F. Collura, E. C. Jacobs, D. S. Braun, and R. C. Burgess, "EView—a workstation-based viewer for intensive clinical electroencephalography," *IEEE Trans. Biomed. Eng.,* vol. 40, no. 8, pp. 736–744, 1993.

[3] P. L. Nunez, *Electric Fields of the Brain.* New York: Oxford University Press, 1981.

[4] R. J. Sclabassi and D. N. Krieger, "Neurophysiological evaluation and monitoring," In *Disorders of the Pediatric Spine,* D. Pang (ed.), New York: Raven Press, 1993.

[5] M. Sun, S. Baumann, and R. J. Sclabassi, "A model-based chracterization of evoked potentials for surgical monitoring," in *Proc. 16th Ann. Int. Conf., IEEE Eng. in Medicine and Biology Soc.,* Baltimore, pp. 195–196, 1994.

[6] M. R. Gaab, M. Ottens, K. Ungersbock and G. Moller, "Computer-aided neuromonitoring: conditions, techniques, and clinical applications." In *Advances in Neurosurgery,* H. Wenker, et al., (ed.), Berlin: Springer-Verlag, pp. 283–294, 1986.

[7] M. Frize, K. B. Taylor, B. G. Nickerson, F. G. Solven, H. Borkar, and V. Dunfield, "A knowledge-based system to assist patient management in an intensive care unit," *Proc. IMA - IFMBE Working Conf. on Biosignal Interpretation,,* Rebild Bakker, Aalborg, Denmark, August 25–27, pp. 156–160, 1993.

[8] D. L. Donoho and I. M. Johnstone, "Ideal spatial adaptation via wavelet shrinkage," *Biometrika,* vol. 81, pp. 425–455, 1994.

[9] S. Mallat and S. Zhong, "Characterization of signals from multiscale edges," *IEEE Trans. Signal Processing,* vol. 14, no. 7, 1992, pp. 710–732.

[10] C. K. Chui, *An Introduction to Wavelets.* New York: Academic Press, 1992.

[11] O. Rioul and M. Vetterli, "Wavelets and signal processing," *IEEE SP Magazine,* October, 1991.

[12] S. Mallat, "Multiresolution approximations and wavelet orthonormal bases of $L^2(R)$," *Trans. Am. Math. Soc.,* vol. 315, no. 1, 1989, pp. 69–87.

[13] M. Sun, C-C. Li, H. Szu, Y. Zhang, and R. J. Sclabassi, "Symmetrical wavelet transforms for edge localization," *Optical Eng.—J. SPIE,* July, 1994.

[14] S. Mallat, "Zero-crossings of a wavelet transform," *IEEE Trans. Inform. Theory,* vol. 37, no. 4, pp. 1019–1033, 1991.

[15] A. Papoulis, *Probability, Random Variables, and Stochastic Processes,* 2nd Ed. New York: McGraw-Hill, 1984.

[16] M. Sun and R. J. Sclabassi, "Discrete instantaneous frequency and its computation," *IEEE Trans. Sig. Proc.,* vol. 41, no. 5, pp. 1867–1880, 1993.

[17] P. Hauri, *The Sleep Disorders,* 2nd Ed. Kalamazoo, MI: The Upjohn Company, 1982.

[18] I. R. Piper, J. D. Miller, N. M. Dearden, J. R. S. Leggate, and I. Robertson, "System analysis of cerebrovascular pressure transmission: an observational study in head-injured patients," *J. Neurosurgery,* vol. 73, pp. 871–880, 1990.

[19] J. M. Tukey, *Exploratory Data Analysis.* Reading, MA: Addison-Wesley, 1977.

[20] W. Freeman, *Mass Action in the Nervous System.* New York: Academic Press, 1975.

Chapter 12

Experiments with Adapted Wavelet De-Noising for Medical Signals and Images*

Ronald R. Coifman, Mladen Victor Wickerhauser

12.1. TIME AND FREQUENCY ANALYSIS

In [1] we describe tools for adapting methods of analysis to various tasks occurring in harmonic and numerical analysis and signal processing. The main point is that by choosing a new coordinate system, in which space and frequency are simultaneously localized, one can more easily separate the signal into coherent structures and incoherent noise. This decomposition is intimately related to efficiency in representation (i.e., compression) and to pattern extraction, or structural understanding.

The separation is done by choosing an appropriate decomposition of the signal into components in the *time-frequency plane*—an abstract two-dimensional (2-D) signal representation in which time and frequency are the horizontal and vertical axes, respectively. A waveform is represented by a rectangle in this plane, as seen in the left half of Fig. 12-1. Let us call such a rectangle an *information cell*. The position in time and the main frequency can be read from the coordinates of the center of the rectangle. The uncertainty in time and the uncertainty in position are given by the width and height of the rectangle, respectively. Heisenberg's inequality, or the *uncertainty principle*, implies that the area of such a rectangle can never be less than 1. The amplitude of a waveform can be encoded by darkening the rectangle in proportion to its waveform's energy.

We will only use time-frequency atoms in our analysis; these are waveforms which are so well localized in both time and frequency that the areas of their infor-

*Research supported by NSF, AFOSR, and the Southwestern Bell Telephone Company.

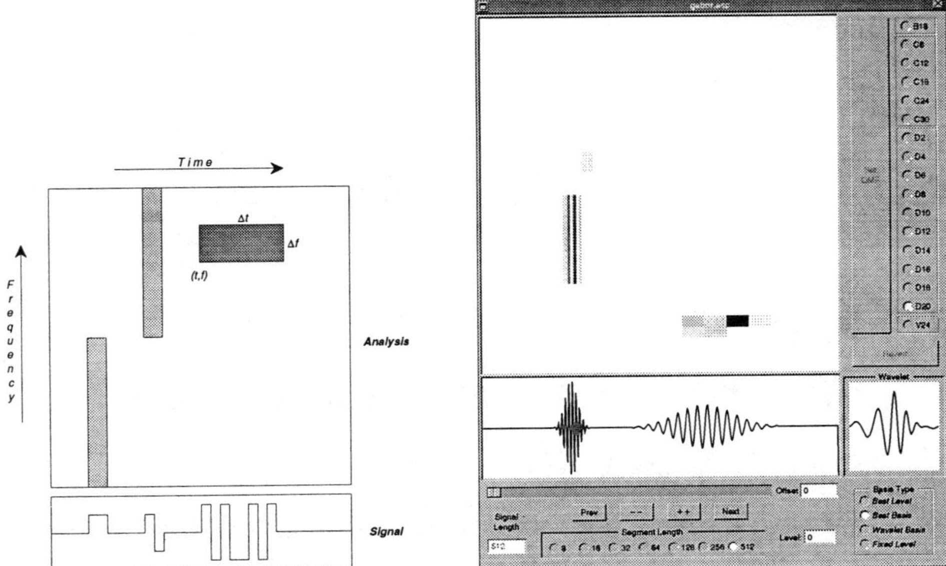

Figure 12-1   Idealized and actual analysis in the time-frequency plane.

mation cells must be close to 1. They may be represented by information cells of equal area in the time-frequency plane. A basis of such atoms corresponds to a covering of the time-frequency plane by rectangles, and we will depict an orthonormal basis by using disjoint rectangles of exactly equal (unit) area. It is well known that only the *Gaussian function* $g(t) = e^{-t^2}$ and its variations have the minimal information cell area. The other time-frequency atoms are not too far off, though, and we will avoid the many restrictions of the Gaussian by relaxing the minimal area condition. The only price we will have to pay is that a single atom might in practice require a few of the approximate atoms. The right half of Fig. 12-1 shows the output of such an approximate analysis on two Gaussian time-frequency plane atoms.

An orthogonal adapted wavelet analysis [1] produces a numerical recipe for the decomposition; the covering partitions are chosen to achieve maximum efficiency with respect to an *information cost function*. Not only does this approach shed light on classical analysis methods, it also suggests new methods of operating on the signals we have analyzed. These include nonstandard forms for linear operators [2,3], which better display the interactions among different signal parts, and discrete time-frequency plane approximations for the evaluation of complicated nonlinear operators [4] and large-scale computation.

12.2. EXAMPLE LIBRARIES OF WAVEFORMS

The appropriate analysis method is chosen from a catalog of tools for application to a particular problem. In an orthogonal adapted waveform analysis, the user begins with a collection of standard libraries of waveforms—called *wavelets, wavelet pack-*

ets, and *windowed trigonometric waveforms*—which can be combined to fit specific classes of signals. All these functions are time-frequency atoms. Examples of such waveforms are displayed in Fig. 12-2.

These libraries are used because they come equipped with fast numerical algorithms. They allow us to perform data compression, de-noising, and diagnostic feature extraction in real time. The process of analysis is usually done by comparing acquired segments of data with stored waveforms. The numerical comparison algorithm itself is fast and perfectly conditioned, always being a factored sparse orthogonal transformation. Then the most efficient orthonormal basis for compression of the signal is selected and used to extract and manipulate relevant features.

Consider first the short-time or windowed Fourier transform. Here the basis functions are exponentials, or maybe sines and cosines, which are enveloped so they oscillate only for a short time before going back to zero. They yield a tiling of the time-frequency plane by congruent rectangles, whose dimensions depend upon the window size. Two choices of window size are shown in Fig. 12-3. It is still necessary to choose a window size appropriate to the analysis. Our measure of quality will be the amount of white space, or negligible waveforms, in the time frequency analysis of a signal. Lots of white space means that most of the components have negligible energy, so that the signal energy is concentrated into just a few waveforms.

Very short windows are most efficient for sharp impulses, while long windows correspond to information cells which spread energy all over the time-frequency plane, as seen in Fig. 12-4. Conversely, long windows are more efficient than short ones for nearly continuous tones, as depicted in Fig. 12-5. Hence it can be useful to

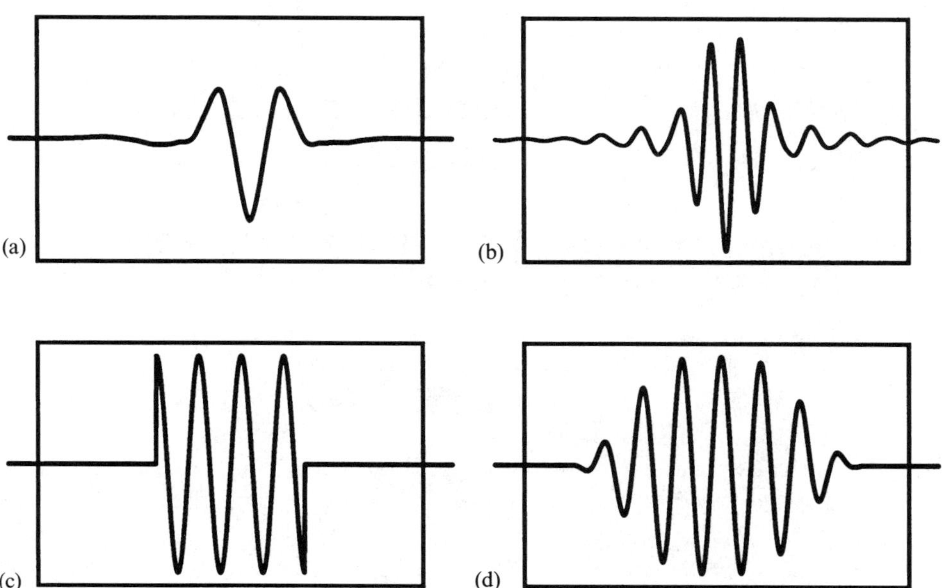

Figure 12-2   Example waveforms: (a) wavelet, (b) wavelet packet, (c) block cosine, and (d) smoothly windowed cosine functions.

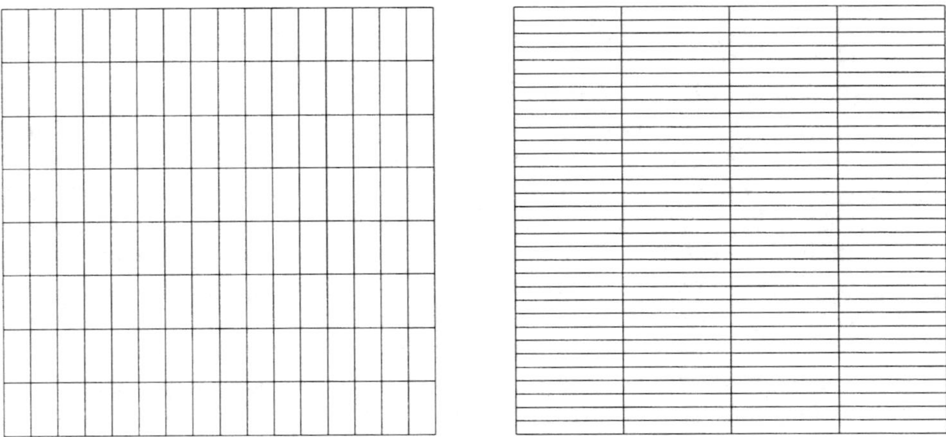

Figure 12-3 Narrow- and wide-windowed Fourier tilings of the time-frequency plane.

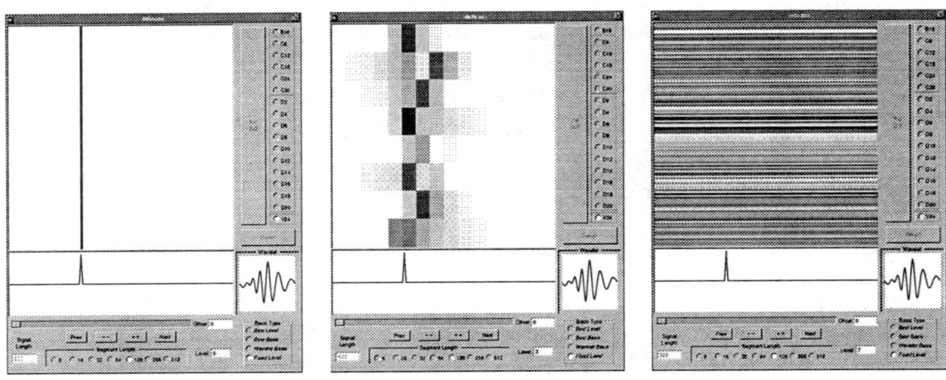

Figure 12-4 Time-frequency analysis of a sharp impulse for increasing window sizes.

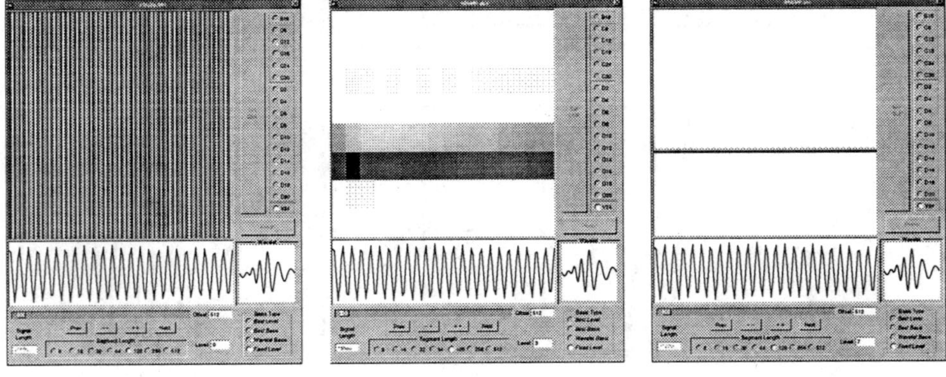

Figure 12-5 Time-frequency analysis of a nearly pure tone for increasing window sizes.

examine the signal in many window sizes at once and then to choose the *best basis* in the sense of efficient representation.

Remark. The Fourier transform is a rotation by 90 degrees in the time-frequency plane. This is most evident in Fig. 12-6, which shows the idealized information cells corresponding to grid-point samples (i.e., *Dirac mass* "waveforms") or sampled pure sine waves as time-frequency atoms. It is also possible to apply the *Hermite semigroup transform* or *angular Fourier transform* to obtain information cells which make arbitrary angles with the time and frequency axes.

Wavelet analysis [5] corresponds to windowing frequency space in "octave" windows. Since the information cells have equal area, they cover the time-frequency plane in the manner depicted in the left half of Fig. 12-7. A natural extension therefore is provided by allowing all dyadic windows in frequency space and adapted window choice. This sort of analysis is equivalent to wavelet packet analysis; one example is shown in the right half of Fig. 12-7.

In the wavelet library, there are both *wavelets*, traditionally denoted $\psi = \psi(t)$, and *scaling functions*, traditionally denoted $\phi = \phi(t)$. The pair of functions satisfy the *two-scale equations*, which in one normalization take the following form:

$$\phi(t) = \sqrt{2} \sum_{k \in Z} h(k)\phi(2t - k) \tag{12-1}$$

$$\psi(t) = \sqrt{2} \sum_{k \in Z} g(k)\phi(2t - k), \qquad g(k) = (-1)^k h(1 - k) \tag{12-2}$$

Here $h = \{h(k) : k \in \mathbf{Z}\}$ is a sequence of coefficients which defines a related sequence of coefficients $g = \{g(k) : k \in \mathbf{Z}\}$. All properties of the wavelet library are determined by h.

The expansion of a signal $u = u(t)$ in the wavelet library starts with the sequence of inner products of the signal with integer translates of the scaling functions:

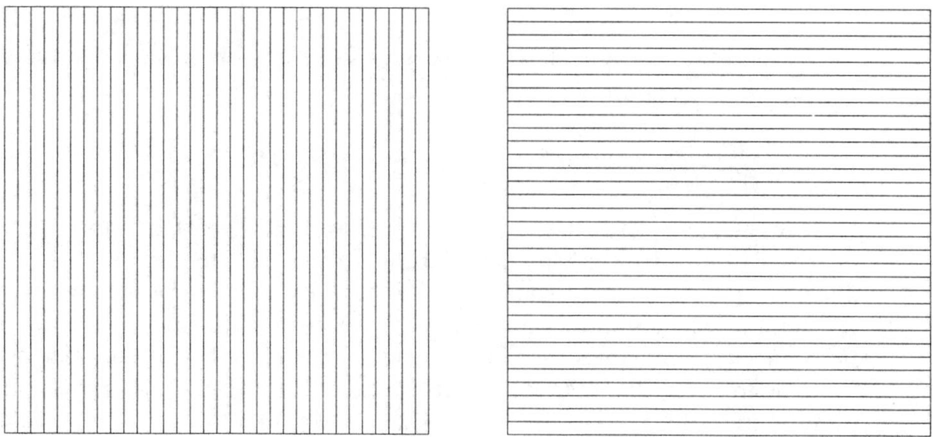

Figure 12-6   Dirac and Fourier decompositions of the time-frequency plane.

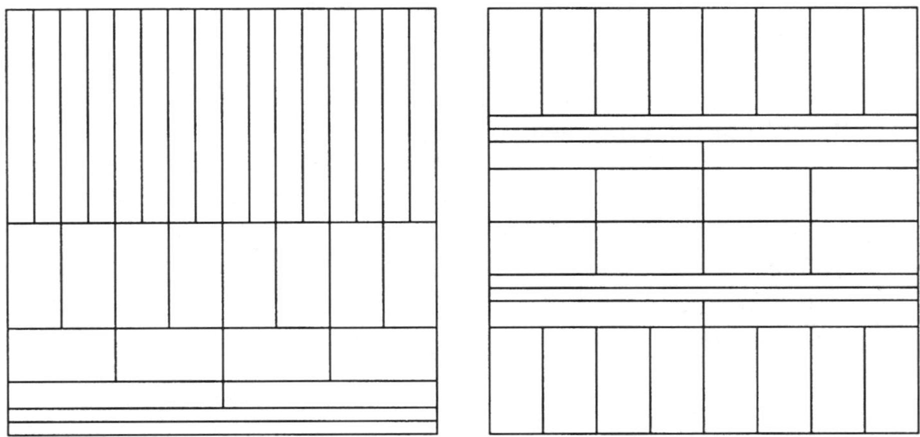

Figure 12-7   Dyadic wavelet and wavelet packet tiling of the time-frequency plane.

$$x(k) \overset{\text{def}}{=} \int_{\mathbf{R}} u(t)\phi(t-k)dt, \qquad k \in \mathbf{Z} \tag{12-3}$$

This amounts to a projection of the original signal, which is an almost arbitrary function of a continuous variable $t \in \mathbf{R}$, onto the much more restricted subspace of functions superposed from integer translates of the nice bump ϕ. Whether this projected image faithfully retains the important properties of the original depends on h and on the other assumptions placed on u. From the new sequence $x = \{x(k) : k \in \mathbf{Z}\}$, the inner products of u with dilated and translated wavelets may be computed with the following pair of *filtering* operations:

$$Hx(n) \overset{\text{def}}{=} \sum_{k \in \mathbf{Z}} h(2n-k)x(k), \qquad Gx(n) \overset{\text{def}}{=} \sum_{k \in \mathbf{Z}} g(2n-k)x(k), \qquad n \in \mathbf{Z} \tag{12-4}$$

This works because wavelets and scaling functions are related by the two-scale equations. For example,

$$\int_{\mathbf{R}} u(t)2^{-j/2}\psi(2^{-j}t-k)dt = GH^{j-1}x(k) \tag{12-5}$$

for any integer $j \geq 0$ and any integer k. All but the last of the inner products are obtained from x by applying a certain number of H filters followed by a single G filter. The procedure is depicted in Fig. 12-8, where the wavelet transform consists of the numbers to be found in the shaded boxes. It is called the *discrete wavelet transform* since it can only recover the discrete approximation x of the original signal u.

The *wavelet packet* library is constructed by recursion of the wavelet algorithm, using more G filters. This library, introduced in [6] and described in detail in [1], contains the wavelet basis, Walsh functions, and smooth versions of Walsh functions called wavelet packets. The basic functions may be denoted by $w_n = w_n(t)$, where

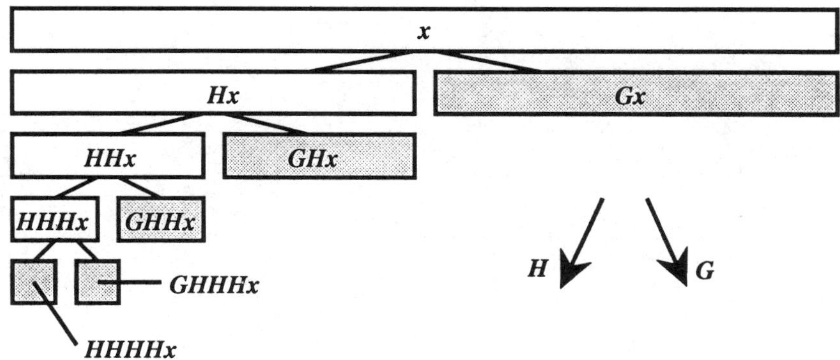

Figure 12-8   Discrete wavelet transform.

$n \geq 0$ is a nominal frequency index. They satisfy a generalization of the two-scale equations:

$$w_{2n}(t) = \sqrt{2} \sum_{k \in \mathbf{Z}} h(k) w_n(2t - k) \qquad (12\text{-}6)$$

$$w_{2n+1}(t) = \sqrt{2} \sum_{k \in \mathbf{Z}} g(k) w_n(2t - k), \qquad n = 0, 1, 2, \ldots \qquad (12\text{-}7)$$

The initial function w_0 is just the scaling function ϕ defined by the first two-scale equation. Likewise, $w_1 = \psi$, and it is easily seen how w_n is developed for all non-negative n. As with wavelets, the properties of all the functions $\{w_n, n \geq 0\}$ are determined by the sequence h.

The expansion of a signal $u = u(t)$ in the wavelet packet library begins, as for wavelets, with the sequence x of inner products with scaling functions defined by Eq. (12-3). Recursive application of the H and G operators defined in Eq. (12-4) then produces inner products with the other wavelet packets. For example,

$$\int_{\mathbf{R}} u(t) 2^{-3/2} w_5(2^{-3} t - k) dt = GHGx(k) \qquad (12\text{-}8)$$

for $k \in \mathbf{Z}$. Here the nominal frequency is 5 and the scale index is 3. The complete procedure for all wavelet packets is depicted in Fig. 12-9. The inner products in Eq. (12-8) will be found in the shaded box labeled *GHGx*, which is number 5 from the left in level 3 from the top since the indexing begins at 0.

The numbers in the shaded boxes in Fig. 12-9 constitute one particular *basis subset*, that is, a minimal set from which the sequence x can be exactly recovered. There are other basis subsets available, for example the one shaded in Fig. 12-10. That is recognizable as a *complete subband decomposition* of the original signal. To get many more examples, the *graph basis theorem* ([1], p. 244) can be used. It shows that the leaves of any binary subtree of the wavelet packet tree form a basis subset.

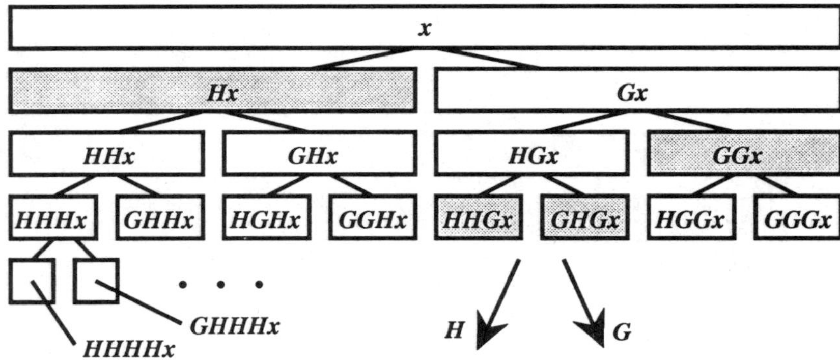

Figure 12-9   Discrete wavelet packet expansion and an example basis.

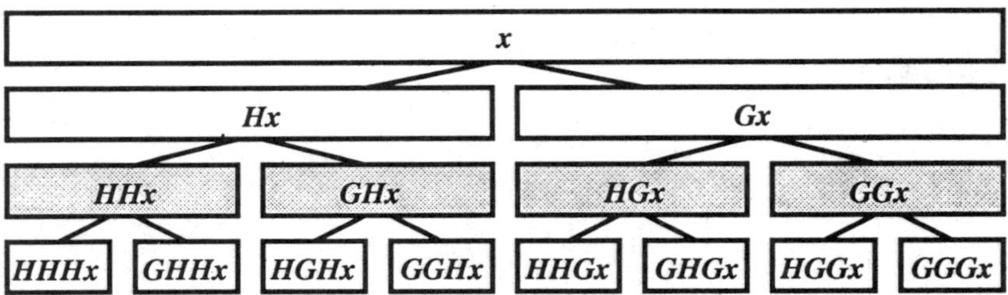

Figure 12-10   Complete level basis of wavelet packets.

Wavelet packet analysis algorithms permit us to perform an adapted Fourier windowing directly in the time domain by successive filtering of a function into different frequency bands. The window size selection algorithm, in this context, gives an adapted subband coding algorithm. It should be mentioned that all of these algorithms have higher dimensional generalizations; in particular, they can be used to analyze still images and movies in the same way that we are analyzing acoustic signals.

To illustrate a complete analysis in a library, we start with a description of an algorithm to compute the windowed Fourier sine expansion of a function on an interval from the Fourier expansion of its restrictions to the left half and the right half. This procedure is depicted in the left half of Fig. 12-11. The main idea is that all the signals under the big window or "bell," which is shown taking negative values, are combinations of signals under the two smaller windows. We see that in order to compute the Fourier expansion on the large interval, we can start with adjacent pairs of small intervals, combine coefficients to obtain the expansion on their union, and continue until we reach the largest interval at the top level. This scheme is depicted in the right half of Fig. 12-11. Along the way we have obtained all dyadic windowed Fourier transforms as intermediate computations. We notice that every disjoint collection of intervals and their orthogonal bases provides us with an orthogonal basis for the union.

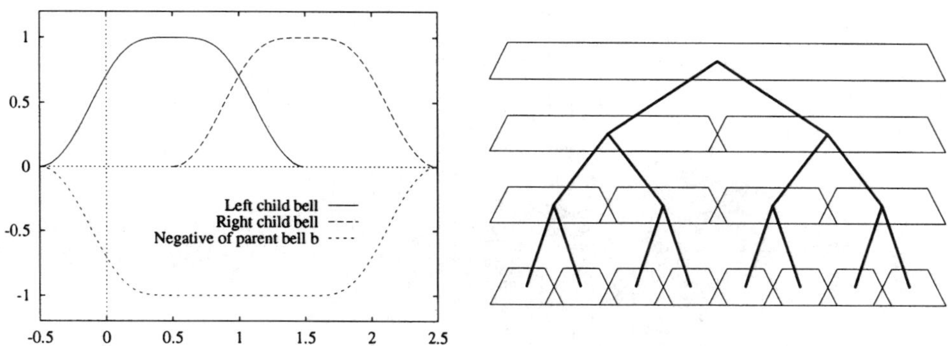

Figure 12-11 Left, big window from two small windows; right, the procedure iterated.

12.3. CHOOSING THE "BEST BASIS"

A natural question that arises in connection with the windowed Fourier transform is how to place the windows. The window selection has a big effect on the number of large coefficients in the expansion. So let us now turn to the question of optimizing the windows to obtain an efficient representation of a function.

The *best-basis* algorithm introduced in [7] fits a time-frequency plane cover to the signal so as to best concentrate the shading into the fewest information cells in the time-frequency plane. This method can use rectangular information cells of all aspect ratios. The *best-level* algorithm, described in [1], fits a cover of congruent rectangles to the signal, so as to best concentrate the shading. The congruence restriction on the rectangles is sometimes useful to avoid undesirable artifacts in partial reconstructions.

We can proceed as follows in our adapted windowed Fourier transform example: we start with the adjacent small intervals and determine the expansion coefficients on each one separately. We then compute the expansion coefficients of the union. Now we can choose that expansion for which the number of coefficients needed to capture 99% of the energy is smallest. Or, we can choose that expansion whose "cost" is smallest: information cost, coding cost, error cost, and so on.

We compare the cost of the chosen expansions on two adjacent unions of pairs to the expansion of their union and again pick the best. We continue until we reach an optimal distribution of time windows.

This algorithm segments a signal into portions that are individually easy to describe. It can be combined with an appropriate recognition criterion to segment continuous speech into voiced and unvoiced segments, as done in [8] and depicted in Fig. 12-12, prior to classification into phonemes.

The adapted wavelet packet decomposition uses the same choice algorithm, only it produces an adapted segmentation of the frequency axis rather than the time axis.

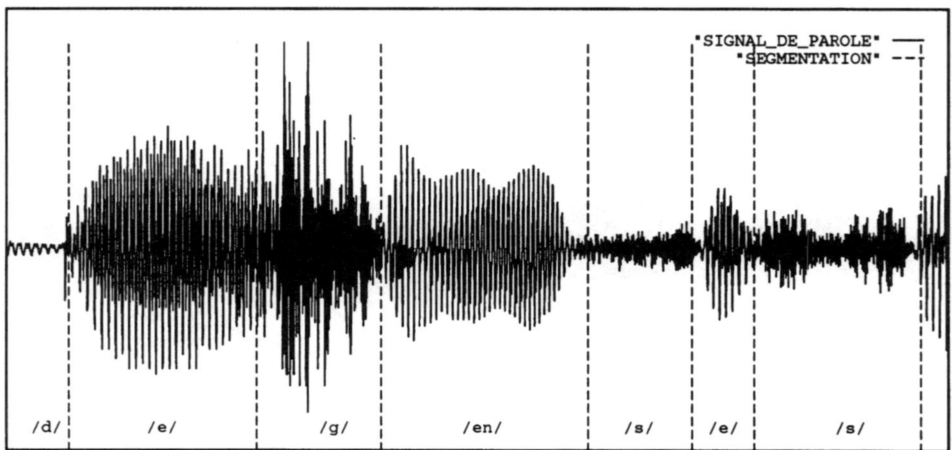

Figure 12-12 Automatic segmentation of a phrase into voiced and unvoiced portions.

12.4. COMPRESSION

The *compressibility* of a sampled signal is the ratio of the total area of time-frequency plane (N, for a signal sampled at N points) divided by the total area of the dark information cells (each of area 1). We may automatically analyze signals by expanding them in the best basis, then drawing the corresponding time-frequency plane representation. The negligible components need not be drawn, as it is not relevant which particular basis is chosen for a subspace containing negligible energy.

As done for the Gaussians in Fig. 12-1, signals can be automatically analyzed in their best wavelet packet bases by a computer program "WPLab"[9]. The user selects a transform by picking an analyzing quadrature filter from a list of 17 on the right. The "mother wavelet" [5] determined by that filter is displayed in the small square window at the lower right, to indicate roughly what the time-frequency atoms look like. The signal is plotted in the rectangular window at bottom, and the time-frequency plane representation is drawn in the large main square window. Examples of canonical signals for this analysis are *chirps* (oscillatory signals with increasing modulation), *spikes* (sharp transients), *whistles* (almost periodic functions), and combinations of all three such as human speech; examples of best basis analyses for these may be seen in [1].

12.5. ADAPTED WAVEFORM "DE-NOISING"

It is also possible to expand a signal in several libraries of waveforms and then to choose the library which best represents it. Or, if no library does particularly well, we can peel off layers of a signal by taking one or a few waveforms out at a time, then re-analyze the remainders. These are examples of *meta-algorithms* which are used at a high level to choose an appropriate analysis for the given signal.

As an application combining these ideas we now describe an algorithm for *de-noising* or, more precisely, *coherent structure extraction*. This is a difficult and ill-defined problem, not least because what is "noise" is not always well defined. We choose instead to view an N-sample signal as being noisy or incoherent relative to a basis of waveforms if it does not correlate well with the waveforms of the basis, i.e., if its entropy is of the same order of magnitude as

$$\log(N) - \epsilon \tag{12-9}$$

with small ϵ. From this notion, we are led to the following iterative algorithm based on the several previously defined libraries of orthonormal bases. We start with a signal f of length N, find the best basis in each library and select from among them the "best" best basis, the one minimizing the cost of representing f. We put the coefficients of f with respect to this basis into decreasing order of amplitude.

The rate at which the coefficients decrease controls the *theoretical dimension N_0*, which is a number between 1 and N describing how many of the coefficients are significant. We can define N_0 in several ways; the simplest is to count the coefficients with amplitudes above some threshold. Another is to exponentiate the entropy of the coefficient sequence, which matches the criterion in Eq. (12-9).

Theoretical dimension is a kind of information cost. We will say that the signal is incoherent if its theoretical dimension is greater than a preset "bankruptcy" threshold $\beta > 0$. The threshold β is chosen to determine if unacceptably bad compression was obtained even with the best choice of waveforms. This condition terminates the iteration when further decompositions gain nothing.

If the signal is not incoherent, then we can pick a fraction $\delta > 0$ and decompose f into $c_1 + r_1$, where c_1 is reconstructed from the δN_0 large coefficients, while r_1 is the remainder reconstructed from the small coefficients. We proceed by using $r_1, r_2, \ldots$ as the signal and iterating the decomposition. The procedure is depicted in Fig. 12-13. We can stop after a fixed number of decompositions, or else we can iterate until we are left with a remainder whose theoretical dimension exceeds β. We then superpose the coherent parts to get the coherent part of the signal. What remains qualifies as noise to us, because it cannot be well represented by any sequence of our adapted waveforms. Thus the adapted waveform de-noising algorithm peels a particular signal into layers; we take as many of the top layers as we want, assured that the bottom layers are not cost-effective to represent.

The two parameters, β and δ, can be adjusted to match an a priori estimate of the signal-to-noise ratio, or can be adjusted by feedback to get the cleanest-looking signal if no noise model is known.

Adapted waveform de-noising is a fast approximate version of the matching pursuit procedure described by Mallat et al. [10]. There the waveforms are Gaussians, and just one component is extracted at each iteration. That procedure always produces the best decomposition, at the cost of many more iterations plus more work for each iteration. Mallat's stopping criterion is to test the amplitude ratio of successive extracted amplitudes; this is a method of recognizing remainders which have the statistics of random noise.

As an example, we start with a mechanical rumble masked by the noise of aquatic life, recorded through an underwater microphone. The calculations were

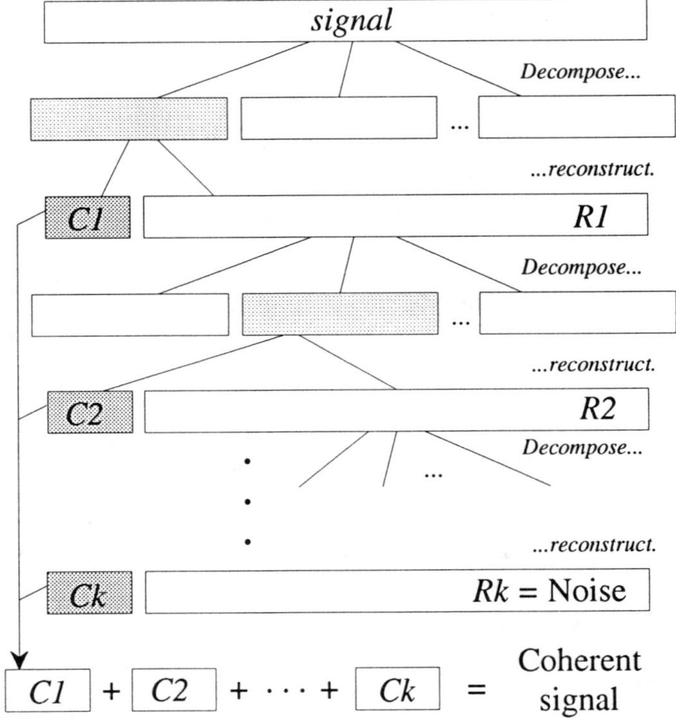

Figure 12-13   Schematic of adapted waveform denoising.

performed by the program "denoise" [11,12], using $\delta = 0.5$ and manually limiting the number of iterations to four. Figure 12-14 shows the original signal paired with its de-noised version. Note that very little smoothing of the signal has taken place. Figs. 12-15, 12-16, 12-17, and 12-18 respectively show the coherent parts and the remainders of the first four iterations. Notice how the total energy in each successive coherent part decreases, while the remainders continue to have roughly the same energy as the original.

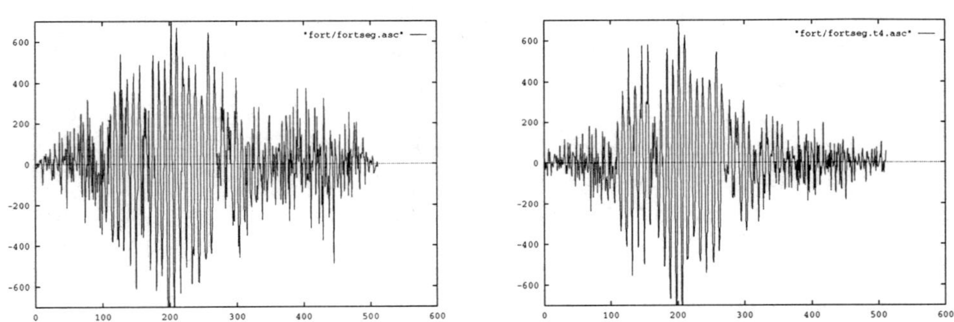

Figure 12-14   Original (left) and denoised (right) signals.

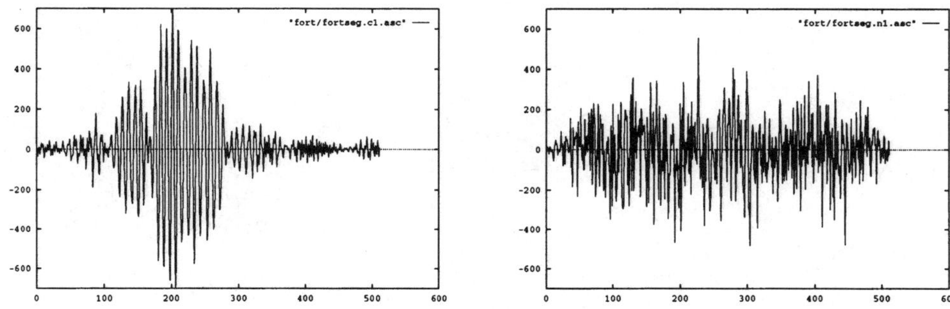

Figure 12-15   First coherent part (left) and first remainder part (right).

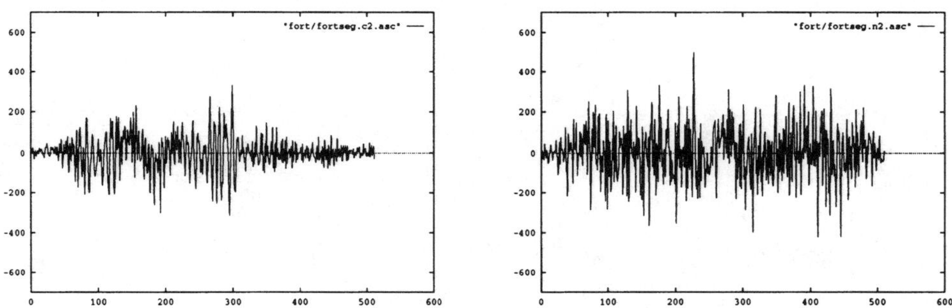

Figure 12-16   Second coherent part (left) and second remainder part (right).

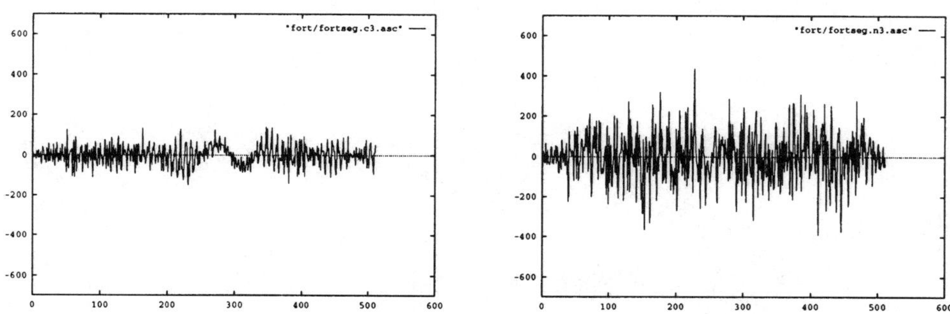

Figure 12-17   Third coherent part (left) and third remainder part (right).

Figures 12-19, 12-20, 12-21, and 12-22 respectively show the successive reconstructions from the coherent parts paired with a plot of the best-basis coefficient amplitudes of the remainders, rearranged into decreasing order. A visual estimate of the theoretical dimension from these plots gives evidence after the fact that little is gained after four iterations.

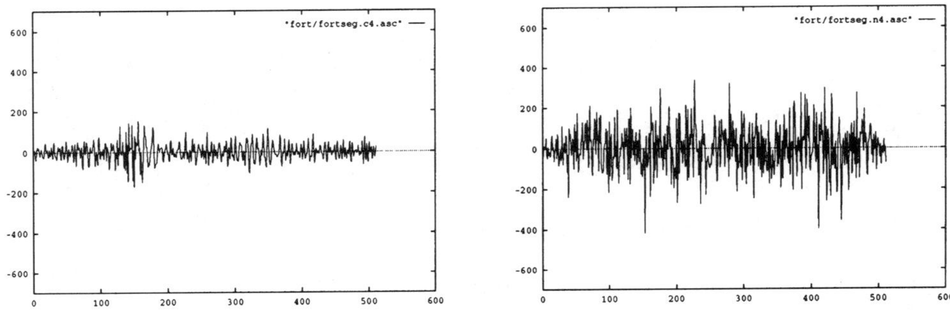

Figure 12-18   Fourth coherent part (left) and fourth noise part (right).

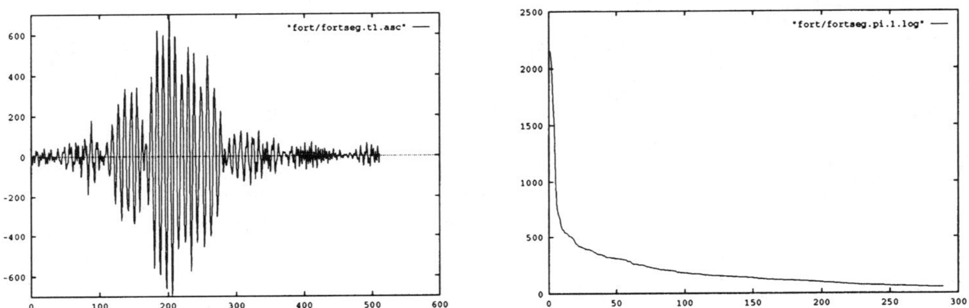

Figure 12-19   First reconstruction (left) and its sorted remainder coefficients (right).

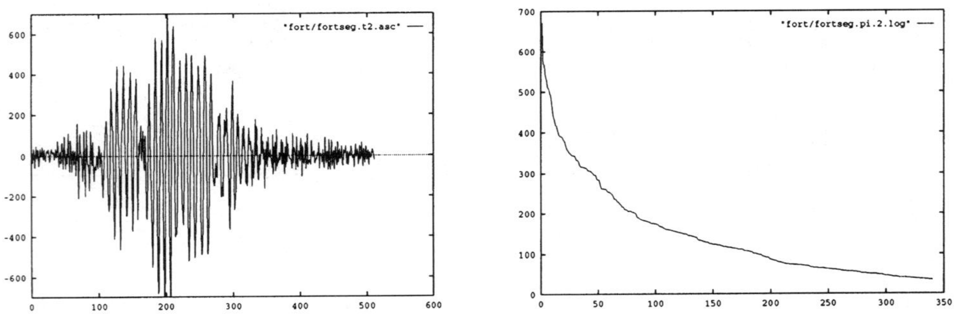

Figure 12-20   Second reconstruction (left) and its sorted remainder coefficients (right).

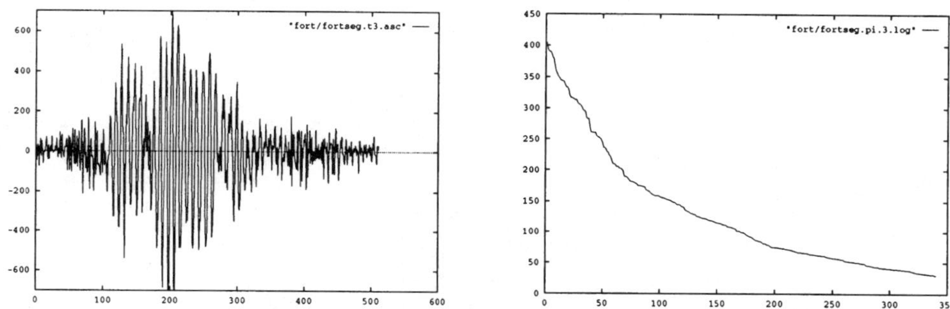

Figure 12-21   Third reconstruction (left) and its sorted remainder coefficients (right).

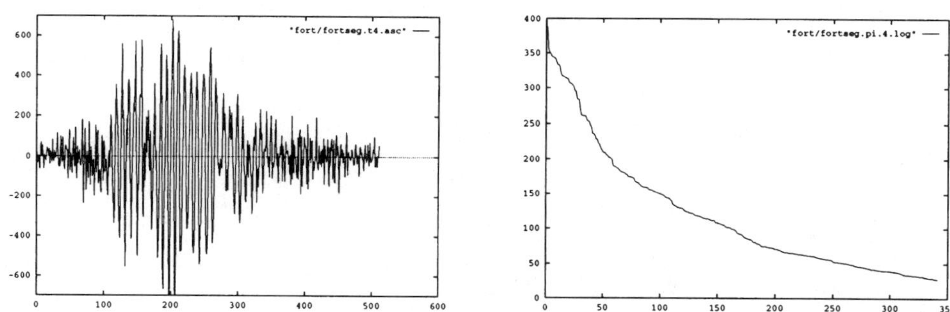

Figure 12-22   Fourth reconstruction (left) and its sorted remainder coefficients (right).

12.6. EXPERIMENTS WITH SNR IMPROVEMENT

12.6.1 Procedure

The software used to perform these experiments is a demonstration package called "denoise" which is maintained and distributed by FMA&H Corporation. It was modified by the authors from the version available by anonymous FTP [12]. It consists of scripts and executables for one- and two-dimensional adapted wavelet denoising, together with some utility functions for signal formatting, adding noise, and computing the SNR in dB.

We began with a 512-point sampled sine function of amplitude 64, biased to the range [64, 196], with seven cycles in the 512-point period. This is plotted in Fig. 12-23. We then added pseudo-random numbers to each sample, approximating independent uniformly distributed random noise with the right variance to produce the files `sine+0db.asc`, `sine+2db.asc`, etc. The filename indicates the nominal signal-to-noise ratio in decibels. We then ran `denoise -i4 -m9 -t0.2 sine+2dB.asc`, etc., to produce the superposition of the four coherent parts `sine+2db.t4.asc`. This is the denoised signal.

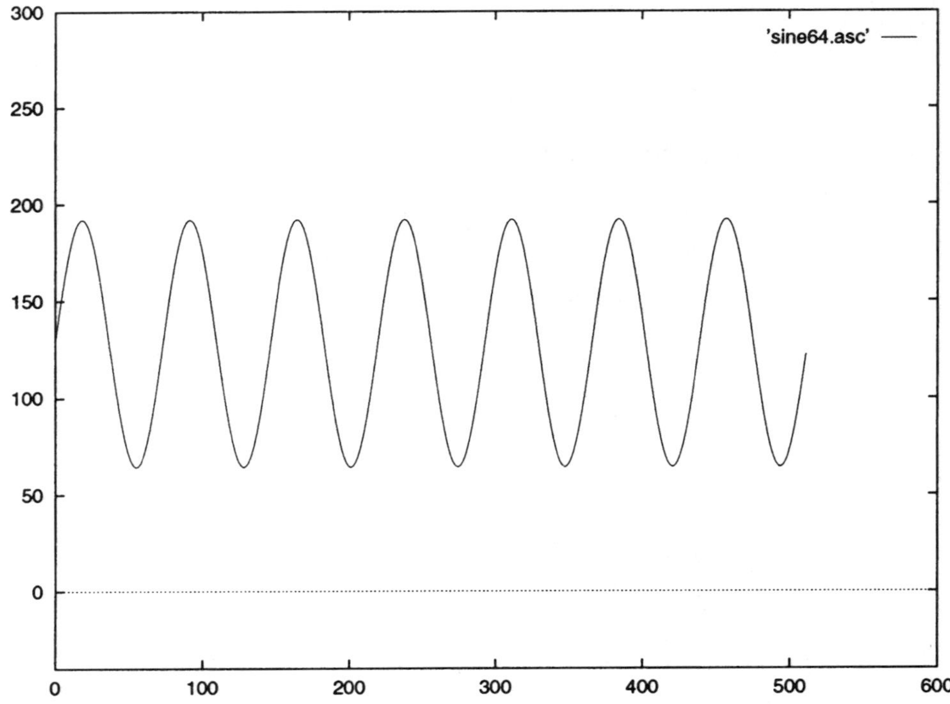

Figure 12-23   Original sampled sine function.

Since the original signal was known, we could compare the actual SNRs of the noised and de-noised sines using `snr sine64.asc sine+2db.asc`, etc., and `snr sine64.asc sine+2db.t4.asc`.

12.6.2 Results

The actual SNR and the gain obtained by denoising are listed in Table 12-1. The graphs of the functions that produced these results are shown in Figs. 12-24–12-31.

We used four iterations and a high threshold, chosen to make the denoised graphs look best. We tried four combinations of "-t" and "-i" parameter values. Further experiments would yield better results for a target SNR.

The noised and de-noised SNRs are computed from the middle half of the signal (samples 128 to 384) in order to avoid edge effects. If edges are included, the gain is lower.

12.7. CONCLUSION

We have analyzed one-dimensional sampled continuous waveforms by decomposing them into building blocks or atoms. The main desirable feature is that the atoms be well localized both in the time domain and in the frequency domain, in the manner of

TABLE 12-1   SNR Improvement from Applying "Denoise" to a Noisy Sine Function

SNR(dB):	Nominal	Actual	De-noised	Gain	Noised and De-noised Filenames
	−4	−3.5	6.5	10.0	sine-4db.asc, sine-4db.t4.asc
	−2	−1.8	7.0	8.8	sine-2db.asc, sine-2db.t4.asc
	+0	0.1	12.2	12.1	sine+0db.asc, sine+0db.t4.asc
	+2	2.1	10.4	8.3	sine+2db.asc, sine+2db.t4.asc
	+4	4.1	11.1	7.0	sine+4db.asc, sine+4db.t4.asc
	+8	8.2	14.2	6.0	sine+8db.asc, sine+8db.t4.asc
	+16	16.4	17.4	1.0	sine+16db,asc, sine+16db.t4.asc
	+24	25.0	28.1	3.1	sine+24db.asc, sine+24db.t4.asc

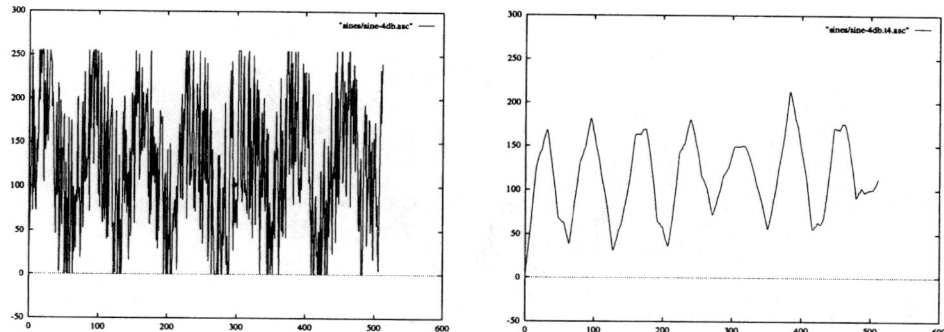

Figure 12-24   Four-iteration reconstruction of sine with −4 dB SNR. Left, noisy sinusoidal signal; right, reconstructed sinusoidal signal.

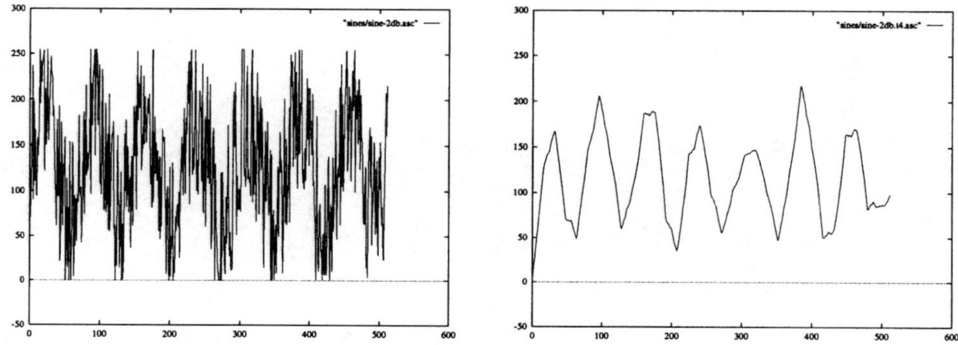

Figure 12-25   Four-iteration reconstruction of sine with −2 dB SNR. Left, noisy sinusoidal signal; right, reconstructed sinusoidal signal.

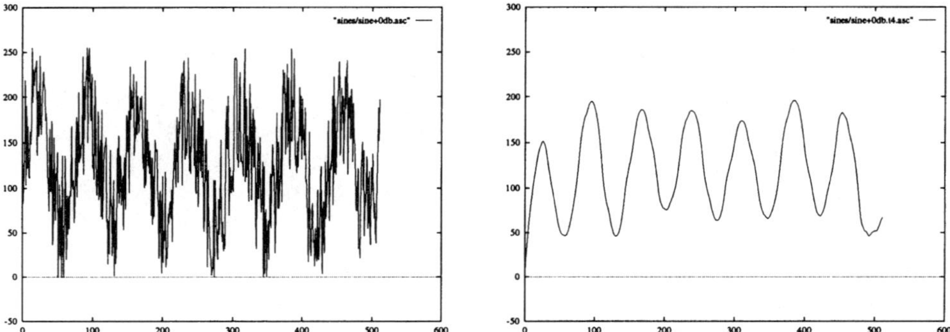

Figure 12-26   Four-iteration reconstruction of sine with 0 dB SNR. Left, noisy sinusoidal signal; right, reconstructed sinusoidal signal.

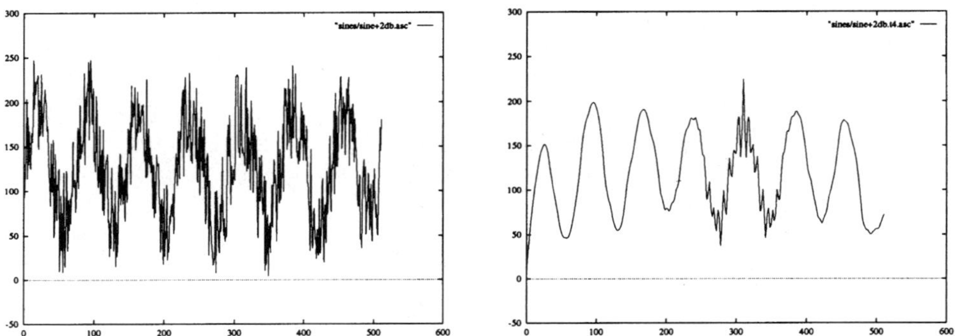

Figure 12-27   Four-iteration reconstruction of sine with +2 dB SNR. Left, noisy sinusoidal signal; right, reconstructed sinusoidal signal.

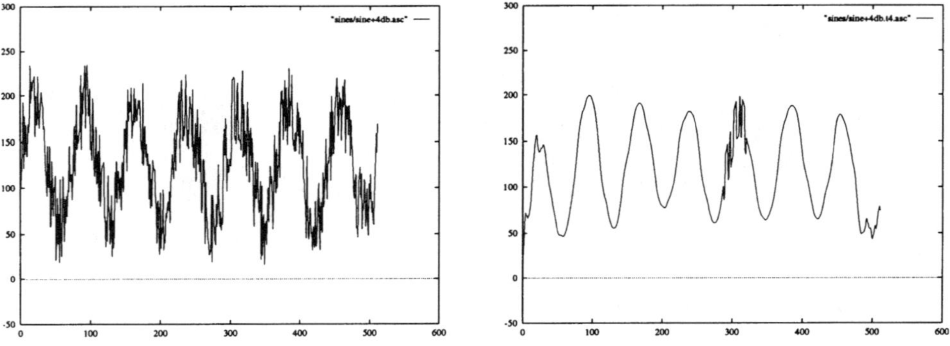

Figure 12-28   Four-iteration reconstruction of sine with +4 dB SNR. Left, noisy sinusoidal signal; right, reconstructed sinusoidal signal.

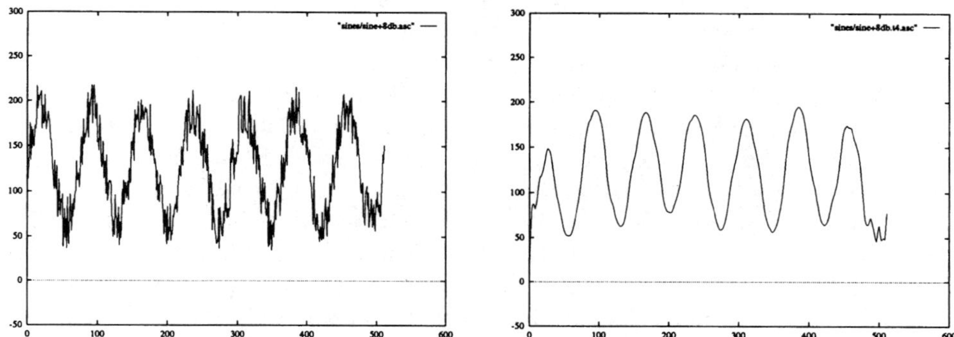

Figure 12-29   Four-iteration reconstruction of sine with $+8\,$dB SNR. Left, noisy sinusoidal signal; right, reconstructed sinusoidal signal.

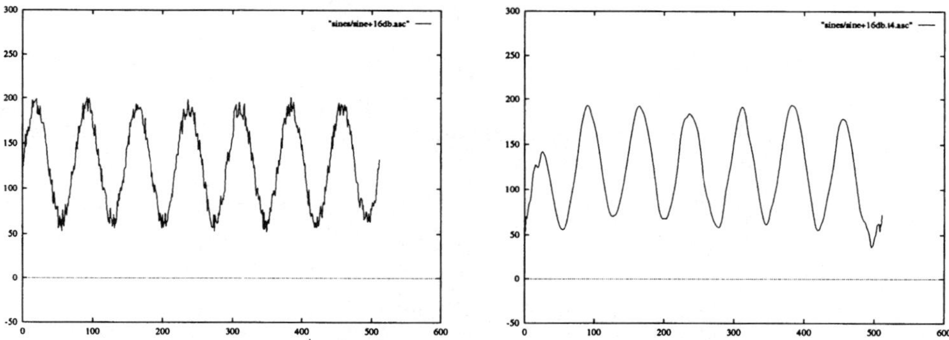

Figure 12-30   Four-iteration reconstruction of sine with $+16\,$dB SNR. Left, noisy sinusoidal signal; right, reconstructed sinusoidal signal.

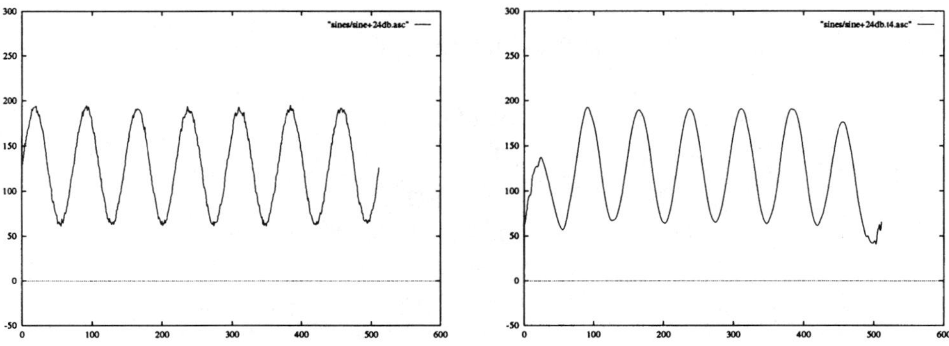

Figure 12-31   Four-iteration reconstruction of sine with $+24\,$dB SNR. Left, noisy sinusoidal signal; right, reconstructed sinusoidal signal.

windowed sines and cosines or Gabor functions. Wavelets and related functions like wavelet packets and local cosines can also be used in such decompositions. They have the added advantage that the resulting expansions are orthogonal or energy-preserving, allowing us to compare and adapt expansions to signals in order to minimize the cost of representation. Such adapted decompositions perform compression and analysis simultaneously. We have designed an idealized graphical presentation of the time-frequency information obtained by such a best-adapted waveform analysis, and from such presentations we can recognize and extract transient features such as parts of speech. Finally, we have shown that the negligibly small components in the analysis may be treated as noise and discarded, and we have designed an iterative algorithm for extracting coherent signals from such noise. Adapted wavelet analyses are practical for realistic signal sizes because the underlying algorithms have low computational complexity.

12.A INSTRUCTIONS AND SAMPLE OUTPUT FOR THE PROGRAM "DENOISE"

12.A.1 Summary of the Algorithm

Denoise takes as input a one-dimensional signal such as a soundfile, and:

1. Breaks it into windows of desired length (a power of two).
2. In each window, tries a wavelet packet transform with a number of filters, keeps the one which has the lowest entropy for its best basis, and sorts the coefficients by decreasing amplitude.
3. In each window, eliminates the coefficients that have an amplitude less than a certain energy threshold.
4. In each window, evaluates a cost function of the coefficients repeatedly until the cost is higher than a certain cost threshold. The coefficients not considered are discarded.
5. Reconstructs a new signal from the remaining coefficients.

The parameters that can be modified are:

1. The cost function which is used in step 4 above.
2. The energy threshold which is used in step 3 above.
3. The threshold for the cost function.
4. The number of times this process is iterated.

12.A.2 Manual page

```
DENOISE(1)          UNIX Programmer's Manual          DENOISE(1)
```

```
NAME
    denoise - denoise a one-dimensional ASCII signal file

SYNOPSIS
    denoise [ -wleitmf ] filename

DESCRIPTION
    Denoise takes as input a one-dimensional signal such as a
    soundfile, and:

        - breaks it into windows of desired length (a power of
          two).
        - in each window, tries a wavelet packet transform
          with a number of filters, keeps the one which has
          the lowest entropy for its best basis, and sorts
          the coefficients by decreasing amplitude.
        - in each window, eliminates the coefficients in the
          wavelet packet that have an amplitude less than a
          certain energy threshold.
        - Either:
            - in each window, evaluates a cost function of the
              coefficients repeatedly until the cost is higher
              than a certain cost threshold. The coefficients
              not considered are discarded.
            - does Wiener-filtering, by extrapolating the
              noise distribution (exponential least squares
              fit).
            - does Wiener-extraction (determining the number
              of coefficients to retain from the exponential fit
              of the noise coefficients).
        - reconstructs a new signal from the remaining coefficients

    The parameters that can be modified are:

        - the energy threshold which is used in step 3 above.
        - the threshold for the cost function.
        - the number of times this process is iterated.

OPTIONS
    -l    Produce a log of all wavelet packet coefficients in the
          selected basis (before E-thresholding).

    -w    Use Wiener-filtering. The -t option is meaningless
          when used with Wiener-filtering. Note that an
          E-threshold must be specified otherwise one will be
          provided.
```

    -x   Use Wiener-extraction. The -t option is meaningless
          when used with Wiener-extraction. Note that an
          E-threshold must be specified otherwise one will be
          provided.

    -s   Do   smooth   cuts.   The   rejected   coefficients   are
          modulated by a bell function to make the algorithm less
          brutal.

    -e   threshold
          select the energy-threshold to use. Possible values
          for threshold are of the form number, in which case
          the threshold is assumed to be an absolute value, or
          number%, where the threshhold will be relative to the
          signal window energy.

    -i   number
          set the number of iterations for the algorithm.
          Default is 1.

    -t   threshold
          set the threshold for the cost function. Default is 1.

    -m   number
          set the maximum number of levels in the wavelet packet
          expansion. Default is 9. The window width is 2 to the
          power of the maximum level.

    -f   parse the arguments quietly. This is useful for
          error-checking in shell-scripts.

SEE ALSO
    winfold(1), gnuplot(1), denoise2d(1)

BUGS
    Only ASCII files can be used as input.

12.A.3 Output from denoise -i4 -m9 -t0.2 sine+8db.asc

Copyright (C) 1989-94 Wickerhauser Consulting,
R.R. Coifman, F. Majid, Y. Meyer, M. V. Wickerhauser.
Patents applied for. All rights reserved.
Version 2.0, 11 September 1994.

denoise1d: sound filtering with periodized wavelet packets.
by Fazal Majid, 1992.

```
*** Window 1 ***
:    Iteration 1
:    Best-basis search:
:    using Daubechies 8 -- entropy = 1.368005
:    Reconstruction phase:
:    retained 512 coefficients after E-thresholding
:    retained 7 coefficients after cost-thresholding
:    noise = 2.3% of original energy
:
:    Iteration 2
:    Best-basis search:
:    using Coifman 12 -- entropy = 7.143196
:    Reconstruction phase:
:    retained 512 coefficients after E-thresholding
:    retained 1 coefficients after cost-thresholding
:    noise = 2.2% of original energy
:
:    Iteration 3
:    Best-basis search:
:    using Coifman 12 -- entropy = 7.218138
:    Reconstruction phase:
:    retained 512 coefficients after E-thresholding
:    retained 1 coefficients after cost-thresholding
:    noise = 2.1% of original energy
:
:    Iteration 4
:    Best-basis search:
:    using Coifman 12 -- entropy = 7.253392
:    Reconstruction phase:
:    retained 512 coefficients after E-thresholding
:    retained 1 coefficients after cost-thresholding
:    noise = 2.0% of original energy
:
```

REFERENCES

[1] M. V. Wickerhauser, *Adapted Wavelet Analysis from Theory to Software.* Wellesley, MA: AK Peters, Ltd., 9 May 1994. With optional diskette.

[2] G. Beylkin, R. R. Coifman, and V. Rokhlin, "Fast wavelet transforms and numerical algorithms I," *Comm. Pure Appl. Math.*, vol. XLIV, pp. 141–183, 1991.

[3] M. V. Wickerhauser, Computation with adapted time-frequency atoms. In Meyer and Roques [13], pp. 175–184.

[4] M. V. Wickerhauser, "Large-rank approximate principal component analysis with wavelets for signal feature discrimination and the inversion of complicated maps," (*Proceedings of Math-Chem-Comp 1993, Rovinj, Croatia*). *J. Chemical Information Computer Sci.*, vol. 34, pp. 1036–1046, September/October 1994.

[5] I. Daubechies, *Ten Lectures on Wavelets*, vol. 61 of *CBMS-NSF Regional Conference Series in Applied Mathematics*. Philadelphia: SIAM Press, 1992.

[6] R. R. Coifman, Y. Meyer, S. R. Quake, and M. V. Wickerhauser, "Signal processing and compression with wavelet packets." In Meyer and Roques [13], pp. 77–93.

[7] R. R. Coifman and M. V. Wickerhauser, "Entropy based algorithms for best basis selection," *IEEE Trans. Information Theory*, vol. 32, pp. 712–718, March 1992.

[8] C. D'Alessandro, Xiang Fang, E. Wesfreid, and M. V. Wickerhauser, "Speech signal segmentation via Malvar wavelets." In Meyer and Roques [13], pp. 305–308.

[9] D. Rochberg and M. V. Wickerhauser, WPLab version 3.03 (for NeXT computers). Available by anonymous ftp from [14], 1992.

[10] S. G. Mallat and Zhifeng Zhang, "Matching pursuits with time-frequency dictionaries," *IEEE Trans. Signal Proc.*, vol. 41, pp. 3397–3415, December 1993.

[11] F. Majid, Applications des paquets d'ondelettes au débruitage du signal. Preprint, Department of Mathematics, Yale University, 28 July 1992. Rapport d'Option, Ecole Polytechnique.

[12] R. R. Coifman, F. Majid, and M. V. Wickerhauser, *Denoise.* (Available from Fast Mathematical Algorithms and Hardware Corporation, 1020 Sherman Avenue, Hamden, CT 06514 USA, 1992.)

[13] Y. Meyer and S. Roques, eds. *Progress in Wavelet Analysis and Applications, Proc. Int. Conf. "Wavelets and Applications,"* Toulouse, France, 8–13 June 1992. Observatoire Midi-Pyrénées de l'Université Paul Sabatier, Gif-sur-Yvette, France: Editions Frontières, 1993.

[14] wuarchive. Washington University in St. Louis. InterNet Anonymous File Transfer (ftp) Site wuarchive.wustl.edu [128.252.135.4], 1991.

Chapter 13

Speech Enhancement for Hearing Aids

Janet C. Rutledge

13.1. INTRODUCTION

Hearing loss is one of the most prevalent handicapping disabilities in the United States. It affects more Americans than cancer, heart disease, tuberculosis, blindness, multiple sclerosis, venereal disease, and kidney disease combined [1]. It is estimated that only 15% of individuals who need a hearing aid are actually using one. Current hearing aid technology does not meet the needs of a large group of people who suffer from certain types of hearing losses. More advanced compensation schemes can be implemented in hearing aid devices by using state-of-the-art digital signal-processing techniques. The focus of the compensation processing is on sensorineural hearing losses, especially those accompanied by recruitment of loudness.

Developing schemes which compensate for the loss of hearing has been an ongoing task for several centuries. Early designs were based on tube acoustics to amplify desired sounds. Invention of the telephone and phonograph technologies in the 1870s launched a new era of hearing aid design and testing. Many obstacles have limited the progress of developing hearing aid devices. The mechanics and characteristics of conductive-type hearing losses are fairly well understood. Frequency selective amplifier hearing aids perform well in compensating for this class of losses. Sensorineural-type hearing losses are less well understood physiologically, and mathematical and computer models can only approximate the complex interactions. Traditional amplifier hearing aids have not worked as well with this class of losses.

The signal processing approach to the design problem is to find a preprocessing operator to enhance a signal that will undergo a known distortion. In this case the distortion comes from the effects of the hearing impairment, and the preprocessor forms the hearing aid device. Models of hearing and hearing loss, together with knowledge of speech perception, are used to develop these preprocessing compensation techniques.

There are several concerns that are peripheral to the discussion here, but are very important to hearing aid designers. These concerns include proper design and fit of ear molds, acoustic interaction with the ear canal, battery size and power, and cosmetic issues. Hearing aids have not gained the public acceptance that eyeglasses have, so the stigma attached to wearing a hearing aid, and the problems of size and general cosmetics, complicate the compensation technology issues. Hearing aids started off with the body-worn types. As technology improved and components got smaller, in-the-ear and behind-the-ear devices became possible. Device size is returning to the realm of body-worn devices as technology moves toward digital hearing aids employing microprocessors.

Past research results have indicated that waveform parameterization methods such as the sinusoidal model [2] and wavelets [3] are well suited for use in hearing loss compensation techniques [4,5]. These methods provide more flexibility than band-pass filter approaches, and allow properties of the auditory system to be incorporated.

Background noise is a major problem with hearing aids since amplified noise is distracting and often painful. Many noise-reduction techniques available today require a separate source input for the noise signal. In hearing aids this is not always practical. Waveform parameterization techniques have offered some promise of noise-reduction systems that use properties that differentiate the desired signal from noise. One such method employs wavelet decomposition along several different sets of basis functions. The algorithm [6] selects the basis and coefficients which provide optimum compression of the signal and reduction of the noise.

Wavelet-based denoising techniques can serve as a preprocessor for frequency-dependent compensation for hearing impairments [7]. The beauty of the wavelet approach is that the same coefficients can be used in both stages of the processing. Presented in this chapter is a wavelet-based amplitude compression technique and a novel single-microphone noise-reduction technique that can work together and may provide improved hearing aid performance.

13.2. BACKGROUND

Presented in this section is some background on the characteristics of hearing impairments along with an overview of some existing noise-reduction and hearing loss compensation techniques.

13.2.1 Hearing Impairments

There are four major categories used to classify hearing impairments. The first is conductive loss, which is associated with a pathology of the middle ear, ear drum, or ear canal. Conductive impairments are characterized by a reduced level of sound transmitted to the sense organ of the inner ear. Examples include otosclerosis, a fixation of the stapes in the oval window, and otitis media, an accumulation of fluid, pus or adhesions accompanying infection [8]. Many of these types of impairments can be corrected medically.

A second category, sensorineural, encompasses defects in the cochlea (sense organ), auditory nerve, or both. There are many possible causes of sensorineural impairments including congenital or hereditary factors, disease, tumors, old age, long-term exposure to industrial noise, acoustic trauma or exposure to toxic drugs [8]. Cochlear impairments are often caused by Ménière's syndrome or noise exposure, the latter of which can damage hair cells, often in the region of high-frequency sensitivity. Neural (retro-cochlear) impairments may arise from tumors, hemorrhage, multiple sclerosis, or other causes. The most common cause of impaired hearing is the natural aging process. Effects such as degeneration of hair cells, alterations in cochlear fluids and the loss of neurons in the auditory cortex are collectively called presbycusis. The result is generally a high-frequency loss which adversely affects speech discrimination. Prolonged exposure to noise results in similar types of reduction in sensitivity and is often difficult to distinguish from presbycusis.

Although most hearing impairments fall into one of the above two categories, some fall into the classes of central loss or functional deafness. Central losses are characterized by a reduction in auditory comprehension without a decrease in auditory sensitivity. With this type of loss, individuals have normal or near normal audiograms but are unable to process sounds fully, especially speech. Functional deafness encompasses all other impairments in which there is no known physical involvement. These include cases which are psychological or motivational rather than physiological.

In addition to reduced dynamic range of hearing, many listeners with sensorineural hearing loss experience reduced spectral resolution. The cochlea behaves like a bank of bandpass filters since each region responds to a different characteristic frequency. For normal-hearing individuals, the auditory bandpass filters are sharply tuned to a particular frequency, and a low threshold level of sound at that frequency is required to activate that portion of the basilar membrane [9]. With sensorineural hearing losses of cochlear origin, there is generally a decrease in the slope of the filter skirts, particularly on the low-frequency side [10], and an increase in the threshold for activation. This tendency can be related to the phenomenon of upward spread of masking experienced by these impaired listeners.

Recruitment of loudness is a sensorineural hearing loss of cochlear origin. By our nomenclature, loudness refers to the subjective magnitude of sound; intensity refers to the measured power of the acoustic signal. Equal loudness curves are functions mapping loudness to intensity at each frequency such that any two tones played at the associated intensities along a given curve will sound equally loud. The thresholds of hearing and pain bound the range of sounds that a person can hear. This range is called the *dynamic range of hearing*. For normal listeners, equal increments in sound intensity produce equal increments in loudness perception uniformly across all frequencies of importance to speech signals. If a listener suffers from recruitment of loudness, perceived loudness grows more rapidly with an increase in sound intensity than it does in the normal ear. Since the threshold of hearing is usually raised with recruitment of loudness while the threshold of pain may remain constant or even be lowered, the dynamic range of hearing is reduced. Thus, the loudness curves are compressed together, causing relatively small changes in intensity to give larger corresponding changes in perceived loudness.

It is possible to simulate the dominant effects of recruitment of loudness in normal-hearing listeners by adding background noise to the audio signal. Models of masking effects often use noise to approximate the contributions made by the inner-ear malfunctions [11]. Thus, compensation algorithms can be tested with normal-hearing listeners by masking the compensated speech with appropriately shaped background noise.

13.2.2 Hearing Loss Compensation Techniques

Compensating for elevated thresholds in all types of hearing impairments involves amplification to raise speech above the threshold of hearing. With sensorineural impairments, there is a reduced dynamic range of hearing, so some form of amplitude limiting is needed to keep speech signals from exceeding the impaired listener's threshold of discomfort. Linear techniques apply gain directly to the amplitude of the incoming signal. Systems which vary with frequency are analogous to frequency equalizers. Nonlinear schemes reduce the amplitude level variations through techniques such as applying the gains exponentially to the amplitude of the signal. Linear techniques work well for conductive hearing losses. For sensorineural hearing losses with severely restricted dynamic ranges, linear processing has limitations [12]. Although linear processing techniques have generally performed better in subject testing, it is argued by Lippmann et al. [12] and Villchur [13] that the laboratory environment under which the testing was conducted did not represent the true conditions of everyday conversation. The focus of the remaining sections will be on nonlinear processing techniques.

Experimental evidence suggests that increasing the power of consonants relative to that of the vowels increases discriminability of speech in a noisy environment [14]. Amplitude compression, like peak clipping, allows fast adjustments to the gain of a speech signal to ensure amplification of low-level components without having high-amplitude components surpass the threshold of discomfort. Furthermore, since amplitude expansion plus attenuation has been shown by Villchur [15] to closely model the effects of recruitment of loudness, it seems natural that the inverse processing, amplitude compression plus equalization gain, should be applied for recruitment compensation.

Amplitude Compression.   Researchers have developed various types of compression systems. These methods include wideband and multiband compression, as discussed in detail by Nábělek [16]. Multiband syllabic compression systems reduce the variation in speech level in each frequency band according to the subject's reduced dynamic range in that band. Single-channel (wideband) systems process the entire speech signal on the basis of overall level. Although wideband processing cannot match a person's hearing profile as well as multiband processing can, wideband processing does not distort the short-term spectral shape. The wideband and multiband compression systems mostly use digital or analog filters along with equalization gain. With these systems, the parameters remain constant over time, regardless of the input conditions.

A modified version of Villchur's system was developed by Kates [17,18]. The compression gain is derived from the envelope $e(n)$ of the speech signal in each band according to the relation

$$g(n) = e(n)^{[\frac{1}{c}-1]} \tag{13-1}$$

where c is the compression ratio [19]. This gain is then multiplied by the speech on a point-by-point basis to achieve the compressed speech in each band. Post compression equalization is necessary in both cases to make sure that the processed speech properly matches the subject's hearing profile. It should sufficiently raise the compressed speech above the threshold of hearing without allowing it to reach the threshold of discomfort. An upper threshold for the maximum allowable value of the speech amplitude is chosen to ensure that the threshold of discomfort is never reached. The equalization gain for each band is based on this upper threshold and the amount of compression applied in that band. For the bandpass filter approach, the equalization gain is given by

$$\gamma = u^{[1-\frac{1}{c}]} \tag{13-2}$$

where u is the upper threshold and c is the compression ratio in that band. When combined with the compression gain, the resultant gain which multiplies the speech in each band is given by

$$\acute{g}(n) = \gamma g(n) = \left(\frac{e(n)}{u}\right)^{[\frac{1}{c}-1]} \tag{13-3}$$

If the speech amplitude at any point is equal to u, then $\acute{g}(n) = 1$. Any speech amplitude which is above the upper threshold is set equal to the upper threshold. Amplitude values which are more than 30 dB below the upper threshold are likely to be noise rather than speech. Therefore, amplitudes below a lower threshold are given a fixed linear gain so that they are not overemphasized. A gain of 15 dB is applied whenever the envelope is more than 30 dB below the respective upper threshold. Kates provides a preemphasis gain to the consonant portions of speech prior to compression. This gain is proportional to the ratio of energy in a high-frequency band to that in a low-frequency reference band.

Parametric Compression.   Compression of speech parameters can also serve to compensate for recruitment of loudness. An interesting example, and the one that is the basis for the wavelet-based algorithm, is TVFD (time-varying frequency-dependent) processing [5]. This processing is based on a sinusoidal model of speech [2] in which speech is represented as the sum of sinusoids with various amplitudes, frequencies and phases. The components are calculated from the short-time Fourier transform using a 20-ms window. TVFD is a digital multiband compression system in which each sinusoidal component corresponds to a separate frequency band. The sinusoidal amplitude parameters are given compression gain based on an intercomponent masking model of both normal-hearing and hearing-impaired listeners. In both cases, the masking model incorporates the listener's thresholds of hearing and pain. Calculating the gain for each frequency component based on preserving the relative distance above masked threshold allows the algorithm to adjust to the time-varying qualities of the input speech.

13.2.3 Noise Reduction

Background noise is a major problem in hearing aids. As the incoming speech sound of interest is amplified, so are other signals and noise that are present. The effects of sensorineural hearing loss exacerbate the problem of trying to understand speech spoken in a noisy environment. Many existing hearing aids use selective filtering to reduce common background sounds which are restricted to a particular frequency region [20]. Digital signal processing techniques to reduce noise have also been applied to the problem with mixed success.

The most widely used noise cancelation system for general signal processing applications uses adaptive filtering with two microphone inputs [21]. The primary input consists of both the speech and noise. The reference has the noise only. The main problem with this approach is that with a hearing aid, the two microphones would have to be placed so close together that it may not be possible to get a clean reference noise input [22]. The achievable output signal-to-noise ratio is directly dependent on the noise-to-signal ratio in the reference input. Techniques such as using a beamforming or directional microphone for the reference input and an omnidirectional microphone for the primary input provide some spatial cancelation [23]. This method works best in small rooms without much reverberation.

Single microphone systems can take advantage of the spectral differences between speech and noise to do the noise-reduction. These systems suffer from the problem that neither the speech or noise is a stationary signal, and thus may change in between spectral model updates. One single microphone system that has shown success uses waveform parameterization techniques, and then separates the speech parameters from the noise parameters by taking advantage of their differing features in each band [24]. The assumption here is that the noise spectrum changes more slowly than the speech spectrum does. Unfortunately, if the speech energy in a band is low level, and the noise in that same band is strong and quasi-stationary, both may be reduced.

A method of noise-reduction has been proposed by Coifman and Wickerhauser [25] which has the ability to reduce unstructured noise in a signal without eliminating the structured component. This *de-noising* program is an algorithm which chooses the best basis from a library of wavelet packets and local trigonometric functions, and retains only the largest coefficients needed to represent the signal. The guiding principle is that the part of the signal that does not compress in any basis is noise. That is, the energy from the random noise is spread out over many coefficients, whereas music or human speech is concentrated in only a few.

13.2.4 Motivation for Using Wavelets

A wavelet transform is equivalent to implementation of a nonuniform filter bank with finer resolution at low frequencies (and finer time resolution at high frequencies). Therefore wavelet-based multiband compression is a viable alternative to traditional multiband filtering techniques. The structure lends itself quite well to this application since the wavelet decomposition provides access to both time-domain and frequency-domain information. Wavelet packets and local cosine bases are especially well suited to this problem since it is possible to decompose the signal into frequency bands that are as fine as desired. Rather than being confined to the octave band format of the wavelet transform, compensation gain can be calculated more carefully in regions where the threshold of hearing changes rapidly.

Many noise-reduction algorithms often attempt to preserve certain features of the speech waveform and remove features unique to the noise waveform. The wavelet parametric approach lends itself well to solving the problems met by the waveform representation, in that:

- Short-time stationarity (on the order of a frame length) may be assumed, since processing may be modified as needed on a frame-by-frame basis.
- Distortion may be more easily controlled, since the output waveforms are synthesized by the noise-reduction system.
- Parameters derived from distorted or noisy data may be input to intelligent signal-processing algorithms for the purpose of enhancement or restoration.

13.3. WAVELET-BASED COMPRESSION

Wavelet-based compression uses the TVFD-style gain calculation, i.e., gain is calculated separately for each wavelet coefficient based on its intensity level, with the wider bandwidths of traditional multiband amplitude compression. The parameters of this parametric compression algorithm are wavelet coefficients. As a compression technique, the strategy of wavelet-based compression remains to invert the expander/attenuator model of recruitment of loudness.

The wavelet analysis/synthesis in this algorithm is performed via the multiresolution analysis wavelet decomposition/reconstruction algorithm developed by Mallat [3]. It is also called a fast wavelet transform (FWT) [26]. To compensate for recruitment of loudness, compression is applied to each wavelet coefficient c_{mn}, where m is the resolution level and n is the time index. The signal is then reconstructed from a wavelet series with the modified coefficients [4]. Gain is applied in such a way as to amplify the coefficient from a given equal loudness curve in the normal-hearing person's hearing profile to the corresponding equal loudness curve in the hearing-impaired person's profile. Since each wavelet coefficient corresponds to a band of frequencies, the equal loudness curves have been averaged within each band to also give one value per band.

Gain is calculated for each wavelet coefficient in each band such that the ratio of log intensity above hearing threshold to dynamic range of hearing is the same for the hearing-impaired listener as the corresponding ratio is for the normal-hearing listener. In other words, the compression gain is found so the desired relation specified in Eq. (13-4) and pictured in Fig. 13-1 holds. Mathematically,

$$\frac{\delta^*}{\delta} = \frac{\Delta^*}{\Delta} \tag{13-4}$$

where $\delta = C_{mn} - T_{\text{nor}}$ = the distance the unamplified wavelet coefficient is above a normal-hearing person's threshold of hearing (in dB)

δ^* = the distance the compensated wavelet coefficient is above a hearing-impaired person's threshold of hearing (in dB)

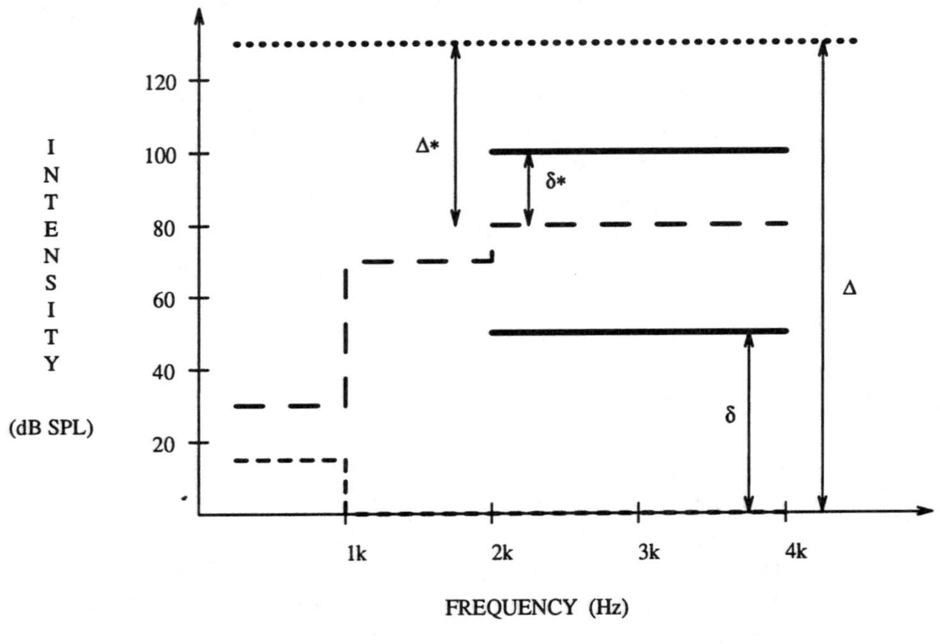

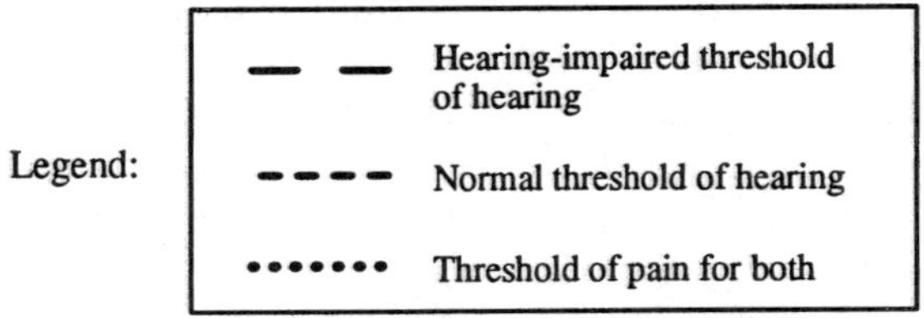

Figure 13-1   Calculation of gains based on the distance the wavelet coefficient is above the approximated threshold of hearing for the normal and impaired listener.

$\Delta = T_{\text{pain}} - T_{\text{nor}}$, the normal-hearing person's dynamic range of hearing (in dB)

$\Delta^* = T_{\text{pain}} - T_{\text{im}}$, the hearing-impaired person's dynamic range of hearing (in dB)

$C_{mn} = 20 \log c_{mn}$

T_{im} = the hearing-impaired person's threshold of hearing (in dB)

T_{nor} = the normal-hearing person's threshold of hearing (in dB)

T_{pain} = the threshold of pain for both (in dB)

With the above-defined relation, compression gain for a coefficient can be derived as follows. Let $C^*_{mn} = 20 \log c^*_{mn}$ be the compensated coefficient in dB. Then

$$C^*_{mn} = T_{\text{im}} + \delta^* = T_{\text{im}} + \delta \frac{\Delta^*}{\Delta} \tag{13-5}$$

Substituting in equation (13-5)

$$C^*_{mn} = T_{\text{im}} + (C_{mn} - T_{\text{nor}}) \frac{\Delta^*}{\Delta} \tag{13-6}$$

Define the thresholds t_{im} and t_{nor} such that $T_{\text{im}} = 20 \log t_{\text{im}}$ and $T_{\text{nor}} = 20 \log t_{\text{nor}}$, and substitute into equation (13-6) to find the compensated coefficient

$$20 \log c^*_{mn} = 20 \log t_{\text{im}} + 20 \frac{\Delta^*}{\Delta} \log c_{mn} - 20 \frac{\Delta^*}{\Delta} \log t_{\text{nor}}$$

Thus, the amplitude compression operator maps the coefficient, c_{mn} to the compressed coefficient, c^*_{mn} as follows:

$$c^*_{mn} = t_{\text{im}} \left(\frac{c_{mn}}{t_{\text{nor}}} \right)^{\Delta^*/\Delta} \tag{13-7}$$

The transformation $c_{mn} \longrightarrow c^*_{mn}$ defined by equation (13-7), could be applied anytime the speech signal is represented as a superposition of basis functions, all of which are reasonably well localized in frequency. For example, this transformation is used in TVFD where speech is modeled as a linear combination of sinusoids. With TVFD the thresholds include the effects of intercomponent masking. The transformation could be applied equally well to a wavelet packet representation of the speech signal or a superposition of local cosine functions.

Reconstruction of the compensated wavelet coefficients gives similar results to amplitude compression. This is illustrated in Fig. 13-2 which shows the original spectrum for the vowel /ʌ/ in the word *other* along with the vowel using the wavelet-based compensation technique and multiband amplitude compression for a flat-loss subject. With both compensation techniques the general formant structure is maintained.

13.3.1 Comparison with Multiband Filter Compression

As discussed in the previous section, this wavelet-based processing technique bears a close resemblance to multiband amplitude compression. Therefore, as a

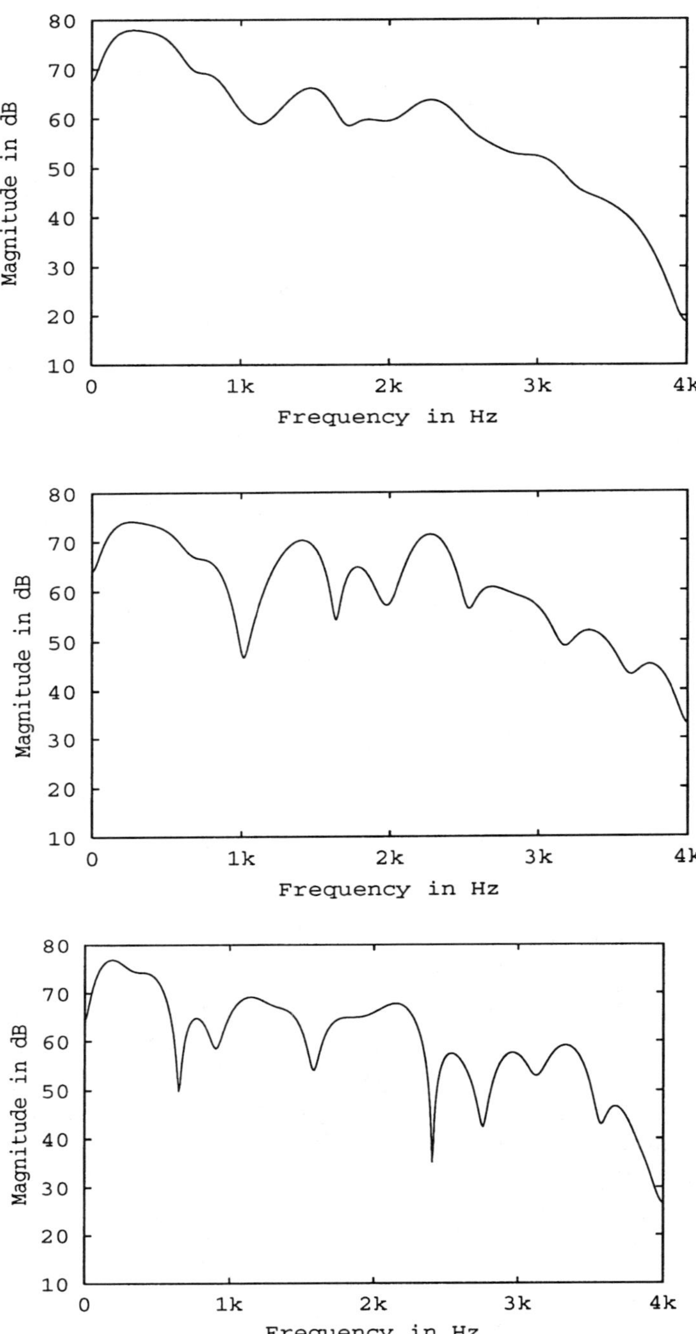

Figure 13-2 (a) Original vowel spectrum based on the FFT magnitude with a 6-ms Hanning window. (b) Wavelet-based compensation vowel spectrum. (c) Multichannel filter amplitude compression vowel spectrum.

preliminary evaluation, this new technique was compared to compression from both a subject-based performance and an implementation perspective. The compression algorithm used 2:1 compression in each band. The bands were the same as those with the wavelet processing: 0–1000 Hz, 1000–2000 Hz, and 2000–4000 Hz. A summary of those results is presented here.

Listening tests were conducted using four normal-hearing subjects between the ages of 30 and 50 with noise masking to simulate a flat-loss hearing impairment. The stimuli were sentences from the 500 Harvard Sentences listed in the *IEEE Recommended Practice for Speech Quality Measurements*. Each sentence has five key words to test for correctness. This provides analysis of the speech reception abilities in conversational speech, where contextual information can be used.

The results show that performance on both multiband amplitude compression and the wavelet-based processing was similar [4]. The average score on the compression processing was 80.6% words correct versus 78.1% words correct for the wavelet-based processing. A comment made by one subject was that he found the compression processing to be a little clearer, while the wavelet processing was more pleasant.

An obvious question in light of the above results is what advantage does wavelet processing have over traditional multiband filtering approaches. From a complexity perspective, both are fairly similar. In the processing stage before applying the compression gain, both mainly involve convolution. With a parallel bank of band-pass filters the amount of information (and thus memory) is increased in proportion to the number of bands used in the system, whereas the wavelet processing has the same number of coefficients as data points. The computation of the gains is similar in complexity for the two techniques. During reconstruction, the wavelet processing requires additional convolutions; the multiband compression merely requires addition of the filtered time waveforms. On a digital signal processing microprocessor, the additional convolutions of wavelet processing would not be as significant as the increased memory required by multiband compression.

An advantage to using wavelet processing is that the variable support of the wavelet in the time domain is dependent upon its scale. Thus, calculation of gain is customized to the characteristics of the speech waveform in each frequency region. The gain for consonants and stops, which contain predominantly high frequencies, is calculated from a smaller window than vowels, which contain more low frequencies. With multiband compression processing, gains are based on the energy envelopes which follow variations in the waveform according to predetermined time constants. With other parameterization techniques the window size is the same for each frequency band.

The calculation of the gains is designed so that each coefficient will be within the impaired listener's residual dynamic range regardless of the intensity level of the input speech. Multiband filter compression systems are time-invariant. They are based on the average intensity level of input speech and may have some reduction in performance for very low-level inputs [12].

13.4. WAVELET-BASED NOISE REDUCTION

The benefit provided by even the most complex signal processing hearing aid can be decimated by the presence of background noise. Much of the difficulty in understanding speech in the presence of noise may be attributed to masking of consonants, which often resemble short-duration bursts of random noise. Numerous signal processing algorithms have been proposed to address this problem. Several of these algorithms have difficulty distinguishing between noise and consonants, and consequently remove both. Furthermore, inaccurate estimates of the noise (which is often assumed to be stationary) can cause some algorithms to create audible artifacts which further mask consonants. Many of the systems succeed at improving the signal-to-noise ratio; they do not, however, provide a corresponding improvement in intelligibility [22,27].

In this section, a novel parametric single-microphone approach is proposed. The new approach reduces noise by projecting noisy speech onto a series of wavelet packet and local trigonometric bases and by retaining the transform coefficients thought to correspond to the original clean speech. Preliminary results indicate that the algorithm may be useful in digital hearing aid applications.

13.4.1 Simultaneous Compression and De-Noising

The new approach uses the minimum description length (or MDL) criterion recently applied by Saito [28] to reduce additive white Gaussian noise in digitized image and geophysical signals. The description length is defined as the length (in bits) of the theoretical binary codeword required to describe both a noisy signal and a model thereof. The length is the sum of the lengths of the individual components:

$$L(x, \lambda) = L(x|\lambda) + L(\lambda) \tag{13-8}$$

The objective here is to find the parameters of the speech from the parameters of the noisy observation. The most probable description of the model for the signal produces the minimum description length for signals which may be constructed from "dictionaries" in the basis library. The observation x, which is an N-length vector speech signal $s(\lambda_k)$ in additive noise n, is expressed as

$$x = s(\lambda_k) + n \tag{13-9}$$

where $s(\lambda_k)$ is assumed to be equal to a linear combination of k basis elements, described by parameter vector λ_k with codeword of length $L(\lambda_k)$. This length, which depends on the value of k, is found using the universal prefix coding method proposed by Rissanen [29]. The Shannon coding scheme is used to find $L(x|\lambda)$, which describes a sample observation of a random process. For zero-mean Gaussian noise with likelihood function

$$p_N(n) = [(2\pi)^N \det(R_n)]^{-\frac{1}{2}} \exp\left[-\tfrac{1}{2}n^T R_n^{-1} n\right] \tag{13-10}$$

the codeword length is given by

$$
\begin{aligned}
L(x|\lambda_k) &= -\log_2 p_{X|\Lambda_k}(x|\lambda_k) \\
&= \varsigma(\ln[(2\pi)^N \det(R_n)] + n^T R_n^{-1} n)|_{n=x-s(\lambda_k)}
\end{aligned}
\tag{13-11}
$$

where R_n is the autocorrelation matrix of the noise, and ς is the constant used in converting from $\log_2$ to ln. $L(x|\lambda)$ is minimized by the maximum likelihood estimate of R_n. For the case of white Gaussian noise (which was considered by Saito [30]), $R_n = \sigma_N^2 I$, and $\hat{\sigma}_N^2 = \frac{1}{N}\|x - s(\lambda_k)\|^2$.

The speech signal $s(\lambda_k)$ is estimated by finding the basis and the model order k giving the minimum description length. Given a library containing M varieties of orthonormal bases (i.e., wavelet packets and local trigonometric functions) with minimum information cost [6], Saito's algorithm selects the basis and coefficients providing optimum compression of the signal and rejection of the noise (which compresses poorly in every basis). For basis m, the wavelet packet transform $W_m^T x$ is calculated and the coefficients are sorted in ascending order. This process is iterated over the model order k, calculating an approximation to the minimum description length (AMDL) as

$$
\frac{3k}{2}\log_2 N + \frac{N}{2}\ln \|(I - \Omega^{(k)})W_m^T x\|^2 + C
\tag{13-12}
$$

Here $\Omega^{(k)}$ is a rank-k thresholding matrix which selects the largest k coefficients of the transform matrix $W_m^T x$, and C is a constant independent of basis and order selection. The values k and m, such that $0 < k < \frac{N}{2}$ and $1 \leq m \leq M$, are selected which minimize the description length, assuming equal probability of basis selection. Demonstrations of the algorithm on image and geophysical signals have been quite successful [30].

13.4.2 Adaptive Multi-band MDL

Application of the MDL algorithm to speech enhancement confirms its capability for robust, autonomous calibration; a desirable property that was not a feature of previous approaches. However, like earlier approaches, the algorithm tends to remove consonants in the presence of noise, and imposes mild distortion on speech (particularly consonants) in the absence of noise. Furthermore, for the short frame lengths appropriate to real-time processing of speech, the additive noise tends to compress efficiently onto a few basis elements. These retained coefficients produce audible artifacts similar to those produced by previous approaches.

Multiband MDL. Several modifications are proposed for adaptation of the MDL algorithm to use with speech [31]. First, the incoming signal is split into two bands: a low-frequency band dominated by vowels and nasal consonants, and a high-frequency band dominated by fricative consonants and plosive bursts. The MDL algorithm is then separately applied to the low- and high-frequency signals. The split-band approach allows salient features of consonants to be reproduced accurately, reducing distortion produced by the original algorithm in the absence of noise.

Unfortunately, the multiband method described above shows reduced capability for noise-reduction, largely because the noise is also filtered. A more efficient approach uses a power-symmetric quadrature mirror filter bank, which maintains the lack of correlation in the noise required by the original MDL. This allows use of multiband MDL without any need for inverse filtering.

Adaptive Processing for Consonants.   Additional modifications are motivated by a relationship between changes in AMDL values and changes in the envelope of the speech waveform [31]. Observed AMDL values have a lower bound dependent on the minimum amplitude of the speech signal, and provide a reliable indication of whether the signal is above or below the noise floor. An example of MDL statistics for noisy speech is shown in Fig. 13-3.

Inspection of Fig. 13-3 indicates that at (or below) the noise floor, the dominant coefficients (i.e., the ones retained) may be attributed to noise. As a result, coefficient retention is unaffected by the presence of low-level speech, and application of MDL produces audible artifacts.

When the signal is below the noise floor, a tracking algorithm which monitors AMDL values is used to adaptively disable MDL processing at high frequencies in

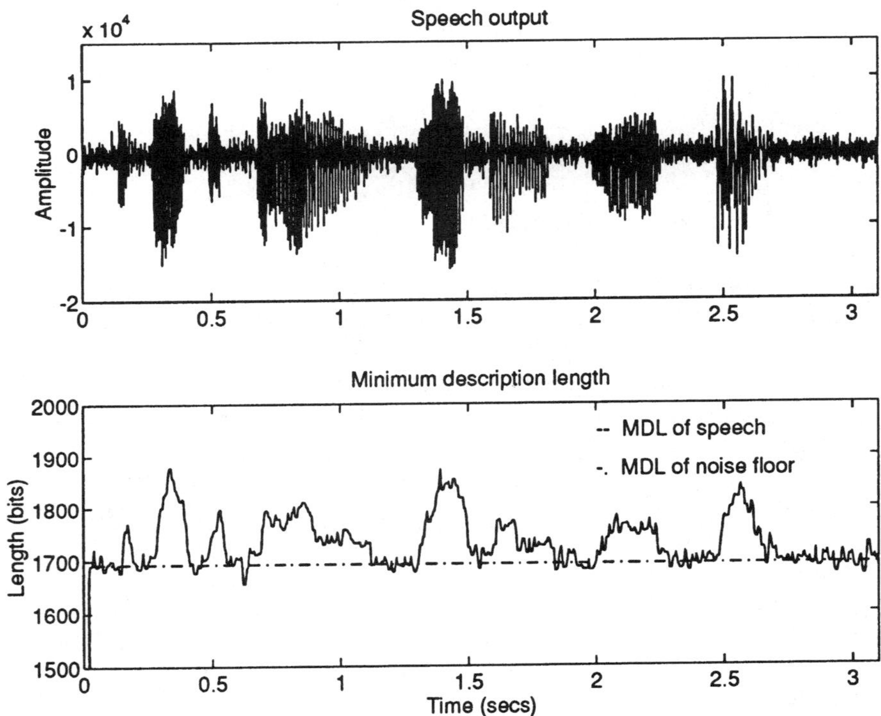

Figure 13-3   MDL statistics for noisy speech.

favor of alternative processing that reduces audible artifacts. Here, a form of power spectrum subtraction using local trigonometric bases is employed. A running average of spectra derived from discarded coefficients in the local trigonometric basis is used to construct an estimate of the noise.

13.4.3 Preliminary Results

A preliminary comparison of the capabilities of original and modified MDL approaches was conducted. An utterance of the sentence, "That hose can wash her feet," was sampled at 8 kHz, digitized to 16 bits, and added to each of three white Gaussian noise sequences to produce waveforms with overall signal-to-noise ratios (SNRs) of 0, 5, and 10 dB. Successive frames of the speech signals (256 samples, 50% overlap) were processed by each of three algorithms: original MDL, multiband MDL with QMFs, and adaptive multiband MDL with QMFs. Root mean square (rms) levels of the /o/ phoneme in "hose" and the closure preceding /t/ in "feet" were used to obtain relative measures of SNR for each of the three methods. (For sentences with 0, 5, and 10 dB average SNRs, vowel-to-silence SNRs were 8.77, 13.78, and 18.75 dB respectively.) The observed SNR increases are presented in Table 13-1.

At all noise levels, the proposed algorithm reduces the "musical noise" produced by the original MDL algorithm. This difference is reflected in the higher SNRs of the proposed algorithm. Informal listening indicates that the proposed algorithm improves perception of consonants at vowel-to-silence ratios of 18.8 dB SNR. At lower SNRs, where the original algorithm produces substantial amounts of "musical noise," spectral subtraction tends merely to attenuate the consonants. The severity of the attenuation is likely due to the spectrum of the noise, which, being flat, is substantially higher in level than the spectrum of the consonants at high frequencies.

13.4.4 Discussion

Reduction of Colored Noise.   The preliminary investigations cited above point to two areas where improvement is required. The first concerns the spectrum of interfering noise in typical speech communication systems, which is generally not white or Gaussian. Unfortunately, the present formulation of MDL has not been extended to colored or non-Gaussian noise.

TABLE 13-1   SNR Improvements (in dB)

Original Vowel-to-Silence SNR	18.8 dB	13.8 dB	8.8 dB
Algorithm under Test	Improved Vowel-to-Silence SNRs		
Original MDL	29.8	23.9	17.7
Multiband MDL	31.0	23.2	17.1
Multiband MDL w/HF alternate	32.6	25.7	20.2

Saito [30] reports a method of removing colored noise (with known parameters) which relies on a pattern classification method using local discrimination bases (LDBs). The method selects for every frame a set of basis functions which optimizes classification of the frame as either "noise" of known spectrum or "signal + noise." Selection relies on the statistics of noise, which may be estimated from discarded elements in the adaptive multiband MDL.

Application of the MDL to the reduction of colored noise requires maximization of the likelihood function given in Eq. (13-11) with respect to arbitrary R_n, which is difficult. Burg et al. [32] proposed an iterative ML estimator which converges to the nonsingular Toeplitz (assuming the noise is stationary) matrix R_n if the number of observations is greater than N [33]. For one observation, Fuhrmann and Miller [34] showed that the iterative approach was likely to fail. One of their proposed alternatives is to model the noise as an autoregressive (AR) time series. Several maximum liklihood (ML) estimation approaches have been developed for this purpose [35–38]. The AR approaches will trade flexibility for computational efficiency since a new AR model must be fit to the data in the time domain for every value of k. Current research focuses on optimizing the efficiency of the AR approach.

Given a compatible statistical model of speech, it might be possible to estimate the coefficients of the noise signal through conventional parameter estimation methods. Hence, a statistical model of speech, suitable for use in parameter estimation, would first be developed. ML estimation methods would then be used in estimating the noise spectrum for each frame. The estimates, in turn, would be used by the LDB-based approach to enhance the speech. Should this approach prove infeasible, a second LDB-based approach using a library of noise classes (one for each type of noise) could be implemented.

Reduction of Residual Noise.   A second area of improvement concerns the residual "musical" noise produced by retention of unwanted coefficients. This residual noise tends to increase in level as the window length decreases. This characteristic creates a challenge for real-time processing of speech, in which short window lengths are generally required.

Berger, et al. [39] have recently proposed an approach for reducing both residual noise and Gibbs-effect artifacts caused by the truncation of the wavelet series. Their algorithm averages together de-noised versions of the noisy speech, with each version delayed by a small number of samples before processing (and shifted back to its origin after processing). The method exploits the fact that the wavelet-packet and local cosine transforms are not shift-invariant, and that best-basis representations of the noise will generally be less regular than the representations of the signal.

Figure 13-4(a) and (b) compare the performance of the original MDL algorithm with that of an MDL algorithm using the shift-de-noise-average method at 10 dB SNR. Here, the signal of Fig. 13-4 (a) is combined with a second version shifted forward by four samples, de-noised, and then shifted back to its original position. The algorithm does significantly reduce the level of audible artifacts; however, it also has the effect of reducing the level of some consonants. In the example given above, the phoneme /t/ is almost completely removed by the latter processor.

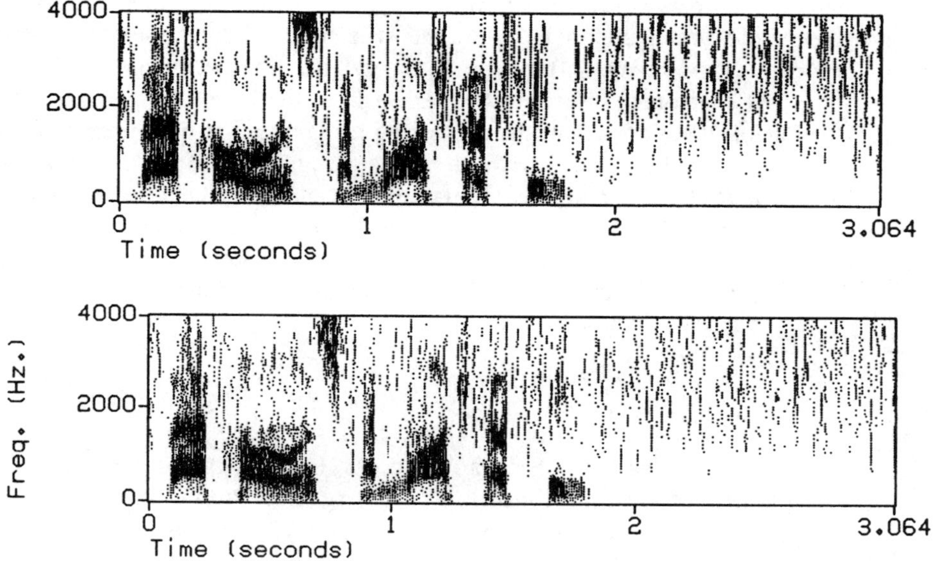

Figure 13-4   (a) Output of MDL (10 dB SNR). (b) Output of MDL with shift-de-noise-averaging (10 dB SNR).

Adaptability for Other Techniques.   As stated previously, one speech-enhancement system that has been shown to improve intelligibility under certain test conditions is the two-microphone least mean square (LMS) filter-based noise canceling system evaluated by Levitt et al [22] for application in digital hearing aids. It is conceivable that an LMS filter-based system could be implemented with one microphone, given a suitable noise reference. The adaptive MDL system constructs such a reference estimate of the noise spectrum. Currently, the estimate is used by the spectral subtraction algorithm. It may also be possible to use this noise estimate in lieu of the reference noise signal required by the two-channel adaptive-filter-based system.

13.5. CONCLUDING REMARKS

There has been a lot of progress in the comfort and performance of hearing aids over the last decade. With the advent of digital signal processing microprocessors, digital hearing aids are becoming a reality. These new digital aids will allow the kind of processing described in this chapter.

The wavelet approach to the combined problem of noise-reduction and amplitude compression offers high quality and flexibility since the parameters can be modified to fit the individual hearing loss and characteristics of the noise. The compression gains are calculated based on the impaired listener's thresholds of hearing and discomfort, and adapt to the characteristics of the incoming speech signal. The noise-reduction algorithm decides whether the wavelet coefficients in a given band are likely to be from the speech signal or noise, and reduces them accordingly. More work remains to extend the noise-reduction technique to the variety of environmental sounds experienced in the course of everyday life.

Performance of each of the processing techniques has been presented separately. Future work will include evaluation of adaptive multiband MDL as a pre-processor for the wavelet-based compression method. In addition, the noise estimate generated by the modified MDL algorithm will be evaluated for use as the reference input for a two-microphone adaptive filtering noise-reduction system. These evaluations will be done using both objective measures and human subject listening tests.

REFERENCES

[1] V. B. Van Hasselt, P. S. Strain, and M. Hersen, *Handbook of Development and Physical Disabilities*. Elmsford, NY: Pergamon Press, 1988.

[2] R. J. McAulay and T. F. Quatieri, "Speech analysis/synthesis based on a sinusoidal representation," *IEEE Trans. Acoustics, Speech Signal Proc.*, vol. ASSP-34, pp. 744–754, 1986.

[3] S. G. Mallat, "A theory for multiresolution signal decomposition: The wavelet representation," *IEEE Trans. Pattern Machine Intelligence*, vol. 11, no. 7, pp. 674–693, 1989.

[4] L. A. Drake, J. C. Rutledge, and J. Cohen, "Wavelet analysis in recruitment of loudness compensation," *IEEE Trans. Signal Proc.*, vol. 41, no. 12, pp. 3306–3312, 1993.

[5] J. C. Rutledge and M. A. Clements, "Compensation for recruitment of loudness in sensorineural hearing impairments using a sinusoidal model of speech." In *Proc. Int. Conf. Acoustics, Speech and Signal Processing*. New York: IEEE, 1991.

[6] R. Coifman and V. Wickerhauser, "Entropy-based algorithms for best basis selection," *IEEE Trans. Information Theory*, vol. 38, pp. 713–718, 1992.

[7] N. A. Whitmal and J. C. Rutledge, "Noise reduction algorithms for digital hearing aids." In *Proc. Int. Conf. Engineering in Medicine and Biology*. New York: IEEE, 1994.

[8] H. Levitt, J. M. Pickett, and R. A. Houde, *Sensory Aids for the Hearing Impaired*. New York: IEEE, pp. 3–27, 1980.

[9] N. Y-S. Kiang, T. Watanabe, E. C. Thomas, and L. F. Clarke, *Discharge Patterns of Single Fibers in the Cat's Auditory Nerve*. Cambridge, MA: MIT Press, 1965.

[10] E. F. Evans, "Peripheral auditory processing in normal and abnormal ears: Physiological considerations for attempts to compensate for auditory deficits by acoustic and electrical prostheses." In *Scandinavian Audiology, Supplement 6*. C. Ludvigsen and J. Barfod (eds.). Stockholm: Almqvist & Wiksell Periodical Company, pp. 9–47, 1978.

[11] L. E. Humes, B. Espinoza-Varas, and C. S. Watson, "Modeling sensorineural hearing loss. I. Model and retrospective evaluation," *J. Acoust. Soc. Am.*, vol. 83, pp. 188–202, 1988.

[12] R. P. Lippmann, L. D. Braida, and N. I. Durlach, "Study of multichannel amplitude compression and linear amplification for persons with sensorineural hearing loss," *J. Acoust. Soc. Am.*, vol. 69, pp. 524–534, 1981.

[13] E. Villchur, "Comments on the negative effect of amplitude compression in multichannel hearing aids in the light of the modulation-transfer function," *J. Acoust. Soc. Am.*, vol. 86, pp. 425–428, 1989.

[14] J. C. R. Licklider, "Effects of amplitude distortion upon the intelligibility of speech," *J. Acoust. Soc. Am.*, vol. 29, pp. 429–434, 1946.

[15] E. Villchur, "Simulation of the effect of recruitment on loudness relationships in speech," *J. Acoust. Soc. Am.*, vol. 56, pp. 1601–1611, 1974.

[16] I. V. Nábělek, "Performance of hearing-impaired listeners under various types of amplitude compression," *J. Acoust. Soc. Am.*, vol. 74, pp. 776–791, 1983.

[17] J. M. Kates, "Speech intelligibility enhancements." *U.S. Patent 4,454,609*, June 1984.

[18] J. M. Kates, "Adaptive speech enhancement for the hearing impaired." In *Proc. Digital Signal Processing Workshop*, Chatham, MA, pp. 4.4.1–4.4.2, October, 1984.

[19] S. DeGennaro, K. R. Krieg, L. D. Braida, and N. I. Durlach, "Third-octave analysis of multichannel amplitude compressed speech." In *Proc. Int. Conf. on Acoustics, Speech and Signal Processing*, New York: IEEE, 1981.

[20] M. Weiss and A. C. Neuman, "Noise reduction in hearing aids." In *Acoustical Factors Affecting Hearing Aid Performance*. G. A. Studebaker and I. Hochberg (eds.). Rockleigh, NJ: Allyn and Bacon, 1993.

[21] B. Widrow et al., "Adaptive noise canceling: Principles and applications." *Proc. IEEE*, vol. 63, pp. 1692–1716, December 1975.

[22] H. Levitt, M. Bakke, J. Kates, A. Neuman, T. Schwander, and M. Weiss, "Signal processing for hearing impairment," *Scand. Audiology*, vol. suppl. 38, pp. 7–19, 1993.

[23] J. E. Greenberg and P. M. Zurek, "Evaluation of an adaptive beamforming method for hearing aids," *J. Acoust. Soc. Am.*, vol. 91, no. 3, pp. 1662–1676, 1992.

[24] D. Graupe and J. K. Grosspietsch, "A single-microphone-based self-adaptive filter of noise from speech and its performance evaluation," *J. Rehab. Res. Dev.*, vol. 24, pp. 119–126, 1987.

[25] R. R. Coifman and M. V. Wickerhauser, "Wavelets and adapted waveform analysis." In *Wavelets: Mathematics and Applications*. J. Benedetto and M. Frazier (eds.), Cleveland, OH: CRC Press, 1993.

[26] M. A. Cody, "The fast wavelet transform," *Dr. Dobb's J.*, pp. 16–28, April 1992.

[27] H. Dillon and R. Lovegrove, "Single-microphone noise reduction systems for hearing aids: A review and evaluation." In *Acoustical Factors Affecting Hearing Aid Performance*. G. A. Studebaker and I. Hochberg (eds.), Rockleigh, NJ: Allyn and Bacon, 1993.

[28] N. Saito, "Simultaneous noise suppression and signal compression using a library of orthonormal bases and the minimum description length criterion," In *Wavelets in Geophysics*. E. Foufoula-Georgiou and P. Kumar (eds.). New York: Academic Press, Inc., 1994.

[29] J. Rissanen, "Modeling by shortest data description," *Automatica*, vol. 14, pp. 465–471, 1978.

[30] N. Saito, Local feature extraction and its application using a library of bases. *PhD Thesis*, Yale University, 1994.

[31] N. A. Whitmal, J. C. Rutledge, and J. Cohen, "Wavelet-based noise reduction." In *Proc. Int. Conf. Acoust., Speech, Signal Proc.* New York: IEEE, 1995.

[32] J. P. Burg, D. G. Luenberger, and D. L. Wenger, "Estimation of structured covariance matrices," *Proc. IEEE*, vol. 70, pp. 963–974, 1982.

[33] Nguyen Quang A, "On the uniqueness of the maximum-likelihood estimate of structured covariance matrices" *IEEE Trans. Acoust., Speech, Signal Proc.*, vol. 32, no. 6, pp. 1249–1251, 1984.

[34] D. R. Fuhrmann and M. I. Miller, "On the existence of positive-definite maximum-likelihood estimates of structured covariance matrices," *IEEE Trans. Information Theory*, vol. 34, pp. 722–729, 1988.

[35] S. M. Kay, "Recursive maximum likelihood estimation of autoregressive processes," *IEEE Trans. Acoust. Speech, Signal Proc.*, vol. 31, no. 1, pp. 56–65, 1983.

[36] P. D Tuan, "Maximum likelihood estimation of the autoregressive model by relaxation on the reflection coefficients," *IEEE Trans. Acoust., Speech, Signal Proc.*, vol. 36, no. 8, pp. 1363–1367, 1988.

[37] B. Armour and S. Morgera, "An exact forward-backward maximum likelihood autoregressive parameter estimation method," *IEEE Trans. Signal Proc.*, vol. 39, no. 9, pp. 1985–1993, 1991.

[38] M. L. Vis and L. L. Scharf, "A note on recursive maximum likelihood for autoregressive modeling," *IEEE Trans. Signal Proc.*, vol. 42, no. 10, pp. 2881–2883, 1994.

[39] J. Berger, R. Coifman, and M. Goldberg, "Removing noise from music using local trigonometric basis and wavelet packets," *J. Audio Eng. Soc.*, pp. 808–818, December 1994.

Chapter 14

From Continuous Wavelet Transform to Wavelet Packets: Application to the Estimation of Pulmonary Microvascular Pressure

Mohsine Karrakchou, Murat Kunt

14.1. INTRODUCTION

It is well known that the Fourier transformation is a perfect tool to analyze periodic functions and represent them as a superposition of pure frequencies. However, this only makes sense if the signal is stationary. If not, the signal should be considered locally, which means performing a local Fourier transform or other local transforms in the same way as a musician would describe a piece of music as a sequence of pure tones well localized in time. This has led to the development of different time-frequency representations like the short time Fourier transform, the Wigner–Ville, or the Gabor transform. Wavelets have emerged in the last few years as a powerful alternative to these transforms for time-frequency analysis. They have since generated much interest in various applied and theoretical areas.

This chapter is devoted to the presentation of some of the principal properties of wavelets ranging from the continuous wavelet transform to wavelet packets. Since wavelet "history" becomes more and more complicated, writing a self-contained, complete and comprehensive overview of wavelets is inevitably nearly impossible. Subsequent papers and books which throw more light on wavelets and their applications, and give excellent reviews on the topic are thus recommended [1–5].

In addition to the various theoretical developments, wavelet representation is now used in various fields ranging from signal and image compression to the analysis of fractals and multifractals, to name a few. In the biomedical field, the wavelet transform has been used to analyze electrocardiogram signals for detecting P and T waves. Different approaches have been investigated using either the continuous wavelet transform [6–9] or the orthogonal wavelet transform [8,10]. Where electro-encephalogram signals and particularly the extraction of evoked potentials are concerned, successful attempts have been carried out using the discrete wavelet transform [11]. An analysis of complex changes in the shape of evoked potential signals that appear in the case of neurologic injury has also been performed with the aid of wavelets to model such responses [12]. Another application of the multiresolution decomposition consists in the extraction of nerve motor potentials from needle electromyogram signals [13]. These applications show that wavelet transformation does indeed constitute a useful tool to deal with nonstationary signals, and most medical signals are of this type. In this chapter, we focus on one particular application, i.e., the estimation of pulmonary capillary pressure and show that wavelets successfully solve such a problem.

14.2. WAVELET PACKETS

Wavelet packets are a generalization of the concept of wavelet transforms in which arbitrary time-frequency resolution can be chosen according to the signal [14]. This will be done, of course, within the bounds of the Heisenberg uncertainty principle. The idea is to obtain an adaptive partitioning of the time-frequency plane depending on the signal of interest.

This section consists in a short review of wavelet packets and a presentation of their most interesting properties. All subsequent information can be found in deeper detail in [15,16].

Let $h[n]$ and $g[n]$ be the impulse response of two analysis filters, and let $A_{n,k+1}$ be one of the vector spaces at level $k + 1$ onto which signals are projected. The filtering operations performed by $h[n]$ and $g[n]$ split the vector space into two subspaces ($A_{2n,k}$ and $A_{2n+1,k}$). This is illustrated in Fig. 14-1. It can be shown [15] that these two subspaces are orthogonal and that $A_{n,k+1}$ is the direct sum of these two subspaces, namely:

$$A_{n,k+1} = A_{2n,k} \oplus A_{2n+1,k} \tag{14-1}$$

Using this scheme, a tree of depth $\log L$ (where L is the length of the signal being analyzed) can be built by iterating this decomposition at each of the newly created nodes. The idea of wavelet packets is to allow any orthogonal decomposition to be performed over this entire tree, instead of choosing a rigid decomposition of a signal. The property (14-1) ensures that pruning any subtree still yields an orthonormal basis onto which the signal can be decomposed. Figure 14-2 shows the entire tree.

The tree contains several admissible bases, one of which is the wavelet basis itself (Fig. 14-3). As there is a large but finite library of bases, an ordering of this library can be performed with respect to some criterion. Hence it is possible to extract the best basis relative to the criterion considered. This is known as the best-basis method [14–19]. The basis can be any subtree of the initial entire tree, and the best-basis method will yield this subtree by a pruning scheme performed on the complete tree.

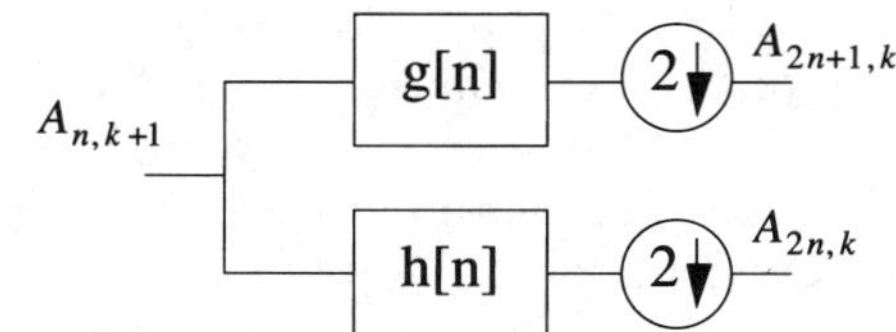

Figure 14-1 One-stage decompositon of a signal.

Figure 14-2 Sketch of an entire tree obtained by the decomposition of wavelet packets.

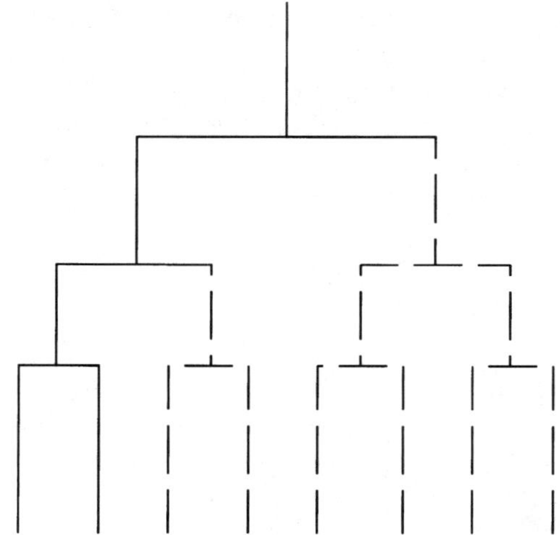

Figure 14-3 The wavelet basis obtained from the entire tree by pruning the dashed subtree.

14.2.1 The Best-Basis Method

Let $M(x)$ be a real-valued functional on the signal sequences x_i. The purpose of the method is to find the optimal basis that minimizes this functional on the manifold of orthonormal bases. The search for the best basis is straightforward: the entire decomposition tree is spanned. A bottom-up pruning scheme is applied in which, going from leaves to the root, the values of the functional on the two possible bases (the node and its two children) are compared (see Fig. 14-4).

Let $A_{2n,k}$ and $A_{2n+1,k}$ be the bases chosen at level k. Let $B_{n,k+1}$ be the parent of the latter two, and let $M(B_{l,m}x)$ be the value of the functional computed on the projection of the signal x onto some basis $B_{l,m}$. The optimal basis, $A_{n,k+1}$, is going to be:

$$A_{n,k+1} = \begin{cases} B_{n,k+1} & \text{if } M(B_{n,k+1}) \leq (M(A_{2n,k}) + M(A_{2n+1,k})) \\ A_{2n,k} \oplus A_{2n+1,k} & \text{if } M(B_{n,k+1}) > (M(A_{2n,k}) + M(A_{2n+1,k})) \end{cases}$$

and it is assigned the minimum of the two functional values $M(B_{n,k+1}x)$ and $[M(A_{2n,k}x) + M(A_{2n+1,k}x)]$ for further comparisons.

14.2.2 Criteria for the Selection of the Best Basis

The criteria for the best-basis selection that have already been used in the literature are summarized in this section. It should be noticed that most of them concern only signal compression. Another criterion has been proposed for high-resolution spectral estimation and a new one will be detailed within this chapter for adaptive filtering purposes.

When applying a dyadic wavelet transform to a speech signal or an image, the wavelet transform partitions those signals into octave bands in the frequency domain. A similar operation seems to be performed both in the hearing system and the visual cortex. The other main feature of the octave band partitioning is that it allows a multiresolution representation of the data. From an energy compaction point of view, however, there is no evidence that an octave band partitioning in the frequency domain is the best choice [20]. Decomposition of the wavelet packets is performed in order to compact the energy in a small number of coefficients.

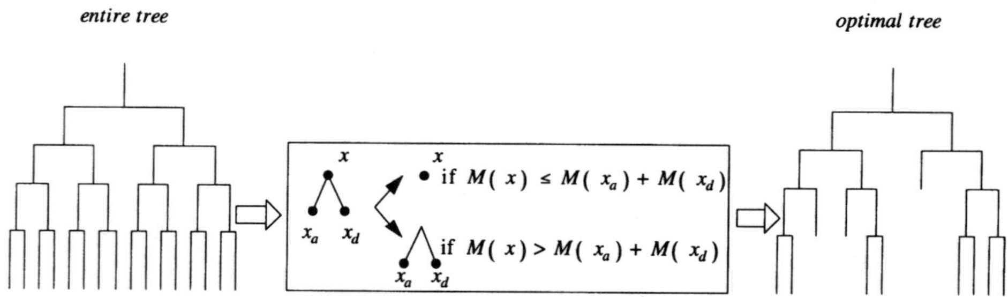

Figure 14-4   Illustration of the best-basis selection principle.

Consequently, to represent the signal with a good accuracy and using a small number of high-energy coefficients, the choice of the decomposition criterion (defining the best partitioning) is of particular importance.

Threshold Criterion.   This involves setting an arbitrary threshold ϵ and counting the elements in the sequence x whose absolute value exceeds ϵ. This gives the number of coefficients needed to transmit the signal with a precision ϵ.

Entropy Criterion.   This involves taking the non-additive Shannon–Weaver entropy criterion expressed by

$$\epsilon_X = -\sum_i \frac{|x_i|^2}{|X|^2} \log_2 \frac{|x_i|^2}{|X|^2} \qquad \text{where} \quad |X|^2 = \sum_i |x_i|^2 \qquad (14\text{-}2)$$

This choice is motivated by the fact that the entropy criterion is computationally simple and permits a good estimation of the bit-rate needed for sequence coding.

Rate–Distortion Criterion.   Ramchandran and Vetterli proposed in [21] to use a criterion for image compression that includes both rate and distortion in order to minimize the global distortion for a given bit budget. This criterion based on rate–distortion theory, although yielding better results, is considered too complex for real-time implementation.

Minimizing one of these three criteria leads to a decomposition using as few coefficients as possible. This is most advantageous for signal or image compression [17]. However, for other applications, specific criteria should be designed. This is the case for instance when high-resolution spectral estimation is of concern.

Minimum Description Length.   The motivation for using subband decomposition techniques for sinusoid parameter estimation is not only because of the mentioned advantages of subbands but more because of the importance of the isolation of the different modes present in the signal. Indeed, it is well known that the lower the number of the present modes, the better is the limit of Cramer–Rao. However, performing a subband decomposition would lead to the attenuation of some modes or even their complete disappearance. For this reason a signal-dependent decomposition is of major interest. Such decomposition will not only avoid the attenuation of these modes but will also isolate as far as possible each mode in a proper frequency band. This will lead to a better quality of the estimation due to the reduction of the interference effect between different modes. For the present problem, the selection of the optimal decomposition is made by maximizing the number of modes on the whole decomposition. This allows us to stop the decomposition as soon as one mode disappears by a further decomposition. For this purpose the minimum description length [22] has been used in [23,24].

14.3. ESTIMATION OF PULMONARY CAPILLARY PRESSURE

14.3.1 The Clinical Importance of Effective Pulmonary Capillary Pressure

The so-called adult respiratory distress syndrome (ARDS) is frequently encountered in intensive care patients. It usually arises one to three days after one or several of many associated and possibly causative conditions, such as sepsis, massive trauma, various lung infections, near drowning, or surface burns [24]. It is characterized physiologically by diffuse damage to the alveolar capillary membrane, with subsequent interstitial edema. Its most important clinical characteristic is progressive hypoxemia despite increased levels of inspired oxygen. Although a vast amount of experimental and clinical research which has been devoted to ARDS (for a review see [26,27]), the mortality has remained high (40–70%) [28]. It has not yet been possible to successfully interrupt the chain of pathophysiological events injuring the lung, so that therapy is still largely based on supportive measures aimed at improving tissue oxygenation, such as mechanical ventilation and drug support of the cardiovascular system.

One key issue in the management of ARDS is interstitial edema of the lung. There is ample experimental evidence that interventions aimed at minimizing the accumulation of interstitial fluid improve gas exchange [29]. Recent clinical studies even suggest that control of extravascular lung water may improve outcome [30–32].

The rate of pulmonary edema formation is the difference between the rate of fluid efflux through the injured microvascular walls and the rate at which fluid is removed from the interstitium by the lymphatic system [33]. The ability to therapeutically influence lung lymph flow has been limited so far, even though it is now known that a high central venous pressure impedes the lymphatic drainage of the lung [33,34]. The rate of fluid efflux from the microvessels has proven more amenable to therapeutic control. This rate depends on the microvascular pressure P_{mv} within the capillaries.

The hallmark of ARDS is an increase in the permeability of the capillaries to both plasma protein and fluid [29,33,35]. The pulmonary edema occurring in this condition has been termed acute permeability edema. In permeability edema, transvascular fluid flux becomes highly sensitive to microvascular pressure [36]. Moreover, P_{mv} has been found to be elevated in a variety of experimental models of lung injury; this was explained by the vasoconstriction of pulmonary veins induced by various mediators of inflammation [37–39]. These experimental results suggest that increased P_{mv} may contribute to the pulmonary edema of ARDS. In summary, acute permeability pulmonary edema is extremely sensitive to the pressure in the microvessels of the lung (P_{mv}), which may be elevated in this condition. Therefore, an essential therapeutic goal in the clinical management of ARDS is to lower P_{mv} as much as possible [40]. It is therefore important to monitor P_{mv}, the "effective" pulmonary capillary pressure, in these patients.

The arterial occlusion (AO) technique has been proposed as a convenient means for the *in vivo* estimation of P_{mv} [41–46]. This technique relies on the analysis of the pressure transient observed in a pulmonary arterial branch abruptly occluded by the inflation of a Swan–Ganz catheter balloon. The sudden inflation of the balloon causes an interruption of the blood flow, yielding a sudden drop in the measured pressure. P_{mv} is then estimated from an analysis of the post-occlusion pressure transient (POPT) observed after inflating the balloon. A schematic representation of the cardio-pulmonary system is given in Fig. 14-5, while a typical apneic signal is shown in Fig. 14-6.

Pulmonary arterial occlusion pressure (PAOP) as measured with a Swan–Ganz catheter, is widely used as an estimate of P_{mv}. In fact, PAOP estimates the pressure within the large pulmonary veins or left atrium (P_{lv}). In order for blood to flow forward through the lungs, P_{mv} must exceed P_{lv}. Under normal conditions, the vascular pressure gradient across the pulmonary circulation is so small [47] that P_{mv} or PAOP are very close to P_{lv}. Under conditions of increased pulmonary vascular resistance (PVR), however, P_{mv} may differ considerably from the PAOP because it depends on the relative magnitudes of vascular resistances on the arterial (R_a) and on the venous (R_v) sides of the capillary bed. Discrepancies between P_{mv} and PAOP are expected in ARDS, because of the frequently associated pulmonary hypertension [48]. Indeed, it is clear that elevated PVRs are intimately related to the pathogenesis of ARDS.

14.3.2 Arterial Occlusion (AO)

Arterial occlusion is another way to estimate microvascular pressure. A balloon-tipped Swan–Ganz catheter is placed in the pulmonary artery. When the balloon is suddenly inflated, an abrupt interruption of flow occurs and the pressure measured at the catheter tip drops within a few seconds to a new stationary value PAOP. The AO technique takes advantage of the special shape of the pressure transient resulting from the sudden occlusion of a medium-sized pulmonary arterial branch. After the

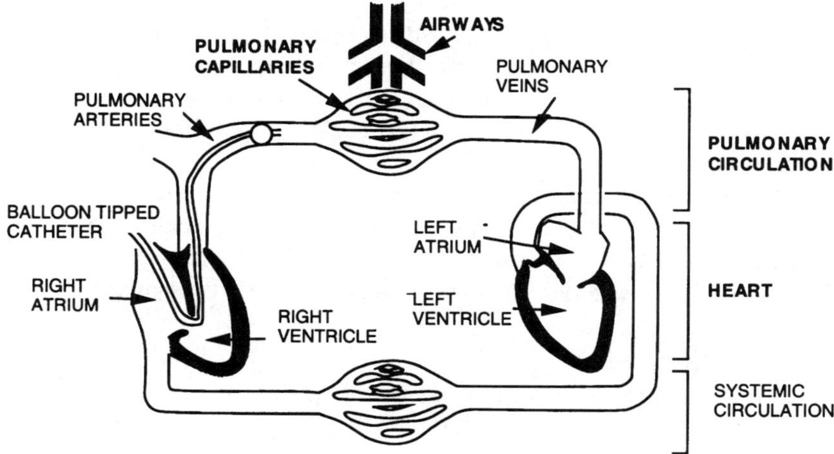

Figure 14-5 A schematic representation of the cardio-pulmonary system, with a balloon-tipped catheter inserted in a pulmonary artery.

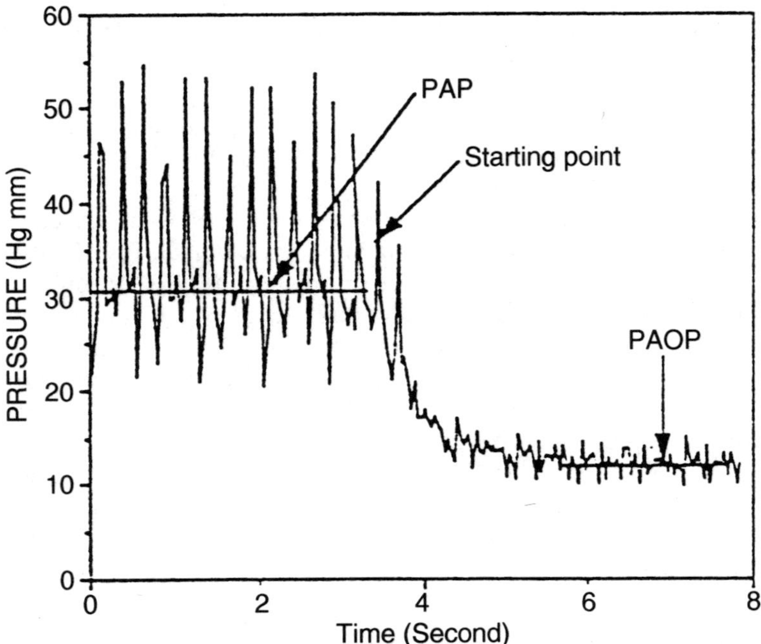

Figure 14-6   A typical signal of effective pulmonary pressure.

onset of occlusion (starting point), there is a rapid initial drop (fast transient), and pressure then decays exponentially to PAOP (slow transient). The point where the two transients meet is angular. It was shown from animal experimentation [41,49] that the pressure obtained by back-extrapolation of the slow transient from the inflection point to the time of occlusion is a good estimate of P_{mv}. Several attempts to estimate P_{mv} using this principle have been made in intensive-care patients. However, none of the described methods appears to be very robust nor lends itself easily to automation in an intensive-care environment. The search for the inflection point by means of visual inspection, even feasible in nonpulsatile flow conditions [41,50,51], is subject to undocumented inter-observer variation. Collee et al. [43] based their estimate of P_{mv} on the back-extrapolation to time zero (the time of occlusion) of an exponential fit to the slow part of the transient. Their estimated values, however, might be very sensitive to the choice of both inflection and occlusion points. D'orio et al. [52] and Siegel et al. [53] modeled the transient as a sum of two exponential processes that were fitted by least-square fitting algorithms. The post-occlusion arterial pressure transient is thus represented as:

$$P(t) = A_0 e^{-\alpha_0 t} + B_0 e^{-\beta_0 t} + C_0 \qquad (14\text{-}3)$$

where α_0 is the time constant of the slow exponential, β_0 is the time constant of the fast exponential, and where the time origin is assumed to be at the occlusion point that corresponds to the inflation moment. The asymptotic value C_0 was usually assumed to be the estimation of the capillary pressure, but has been shown to underestimate P_{mv}.

14.3.3 Limitations of the Arterial Occlusion to Apneic Transients

The patient should be relaxed. He/she should be mechanically ventilated with a properly positioned pulmonary artery catheter and the ventilation should be held during balloon inflation. Indeed, even a slight contamination of the transient by pressure fluctuations due to respiratory movements will make any of the above-described methods difficult to apply. An illustration of this contamination is given in Fig. 14-7 where different transients recorded in non-apneic conditions are shown. This clearly shows the difficulty of direct modeling of such transients.

Therefore, the *in vivo* determination of P_{ao} has so far been restricted to para-lyzed, mechanically ventilated subjects in whom the pressure transient can be recorded during a brief (20 s) period of apnea. Although transients heavily distorted by respiratory artifacts may be beyond the technique's possibilities, the practical use of the AO technique would be greatly improved if, at least, some respiratory fluctua-tions could occur without compromising the estimation of P_{mv}.

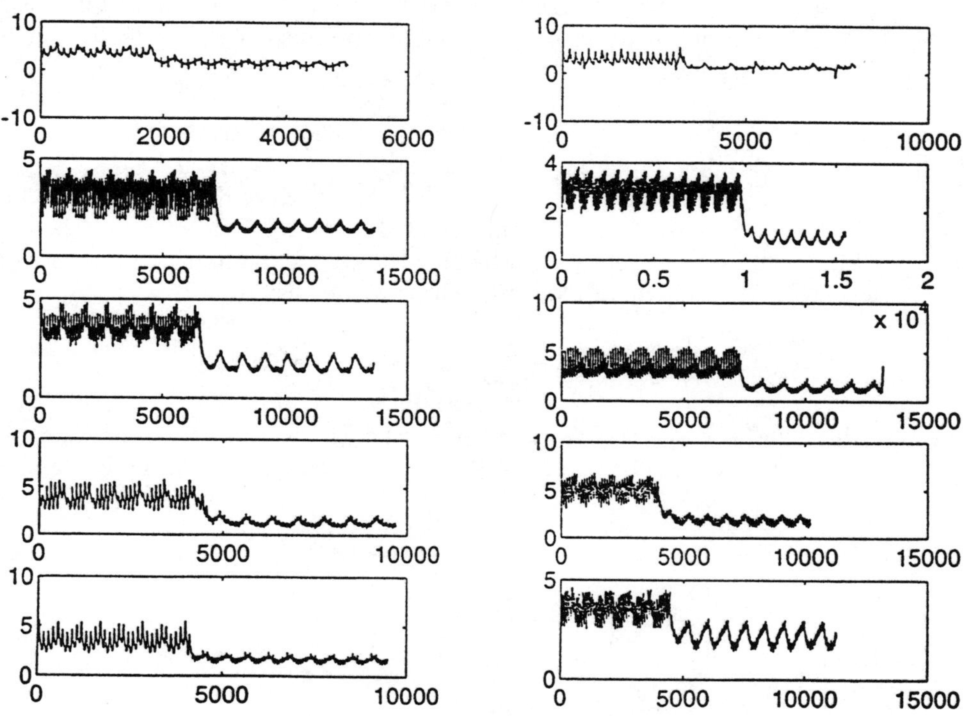

Figure 14-7   An illustration of various transients perturbed by respiratory interfer-ence.

14.4. HOW WAVELETS CAN HELP TO SOLVE THE PROBLEM

In this section, we present new signal-processing algorithms which are useful for interference canceling. Multirate adaptive filtering is discussed and new structures based on what we call mutual wavelet packets are described. These algorithms are applied for pulmonary capillary pressure transients recorded in patients.

14.4.1 Classical Finite Impulse Response Adaptive Filtering

Most everyday signal-processing tasks tend to be done in an unknown environment, which prevents engineers designing context-dependent filters. Since the pioneering work of Widrow and coworkers [54,55] adaptive filtering has emerged as a powerful and intensively used tool in many applications such as control, system identification, and echo cancellation.

The basic idea of adaptive-filtering algorithms can be illustrated by the block diagram of Fig. 14-8. The unknown system characterized by the transfer function $H(z)$ between the input signal $x[n]$ and the desired signal $d[n]$ has to be identified. The identification is performed using an estimate of $H(z)$, namely $\hat{H}(z)$. The error $e[n]$ between the desired response $d[n]$ and its estimate $\hat{d}[n]$ is used by an algorithm which adjusts the coefficients of $\hat{H}(z)$ in order to minimize some function of the error.

A frequently used error criterion is the mean square error (MSE), defined as:

$$\varepsilon = E\{e^2[n]\} \tag{14-4}$$

where

$$e[n] = d[n] - \hat{d}[n] \tag{14-5}$$

Assuming that $H(z)$ can be approximated using an finite impulse response (FIR) filter, then it can be represented by an FIR filter whose impulse response is denoted by $w[n]$. The estimate of $d[n]$ can then be written as:

$$\hat{d}[n] = \sum_{i=0}^{N-1} w_i[n]x[n-i] = \mathbf{w}_N^T[n]\mathbf{x}_N[n] \tag{14-6}$$

where the bold symbols represent N-component vectors. It can be shown [56] that substituting Eqs. (14-6) and (14-5) in (14-4) yields the following expression of the MSE:

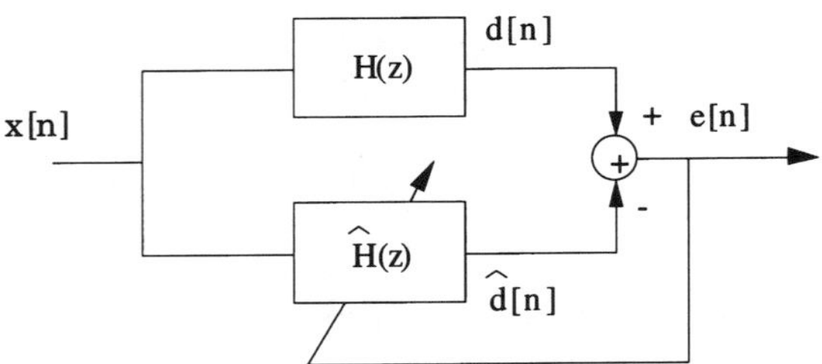

Figure 14-8   An adaptive filter in the system identification configuration.

$$\varepsilon = E\{e^2[n]\} = \sigma_d^2 - 2\mathbf{w}_N^T[n]\mathbf{p}_N + \mathbf{w}_N^T[n]\mathbf{R}_{NN}\mathbf{w}_N[n] \tag{14-7}$$

where

$$\mathbf{R}_{NN} = E\{\mathbf{x}_N[n]\mathbf{x}_N^T[n]\}, \qquad \mathbf{p}_N = E\{d[n]\mathbf{x}_N[n]\} \tag{14-8}$$

and σ_d^2 is the variance of the desired signal. The MSE is thus a quadratic function of the predictive filter coefficients. It can easily be shown that this performance surface is concave and, hence, has one global minimum. The optimal filter in a Wiener sense is the one minimizing the MSE. This is obtained by differentiating Eq. (14-4) with respect to the filter weights:

$$\frac{\partial \varepsilon}{\partial \mathbf{w}_N} = 0 \tag{14-9}$$

yielding:

$$\mathbf{R}_{NN}\mathbf{w}_N^* = \mathbf{p}_N \tag{14-10}$$

Equation (14-10) is called the normal equation. Although this equation is usually solvable, it is not solved as such due to the complexity of the problem. Moreover, gathering the long-term statistics of the signals may also be too time-consuming. Real-world applications mostly use the well known and simple least mean square (LMS) algorithm which tries to iteratively solve this normal equation using an approximation of the long-term statistics [55,56]. This algorithm belongs to the family of stochastic methods since it minimizes a statistical error measure.

It should be noted that this general introduction to adaptive filtering is made in the context of system identification. The same remains valid for interference canceling. The general principle of interference cancelling is the following. A first sensor receives the composite signal $x = s + n_1$ with s the signal of interest and n_1 the interference (or "noise") which is supposed to be uncorrelated with s. A second sensor receives a reference input signal $y = n_2$, where n_2 is uncorrelated with s but correlated with n_1. The noise n_2 is then filtered by a filter H to produce an output as close as possible to n_1. This output is finally subtracted from $s + n_1$. This scheme is illustrated in Fig. 14-9. Due to the assumptions made about the correlations of the

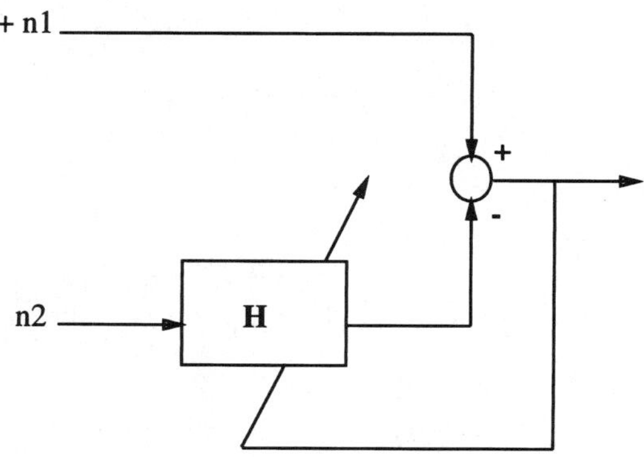

Figure 14-9   A simple diagram of the canceller.

different signals, minimizing the error power by adjusting the filter H will correspond to estimating the signal s.

While in most applications, H is supposed be linear and simple adaptive algorithms are used, there are many ways to improve the performance of such a simple scheme. Multirate adaptive filtering is one of these solutions, and is described in the next section.

14.4.2 Fundamentals of Adaptive Filtering in Subbands

Subband adaptive filtering has been suggested as an alternative to classical adaptive filtering to increase computation speed and improve convergence of the adaptation process. It was first introduced by Furukawa [57] and Kellermann [58]. A general subband echo canceler is depicted in Fig. 14-10, where a noisy signal ($x = s + n_1$) composed of a desired signal s and an interfering noise n_1 is decomposed through the analysis filter bank along with an auxiliary noise signal $y = n_2$. The purpose of echo cancellation is to recover the desired signal s from the noisy one using an adaptive filter.

The subband approach consists in performing the subband decomposition of both signals yielding M pairs of sequences decimated by a factor R. Regular adaptive filtering can be performed in each subband before reconstructing the sequence of output signals. Besides the usual advantages of subband decomposition consisting in the possibility of parallel processing and the reduction in the number of data to be processed, other benefits are achieved, namely the reduction of the adaptive filter length and the speed up of convergence.

However, there is one important drawback to subband adaptive filtering due to the non ideal nature of the analysis filters. Since rejection in the stopband is not infinite, when performing one-bandpass filtering some frequency components of the adjacent bands will be included in the output signal of that bandpass filter. The problem is that these components will be aliased at the downsampling stage.

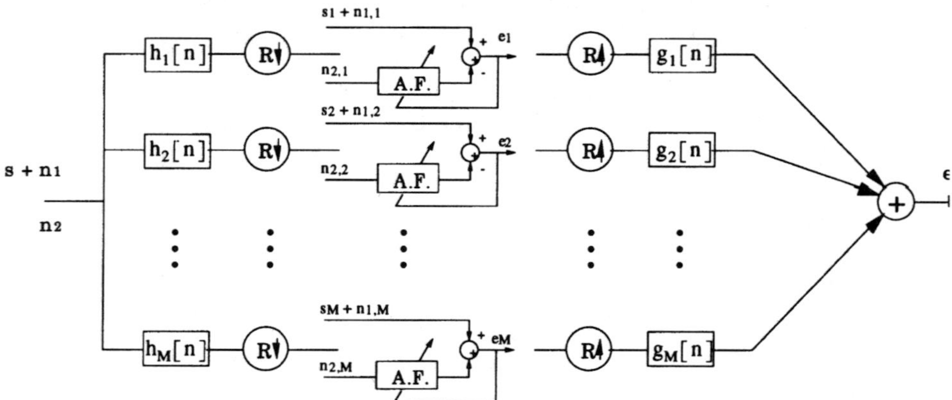

Figure 14-10   Adaptive filtering purformed with an M-band filter bank.

Remedies of this situation are various. Of course, the use of sharp analysis filters will reduce this effect. Unfortunately this requires longer analysis filters. Another solution is to have a decimation factor lower than the number of subbands. Gilloire and Vetterli proposed in [59] a solution consisting in performing an additional adaptation with the reference input of the contiguous bands.

Another drawback concerns the rigid nature of the decomposition, which is totally independent of the spectral content of the signal. For instance, in the design of an adaptive notch filter or of an adaptive line enhancer, the sinusoids of interest might fall into the gap regions of the filter banks and thus might not be enhanced or canceled by the adaptive system. A signal-dependent partitioning of the frequency axis may solve the problem. This can be accomplished in the framework of the wavelet packets. The concern is to obtain a nonuniform filter bank for subband decomposition that performs a signal-dependent partitioning of the frequency axis in order to overcome the intrinsic limitation of classical signal-independent subband adaptive filtering. This section shows how, using a new mutual wavelet packets scheme along with a new decomposition criterion, a nonuniform signal-dependent filter bank is obtained [23,60].

14.4.3 The Decomposition of Mutual Wavelet Packets

As described in section 14-2, wavelet packets can be regarded as a powerful tool to obtain an optimal subband decomposition with respect to some criterion. This is achieved by splitting and/or merging subbands obtained by successive application of analysis filters.

The concern here is to favor adaptive filtering. Considering Eq. (14-10), this can be obtained by somehow maximizing the magnitude of either the vector $\mathbf{p}_N$ or some of its relevant components. This corresponds to maximizing some measure of the cross-correlation between the two signals considered in the adaptive process. As an illustration of the influence of the cross-correlation vector on the adaptation, consider for instance the case of two uncorrelated signals. Since the vector $\mathbf{p}_N$ will be equal to zero, the optimal filter will be zero as well. This is quite a reasonable result since there is no way to extract one signal from the other. Intuitively, maximizing the cross-correlation between the two signals will maximize their similarity and favor the performance of the adaptation. Therefore the criterion suggested for the decomposition of the mutual wavelet packets is based on the maximization of the sum of the magnitude of the cross-correlation samples and is given by:

$$M(x_i[n], y_i[n]) = \sum_{k=0}^{N-1} |\phi_{x_i y_i}[k]| \qquad (14\text{-}11)$$

where $x_i[n]$ is the signal to be filtered at a given node of the decomposition, $y_i[n]$ is the reference signal at the same node, N is the number of coefficients of the local adaptive filter within the node, and $\phi_{xy[k]}$ is the cross-correlation sequence of $x_i[n]$ and $y_i[n]$. Each of these cross-correlation samples is estimated, as usually, with the biased correlation estimate of an L-sample sequence given by:

$$\hat{\phi}_{x_i y_i}[k] = \frac{1}{L} \sum_{n=0}^{L-1} x_i[n] y_i[n+k]$$

The choice of computing the criterion using N samples of the cross-correlation sequence is justified by the fact that this number corresponds to the number of components of the vector $\mathbf{p}_N$. Moreover, this could be reduced to some smaller constant when speed of the best-basis selection is of major concern. Compared with the usual wavelet packets used for signal analysis or signal compression, the decomposition of the mutual wavelet packets simultaneously decomposes two signals onto the same basis with respect to a common criterion, as illustrated in Fig. 14-11. The best basis is thus optimal for the two signals instead of being adapted to only one of them.

The purpose of the method is to find the optimal basis that maximizes this functional on the manifold of orthonormal bases. The search for the best basis is performed exactly as explained for the decomposition of one signal, except that in the present case, two binary trees will be spanned instead of one. This works as follows: the two decomposition trees are completely spanned. A bottom-up pruning scheme is applied simultaneously for both of them. When going from leaves to the root, a correspondence is made between the nodes and the values of the functional on the two possible bases (the nodes basis and their two children bases) are compared. The optimal basis is going to be the one that maximizes the functional. The parent nodes are assigned the maximum of the two functional values for further comparisons.

Compared to regular subband adaptive filtering, the main advantage of this new scheme consists in creating a nonuniform filter bank adapted to the spectral content of the signals. The decomposition is chosen so that the signals are maximally correlated in each of the subbands, which assures a better adaptation process. The performance of this new scheme will mostly show up in the presence of structured and colored noise or interferences, where the band selection is critical. Performance in the presence of wideband signals is expected to be close to regular subband filtering since the optimal decomposition is not going to change significantly the filterbank structure.

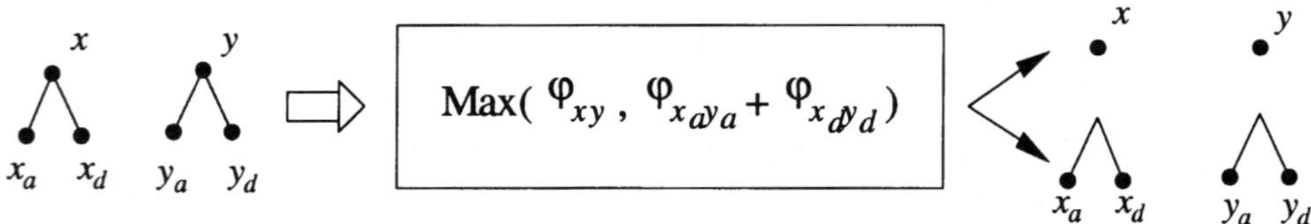

Figure 14-11   An illustration of the best-basis selection for the decomposition of the mutual wavelet packets.

14.4.4 Implementation Scheme

Once the best basis has been selected and the nonuniform filter-bank decomposition performed, the adaptation remains to be done. Recent works in classical subband adaptive filtering [61–63] propose structures that differ in their adaptation criteria. The two principal structures are investigated in this section and applied to the decomposition of the mutual wavelet packets. The first one is called the synthesis-independent structure and the whole adaptation process is performed in the subbands. The second one, called the synthesis-dependent structure, computes its error criterion on the reconstructed error signal.

For both schemes, the size of the filter in each of the subbands is derived from the following equation given in [59]:

$$L_S = \frac{L_F}{2} + L_A \qquad (14\text{-}12)$$

where L_A is the length of the analysis filter, L_F the length of the adaptive filter used for the parent and L_S the length of the adaptive filter used for the children. At every stage of the decomposition process, when a node is split into two children, the filter lengths of the children nodes are deduced from that of their parent according to the above equation.

14.4.5 Experimental Results

Although the methods proposed in the literature for the estimation of P_{mv} have been validated on some animals, their use for clinical studies is considerably limited. In fact, strict apnea is required during the recording of the signal to avoid respiratory artifacts. However, we have proposed a method to suppress respiratory interference, so as to make the method more generally applicable by overcoming the requirement of strict apnea for the *in vivo* measurement of P_{mv}.

Our approach consists in using an interference canceler, where the auxiliary signal used is the right atrial pressure signal. This signal can be recorded simultaneously with the POPT using the same Swan–Ganz catheter. Also, the right atrial pressure signal is well correlated with the respiratory interference without being correlated with the pressure transient. The proposed mutual wavelet packets adaptive-filtering algorithm is used for the cancellation of respiratory interference in pulmonary capillary transients [64]. Two respiratory periods are selected in the last part of each signal to adapt the filters. After convergence, the filters are applied to the entire signal. This gives a measure of the average magnitude of the respiratory interference before and after cancellation.

The basic adaptive-filtering scheme, performed on the original signals and denoted by full-band implementation, is compared to the mutual wavelet packets scheme. The basic analysis wavelet used for the decomposition is the D_{10} Daubechies wavelet [65].

A typical result is illustrated in Fig 14-12. It can be seen that the respiration is much more attenuated when using the mutual wavelet packets scheme. We have performed the simulations on 115 different transients recorded on different patients.

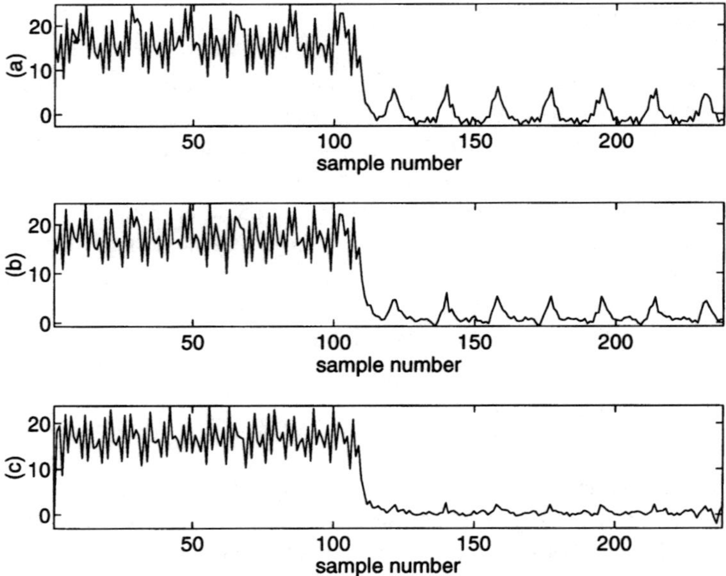

Figure 14-12 Typical result for respiratory interference canceling on the pulmonary pressure transients: (a) before interference canceling; (b) full-band implementation; (c) the mutual wavelet packets scheme.

In every experiment, the performance of the mutual wavelet packets scheme has been contrasted with the performance of the fullband scheme. The adaptive algorithm used in both cases is the classical least mean square algorithm. The metric used to estimate their performance is an estimate of the average magnitude of the residual error. This average magnitude has been computed on the last part of the signal where the influence of the transient is negligible. The results of these experiments are reported in Fig. 14-13. It appears that the performance achieved using the mutual wavelet packets is better than that of the fullband scheme. Indeed the respiratory artifacts are more efficiently canceled in all cases.

The main advantage of the proposed method is of course an adaptive subband decomposition of the signals to be filtered, adaptively reducing the aliasing gaps of the filter bank at the sensitive locations on the frequency axis. To favor the adaptation process, the similarity between the signals to be filtered has to be maximized in every subband. The sought application being adaptive filtering—a method trying to be as independent of the environment as possible—the decomposition process should be adaptive itself. The mutual wavelet packets framework offers such an adaptive decomposition of the signals and has proved to be advantageous.

14.5. CONCLUSION

This chapter has illustrated the fact that the Fourier transform is not sufficient for the analysis of nonstationary signals, from the viewpoint of balancing between time

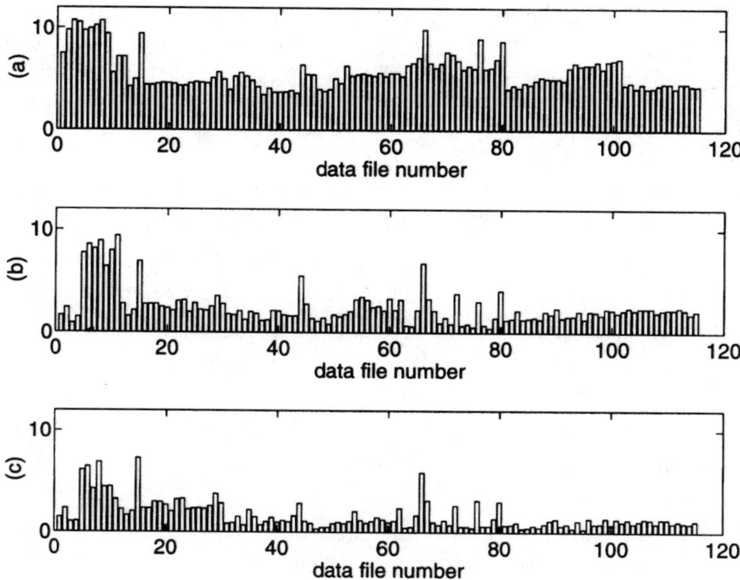

Figure 14-13   Performance comparison for interference canceling: (a) before interference canceling; (b) full-band implementation; (c) mutual wavelet packets scheme. For each data file, the average magnitude of the residual error is reported.

and frequency localization. The principle of continuous wavelet transformation has been summarized. The orthogonal wavelet transform has been presented in the general context of the multiresolution representation. Some of the important properties of wavelets such as orthogonality, regularity, or compact support have been described.

These properties allow us, at least from a theoretical point of view, to distinguish wavelets from filter banks while their numerous similarities have been highlighted from a practical viewpoint. The generalization to the concept of wavelet packets has been introduced together with the best-basis method. The different criteria used for the selection of the best basis and already used in the literature have been described. Finally, the estimation of pulmonary capillary pressure has been detailed and a new mutual wavelet packets scheme has been proposed for subband adaptive filtering. Simulation results proved that the proposed method is more efficient than classical methods.

REFERENCES

[1] C. K. Chui, *An Introduction to Wavelets*. New York: Academic Press, 1992.

[2] C. K. Chui, *Wavelets: A Tutorial in Theory and Applications*. New York: Academic Press, 1992.

[3] J. M. Combes, A. Grossman, and Ph. Tchamitchian, *Wavelets: Time–Frequency Methods and Phase Space. Inverse Problems and Theoretical Imaging*. Berlin: Springer Verlag, 1989.

[4] Y. Meyer, *Ondelettes et Opérateurs, I*. Paris: Hermann, 1990.

[5] Y. Meyer, *Ondelettes, II*. Paris: Hermann, 1990.

[6] F. B. Tuteur, "Wavelet transformations in signal detection," *IEEE Int. Conf. Acoust., Speech, Signal Proc.*, vol. CH2561-9, pp. 1435–1438, 1988.

[7] O. Meste, H. Rix, R. Jane, and P. Caminal, "Detection of late potentials by means of wavelet transform," *IEEE/EMBS*, November 1989.

[8] O. Meste, H. Rix, R. Jane, and P. Caminal, "Détection battement par battement de potentiels tardifs dans les électrocardiogrammes à haute amplification," *ITBM*, 1991.

[9] L. Senhadji, G. Carrault, J. J. Bellanger, and G. Passariello, "Quelques nouvelles applications de la transformée en ondelettes," *Innovation Tecnologique en Biologie et Medecine*, 1993.

[10] M. S. Fuller, T. Dustman, and R. Freeman, "Wavelet analysis of the signal averaged electrocardiogram," *IEEE/EMBS*, November 1991.

[11] L. Senhadji, G. Carrault and J. J. Bellanger, "Détection et cartographie multi-échelles en EEG." *Proc. Int. Conf. on Wavelets and Applications*, Toulouse, France, S. Roques ed. Berlin: Springer Verlag, 1993.

[12] N. V. Thakor, G. Xin-rong, S. Yi-chun, and D. Hanley, "Multiresolution wavelet analysis of evoked potentials," *IEEE Trans. on Biomedical Eng.*, November 1993.

[13] J. C. Smerek and H. Debruin, "Multiresolution decomposition of needle EMG," *IEEE/EMBS*, November 1991.

[14] P. Reynaud and B. Torresani, "Paquets Continus d'Ondelettes et Décomposition Optimale." *Treizième colloque Gretsi*, Juan–les–Pins, September 1991.

[15] M. V. Wickerhauser, "INRIA lectures on wavelet packet algorithms," *INRIA Lectures on Wavelet Packet Algorithms*, March 1991.

[16] R. R. Coifman and M. V. Wickerhauser, "Entropy-based algorithms for best basis selection," *IEEE Trans. Inform. Theory*, vol. 38, pp. 713–718, March 1992.

[17] R. Coifman, Y. Meyer, D. Quacke, and M. Wickerhauser, "Acoustic signal compression with wave packets." *Wavelet Workshop*, Marseilles, October 1990.

[18] R. R. Coifman and M. V. Wickerhauser, *Best Adapted Wave Packet Bases*. 1990.

[19] Y. Meyer, "Méthodes temps-fréquence et méthodes temps-échelle en traitement du signal et de l'image," *INRIA Lectures on Wavelet Packet Algorithms*, March 1991.

[20] B. Rouchouze, M. Karrakchou, A. Geurtz, and M. Kunt, "On the use of wavelet packets for image sequence coding." *Int. Symp. on Fiber Optic Networks and Video Communications*, Berlin, April pp. 316–327, 1993.

[21] K. Ramchandran and M. Vetterli, "Best wavelet packets in a rate-distortion sense," *IEEE Trans. Image Proc.*, vol. 2, pp. 160–175, 1993.

[22] M. Wax and T. Kailath, "Detection of signals by information theoretic criteria," *IEEE Trans. Acoust., Speech, Signal Proc.*, vol. 33 April 1985.

[23] M. Karrakchou and C. van den branden Lambrecht, "New issues for the use of wavelet packets." *Workshop on Time Frequency Analysis, Lyon*, March, 1994, pp. 26.1–26.4, 1994.

[24] C. van den branden Lambrecht and M. Karrakchou, "Wavelet packets for the estimation of parameters of localized sinusoids." *European Conf. Signal Processing*, Edinburgh, August, 1994, pp. 653–657, 1994.

[25] P. E. Pepe, "The clinical entity of adult respiratory distress syndrome: definition, prediction and prognosis," *Critical Care Clinics*, vol. 2, 1986.

[26] A. Fein, J. P. Wiener-Kronich, M. Nieder, and M. A. Matthay, "Pathophysiology of the adult respiratory distress syndrome. What have we learned from human studies?," *Critical Care Clinics*, vol. 2, 1986.

[27] M. R. Flick, "Mechanisms of lung injury. What have we learned from experimental animal models?," *Critical Care Clinics*, vol. 2, 1986.

[28] D. R. McCaffree, "Adult respiratory distress syndrome." In *Cardiopulmonary Critical Care*, D. R. Dantzker ed., New York: Grune-Stratton, 1986.

[29] R. M. Prewitt, J. McCarthy, and D. H. Wood, "Treatment of acute low pressure pulmonary edema in dogs," *J. Clin. Invest.*, vol. 67, 1981.

[30] P. R. Eisenberg, J. R. Hansbrough, D. Anderson, and D. P. Schuster, "A prospective study of lung water measurements during patient management in intensive care unit," *Am. Rev. Respir. Dis.*, vol. 136, 1987.

[31] R. S. Simmons, G. G. Berdine, J. J. Seidenfeld, T. J. Prihoda, G. D. Harris, J. D. Smith, T. J. Gilbert, E. Mota, and W. G. Johanson, "Fluid balance and the adult respiratory distress syndrome," *Am. Rev. Respir. Dis.*, vol. 135, 1987.

[32] M. A. Mattay and J. P. Wiener-Kronish, "Intact epithelial barrier function is critical for the resolution of alveolar edema," *Am. Rev. Respir. Dis.*, vol. 142, 1990.

[33] S. J. Allen, R. E. Drake, J. P. Williams, G. A. Williams, and J. C. Gabel, "Recent advances in pulmonary edema," *Critical Med.*, vol. 15, 1987.

[34] G. A. Laine, S. J. Allen, and J. Katz, "Effect of systemic venous pressure elevation in lymph flow and lung edema formation," *J. Appl. Physiol*, vol. 61, 1986.

[35] R. H. Simmons. "Mechanisms of lung injury." In *Cardio-Pulmonary Critical Care*. D. R. Dantzker ed., New York, 1986.

[36] A. E. Taylor, J. W. Barnard, S. A. Barman, and W. Keith Adkins, "Fluid Balance." In *The Lung: Scientific Foundations*. R. G. Crystal and J. B. West eds., New York: Raven Press, 1991.

[37] C. E. Patterson, J. W. Barnard, J. E. Lafuze, M. T. Hull, S. J. Baldwin, and R. A. Rhoades, "The role of activation of neutrophils and microvascular pressure in acute pulmonary edema," *Am. Rev. Respir. Dis.*, vol. 140, 1989.

[38] D. C. Hocking, P. G. Phillips, T. J. Ferro, and A. Johnson, "Mechanisms of pulmonary edema induced by tumor necrosis factor alpha," *Circul. Res.*, vol. 67, 1990.

[39] C. R. Chen, N. F. Voelkel, and S. W. Chang, "PAF potentiates protamine-induced lung edema: role of pulmonary venoconstriction," *J. Appl. Physiol.*, vol. 68, pp. 1059–1068, 1990.

[40] N. C. Staub, "The hemodynamics of pulmonary edema," *Bull. Eur. Physiopathol. Respir.*, vol. 22, 1986.

[41] H. Holloway, M. Perry, J. Downey, J. Parker and A. Taylor, "Estimation of effective pulmonary capillary pressure in intact lungs," *Am. Physiol. Soc.*, vol. 54, pp. 846–851, 1983.

[42] Y. Yamada, M. Suzukawa, M. Chinzei, T. Chinzei, N. Kawahara, K. Suwa, and K. Numata, "Phasic capillary pressure determined by arterial occlusion in intact dog lung lobes," *Am. Physiol. Soc.*, vol. 67, pp. 2205–2211, 1989.

[43] G. G. Collee, K. E. Lynch, R. D. Hill, and W. M. Zapol, "Bedside measurement of pulmonary capillary pressure in patients with acute respiratory failure," *Anesthesiology*, vol. 66, pp. 614–620, May 1987.

[44] S. Audi, C. A. Dawson, D. A. Rickaby, and J. H. Linehan, "Localization of the sites of pulmonary vasomotion by use of arterial and venous occlusion," *Am. Physiol. Soc.*, 2127–2136, 1991.

[45] C. A. Dawson, T. A. Bronikowski, J. H. Linehan, S. T. Haworth, and D. A. Rickaby, "On the estimation of pulmonary capillary pressure from arterial occlusion," *Am. Rev. Respir. Dis.*, vol. 140, pp. 1228–1236, 1989.

[46] T. S. Hakim, J. M. I. Maarek, and H. K. Chang, "Estimation of pulmonary capillary pressure in intact dog lungs using the arterial occlusion technique," *Am. Rev. Respir. Dis.*, vol. 140, pp. 217–224, 1989.

[47] C. A. Dawson, "Pulmonary circulation." In *American Handbook of Physiology. Section 2. Respiration. Part I.* A. P. Fishmans (ed.), Bethesda, 1986.

[48] W. M. Zapol and M. T. Snider, "Pulmonary hypertension in severe acute respiratory failure," *New. Engl. J. Med.*, vol. 96, 1977.

[49] T. S. Hakim, R. R. J. Michel, and H. K. Chang, "Partitioning of pulmonary vascular resistance in dogs by arterial and venous occlusion," *J. Appl. Physiol.*, vol. 52, 1982.

[50] D. K. Cope, R. C. Allison, J. L. Parmentier, J. N. Miller, and A. E. Taylor, "Measurement of effective pulmonary capillary pressure using the pressure profile after pulmonary artery occlusion," *Critical Care Med.*, vol. 14, pp. 16–22, 1986.

[51] F. Grimbert, "Effective pulmonary capillary pressure in pulmonary edema." In *Update in Intensive Care and Emergency Medicine.* J. L. Vincent (ed.) Berlin: Springer, 1986.

[52] V. D'Orio, J. Halleux, L. M. Rodriguez, C. Wahlen, and R. Marielle, "Effects of *Escherischia coli* endotoxin on pulmonary vascular resistance in intact dogs," *Critical Care Med.*, vol. 14, 1986.

[53] L. C. Siegel, R. G. Pearl, and S. L. Shafer, "The longitudinal distribution of pulmonary vascular resistance during unilateral hypoxia," *Anesthesiology*, 1989.

[54] B. Widrow et al., "Adaptive noise cancelling: principles and applications," *Proc. IEEE*, December 1975.

[55] B. Widrow and S. D. Stearns, *Adaptive Signal Processing*. Englewood Cliffs, NJ: Prentice-Hall, 1985.

[56] S. T. Alexander, *Adaptive Signal Processing*. Berlin: Springer Verlag, 1986.

[57] I. Furukawa, "A design of canceller of broad band acoustic echo." *Int. Teleconferencing Symp*, Tokyo, 1984.

[58] W. Kellerman, "Kompensation akusticher Echos in Frequenzteil-bändern. *Aachener Kolloquium*, Aachen, FRG, 1984.

[59] A. Gilloire and M. Vetterli, "Adaptive filtering in sub-bands." *IEEE Int. Conf. on Acoustics, Speech and Signal Processing*, 1988.

[60] M. Karrakchou, C. van den branden Lambrecht, and M. Kunt, "Mutual wavelet packets adaptive filtering for the analysis of pulmonary capillary pressure." Invited paper for the *IEEE Mag. Eng. Med. Biol. Soc.*, 1995.

[61] W. Kellermann, "Analysis and design of multirate systems for cancellation of acousticals echoes." *IEEE Int. Conf. Acoustics, Speech and Signal Processing Conf.*, 1988.

[62] S. Furui and M. M. Sondhi, *Advances in Speech Signal Processing*. New York: Marcel Dekker, 1992.

[63] M. Petraglia, "Efficient adaptive filtering structures based on multirate techniques." *Ph.D. Thesis*, University of California at Santa Barbara, December 1991.

[64] M. Karrakchou and M. Kunt, "Wavelet packets for interference cancelling in biomedical systems." Invited paper for the *Int. Conf. Artificial Neural Networks in Engineering*, St Louis, MI, November, 1994, pp. 514–518, 1994.

[65] I. Daubechier, "Orthonormal bases of compactly supported wavelets". *Comm. Pure Appl. Math.*, vol.41, pp. 909–996, 1988.

Chapter 15

In Pursuit of Time-Frequency
Representation of Brain Signals

P. J. Durka, K. J. Blinowska

15.1. INTRODUCTION

Most traditional methods of signal analysis are based on the assumption of the ergodicity of the time series which requires stationarity of the signals. Physiological time series comply with that assumption only for very limited time intervals and quite often the variation of the signal in time is of primary interest. This is especially the case for brain electrical activity. According to the present understanding, the processing of information by the brain is reflected in dynamical changes of electrical activity in time, frequency, and space. Therefore, the study of these processes requires methods which can describe the variation of the signal in time and frequency in a quantitative way.

The traditional windowed Fourier transform is not sufficient in this case. It is subject to high statistical errors, and it is severely biased as a consequence of the unfulfilled assumption that the signal is either infinite or periodic outside the measurement window. Parametric methods like the autoregressive (AR) model are free from the "windowing" effect since no assumptions about the signal outside the measurement window are needed, but still stationarity of the signal is required and signal structures of duration shorter than the measurement window cannot be identified. Wavelet analysis and related methods such as matching pursuit (MP) have brought essential progress in this respect.

The wavelet transform introduced by Grossman and Morlet [1] very quickly found applications in different research areas where signal and image analysis is applied, e.g., in biomedical engineering. In the study of brain signals it was applied, e.g., by Bartnik and Blinowska [2] and Thakor et al. [3] to the evaluation of evoked potentials and by Akay et al. [4] in the analysis of electrocortical activity.

Evoked potentials are signals of amplitude an order of magnitude smaller than the ongoing electroencephalogram (EEG) activity. The most commonly applied method of evoked potentials (EP) evaluation is based on averaging single EPs triggered by a repeating stimulus. This method relies on unrealistic assumptions concerning the purely deterministic and repeatable character of the EP, the purely stochastic character of the on-going EEG, and the independence of both signals. Wavelet analysis offers an alternative method of EP evaluation which will be described below. It is based on an important feature of the method—its ability to discriminate between signals lying in the same frequency range but with different temporal localizations.

Nevertheless, wavelet analysis is itself subject to certain limitations connected with the fact that the bandwidth is inversely proportional to the time scale, which effectively limits the resolution. Another drawback is the sensitivity of the representation to the shift in time of the analyzed window. These limitations can be overcome by the MP technique, which may be considered a generalization of the wavelet representation. The application of this method to EEG signal transients will be described in the second part of this chapter.

15.2. APPLICATION OF THE WAVELET TRANSFORM TO EVOKED-POTENTIAL ANALYSIS

15.2.1 Method

The wavelet transform describes signals in terms of coefficients representing the energy content in a specified time-frequency region. This representation is constructed by decomposition of a signal over a set of functions generated by translating and scaling one function (wavelet ψ).

The choice of ψ and dilation by powers of two yields an orthogonal multiresolution decomposition as proposed by Mallat [5]. It can be viewed as a recursive approximation of a time series $x(t)$ at resolutions changing as powers of two. If we denote the approximation of signal x at scale 2^j as $A_{2^j}x$, then obviously between scale 2^{j+1} and a coarser scale 2^j some information is lost. It can be retrieved in a "detail signal" $D_{2^j}x$. Both operations (approximation and extracting the difference) are orthogonal projections on subspaces of $L^2(\mathbb{R})$, and are denoted, respectively, as V_{2^j} and O_{2^j}, where $O_{2^j} \oplus V_{2^j} = V_{2^{j+1}}$. Orthogonal bases of both spaces are generated by dilating and translating a scaling function ϕ (for approximations) and wavelet ψ (for the detail signal). So if we denote $\psi_{2^j}(t) = 2^j \psi(2^j t)$, then $[\sqrt{2^{-j}}\phi_{2^j}(t - 2^{-j}n)]_{n \in Z}$ and $[\sqrt{2^{-j}}\psi_{2^j}(t - 2^{-j}n]_{n \in Z}$ form orthonormal bases of V_{2^j} and O_{2^j}, respectively. Finally, a set of wavelets

$$\left[\sqrt{2^{-j}}\psi_{2^j}(t - 2^{-j}n)\right]_{(n,j)\in Z^2} \tag{15-1}$$

is an orthonormal basis of $L^2(\mathbb{R})$. The signal $x(t)$ is fully characterized by (and can be reconstructed from) wavelet coefficients:

$$D_{2^j}^n(x) = \langle x(t), \psi_{2^j}(t - 2^{-j}n) \rangle \tag{15-2}$$

The scale j corresponds to an octave of the signal bandwidth. If we denote the Nyquist frequency as f_N, then scale 0 covers frequencies from $f_N/2$ to f_N scale 1 frequencies from $f_N/4$ to $f_N/2$ and so on (Fig. 15-1).

This approach yields a very efficient pyramidal algorithm for calculating the $D_{2^j}^n$ coefficients, based on quadrature mirror filters (Fig. 15-1, lower part). The approximation of a signal at scale 2^j contains all the information necessary to compute a coarser approximation at scale 2^{j+1} as well as their difference. Decomposition consists of applying low-pass (for $A_{2^j}(x)$) and bandpass (for $D_{2^j}(x)$) filters followed by downsampling (keeping every second sample). The original signal can be retrieved by an inverse procedure. By keeping only those $D_{2^j}^n$ coefficients which correspond to specified components, we can reconstruct any desired part of the signal variance.

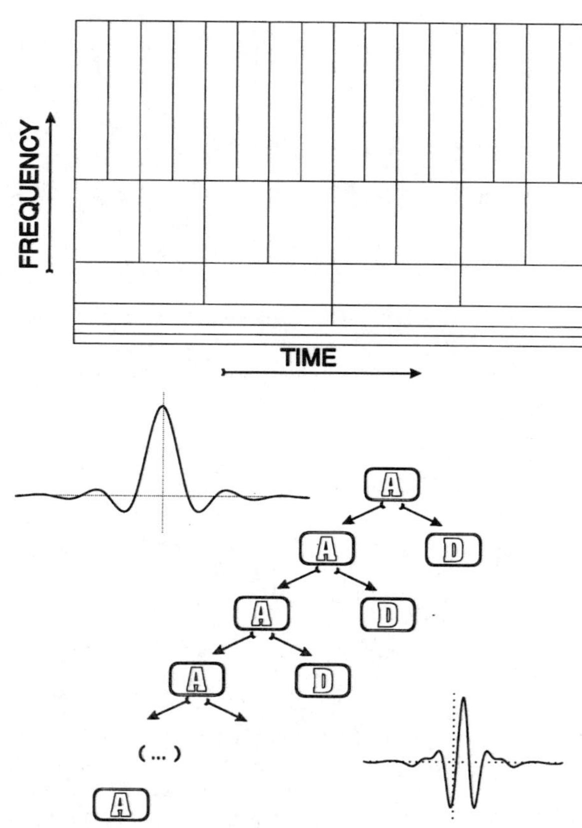

Figure 15-1   The scheme of multiresolution decomposition: upper part—a symbolic division of the time-frequency plane; lower part—pyramid algorithm; top left—φ; bottom right of lower part—ψ.

15.2.2 Application to EP Analysis

We have evaluated auditory EPs from scalp electrodes and somatosensory evoked potentials (SEPs) recorded by electrodes placed directly on the skull of experimental animals. In the first of these studies the stimuli were given in irregular intervals ranging from 2.5 to 4 s. The only assumption made was that the EP occurs in the first second after the stimulus, and between the first and second seconds we have only the ongoing EEG. The multiresolution decomposition was performed on signal segments 1 s in length. The obtained wavelet coefficients served as the input values to discriminant analysis procedures which were used to find D_j parameters optimally distinguishing the EP from the EEG. It was found that only five parameters were needed to obtain a level of significance $< 10^{-5}$ for the null hypothesis of no difference between the signals. These five parameters were used for reconstruction of single potentials. Details of the statistical procedures used are given in [2] and

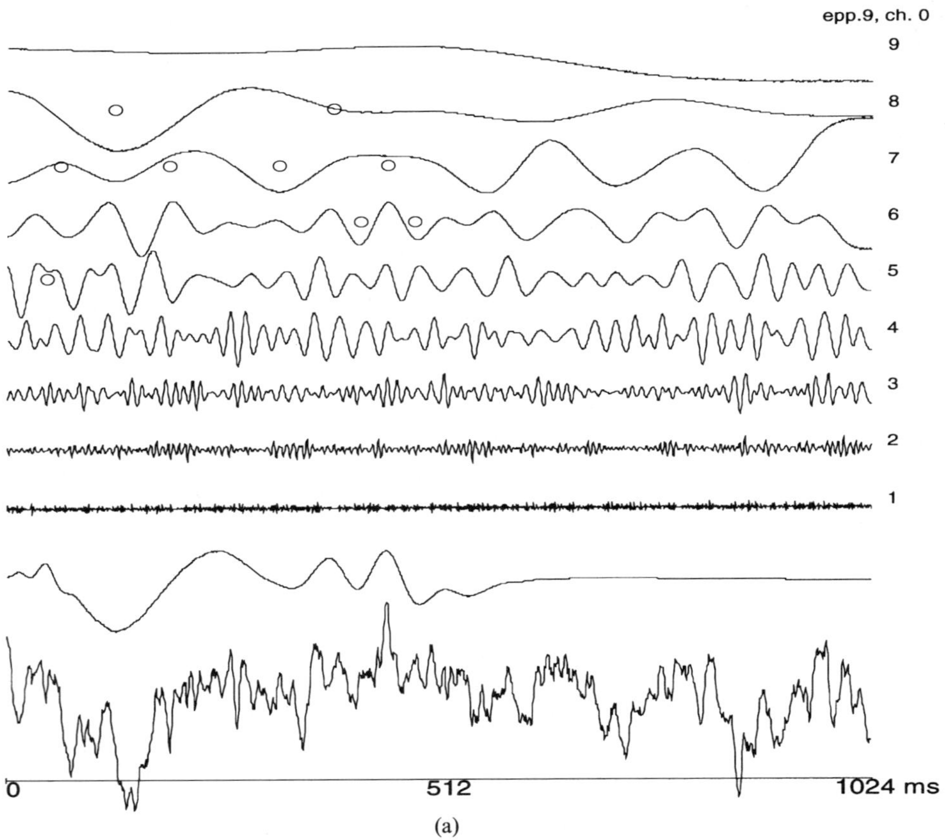

Figure 15-2   Reconstructions of a single evoked potential (a) and an average potential (b). From the bottom to the top: experimental signal, reconstructed signal, detailed continuous signal at different resolutions. The wavelet components taken into account in the reconstruction are marked by small circles.

reconstructions of single EP are presented in [6]. In Fig. 15-2, examples of the reconstruction of a single trial and the average EP from nine D_j coefficients are shown. The basic features of the EP are represented with good accuracy.\over \over

In all studied cases it was observed that the components most significant for the discrimination were the early ones. That could have been expected, since the EP occurs in the first 500 ms after the stimulus, nevertheless our assumption was much more general, namely that the EP appears in the first second after the stimulus. From our study follows a useful method of parameterization of EPs—an alternative to the conventionally used description by latencies and amplitudes based on averaging. For EPs changed pathologically, it is sometimes difficult to find latencies corresponding to the control group and to decide whether they were shifted or missing or whether there are some additional components. Wavelet analysis overcomes this difficulty, offering a universal time-frequency scale for all potentials.

Parameterization based on wavelet analysis is assumption-free, objective, and it provides a high rate of information reduction. By means of only a few parameters it

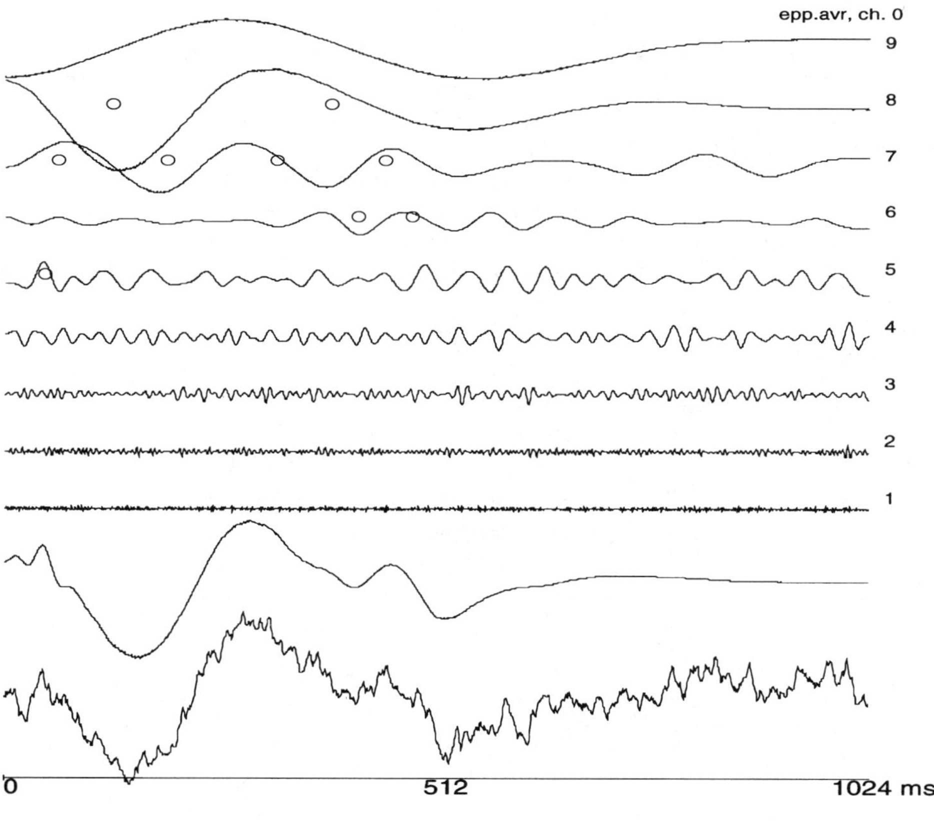

(b)

is possible to reconstruct the shape of the EP with reasonable fidelity. The high rate of signal compression might be useful for storing large amounts of patient data.

The convenient parameterization by means of wavelet analysis might be very helpful in neurophysiological studies. An example is a study of the cat somatosensory EP before and after a lesion in part of the cerebellum [7]. In this study signals from 16 electrodes were recorded; the procedure was similar to the one described above. First, multiresolution decomposition was performed on the data segments containing EPs and on those representing ongoing EEGs. Then the Mann–Whitney test was performed on each channel of the data set in order to identify parameters characterizing the SEP in contrast with the background. In this study the time course of the EP signals was more complicated than in the clinical study. In the higher-frequency bands 250–125 Hz, 125–62 Hz, and 62–31 Hz, early components were more important, and for the 31–16 Hz frequency band some later components were also significant. Neighboring electrodes had similar coefficients D_j. There were pronounced differences between the left and right sides of the cortex, depending on which paw was stimulated. In the hemisphere contralateral to the stimulus, high-frequency components were more important. After removal of part of the cerebellum the multiresolution decomposition was again performed and the Mann–Whitney test used to discriminate between the SEP and EEG, and SEP before and after the lesion. It was found that the SEP components which changed after the lesion were located in the time-frequency region which also best distinguished the SEP from the EEG (Fig. 15-3). This shows, without any prior assumptions, that the lesion affected the event-related part of the brain activity—not the ongoing EEG. The changes were rather subtle and they were not easy to identify visually or to quantify by other methods. In Fig. 15-4, an example of the multiresolution decomposition of the SEP before and after the lesion is shown for one of the channels where the difference was most pronounced. In this experiment wavelet analysis provided the basis of an objective parameterization and distinction between EPs arising under differing experimental conditions.

15.2.3 Discussion

The previously described studies demonstrated the new possibilities opened up by wavelet analysis in the study of time-locked phenomena such as EPs. Wavelet analysis provides a frame of much finer time-frequency resolution than previously used methods. By means of this frame signals can be conveniently parameterized and time-frequency components characterizing the signals can be chosen. This kind of signal description is very effective and offers the possibility of an important reduction of information—often very few components are needed to describe the signal with good fidelity. Wavelet analysis can be used for extraction of the signals from a noisy background. In the process of signal synthesis from the multiresolution wavelet coefficients, background components can be effectively suppressed, on the basis of objective statistical tests. Wavelet analysis has proved to be useful in characterizing changes in brain activity, e.g., during hypoxic injury and during maturation. In the study of cerebral hypoxia [3], the detailed components of multiresolution decom-

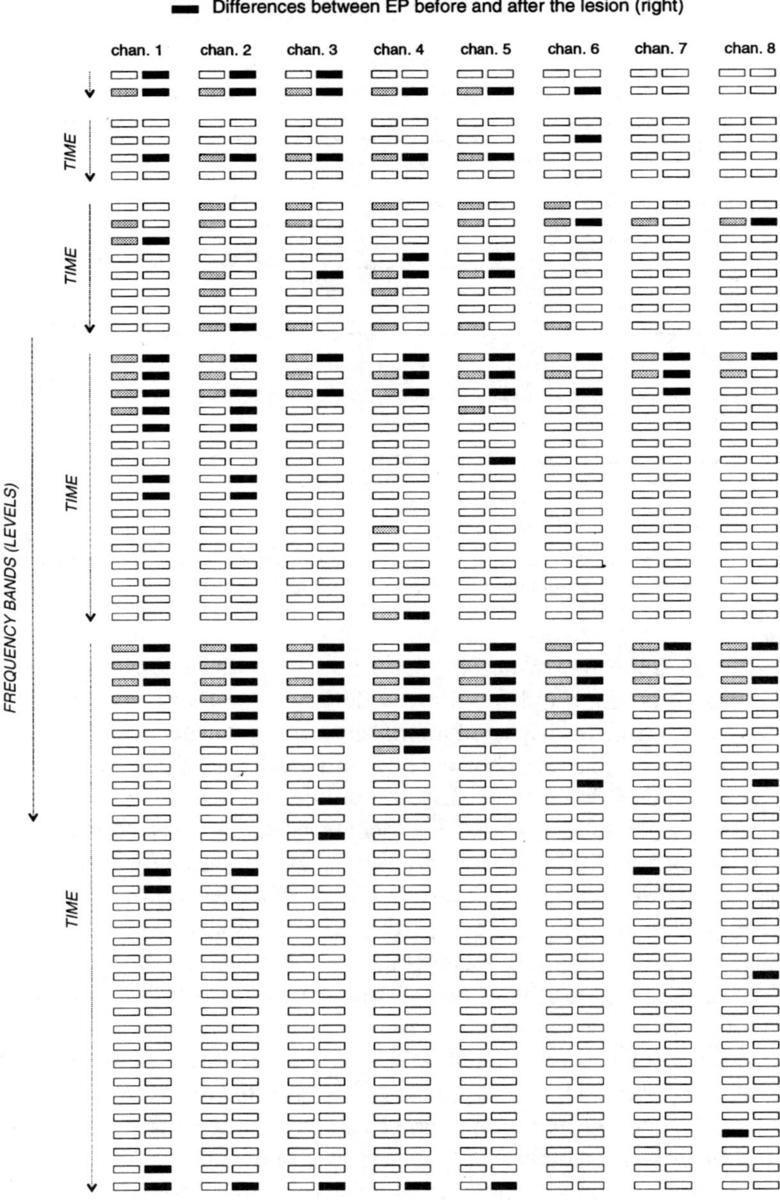

Figure 15-3   The results of the Mann–Whitney test applied to wavelet coefficients in order to discriminate between an EEG and an EP (gray rectangles) and an EP before and after the lesion (black rectangles). The blank rectangles mean that no difference on the significance level < 0.01 for given D_j was found. The dyadic frequency bands starting from the lowest (18–16 Hz) band are separated by spaces. In each block the time sequence of the coefficients is marked by an arrow. The columns show the different channels.

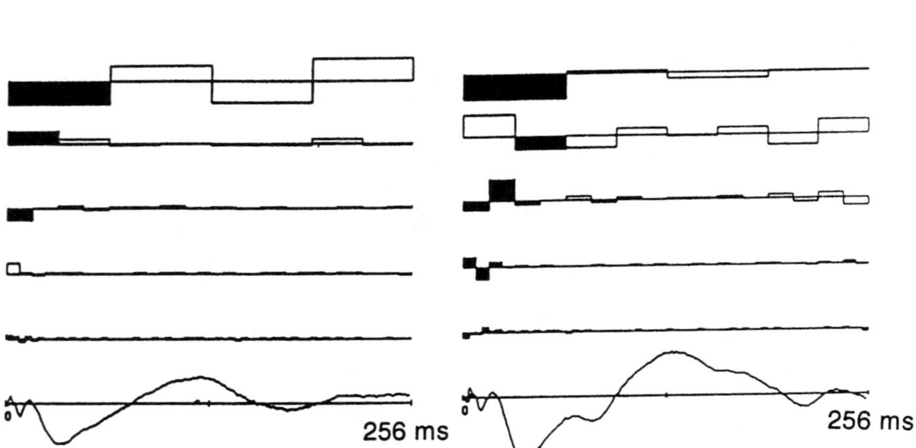

Figure 15-4   The sensory EPs and their decomposition into discrete detail function components before and after the lesion.

positions appeared to be of diagnostic value—the fine resolution coefficients characterized an early and more rapid decline in response to hypoxic injury while the coarse component displayed a recovery upon reoxygenation. In [4], wavelet analysis was used to confirm the presence of statistically different frequency patterns in three age groups of fetal lambs. In both studies mentioned, comparison of the wavelet transform with short-term Fourier transforms revealed the much better time-frequency resolution and discrimination power of the wavelet transform. However, the wavelet analysis also has certain limitations. The most important of these is the fact that the bandwidth changes in steps (usually by a factor of two) and therefore expansion coefficients in a wavelet frame do not provide precise frequency estimates of waveforms whose Fourier transform is well localized, especially at high frequencies.

There have been different attempts to overcome this limitation. When the orthogonality assumption is relaxed, finer tuning of the time-frequency scale is possible [8]; however, the computational procedures are much more complicated in this case. Wavelet packet analysis [9] generalizes the compactly supported wavelets and provides an optimal orthonormal basis for a given signal.

The wavelet representation depends also on the data window setting and is sensitive to the time shift of the window. This limitation is not critical in the case of time-locked phenomena like EPs. In such cases the wavelet transform offers a universal and convenient parameterization which can be used for comparison of signals, but this kind of representation is much less suitable for transients which occur more or less randomly in the signal (e.g., spindles, K-complexes, etc.). In Fig. 15-5 (a and b) the multiresolution decomposition by orthogonal wavelets of a signal

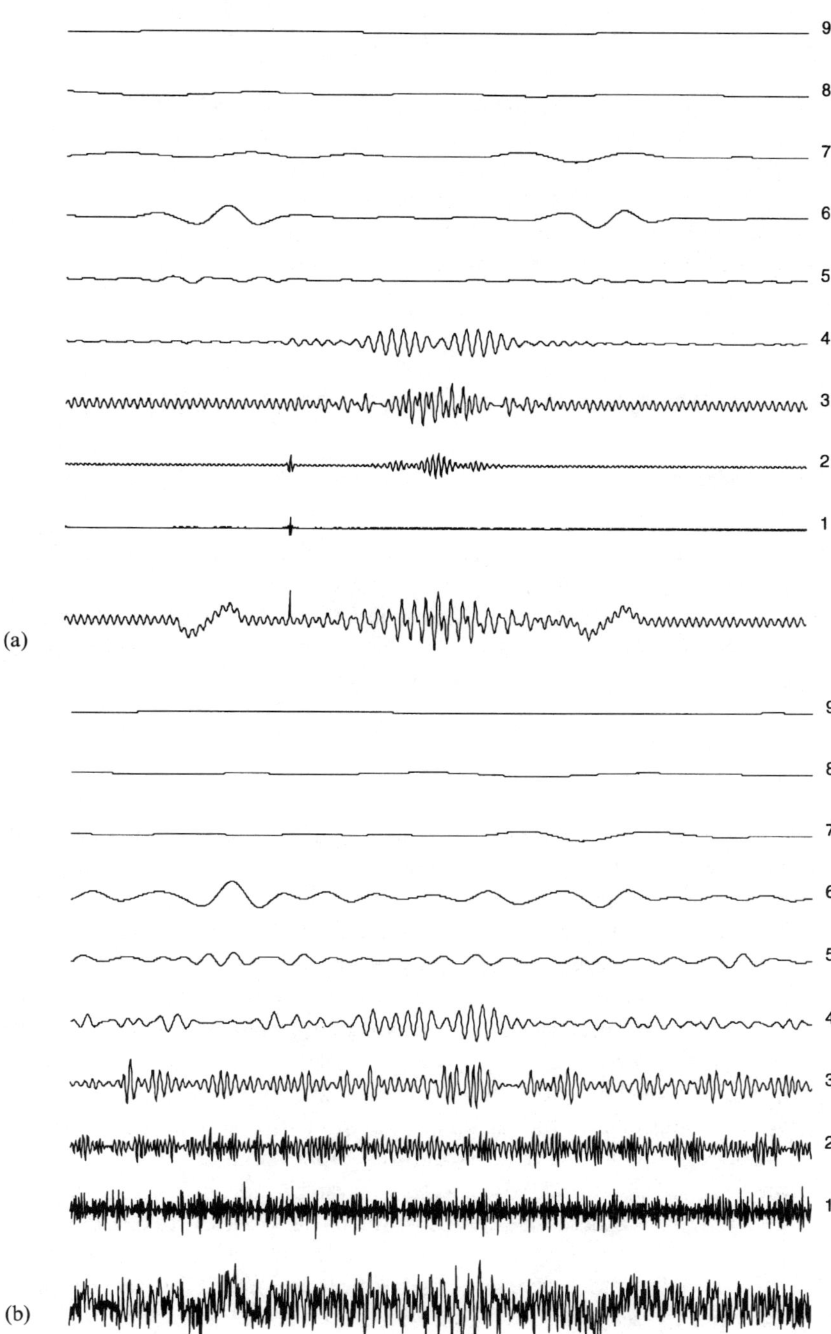

Figure 15-5 (a) Multiresolution decomposition of the simulated signal shown below. The components of this signal are shown in Fig. 15-6(a). (b) The same as (a), except that the noise component of amplitude similar to the signal was added [compare Fig. 15-6(b)].

synthesized from a sinusoid and various kinds of transients is shown. [The construction of the signal is presented in Fig. 15-6(a).] In the absence of noise, signal structures can be identified, although generally data structures are decomposed into multiple components belonging to different frequency bands. It is difficult to tell how many spindles are present in the data. When the signal is perturbed by noise [see Fig. 15-6(b) lower part], the picture becomes quite complex and it is more difficult to recognize data structures. A much more convenient description of this kind of signals is offered by the MP approach, in which the basic functions are fitted to local signal features.

15.3. MATCHING PURSUIT METHOD AND ITS APPLICATIONS

15.3.1 Method

In MP, the repertoire of waveforms used for the decomposition of signals is very broad and redundant. From this large dictionary of possible functions, a subset is chosen in such a way as to match optimally the local signal structures.

A general family of time-frequency atoms can be generated by scaling, translating, and modulating a window function $g(t)$:

$$g_I(t) = \frac{1}{\sqrt{s}} g\left(\frac{t-u}{s}\right) e^{i\xi t} \tag{15-3}$$

where $s > 0$ is the scale, ξ is the frequency modulation, and u is the translation. The index $I = (s, \xi, u)$ describes the chosen set of parameters. $g(t)$ is usually even and its energy is mostly concentrated around u in a time domain proportional to s. In the frequency domain, energy is mostly concentrated around ξ with a spread proportional to $1/s$. The minimum of time-frequency variance is obtained when $g(t)$ is Gaussian. The windowed Fourier transform and wavelet transform (WT) can be considered as particular cases of MP corresponding to restrictions concerning the choice of parameters.

In the case of the windowed Fourier transform, the scale s is constant—equal to the window length—and the parameters ξ and u are uniformly sampled; therefore, it is not appropriate to describe structures much smaller or much larger than the window.

The WT overcomes this limitation since it allows for a change of scale, decomposing the signal over atoms of varying time-frequency coordinates. Nevertheless, in the case of WTs, the frequency modulation is limited by the restriction on the frequency parameter $\xi = \xi_0/s$. Therefore, the wavelet frame does not provide precise estimates of the frequency content of waveforms well localized in the frequency domain and time localization of low-frequency structures.

In MP, all the parameters defined in Eq. (15-3) can vary freely (in practice certain restrictions can be imposed to speed up the computations); all that remains is to choose the time-frequency atoms in an optimal way. The method of iterative

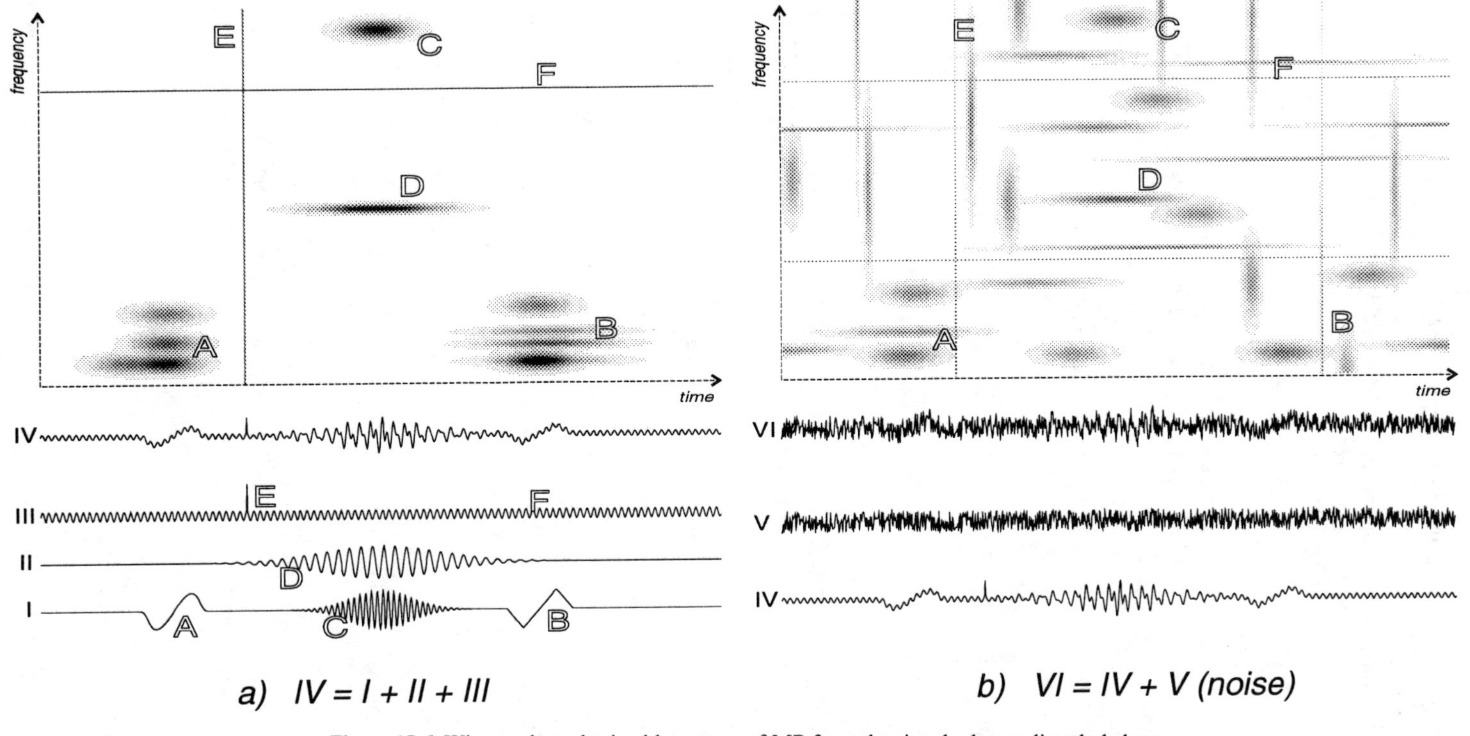

Figure 15-6 Wigner plots obtained by means of MP from the signals shown directly below. Letters mark signal structures and corresponding atoms or groups of atoms: F—sinusoid; E—delta function; D, C—sinusoids modulated by Gauss. (a) The analyzed signal IV is a sum of I, II and III. (b) The analyzed signal VI is a sum of signal IV (the same as IV in (a)) and signal V which is white noise of an amplitude similar to signal VI.

adaptive selection of functions best approximating the signal was proposed by Mallat and Zhang [10].

In the first step of the iteration procedure, the function g_{I_0} is chosen which gives the biggest product with the signal $x(t)$.

$$x = \langle x, g_{I_0} \rangle g_{I_0} + R^1 x \tag{15-4}$$

Then the residual vector $R^1 x$ obtained after approximating $x(t)$ in the direction g_{I_0} is decomposed in a similar way. The iterative procedure is repeated on the following obtained residues:

$$R^n x = \langle R^n x, g_{I_n} \rangle g_{I_n} + R^{n+1} x \tag{15-5}$$

In this way the signal x is decomposed into a sum of time-frequency atoms, which are chosen to match optimally its residues:

$$x = \sum_{n=0}^{m} \langle R^n x, g_{I_n} \rangle g_{I_n} + R^{n+1} x \tag{15-6}$$

Although the procedure converges to $x(t)$, we have to stop it at some point. We can define a magnitude $\lambda(n)$ as the proportion of the energy of residue $R^n x$ explained by g_{I_n}:

$$\lambda(n) = \frac{\langle R^n x, g_{I_n} \rangle}{\| R^n(x) \|} \tag{15-7}$$

It converges to a constant value depending on the size of the signal, which corresponds to a situation in which there are no more structures in the residuum coherent with the dictionary. Recent research shows that residua converge to a chaotic attractor of a process called "dictionary noise" [11].

Energy conservation:

$$\| x \|^2 = \sum_{n=0}^{m-1} |\langle R^n x, g_{I_n} \rangle|^2 + \| R^m x \|^2 \tag{15-8}$$

allows us to conveniently visualize its density in the time-frequency plane in the form of a Wigner distribution. Unlike the Wigner or Cohen class distributions, MP representation does not include interference terms and thus provides a clear picture in time-frequency space.

15.3.2 Results and Discussion

In Fig. 15-6(a), a Wigner plot of the same signals as presented in Fig. 15-5 is shown. In this case, in the time-frequency plane, data structures such as the sinusoid, delta function, and spindles are described by only one waveform—one atom: the rhythmic component by a horizontal line, the delta function by a vertical line, and spindles by ellipsoidal "blobs." Spindles of different frequencies but appearing at the same time are well resolved. The Gabor dictionary is especially well suited to describe this kind of structures. Complex transients such as triangular waves are coded by few atoms.

Addition of noise degrades the picture only superficially. Atoms describing the signal structures are not so clean-cut, but in fact the parameters characterizing these atoms are not changed and the resolution is not really degraded. Each atom is characterized quantitatively by four parameters: time and frequency coordinates, time span, and intensity. Resolution is better than in any other method known at present, indeed it is close to the theoretical limit [10].

In Fig. 15-7(a), an example of a Wigner map for an EEG signal (sleep stage 2) is shown. It is easy to observe that several rhythmic components of frequency below 7 Hz are present. Sharp, spike-like features of the signal are visible as vertical lines. The structure marked with an arrow corresponds to the sleep spindle. In Fig. 15-7(a), 50 waveforms corresponding to 94.25% of the energy of signal are shown. In Fig. 15-7(b) and (c), 100 atoms (97.7%) of energy and 200 atoms (99.32% of energy) for the same data segment are presented. Compared with Fig. 15-7(a), in Fig. 15-7(b) some higher-frequency rhythmical components and more spindle-like atoms appear, and in the next Fig. 15-7(c) the picture becomes still more complicated. While inspecting Fig. 15-7(a), (b), and (c) we have encountered the problem of determining how many atoms are sufficient to characterize the signal. Where should the threshold be set in order to discriminate against noise without information loss? The λ function (Fig. 15-8) provides some help in this respect, since it characterizes coherence of the signal with respect to the dictionary of functions used in the approximation. Studying the behavior of λ as a function of the number of atoms (i.e., the algorithm's iterations) we can observe that for a number of the atoms higher than 100, its decrease becomes very slow and at 150 atoms it approaches an asymptotic value $\lambda \approx 0.1$. λ^e is an average level of λ (for 2048 signal points) for a Gabor dictionary of functions fitted to white noise.

These considerations indicate that in order to obtain a clear Wigner plot, for our signal consisting of 2048 points, 100–150 atoms should be taken into account. In general this number of atoms depends on the amount of information contained in the signal and the coherence with the chosen dictionary. Addition of more atoms increases mainly noise components. If very weak components of the signal are not of particular interest, it is better not to use too many atoms in the construction of time-frequency representations. In Fig. 15-9 only 60 atoms were used to construct the time-frequency plot (see also Table 15-1).

We have used MP for detection of spindles in the EEG of overnight sleep. The spindles were identified according to the following criteria: frequency 12–15 Hz, time span 64 points. The setting of the intensity threshold merits some attention. In Fig. 15-10, the amplitude of the detected spindles is shown as a function of time (in hours). The time intervals where very few or no spindles were observed corresponded to periods of REM (rapid eye movement) sleep, where, according to present knowledge, no spindles occur. More accurately, the REM epochs correspond to time

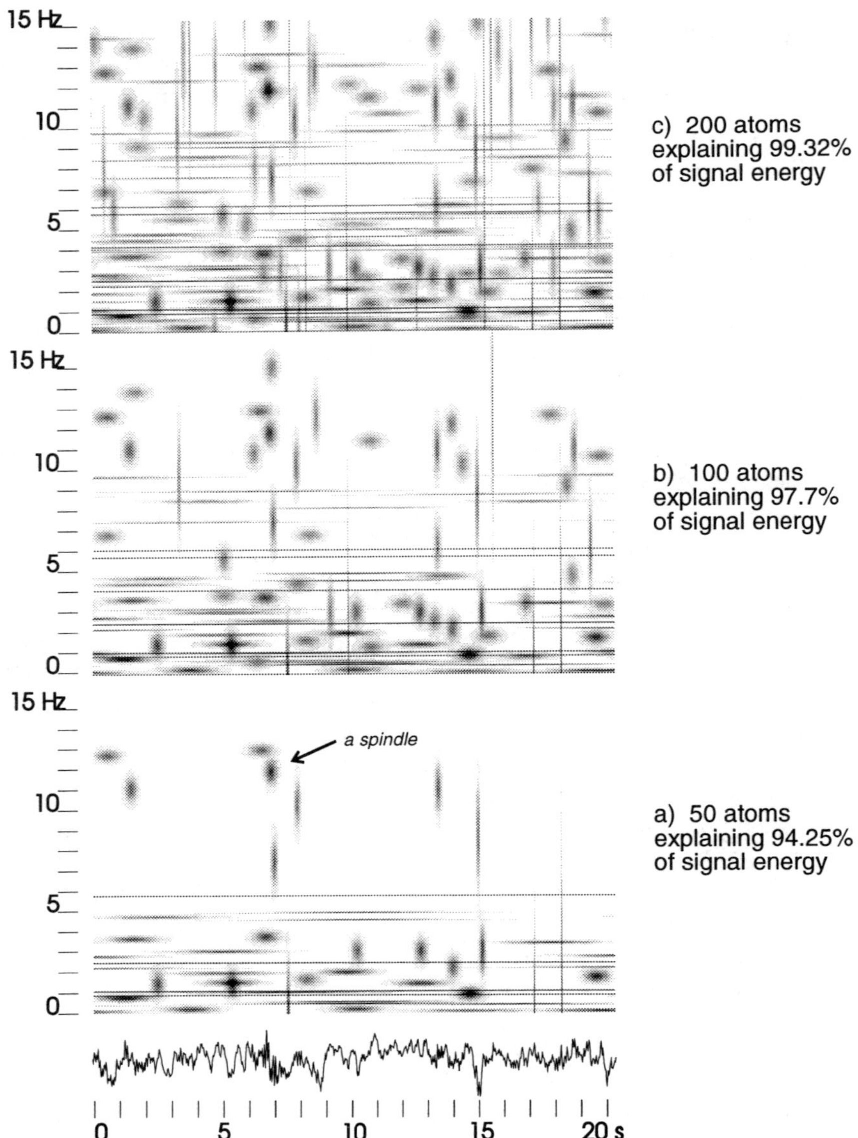

Figure 15-7   Wigner maps of the signal shown below. 2048 data points sampled at 102.4 Hz. Number of atoms shown: (a) 50, (b) 100, (c) 200.

intervals where no spindles of intensity higher than 400 occur. It is interesting that this threshold is connected with the possibility of visual spindle detection. In Fig. 15-9, spindle E has intensity $26\,\mu V$ and lies on the verge of visual identification. It seems that MP provides an extremely sensitive tool for spindle identification which might be helpful in the investigation of mechanisms of their generation and propagation in brain structures.

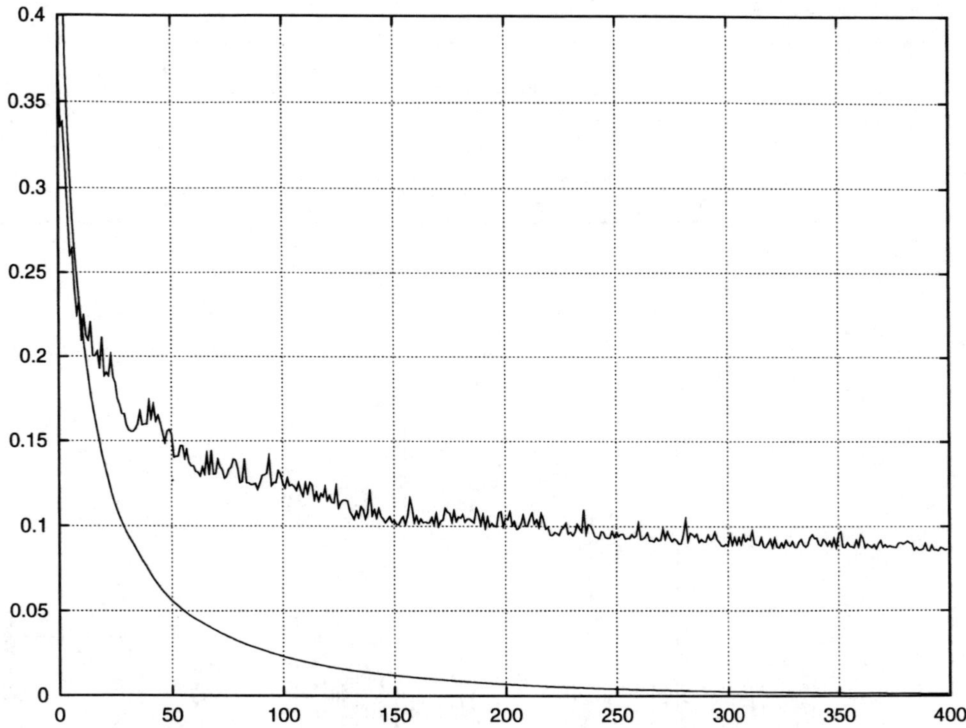

Figure 15-8   Top line—λ as a function of number of atoms taken into account. Bottom line—the amount of energy not accounted for (total energy normalized to 1).

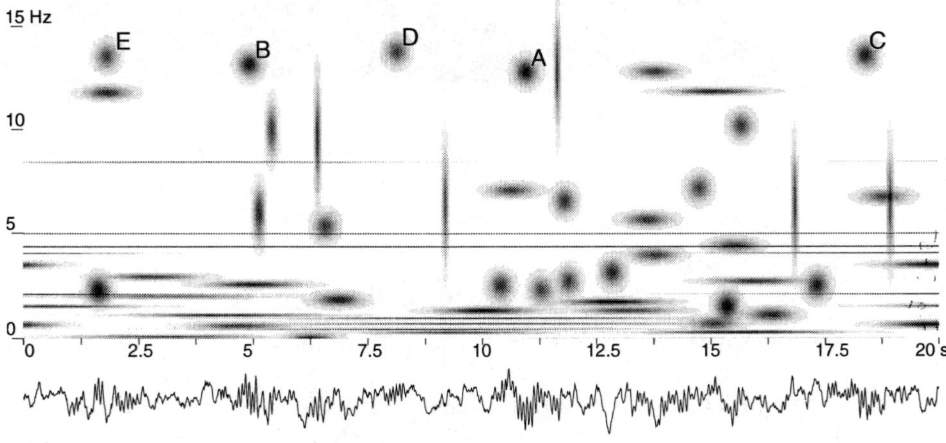

Figure 15-9   A time-frequency analysis obtained by means of MP for the signal shown below (sleep stage 2). The spindles are marked as A, B, C, D, E.

TABLE 15-1   Parameters of spindles presented in Fig. 15-9

	A	B	C	D	E
Amplitude (μV)	82.4	68.4	53.5	30.9	26
Time span (s)	0.59	0.59	0.59	0.59	0.59
Frequency (Hz)	12.8	13.2	13.6	13.8	13.6
Position in time (s)	10.94	4.92	18.36	8.13	1.88

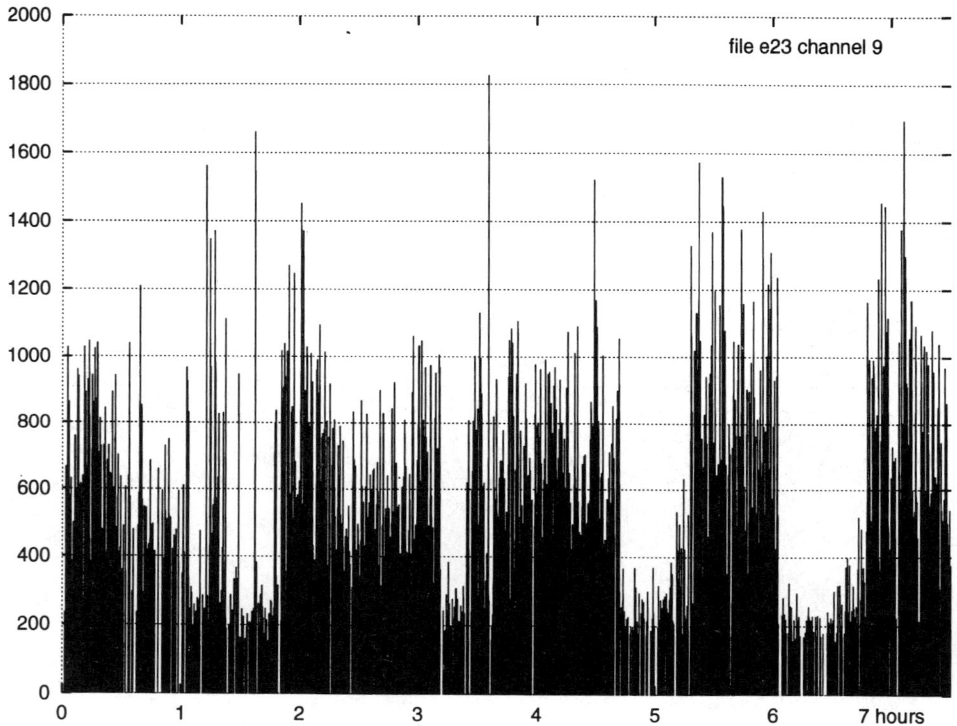

Figure 15-10   Spindles occurring during overnight sleep. Each spindle is marked as a vertical line of length corresponding to its intensity. Horizontal axis—time in hours.

15.4. CONCLUSIONS

The aim of this chapter was to show the possibilities offered by wavelet analysis and MP. The former method provided the possibility of simultaneous description of signals in terms of time and frequency. However, some restrictions connected with the relation between the time and frequency scales effectively limited the resolution. MP overcomes this limitation, offering a maximally adaptive approach and very fine resolution close to the theoretical limit. MP fits the waveforms to each data structure; sometimes this kind of extremely adaptive approach is not desired; e.g., for a comparison of datasets, we would like to have the same basis functions and the same frame, as is the case in WT. Another feature which makes the WT attractive is the simplicity of the algorithm (at least for orthogonal wavelets), keeping the cost of computation low. The MP algorithm requires longer computation time. Wavelet analysis seems to be a good method for time-locked phenomena, but for tracking transients, MP seems to be a better choice.

Another interesting approach to the problem of finding an optimal signal representation is wavelet networks [12]. Together with MP they represent a very promising trend of signal-adaptive techniques which in the near future should yield major advances in nonstationary time series analysis.

ACKNOWLEDGMENTS

We gratefully acknowledge S. Mallat, Z. Zhang, and G. Davis for the "mpp" software package, available through anonymous ftp from host `cs.nyu.edu`. We are grateful to W. Szelenberger for allowing us to use his data and for consultations concerning sleep EEG. We thank E. Kelly for help in preparation of the manuscript. This work was partially supported by KBN grants 3 P401 003 07 and 8T11E01209.

REFERENCES

[1] A. Grossman and J. Morlet, "Decomposition of Hardy functions into square integrable wavelets of constant shape," *SIAM J. Math.* vol. 15, pp. 723–736, 1984.

[2] E. A. Bartnik, and K. J. Blinowska, "Wavelets—A new method of evoked potential analysis," *Med. & Biol. Eng. & Comput.* 30:125–126, 1992.

[3] N. V. Thakor, G. Xin-Rong, S. Yi-Chun, and D. F. Hanley, "Multiresolution wavelet analysis of evoked potentials," *IEEE Trans. BME*, vol. 40, pp. 1085–1093, 1993.

[4] M. Akay, Y. M. Akay, P. Cheng, and H. H. Szeto, "Time frequency analysis of electrocortical activity during maturation using wavelet transform," *Biol. Cybern.* vol. 71, pp. 169–176, 1994.

[5] S. G. Mallat, "A theory of multiresolution signal decomposition: the wavelet representation," *IEEE Trans. Pattern Anal. Machine Intell.*, vol. 11, pp. 674–693, 1989.

[6] E. A. Bartnik, K. J. Blinowska, and P. J. Durka, "Single evoked potential reconstruction by means of wavelet analysis," *Biol. Cybern.* vol. 67, pp. 175–181, 1992.

[7] K. J. Blinowska, P. J. Durka, A. Kołodziejak, and R. Tarnecki, "Application of wavelet transform to the single evoked potentials analysis and reconstruction," *Techn. Health Care*, vol. 1, pp. 344–345, 1993.

[8] I. Daubechies, "The wavelet transform, time-frequency localization and signal analysis," *IEEE Trans. Inform. Theor.*, vol. 36, pp. 961–1005, 1990.

[9] R. R. Coifman, Y. Meyer, S. Quake, and M. V. Wickerhause, "Signal Processing and compression with wavelet packets." In *Progress in Wavelet Analysis and Applications*. Y. Mayer and S. Rouges (ed.), Gif-sur-Yvette: Edition Frontières, 1993.

[10] S. G. Mallat, and Z. Zhang, "Matching pursuit with time-frequency dictionaries," *IEEE Trans. Signal Proc.*, vol. 41, pp. 3397–3415, 1993.

[11] G. Davis, S. Mallat and Z. Zhang, "Adaptive time-frequency decomposition with matching pursuit." In *Wavelets Theory, Algorithms and Applications*. C. Chui, L. Montefusco, and L. Pucio (ed.) Boston: Academic Press, 1994.

[12] Q. Zhang, and A. Benveniste, "Wavelet networks," *IEEE Trans. Neural Networks*, vol. 3, pp. 889–898, 1992.

Chapter 16

EEG Spike Detectors Based on Different Decompositions: A Comparative Study

L. Senhadji, J. J. Bellanger, G. Carrault

16.1. INTRODUCTION

The electroencephalogram (EEG) signal of epileptic patients highlights particular events, the most relevant of which are the seizures. Between seizures, the EEG is characterized by occasional epileptiform transients such as spikes and sharp waves. Because seizures do not occur frequently, the recording might require long-term EEG monitoring. The detection of the interictal events is therefore of particular importance in the characterization of epilepsy and may have a high significance in terms of localization of epileptic foci. The interpretation of the underlying process (propagation routes, synchronism between brain regions) depends on the detection quality of the EEG segments where these transients are present. Gotman [1] gave an overview of the methods designed to recognize and quantify spikes, sharp-waves, and spike-waves. Recent approaches have increased the detection performances by making use of the spatial and temporal context of the EEG [2–4]. Although an effort to automate the detection of epileptiform transients was undertaken, a complete solution has not been found yet. This is mainly due to the wide variety of shapes of these transient signals, their similarities to waves that are part of the background activity, and to impulsive artifacts.

EEG signals observed over a period of time $[0, T]$ can be described, after sampling, by a random process $X(k)$ whose form is

$$X(k) = F(k) + B(k) + \sum_{i=1}^{n_p} P_i(k - \theta_{p_i}) + \sum_{j=1}^{n_a} A_j(k - \theta_{A_j}) + B(k); \, k \in \{0, 1, \ldots, T\}$$

$$= (F(k) + B(k)) + P(k) + A(k) \tag{16-1}$$

This relation depicts relevant activities (elementary waves, background activity, noise, artifacts, etc.) which constitute the signal. $F(k)$ may be considered as a piece-wise stationary signal present either casually or over the whole duration of the observation; for each i, P_i represents a brief duration potential, with time occurrence θ_{P_i}, and corresponds to an abnormal neural discharge; the A_j terms may be related to artifacts occurring at times θ_{A_j}; finally, measurement noise which can be considered as stationary over the observation duration is described by $B(k)$. Over the period of observation, the entities n_p and n_a represent respectively the number of temporal occurrences of brief useful events and artifact transient signals.

Cerebral background activity includes basic activities (alpha, beta, delta, gamma) as well as ictal stationary periods of time (recruitment phase during an epileptic seizure for example) and is modeled here by $F(k)$. The distinction between the A_j and P_j components depends on the goals of the study: in our case, the epileptic events to detect are described by the P_j terms; accordingly, transitory waves associated to sleep, vertex sharp transients or K complexes, belong to the set of artifacts. Whatever the clinical objectives, the transient signals generated by eye movements or electrode shifts are represented by A_j terms.

From a signal-processing point of view, the detection of spikes and sharp-waves can be seen as a classical detection problem where, at each point in time, the hypothesis "presence of spike" is confronted with its opposite hypothesis. The difficulties encountered here come from the time-varying characteristics of both relevant signals (not perfectly known signals) and noises (superposition of transient artifacts and locally stationary activities), and the unknown firing rate of the interictal events. To face the composite structure of the noises and the inherent nonstationary character of the spikes and artifacts, we proposed in [5,6] a two-stage detection scheme based on a time-scale representation of the observation X under the following hypothesis:

- $F + B$ is approximately stationary on $\{0, \ldots, T\}$ and is a zero mean gaussian signal with unknown covariance matrix.
- The frequency range of the artifacts (A_j) is statistically higher than the frequency band of the spikes (P_i).
- The transients being sparse events in the observation, first- and second-order statistics of $F + B$ can be learned using X.
- The shapes of A_j and P_i are random, only their duration is approximately known.

The proposed detection structure in [5] is based on two decision stages N_1 and N_2: the first one is a quadratic imposed structure aimed at the detection of P_i and the second one is aimed at the rejection of false alarms (of N_1) due to A_j. N_1 and N_2 use a wavelet filter bank determined heuristically.

In this chapter we compare the stage N_1 of the proposed detector with other quadratic detectors (with imposed structure or not) that make more or less use of information on transients and background activity. The problem is stated in the next section and the solution proposed in [5] is briefly presented. The new solutions for N_1 are introduced in section 16.3. Experimental results and discussions are then reported.

16.2. PROBLEM STATEMENT

When dealing with the detection, over an observation time interval $[0, T]$, of an unknown number of transient signals whose supports are disjoined and arrival times are unknown, a classical sub-optimal solution, which we use here, is to consider the detection task as a sequence of elementary detection problems, each consisting in the detection of the presence of one transient in a short observation window covering the time support of the expected event. A series of tests of the same type are conducted on a set of observation windows with equal length L, the union of them being the whole observation time interval. More precisely:

1. We define the following vectors

$$\mathbf{X}(k) = [X(k-L+1), \ldots, X(k)]^t, k \in \{L-1, \ldots, T\}$$
$$\mathbf{F}(k) + \mathbf{B}(k) = [F(k-L+1) + B(k-L+1), \ldots, F(k) + B(k)]^t$$
$$\mathbf{P}(k) = [P(k-L+1), \ldots, P(k)]^t$$
$$\mathbf{A}(k) = [A(k-L+1), \ldots, A(k)]^t$$

All these vectors are built, for each k, in the same way, with L consecutive samples extracted respectively from the observation X, the sum of instrumental noise B and background activity F, the signal describing the spikes and the artifacts

2. For each k, we have to decide between two hypotheses:

$$H_{0,k} : \mathbf{X}(k) = \mathbf{F}(k) + \mathbf{B}(k) \qquad \text{or } \mathbf{X}(k) = \mathbf{F}(k) + \mathbf{B}(k) + \mathbf{A}(k)$$
$$H_{1,k} : \mathbf{X}(k) = \mathbf{F}(k) + \mathbf{B}(k) + \mathbf{P}(k)$$

which are mutually exclusive because we neglect the probability of $\mathbf{A}(k) \neq 0$ and $\mathbf{P}(k) \neq 0$ on $\{k-L+1, \ldots, k\}$. The hypothesis $H_{0,k}$ can be decomposed in $H_{0,k} = H'_{0,k} \cup H''_{0,k}$ such that:

$$H'_{0,k} : \mathbf{X}(k) = \mathbf{F}(k) + \mathbf{B}(k)$$
$$H''_{0,k} : \mathbf{X}(k) = \mathbf{F}(k) + \mathbf{B}(k) + \mathbf{A}(k)$$

When assuming that the distribution of $F + B$ is known (i.e., given or estimated with enough precision), the hypothesis $H'_{0,k}$ is simple while the hypotheses $H''_{0,k}$ and $H_{1,k}$ are composite because the shape of the transients and their arrival times are governed by a priori unknown distributions.

3. We introduce here a series of tests T_k between the hypothesis $H_{0,k}$ and $H_{1,k}$ and the corresponding series of decision variables δ_k which are set to 0 if $H_{0,k}$ is true and to 1 otherwise.

4. A "fusion" procedure on δ_k is then used to ensure that: (1) one transient signal leads at most to one detection; (2) the number of false alarms is not greater than 1 for a series $\{k1, \ldots, k2\}$ of decision instants such that $k2 - k1$ is less than a given value (the minimal time duration of the spikes).

16.3. DESCRIPTION OF THE TEST T_1

The heuristic detection algorithm proposed in [5] avoids the direct classical construction of the tests T_k. It uses a wavelet filter bank, the outputs of which are processed by a two-stage decision scheme. The first stage N_1 uses a series of identical tests denoted T_1 between the hypothesis $H_{0,k}'$ and $H_{1,k}$ by ignoring the hypothesis $H_{0,k}''$. For each k we have $\delta_k = 0$ ($H_{0,k}'$ true) if $S_1(\mathbf{X}(k))\,\lambda_1 <$ otherwise $\delta_k = 1(H_{1,k}$ true). S_1 is a quadratic form and λ_1 is a threshold determined adaptively based on the empirical distribution of $S_1(\mathbf{X}(k))$, k belonging to $\{L, \ldots, T\}$. The second stage, N_2, is aimed at the elimination of the false alarms due to the presence, at the input of N_1, of artifacts A_j. This stage has already been reported and will not be presented in this chapter. It has been shown experimentally (for detailed description see [5,6]) that most of the artifacts are removed through N_2.

The decision statistic, associated to the proposed detector is $S_1(\mathbf{X}(k)) = \|\, \mathbf{FX}(k)\|^2$ where $\mathbf{F}$ is a matrix such that $\mathbf{F} = [\mathbf{F}_1^t, \mathbf{F}_2^t, \ldots, \mathbf{F}_M^t]^t$ (t denotes the matrix transposition) has M rows $\mathbf{F}_i, i = 1, \ldots, M$ which are finite impulse response (FIR) filters of the same length L and where $\|\, \mathbf{Z}\, \|^2$ is equal to $\mathbf{Z}^t\mathbf{Z}$ for $\mathbf{Z}$ vector belonging to $\mathbb{R}^L$. The detection test T_1 built in N_1 is of quadratic form (Fig. 16-1) i.e., $\|\mathbf{Y}(k) = \mathbf{FX}(k)\|^2 \gtrless \lambda_1$.

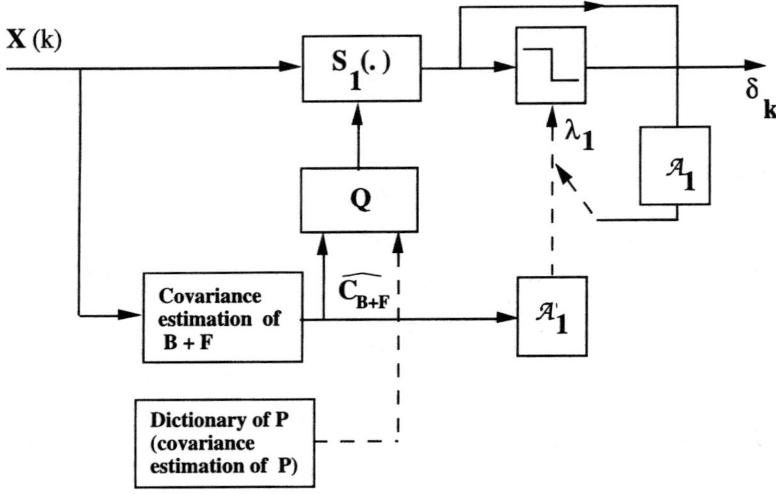

Figure 16-1   Block diagram of the quadratic detection structure.

As the transients P_i and A_j are brief and occur in the observation as "details," well localized in time-scale space, the matrix $\mathbf{F}$ was built based on a wavelet transform [7]. The $\mathbf{F}_i$ are then the sampled versions of the functions $\psi_{a_i}(t) = \frac{1}{\sqrt{a_i}} \psi\left(\frac{t}{a_i}\right)$, $i = 1, \ldots, M$, where a_i are scale values and ψ is a complex analyzing wavelet defined by:

$$\psi(t) = C \cdot (1 + \cos 2\pi f_0 t) \cdot e^{2i\pi k f_0 t}, \ |t| \leq 1/2f_0, \ k \text{ integer} \notin \{-1, 0, 1\}$$

The parameter k sets up the number of oscillations of the complex part (admissibility conditions are verified for k different from $-1, 0, 1$), f_0 is the normalized frequency and C is a normalization coefficient ($\|\psi\| = 1$). The wavelet transform is then exploited as a linear feature extraction procedure to obtain relevant time-scale atoms where the spikes P_i are located. The $\mathbf{F}_i$ were determined in [5] based on visual inspection of a large amount of interictal EEG data processed by wavelets (different patients cand different interictal periods). The adaptive strategy presented in [5] allows a good control of the false alarm probability of T_1 both on simulated and real data. $S_1(\mathbf{X}(k))$ can be expressed by the quadratic form $\mathbf{X}(k)^t \mathbf{Q} \mathbf{X}(k)$ where $\mathbf{Q} = \mathbf{Q_o} = \mathbf{F}^t\mathbf{F}$. The associated test is then denoted as $T_{1,o}$. Obviously, such construction of the decision test $T_{1,o}$ exploits heuristically the morphology of P_i to determine the matrix $\mathbf{F}$ and does not make full use of the statistical properties of $F + B$ that can be achieved by learning on $\mathbf{X}$. The main concern of this contribution is to propose alternative forms of S_1, in other words new matrices $\mathbf{Q}$ by introducing more prior or learned information. For this purpose, a set of N_E sampled spikes $\hat{P}_r \in \mathbb{R}^L$, $r = 1, \ldots, N_E$ was selected manually from a real digitized EEG channel to build a dictionary of sampled spikes which is used to estimate the covariance matrix $\mathbf{C_p}$ of P_i.

16.4. VARIATIONS OF S_1

All the proposed forms of S_1 are expressed by a quadratic function $S_1(\mathbf{X}(k)) = \mathbf{X}(k)^t \mathbf{Q} \mathbf{X}(k)$, where $\mathbf{Q}$ is set according to the selected criteria (each criterion gives to the test $S_1(\mathbf{X}(k)) \gtrless \lambda_1$ a statistical signification). In the following, $F + B$ will be replaced by B for simplification and the distribution assumed to be Gaussian with the estimated covariance matrix $\widehat{\mathbf{C}_B}$.

16.4.1 Detectors Built Without Using the Spike Waveform

Null Hypothesis Rejection Test : $T_{1,\mathrm{N}}$. The only required information on the signal is its duration. The test is based on the distance between the observation $X(k)$ and a null reference for the metric associated to $\widehat{\mathbf{C_B}}^{-1}$. $\widehat{\mathbf{C_B}}$ is an estimate of the expectation of $\mathbf{X}(k)\mathbf{X}(k)^t$ (i.e., $\mathbb{E}[\mathbf{X}(k)\mathbf{X}(k)^t]$) conditionally to the hypothesis $H_{0,k}'$. The resulting quadratic form is obtained by setting $\mathbf{Q}$ to $\mathbf{Q_N} = \widehat{\mathbf{C_B}}^{-1}$,

Null Hypothesis Rejection Test Applied to $Y(k) : T_{1,o}'$. Q is set to $\mathbf{F}^t \widehat{\mathbf{C}_y}^{-1} \mathbf{F}$ where $\widehat{\mathbf{C_y}}$ is an estimate of the expectation of $\mathbf{Y}(k)\mathbf{Y}(k)^t$ under the null hypothesis $H_{0,k}'$. $T_{1,o}'$ may be interpreted as $T_{1,\mathrm{N}}$, with $\mathbf{Y}$ instead of $\mathbf{X}$.

16.4.2 Detectors Based on Objective Knowledge on P_i (Other Than Their Time Duration)

In this section, the noise B and spikes P_i are supposed to be independent. With the help of the dictionary, the covariance matrix $\mathbf{C_p}$ of the transients P_i is estimated according to:

$$\widehat{\mathbf{C_p}} = \frac{1}{N_E} \sum_{r=1}^{N_E} \tilde{\mathbf{P}}_r \tilde{\mathbf{P}}_r^t$$

Neyman–Pearson Detector: $T_{1,\mathrm{NP}}$. This detector is based on the likelihood ratio calculated assuming that the spikes are normally distributed with the estimated covariance matrix $\widehat{\mathbf{C_p}}$. The log-likelihood ratio is equal to $\mathbf{X}'(k)\mathbf{Q_{NP}}\mathbf{X}(k)$ (up to an additive constant) where $\mathbf{Q_{NP}} = \widehat{\mathbf{C_B}}^{-1} - \left(\widehat{\mathbf{C_B}} + \widehat{\mathbf{C_p}}\right)^{-1}$.

Deflection Criteria Detector: $T_{1,\mathrm{DC}}$. The deflection associated to $S_1(\mathbf{X}(k))$ is defined by:

$$D(S_1) = \frac{(\mathbb{E}_1[S_1] - \mathbb{E}_0[S_1])^2}{\mathrm{var}_0[S_1]}$$

where $\mathbb{E}_i[S_1]$ and $\mathrm{var}_i[S_1]$ are the expectation and the variance of S_1 conditionally to the hypothesis $H_{i,k}$; $i = 0, 1$ respectively. The matrix $\mathbf{Q}$ is obtained by maximizing the deflection $D(S_1)$. Consequently $\mathbf{Q}$ is equal to $\mathbf{Q_{DC}} = \mathbf{C_B}^{-1}\widehat{\mathbf{C_P}}\mathbf{C_B}^{-1}$.

Generalized Likelihood Ratio Detector: $T_{1,\mathrm{GLR}}$. The detector relies on the composite hypothesis $H_{1,k}$ which is defined by "P_i is present and is localized in a given d-dimensional subspace, denoted E_p, of $\mathbb{R}^L$." The generalized log-likelihood ratio between $H_{0,k}$ and $H_{1,k}$ leads to the statistic $S_1(\mathbf{X}(k))$ where the matrix $\mathbf{Q}$ is set to:

$$\mathbf{Q} = \mathbf{Q_{GLR}} = \widehat{\mathbf{C_B}}^{-1}\mathbf{G}\left[\mathbf{G}^t\widehat{\mathbf{C_B}}^{-1}\mathbf{G}\right]^{-1}\mathbf{G}^t\widehat{\mathbf{C_B}}^{-1}$$

where $\mathbf{G}$ is a matrix, of d columns and L rows, such that the column vectors span the space E_p. The hypothesis that the spikes are elements of E_p is approximately verified if we suppose that the whole energy of the transient signals is localized in the subspace spanned by d ($d < L$) eigenvectors corresponding to the d largest eigenvalues of the covariance matrix $\mathbf{C_p}$.

Stochastic Extension of the Matched Filter. The method described in [8] considers the signal to be detected as random and relies on a quadratic detection statistic $S_1(\mathbf{X}(k))$ with imposed form. Moreover, $S_1(\mathbf{X}(k)) = \|\mathbf{F}'\mathbf{X}(k)\|^2$ where $\mathbf{F}' = [\mathbf{F}_1'^t, \mathbf{F}_2'^t, \ldots, \mathbf{F}_{M'}'^t]^t$ is a bank of M' filters defined by the M' eigenvectors associated to the M' greatest eigenvalues of the matrix $\mathbf{C_B}^{-1}\mathbf{C_P}$. For $M' = 1$, $\mathbf{F}'\mathbf{X}(k)$ is the linear form which maximizes a deflection criterion. When $M' > 1$ and assuming that the spike distribution is Gaussian, the authors argue that $S_1(\mathbf{X}(k))$ is an approximation of the likelihood ratio. As conducted, the method supposes that the transients P_i

are stationary upon the time support of $\mathbf{X}(k)$ and that their time locations are uniformly distributed over the time interval $[0, T]$, which lead to covariance matrices of stationary structures. The estimated covariance matrix is then a Toeplitz matrix noted $\widehat{\mathbf{C}_{\mathbf{PS}}}$. According to our notations, $\widehat{\mathbf{C}_{\mathbf{PS}}}$ is proportional to the square matrix of dimension L whose elements are given by $\widehat{\mathbf{C}_{\mathbf{PS}}}(i,j) = \sum_{r=1}^{N_E} \sum_k \widetilde{\mathbf{P}}_{\mathbf{r}}(k)\widetilde{\mathbf{P}}_{\mathbf{r}}(k - (i - j))$. The above method was implemented and a modified version was proposed based on the following remarks: for the detection problem, we introduced the hypothesis $H_{1,k}$ which assumes implicitly that the whole energy of the spike is concentrated around the midpoint of $\mathbf{X}(k)$. In other words, the spike is centered over the time support of $\mathbf{X}(k)$. The spikes are then nonstationary random signals and the estimate $\widehat{\mathbf{C}_{\mathbf{PS}}}$ is replaced by $\widehat{\mathbf{C}_{\mathbf{P}}}$. The original and modified statistics are associated to the tests $T_{1,\mathrm{SMF}}$ and $T'_{1,\mathrm{SMF}}$, respectively.

16.5. EXPERIMENTATION AND PERFORMANCE EVALUATION

The performance of the tests described above has been evaluated and compared for simulated data using receiver operating curves (ROCs) and modified ROC curves [9]. For each value of the decision threshold λ_1, the associated ROC point represents the probability PD for detecting a signal P_i, which is present in an observation window of length L, versus the false alarm probability (FAP) defined by: $Pr[S_1(\mathbf{X}(k)) > \lambda_1/H_{0,k}]$, while the modified ROC curves represent the evolution of PD versus the false alarm rate $FA\tau$ which corresponds, for large values of λ_1, to the mean value of the number of λ_1 up crossing of $S_1(\mathbf{X}(k))$ per time unit.

A dictionary of $N_E = 100$ sampled spikes, extracted manually from an interictal EEG signal recorded using the standard protocol 10–20 and a sampling frequency equal to 200 Hz, was built and the covariance estimates were computed. According to the time duration of the spikes and to the frequency sampling, the observation vector dimension L was set to 60 for all the statistics except for those based on wavelets ($T_{1,0}$ and $T'_{1,0}$) where the retained a_i's led to $L = 30$. For the test $T_{1,\mathrm{GLR}}$, the dimension of the space generated by the covariance matrix of the spikes was set to $d = 4$. Hence the number of filters used for both original and modified stochastic extension of the matched filter is $M' = 4$. These choices were a compromise between the calculus complexity of $T_{1,k}$ at each step k and the *a priori* information on the signal and noise used (number of eigenvectors selected for $\widehat{\mathbf{C}_{\mathbf{P}}}$ and $\widehat{\mathbf{C}_{\mathbf{B}}^{-1}\mathbf{C}_{\mathbf{P}}}$).

Before presenting experimental results, we must emphasize that the validation of detection methods like those mentioned above is a very difficult task. The main problems encountered are as follows.

1. The definition of a set of EEG recordings to build a database with clinical significance. Signals are generally recorded over a long time period (several days) with a large number of electrodes, and they depend on the relative positions between cortical sources and electrodes. On the other hand, the shapes of epileptiform transients vary with patients and may change from one interictal period to another.

2. The labeling of the signal components involves the detection of epileptiform transients by a visual inspection of the recordings and validation by different phy-

sicians. Such a procedure is subjective and expert-dependent. It may discard some transients which could be detected in other circumstances or by another group of experts.

A complete evaluation of the above-presented detectors for real data is far from being straightforward. Performance was studied mainly using simulation (artificially generated EEG) and preliminary results were obtained on real data.

16.6. RESULTS AND DISCUSSION

For performance analysis, artificial EEG signals were generated based on the following method: background activities were generated using autoregressive (AR) modeling of real EEG stationary periods without transients. 24 models were estimated based on classical criteria [10]. Abnormal EEGs were obtained by superimposing spikes periodically, randomly selected over the dictionary. An example of the simulated signal is presented in Fig. 16-2. Based on the AR model's coefficient the matrices C_B were calculated (one model corresponds to one matrix). Both FAP and false alarm rate were measured on simulated background activities (204 800 samples for each model). Two representative models were selected to illustrate the behavior of the detectors. Both ROCs and modified ROCs exhibit a similar hierarchy between the detection statistics. Over all the models, three groups can be distinguished. The best results are obtained for the Neyman–Pearson test (the best), the deflection criteria test, and the generalized likelihood test. The performance decreases when imposed quadratic structures (i.e., wavelet and stochastic matched filter) are used. The null hypothesis test leads to an intermediate level of performance. Among all the models, the modified stochastic matched filter extension detector as well as the modified wavelet transform detector performed better than their associated original versions (Figs 16-3(a) and (b) and 16-4(a) and (b)).

The tests on real data were conducted on the EEG signal from which the spike dictionary was built. C_B was estimated on a spike-free segment of the signal, while the estimates of $\widehat{C_P}$ and $\widehat{C_{PS}}$ are those used on synthetic EEG data. Figure 16-5 depicts the behavior of the detection statistics for an interictal EEG period of 10 s. The performance evaluations were made only through ROCs. Figure 16-6 shows the behavior of the studied tests. The results indicate a new performance-based hierarchy. The Neyman–Pearson test and the null hypothesis rejection test are the worst methods. None of the tests perform better than the others for all FAP values. However, two main points can be emphasized. Firstly, the introduction of the estimated covariance matrix $\widehat{C_y}$ in the detector using wavelets (T_{10}) significantly decreases the performance especially for small values of FAP, while the use of $\widehat{C_P}$ in place of $\widehat{C_{PS}}$ improves the performance of the stochastic extension of the matched filter ($T'_{1,SMF}$). Secondly, the test $T_{1,o}$ produces satisfactory results regarding the required *a priori* information (i.e., time-scale atoms where the spikes are located).

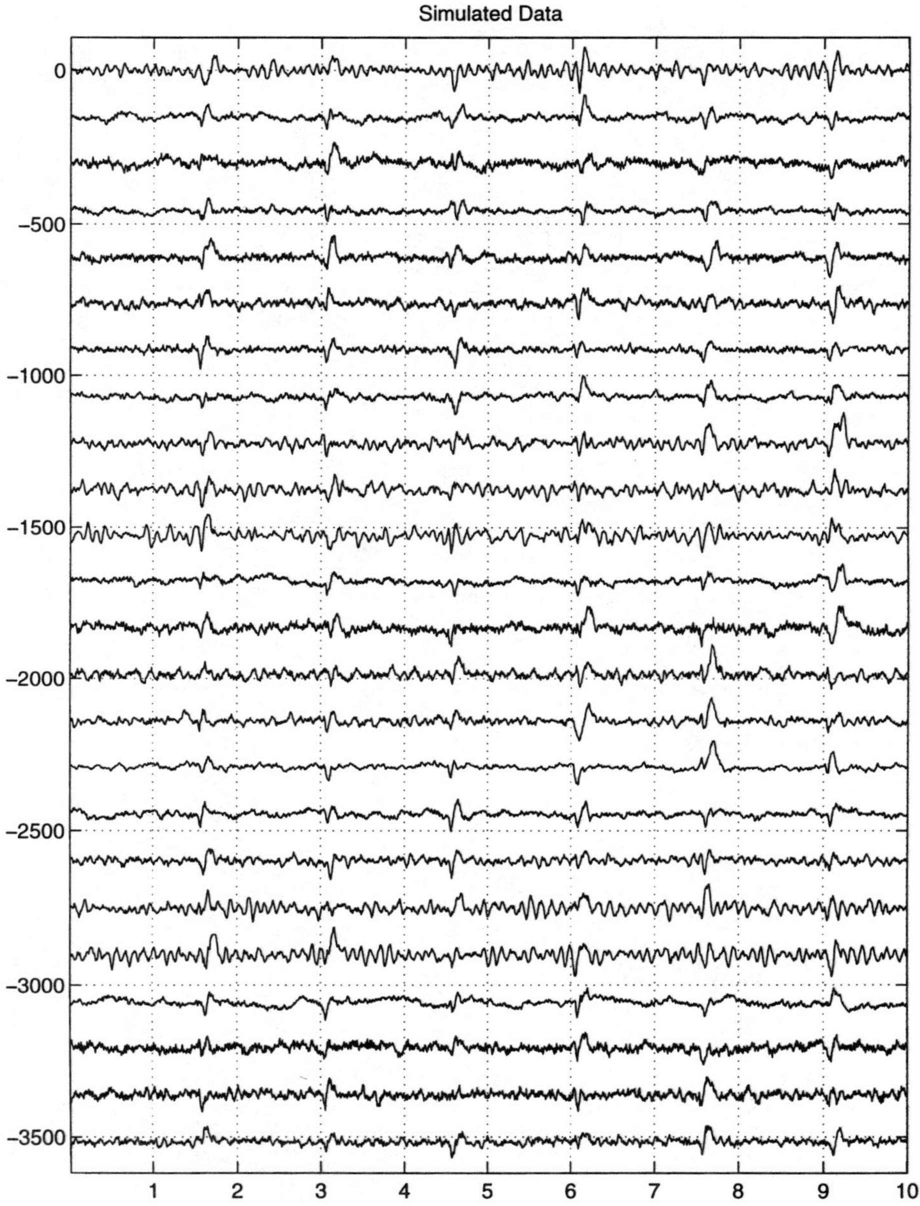

Figure 16-2   Examples of 10 s simulated EEG signals: from top to bottom, 24 signals corresponding to 24 AR models, the spikes were added periodically (every 300 samples).

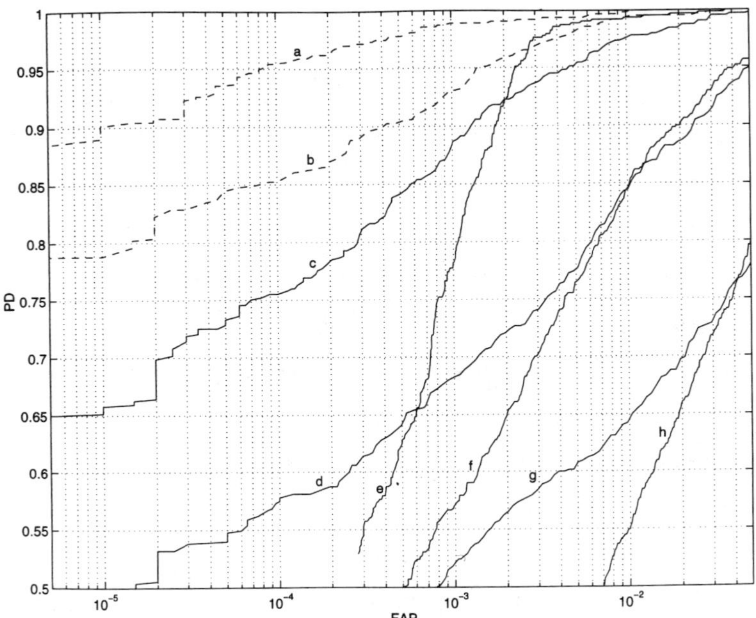

Figure 16-3(a) ROCs for the model 11: (a) $T_{1,\mathrm{NP}}$ (b) $T_{1,\mathrm{DC}}$; (c) $T_{1,\mathrm{GLR}}$; (d) $T'_{1,\mathrm{SMF}}$; (e) $T_{1,N}$; (f) $T'_{1,\mathrm{o}}$; (g) $T_{1,\mathrm{SMF}}$; (h) $T_{1,\mathrm{o}}$.

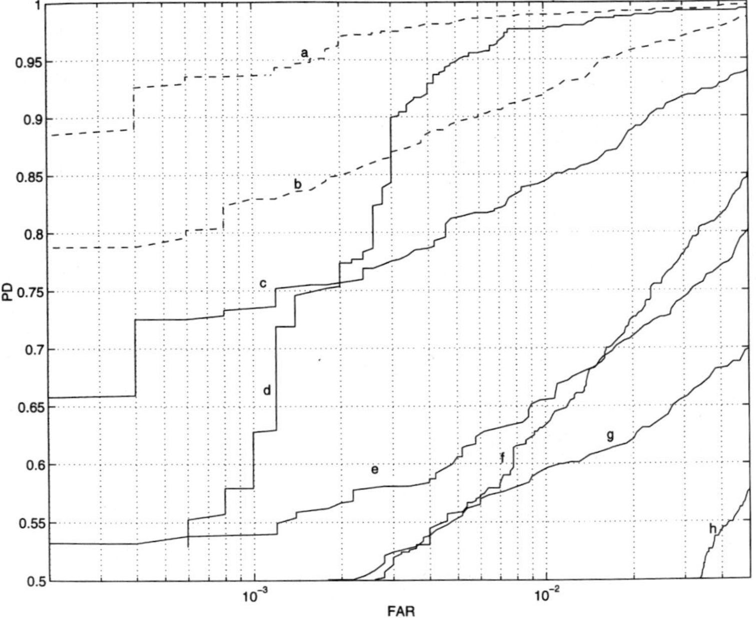

Figure 16-3(b) Modified ROCs for the model 11: (a) $T_{1,\mathrm{NP}}$; (b) $T_{1,\mathrm{DC}}$; (c) $T_{1,\mathrm{GLR}}$; (d) $T_{1,\mathrm{N}}$;(e) $T'_{1,\mathrm{SMF}}$; (f) $T'_{1,\mathrm{o}}$; (g) $T_{1,\mathrm{SMF}}$; (h) $T_{1,\mathrm{o}}$.

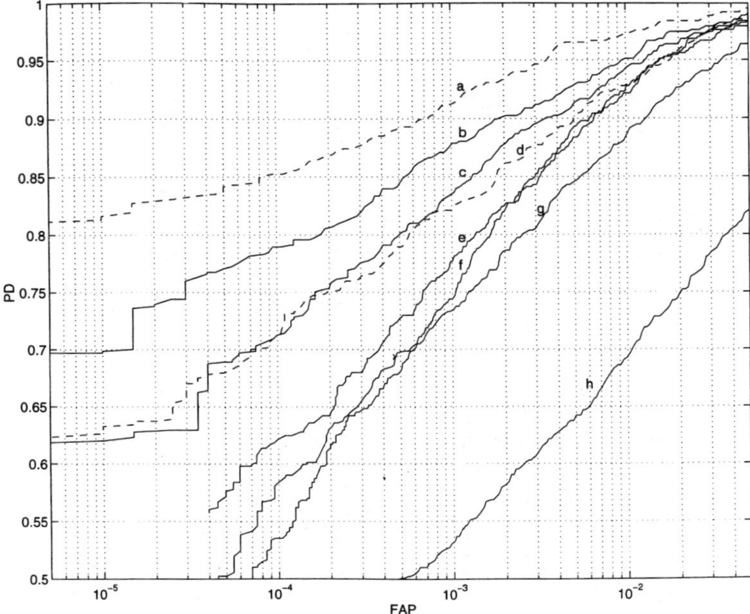

Figure 16-4(a)   ROCs for the model 14: (a) $T_{1,\text{NP}}$; (b) $T_{1,\text{GLR}}$; (c) $T_{1,\text{N}}$; (d) $T_{1,\text{DC}}$; (e) $T'_{1,\text{o}}$; (f) $T_{1,\text{o}}$; (g) $T'_{1,\text{SMF}}$; (h) $T_{1,\text{SMF}}$.

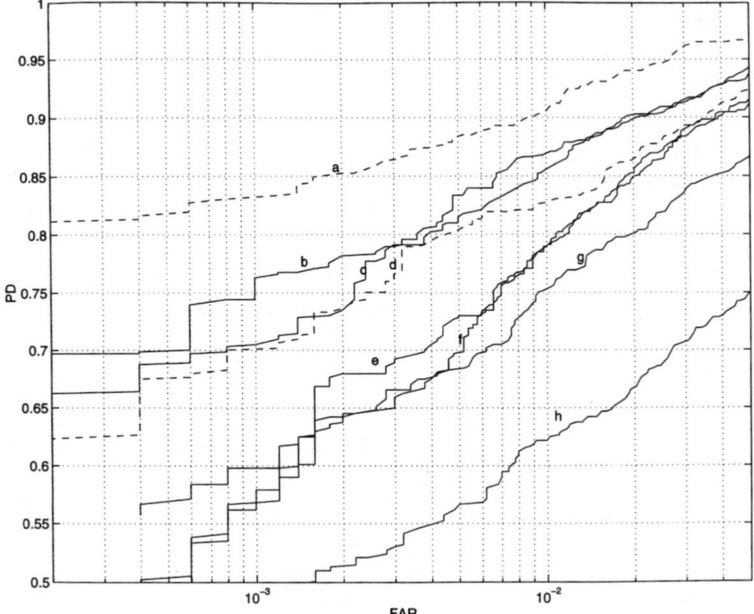

Figure 16-4(b)   Modified ROCs for the model 14: (a) $T_{1,\text{NP}}$; (b) $T_{1,\text{GLR}}$; (c) $T_{1,\text{N}}$ (d) $T_{1,\text{DC}}$; (e) $T'_{1,\text{o}}$, (f) $T_{1,\text{o}}$; (g) $T'_{1,\text{SMF}}$; (h) $T_{1,\text{SMF}}$.

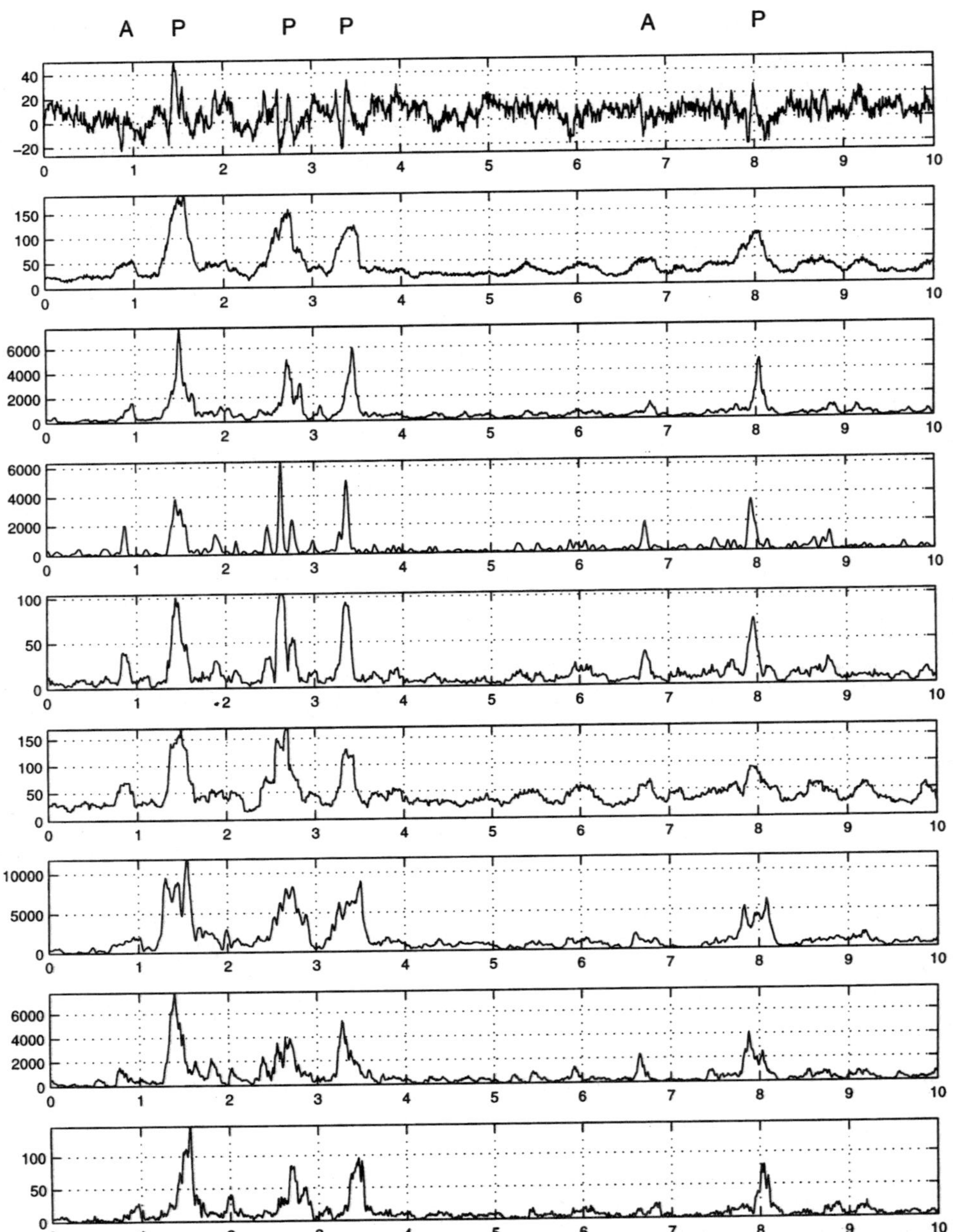

Figure 16-5 Detection statistics behavior on real EEG signal: A—artifact, P—spike. From top to bottom: raw EEG signal, $T_{1,NP}$, $T_{1,DC}$, $T_{1,o}$, $T'_{1,o}$ $T_{1,N}$, $T_{1,SMF}$, $T'_{1,SMF}$ and $T_{1,GLR}$.

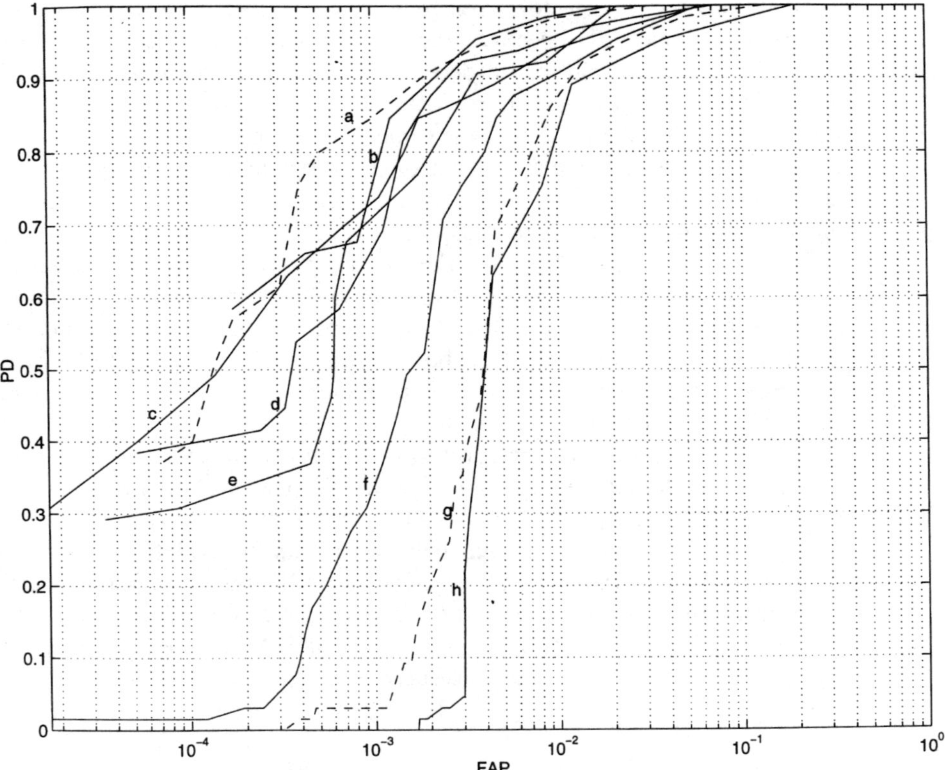

Figure 16-6   ROCs obtained on a real EEG channel (a) $T_{1,\mathrm{DC}}$; (b) $T'_{1,\mathrm{SMF}}$; (c) $T_{1,\mathrm{o}}$; (d) $T_{1,\mathrm{GLR}}$; (e) $T_{1,\mathrm{SMF}}$; (f) $T'_{1,\mathrm{o}}$; (g) $T_{1,\mathrm{NP}}$; (h) $T_{1,\mathrm{N}}$.

16.7. CONCLUSION

Different quadratic methods for detecting transients embedded in stationary background EEG activity were compared using simulated EEG. Some methods make use of objective information on the morphology of the waves to be detected (based on a dictionary of transient prototypes). All of the proposed structures, except the one proposed in [5], require an estimated covariance matrix of the background activity. Performance analysis in terms of both ROCs and modified ROCs, conducted on artificial interictal EEGs shows that a significant improvement of the quadratic tests performance is achieved when using the nonstationary estimate of the spike covariance matrix. The original version of the stochastic extension of the matched filter cannot be used as it is. The appropriate modification of the spike covariance estimate leads to better result. However, its performance remains lower than those obtained by means of the tests $T_{1,\mathrm{NP}}$, $T_{1,\mathrm{GLR}}$ and $T_{1,\mathrm{DC}}$. The behavior of the wavelet tests $T_{1,\mathrm{o}}$, $T'_{1,\mathrm{o}}$ is similar to $T'_{1,\mathrm{SMF}}$. The first attempt made on real EEG data shows some modifications in the detectors hierarchy. However, we note that the best methods require an estimate of the covariance matrices of both background activity and spikes. Thus, the wavelet-based detector $T_{1,0}$ is more attractive because such estimates are not required and its performance level is satisfactory. Furthermore, the procedure developed in [5] using the same wavelet decomposition scales to reject the artifacts detected due to the presence of A_j and which leads to good results, cannot be easily replaced by classical detectors because, in practice, the morphological information on artifacts at our disposal is not precise enough. On real EEG data, the Neyman–Pearson detector and the null hypothesis rejection test exhibit the worst performances. Both tests assume that the background activity is stationary and normally distributed and the first ($T_{1,\mathrm{NP}}$) supposes that the P_i are Gaussian. Real signals may not comply with the above assumptions and it would be interesting to investigate more extensively the domain of validity of such hypotheses and to study the robustness under non-Gaussianity.

The determination of threshold (λ_1) has not been discussed in this chapter. It can be evaluated adaptively, as presented in [5], to control the FAP based on the $T_{1,.}$ samples (algorithm $\mathcal{A}_1$ in Fig. 16-1). Another approach is possible if the probability density law of $T_{1,.}$ under the null hypothesis, can be analytically determined as a function of the covariance matrix $\mathbf{C_B}$. In fact, using the analytical expression of the law, a threshold value is computed for a given FAP (algorithm $\mathcal{A}'_1$ in Fig. 16-1). To control the false alarm rate, the algorithms $\mathcal{A}_1$ and $\mathcal{A}'_1$ require the joint density of $T_{1,.}$ and its variation between two consecutive samples.

REFERENCES

[1] J. Gotman, "Computer analysis of the EEG in epilepsy." In *Handbook of Electroencephalography and Clinical Neurophysiology* (revised series), vol. 2, F. H. Lopes Da Silva, W. Storm Van Leeuwen and A. Rémond (ed.) Amsterdam: Elsevier, pp. 171–204, 1985.

[2] R. J. Glover, N. Raghavam, P. Y. Ktonas, and J. D. Frost, "Context-based automated detection of epileptogene sharp transient in the EEG: elimination of false positives." *IEEE Trans. BME*, vol. 36, pp. 519–527, 1989.

[3] J. Gotman and L. Y. Wang, "State-dependent spike detection: concepts and preliminary results." *Electroenceph. Clin. Neurophysiol.*, vol. 79, pp. 11–19, 1991.

[4] A. A. Dingle, R. D. Jones, G. J. Carroll, and W. R. Fright, "A multistage system to detect epileptiform activity in the EEG," *IEEE. Trans. BME.* vol. 40, pp. 1260–1268, 1993.

[5] L. Senhadji, J. J. Bellanger, and G. Carrault, "Détection temps-echelle d'événements paroxystiques intercritiques en eléctroencéphalographie," *Traitement du Signal*, vol. 12, pp. 357–371, 1995.

[6] L. Senhadji, J. L. Dillenseger, F. Wendling, C. Rocha, and A. Kinie, "Wavelet analysis of EEG for three-dimensional mapping of epileptic events." *Ann. Biomed. Eng.*, vol. 23, pp. 543–551, 1995.

[7] R. K. Martinet, J. Morlet, and A. Grossmann, "Analysis of sound patterns through wavelet transform." *Jour. Pat. Rec. Art. Intel.*, vol. 1, pp. 273–301, 1987.

[8] J. F. Cavassilas, and B. Xerri, "Extension de la notion de filtre adapté. Contribution à la détection de signaux courts en présence de termes perturbateurs." *Traitement du Signal*, vol. 10, pp. 215–221, 1993.

[9] P. Y. Arques, "Décisions en traitement du signal." *Collection Cnet/Enst*, Paris: Masson, 1979.

[10] S. L. Marpel, "Digital spectral analysis with applications," A. V. Oppenheim (ed.), Englewood Cliffs, NJ: Prentice-Hall, 1987.

PART III

Wavelets and Medical Imaging

In this part, we will focus on the two-dimensional (2-D) and three-dimensional (3-D) wavelet analysis methods and their applications to medical images.

Chapter 17 by Koren and Laine reviews the one-dimensional and multidimensional discrete dyadic wavelet transform methods with their implementations and applications to digital mammography.

Chapter 18 by Schuler and Laine discusses the design and implementations of hexagonal quadrature mirror filters banks and wavelets with their medical applications.

Chapter 19 by Sahiner and Yagle reviews the inversion of the Radon transform under wavelet constraints. It also addresses the important problems related to the inversion of the Radon transform.

Chapter 20 by Richardson summarizes data compression and teleradiology, feature enhancement and classification, fractal dimension estimation and de-noising using the wavelet transform methods, and their applications to mammograms.

Chapter 21 by Clarke et al. discusses the theory and application of the hybrid wavelet transform method for computer-assisted diagnosis and telemedicine applications.

Chapter 22 by Saipetch et al. describes the medical image enhancement method based on the wavelet transform and arithmetic coding with its application to chest radiography.

Chapter 23 by Healy et al. presents techniques for signal acquisition in magnetic resonance imaging based on adapted wavelet theory.

Chapter 24 by Olson gives a tutorial overview of a stabilization algorithm for limited angle tomography.

Chapter 25 by Manduca discusses the medical image compression using the wavelet transform.

Chapter 17

A Discrete Dyadic Wavelet Transform for Multidimensional Feature Analysis

Iztok Koren, Andrew Laine

17.1. INTRODUCTION

Discrete nonredundant wavelet transforms have been successfully applied previously in image compression applications [1–3]. However, the lack of translation invariance and aliasing present after the decomposition stage [4] may introduce undesirable artifacts for the analysis of medical signals and images, and can justify the use of a redundant wavelet representation.

The discrete dyadic wavelet transform is one example of a redundant representation. As originally proposed, the wavelet was a first derivative of a smoothing function, and was used as a multiscale edge detector to obtain a translation-invariant parsimonious representation consisting of edges [5]. A reconstruction algorithm to approximate an original signal from its multiscale edge coefficients alone was devised in [5,6].

On the other hand, previous applications described in [7–10] made no attempt to obtain a parsimonious representation from a discrete dyadic wavelet transform. Rather, the transform intentionally remained highly redundant. This redundancy was exploited for image enhancement by first modifying transform coefficients in some nonlinear fashion and reconstructing. Here, we continue this theme and present a discrete dyadic wavelet transform as a redundant representation which can be implemented efficiently and is well matched for quantification problems in the analysis of medical images.

The discrete dyadic wavelet transform was originally proposed in one and two dimensions. However, in medical imaging, there is a more general need for signal processing in more than two dimensions. In this chapter, we extend the discrete dyadic wavelet transform to multiple dimensions and describe an efficient implementation within a fast hierarchical digital filtering scheme.

When digital filtering of a finite-duration discrete signal is performed via circular convolution the filter will act on both ends of a signal simultaneously. This may lead to artifacts near both ends of the result. In image processing, mirror extension of an input signal is a popular method for alleviating such boundary effects. We present a fast filter bank implementation of the discrete dyadic wavelet transform which takes advantage of the fact that the input signal to the filter bank is mirror extended.

The chapter is organized as follows: section 17.2 presents a discrete dyadic wavelet transform in one dimension. Next, section 17.3 describes the transform for higher dimensions. Section 17.4 presents sample applications of the transform to problems in medical imaging for two- and three-dimensional modalities. Finally, section 17.5 summarizes along with concluding remarks.

17.2. ONE-DIMENSIONAL DISCRETE DYADIC WAVELET TRANSFORM

In this section we shall describe a one-dimensional discrete dyadic wavelet transform. We first formulate the transform in section 17.2.1. Next, section 17.2.2 describes a fast implementation. Section 17.2.3 then concludes with remarks on shift invariance, initialization procedure, connection to scale-space filtering, and further extensions of the transform.

17.2.1 Wavelet Transform

A discrete wavelet transform is obtained from a continuous representation by discretizing dilation and translation parameters such that the resulting set of wavelets constitutes a frame. The dilation parameter is typically discretized by an exponential sampling with a fixed dilation step and the translation parameter by integer multiples of a dilation-dependent step [4]. Unfortunately, the resulting transform is variant under translations, a property which makes it less attractive for the analysis of nonstationary signals.

Sampling the translation parameter with the same sampling period as the input function to the transform results in a translation-invariant, but redundant representation. The dyadic wavelet transform proposed by Mallat and Zhong [5] is one such representation. Let us begin with a brief review of properties of the dyadic wavelet transform as described in [5], but included here for completeness.

The dyadic wavelet transform of a function $s(x) \in L^2(R)$ is defined as a sequence of functions

$$\{W_m s(x)\}_{m \in Z}$$

where $W_m s(x) = s * \psi_m(x) = \int_{-\infty}^{\infty} s(t)\psi_m(x - t)dt$, and $\psi_m(x) = 2^{-m}\psi(2^{-m}x)$ is a wavelet $\psi(x)$ expanded by a dilation parameter (or scale) 2^m.

To ensure coverage of the frequency axis the requirement on the Fourier transform of $\psi_m(x)$ is the existence of $A_1 > 0$ and $B_1 < \infty$ such that

$$A_1 \le \sum_{m=-\infty}^{\infty} |\hat{\psi}(2^m\omega)|^2 \le B_1$$

is satisfied almost everywhere. The constraint on the Fourier transform of the (nonunique) reconstructing function $\chi(x)$ is

$$\sum_{m=-\infty}^{\infty} \hat{\psi}(2^m\omega)\hat{\chi}(2^m\omega) = 1$$

A function $s(x)$ can then be completely reconstructed from its dyadic wavelet transform using the identity

$$s(x) = \sum_{m=-\infty}^{\infty} W_m s * \chi_m(x)$$

where $\chi_m(x) = 2^{-m}\chi(2^{-m}x)$.

In numerical applications, processing is performed on discrete rather than continuous functions. When the function to be transformed is in the discrete form, the scale 2^m can no longer vary over all $m \in Z$. Finite sampling rate prohibits the scale from being arbitrarily small, while computational resources restrict the use of an arbitrarily large scale. Let the finest scale be normalized to 1 and the coarsest scale be set to 2^M, where $M \in N$ denotes the number of analysis levels.

The smoothing of a function $s(x) \in L^2(R)$ is defined as

$$S_m s(x) = s * \phi_m(x)$$

where $\phi_m(x) = 2^{-m}\phi(2^{-m}x)$ with $m \in Z$, and $\phi(x)$ is a smoothing function (i.e., its integral is equal to 1 and $\phi(x) \to 0$ as $|x| \to \infty$).

In Mallat and Zhong [5], a real smoothing function $\phi(x)$ was selected, whose Fourier transform satisfied

$$|\hat{\phi}(\omega)|^2 = \sum_{m=1}^{\infty} \hat{\psi}(2^m\omega)\hat{\chi}(2^m\omega) \tag{17-1}$$

In addition, it was shown that any discrete function of finite energy ($s(n) \in l^2(Z)$) can be written as the uniform sampling of some function smoothed at scale 1, i.e., $s(n) = S_0 t(n)$, where $t(x) \in L^2(R)$ is not unique. Thus, the discrete dyadic wavelet transform of $S_0 t(n)$ for any coarse scale 2^M is defined as a sequence of discrete functions

$$\{S_M t(n+s), \{W_m t(n+s)\}_{m\in[1,M]}\}_{n\in Z}$$

where s is a $\psi(x)$ dependent sampling shift.

For a certain choice of wavelets the discrete dyadic wavelet transform can be implemented within a fast hierarchical digital filtering scheme. Next, we shall summarize the relations between filters, wavelets, and smoothing functions.

The Fourier transform of $\phi(x)$ must satisfy [5]

$$\hat{\phi}(\omega) = e^{-j\omega s} \prod_{k=1}^{\infty} H(2^{-k}\omega) \tag{17-2}$$

where j stands for $\sqrt{-1}$, the low-pass filter frequency response $H(\omega)$ is differentiable, and

$$|H(\omega)|^2 + |H(\omega + \pi)|^2 \le 1 \qquad \text{with} \qquad |H(0)| = 1$$

Computing Eq. (17-1) for the finest two scales shows that

$$\hat{\psi}(2\omega)\hat{\chi}(2\omega) = |\hat{\phi}(\omega)|^2 - |\hat{\phi}(2\omega)|^2 \tag{17-3}$$

while computing Eq. (17-2) for $\hat{\phi}(2\omega)$ yields

$$\hat{\phi}(2\omega) = e^{-j\omega s} H(\omega)\hat{\phi}(\omega) \tag{17-4}$$

If we choose

$$\hat{\psi}(2\omega) = e^{-j\omega s} G(\omega)\hat{\phi}(\omega) \tag{17-5}$$

and

$$\hat{\chi}(2\omega) = e^{j\omega s} K(\omega)\hat{\phi}^*(\omega) \tag{17-6}$$

where $G(\omega)$ and $K(\omega)$ are digital filter frequency responses, "*" denotes complex conjugation, and insert Eqs. (17-4–17-6) into Eq. (17-3) we observe a relation between the filter frequency responses [5],

$$|H(\omega)|^2 + G(\omega)K(\omega) = 1 \tag{17-7}$$

Now we are in a position to choose filters that will give rise to wavelets and scaling functions for a discrete dyadic wavelet transform. Filters that are associated with a compactly supported orthonormal wavelet basis are certainly a possible choice. However, we suggest that other choices, as described below, may provide distinct advantages for the analysis of medical images.

Suppose we seek a wavelet that is compactly supported, antisymmetric or symmetric, exhibits good edge detection capability (i.e., equivalent to a first or a second derivative of some smoothing function), and is as regular as possible. With these additional constraints orthonormal wavelet bases are completely ruled out.

In [8], Laine et al. proposed an extension to the family of filters described in [5]. In this design, the wavelet could be either antisymmetric and equal to the first derivative of some smoothing function $\theta(x)$, or symmetric and equal to the second derivative of $\theta(x)$. When a wavelet $\psi(x)$ is antisymmetric around zero (i.e., an odd function) $e^{-j\omega s}G(\omega)$ is an odd imaginary function, and when a wavelet $\psi(x)$ is symmetric around zero (i.e., an even function) $e^{-j\omega s}G(\omega)$ is an even real function. The function $e^{-j\omega s}H(\omega)$ is even and real in both cases.

For a wavelet $\psi(x)$ to be a first (second) derivative of some smoothing function $\theta(x)$, $\hat{\psi}(\omega)$ must have a first-order (second-order) zero at $\omega = 0$, and, therefore, $G(\omega)$ must have a first-order (second-order) zero at $\omega = 0$.

However, even after satisfying all of these constraints, there remains a large number of possible choices for $H(\omega)$. Here, we extend the class of filters from [8] by choosing

$$H(\omega) = e^{j\omega s}\left[\cos\left(\frac{\omega}{2}\right)\right]^{p+1} \tag{17-8}$$

where p is a nonnegative integer and $s = (p+1)\bmod 2/2$. Once $H(\omega)$ is chosen, the product $G(\omega)K(\omega)$ is constrained by Eq. (17-7). For example, choosing

$$G(\omega) = e^{j\omega s}\left[4j\sin\left(\frac{\omega}{2}\right)\right]^{r} \tag{17-9}$$

where $r \in \{1, 2\}$ and $s = r\bmod 2/2$, determines

$$K(\omega) = -\frac{1}{16}\left[e^{-j\omega}G(\omega)\right]^{r\bmod 2}\sum_{l=0}^{2}\left[\cos\left(\frac{\omega}{2}\right)\right]^{2l} \tag{17-10}$$

Note that $H(\omega)$ is a low-pass filter, $G(\omega)$ a high-pass filter, and $K(\omega)$ a high-pass filter for $r = 1$ and a low-pass filter when $r = 2$ and $p > 0$.

The filters described in Eqs. (17-8) through (17-10) are finite impulse response (FIR) filters. For implementation, Tables 17-1 and 17-2 list the filter coefficients for the cases $r \in \{1, 2\}$ and $p \in \{0, 1, 2, 3\}$.

TABLE 17-1   Impulse Responses of Filters $H(\omega)$, $G(\omega)$, and $K(\omega)$ for $r = 1$, $p \in \{0, 1, 2, 3\}$

n	$g(n)$	$p = 0$		$p = 1$		$p = 2$		$p = 3$	
		$h(n)$	$k(n)$	$h(n)$	$k(n)$	$h(n)$	$k(n)$	$h(n)$	$k(n)$
−3									−0.001 953 125
−2						0.125	−0.007 8125	0.0625	−0.017 578 125
−1	2	0.5		0.25	−0.031 25	0.375	−0.054 6875	0.25	−0.072 265 625
0	−2	0.5	−0.125	0.5	−0.156 25	0.375	−0.171 875	0.375	−0.181 640 625
1			0.125	0.25	0.156 25	0.125	0.171 875	0.25	0.181 640 625
2					0.031 25		0.054 6875	0.0625	0.072 265 625
3							0.007 8125		0.017 578 125
4									0.001 953 125

TABLE 17-2   Impulse Responses of Filters $H(\omega)$, $G(\omega)$, and $K(\omega)$ for $r = 2$, $p \in \{0, 1, 2, 3\}$

n	$g(n)$	$p = 0$		$p = 1$		$p = 2$		$p = 3$	
		$h(n)$	$k(n)$	$h(n)$	$k(n)$	$h(n)$	$k(n)$	$h(n)$	$k(n)$
−3									−0.000 976 5625
−2						0.125	−0.003 906 25	0.0625	−0.009 765 625
−1	4	0.5		0.25	−0.015 625	0.375	−0.031 25	0.25	−0.045 898 4375
0	−8	0.5	−0.0625	0.5	−0.093 75	0.375	−0.117 1875	0.375	−0.136 718 75
1	4			0.25	−0.015 625	0.125	−0.031 25	0.25	−0.045 898 4375
2							−0.003 906 25	0.0625	−0.009 765 625
3									−0.000 976 5625

By inserting Eq. (17-8) into Eq. (17-2) and using $\prod_{k=1}^{\infty} \cos(2^{-k}\omega) = \sin\omega/\omega$ we obtain

$$\hat{\phi}(\omega) = \left(\frac{\sin(\omega/2)}{\omega/2}\right)^{p+1}$$

while by applying Eq. (17-9) and Eq. (17-5) we see that

$$\hat{\psi}(\omega) = (j\omega)^r \left(\frac{\sin(\omega/4)}{\omega/4}\right)^{p+r+1}$$

Thus, the wavelet $\psi(x)$ is a first ($r = 1$) or a second ($r = 2$) derivative of a smoothing function $\theta(x)$, whose Fourier transform is

$$\hat{\theta}(\omega) = \left(\frac{\sin(\omega/4)}{\omega/4}\right)^{p+r+1}$$

Note that $\theta(x)$ is a spline function of degree $p + r$. By increasing its degree, $\hat{\theta}(\omega)$ becomes more localized in the frequency domain and has larger support in the spatial domain.

For exposition, Fig. 17-1 shows $\theta(x)$ for $p + r \in \{1, 2, 3, 4, 5\}$ and the corresponding wavelets $\psi(x)$ for $r \in \{1, 2\}$. Wavelets from this family have a support of length $(p + r + 1)/2$, regularity order p (i.e., $\psi(x) \in C^p$), and are either symmetric ($r = 2$) or antisymmetric ($r = 1$).

17.2.2 Implementation

Similar to orthogonal and biorthogonal discrete wavelet transforms [4], the discrete dyadic wavelet transform can be implemented within a hierarchical filtering scheme. Using the definition of the discrete dyadic wavelet transform along with Eqs. (17-4) and (17-5) we can formulate the analysis section of such a filter bank. The synthesis section simply follows from Eq. (17-7). Suppose

$$F_s(\omega) = e^{-j\omega s} F(\omega) \tag{17-11}$$

where $F(\omega)$ is either $H(\omega)$, $G(\omega)$, or $K(\omega)$ [Eqs. (17-8) through (17-10)]. We may then construct a filter-bank implementation of the discrete dyadic wavelet transform as shown in Fig. 17-2.

Filters referred to in Eqs. (17-8) through (17-10) at level $m + 1$ (i.e., filters applied at some scale 2^m) become $F(2^m\omega)$, where $F(\omega)$ denotes any of the three filters at level 1. In the spatial domain this is equivalent to up-sampling the filter impulse response by 2^m (i.e., inserting $2^m - 1$ zeros between subsequent filter coefficients at level 1). Noninteger shifts at level 1 are rounded to the nearest integer. An implementation with up-sampling of filter impulse responses (called "algorithme à trous") was first proposed by Holschneider et al. [11]. The complexity of such a filter-bank implementation increases linearly with the number of levels.

Let us refer to the filters at level 1 [Eqs. (17-8) through (17-10)] as "original filters," to distinguish them from their up-sampled versions. Let an input signal $x(n)$ be real, $x(n) \in l^1(\mathbf{Z})$, $n \in [0, N - 1]$, and let $X(\omega)$ be its Fourier transform.

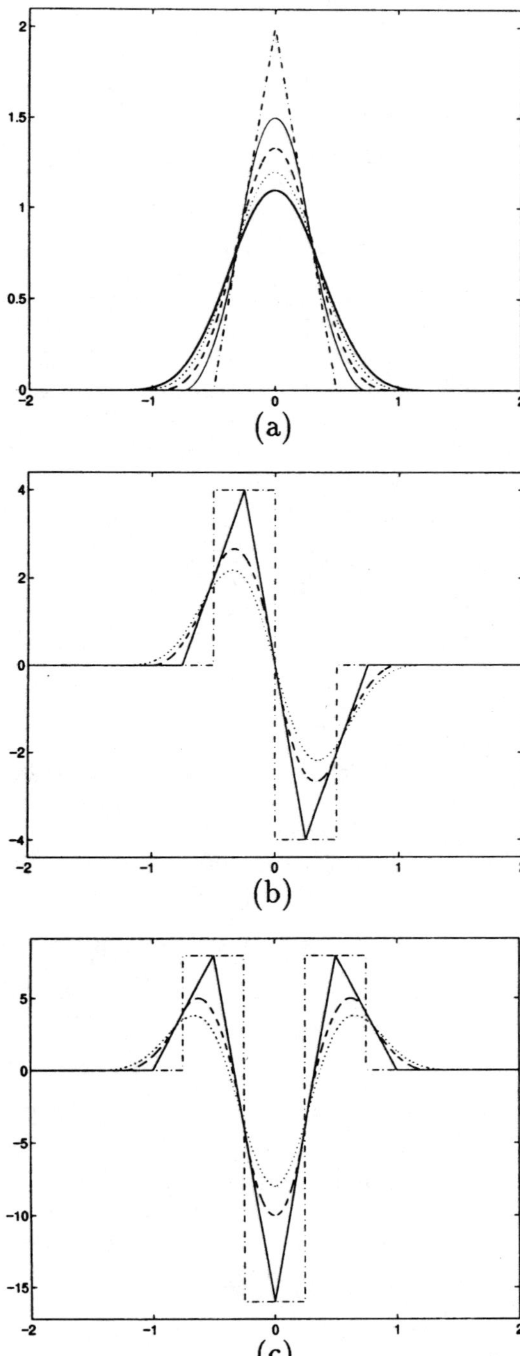

Figure 17-1 (a) Primitives $\theta(x)$: piecewise linear spline (dash-dotted), quadratic spline (thin solid), cubic spline (dashed), quartic spline (dotted), and quintic spline (thick solid); (b) wavelets $\psi(x) = d\theta(x)/dx$: the first derivative of the piecewise linear spline (dash-dotted), of the quadratic spline (solid), of the cubic spline (dashed), and of the quartic spline (dotted); (c) wavelets $\psi(x) = d^2\theta(x)/dx^2$: the second derivative of the quadratic spline (dash-dotted), of the cubic spline (solid), of the quartic spline (dashed), and of the quintic spline (dotted).

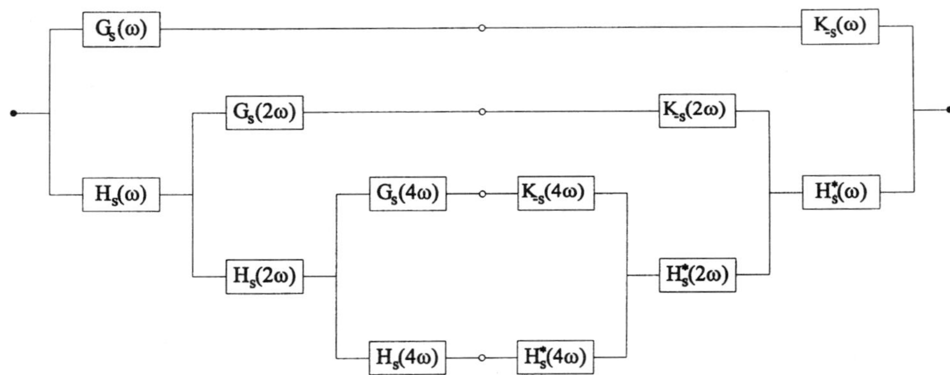

Figure 17-2   Filter-bank implementation of a one-dimensional discrete dyadic wavelet transform decomposition (left) and reconstruction (right) for three levels of analysis. $H_s^*(\omega)$ denotes the complex conjugate of $H_s(\omega)$.

Depending on the length of each filter impulse response, filtering an input signal may be computed either by multiplying $X(\omega)$ by a filter's frequency response or by circularly convolving $x(n)$ with a filter's impulse response. Of course, such a periodically extended signal may change abruptly at the boundaries, causing artifacts. A common remedy for such a problem is realized by constructing a mirror extended signal

$$x_{me}(n) = \begin{cases} x(-n-1) & \text{if } n \in [-N, -1] \\ x(n) & \text{if } n \in [0, N-1] \end{cases} \tag{17-12}$$

where we chose the signal $x_{me}(n)$ to be supported in $[-N, N-1]$. It will become evident shortly, that mirror extension is particularly elegant in conjunction with symmetric/antisymmetric filters.

Let us first classify symmetric/antisymmetric real even-length signals into four types [2]:

$$\textbf{Type I} \quad f(n) = f(-n)$$
$$\textbf{Type II} \quad f(n) = f(-n-1)$$
$$\textbf{Type III} \quad f(n) = -f(-n)$$
$$\textbf{Type IV} \quad f(n) = -f(-n-1)$$

where $n \in [-N, N-1]$. Note that for Type I signals the values at $f(0)$ and $f(-N)$ are unique, and that for Type III signals the values at $f(0)$ and $f(-N)$ are equal to zero.

Using properties of the Fourier transform, it is easy to show that the convolution of symmetric/antisymmetric real signals results in a symmetric/antisymmetric real signal. If a symmetric/antisymmetric real signal has an even length, then there always exists an integer shift such that the shifted signal belongs to one of the four types.

Next, we examine the filter-bank implementation of a one-dimensional discrete dyadic wavelet transform (Fig. 17-2) with filters derived from Eqs. (17-8) through (17-10) driven by a mirrored signal $x_{me}(n)$ at the input. Let the number of levels M be restricted by

$$M \leq 1 + \log_2 \frac{N-1}{L_{\max}-1} \tag{17-13}$$

where $L_{\max}$ is the length of the longest original filter impulse response.

Each block in the filter bank consists of a filter and a circular shift operator [Eq. (17-11)]. Equation (17-13) guarantees that the length of the filter impulse response does not exceed the length of the signal at any block.

Since our input signal $x_{me}(n)$ is of Type II and noninteger shifts at level 1 are rounded to the nearest integer, it follows that a processed signal at any point in the filter bank belongs to one of the types defined above. This means that filtering a signal of length $2N$ can be reduced to filtering a signal of approximately one half of its length. (For Types I and III, $N+1$ samples are needed. However, for Type III one needs to store only $N-1$ values because zero values are always present at the zeroth and $(-N)$th sample position.)

Implementation is particularly simple for filters designed with $r=2$ and p odd [Eqs. (17-8) through (17-10)]. Filters are of Type I in this case, so the signal at any point of the filter bank will be of Type II. A block from the filter bank shown in Fig. 17-2 [Eq. (17-11)] can therefore be implemented by

$$F_{s,m}u(n) = f(0)u_{II}(n) + \sum_{l=1}^{\frac{L-1}{2}} f(l)[u_{II}(n - 2^m l) + u_{II}(n + 2^m l)], \qquad n \in [0, N-1]$$

$$\tag{17-14}$$

where

$$u_{II}(n) = \begin{cases} u(-n-1) & \text{if } n \in [-\frac{N}{2}, -1] \\ u(n) & \text{if } n \in [0, N-1] \\ u(2N-n-1) & \text{if } n \in [N, \frac{3N}{2}] \end{cases} \tag{17-15}$$

$u(n)$ is an input signal to a block, $m+1$ denotes a level (corresponding to scale 2^m), $f(n)$ is an impulse response of some original filter, L is the length of the filter, and N is the length of an input signal $x(n)$ to the filter bank.

A filter bank with the above implementation of blocks and signal $x(n)$ at the input yields equivalent results as circular convolution for $x_{me}(n)$ as defined by Eq. (17-12). In addition to requiring one half the amount of memory, the computational savings over a circular convolution implementation of blocks are, depending on the original filter length, three to four times fewer multiplications and one half as many additions.

A similar approach can be used for other filters. However, things get slightly more complicated in this case, because the filters are not of the same type and the signal components within the filter bank are of distinct types. As a consequence, an implementation of blocks that use distinct original filters may not be the same, and the implementation of blocks at level 1 may differ from the implementation of blocks at other levels of analysis.

The decomposition blocks at level 1 can be implemented by

$$G_{s,0}u(n) = g(0)[u_{II}(n-1) - u_{II}(n)], \qquad n \in [1, N-1]$$

for $r = 1$, Eq. (17-14) for $r = 2$,

$$H_{s,0}u(n) = \sum_{l=0}^{\frac{L}{2}-1} h(l)[u_{II}(n - l - 1) + u_{II}(n + 1)], \qquad n \in [0, N]$$

for p even, and Eq. (17-14) for p odd, where $u_{II}(l)$ is defined by Eq. (17-15), $g(n)$ and $h(n)$ are impulse responses of the filters computed from Eqs. (17-9) and (17-8), respectively, and L is the length of the corresponding impulse response.

The output from a block $G_s(\omega)$ at level 1 is of Type III for $r = 1$ and of Type II for $r = 2$, while the output from $H_s(\omega)$ at the same level is of Type 1 for p even and of Type II for p odd.

The decomposition blocks at subsequent levels $m \in [1, M - 1]$ can be implemented by

$$G_{s,m}u(n) = g(0)[u_I(n - 2^m s) - u_I(n + 2^m s)], \qquad n \in [1, N - 1]$$

for $r = 1$ and p even,

$$G_{s,m}u(n) = g(0)[u_{II}(n - 2^m s) - u_{II}(n + 2^m s)], \qquad n \in [0, N - 1]$$

for $r = 1$ and p odd,

$$F_{s,m}u(n) = f(0)u_I(n) + \sum_{l=1}^{\frac{L-1}{2}} f(l)[u_I(n - 2^m l) + u_I(n + 2^m l)], \qquad n \in [0, N] \quad (17\text{-}16)$$

with $f(n) = g(n)$ for $r = 2$ and p even,

$$H_{s,m}u(n) = \sum_{l=0}^{\frac{L}{2}-1} h(l)[u_I(n - 2^m(l + s)) + u_I(n + 2^m(l + s))], \qquad n \in [0, N] \quad (17\text{-}17)$$

for p even, and Eq. (17-14) for p odd, where

$$u_I(n) = \begin{cases} u(-n) & \text{if } n \in [-\frac{N}{2}, -1] \\ u(n) & \text{if } n \in [0, N] \\ u(2N - n) & \text{if } n \in [N + 1, \frac{3N}{2}] \end{cases} \quad (17\text{-}18)$$

The outputs from blocks $G_s(2^m \omega)$ are of Type III for $r = 1$ and p even, of Type IV for $r = 1$ and p odd, and of Type I for $r = 2$ and p even, whereas the outputs from $H_s(2^m \omega)$ are of Type I for p even and of Type II for p odd.

Next, the reconstruction blocks at level 1 can be implemented by

$$K_{-s,0}u(n) = \sum_{l=1}^{\frac{L}{2}} k(l)[u_{III}(n - l + 1) - u_{III}(n + l)], \qquad n \in [0, N - 1]$$

for $r = 1$, Eq. (17-14) for $r = 2$,

$$H_{s,0}^* u(n) = \sum_{l=1}^{\frac{L}{2}} h(-l)[u_I(n - l + 1) + u_I(n + l)], \qquad n \in [0, N - 1]$$

for p even, and Eq. (17-14) for p odd, where

$$u_{III}(n) = \begin{cases} -u(-n) & \text{if } n \in [-\frac{N}{2}, -1] \\ 0 & \text{if } n = 0 \\ u(n) & \text{if } n \in [1, N-1] \\ 0 & \text{if } n = N \\ -u(2N-n) & \text{if } n \in [N+1, \frac{3N}{2}] \end{cases} \qquad (17\text{-}19)$$

$u_I(n)$ is as defined by Eq. (17-18) and $k(n)$ is an impulse response of the filter from Eq. (17-10). Note that both outputs from blocks $K_{-s}(\omega)$ and $H_s^*(\omega)$ are of Type II. The reconstruction blocks at subsequent levels can be implemented by

$$K_{-s,m}u(n) = \sum_{l=0}^{\frac{L}{2}-1} k(l+1)[u_{III}(n - 2^m(l+s)) - u_{III}(n + 2^m(l+s))], \qquad n \in [0, N]$$

for $r = 1$ and p even, Eq. (17-16) with $f(n) = k(n)$ for $r = 2$ and p even,

$$K_{-s,m}u(n) = \sum_{l=0}^{\frac{L}{2}-1} k(l+1)[u_{IV}(n - 2^m(l+s)) - u_{IV}(n + 2^m(l+s))], \qquad n \in [0, N-1]$$

for $r = 1$ and p odd,

$$H_{s,m}^* u(n) = H_{s,m} u(n)$$

for p even, and Eq. (17-14) for p odd, where $u_{III}(l)$ is given by Eq. (17-19),

$$u_{IV}(n) = \begin{cases} -u(-n-1) & \text{if } n \in [-\frac{N}{2}, -1] \\ u(n) & \text{if } n \in [0, N-1] \\ -u(2N-n-1) & \text{if } n \in [N, \frac{3N}{2}] \end{cases}$$

and $H_{s,m}u(n)$ is given by Eq. (17-17). We observe that the outputs from blocks $K_{-s}(2^m\omega)$ and $H_s^*(2^m\omega)$, $m \in [1, M-1]$, are of Type I for p even, and of Type II for p odd.

When we compare the above implementation of blocks with circular convolution driven by a mirrored signal $x_{me}(n)$ at the input, we observe that approximately twofold less memory space, three to four times fewer multiplications and one half as many additions are required. (For Type I signals an additional sample has to be saved because two values are without a pair.)

The implementation presented in this section performs all operations in the spatial domain. However, one could also implement the structure shown in Fig. 17-2 with an input signal $x_{me}(n)$ [Eq. (17-12)] in the frequency domain. For short filter impulse responses, such as those given in Tables I and II, the spatial implementation described in this section is certainly more efficient. For long filter impulse responses, however, filtering is faster if implemented in the frequency domain. Indeed, filtering with $G(\omega)$, which has never more than three nonzero coefficients, in the spatial domain is more practical. Additional details on alternative implementation strategies can be found in [13].

17.2.3 Remarks

1. As mentioned earlier, the translation invariance property of the discrete dyadic wavelet transform is due to the fact that the translation parameter is sampled with the same sampling period as the input signal, over all scales. When comparing Fig. 17-2 to traditional filter-bank implementations of an orthogonal and biorthogonal wavelet transform, we observe that each subband shown in Fig. 17-2 is retained at its original density rather than being critically sampled. Note that the filters at distinct scales are not the same as in orthogonal or biorthogonal cases of analysis [4].

Thus, the discrete dyadic wavelet transform is highly redundant. To obtain a more parsimonious representation we may sample the translation parameter in a translation invariant manner (e.g., sampling a function at its extrema or the extrema of its first few derivatives). Of importance is whether it remains possible to reconstruct an original signal from this subset of transform coefficients.

In [5] and [6], a reconstruction algorithm is presented that reconstructs an approximation of a signal given the position of local maxima of the wavelet coefficient's moduli and the values of wavelet coefficients at each corresponding location. Wavelets with $r = 1$ were used so that the locations of maxima corresponded to inflection points of the original signal smoothed at dyadic scales.

2. The discrete dyadic wavelet transform was derived from a continuous representation by setting $s(n) = S_0 t(n)$ [5]. Such a discrete wavelet transform is equal to the sampling of its continuous counterpart only when $t(n) = s(n)$ or, equivalently, $s(n) = S_0 s(n)$. Approximation by sampling of the continuous transform improves with larger scales; however, a better approximation can be achieved by appropriate initialization of the transform. For more details on this subject please refer to [14] and [15].

3. In section 17.2.1 the wavelet $\psi(x)$ was chosen to be the first ($r = 1$) or second ($r = 2$) derivative of a smoothing function $\theta(x)$. For these cases, the dyadic wavelet transform of a function $s(x) \in L^2(R)$ can be written as

$$W_m s(x) = s * \left(2^{mr} \frac{d^r \theta_m}{dx^r} \right)(x) = 2^{mr} \frac{d^r}{dx^r}(s * \theta_m)(x) \qquad m \in Z, \ r \in \{1, 2\}.$$

Depending on the wavelet selected for analysis, the wavelet transform $W_m s(x)$ is proportional either to the first ($r = 1$) or second ($r = 2$) derivative of $s(x)$ smoothed by θ_m. By recording the zero-crossings of $W_m s(x)$ for $r = 2$, a scale-space image of the representation similar to [16] can be obtained. The only differences are that curves in the scale-space plane are computed for dyadic scales only and that θ_m is a close approximation of a Gaussian instead of a true Gaussian.

4. In this chapter, we discuss first and second derivative of smoothing function wavelets; however, it is possible to choose wavelets that are higher-order derivatives of $\theta(x)(r > 2)$ as well. In such a case, a smoothing function $\phi(x)$ can be used for the decomposition only, while for the reconstruction another (nonunique) function must be introduced. In terms of the filter bank from Fig. 17-2 this means that a fourth filter, which replaces $H_s^*(\omega)$, needs to be computed. A discrete dyadic wavelet trans-

form that employs wavelets with $r > 2$ in addition to those discussed so far, is presented in [15].

17.3. MULTIDIMENSIONAL DISCRETE DYADIC WAVELET TRANSFORM

In this section we shall extend the one-dimensional transform described in section 17.2 for applications of higher dimension. The transform in higher dimensions is first formulated in section 17.3.1. Then, section 17.3.2 shows how to implement the transform in a fast and efficient fashion. Finally, a connection to traditional Canny and Marr–Hildreth edge detectors is made in section 17.3.3.

17.3.1 Wavelet Transform

In this section a two-dimensional discrete dyadic wavelet transform [5] is generalized for processing signals of multiple dimensions. First, we define the dyadic wavelet transform of a function $s(x_1, x_2, \ldots, x_d) \in L^2(R^d)$ as a set of functions

$$\{W_m^1 s(x_1, x_2, \ldots, x_d), W_m^2 s(x_1, x_2, \ldots, x_d), \ldots, W_m^d s(x_1, x_2, \ldots, x_d)\}_{m \in Z}$$

where

$$W_m^l s(x_1, x_2, \ldots, x_d) = s * \psi_m^l(x_1, x_s, \ldots, x_d)$$

$$= \int \int \cdots \int_{-\infty}^{\infty} s(t_1, t_2, \ldots, t_d) \psi_m^l(x_1 - t_1, x_2 - t_2, \ldots, x_d - t_d) dt_1 dt_2 \ldots dt_d$$

$$\text{for } l = 1, 2, \ldots, d$$

and $\psi_m^l(x_1, x_2, \ldots, x_d) = 2^{-dm} \psi^l(2^{-m}x_1, 2^{-m}x_2, \ldots, 2^{-m}x_d)$ are wavelets $\psi^l(x_1, x_2, \ldots, x_d)$ expanded by a dilation parameter 2^m.

To ensure coverage of the frequency space there must exist an $A_d > 0$ and $B_d < \infty$ such that

$$A_d \leq \sum_{m=-\infty}^{\infty} \sum_{l=1}^{d} |\hat{\psi}^l(2^m \omega_1, 2^m \omega_2, \ldots, 2^m \omega_d)|^2 \leq B_d$$

is satisfied almost everywhere. If (nonunique) functions $\chi^1(x_1, x_2, \ldots, x_d)$, $\chi^2(x_1, x_2, \ldots, x_d), \ldots, \chi^d(x_1, x_2, \ldots, x_d)$ are chosen such that their Fourier transforms satisfy

$$\sum_{m=-\infty}^{\infty} \sum_{l=1}^{d} \hat{\psi}^l(2^m \omega_1, 2^m \omega_2, \ldots, 2^m \omega_d) \hat{\chi}^l(2^m \omega_1, 2^m \omega_2, \ldots, 2^m \omega_d) = 1$$

the function $s(x_1, x_s, \ldots, x_d)$ may be reconstructed from its dyadic wavelet transform by

$$s(x_1, x_2, \ldots, x_d) = \sum_{m=-\infty}^{\infty} \sum_{l=1}^{d} W_m^l s * \chi_m^l(x_1, x_2, \ldots, x_d)$$

where $\chi_m^l(x_1, x_2, \ldots, x_d) = 2^{-dm} \chi^l(2^{-m}x_1, 2^{-m}x_2, \ldots, 2^{-m}x_d)$.

However, when processing discrete functions the scale 2^m may no longer vary over all $m \in \mathbf{Z}$. Let the finest scale be normalized to 1 and the coarsest scale set to be 2^M. Let us introduce a real smoothing function $\phi(x_1, x_2, \ldots, x_d)$ such that its Fourier transform satisfies

$$|\hat{\phi}(\omega_1, \omega_2, \ldots, \omega_d)|^2 = \sum_{m=1}^{\infty} \sum_{l=1}^{d} \hat{\psi}^l(2^m\omega_1, 2^m\omega_2, \ldots, 2^m\omega_d)\hat{\chi}^l(2^m\omega_1, 2^m\omega_2, \ldots, 2^m\omega_d)$$

$$(17\text{-}20)$$

Here, as in one dimension, a finite energy discrete function $(s(n_1, n_2, \ldots, n_d) \in l^2(\mathbf{Z}^d))$ can be written as the uniform sampling of some function smoothed at scale 1: $s(n_1, n_2, \ldots, n_d) = S_0 t(n_1, n_2, \ldots, n_d)$, where $t(x_1, x_2, \ldots, x_d) \in L^2(\mathbf{R}^d)$ is not unique, and $S_m t(x_1, x_s, \ldots, x_d) = t * \phi_m(x_1, x_2, \ldots, x_d)$. Thus, the discrete dyadic wavelet transform of $S_0 t(n_1, n_2, \ldots, n_d)$ for any coarse scale 2^M is defined as

$$\{S_M t(n_1 + s, n_2 + s, \ldots, n_d + s),$$
$$\{W_m^1 t(n_1 + s, n_2 + s, \ldots, n_d + s),$$
$$W_m^2 t(n_1 + s, n_2 + s, \ldots, n_d + s), \ldots,$$
$$W_m^d t(n_1 + s, n_2 + s, \ldots, n_d + s)\}_{m \in [1,M]}\}_{(n_1, n_2, \ldots, n_d) \in \mathbf{Z}^d}$$

where s is a wavelet dependent sampling shift.

To implement a multidimensional discrete dyadic wavelet transform within a fast hierarchical digital filtering scheme the wavelets were chosen to be separable products of one-dimensional functions:

$$\psi^l(x_1, x_2, \ldots, x_d) = \psi(x_l)2^{d-1} \prod_{\substack{n=1 \\ n \neq l}}^{d} \phi(2x_n), \qquad l \in [1, d] \qquad (17\text{-}21)$$

where $\phi(x)$ and $\psi(x)$ were chosen as described in section 17.2.1 (recall that the Fourier transforms $\hat{\phi}(\omega)$ and $\hat{\psi}(2\omega)$ were defined by Eqs. (17-2) and (17-5), respectively).

From Eqs. (17-21), (17-5), and (17-2) we may write

$$\hat{\psi}^l(2\omega_1, 2\omega_2, \ldots, 2\omega_d) = e^{-j\omega_l s} G(\omega_l) \prod_{n=1}^{d} \hat{\phi}(\omega_n), \qquad l \in [1, d] \qquad (17\text{-}22)$$

where $G(\omega)$ is the frequency response of a digital filter. Choosing

$$\hat{\chi}^l(2\omega_1, 2\omega_2, \ldots, 2\omega_d) = e^{j\omega_l s} K(\omega_l) L_d(\omega_1, \ldots, \omega_{l-1}, \omega_{l+1}, \ldots, \omega_d) \prod_{n=1}^{d} \hat{\phi}^*(\omega_n),$$

$$l \in [1, d] \qquad (17\text{-}23)$$

where $K(\omega)$ and $L_d(\omega_1, \ldots, \omega_{l-1}, \omega_{l+1}, \ldots, \omega_d)$ are digital filter frequency responses, we may compute Eq. (17-20) for the finest two scales by

$$\sum_{l=1}^{d} \hat{\psi}^l(2\omega_1, 2\omega_2, \ldots, 2\omega_d)\hat{\chi}^l(2\omega_1, 2\omega_2, \ldots, 2\omega_d)$$

$$= |\hat{\phi}(\omega_1, \omega_2, \ldots, \omega_d)|^2 - |\hat{\phi}(2\omega_1, 2\omega_2, \ldots, 2\omega_d)|^2 \tag{7-24}$$

Inserting the terms defined by Eqs. (17-23), (17-22), and (17-4) with $\hat{\phi}(\omega_1, \omega_2, \ldots, \omega_d) = \prod_{n=1}^{d} \hat{\phi}(\omega_n)$ into Eq. (17-24) results in

$$\sum_{l=1}^{d} K(\omega_l)G(\omega_l)L_d(\omega_1, \ldots, \omega_{l-1}, \omega_{l+1}, \ldots, \omega_d) + \prod_{l=1}^{d} |H(\omega_l)|^2 = 1 \tag{17-25}$$

Equation (17-25) represents a relation between the frequency responses of the digital filters used to implement a multidimensional discrete dyadic wavelet transform and is a multidimensional analog to Eq. (17-7). Note that with the exception of $L_d(\omega_1, \ldots, \omega_{l-1}, \omega_{l+1}, \ldots, \omega_d)$ each filter is one-dimensional.

Solving Eq. (17-25) for $L_d(\omega_1, \ldots, \omega_{l-1}, \omega_{l+1}, \ldots, \omega_d)$ by substituting $K(\omega_l)$ $G(\omega_l)$ from Eq. (17-7) yields the closed formulae

$$L_2(\omega) = \frac{1}{2}(1 + |H(\omega)|^2)$$

$$L_3(\omega_{n_1}, \omega_{n_2}) = \frac{1}{3}\left(1 + \binom{2}{1}^{-1}(|H(\omega_{n_1})|^2 + |H(\omega_{n_2})|^2) + |H(\omega_{n_1})H(\omega_{n_2})|^2\right)$$

$$L_4(\omega_{n_1}, \omega_{n_2}, \omega_{n_3}) = \frac{1}{4}\left(1 + \binom{3}{1}^{-1}(|H(\omega_{n_1})|^2 + |H(\omega_{n_2})|^2 + |H(\omega_{n_3})|^2)\right.$$

$$+ \binom{3}{2}^{-1}(|H(\omega_{n_1})H(\omega_{n_2})|^2 + |H(\omega_{n_1})H(\omega_{n_3})|^2 + |H(\omega_{n_2})H(\omega_{n_3})|^2)$$

$$\left. + |H(\omega_{n_1})H(\omega_{n_2})H(\omega_{n_3})|^2\right) \tag{17-26}$$

Higher-dimensional $L_d(\omega_1, \ldots, \omega_{l-1}, \omega_{l+1}, \ldots, \omega_d)$ follow in the same manner as above.

Note that except for the case of $d = 2$, the functions $L_d(\omega_1, \ldots, \omega_{l-1}, \omega_{l+1}, \ldots, \omega_d)$ are nonseparable. However, all are sums of separable functions.

For implementation, $L_3(\omega_{n_1}, \omega_{n_2})$ can be reformulated as

$$L_3(\omega_{n_1}, \omega_{n_2}) = \frac{1}{3}\left(L_3'(\omega_{n_1})L_3'(\omega_{n_2}) + \frac{3}{4}\right) \tag{17-27}$$

where $L_3'(\omega) = |H(\omega)|^2 + \frac{1}{2}$. In Tables 17-3 and 17-4 we provide the filter coefficients for $L_2(\omega)$ from Eq. (17-26) and $L_3'(\omega)$ computed from Eq. (17-27) for $p \in \{0, 1\}$ and $p \in \{2, 3\}$, respectively.

TABLE 17-3   Impulse Responses of Filters $L_2(\omega)$ and $L_3'(\omega)$ for $p \in \{0, 1\}$

	$p = 0$		$p = 1$	
n	$l_2(n)$	$l_3'(n)$	$l_2(n)$	$l_3'(n)$
-2			0.031 25	0.0625
-1	0.125	0.25	0.125	0.25
0	0.75	1	0.6875	0.875
1	0.125	0.25	0.125	0.25
2			0.031 25	0.0625

TABLE 17-4   Impulse Responses of Filters $L_2(\omega)$ and $L_3'(\omega)$ for $p \in \{2, 3\}$

	$p = 2$		$p = 3$	
n	$l_2(n)$	$l_3'(n)$	$l_2(n)$	$l_3'(n)$
-4			0.001 953 125	0.003 906 25
-3	0.007 8125	0.015 625	0.015 625	0.031 25
-2	0.046 875	0.093 75	0.054 6875	0.109 375
-1	0.117 1875	0.234 375	0.109 375	0.218 75
0	0.656 25	0.8125	0.636 718 75	0.773 4375
1	0.117 1875	0.234 375	0.109 375	0.218 75
2	0.046 875	0.093 75	0.054 6875	0.109 375
3	0.007 8125	0.015 625	0.015 625	0.031 25
4			0.001 953 125	0.003 906 25

17.3.2 Implementation

As in the one-dimensional case, a multidimensional discrete dyadic wavelet transform can be implemented as a fast hierarchical filtering scheme. The filter-bank implementation follows from Eqs. (17-22), (17-4), and (17-25), and is shown in Fig. 17-3.

The only nonseparable filter in the filter-bank implementation of a multidimensional discrete dyadic wavelet transform is $L_d(\omega_1, \ldots, \omega_{l-1}, \omega_{l+1}, \ldots, \omega_d)$. Filtering with $L_d(\omega_1, \ldots, \omega_{l-1}, \omega_{l+1}, \ldots, \omega_d)$ can be accomplished as a sum of separable filters or by computing the Fourier transform of the nonseparable filter.

Implementation of filtering with $L_d(\omega_1, \ldots, \omega_{l-1}, \omega_{l+1}, \ldots, \omega_d)$ in two and three dimensions (Tables 17-3 and 17-4) is straightforward. We note that $d = 2$ is the only case when the filter is separable and the implementation for $d = 3$ using Eq. (17-27) is almost as fast as computing a separable $L_3(\omega_{n_1}, \omega_{n_2})$. In addition to filtering by l_3' along two dimensions, we need only add three fourths of the input signal, and then scale the result by a factor of one third.

Suppose our multidimensional structure is of size $N_1 \times N_2 \times \cdots \times N_d$ and its mirror extension handles boundary effects. The number of levels remains restricted

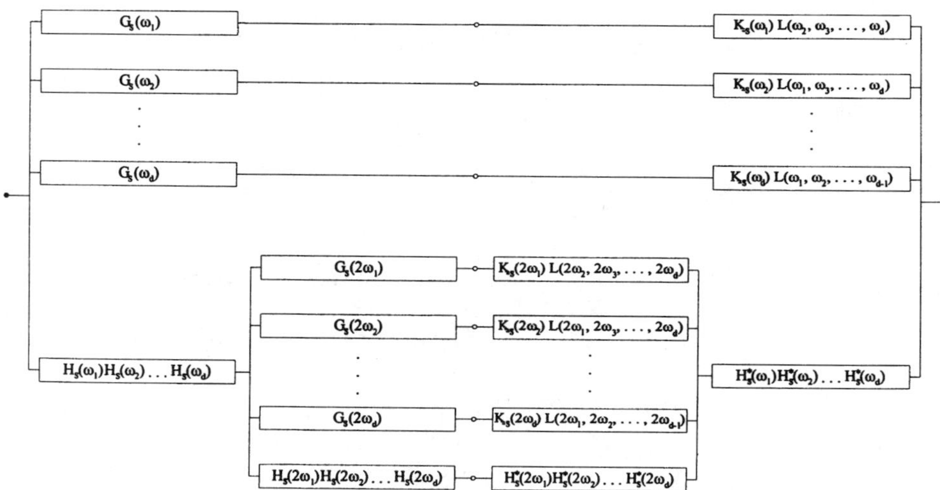

Figure 17-3   Filter-bank implementation of a miltidimensional discrete dyadic wavelet transform decomposition (left) and reconstruction (right) for two levels of analysis. $H_s^*(\omega)$ denotes the complex conjugate of $H_s(\omega)$.

by Eq. (17-13), where $N = \min\{N_l; \ l \in [1, d]\}$ and L_{max} is the length of the impulse response of $|H(\omega)|^2$.

If only separable filters are included in the filter-bank implementation, we may then use the one-dimensional implementation as described in section 17.2.2. Equations for implementing blocks $G_s(\omega)$, $H_s(\omega)$, $K_{-s}(\omega)$, and $H_s^*(\omega)$ at distinct levels and values of p and r are also straightforward in this case.

The one-dimensional filters comprising $L_d(\omega_1, \ldots, \omega_{l-1}, \omega_{l+1}, \ldots, \omega_d)$ can be realized by Eq. (17-14) for p odd or $m = 0$ and Eq. (17-16) otherwise ($f(n)$ is an impulse response of $L_2(\omega)$, $L_3'(\omega)$, or a similar zero-phase filter).

As previously mentioned in section 17.2.2, the computational savings for this implementation over circular convolution and a mirror-extended signal are (depending on the original filter length) approximately three to four times fewer multiplications and one half as many additions. As in one dimension, for long filters a frequency domain implementation may be more efficient.

Frequency domain filtering with $L_d(\omega_1, \ldots, \omega_{l-1}, \omega_{l+1}, \ldots, \omega_d)$, however, becomes more efficient for larger d even with short filters. The drawback of using the Fourier transform of a nonseparable filter $L_d(\omega_1, \ldots, \omega_{l-1}, \omega_{l+1}, \ldots, \omega_d)$ is that the multidimensional structure has to be mirror extended in all directions. This requires 2^d times more storage than needed for the original structure.

17.3.3 Remarks

Knowing that a wavelet behaves as either a first ($r = 1$) or a second ($r = 2$) derivative of a smoothing function $\theta(x)$, Eq. (17-21) may be rewritten as

$$\psi^l(x_1, x_2, \ldots, x_d) = \frac{\partial^r \theta^l(x_1, x_2, \ldots, x_d)}{\partial x_l^r}, \qquad l \in [1, d], \qquad r \in \{1, 2\}$$

where

$$\theta^l(x_1, x_2, \ldots, x_d) = \theta(x_l) 2^{d-1} \prod_{\substack{n=1 \\ n \neq l}}^{d} \phi(2x_n)$$

Let us denote $W_m s(x_1, x_2, \ldots, x_d) = (W_m^1 s(x_1, x_2, \ldots, x_d), W_m^2 s(x_1, x_2, \ldots, x_d), \ldots,$ $W_m^d s(x_1, x_2, \ldots, x_d))$, $\mathbf{V} = \left(\frac{\partial}{\partial x_1}, \frac{\partial}{\partial x_2}, \ldots, \frac{\partial}{\partial x_d}\right)$, $\Delta = \mathbf{V}^2 = \left(\frac{\partial^2}{\partial x_1^2} + \frac{\partial^2}{\partial x_2^2} + \cdots + \frac{\partial^2}{\partial x_d^2}\right)$, and assume that $\theta^l(x_1, x_2, \ldots, x_d)$ can be approximated by $\theta(x_1, x_2, \ldots, x_d)$ for all $l \in [1, d]$.

For $r = 1$ it then follows that

$$W_m s(x_1, x_2, \ldots, x_d) = 2^m \mathbf{V}(s * \theta_m)(x_1, x_2, \ldots, x_d) \tag{17-28}$$

Thus for $r = 2$ we can write

$$\sum_{l=1}^{d} W_m^l s(x_1, x_2, \ldots, x_d) = 2^{2m} \Delta(s * \theta_m)(x_1, x_2, \ldots, x_d) \tag{17-29}$$

With $\theta(x_1, x_2, \ldots, x_d)$ being a Gaussian, finding local extrema of Eq. (17-28) in the direction of gradient $\mathbf{V}$ corresponds to the filtering stage of a Canny edge detector [17], and finding zero-crossings of Eq. (17-29) with $d = 2$ corresponds to the filtering carried out with a Marr–Hildreth edge detector (Laplacian or Gaussian) [18]. (Note that both edge detectors involve postprocessing.) Edge detection based on finding local extrema of $W_m s(x_1, x_2, \ldots, x_d)$ or zero-crossings of $\sum_{l=1}^{2} W_m^l s(x_1, x_2)$ is therefore an approximation to the Canny or the Marr–Hildreth edge detector over a range of dyadic scales. The differences stem from the fact that $\theta(x_1, x_2, \ldots, x_d)$ is neither a Gaussian nor is $\theta^l(x_1, x_2, \ldots, x_d)$ equal to $\theta(x_1, x_2, \ldots, x_d)$.

17.4. APPLICATIONS

In this section we present two applications of the discrete dyadic wavelet transform to problems in medical imaging. An example of nonlinear contrast enhancement via a two-dimensional discrete dyadic wavelet transform is given in section 17.4.1. In addition, a three-dimensional discrete dyadic wavelet transform is applied for edge detection in echocardiographic image sequences in section 17.4.2. Finally, in section 17.4.3 two applications of the transform using the reconstruction algorithm from multiscale edges are mentioned [5, 6].

17.4.1 Contrast Enhancement in Digital Mammography

In mammography, early detection of breast cancer relies upon the ability to distinguish between malignant and benign mammographic features. The detection of small malignancies and subtle lesions is often difficult. Contrast enhancement can make more obvious unseen or barely seen features of a mammogram without requiring additional radiation.

Within a discrete dyadic wavelet transform, a framework for contrast enhancement was achieved by applying a (possibly nonlinear) function (referred to as an "enhancement function") to wavelet coefficients $\{\{W_m^1 t(n_1 + s, n_2 + s), W_m^2 t(n_1 + s, n_2 + s)\}_{m\in[1,M]}\}_{(n_1,n_2)\in Z^2}$ and then reconstructing an enhanced image with modified coefficients [8].

In [7] it was shown that multiscale contrast enhancement techniques based on three multiscale representations: (1) discrete dyadic wavelet transform, (2) φ-transform, and (3) hexagonal wavelet transform, can out-perform traditional contrast enhancement techniques such as histogram equalization and unsharp masking [19] when applied to subtle features of importance in digital mammography. Furthermore, in [9], it was shown that unsharp masking with a Gaussian low-pass filter can be formulated as a special case of contrast enhancement via a discrete dyadic wavelet transform.

The enhancement function was chosen such that: (1) low-contrast areas were treated more aggressively than existing areas of high contrast, (2) edges were not blurred, (3) it was monotonically increasing and (4) antisymmetric (i.e., odd). A simple function that satisfies the above conditions is [9]

$$E(x) = \begin{cases} x - (K-1)T & \text{if } x < -T \\ Kx & \text{if } |x| \le T \\ x + (K-1)T & \text{if } x > T \end{cases} \qquad (17\text{-}30)$$

where $K > 1$. Figure 17-4 shows the enhancement function from Eq. (17-30) for parameter values $K = 20$ and $T = 1$.

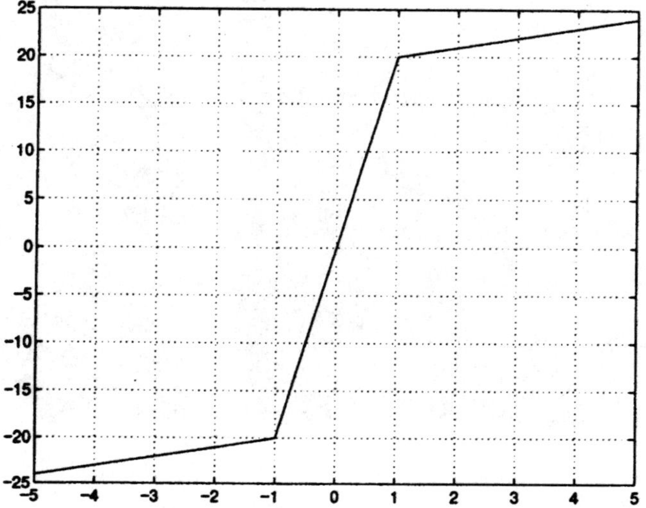

Figure 17-4 The enhancement function (Eq. (17-30) with $K = 20$ and $T = 1$).

Filters with $r = 2$ [Eqs. (17-8) through (17-10)] were found to be more suitable for contrast enhancement than wavelet filters designed with $r = 1$. In addition, zero-phase filters are preferred, which led to the choice of filters with $r = 2$ and p odd. We point out that changed positions of edges and the generation of false edges were artifacts observed after employing a nonlinear enhancement function with improperly chosen filters [10].

Figure 17-5 shows an original mammographic image and its corresponding enhancement obtained by using filters and wavelets with $r = 2$, $p = 3$, and an enhancement function defined by Eq. (17-30) with $K = 30$ and $T_m = 0.12 \max\{|W_m^l t(n_1 + s, n_2 + s)|; \ l \in \{1, 2\}, \ (n_1, n_2) \in \mathbf{Z}^2\}, \ m \in \{1, M\}$. The image matrix size was 775×436 pixels, 210 micron spot size, and the analysis was performed up to the fifth level (i.e., $M = 5$). Note the improved visibility of the spicular boundary around the mass.

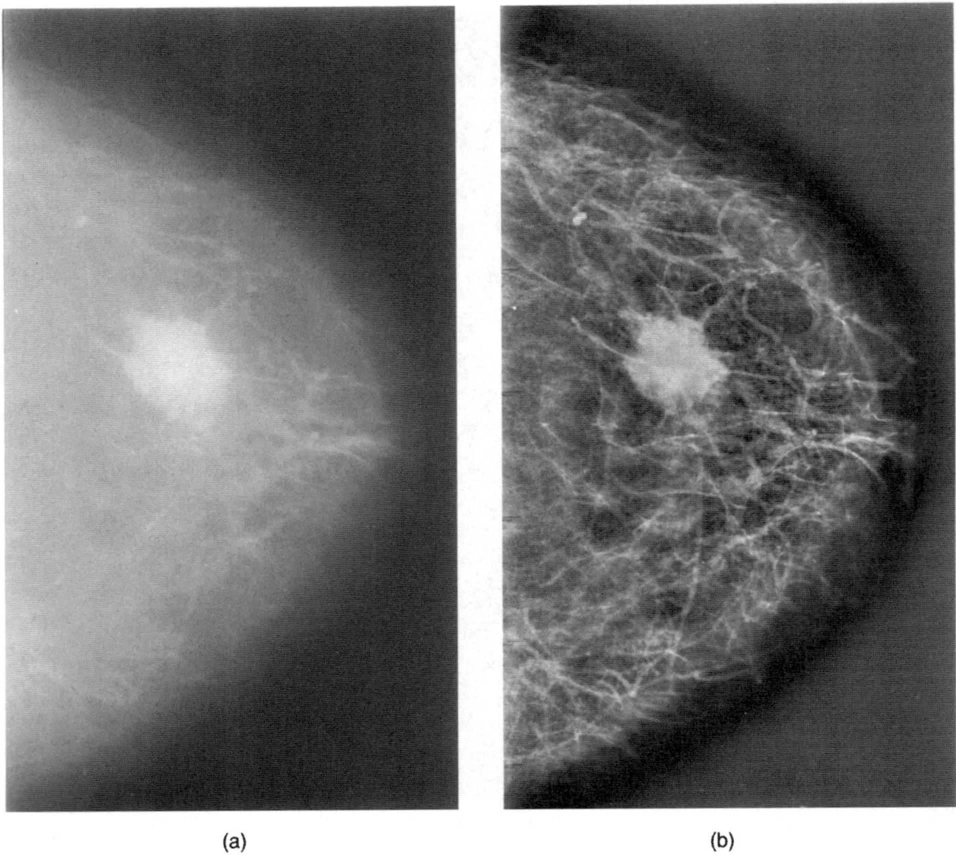

(a)   (b)

Figure 17-5   (a) An original mammographic image containing a spicular mass. (b) An enhanced image with spicular borders well delineated.

Enhancement of a suspicious area of a mammogram using wavelets with $r = 2$, $p = 1$, and an enhancement function given by Eq. (17-30) is demonstrated in Fig. 17-6. The parameters were: $K = 20$, $T_m = 0.6\max\{|W_m^l t(n_1 + s, n_2 + s)|; \ l \in \{1, 2\}, (n_1, n_2) \in \mathbf{Z}^2\}$, $m \in \{1, M\}$, and $M = 6$. The image and enhanced area matrix sizes were 655×615 pixels and 175×165 pixels, respectively.

Additional details on contrast enhancement by a discrete dyadic wavelet transform including incorporation of de-noising into the enhancement scheme, can be found in [8–10].

17.4.2 Edge Detection in Echocardiographic Image Sequences

In echocardiography, reliable detection of heart wall boundaries is of great clinical importance. Diagnostic parameters for quantification, such as time variations of heart wall motion, thickness, and enclosed area, can be evaluated once boundaries of the heart wall are determined.

Echocardiography is attractive in that it is convenient and less expensive than nuclear or X-ray imaging modalities. Automatic processing of echocardiographic images, however, is hampered by poor image quality. Specifically, low contrast, dropouts, nonlinear noise, blur, and distortions are frequently encountered in practice.

In [20], temporal information was included in analysis to make edge detection more reliable. A sequence of images was treated as a volume and processed by a three-dimensional discrete dyadic wavelet transform. Similar to the work of Wilson et al. [21], heart wall boundaries were detected along the posterior and anterior walls; a good starting point for extracting complete heart wall boundaries.

Edge detection was performed by wavelet filters with $r = 1$ as defined in Eqs. (17-8) through (17-10). Wavelet filters with $r = 1$ were more appropriate for this application than filters designed with $r = 2$. Since gradient vector extrema carry information about edge strength, they allow for application of a simple heuristic, which is easy to implement (e.g., the fact that a cavity tends to be darker than surrounding muscle since there is little tissue to reflect the ultrasound waveform within a cavity).

The search for the epicardial boundary along the posterior wall was performed first. The detected boundary served as a reference for subsequent searches [21]. The search for endocardial boundaries was performed next and used both the detected epicardial boundary and the fact that the endocardial boundaries enclose the cavity. Note that from the direction of the gradient vector at its modulus maxima the direction of variation was determined. Finally, the epicardial boundary along the anterior wall was found.

Figure 17-7 shows a sample of an original 256×256 echocardiographic image frame (a), the detected boundaries (b), and edges at scales 2^3 (c) and 2^4 (d). Note that in this application reconstruction from wavelet coefficients was not used. A three-dimensional discrete dyadic wavelet transform was employed in this analysis for the detection of edge features.

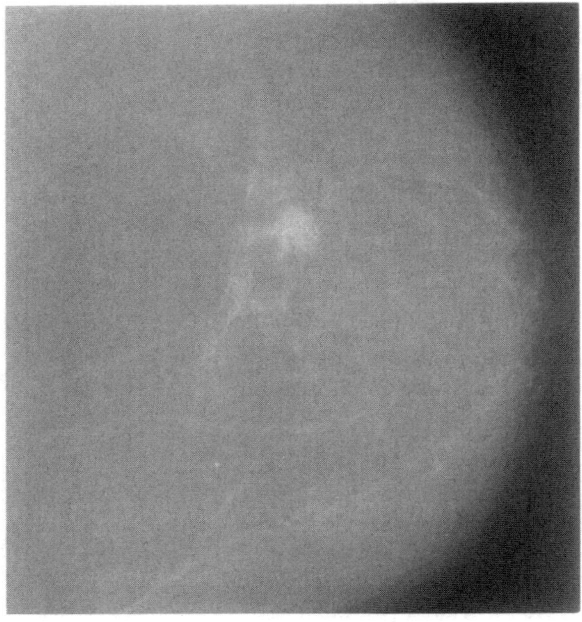

(a)

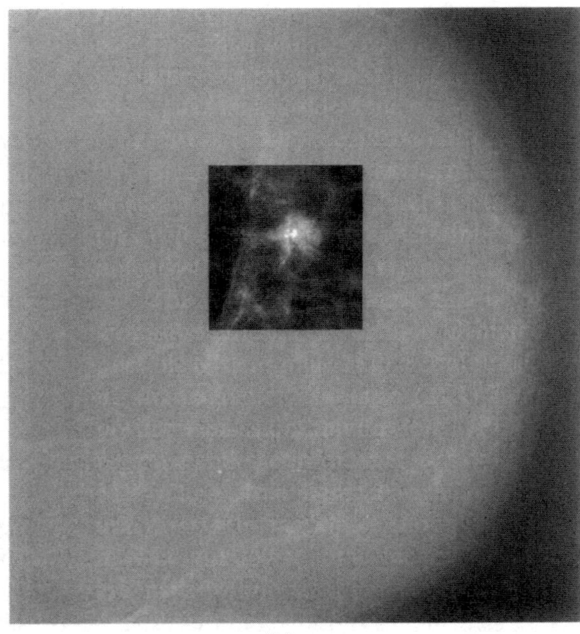

(b)

Figure 17-6 (a) An original mammographic image containing an ill-defined mass. (b) Enhanced area of a suspicious region.

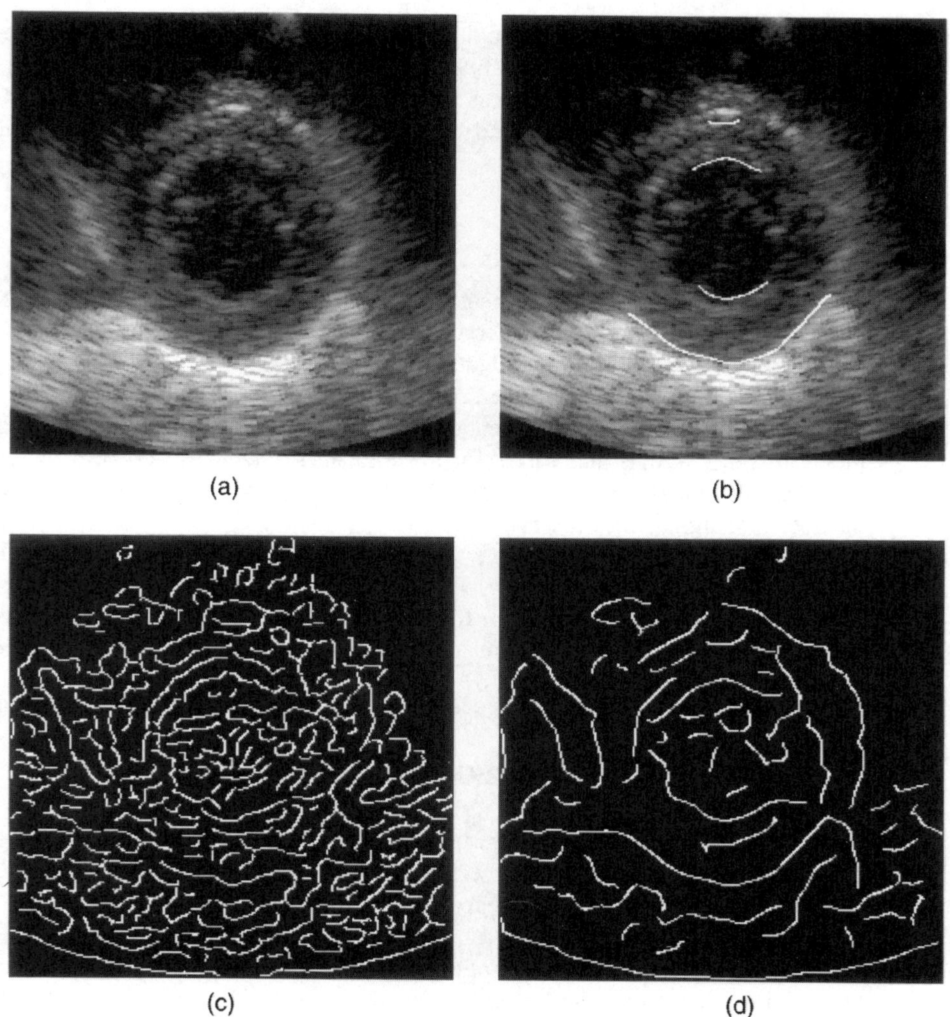

(a)

(b)

(c)

(d)

Figure 17-7   (a) An original echocardiographic image frame. (b) Original image with detected boundaries superimposed. (c) Detected edges obtained at scale 2^3. (d) Detected edges obtained at scale 2^4

17.4.3 Remarks

An attractive property of the discrete dyadic wavelet transform as originally proposed was that reconstruction from edges alone could be accomplished. The reconstruction from local maxima mentioned in section 17.2.3 was extended for two dimensions and used for compression in [5]. By selecting and encoding only the most prominent edges high compression ratios were achieved.

An example of the removal of white noise by reconstruction from edges was shown in [6]. The algorithm was based on the idea that noise singularities behave differently across scales than signal singularities. Singularities were characterized by use of Lipschitz exponents, which were estimated from the propagation of wavelet transform modulus maxima across scales. Modulus maxima were traced through a three-dimensional scale-space and coefficients considered to belong to noise were removed prior to reconstruction.

17.5. CONCLUSION

We have reviewed the one-dimensional discrete dyadic wavelet transform using wavelets that corresponded either to the first or second derivative of an arbitrary order spline function and extended it to higher dimensions.

A mirror extended input signal to a filter-bank implementation of the discrete dyadic wavelet transform enabled us to take advantage of the symmetry/antisymmetry of both signals and filters for an efficient implementation of a dyadic transform. Similar ideas can be used for any scheme employing symmetric/antisymmetric discrete real signals and filters.

We computed the coefficients for filters associated with low-order splines used as smoothing functions in a fast hierarchical digital filtering implementation of the discrete dyadic wavelet transform for one-, two-, and three-dimensional processing. Implementation of the transform algorithm in the "C" programming language is available via the Internet at `http://www.iprg.cise.ufl.edu/`.

ACKNOWLEDGMENT

Original digitized mammograms shown in Figs. 17-5 and 17-6 were provided courtesy of the Center for Engineering and Medical Image Analysis and the H. Lee Moffitt Cancer Center and Research Institute at the University of South Florida, Tampa.

The digitized echocardiographic image shown in Fig. 17-7 was provided courtesy of the Division of Cardiology, Department of Medicine and the J. Hillis Miller Health Science Center at the University of Florida, Gainesville.

REFERENCES

[1] R. R. Coifman and M. V. Wickerhauser, "Entropy-based algorithms for best basis selection," *IEEE Trans. Inform. Theory*, vol. 38, pp. 713–718, 1992.

[2] B. Jawerth and W. Sweldens, "Overview of wavelet based multiresolution analyses," *SIAM Rev.*, vol. 36, pp. 377–412, 1994.

[3] K. Ramchandran, M. Vetterli, and C. Herley, "Wavelets, subband coding, and best bases," *Proc. IEEE*, vol. 84, pp. 541–560, 1996.

[4] I. Daubechies, *Ten Lectures on Wavelets*, Philadelphia, PA: SIAM, 1992.

[5] S. Mallat and S. Zhong, "Characterization of signals from multiscale edges," *IEEE Trans. Pattern Anal. Mach. Intell.*, vol. 14, pp. 710–732, 1992.

[6] S. Mallat and W. L. Hwang, "Singularity detection and processing with wavelets," *IEEE Trans. Inf. Theory*, vol. 38, pp. 617–643, 1992.

[7] A. F. Laine, S. Schuler, J. Fan, and W. Huda, "Mammographic feature enhancement by multiscale analysis," *IEEE Trans. Med. Imaging*, vol. 13, pp. 725–740, 1994.

[8] A. Laine, J. Fan, and S. Schuler, "A framework for contrast enhancement by dyadic wavelet analysis." In *Digital Mammography*, A. G. Gale et al. (eds.). Amsterdam: Elsevier, pp. 91–100, 1994.

[9] A. Laine, J. Fan, and W. Yang, "Wavelets for contrast enhancement of digital mammography," *IEEE Eng. Med. Biol. Mag.*, vol. 14, pp. 536–550, 1995.

[10] J. Fan and A. Laine, "Multiscale contrast enhancement and denoising in digital radiographs." In *Wavelets in Medicine and Biology*. A. Aldroubi and M. Unser, (eds.), Boca Raton, FL: CRC Press, pp. 163–189, 1996.

[11] M. Holschneider, R. Kronland-Martinet, J. Morlet, and Ph. Tchamitchian, "A real-time algorithm for signal analysis with the help of the wavelet transform." In *Wavelets, Time-Frequency Methods and Phase Space*, J. M. Combes, A. Grossmann, and Ph. Tchamitchian (eds.), Berlin: Springer Verlag, pp. 286–297, 1989.

[12] A. V. Oppenheim and R. W. Schafer, *Discrete-Time Signal Processing*. Englewood Cliffs, NJ: Prentice-Hall, 1989.

[13] O. Rioul and P. Duhamel, "Fast algorithms for discrete and continuous wavelet transforms," *IEEE Trans. Inf. Theory*, vol. 38, pp. 569–586, 1992.

[14] M. Unser, A. Aldroubi, and S. J. Schiff, "Fast implementation of the continuous wavelet transform with integer scales" *IEEE Trans. Signal Process.*, vol. 42, pp. 3519–3523, 1994.

[15] I. Koren, "A multiscale spline derivative-based transform for image fusion and enhancement. *Ph.D. Thesis*, Department of Electrical and Computer Engineering, University of Florida, Gainesville, FL, 1996.

[16] A. Witkin, "Scale space filtering." *Proc. Int. Joint Conf. Artif. Intell.*, Karlsruhe, Germany, 1983, pp. 1019–1022, 1983.

[17] J. Canny, "A computational approach to edge detection," *IEEE Trans. Pattern Anal. Mach. Intell.*, vol. 8, pp. 679–698, 1986.

[18] D. Marr and E. Hildreth, "Theory of edge detection," *Proc. R. Soc. London Series B*, vol. 207, pp. 187–217, 1980.

[19] J. Lim, *Two-Dimensional Signal and Image Processing*. Englewood Cliffs, NJ: Prentice-Hall, 1990.

[20] I. Koren, A. F. Laine, J. Fan, and F. J. Taylor, "Edge detection in echocardiographic image sequences by 3-D multiscale analysis." In *Proc. IEEE Int. Conf. Image Process.*, Austin, TX, November 1994, vol. 1, pp. 288–292, 1994.

[21] D. C. Wilson, E. A Geiser, and J.-H. Li, "Feature extraction in two-dimensional short-axis echocardiographic imager." *J. Math. Imaging Vision*, vol.3, pp. 285–298, 1993.

Chapter 18

Hexagonal QMF Banks
and Wavelets

Sergio Schuler, Andrew Laine

18.1. INTRODUCTION

In this chapter we shall lay bare the theory and implementation details of hexagonal sampling systems and hexagonal quadrature mirror filters (HQMF). Hexagonal sampling systems are of particular interest because they exhibit the tightest packing of all regular two-dimensional (2-D) sampling systems and for a circularly band-limited waveform, hexagonal sampling requires 13.4% fewer samples than rectangular sampling [1]. In addition, hexagonal sampling systems also lead to non-separable quadrature mirror filters in which all basis functions are localized in space, spatial frequency, and orientation [2]. This chapter is organized in two main sections. Section 18.2 describes the theoretical aspects of hexagonal sampling systems while section 18.3 covers important implementation details.

18.2. HEXAGONAL SAMPLING SYSTEM

This section presents the theoretical foundation of hexagonal sampling systems and HQMFs. Most of this material has appeared elsewhere in [1–6] but is described here under a unified notation for completeness. In addition, it will provide continuity and a foundation for the original material that follows in section 18.3. The rest of the section is organized as follows. Section 18.2.1 covers the general formulation of a hexagonal sampling system. Section 18.2.2 introduces up-sampling and down-sampling in hexagonal sampling systems. Section 18.2.3 reviews the theory of

HQMFs. Section 18.2.4 describes redundant analysis/synthesis filter banks in hexagonal systems. Finally, section 18.2.5 covers the formulation of the discrete Fourier transform in hexagonal sampling systems.

18.2.1 Hexagonal Systems

Let $x_a(\mathbf{t}) = x_a(t_1, t_2)$ be a 2-D analog waveform, then a sampling operation in 2-D can be represented by a lattice formed by taking all integer linear combinations of a set of two linearly independent vectors $\mathbf{v}_1 = [v_{11}\ v_{21}]^T$ and $\mathbf{v}_2 = [v_{12}\ v_{22}]^T$. Using vector notation we can represent the lattice as the set of all vectors $\mathbf{t} = [t_1\ t_2]^T$ generated by

$$\mathbf{t} = \mathbf{Vn} \tag{18-1}$$

where $\mathbf{n} = [n_1\ n_2]^T$ is an integer-valued vector and $\mathbf{V} = [\mathbf{v}_1\ \mathbf{v}_2]$ is a 2×2 matrix, known as the *sampling matrix*. Because $\mathbf{v}_1$ and $\mathbf{v}_2$ are chosen to be linearly independent, the determinant of $\mathbf{V}$ is nonzero. Note that $\mathbf{V}$ is not unique for a given sampling pattern and that two matrices representing the same sampling process are related by a linear transformation represented by a unimodular matrix [7].

Sampling an analog signal $x_a(\mathbf{t})$ on the lattice defined by (18-1) produces the discrete signal

$$x(\mathbf{n}) = x_a(\mathbf{Vn})$$

Figure 18-1(a) shows a hexagonal sampling lattice defined by the pair of sampling vectors

$$\mathbf{v}_1 = \begin{bmatrix} 2T_1 \\ 0 \end{bmatrix} \quad \text{and} \quad \mathbf{v}_2 = \begin{bmatrix} -T_1 \\ T_2 \end{bmatrix} \tag{18-2}$$

where $T_1 = \frac{1}{2}$ and $T_2 = \frac{\sqrt{3}}{2}$. The lattice is hexagonal since each sample location has exactly six nearest neighbors when $T_2 = T_1\sqrt{3}$.

Let the Fourier transform of $x_a(\mathbf{t})$ be defined by

$$X_a(\mathbf{\Omega}) = \int_{-\infty}^{+\infty} x_a(\mathbf{t}) \exp(-j\mathbf{\Omega}^T\mathbf{t})dt$$

where $\mathbf{\Omega} = [\Omega_1\ \Omega_2]^T$. Similarly, let the Fourier transform of the sequence $x(\mathbf{n})$ be defined as

$$X(\omega) = \sum_{\mathbf{n}} x(\mathbf{n}) \exp(-j\omega^T\mathbf{n}) \tag{18-3}$$

where $\omega = [\omega_1\ \omega_2]^T$. Mersereau and Dudgeon [4] showed that the spectrum of the sequence $x(\mathbf{n})$ and the spectrum of the signal $x_a(\mathbf{t})$ are related by

$$X(\omega) = \frac{1}{|\det \mathbf{V}|} \sum_{\mathbf{k}} X_a(\mathbf{V}^{-T}(\omega - 2\pi\mathbf{k})) \tag{18-4}$$

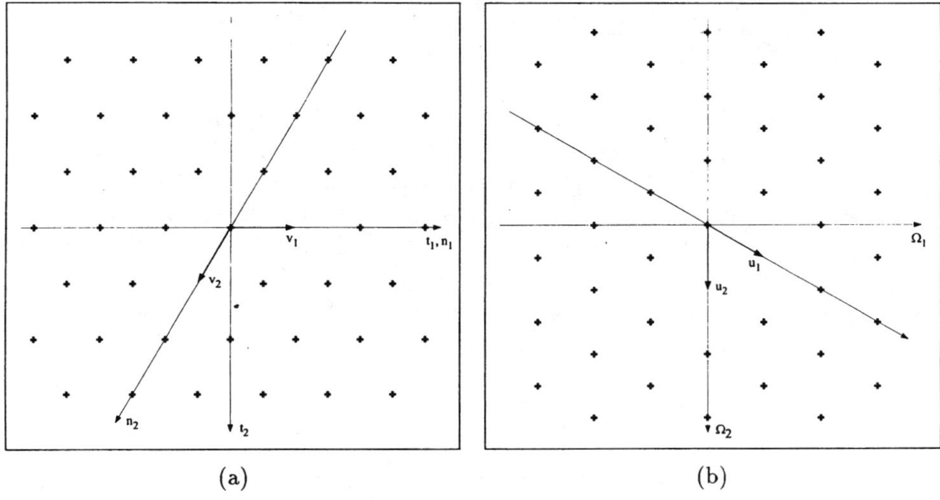

Figure 18-1   (a) A hexagonal sampling lattice in the spatial domain. (b) A hexagonal sampling lattice in the frequency domain.

where $\mathbf{k}$ is an integer-valued vector and $\mathbf{V}^{-T}$ denotes $(\mathbf{V}^{-1})^T$.

Alternatively, we can define the Fourier transform of the sequence $x(\mathbf{n})$ as

$$X_V(\mathbf{\Omega}) = \sum_{\mathbf{n}} x(\mathbf{n}) \exp(-j\mathbf{\Omega}^T \mathbf{V} \mathbf{n})$$

$$= X(\mathbf{V}^T \mathbf{\Omega}) \tag{18-5}$$

then Eq. (18-4) may be written as

$$X_V(\mathbf{\Omega}) = \frac{1}{|\det \mathbf{V}|} \sum_{\mathbf{k}} X_a(\mathbf{\Omega} - \mathbf{U}\mathbf{k}) \tag{18-6}$$

where

$$\mathbf{U} = 2\pi \mathbf{V}^{-T} \tag{18-7}$$

Thus, Eq. (17-6) can be interpreted as a periodic extension of $X_a(\mathbf{\Omega})$ with periodicity vectors $\mathbf{u}_1 = [u_{11}\, u_{21}]^T$ and $\mathbf{u}_2 = [u_{12}\, u_{22}]^T$, where $\mathbf{U} = [\mathbf{u}_1\, \mathbf{u}_2]$. The set of all vectors $\mathbf{\Omega}$ generated by $\mathbf{\Omega} = \mathbf{U}\mathbf{n}$ defines a lattice in the frequency domain known as the modulation or reciprocal lattice. Thus, the spectrum of a sequence $x(\mathbf{n})$ can be viewed as the convolution of the spectrum of $x_a(\mathbf{t})$ with a modulation lattice defined by $\mathbf{U}$.

Figure 18-1(b) shows the reciprocal lattice corresponding to the sampling vectors defined in Eq. (18-2), i.e., the lattice defined by the pair of modulation vectors

$$\mathbf{u}_1 = \begin{bmatrix} \frac{\pi}{T_1} \\ \frac{\pi}{T_2} \end{bmatrix} \quad \text{and} \quad \mathbf{u}_2 = \begin{bmatrix} 0 \\ \frac{2\pi}{T_2} \end{bmatrix}$$

18.2.2 Up-Sampling and Down-Sampling in Hexagonal Systems

Let Λ denote the integer lattice defined by the set of integer vectors $\mathbf{n}$, and let Λ_K denote the sampling sublattice generated by the subsampling matrix $\mathbf{K}$, that is the set of integer vectors $\mathbf{m}$ such that $\mathbf{m} = \mathbf{Kn}$. Note that in order to properly define a sublattice of Λ, a subsampling matrix must be nonsingular with integer-valued entries. In general, a sublattice of Λ is called separable if it can be represented by a diagonal matrix $\mathbf{K}$, otherwise it is called nonseparable. Figure 18-2 shows an integer sampling lattice Λ and a sampling sublattice Λ_K, for the separable sub-sampling matrix

$$\mathbf{K} = \begin{bmatrix} 2 & 0 \\ 0 & 2 \end{bmatrix} \tag{18-8}$$

With Λ and Λ_K defined in this way, we can view the operations of up-sampling and down-sampling as described below.

The process of up-sampling maps a signal on Λ to a new signal that is nonzero only at points on the sampling sublattice Λ_K. The output of an up-sampler is related to the input by

$$y(\mathbf{n}) = \begin{cases} x(\mathbf{K}^{-1}\mathbf{n}), & \text{if } \mathbf{K}^{-1}\mathbf{n} \in \Lambda \\ 0, & \text{otherwise} \end{cases}$$

It is easy to show [5] that the Fourier transform relates the output and input of an up-sampler by

$$Y(\omega) = X(\mathbf{K}^T\omega)$$

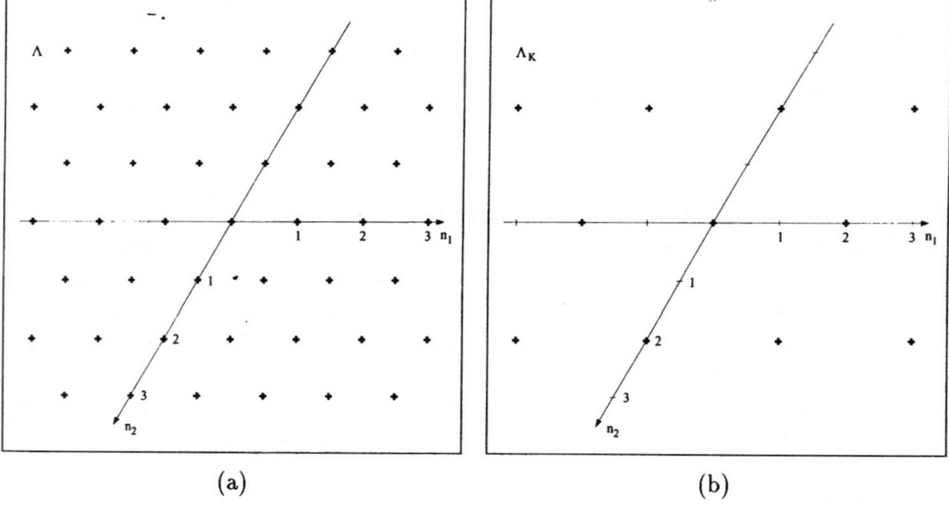

(a)   (b)

Figure 18-2   (a) An integer sampling lattice. (b) A sampling sublattice.

where $X(\omega)$ is defined by Eq. (18-3). Figure 18-3 shows the block diagram of an up-sampler and the process of up-sampling for the subsampling matrix defined by Eq. (18-8).

The process of down-sampling maps points on the sublattice Λ_K to Λ according to

$$y(\mathbf{n}) = x(\mathbf{Kn}) \tag{18-9}$$

and discards all other points.

The Fourier transform relation between the output and input of a down-sampler can be derived by introducing the concept of a sampling function $s_K(\mathbf{n})$ associated with the sampling matrix $\mathbf{K}$ [5], that is,

$$s_K(\mathbf{n}) = \begin{cases} 1, & \text{if } \mathbf{n} \in \Lambda_K \\ 0, & \text{otherwise} \end{cases}$$

Since $s_K(\mathbf{n})$ can be interpreted as a periodic sequence, with periodicity matrix $\mathbf{K}$, i.e., $s_K(\mathbf{n}) = s_K(\mathbf{n} + \mathbf{Km})$, it may be expressed as a Fourier series

$$s_K(\mathbf{n}) = \frac{1}{|\det \mathbf{K}|} \sum_{l=0}^{|\det \mathbf{K}|-1} \exp(-j2\pi \mathbf{k}_l^T \mathbf{K}^{-1} \mathbf{n}) \tag{18-10}$$

where each of the $|\det \mathbf{K}|$ vectors $\mathbf{k}_l = [k_{l_1}\ k_{l_2}]^T$ is associated with one of the cosets of $\mathbf{K}^T$. Notice that a coset of a sublattice Λ_K is defined as the set of points obtained by shifting the entire sublattice by an integer shift vector $\mathbf{k}$. There are exactly $|\det \mathbf{K}|$ distinct cosets of Λ_K, and their union is the integer lattice Λ. Each shift vector $\mathbf{k}_l$ associated with a certain coset is known as a coset vector. For example, one choice for the $\mathbf{k}_l$ given the sampling sublattice defined by Eq. (18-8) is

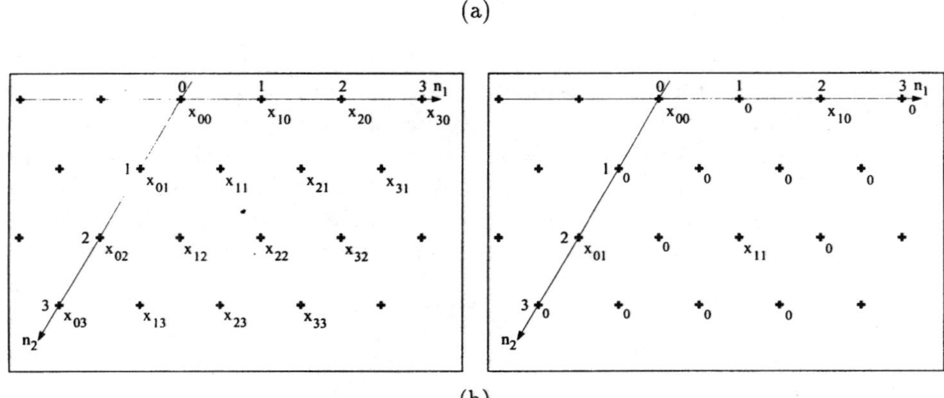

(a)

(b)

Figure 18-3   (a) An up-sample operator. (b) A mapping of samples under up-sampling: left—input signal; right—output signal.

$$\mathbf{k}_0 = \begin{bmatrix} 0 \\ 0 \end{bmatrix}, \qquad \mathbf{k}_1 = \begin{bmatrix} 1 \\ 0 \end{bmatrix}, \qquad \mathbf{k}_2 = \begin{bmatrix} 0 \\ 1 \end{bmatrix}, \qquad \text{and} \qquad \mathbf{k}_3 = \begin{bmatrix} 1 \\ 1 \end{bmatrix} \tag{18-11}$$

Let

$$w(\mathbf{n}) = x(\mathbf{n})s_{\mathbf{K}}(\mathbf{n}) \tag{18-12}$$

then it is easy to see from (18-9) that a down-sampled signal $y(\mathbf{n})$ can be written as

$$y(\mathbf{n}) = w(\mathbf{Kn})$$

since $w(\mathbf{n})$ equals $x(\mathbf{n})$ on $\Lambda_{\mathbf{K}}$. Therefore, the Fourier transform of the sequence $y(\mathbf{n})$ may be written as

$$Y(\omega) = \sum_{\mathbf{n}} w(\mathbf{Kn}) \exp(-j\omega^{\mathrm{T}}\mathbf{n}) \tag{18-13}$$

Since $w(\mathbf{n})$ is zero for $\mathbf{n}$ not in $\Lambda_{\mathbf{K}}$ we may write (18-13) as

$$Y(\omega) = \sum_{\mathbf{n}} w(\mathbf{n}) \exp(-j\omega^{\mathrm{T}}\mathbf{K}^{-1}\mathbf{n})$$
$$= W(\mathbf{K}^{-\mathrm{T}}\omega)$$

where $W(\omega)$ is the Fourier transform of the sequence $w(\mathbf{n})$. From (18-10) and (18-12) it is easy to show that

$$W(\omega) = \frac{1}{|\det \mathbf{K}|} \sum_{l=0}^{|\det \mathbf{K}|-1} X(\omega + 2\pi\mathbf{K}^{-\mathrm{T}}\mathbf{k}_l)$$

therefore, the Fourier transform relation between the output and input of a down-sampler is given by

$$Y(\omega) = \frac{1}{|\det \mathbf{K}|} \sum_{l=0}^{|\det \mathbf{K}|-1} X(\mathbf{K}^{-\mathrm{T}}(\omega + 2\pi\mathbf{k}_l))$$

Figure 18-4 shows the block diagram of a down-sampler and the process of down-sampling for the subsampling matrix defined in (18-8).

Note that the relations derived above are based on the Fourier transform defined in Eq. (18-3). However, a more general definition is described in Eq. (18-5). This formulation takes into account the lattice structure used to sample the original 2-D analog waveform and allows the Fourier transform relation between the input and output of an up-sampler and a down-sampler to be written as

$$Y_{\mathbf{V}}(\Omega) = X(\mathbf{K}^{\mathrm{T}}\mathbf{V}^{\mathrm{T}}\Omega) \tag{18-14}$$

and

$$Y_{\mathbf{V}}(\Omega) = \frac{1}{|\det \mathbf{K}|} \sum_{l=0}^{|\det \mathbf{K}|-1} X(\mathbf{K}^{-\mathrm{T}}(\mathbf{V}^{\mathrm{T}}\Omega + 2\pi\mathbf{k}_l)) \tag{18-15}$$

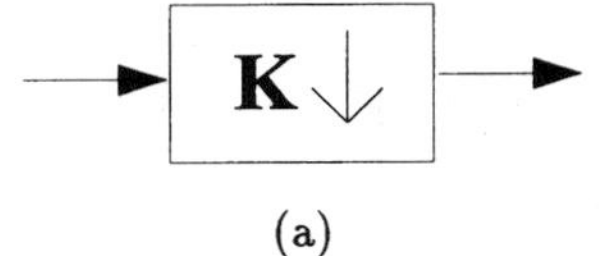

(a)

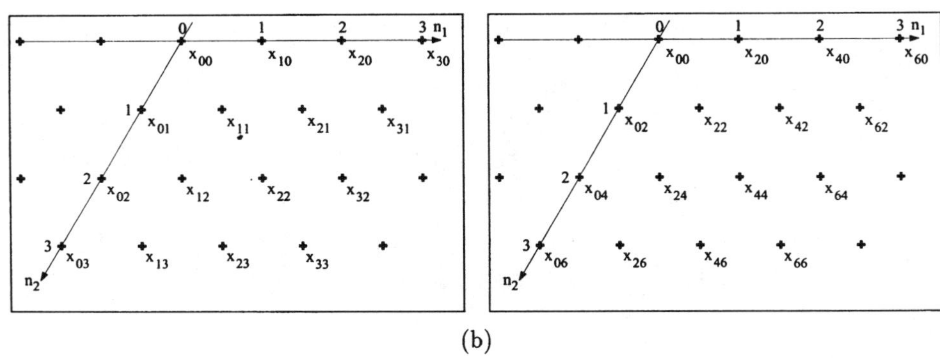

(b)

Figure 18-4   (a) A down-sample operator. (b) A mapping of samples under down-
sampling: left—input signal; right—output signal.

Therefore, if we assume $\mathbf{K}$ as defined in (18-8) we may write Eqs. (18-14) and
(18-15) as

$$Y_{\mathrm{V}}(\mathbf{\Omega}) = X_{\mathrm{V}}(\mathbf{K}^{\mathrm{T}}\mathbf{\Omega}) \tag{18-16}$$

and

$$Y_{\mathrm{V}}(\mathbf{\Omega}) = \frac{1}{|\det \mathbf{K}|} \sum_{l=0}^{|\det \mathbf{K}|-1} X_{\mathrm{V}}(\mathbf{K}^{-\mathrm{T}}\mathbf{\Omega} + \tilde{\mathbf{k}}_l) \tag{18-17}$$

respectively, where

$$\tilde{\mathbf{k}}_l = \mathbf{U}\mathbf{K}^{-\mathrm{T}}\mathbf{k}_l \tag{18-18}$$

18.2.3 Analysis/Synthesis Filter Banks in Hexagonal Systems

This section focuses on perfect reconstruction filter banks in hexagonal sampling systems and wavelets that can be obtained by iterating such filter banks. Parts
of this material are described in Simoncelli and Adelson [2], but are reviewed here for
completeness of presentation.

There are a wide variety of analysis/synthesis (A/S) filter banks for 2-D systems.
We restrict our focus to A/S filter banks in which each channel shares the same
subsampling matrix $\mathbf{K}$ and the number of channels equals $|\det \mathbf{K}|$. Figure 18-5 shows
a four-channel A/S filter bank. We further restrict our study to the separable sublattice defined in Eq. (18-8) since this choice will enable us to apply the A/S filter

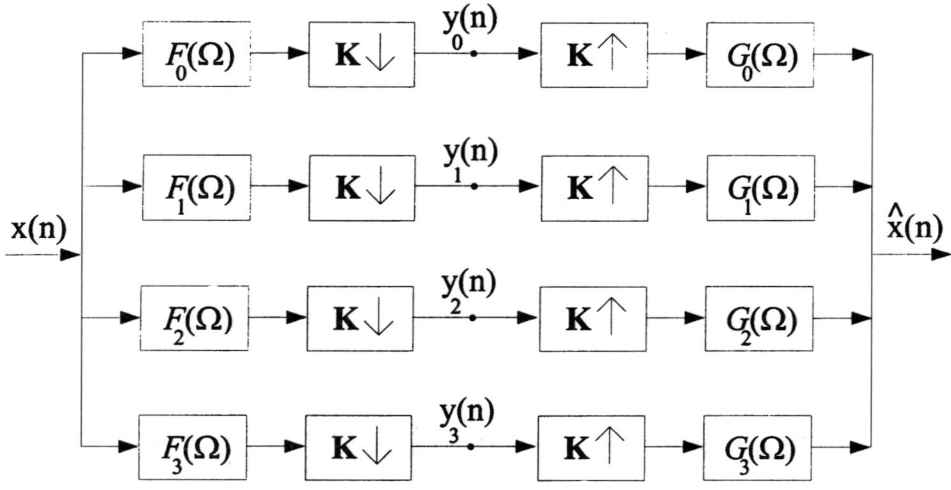

Figure 18-5 A 2-D four-channel A/S filter bank.

bank recursively to each of the subband signals $y_i(\mathbf{n})$ shown in Fig. 18-5 as described in [2].

Consider a four-channel A/S filter bank with $\mathbf{K}$ defined by Eq. (18-8), then using (18-17) we can show that the Fourier transform of $y_i(\mathbf{n})$ may be written as

$$Y_i(\mathbf{\Omega}) = \frac{1}{|\det \mathbf{K}|} \sum_{l=0}^{|\det \mathbf{K}|-1} F_i(\mathbf{K}^{-T}\mathbf{\Omega} + \tilde{\mathbf{k}}_l) X(\mathbf{K}^{-T}\mathbf{\Omega} + \tilde{\mathbf{k}}_l) \tag{18-19}$$

where the subindex V has been suppressed for simplicity. Similarly, using (18-16) we have that the Fourier transform of $\hat{x}(\mathbf{n})$ is given by

$$\hat{X}(\mathbf{\Omega}) = \sum_{i=0}^{|\det \mathbf{K}|-1} G_i(\mathbf{\Omega}) Y_i(\mathbf{K}^{T}\mathbf{\Omega}) \tag{18-20}$$

Therefore, combining Eqs. (18-19) and (18-20) we obtain an overall filter-bank response of

$$\hat{X}(\mathbf{\Omega}) = \frac{1}{|\det \mathbf{K}|} \sum_{i=0}^{|\det \mathbf{K}|-1} G_i(\mathbf{\Omega}) \sum_{l=0}^{|\det \mathbf{K}|-1} F_i(\mathbf{\Omega} + \tilde{\mathbf{k}}_l) X(\mathbf{\Omega} + \tilde{\mathbf{k}}_l)$$

$$= \frac{1}{|\det \mathbf{K}|} \sum_{l=0}^{|\det \mathbf{K}|-1} X(\mathbf{\Omega} + \tilde{\mathbf{k}}_l) \left[\sum_{i=0}^{|\det \mathbf{K}|-1} G_i(\mathbf{\Omega}) F_i(\mathbf{\Omega} + \tilde{\mathbf{k}}_l) \right] \tag{18-21}$$

Combining Eqs. (18-7), (18-11), and (18-18) for the values of T_1 and T_2 in Eq. (18-2) yields the following set of vectors $\tilde{\mathbf{k}}_l$

$$\tilde{\mathbf{k}}_0 = \begin{bmatrix} 0 \\ 0 \end{bmatrix}, \qquad \tilde{\mathbf{k}}_1 = \begin{bmatrix} \pi \\ \frac{\pi}{\sqrt{3}} \end{bmatrix}, \qquad \tilde{\mathbf{k}}_2 = \begin{bmatrix} 0 \\ \frac{2\pi}{\sqrt{3}} \end{bmatrix}, \qquad \text{and} \qquad \tilde{\mathbf{k}}_3 = \begin{bmatrix} \pi \\ \frac{3\pi}{\sqrt{3}} \end{bmatrix}. \tag{18-22}$$

From Eq. (18-22) it is clear that one term of the sum in Eq. (18-21) corresponds to the linear shift invariant (LSI) system response, and the remaining terms correspond to the system alias. The A/S filter bank for which the system aliasing terms in Eq. (18-21) are canceled is generally known as a quadrature mirror filter (QMF) bank.

We can choose the filters to eliminate the aliasing terms in Eq. (18-21) as follows

$$
\begin{aligned}
F_0(\mathbf{\Omega}) &= G_0(-\mathbf{\Omega}) = H(\mathbf{\Omega}) = H(-\mathbf{\Omega}), \\
F_1(\mathbf{\Omega}) &= G_1(-\mathbf{\Omega}) = \exp(j\mathbf{\Omega}^\mathrm{T} \mathbf{s}_1)H(\mathbf{\Omega} + \tilde{\mathbf{k}}_1) \\
F_2(\mathbf{\Omega}) &= G_2(-\mathbf{\Omega}) = \exp(j\mathbf{\Omega}^\mathrm{T} \mathbf{s}_2)H(\mathbf{\Omega} + \tilde{\mathbf{k}}_2) \\
F_3(\mathbf{\Omega}) &= G_3(-\mathbf{\Omega}) = \exp(j\mathbf{\Omega}^\mathrm{T} \mathbf{s}_3)H(\mathbf{\Omega} + \tilde{\mathbf{k}}_3)
\end{aligned}
\tag{18-23}
$$

where $\mathbf{s}_1$, $\mathbf{s}_2$ and $\mathbf{s}_3$ must satisfy the following equations

$$
\begin{aligned}
1 + e^{j\tilde{\mathbf{k}}_1^\mathrm{T}\mathbf{s}_1} = 0, && e^{j\tilde{\mathbf{k}}_1^\mathrm{T}\mathbf{s}_2} + e^{j\tilde{\mathbf{k}}_1^\mathrm{T}\mathbf{s}_3} = 0 \\
1 + e^{j\tilde{\mathbf{k}}_2^\mathrm{T}\mathbf{s}_2} = 0, && e^{j\tilde{\mathbf{k}}_2^\mathrm{T}\mathbf{s}_1} + e^{j\tilde{\mathbf{k}}_2^\mathrm{T}\mathbf{s}_3} = 0 \\
1 + e^{j\tilde{\mathbf{k}}_3^\mathrm{T}\mathbf{s}_3} = 0, && e^{j\tilde{\mathbf{k}}_3^\mathrm{T}\mathbf{s}_1} + e^{j\tilde{\mathbf{k}}_3^\mathrm{T}\mathbf{s}_2} = 0
\end{aligned}
$$

Therefore, a suitable choice for the vectors $\mathbf{s}_l$ given the vectors $\tilde{\mathbf{k}}_l$ in Eq. (18-22) is

$$
\mathbf{s}_1 = \begin{bmatrix} 1 \\ 0 \end{bmatrix}, \qquad
\mathbf{s}_2 = \begin{bmatrix} 1/2 \\ \sqrt{3}/2 \end{bmatrix}, \qquad \text{and } \mathbf{s}_3 = \begin{bmatrix} 1/2 \\ -\sqrt{3}/2 \end{bmatrix}
\tag{18-24}
$$

After canceling all of the aliasing terms in Eq. (21) the remaining LSI system response becomes

$$
\begin{aligned}
\hat{X}(\mathbf{\Omega}) &= \frac{1}{4} X(\mathbf{\Omega}) \sum_{i=0}^{3} G_i(\mathbf{\Omega})F_i(\mathbf{\Omega}) \\
&= \frac{1}{4} X(\mathbf{\Omega}) \sum_{i=0}^{3} H(\mathbf{\Omega} + \tilde{\mathbf{k}}_i)H(-\mathbf{\Omega} + \tilde{\mathbf{k}}_i) \\
&= \frac{1}{4} X(\mathbf{\Omega}) \sum_{i=0}^{3} |H(\mathbf{\Omega} + \tilde{\mathbf{k}}_i)|^2
\end{aligned}
$$

Note that the aliasing cancellation is exact, and independent of the choice of $H(\mathbf{\Omega})$, and the design problem is reduced to finding a filter satisfying the constraint

$$
\sum_{i=0}^{3} |H(\mathbf{\Omega} + \tilde{\mathbf{k}}_i)|^2 = 4
\tag{18-25}
$$

A low-pass solution for $H(\mathbf{\Omega})$ in the above equation results in a band-splitting system which may be cascaded hierarchically through the low-pass band of the QMF bank to produce a multiresolution decomposition in two dimensions. Simoncelli and Adelson [2] describe a simple frequency-sampling design method that produces hexagonally symmetric QMFs with small regions of support for which perfect reconstruction is well approximated. Figure 18-6 shows the region of support of a five-ring

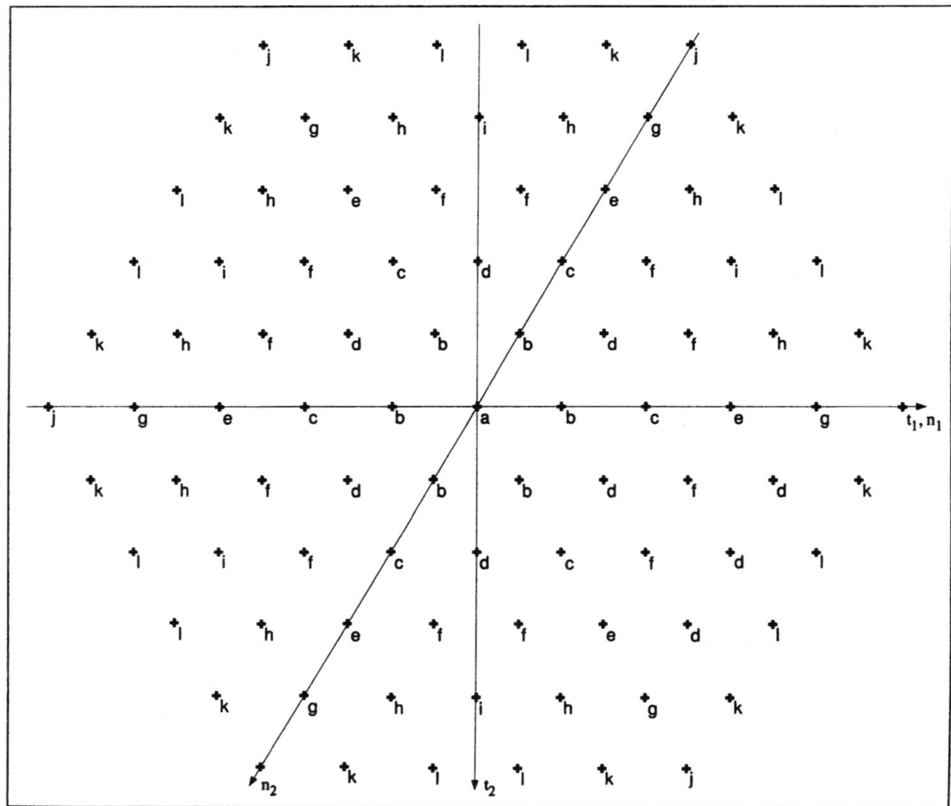

Figure 18-6   Region of support of a five-ring hexagonally symmetric filter. The parameters a through l refer to the low-pass filter coefficients $h(\mathbf{n})$.

hexagonally symmetric filter. Notice that the size of the filter is measured in terms of the number of hexagonal rings it contains. The parameters a through l in Fig. 18-6 refer to the filter coefficients of the low-pass solution $h(\mathbf{n})$ computed in [2]. Figure 18-7(a) shows an idealized diagram of the partition of the frequency domain resulting from a two-level hexagonal multiresolution decomposition.

18.2.4 Redundant Analysis/Synthesis Filter Banks in Hexagonal Systems

In this section we discuss the mathematical formulation of redundant A/S filter banks in hexagonal systems. In particular, we would like to find equivalent filters for the ith stage of the traditional A/S system shown in Fig. 18-7(b).

It can be easily shown that subsampling by $\mathbf{K}$ followed by filtering with $F_0(\boldsymbol{\Omega})$ is equivalent to filtering by $F_0(\mathbf{K}\boldsymbol{\Omega})$ followed by subsampling. Hence, the first two steps

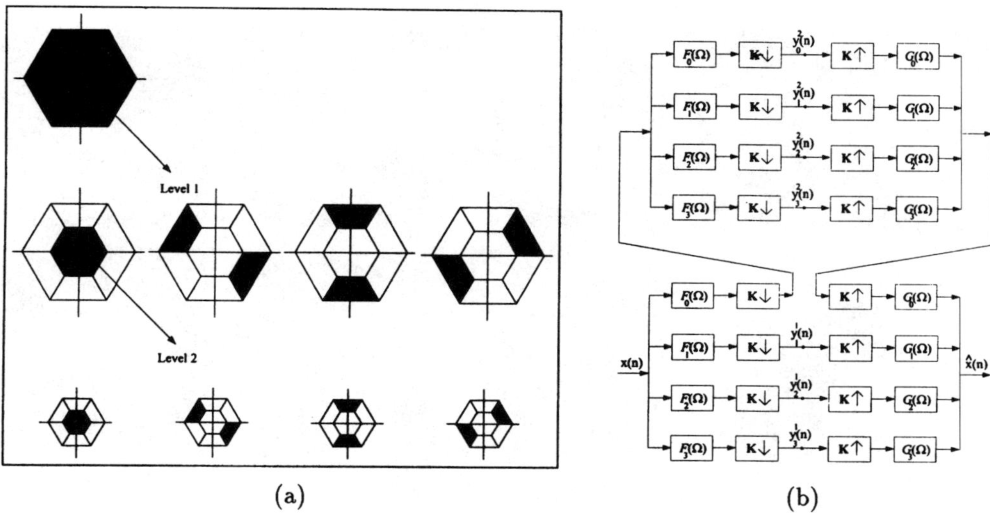

Figure 18-7   (a) Partitions of the frequency domain resulting from a two-level multi-resolution decomposition of hexagonal filters. The upper left frequency diagram represents the spectrum of the original image. (b) A two-stage four-channel A/S filter bank.

of low-pass filtering in Fig. 18-7(b) can be replaced by a filter with Fourier transform $F_0(\Omega)F_0(\mathbf{K}\Omega)$, followed by subsampling by $\mathbf{K}^2$.

In general, equivalent filters for the tth stage ($i \geq 1$) of a cascade of analysis filters are given by

$$F_0^i(\Omega) = F_0(\mathbf{K}^{i-1}\Omega) \prod_{l=0}^{i-2} F_0(\mathbf{K}^l\Omega)$$

$$F_1^i(\Omega) = F_1(\mathbf{K}^{i-1}\Omega) \prod_{l=0}^{i-2} F_0(\mathbf{K}^l\Omega)$$

$$F_2^i(\Omega) = F_2(\mathbf{K}^{i-1}\Omega) \prod_{l=0}^{i-2} F_0(\mathbf{K}^l\Omega) \qquad (18\text{-}26)$$

$$F_3^i(\Omega) = F_3(\mathbf{K}^{i-1}\Omega) \prod_{l=0}^{i-2} F_0(\mathbf{K}^l\Omega)$$

followed by subsampling by $\mathbf{K}^i$. The synthesis filters are obtained in a similar way.

By removing the operations of down-sampling and up-sampling from the resulting equivalent A/S system we obtain an overcomplete hexagonal multiresolution representation. From Eq. (18-25), perfect reconstruction is also accomplished in this case. Figure 18-8 shows the magnitude of the equivalent hexagonal filters F^i for levels 1 and 2 for the four-ring filter coefficients computed in [2].

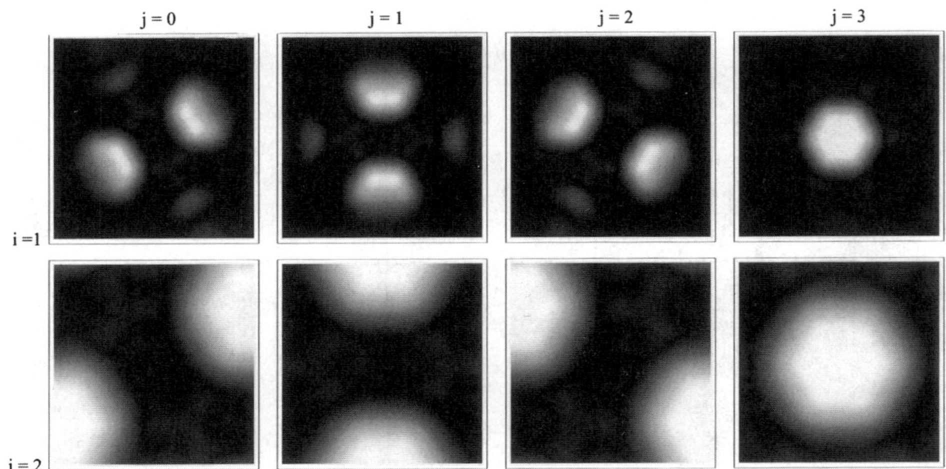

Figure 18-8   Analyzing filters F_j^i for levels 1 and 2.

18.2.5 The Discrete Fourier Transform in Hexagonal Systems

Let $\tilde{x}(\mathbf{n})$ be a periodic sequence with periodicity matrix $\mathbf{N}$, i.e., $\tilde{x}(\mathbf{n}) = \tilde{x}(\mathbf{n} + \mathbf{Nr})$ for any integer vector $\mathbf{r}$. Then it is easy to verify [4] that the following Fourier series relations hold,

$$\tilde{x}(\mathbf{n}) = \frac{1}{|\det \mathbf{N}|} \sum_{\mathbf{k} \in J} \tilde{X}(\mathbf{k}) \exp(j\mathbf{k}^{\mathrm{T}}(2\pi \mathbf{N}^{-1})\mathbf{n}) \tag{18-27}$$

and,

$$\tilde{X}(\mathbf{k}) = \sum_{\mathbf{n} \in I} \tilde{x}(\mathbf{n}) \exp(-j\mathbf{k}^{\mathrm{T}}(2\pi \mathbf{N}^{-1})\mathbf{n}) \tag{18-28}$$

where I and J denote finite-extent regions (consisting of $|\det \mathbf{N}|$ samples) in the $\mathbf{n}$-domain and $\mathbf{k}$-domain, respectively.

Let $x(\mathbf{n})$ be any finite-extent sequence confined to a region I containing S samples. We say that the sequence $x(\mathbf{n})$ admits a periodic extension $\tilde{x}(\mathbf{n})$ with periodicity matrix $\mathbf{N}$ if

$$\tilde{x}(\mathbf{n}) = \tilde{x}(\mathbf{n} + \mathbf{Nr})$$
$$\tilde{x}(\mathbf{n}) = x(\mathbf{n}), \text{ for } \mathbf{n} \in I$$

and $S = |\det \mathbf{N}|$.

If $x(\mathbf{n})$ admits a periodic extension with periodicity matrix $\mathbf{N}$ then the Fourier series relation in Eq. (18-28) can be used to define its discrete Fourier transform (DFT) as follows,

$$\hat{X}(\mathbf{k}) = \tilde{X}(\mathbf{k}), \qquad \mathbf{k} \in J$$

$$= \sum_{\mathbf{n} \in I} x(\mathbf{n}) \exp(-j\mathbf{k}^{\mathrm{T}}(2\pi \mathbf{N}^{-1})\mathbf{n}), \qquad \mathbf{k} \in J \tag{18-29}$$

Similarly, the Fourier series relation in Eq. (18-27) can be used to recover the sequence $x(\mathbf{n})$ from its DFT as follows,

$$x(\mathbf{n}) = \tilde{x}(\mathbf{n}), \qquad \mathbf{n} \in I$$

$$= \frac{1}{|\det \mathbf{N}|} \sum_{\mathbf{k} \in J} \hat{X}(\mathbf{k}) \exp(j\mathbf{k}^{\mathrm{T}}(2\pi \mathbf{N}^{-1})\mathbf{n}), \qquad \mathbf{n} \in I. \tag{18-30}$$

Suppose $x(\mathbf{n})$ is a hexagonally sampled signal with support confined to a region I containing $(2N_1 + N_2)N_2$ samples. In addition suppose that $x(\mathbf{n})$ admits a periodic extension with periodicity matrix

$$\mathbf{N} = \begin{bmatrix} N_1 + N_2 & N_2 \\ N_2 & 2N_2 \end{bmatrix} \tag{18-31}$$

Then, we say that $x(\mathbf{n})$ admits a hexagonally periodic extension. Notice that $x(\mathbf{n})$ may admit more than one periodic extension and that each periodic extension defines a different DFT. It is easy to show that the DFT of each admissible periodic extension of $x(\mathbf{n})$ corresponds to a sampled version of its Fourier transform, and that the sampling lattice is controlled through the periodicity matrix $\mathbf{N}$. This result is a consequence of the following relation between the Fourier transform (18-3) and the DFT (18-29) of $x(\mathbf{n})$,

$$\hat{X}(\mathbf{k}) = X(\omega)|_{\omega = 2\pi \mathbf{N}^{-\mathrm{T}} \mathbf{k}}$$

Using the more general definition of the Fourier transform (18-5) that accounts for the sampling lattice used to sample the original 2-D analog signal we can write

$$\hat{X}(\mathbf{k}) = X_{\mathbf{V}}(\mathbf{\Omega})|_{\mathbf{\Omega} = 2\pi(\mathbf{VN})^{-\mathrm{T}} \mathbf{k}} \tag{18-32}$$

It then follows from Eqs. (18-31) and (18-32) that for $N_1 = N_2 = N$, i.e., for

$$\mathbf{N} = \begin{bmatrix} 2N & N \\ N & 2N \end{bmatrix} \tag{18-33}$$

the DFT of $x(\mathbf{n})$ corresponds to a hexagonal sampled version of its Fourier transform. In this case Eq. (18-29) is referred to as the hexagonal discrete Fourier transform (HDFT) of $x(\mathbf{n})$. It is easy to verify from (18-29) and (18-33) that the HDFT of $x(\mathbf{n})$ is given by

$$\hat{X}(k_1, k_2) = \sum_{(n_1, n_2) \in I} x(n_1, n_2) \exp\left[-j\frac{2\pi}{3N}((2n_1 - n_2)k_1 + (2n_2 - n_1)k_2) \right]$$

$$= \sum_{(n_1, n_2) \in I} x(n_1, n_2) \exp\left[-j\left(\frac{\pi}{3N}(2n_1 - n_2)(2k_1 - k_2) + \frac{\pi}{N} n_2 k_2 \right) \right] \tag{18-34}$$

Mersereau and Dudgeon [4] showed that efficient algorithms for the implementation of the discrete Fourier transform (18-29) exist if the periodicity matrix $\mathbf{N}$ is composite (i.e., if $\mathbf{N}$ can be factored into a nontrivial product of integer matrices). For $N = 2^l, l \geq 0$ we observe that the periodicity matrix $\mathbf{N}$ in (18-33) can be factored as,

$$\mathbf{N} = \begin{bmatrix} 2 & 1 \\ 1 & 2 \end{bmatrix} \begin{bmatrix} 2 & 0 \\ 0 & 2 \end{bmatrix}^l \tag{18-35}$$

This factorization leads to an efficient implementation of (18-34) known as the hexagonal fast Fourier transform (HFFT). Implementation details of the HFFT can be found in [1].

18.3. IMPLEMENTATION

Next, we present implementation details of hexagonal multiresolution representations. Section 18.3.1 describes the selection of image support for efficient signal processing with hexagonal systems. Section 18.3.2 describes filtering in hexagonal sampling systems using a HFFT-based strategy and the computation of hexagonal multiresolution representations. Finally, section 18.3.3 describes the computation of overcomplete hexagonal multiresolution representations. A listing of MATLAB functions implementing hexagonal multiresolution representations is available at `http://www.iprg.cise.ufl.edu/`.

18.3.1 Image Support in Hexagonal Systems

Efficient discrete signal processing in hexagonal sampling systems can be achieved by using the HFFT but at the cost of restricting sequences to those that admit a hexagonally periodic extension with $N_1 = N_2 = N$ and $N = 2^l, l \geq 0$. Systematic application of the HFFT to image processing further restricts sequences to rectangular regions. It is easy to verify that for any $N = 2^l, l \geq 1$ it is possible to find a rectangular region I_l such that any sequence with support confined to I_l admits a hexagonally periodic extension. Figure 18-9 shows such a sequence confined to region I_1 and its hexagonally periodic extension. For any $l \leq 1$, region I_{l+1} may be obtained from I_l by doubling the number of rows and the number of samples per rows in I_l. Notice that given the periodic sequence shown in Fig. 18-9 there exists more than one fundamental period such that when extended periodically in a hexagonal fashion the result is the same periodic sequence. In particular, Fig. 18-9 shows a parallelogram P_1 containing an alternative set of samples that could be used to define the same periodic sequence but greatly simplifies the HFFT computation. The simplification comes from the fact that samples contained in the parallelogram can be stored in an array of size $N \times 3N$ where the indices of the array directly correspond to the coordinates (n_1, n_2) of the samples. It is straightforward to verify that for any sequence confined to $I_l, l \geq 1$ there exists a parallelogram P_l containing an alternative set of samples that defines the same periodic sequence. For any $l \leq 1$, P_{l+1} is obtained from P_l in the same way I_{l+1} was obtained from I_l. It follows that

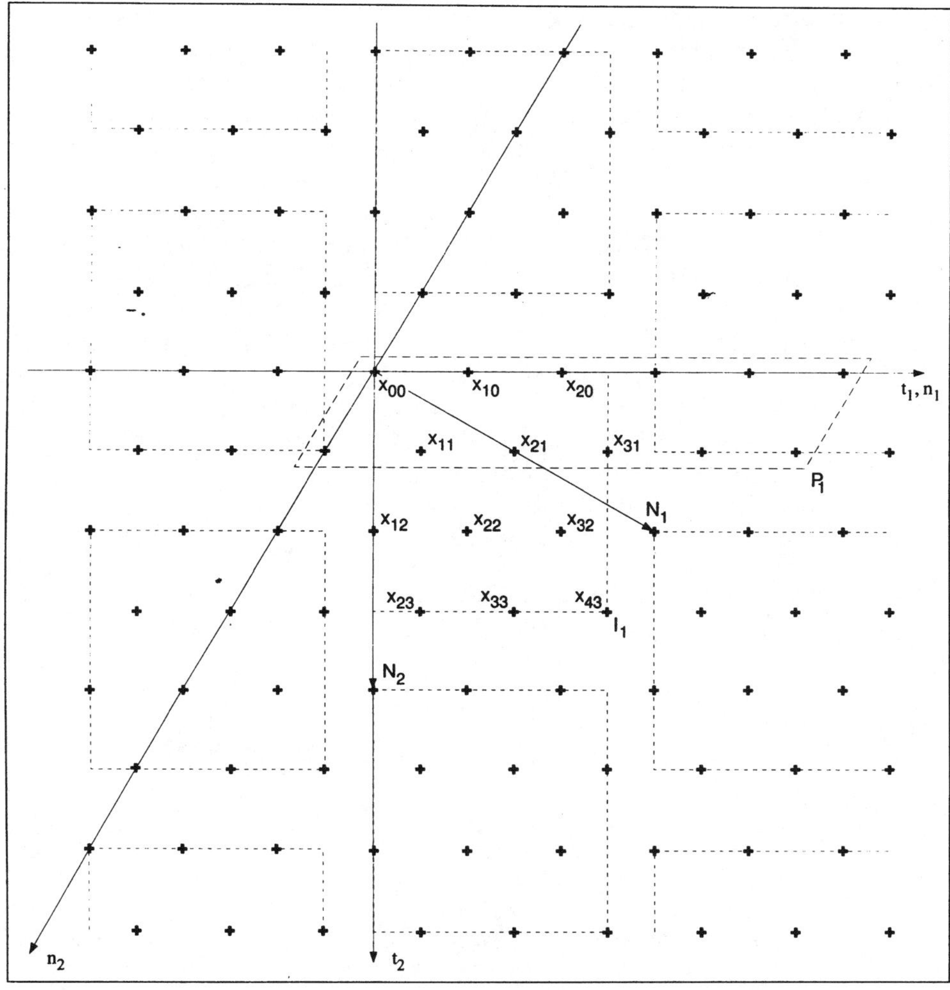

Figure 18-9 A hexagonally sampled sequence confined to a rectangular region I_1 and its hexagonally periodic extension. Parallelogram P_1 contains an alternative set of samples that defines the same periodic sequence.

computation of the HFFT for a sequence confined to I_l can be accomplished by using Eq. (18-34) where I corresponds to a region defined by P_l.

Although there are certain image-processing applications in which image samples are acquired in a hexagonal fashion, most digital detectors (e.g., charge-coupled devices, CCDs) sample images rectangularly with the same sampling rate along both directions. Processing a rectangularly sampled image with a hexagonal sampling system requires the rectangularly sampled image to be mapped into a hexagonal sampling lattice [2, 3]. For square images whose size is a power of two we describe a strategy that maps a rectangularly sampled image confined to a square region consisting of $2N \times 2N$ samples ($N = 2^l$, $l \geq 1$) into a hexagonally sampled image confined to a rectangular region for which it is possible to compute its HFFT. This mapping may be accomplished in two distinct ways. If the original image is oversampled, we first interpolate horizontally by a factor of 3 and then mask the result with the masking function $M_1(n_h, n_v) = \frac{1}{4}(1 + (-1)^{n_h})(1 + (-1)^{n_h/2 + n_v})$ as shown in Fig. 18-10. The resulting image is confined to region I_l and consists of $3N^2$ samples (25% fewer samples than the original). In this case, it is assumed that oversampling of the original image accounts for the reduction of the number of samples in the resulting image. Alternatively, if the original image is critically sampled, we interpolate horizontally by a factor of 3 and vertically by a factor of 2 and mask the result with the masking function $M_2(n_h, n_v) = \frac{1}{2}(1 + (-1)^{n_h + n_v})$. In this case the resulting image is confined to the region I_{l+1} and consists of $3(2N)^2$ samples. In each case the resulting sampling lattice gives a reasonable geometric *approximation* to a hexagonal sampling lattice. Note that the oversampled method or the critically sampled method, together with the equivalence between I_l and P_l can be used to map a $2N \times 2N$ rectangularly sampled image into P_l or P_{l+1}, respectively. Figure 18-11 shows a 512×512 rectangularly sampled image and its mapping into P_9 using the critically sampled strategy described above.

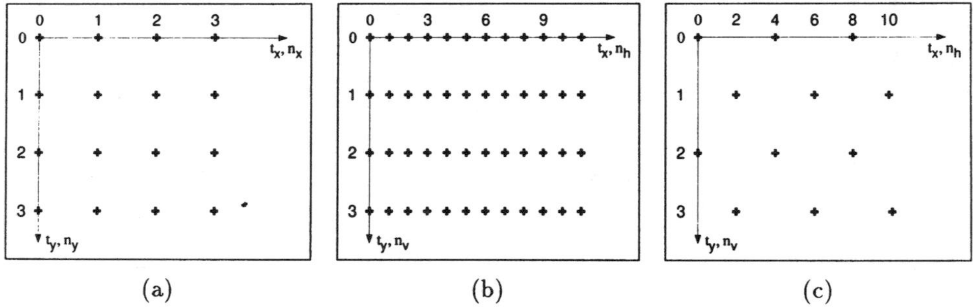

Figure 18-10	Mapping a rectangularly sampled image into a hexagonal sampling lattice: (a) rectangular lattice; (b) intermediate lattice with interpolated samples; (c) sampling function.

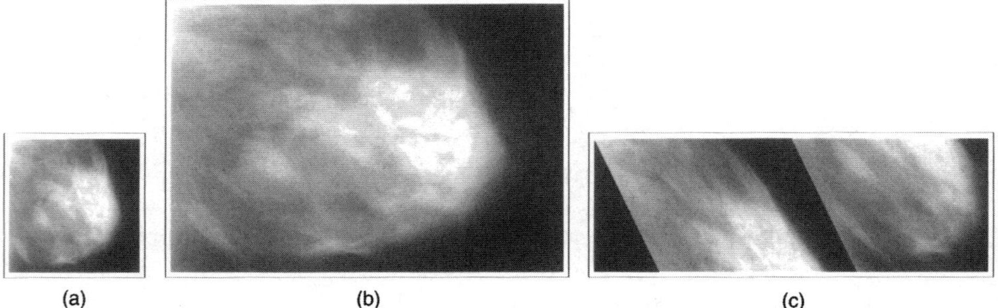

Figure 18-11	(a) A 512×512 rectangularly sampled radiograph of the breast. (b) An image with interpolated samples. (c) A hexagonally sampled image confined to P_9—axes n_1 and n_2 are shown at 90 degrees.

18.3.2 Multiresolution Representations in Hexagonal Systems

Filtering an image with a hexagonal filter may be accomplished by computing the product of the HFFTs of the image and the filter kernel and taking the inverse hexagonal fast Fourier transform (IHFFT) of the result. This approach requires that both the image and the filter kernel be supported in the same region I_l. Again, computation of the HFFT can be accomplished using P_l instead of I_l due to periodicity. Figure 18-12 shows a block diagram of the filtering process described above for a rectangularly sampled signal confined to a square region of size $2N \times 2N$, $N = 2^l$.

Here, we are interested in filtering an image with the quadrature mirror filters in Eq. (18-23) given the low-pass solution $h(n_1, n_2)$ computed in [2]. Replacing equations (18-22) and (18-24) in (18-23) and using relations (18-5) and (18-32) we obtain the following HDFT relations for the filters,

$$\hat{F}_0(\mathbf{k}) = \hat{G}_0(-\mathbf{k}) = \sum_{\mathbf{n} \in I} h_0(\mathbf{n}) \exp(-j\mathbf{k}^{\mathrm{T}}(2\pi\mathbf{N}^{-1})\mathbf{n})$$

$$\hat{F}_1(\mathbf{k}) = \hat{G}_1(-\mathbf{k}) = \exp\left(j\frac{2\pi}{3N}(2k_1 - k_2)\right) \sum_{\mathbf{n} \in I} h_1(\mathbf{n}) \exp(-j\mathbf{k}^{\mathrm{T}}(2\pi\mathbf{N}^{-1})\mathbf{n})$$

$$\hat{F}_2(\mathbf{k}) = \hat{G}_2(-\mathbf{k}) = \exp\left(j\frac{2\pi}{3N}(k_1 + k_2)\right) \sum_{\mathbf{n} \in I} h_2(\mathbf{n}) \exp(-j\mathbf{k}^{\mathrm{T}}(2\pi\mathbf{N}^{-1})\mathbf{n}) \qquad (18\text{-}36)$$

$$\hat{F}_3(\mathbf{k}) = \hat{G}_3(-\mathbf{k}) = \exp\left(j\frac{2\pi}{3N}(k_1 - 2k_2)\right) \sum_{\mathbf{n} \in I} h_3(\mathbf{n}) \exp(-j\mathbf{k}^{\mathrm{T}}(2\pi\mathbf{N}^{-1})\mathbf{n})$$

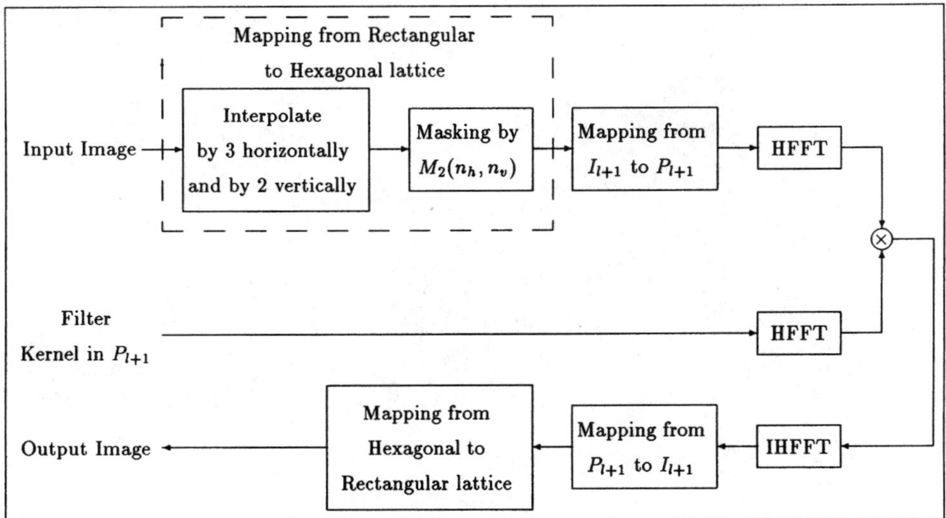

Figure 18-12   Filtering a rectangularly sampled image of size $2N \times 2N$, $N = 2^l$ with a hexagonally sampled system by means of the HFFT. Critically sampled case shown.

where $h_0(\mathbf{n}) = h(n_1, n_2)$, $h_1(\mathbf{n}) = (-1)^{n_1} h(n_1, n_2)$, $h_2(\mathbf{n}) = (-1)^{n_2} h(n_1, n_2)$, $h_3(\mathbf{n}) = (-1)^{n_1 + n_2} h(n_1, n_2)$, and I is the region of support of the filter kernels. It follows from the equations above that the HFFT of filters $f_k(\mathbf{n})$ can be obtained by modulating the HFFTs of the kernels $h_k(\mathbf{n})$ with complex exponentials. Notice that filtering an image confined to a region I_l with an r-ring filter kernel $f_k(\mathbf{n})$ using the HFFT-filtering strategy described above requires that the region of support of the kernel be confined to I_l (or equivalently to P_l). Indeed, this requirement is satisfied as long as $l \geq [\log_2(2 + r_1)] - 1$.

A one-level multiresolution decomposition of an image can be obtained by filtering the image with the filters kernels $f_k(\mathbf{n})$ followed by down-sampling (as shown in Fig. 18-5.) This can be accomplished using the HFFT-filtering strategy described above if the image is confined to I_l or P_l. Notice that if we work with sequences confined to P_l then the down-sampling operation is equivalent to taking every other row and every other column of the array storing the sequence. An $(L + 1)$-level multiresolution decomposition can then be obtained recursively by cascading an analysis section through the low-pass branch of an L-level multiresolution decomposition. Notice that the maximum number of levels is limited by the smallest P_l supporting the filter kernels. Figure 18-13 shows a two-level hexagonal multiresolution decomposition and reconstruction using the three-ring low-pass solution to (18-23) computed in [2]. An algorithm for the multiresolution reconstruction follows directly from its decomposition.

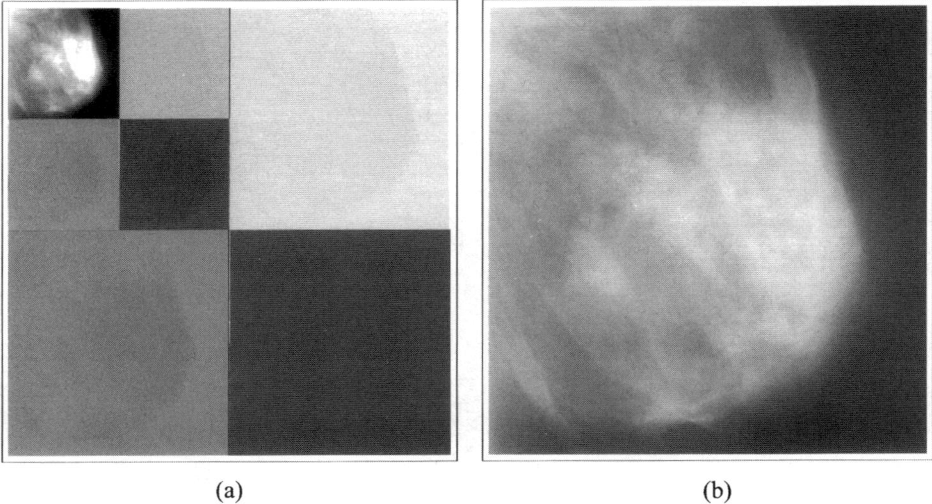

(a)   (b)

Figure 18-13 A two-level hexagonal multiresolution representation: (a) decomposition; (b) reconstruction. Both images are displayed on their original sampling lattice.

18.3.3 Overcomplete Multiresolution Representations in Hexagonal Systems

An overcomplete hexagonal multiresolution representation is computed by filtering an image with the equivalent filters introduced in section 18.2.4. Using the HFFT-filtering strategy described previously it follows that the equivalent filters F_k^i can be computed taking the product of the HFFTs of filter kernels $f_{k,l}(\mathbf{n})$ where $F_{k,l}(\mathbf{\Omega}) = F_k(\mathbf{K}^l\mathbf{\Omega})$. It is easy to show that the following HDFT relation holds for $f_{k,l}(\mathbf{n})$,

$$\hat{F}_{k,l}(\mathbf{k}) = \hat{F}_k(\mathbf{K}^l\mathbf{k})$$

This result combined with equation (18-36) leads to the following HDFT relations for the filters,

$$\hat{F}_{0,l}(\mathbf{k}) = \hat{G}_{0,l}(-\mathbf{k}) = \sum_{\mathbf{n}\in I} h_{0,l}(\mathbf{n}) \exp(-j\mathbf{k}^{\mathrm{T}}(2\pi\mathbf{N}^{-1})\mathbf{n})$$

$$\hat{F}_{1,l}(\mathbf{k}) = \hat{G}_{1,l}(-\mathbf{k}) = \exp\left(j\frac{2^{l+1}\pi}{3N}(2k_1 - k_2)\right) \sum_{\mathbf{n}\in I} h_{1,l}(\mathbf{n}) \exp(-j\mathbf{k}^{\mathrm{T}}(2\pi\mathbf{N}^{-1})\mathbf{n})$$

$$\hat{F}_{2,l}(\mathbf{k}) = \hat{G}_{2,l}(-\mathbf{k}) = \exp\left(j\frac{2^{l+1}\pi}{3N}(k_1 + k_2)\right) \sum_{\mathbf{n}\in I} h_{2,l}(\mathbf{n}) \exp(-j\mathbf{k}^{\mathrm{T}}(2\pi\mathbf{N}^{-1})\mathbf{n})$$

$$\hat{F}_{3,l}(\mathbf{k}) = \hat{G}_{3,l}(-\mathbf{k}) = \exp\left(j\frac{2^{l+1}\pi}{3N}(k_1 - 2k_2)\right) \sum_{\mathbf{n}\in I} h_{3,l}(\mathbf{n}) \exp(-j\mathbf{k}^{\mathrm{T}}(2\pi\mathbf{N}^{-1})\mathbf{n})$$

$$(18\text{-}37)$$

where

$$h_{k,l}(\mathbf{n}) = \begin{cases} h\left((\mathbf{K}^l)^{-1}\mathbf{n}\right), & \text{if } (\mathbf{K}^l)^{-1}\mathbf{n} \text{ is an integer vector} \\ 0, & \text{otherwise} \end{cases}$$

Note that in the above derivation we used the fact that $\mathbf{K}$ defined a separable subsampling matrix. It follows from the equation above that the HFFT of the filter kernels $f_{k,l}(\mathbf{n})$ can be obtained by modulating the HFFTs of the kernels $h_{k,l}(\mathbf{n})$ with complex exponentials. The filters $h_{k,l}(\mathbf{n})$ can be constructed by up-sampling $h_k(\mathbf{n})$ with sampling matrix $\mathbf{K}^l$. The filters F_k^i can then be computed following equation (18-26) with the filters given in (18-37). Notice that it is possible to discard the complex exponentials and still obtain perfect reconstruction. However, this will define a different set of filters. A hexagonal overcomplete multiresolution representation of an image can be obtained by filtering the image with the filters f_k^i derived above. Figure 18-14 shows an example of hexagonal overcomplete multiresolution representations applied to a digitized radiograph of the breast. Contrast enhancement was accomplished adaptively, based on the location of multiscale edges derived

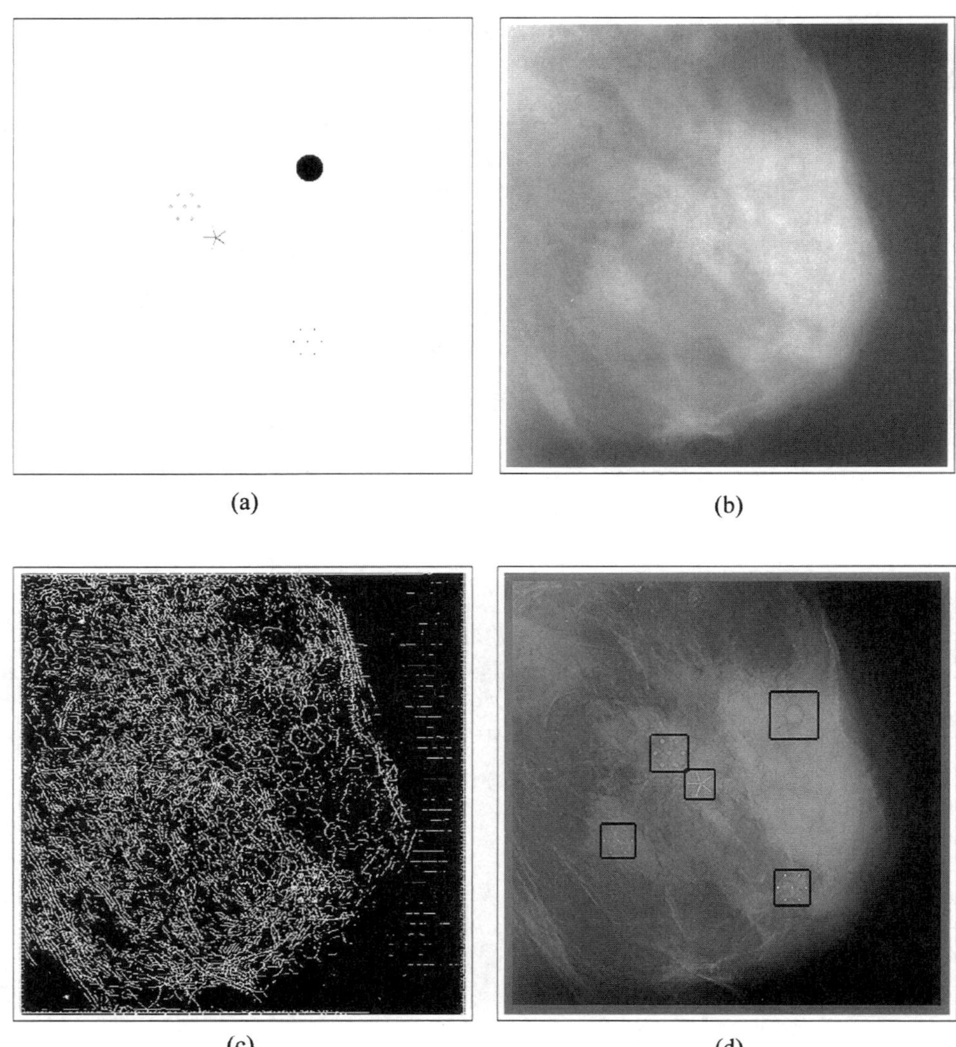

(a)   (b)

(c)   (d)

Figure 18-14   (a) A mathematical phantom. (b) A mammogram blended with phantom. (c) Combined orientations of hexagonal edges obtained from level 3 coefficients. (d) Contrast enhancement by multiscale edges obtained from a hexagonal overcomplete multiresolution representation.

from the hexagonal overcomplete multiresolution representation. Figure 18-15 shows another example of this technique applied to a region of interest of a digitized radiograph of the chest. Please refer to [8] for a complete description of this enhancement algorithm and other possible methods of enhancement.

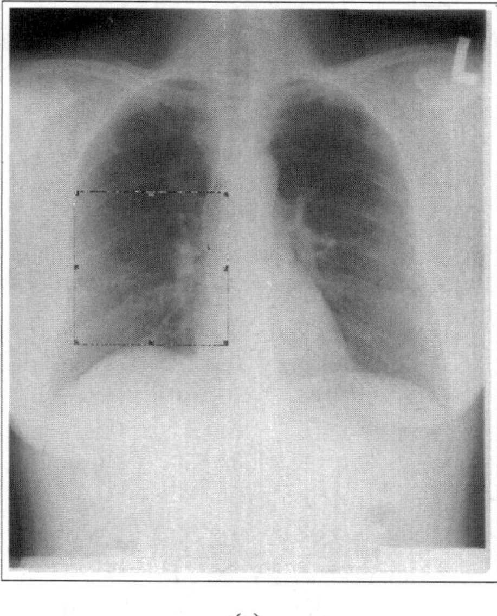

(a)

(b)   (c)

Figure 18-15   (a) Original image. (b) Region of interest. (c) Enhanced region of
interest via multiscale analysis.

ACKNOWLEDGMENTS

The original digitized mammogram shown in Fig. 18-11 was provided courtesy
of the Center for Engineering and Medical Image Analysis and the H. Lee Moffitt
Cancer Center and Research Institute at the University of South Florida, Tampa.

The original digitized radiograph shown in Fig. 18-15 was provided courtesy of the Department of Radiology and the J. Hillis Miller Health Science Center at the University of Florida, Gainsville.

REFERENCES

[1] R. M. Mersereau, "The processing of hexagonally sampled two-dimensional signals." *Proc. IEEE*, vol. 67, pp. 930–949, 1979.

[2] E. P. Simoncelli and E. H. Adelson, "Non-separable extensions of quadrature mirror filters to multiple dimensions." *Proc. IEEE*, vol. 78, pp. 652–663, 1990.

[3] R. M. Mersereau and T. C. Speake, "The processing of periodically sampled multidimensional signals." *IEEE Trans. Acoustics, Speech, and Signal Processing*, vol. 31, pp. 188–194, 1983.

[4] R. M. Mersereau and D. E. Dudgeon, *Multidimensional Digital Signal Processing*, Englewood Cliffs, NJ: Prentice-Hall, 1984.

[5] E. Viscito and J. P. Allebach, "The analysis and design of multidimensional FIR perfect reconstruction filter banks for arbitrary sampling lattices," *IEEE Trans. Circuits Systems*, vol. 38, pp. 29–41, 1981.

[6] A. Laine and S. Schuler, "Hexagonal wavelet processing of digital mammography." In *Medical Imaging 1993*, Newport Beach, February 1993. (Part of SPIE's Thematic Applied Science and Engineering Series.)

[7] M. Newman, *Integral Matrices*, New York: Academic Press, 1972.

[8] A. Laine, S. Schuler, J. Fan, and W. Huda, "Mammographic feature enhancement by multiscale analysis," *IEEE Trans. Medical Imaging*, vol. 13, pp. 725–740, 1994.

Chapter 19

Inversion of the Radon Transform under Wavelet Constraints

Berkman Sahiner, Andrew E. Yagle

19.1. INTRODUCTION

In many problems in the field of medical imaging, one needs to reconstruct a two-dimensional (2-D) object or image from its projections, which amounts to computing the inverse Radon transform. In this chapter, we investigate two important problems encountered in the inversion of the Radon transform [1–4].

The first problem is reconstruction from noisy data. The most common method for computing the inverse Radon transform is the filtered backprojection (FBP) method, which involves filtering the projection data with a ramp filter ($Q(w) \sim |w|$ in (19-2) below). A problem with this filter is that it amplifies the high-frequency components of the noise. Since noise usually dominates at high frequencies, it is common practice to use a low-pass filter in conjunction with the ramp filter to improve the signal-to-noise ratio (SNR). However, the SNR improvement obtained by using a low-pass filter comes at the expense of degraded image resolution, since high-resolution features in the image will also be smoothed. It is desirable to reduce the noise energy in the reconstructed image over regions where high-resolution features are not present, by using spatially-varying filtering.

In the first part of this chapter, we use wavelets to perform this desired localized low-pass filtering. We show how thresholding can be used to determine the regions in the wavelet domain where wavelet coefficients may be set to zero, affecting spatially-varying filtering, and provide a statistical justification for it. Alternatively, *a priori* information about the image can be used to identify such regions. We then use these zero wavelet coefficients as constraints, and compute the minimum mean-squared error image which satisfies these constraints.

The second problem that we investigate is the limited-angle reconstruction problem, which occurs when projection data are missing over a range of angles. Without *a priori* information, the object cannot be uniquely reconstructed from its limited angle projections [5]. For this reason, many researchers have attempted to incorporate *a priori* knowledge about the object into the limited-angle problem [6–8]. The *a priori* information that was assumed in these publications included: (1) upper and lower bounds on reconstructed pixel values; (2) finite support; (3) bounded energy; and (4) closeness to a reference function [6].

In the second part of this chapter, we first reinterpret the missing-angle problem using the wavelet framework, and explain why only certain wavelet coefficients are affected by the missing views. We then add a new item to the *a priori* information list in the previous paragraph. The new item is the knowledge of edges that lie parallel or almost parallel to the missing projection angles, which is readily translated into the wavelet transform language. The affected high-resolution wavelet images are restored using this *a priori* knowledge. The low-resolution wavelet-domain image is restored by interpolating some of the low-wavenumber components of the image, which were reconstructed using the given projections. The wavelet transform is then used to combine the low-resolution and high-resolution images.

This chapter is organized as follows. In section 19.2, we review the inverse Radon transform and the wavelet transform. In section 19.3, we first discuss how the wavelet constraints for spatially-varying filtering may be obtained. We then define and solve the filtering problem, first for an important special case in section 19.3.2, and then for a more general case in section 19.3.3. We conclude section 19.3 with some illustrative examples. In section 19.4, we define the limited-angle problem, give the wavelet interpretation, and discuss the possible sources of *a priori* edge information that we use to solve this problem. We present our algorithm in section 19.4.3, and give numerical examples in section 19.4.4. We conclude with a summary in section 19.5.

19.2. INVERSE RADON TRANSFORMS AND DISCRETE WAVELET TRANSFORMS

19.2.1 The Inverse Radon Transform

The basic reconstruction from projections or inverse Radon transform problems is to reconstruct an image $\mu(x, y)$ from its projections $p(r, \theta)$ where

$$p(r, \theta) = \mathcal{R}\{\mu(x, y)\} = \int_{-\infty}^{\infty} \int_{-\infty}^{\infty} \mu(x, y)\delta(r - x\cos\theta - y\sin\theta)dxdy \qquad (19\text{-}1)$$

is the Radon transform of $\mu(x, y)$. The image is reconstructed from its projections using the inverse Radon transform

$$\mu(x, y) = \mathcal{R}^{-1}\{p(r, \theta)\} = \frac{1}{4\pi^2} \int_0^\pi \int_{-\infty}^\infty P(w, \theta)e^{jw(x\cos\theta + y\sin\theta)}Q(w)\,dw\,d\theta \qquad (19\text{-}2)$$

where $P(w, \theta) = \mathcal{F}_{r \to w}\{p(r, \theta)\}$ is the Fourier transform in the r variable of $p(r, \theta)$, and $Q(w) = |w|$. In practice, $Q(w)$ is a real and symmetric function of w that approximates the ideal Radon kernel $|w|$ [9,10].

Equation (19-2) can be sampled in the image domain to yield a discrete image $x(n, m)$:

$$x(n, m) = \mathcal{R}_s^{-1}\{p(r, \theta)\} = \frac{1}{4\pi^2} \int_0^\pi \int_{-\infty}^\infty P(w, \theta)e^{jw(n\cos\theta + m\sin\theta)}Q(w)\,dw\,d\theta \qquad (19\text{-}3)$$

In practical problems, we have only samples $p(r_j, \theta_i)$ of $p(r, \theta)$, where $\theta_i = (i + 0.5)\pi/N$, $i = 0, 1, \ldots, N - 1$, and $r_j = \Delta_r(j - M/2)$, $j = 0, 1, \ldots, M - 1$. N is the number of angular samples and M is the number of radial samples in the projections. Given samples of its Radon transform, samples $x(n, m)$ of the image $\mu(x, y)$ are obtained by discretizing (19-2) into [9]

$$x(n, m) = \mathcal{R}_d^{-1}\{p(r_j, \theta_i)\} = \frac{1}{2N}\sum_{i=0}^{N-1}\sum_{j=-\infty}^{\infty}\sum_{l=-\infty}^{\infty} q(l - j)p(r_j, \theta_i)h(n\cos i\Delta_\theta + m\sin i\Delta_\theta - l)$$

$$(19\text{-}4)$$

where $\Delta_\theta = \pi/N$, $h(x)$ is an interpolation function, $q(m)$ is a discrete filter that approximates $q(r) = \mathcal{F}^{-1}\{Q(w)\}$, and $\mathcal{R}_d^{-1}$ defines the discrete version of the inverse Radon transform operator. It is assumed that Δ_r, M, and N have been chosen appropriately to obtain an accurate reconstruction $x(n, m)$.

19.2.2 The Discrete Wavelet Transform

The basic idea of the orthogonal discrete wavelet transform (DWT) is to represent a sequence $x(n)$ as a superposition of translations and dilations of a wavelet $g(n)$. The recursive formula for the wavelet decomposition $W_{2^l}x(n)$ of $x(n)$ is [11]

$$x_l(n) = \sum_k h(2n - k)x_{l-1}(k)$$
$$W_{2^l}x(n) = \sum_k g(2n - k)x_{l-1}(k) \qquad (19\text{-}5)$$

where, for a fixed scale l_0, $W_{2^{l_0}}x(n)$ is called the detail signal at scale l_0, and $x_{l_0}(n)$ is called the average signal at scale l_0. The recursion is started with $x_0(n) = x(n)$. The sequences $h(n)$ and $g(n)$ are called the scaling function and the wavelet, respectively, and satisfy

$$g(n) = (-1)^n h(1 - n) \qquad (19\text{-}6)$$

The scaling function is usually a low-pass filter, and (19-6) ensures that the wavelet is a high-pass filter. At each scale l, the average signal from the previous scale is convolved with the high-pass filter $g(n)$, and one sample out of two is retained. Note that if we start with a signal $x(n)$ which is nonzero over an interval of M samples, then the wavelet representation $\{W_{2^1}x(n), W_{2^2}x(n)\ldots, W_{2^L}x(n), x_l(n)\}$ will also have M nonzero samples (ignoring end effects due to the nonzero length of $h(n)$ and $g(n)$) for any L. Thus, the DWT is not redundant.

The recursive formula for the reconstruction of $x(n)$ from its wavelet transform $W_{2^l}x(n)$ is

$$x_{l-1}(n) = \sum_k h(2k - n)x_l(k) + g(2k - n)W_{2^l}x(k) \tag{19-7}$$

The conditions that $h(n)$ and $g(n)$ must satisfy for the orthogonal decomposition-reconstruction of (19-5) and (19-7) to work are derived in [12].

For a 2-D sequence $x(n, m)$, the separable, orthogonal 2-D wavelet transform $W_{2^l}x(n, m)$ is defined recursively as

$$x_l(n, m) = \sum_{k_1}\sum_{k_2} h(2n - k_1)h(2m - k_2)x_{l-1}(k_1, k_2)$$

$$W_{2^l}^{(z)}x(n, m) = \sum_{k_1}\sum_{k_2} g^{(z)}(2n - k_1, 2m - k_2)x_{l-1}(k_1, k_2), \qquad z = 1, 2, 3 \tag{19-8}$$

where $g^{(z)}$, $z = 1, 2, 3$ are called sub-wavelets and are defined by $g^{(1)}(n, m) = g(m)h(n)$, $g^{(2)}(n, m) = g(n)h(m)$, and $g^{(3)}(n, m) = g(n)g(m)$ [11]. The signals $W_{2^l}^{(z)}x$, $z = 1, 2, 3$ represent details of x in the m, n, and diagonal directions, respectively.

As opposed to the recursive formula (19-8), the detail signals $W_{2^l}^{(z)}x(n, m)$ and the average signal $x_l(n, m)$ can also be computed directly from $x(n, m)$ for any given l. The direct decomposition formula is

$$x_l(n, m) = \sum_{k_1}\sum_{k_2} h_l(2^l n - k_1)h(2^l m - k_2)x(k_1, k_2)$$

$$W_{2^l}^{(z)}x(n, m) = \sum_{k_1}\sum_{k_2} g_l^{(z)}(2^l n - k_1, 2^l m - k_2)x(k_1, k_2) \tag{19-9}$$

Using induction, it can easily be shown that the filters $h_l(n, m)$ and $g_l^{(z)}(n, m)$ satisfy

$$h_{l+1}(n) = \sum_k h_1(k)h_l(n - 2^l k)$$

and

$$g_{l+1}^{(z)}(n, m) = \sum_{k_1}\sum_{k_2} g_l^{(z)}(k_1, k_2)h_l(n - 2^l k_1)h_l(m - 2^l k_2) \tag{19-10}$$

where $h_1(n) = h(n)$ and $g_1^{(z)}(n, m) = g^{(z)}(n, m)$.

As a direct consequence of the orthogonality of the wavelets, it can be shown that properly translated versions of the filters $g_l^{(z)}(n, m)$ are orthogonal, i.e.,

$$\sum_{k_1}\sum_{k_2} g_l^{(z)}(2^l n - k_1, 2^l m - k_2)g_{l'}^{(z')}(2^{l'} n' - k_1, 2^{l'} m' - k_2)$$

$$= \delta(l - l')\delta(z - z')\delta(n - n')\delta(m - m') \tag{19-11}$$

Given the detail signals for $j = 1, \ldots, L$ and the average signal $x_L(n, m)$, the original signal $x(n, m)$ can be recovered either using a recursive formula analogous to (19-7) [11], or using the direct formula

$$x(n, m) = \sum_{z=1}^{3}\sum_{l=1}^{L}\sum_{k_1}\sum_{k_2} g_l^{(z)}(2^l k_1 - n, 2^l k_2 - m) W_{2^l}^{(z)} x(k_1, k_2)$$

$$+ \sum_{k_1}\sum_{k_2} h_L(2^L k_1 - n)h_L(2^L k_2 - m)x_L(k_1, k_2) \tag{19-12}$$

19.2.3 The Unsubsampled Wavelet Transform

The DWT described in section 19.2.2 has the advantage that the filters $h_l(n)$ and $g_l(n)$ are orthogonal. As a result of this orthogonality, the DWT is nonredundant. Although nonredundancy is an advantage when dealing with large sets of data, there are cases where it may become a disadvantage. A primary weakness of the orthogonal DWT is that it is translation-varying [13]. When the signal $x(n)$ is translated, its wavelet coefficients, defined by (19-5), change dramatically within and across scales [14]. This is a major problem when wavelets are used to extract or process edges [15].

To overcome difficulties caused by the translation-variance of the orthogonal wavelet transform, a redundant form of the wavelet transform is proposed in [16], which we call the unsubsampled wavelet transform (UWT) in this chapter. In [16], this transform is called the "wavelet transform," and the DWT is not used at all; in this chapter, since we use both transforms, we choose to give a new name to this transform to avoid confusion. As opposed to DWT, no subsampling is performed in the UWT after convolution with the wavelet.

As in DWT, a low-pass filter $h(n)$ and a high-pass filter $g(n)$ are used to define the UWT recursively. Let $d_{2^l}x(n)$ and $r_{2^l}x(n)$ denote the difference and average signals at scale l. These signals are computed recursively from $r_1(n) = x(n)$ as

$$d_{2^{l+1}}x(n) = r_{2^l}x(n) * g_l(n), \qquad 0 \le l \le L$$
$$r_{2^{l+1}}x(n) = r_{2^l}x(n) * h_l(n), \qquad 0 \le l \le L \tag{19-13}$$

where $*$ is convolution and $h_l(n)$ and $g_l(n)$ are obtained by inserting $2^l - 1$ zeros between the coefficients of $h(n)$ and $g(n)$. The Fourier transforms $\hat{H}_l(w)$ and $\hat{G}_l(w)$ of $h_l(n)$ and $g_l(n)$ satisfy

$$\hat{H}_l(w) = \hat{H}(2^l w) \qquad \text{and} \qquad \hat{G}_l(w) = \hat{G}(2^l w) \tag{19-14}$$

Let $\hat{K}(w)$ be a filter that satisfies $\hat{K}(w)\hat{G}(w) + |\hat{H}(w)|^2 = 1$. Then, a perfect reconstruction from the UWT is given by

$$r_{2^{l-1}}x(n) = W_{2^l}x(n) * k_{l-1}(n) + r_{2^l}x(n) * \tilde{h}_{l-1}(n) \tag{19-15}$$

where $\tilde{h}_{l-1}(n) = h_{l-1}(-n)$ and $k_{l-1}(n)$ is obtained by inserting $2^{l-1} - 1$ zeros between the coefficients of $k(n)$.

In this chapter, we choose $g(n) = (-1)^n h(1 - n)$ (as in section 19.2.2), and $k(n) = \tilde{g}(n) = g(-n)$, implying that the condition for perfect reconstruction is

$$|\hat{H}(w)|^2 + |\hat{H}(w + \pi)|^2 = 1 \tag{19-16}$$

Filters that satisfy (19-16) are called quadrature mirror filters, and orthogonal wavelets are a special class of quadrature mirror filters. Therefore, we can use the wavelets defined in section 19.2.2 for the UWT.

For a 2-D signal $x(n, m)$, the UWT is defined similarly as

$$d_{2^{l+1}}^{(1)} x(n, m) = r_{2^l} x(n, m) * g_l(n) * h_l(m) \qquad 0 \leq l \leq L$$

$$d_{2^{l+1}}^{(2)} x(n, m) = r_{2^l} x(n, m) * h_l(n) * g_l(m) \qquad 0 \leq l \leq L$$

$$d_{2^{l+1}}^{(3)} x(n, m) = r_{2^l} x(n, m) * g_l(n) * g_l(m) \qquad 0 \leq l \leq L \qquad (19\text{-}17)$$

$$r_{2^{l+1}} x(n, m) = r_{2^l} x(n, m) * h_l(n) * h_l(m) \qquad 0 \leq l \leq L$$

If $h(n)$ satisfies (19-16), and with the choice of $h(n) = (-1)^n h(1 - n)$, the reconstruction formula is

$$r_{2^{l-1}} x(n, m) = r_{2^l}^{(1)} x(n, m) * \tilde{g}_l(n) * \tilde{h}_l(m) + d_{2^l}^{(2)} x(n, m) * \tilde{h}_l(n) * \tilde{g}_l(m)$$

$$+ d_{2^l}^{(3)} x(n, m) * \tilde{g}_l(n) * \tilde{g}_l(m) + r_{2^l} x(n, m) * \tilde{h}_l(n) * \tilde{h}_l(m) \qquad (19\text{-}18)$$

19.3. FILTERING WITH USE OF DWT CONSTRAINTS

In this section, we consider the problem of reconstructing an image from its projections, given the constraint that some fine-scale wavelet transform values around a region $\mathcal{A}$ of the image are zero. Since fine-scale wavelet transform components represent localized high-resolution features of the image, this constraint means that $\mathcal{A}$ represents a flat or slowly-varying part of the image, or that $\mathcal{A}$ is free of edges. Constraining fine-scale wavelet coefficients in $\mathcal{A}$ to be zero effectively smoothes the image, much as low-pass filtering does. The advantage of using wavelets is that this can be done on a localized basis, smoothing some areas while leaving other areas (such as edges) unaffected.

Since the Radon transform is not unitary, knowledge about one part of the image will improve the quality of the overall reconstructed image. Numerical examples will show that the constraints improve the reconstruction of the entire image, not just the constrained region.

The wavelet constraints may be obtained from measurements using other imaging modalities. For example, if image registration problems are solved, it may be possible to determine $\mathcal{A}$ in a computed tomography (CT) image from the flat regions in an magnetic resonance (MR) image. A similar idea was used in [17], where positron-emission tomography (PET) images are modeled as a number of flat regions, and the information about region boundaries is extracted from segmented MR images. In a medical application, anatomical site information may also be used as *a priori* information. Although many details of the anatomy of a particular patient may be unknown, some general features, such as high-detail and low-detail regions are usually known *a priori*. Since the constraints improve the entire image, the constraints in low-detail regions will improve the reconstruction of high-detail regions.

The constraints may also be obtained directly from a noisy image, by thresholding the absolute value of the wavelet transform of the reconstructed image. The idea of thresholding the time-frequency representation of a signal for spatially-varying filtering has been previously used in the literature [18]. Below, we supply a statistical justification for thresholding the absolute value of DWT.

Detection Problem Formulation.   Assume that $x(n,m)$ is a zero-mean white Gaussian random sequence with power spectral density σ_x^2. Let $W_{2^j}^{(z)}x(n,m)$, $j = 1,\ldots,\infty$, $z = 1,2,3$ be its wavelet transform defined using (19-9). Then, by the orthonormality of the wavelets (19-11), the quadruply indexed random sequence $W_{2^j}^{(z)}x(n,m)$ is uncorrelated and zero-mean with variance σ_x^2. To obtain a random sequence $\bar{x}(n,m) \neq x(n,m)$ whose wavelet coefficients are zero with probability one outside a region D_1, we define [compare to (19-12)]

$$\bar{x}(n,m) = \sum_{l,k_1,k_2,z \in D_1} g_l^{(z)}(2^l k_1 - n, 2^l k_2 - m) W_{2^j}^{(z)} x(k_1,k_2) \tag{19-19}$$

We now state the problem. Given the noisy observations

$$x_\eta(n,m) = \bar{x}(n,m) + \eta(n,m) \tag{19-20}$$

of $\bar{x}(n,m)$, where $\eta(n,m)$ is a zero-mean white Gaussian noise sequence with power spectral density σ_η^2, determine the region D_1.

Detection Problem Solution.   The wavelet transform of $x_\eta(n,m)$ is

$$W_{2^j}^{(z)} x_\eta(n,m) = \begin{cases} W_{2^j}^{(z)} x(n,m) + W_{2^j}^{(z)} \eta(n,m) & \text{if } (l,n,m,z) \in D_1 \\ W_{2^j}^{(z)} \eta(n,m) & \text{otherwise} \end{cases} \tag{19-21}$$

where $W_{2^j}^{(z)} \eta(n,m)$ is the wavelet transform of $\eta(n,m)$. Note that $W_{2^j}^{(z)} x_\eta(n,m)$ is a zero-mean uncorrelated Gaussian random sequence whose variance is $\sigma_\eta^2 + \sigma_x^2$ for (l,n,m,z) inside D_1 and σ_η^2 for (l,n,m,z) outside D_1. Therefore, the decision of whether a point $(l,n,m,z) \in D_1$ decouples from similar decisions for other points. Furthermore, for each point in Z^4, this becomes the well known problem of detection of a Gaussian random variable in Gaussian noise. Its solution is the likelihood ratio test [19]

$$|W_{2^j}^{(z)} x_\eta(n,m)| \begin{array}{c} \in D_1 \\ \gtrless \\ \notin D_1 \end{array} \nu \tag{19-22}$$

where ν is the threshold. This means that we can decide whether (l,n,m,z) is in D_1 simply by thresholding the absolute value of the wavelet transform of the noisy image at that point. The threshold ν can be determined using the Neyman–Pearson criterion.

When the whiteness or orthonormality assumptions are relaxed, the threshold test described above is no longer guaranteed to be optimal. However, the threshold test still seems to be the "natural" approach.

19.3.1 Problem Definition

In this section, we address the following problem. Suppose that we are given noisy observations $p_\eta(r, \theta) = p_a(r, \theta) + \eta(r, \theta)$ of the projections $p_a(r, \theta)$ of an actual image $x_a(r, \theta)$, where $\eta(r, \theta)$ is a zero-mean Gaussian random process in r for each θ, and is uncorrelated in θ. Suppose also that the wavelet transform $W_{2^l}^{(z)} x_a(n, m)$ of $x_a(n, m)$ is known to be zero for several values of n and m on L different scales. The problem is to determine the filtered image $\tilde{x}(n, m)$ such that:

1. $\tilde{x}(n, m)$ satisfies the wavelet constraints and
2. $E\{\sum_n \sum_m (x_a(n, m) - \tilde{x}(n, m))^2\}$ is minimized

For clarity of presentation, we first consider constraints on a single sub-wavelet, and we assume that the additive noise in the projections is white in each slice in r, i.e., $E[\eta(r_1, \theta_1)\eta(r_2, \theta_2)] = \delta(r_1 - r_2)\delta(\theta_1 - \theta_2)$. We generalize to nonwhite noise, and several sub-wavelets, in section 19.3.3.

19.3.2 Constraints on a Single Wavelet

Let $x_\eta(n, m) = \mathcal{R}_s^{-1}\{p_\eta(r, \theta)\}$ and $\epsilon(n, m) = \mathcal{R}_s^{-1}\{\eta(r, \theta)\}$ so that

$$x_\eta(n, m) = x_a(n, m) + \epsilon(n, m) \tag{19-23}$$

We assume that the wavelet transform of $x_a(n, m)$ with respect to the first wavelet $g^{(1)}$ is known to be zero at $C(l)$ points at scale l, where $1 \leq l \leq L$. This knowledge could come either from *a priori* information about the image, or from thresholding the absolute value of the wavelet transform of the image, as discussed at the beginning of this section. We therefore constrain the wavelet transform to be zero at those points. We index each of these points by a pair (c, l), where l denotes the scale and c enumerates the points at each scale. The constraints can then be written as

$$W_{2^l}^{(1)} x_a(i_{c,l}, j_{c,l}) = W_{2^l}^{(1)} \tilde{x}(i_{c,l}, j_{c,l}) = 0, \qquad 1 \leq l \leq L, \qquad 1 \leq c \leq C(l) \tag{19-24}$$

To solve the problem defined at the beginning of this section, we first compute an estimate $\hat{\epsilon}(n, m)$ of $\epsilon(n, m)$ using the given constraints, and then subtract the noise estimates from the noisy image $x_\eta(n, m)$.

By taking the wavelet transform of both sides of (19-23), and using (19-24), we find

$$W_{2^l}^{(1)} \epsilon(i_{c,l}, j_{c,l}) = \sum_n \sum_m \epsilon(2^l i_{c,l} - n, 2^l j_{c,l} - m) g_l^{(1)}(n, m) = W_{2^l}^{(1)} x_\eta(i_{c,l}, j_{c,l}) \tag{19-25}$$

Since $x_\eta(n, m)$ can be computed from the given projections $p_\eta(r, \theta)$, $W_{2^l}^{(1)} x_\eta(i_{c,l}, j_{c,l})$ are known. Our goal is to compute $\hat{\epsilon}(n, m)$ from which the image $\tilde{x}(n, m)$ can be computed

$$\tilde{x}(n, m) = x_\eta(n, m) - \hat{\epsilon}(n, m) \tag{19-26}$$

Since $\eta(r, \theta)$ is zero-mean and jointly Gaussian in r and θ, $\epsilon(n, m)$ (which is a linear combination of $\eta(r, \theta)$) is also zero-mean and jointly Gaussian in n and m. The solution to the problem of finding $\hat{\epsilon}(n, m)$ is the linear minimum mean-square estimate (LMMSE)

$$\hat{\epsilon}(n, m) = \sum_{l=1}^{L} \sum_{c=1}^{C(l)} \beta_{c,l} E[\epsilon(n, m) W_{2^l}^{(1)} \epsilon(i_{c,l}, j_{c,l})] \tag{19-27}$$

where the $\beta_{c,l}$ are computed by solving the matrix equation

$$\mathcal{M}\underline{\beta} = \underline{b} \tag{19-28}$$

In (19-28), $\underline{\beta}$ and $\underline{b}$ are vectors which contain unknowns $\beta_{c,l}$ and knowns $W_{2^m}^{(1)} x(i_{c,l}, j_{c,l})$

$$\underline{\beta} = [\beta_{1,1}, \ldots, \beta_{C(1),1}, \beta_{1,2}, \ldots, \beta_{C(2),2}, \ldots, \beta_{C(L),L}]^{\mathrm{T}}$$

$$\underline{b} = [W_{2^1}^{(1)} \mu(i_{1,1}, j_{1,1}), \ldots, W_{2^1}^{(1)} x(i_{C(1),1}, j_{C(1),1}), W_{2^2}^{(1)} x(i_{1,2}, j_{1,2}), \ldots,$$

$$W_{2^2}^{(1)} x(i_{C(2),2}, j_{C(2),2}), \ldots W_{2^L}^{(1)} x(i_{C(L),L}, j_{C(L),L})]^{\mathrm{T}} \tag{19-29}$$

The system matrix $\mathcal{M}$ consists of L^2 submatrices $\mathcal{M}_{l,m}$,

$$\mathcal{M} = \begin{bmatrix} \mathcal{M}_{1,1} & \mathcal{M}_{1,2} & \cdots & \mathcal{M}_{1,L} \\ \mathcal{M}_{2,1} & \mathcal{M}_{2,2} & \cdots & \mathcal{M}_{2,L} \\ \vdots & \vdots & & \vdots \\ \mathcal{M}_{L,1} & \mathcal{M}_{L,2} & \cdots & \mathcal{M}_{L,L} \end{bmatrix} \tag{19-30}$$

where the (u, v)-th entry of $\mathcal{M}_{l,m}$ is

$$\mathcal{M}_{l,m}(u, v) = E[W_{2^l}^{(1)} \epsilon(i_{u,l}, j_{u,l}) W_{2^m}^{(1)} \epsilon(i_{v,m}, j_{v,m})] \tag{19-31}$$

Let us now define the auto- and cross-correlations $E[W_{2^l}^{(1)} \epsilon(i_{u,l}, j_{u,l}) W_{2^m}^{(1)} \epsilon(i_{v,m}, j_{v,m})]$ and $E[\epsilon(n, m) W_{2^l}^{(1)} \epsilon(i_{c,l}, j_{c,l})]$ that appear in (19-31) and (19-27) in terms of known quantities.

If the noise in the projections is white in r, i.e., if $E[\eta(r_1, \theta_1)\eta(r_2, \theta_2)] = \delta(r_1 - r_2)\delta(\theta_1 - \theta_2)$, then it can be shown [20] that the autocorrelation of $\epsilon(n, m)$ is given by

$$R_\epsilon(n, m) = E[\epsilon(n', m')\epsilon(n + n', m + m')] = \frac{1}{2\pi} \mathcal{R}_s^{-1}\{q(r)\} = \frac{1}{2\pi} R(n, m) \tag{19-32}$$

Defining $y_l^{(1)}$ in terms of the convolution of $g_l^{(1)}$ and R as

$$y_l^{(1)}(n, m) = \frac{1}{2\pi} \sum_{s_1} \sum_{t_1} g_l^{(1)}(s_1, t_1) R(n - s_1, m - t_1) \tag{19-33}$$

we find that

$$E[\epsilon(n, m) W_{2^l}^{(1)} \epsilon(i_{c,l}, j_{c,l})] = y_l^{(1)}(2^l i_{c,l} - n, 2^l j_{c,l} - m) \tag{19-34}$$

The entries of the system matrix $\mathcal{M}$ are computed as

$$E[W^{(1)}_{2^l}\epsilon(i_{u,l}, j_{u,l})W^{(1)}_{2^m}\epsilon(i_{v,l}, j_{v,l})] = \frac{1}{2\pi}\sum_{s_1}\sum_{t_1}\sum_{s_2}\sum_{t_2}$$

$$g^{(1)}_l(s_1, t_1)g^{(1)}_m(s_2, t_2)R(2^m i_{v,m} - 2^l i_{u,l} + s_1 - s_2, 2^m j_{v,m} - 2^l j_{u,l} + t_1 - t_2) \qquad (19\text{-}35)$$

To summarize, the noise estimate $\hat{\epsilon}(n, m)$ is computed using

$$\hat{\epsilon}(n, m) = \sum_{l=1}^{L}\sum_{c=1}^{C(l)}\beta_{c,l}y^{(1)}_l(2^l i_{c,l} - n, 2^l j_{c,l} - m) \qquad (19\text{-}36)$$

where $y^{(1)}_l$ is defined in (19-33), and $\beta_{c,l}$ are solved from (19-28). Then, the filtered image is obtained as the difference between the noisy image and the noise estimate.

Note that if the wavelet $g(n)$ is a finite-length sequence, a constraint on the wavelet transform of the image will involve only a finite number of pixel values. However, the perturbed image will be improved for *all* n and m, since the inverse Radon transform $\mathcal{R}^{-1}_s\{q(r)\}$ will have infinite extent for a general choice of the filter $Q(w)$. This is illustrated in the numerical examples of section 19.3.4.

19.3.3 Constraints on Several Sub-Wavelets

We now generalize to more than one sub-wavelet, and nonwhite noise. Let $Z \leq 3$ be the maximum number of wavelets used, and let $z \in 1, \ldots, Z$ denote the sub-wavelet number [see (19-8)]. We index points by the superscript z and subscripts (c, l), where l denotes the scale and c enumerates the points at each scale. The constraints are

$$W^{(z)}_{2^l}\tilde{x}(i^{(z)}_{c,l}, j^{(z)}_{c,l}) = 0, \qquad 1 \leq l \leq L, \qquad 1 \leq c \leq C(l), \qquad 1 \leq z \leq Z. \qquad (19\text{-}37)$$

Let the autocorrelation of the additive noise in the projections be given by

$$E[\eta(r_1, \theta_1)\eta(r_2, \theta_2)] = t(r_1 - r_2)\delta(\theta_1 - \theta_2) \qquad (19\text{-}38)$$

That is, noise is still uncorrelated between projection angles, but correlated in r at a single angle. Then, using the arguments of the previous subsection, it is easily shown that the noise estimate $\hat{\epsilon}(n, m)$ is given by

$$\hat{\epsilon}(n, m) = \sum_{z=1}^{Z}\sum_{l=1}^{L}\sum_{c=1}^{C(l)}\beta^{(z)}_{c,l}yt^{(z)}_l(2^l i^{(z)}_{c,l} - n, 2^l j^{(z)}_{c,l} - m) \qquad (19\text{-}39)$$

where $yt^{(z)}_l(n, m)$ is defined as

$$yt^{(z)}_l(n, m) = \frac{1}{2\pi}\sum_{s_1}\sum_{t_1}g^{(z)}_l(s_1, t_1)Rt(n - s_1, m - t_1) \qquad (19\text{-}40)$$

and

$$Rt(n, m) = \mathcal{R}^{-1}_s\{q * t\} = \mathcal{R}^{-1}_s\left\{\int_{-\infty}^{\infty}q(\varsigma)t(r - \varsigma)d\varsigma\right\} \qquad (19\text{-}41)$$

$\beta_{c,l}^{(z)}$ are computed by solving $\mathcal{M}\underline{\beta} = \underline{b}$, where

$$\mathcal{M} = \begin{bmatrix} \mathcal{M}_{1,1} & \mathcal{M}_{1,2} & \cdots & \mathcal{M}_{1,L} \\ \mathcal{M}_{2,1} & \mathcal{M}_{2,2} & \cdots & \mathcal{M}_{2,L} \\ \vdots & \vdots & & \vdots \\ \mathcal{M}_{L,1} & \mathcal{M}_{L,2} & \cdots & \mathcal{M}_{L,L} \end{bmatrix} \quad \mathcal{M}_{l,m} = \begin{bmatrix} \mathcal{M}_{l,m}(1,1) & \cdots & \mathcal{M}_{l,m}(1,Z) \\ \vdots & & \vdots \\ \mathcal{M}_{l,m}(Z,1) & \cdots & \mathcal{M}_{l,m}(Z,Z) \end{bmatrix}$$

$$(19\text{-}42)$$

and where the (u,v)-th entry of $\mathcal{M}_{l,m}(z_1, z_2)$ is

$$\frac{1}{2\pi} \sum_{s_1} \sum_{t_1} \sum_{s_2} \sum_{t_2} \Big\{ g_l^{(z_1)}(s_1, t_1) g_m^{(z_2)}(s_2, t_2)$$

$$Rt(2^m i_{v,m}^{(z_1)} - 2^l i_{u,l}^{(z_2)} + s_1 - s_2, \times 2^m j_{v,m}^{(z_1)} - 2^l j_{u,l}^{(z_2)} + t_1 - t_2) \Big\}$$

$$(19\text{-}43)$$

19.3.4 Examples and Discussion

In this section, we present two numerical examples which illustrate the results of the previous subsection. Since the stochastic and deterministic developments lead to the same formula, our examples will be formulated in terms of the stochastic (noisy) case only.

The algorithm developed in the previous subsection was based on complete projections. In practice, only samples $p(r_j, \theta_i)$ of projections are available. This does not present any difficulties, and the only modification to our algorithm is the replacement of (19-41) by

$$Rt(n, m) = \mathcal{R}_d^{-1}\{q * t\} = \mathcal{R}_d^{-1}\left\{ \sum_{t_1} q(t_1) t(m - t_1) \right\} \qquad (19\text{-}44)$$

where $q(n)$ is a discrete filter that approximates $q(r)$, and $t(n)$ is the radial autocorrelation function of the discrete noise in the projections.

EXAMPLE 1

In this example, we use the original Shepp–Logan head phantom [10] as our noiseless image. The noise added to the projections is obtained by passing zero-mean white Gaussian noise with variance 4×10^{-6} through a filter whose discrete-time Fourier transform is $[\sin(w)]^{32}$. The 128×128 noisy image is shown in Fig. 19-1. The wavelets that we use are two sub-wavelets of the Haar basis, which can be regarded as difference operators in the n and m directions (the third Haar sub-wavelet, which can be regarded as a difference operator in the diagonal direction, is not used). We constrain the two finest-scale wavelet coefficients to be zero over a region D of the image. The region D is obtained by the thresholding approach over a region D' in the center of the image. First, the second-finest wavelet coefficients inside D' are set to zero whenever their absolute value is below a threshold. The region which will be affected by the above operation is called D''. Then, inside D'', another threshold is used to set the finest-scale coefficients to zero. We use (19-39) to estimate the noise $\hat{\epsilon}(n, m)$, where the matrix $\mathcal{M}$ is given by (19-42). The resulting MMSE image is shown in Fig. 19-2. The noise power has been reduced by 20.3% in the whole image, while still preserving the edges.

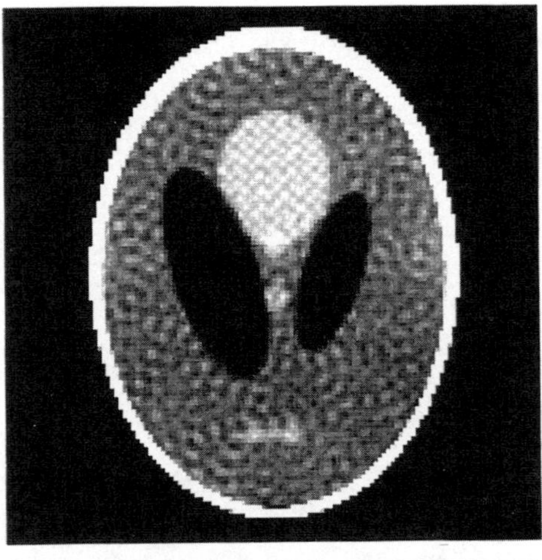

Figure 19-1 The noisy image for Example 1, obtained from the noisy projections of the Shepp–Logan phantom.

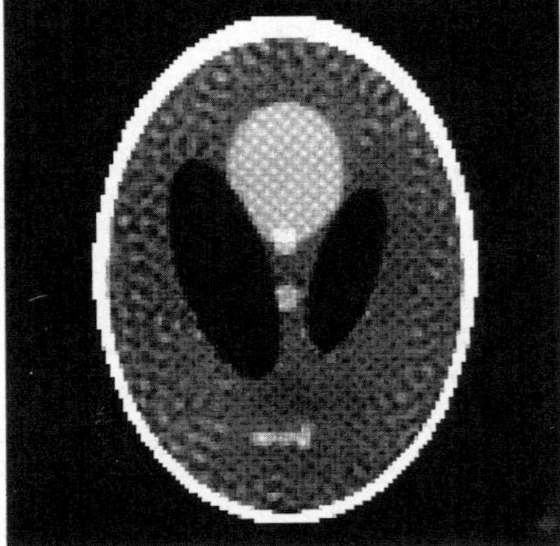

Figure 19-2 MMSE image obtained by constraining the wavelet coefficients of the noisy image; Example 1.

EXAMPLE 2:

The CT image used in this example represents an 8-mm specimen from proximal humerus of a 40–50 year old male, and was obtained at the orthopedic research laboratories of the University of Michigan. Figure 19-3 shows the image reconstructed from its noiseless Radon transform. Figure 19-4 shows the image reconstructed from noisy projections, where the autocorrelation of the additive noise was the same as in Example 1. As in the previous example, the constraints are determined from the noisy image. Figure 19-5 shows the region D' where we set a wavelet constraint when the absolute value of the second-finest wavelet coefficients are below a threshold. The resulting MMSE image is shown in Fig. 19-6. The noise power inside D' has been reduced by 22%, and the noise power in the entire image has been reduced by 15%. We also computed the noise power in a 10 pixel band surrounding D'. Although we do not have any wavelet constraints in this band, the noise power has been reduced by 16%, which shows that constraining wavelet coefficients in a given region improves the reconstruction in other regions. This is because the noise ϵ in the reconstructed image is nonwhite, due to the fact that the Radon transform is nonunitary and the additive noise η on the projections is nonwhite.

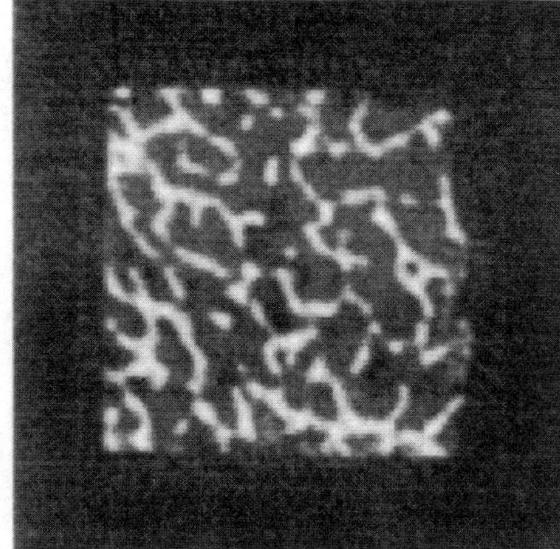

Figure 19-3 The noiseless image for Example 2.

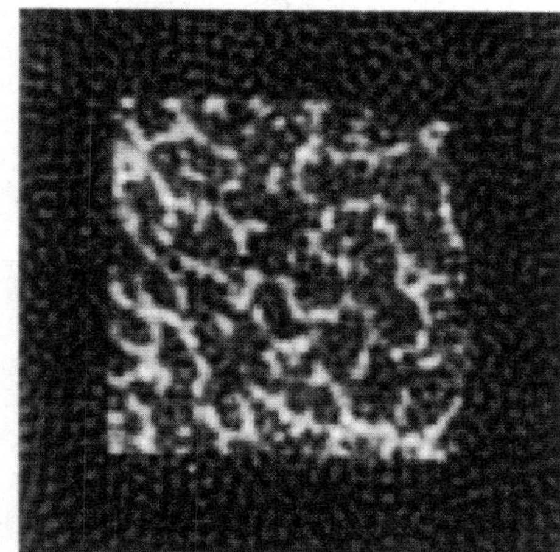

Figure 19-4 The noisy image for Example 2.

Figure 19-5 The region D' where the wavelet constraints are searched.

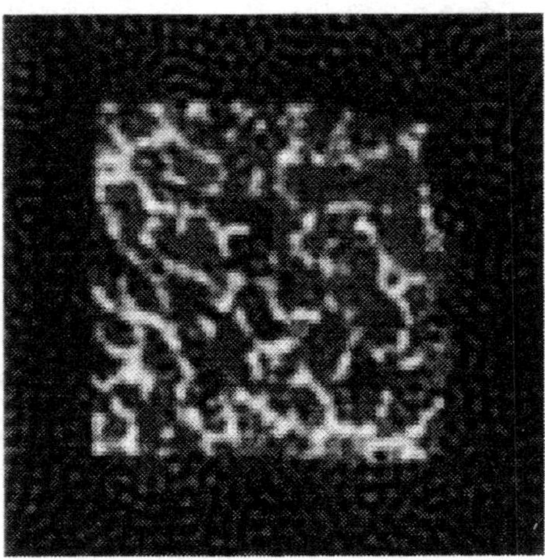

Figure 19-6 MMSE image obtained by constraining the wavelet coefficients of the noisy image; Example 2.

19.4. IMAGE RESTORATION WITH USE OF UWT CONSTRAINTS

In this section, we investigate the application of the UWT to the limited-angle problem, which occurs when projection data are missing over a range of angles. We assume, without loss of generality, that $2L_{\mathrm{miss}}$ views centered around $\pi/2$ are missing in the projection data, corresponding to a missing angle extent of $\theta_{\mathrm{miss}} = 2\pi L_{\mathrm{miss}}/N$. The available data are thus $p(r_j, \theta_i), i = 0, \ldots, \frac{N}{2} - L_{miss} - 1,$ $\frac{N}{2} + L_{miss}, \ldots, N - 1$. The projection-slice theorem states that the Fourier transform $\hat{P}(w, \theta_i)$ of $p(r_j, \theta_i)$ is equal to the Fourier transform $\hat{U}(w_1, w_2)$ of the image along a slice in the Fourier plane that passes through the origin and makes an angle θ_i with the w_1 axis. Thus, the Fourier transform of the image is known on the concentric circles grid shown in Fig. 19.7, except for the bowtie-like region R. The quality of reconstruction from missing views depends on how well we can fill in the missing Fourier transform samples in R. It has recently been shown [21] that the squashing algorithm [22] is equivalent to setting frequency samples in R to zero. Oskoui and Stark have used an interpolation technique to fill R [23]; however, results indicate that the results from their interpolation are not an improvement on those from squashing. Still another method is projections onto convex sets (POCS), where the convex sets are defined by the given projections and the *a priori* knowledge about the image.

In this section, we propose to, in effect, fill in the high-frequency regions of R using some approximate *a priori* knowledge about edges which lie parallel to the n axis, and to fill the low-frequency regions of R using a simple interpolation technique. We characterize the edges, and bring together the low-pass and high-pass regions of R, by using the wavelet transform. Since we want to characterize and restore edges of an image, in this section we use the UWT. The wavelet decomposition formula for the image $x(n, m)$ is (19-17) and the reconstruction formula is (19-18).

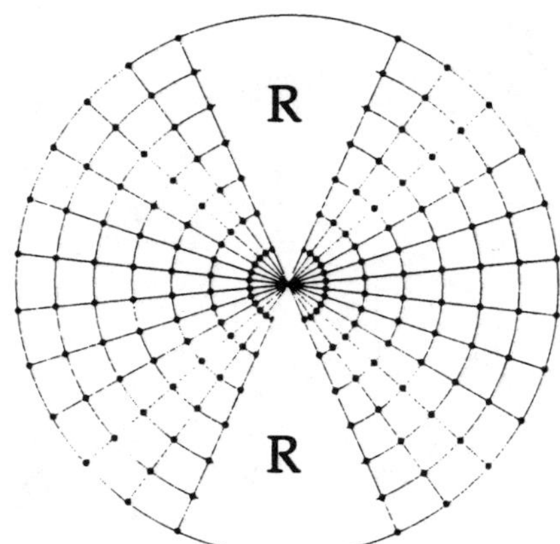

Figure 19-7   The concentric circles grid and the bowtie-like region R over which Fourier transform samples are missing.

A priori edge knowledge that we use does not need to be complete or exact. Since the wavelet transform is local, we need not know *all* of the edge locations parallel to the missing view angles; if some of these edges are unknown, the resulting error will be only local. Also, our simulations indicate that edge magnitudes need not be known exactly—as much as 20% error in the knowledge of edge magnitudes does not result in any significant degradation of the performance of our algorithm, and no *a priori* knowledge at all about edges not parallel to the missing view angles is required.

We now discuss possible sources of *a priori* edge information. As in the case of filtering with DFT constraints, general anatomical site information, such as bone–tissue boundaries, may be used as *a priori* information. A similar idea was used for anatomical localization for PET in [22]. Edge location information may also come from measurements using other imaging modalities. For example, it may be possible to reconstruct parts of the object without 360 degree measurements as in single-positron emission computed tomography (SPECT), or it may be possible to obtain data along missing views using other modalities such as MR or ultrasound. When the missing angle range in CT is relatively small, it may be possible to use the methods in [25] or [17] to align other images with the CT image to obtain edge location information.

19.4.1 Wavelet Interpretation of the Missing Angle Problem

The partitioning of the Fourier plane with ideal low-pass and high-pass filters $\hat{H}$ and $\hat{G}$ is shown in Fig. 19-8. If H and G are ideal, and $\theta_{\text{miss}} \leq 2\tan^{-1}\frac{1}{2} = 53.1°$, then we observe from Fig. 19-8 that $d_{2^l}^{(1)}x(n,m)$ and $d_{2^l}^{(3)}x(n,m)$ will not be affected. This is a new explanation of why edges not parallel to the missing views are relatively unaffected by them.

Edges in the image $x(n,m)$ correspond to high-wavenumber components in the Fourier domain. However, they can also be viewed as localized high-resolution components in the wavelet domain. In a limited-angle tomography problem, if the locations and magnitudes of some of the edges that lie parallel to the x-axis in the image are known, then $d_{2^l}^{(2)}x(n,m)$ can be approximated up to some scale L for values

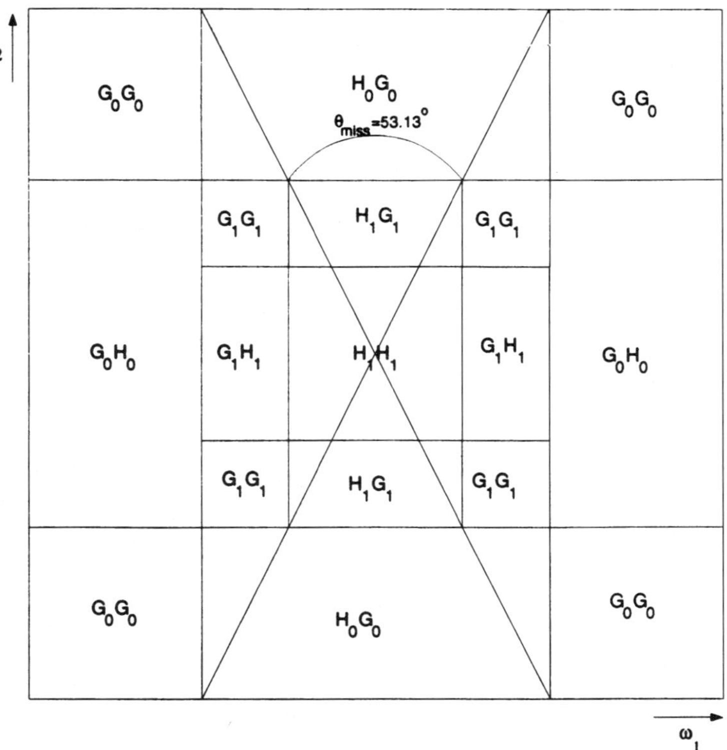

Figure 19-8   The ideal partitioning of the Fourier plane with the filters H and G.

of n and m near the known edges. Since the edges give us little information about the low-resolution behavior of the image, a different method must be used to fill this low-resolution information, which corresponds to low-pass regions in the Fourier plane. A simple algorithm for this is given in the next subsection.

19.4.2 Interpolation of Low-Resolution Missing Data

To understand the relation between low radial frequency components of projection data viewed at different projection angles, we use a result obtained in [26]. Define

$$\mathcal{P}(w, \xi) = \int_{-\infty}^{\infty} \left[\int_{-\infty}^{\infty} p(r, \theta)e^{-jwr}dr \right] e^{-j\xi\theta}d\theta \qquad (19\text{-}45)$$

where $p(r, \theta)$ are the continuous projections. Note that the inner integral in (19-45) is the continuous domain analog of $\hat{P}(w, \theta_i)$, therefore $\mathcal{P}(w, \xi)$ can be thought of as the continuous domain analog of the Fourier transform of $\hat{P}(w, \theta)$ in the θ variable. By considering the bandwidth of Bessel functions of the first kind, Rattey and Lindgren show in [26] that if the image is spatially band-limited to a disk of radius T, then

$$\mathcal{P}(w, \xi) \simeq 0 \qquad \text{for } |\xi| > [T|w|] + 1 \qquad (19\text{-}46)$$

This implies that in the $(w - \xi)$ plane, $\mathcal{P}(w, \xi)$ is effectively support-limited to a bowtie-like region as shown in Fig. 19-9. For small $|w|$, the bandwidth in ξ is also small, which means that low radial frequencies change slowly from projection to projection. As a result, lower frequency samples of the missing projections can be interpolated more accurately than higher frequency samples.

Although a number of interpolation methods could be used, we chose the following simple linear interpolation procedure:

$$
\hat{P}(w_j, \theta_i) = \frac{1}{2L_{\text{miss}} + 1} \left\{ \left(\frac{N}{2} + L_{\text{miss}} - i \right) \hat{P}(w_j, \theta_{N/2-L_{\text{miss}}-1}) \right.
$$
$$
\left. + \left[i - \left(\frac{N}{2} - L_{\text{miss}} - 1 \right) \right] \hat{P}(w_j, \theta_{N/2+L_{\text{miss}}}) \right\}
\tag{19-47}
$$

In (19-47) the Fourier transform of any missing view in the range of missing views is replaced by a linear combination of Fourier transforms of the two views which border the missing view range. Note that since the interpolation is linear, missing projection data can be interpolated simply by taking the inverse Fourier transform of (19-47), i.e., by replacing w_j in (19-47) by r_j.

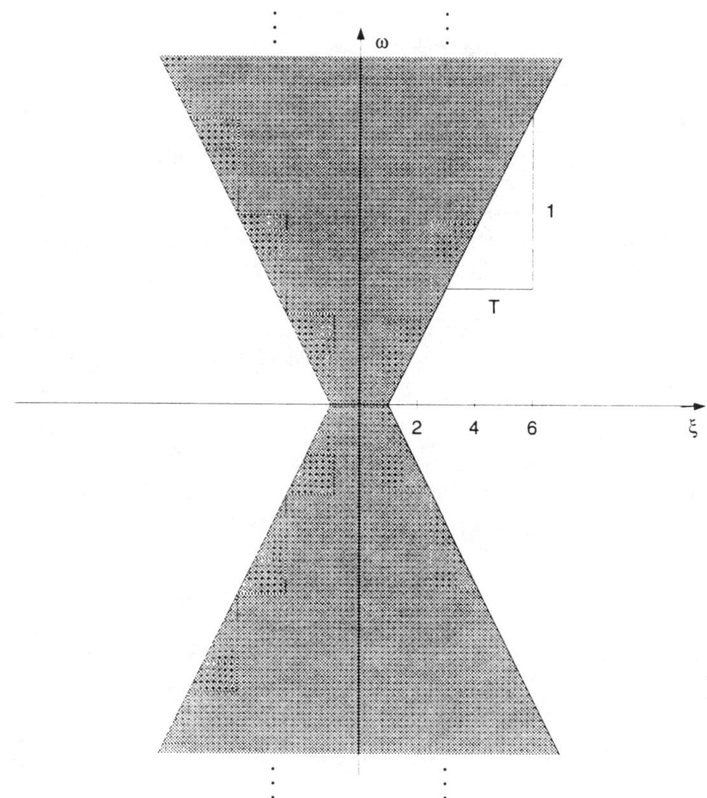

Figure 19-9   Region of support of $\mathcal{P}(w, \xi)$, adopted from [26].

19.4.3 Summary of the Algorithm

We summarize below our algorithm for limited-angle tomographic reconstruction, given the locations and magnitudes of edges that lie parallel to the missing view angles.

1. Interpolate unknown projections from the known ones using (19-47). This interpolation will work well at low frequencies but not so well at high frequencies.
2. Reconstruct an image $x_{pl}(n, m)$ from known and interpolated views using FBP.
3. Compute UWT of $x_{pl}(n, m)$ up to some scale L using (19-17).
4. From the knowledge of edges that lie parallel to the n axis, construct an image $e(n, m)$ that has edges of known magnitude at known locations, and no edges elsewhere. Compute $d_{2^l}^{(2)} e(n, m), l = 1, \ldots, L$.
5. Reconstruct an image $x_{epl}(n, m)$ using the inverse UWT from the images $\{d_{2^l}^{(1)} x_{pl}(n, m), d_{2^l}^{(2)} e(n, m), d_{2^l}^{(3)} x_{pl}(n, m), l = 1, \ldots, L\}$ and $r_{2^L} x_{pl}(n, m)$.
6. Compute projections $p_{epl}(r_j, \theta_i)$ of $x_{epl}(n, m)$ for $\theta_{N/2 - L_{\text{miss}}} \leq \theta_i \leq \theta_{N/2 + L_{\text{miss}} - 1}$ to complete the missing projection data. The completed set of views $\tilde{p}(r_j, \theta_i)$ is

$$\tilde{p}(r_j, \theta_i) = \begin{cases} p_{epl}(r_j, \theta_i) & \text{if } \theta_{N/2 - L_{\text{miss}}} \leq \theta_i \leq \theta_{N/2 + L_{\text{miss}} - 1} \\ p(r_j, \theta_i) & \text{otherwise} \end{cases}. \tag{19-48}$$

7. Finally, use FBP on $\tilde{p}(r_j, \theta_i)$ to obtain the restored image $\tilde{x}(n, m)$.

It is also possible to carry out a similar restoration algorithm without using the wavelet transform. One can replace steps 3 and 4 above with low-pass and high-pass filtering with fixed filters, and step 5 with the combination of the two filtered images. However, such an algorithm (1) would be more rigid, and (2) may be computationally more expensive, than the wavelet transform approach. The flexibility of our algorithm comes from the fact that once steps 3–4 have been performed, one can vary L in step 5 and try to improve the restoration with different effective bandwidths. With fixed low-pass and high-pass filters on the other hand, one would have to repeat steps 3–4 to change the filter bandwidths.

The computational cost comparison of the two approaches depends on the image width and the wavelet used. To give an idea, Haar wavelet requires 16 additions per image point per scale for steps 3–5, whereas low-pass–high-pass filtering using FFT requires approximately $6 \log N$ multiplications and $12 \log N$ additions per image point for an $N \times N$ image. If $N = 256$ and $L = 4$, then the wavelet approach will be more than twice as fast as the low-pass–high-pass filtering approach.

19.4.4 Numerical Examples

We now present numerical examples on two test images. The first image is a geometric phantom. The second image is the Shepp–Logan phantom, in which the gray levels are chosen as in [27]; this is a frequently-used phantom in limited-angle studies [6,23]. The reconstructed images $x(n, m)$ from full views with $N = M = 128$ are shown in Figs. 19-10 and 19-11.

Figure 19-12 shows $x_{\lim}(n, m)$, obtained using FBP when 16 views ($L_{\text{miss}} = 8$, $\theta_{miss} = 22.5°$) are missing from the projections of the geometric phantom. The missing views are replaced by zero in FBP, as in the squashing algorithm [21]. We observe that there is blurring and artifacts around the edges parallel to the n axis. Figure 19-13 shows the locations where we assume *a priori* edge knowledge. Figure 19-14 shows the edge image $e(n, m)$ obtained from these locations and edge magnitudes.

Figure 19-10   The geometric phantom reconstructed from 128×128 projections.

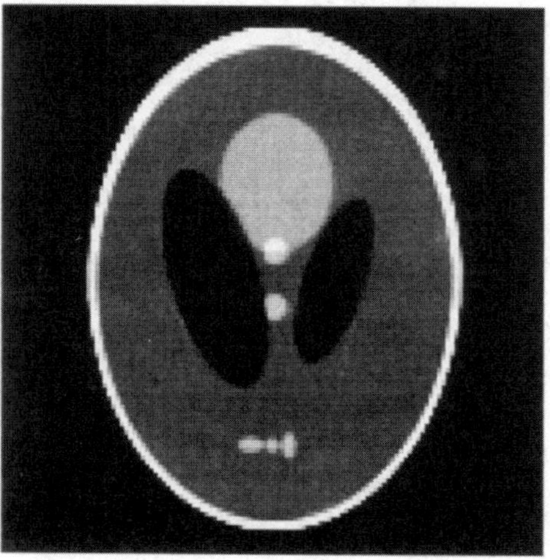

Figure 19-11   The Shepp–Logan phantom reconstructed from 128×128 projections.

Figure 19-12 $\mu_{lim}(n, m)$ obtained from 16 missing view angles ($\theta_{\mathrm{miss}} = 22.5°$).

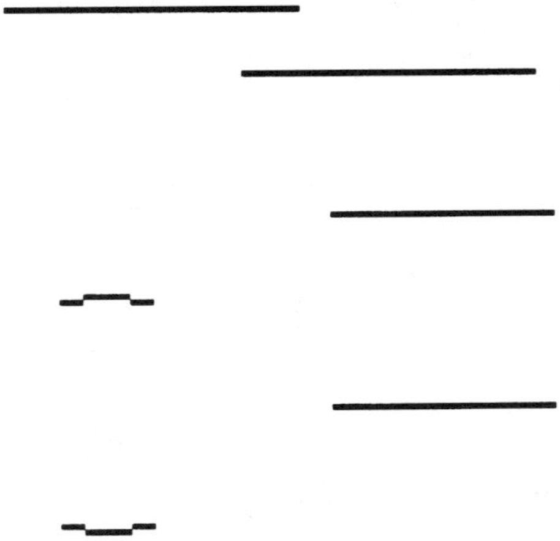

Figure 19-13 Locations of edges that are known to be parallel to the n axis.

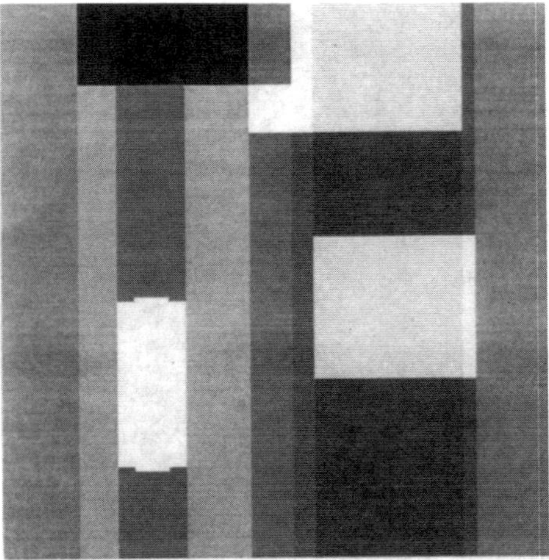

Figure 19-14 The edge image $e(n, m)$.

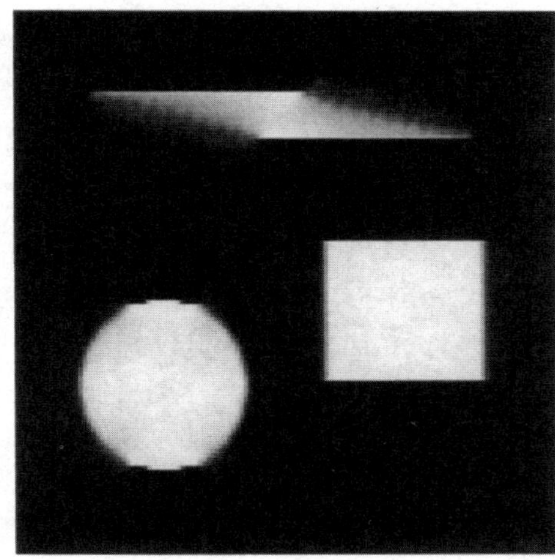

Figure 19-15   The image $x_{epl}(n,m)$.

Figure 19-15 shows $x_{epl}(n, m)$, reconstructed from $\{d_{2^l}^{(1)} x_{pl}(n, m), d_{2^l}^{(2)} e(n, m), d_{2^l}^{(3)} x_{pl}(n, m), l = 1, \ldots, 4\}$ and $r_{2^4} x_{pl}(n, m)$, where the wavelet basis for the decomposition and reconstruction was chosen as the Haar basis. Figure 19-16 shows the final restored image, $\tilde{x}(n, m)$. We note that not only the edges parallel to the n axis are now unblurred (this is natural because we *assumed* that we knew the existence of these edges), but also the artifacts around the edges have been significantly reduced.

Figure 19-17 shows $x_{\lim}(n, m)$ for the Shepp–Logan phantom, with $L_{\mathrm{miss}} = 16$. Figure 19-18 shows the locations where we assume *a priori* edge knowledge, and Figure 19-19 shows the final restored image, $\tilde{x}(n, m)$. Again, artifacts have been eliminated almost completely.

To quantify our results, we use the percent root mean square (rms) error

$$E_{\mathrm{rms}} = (100\%) \frac{\|\tilde{x} - x\|_2}{\|x\|_2} \tag{19-49}$$

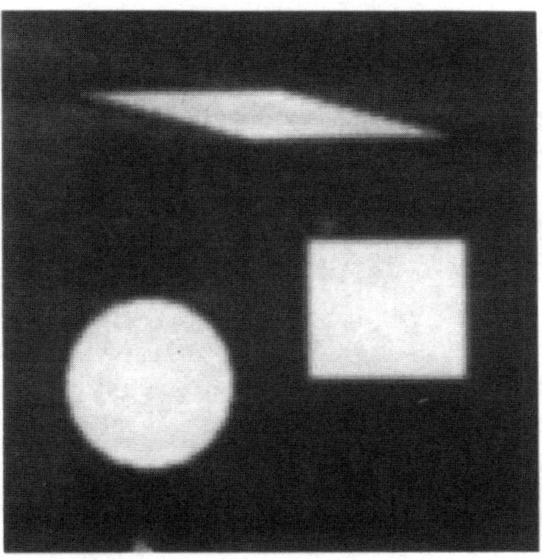

Figure 19-16   The final reconstructed image $\tilde{x}(n, m)$.

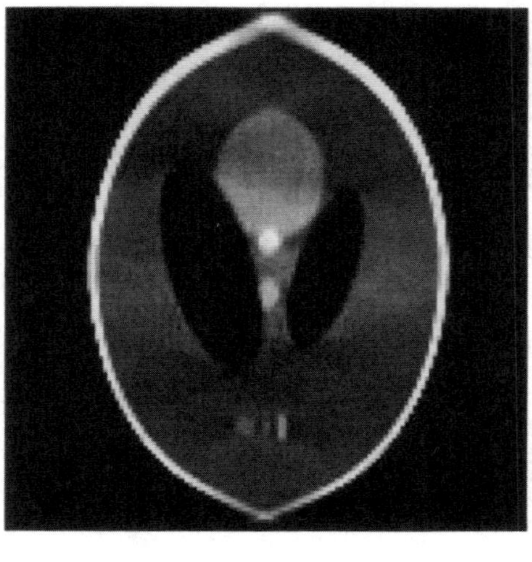

Figure 19-17 $x_{\text{lim}}(n, m)$ obtained from 32 missing view angles ($\theta_{\text{miss}} = 45°$).

Figure 19-18   Locations of edges that are known to be parallel to the n axis.

Table 19-1 summarizes the error for $\tilde{x}$ with three wavelet bases $D2$ (Haar basis), $D4$ and $D6$ defined in [12] and for $L = 3$, 4, 5, and 6. We observe that a short wavelet basis such as $D2$ or $D4$, and a moderate number of scales, e.g., $L = 3$ or $L = 4$, is sufficient to obtain good restoration.

We have also performed two simulations to test the robustness of our algorithm against errors in the edge information. In the first simulation, we assumed that the edge height information was incorrect. We multiplied all the correct edge heights by a number α, and used these erroneous edge heights as our *a priori* information. Table 19-2 shows the rms error after our algorithm for the Shepp–Logan phantom, as α is varied between 0.5 and 2.0. We used the Haar basis with $L = 4$, and $L_{\text{miss}} = 16$. We observe very little change in the rms error when the edge heights are changed by as much as 20% from their original values.

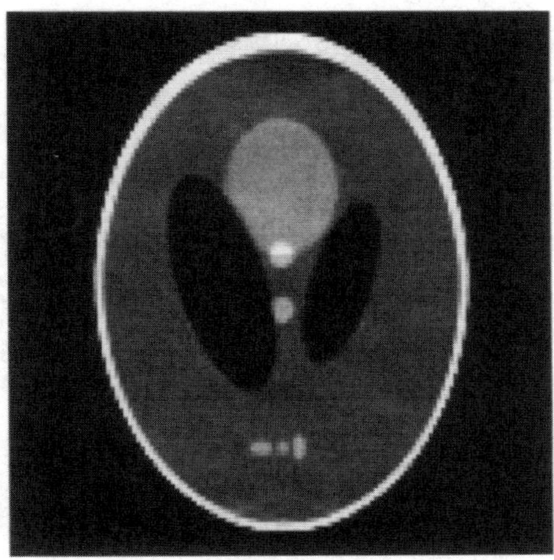

Figure 19-19 The final reconstructed image $\tilde{x}(n,m)$.

Table 19-1 E_{rms} for Different Wavelet Bases and Different J

	Geometric Phantom	Shepp–Logan Phantom
E_{rms} for $f_{\lim}$	38.96	28.91
Haar basis, $J = 3$	11.98	6.98
Haar basis, $J = 4$	10.07	6.21
Haar basis, $J = 5$	11.73	7.09
Haar basis, $J = 6$	12.88	8.88
D4 basis, $J = 3$	12.73	7.29
D4 basis, $J = 4$	9.89	6.33
D4 basis, $J = 5$	11.86	7.03
D4 basis, $J = 6$	13.24	8.90
D6 basis, $J = 3$	13.09	7.41
D6 basis, $J = 4$	9.86	6.42
D6 basis, $J = 5$	11.93	7.09
D6 basis, $J = 6$	13.35	8.93

Table 19-2 E_{rms} for the Shepp–Logan Phantom for Different Values of α

α	0.5	0.8	0.9	1.0	1.1	1.2	2.0
E_{rms}	9.61	6.81	6.34	6.21	6.44	6.99	16.53

In the second experiment, we kept the correct values for the edge heights, but we displaced all the edges in the horizontal or vertical direction by one pixel. The rms error after our algorithm was 6.68% for horizontal displacement, and 11.55% for vertical displacement, which means that good edge location information is more important than good edge height information.

19.5. CONCLUSION

We investigated two applications of the wavelet transforms to the problem of image reconstruction from projections. Our first application was the use of the DWT to perform spatially varying filtering on reconstructed noisy images. We constrained wavelet coefficients of the noisy image to zero in certain regions of the wavelet transform domain. These regions are known either from anatomical site information or from images obtained using other modalities, or are determined by thresholding the DWT of the noisy image. After the constraints were set, we computed the filtered image as the minimum mean squared error image which satisfied the constraints. We showed that even when the additive noise is white, this filtering will affect not just the pixels that contribute to the constrained wavelet coefficients, but the entire image. Using two examples, we demonstrated this property, and displayed the edge-preserving low-pass filtering achieved.

Our second application was the use of the unsubsampled wavelet transform for image reconstruction from limited-angle projections. We gave a new interpolation procedure to obtain a low-resolution image, and observed that when the missing angle range is small, two out of three sets of detail images are largely unaffected. This isolated one set of high-resolution images as the missing information. Any restoration method for the limited-angle problem must use some kind of *a priori* information; in our case, we used the magnitude and location of edges that lie parallel to the missing angles as our *a priori* information. This information was used to restore the affected high-resolution images, and this in turn enabled us to substantially reduce artifacts in the final restored image. In our simulations, we have observed that this approach reduced the mean squared error in the reconstructed image by a factor of 3 to 4 times, compared with the naive reconstruction method which assumed all missing data to be zero.

REFERENCES

[1] S. R. Deans, *The Radon Transform and Some of Its Applications*. New York: John Wiley and Sons, 1983.

[2] B. Sahiner and A. E. Yagle, "Image reconstruction from projections under wavelet constraints," *IEEE Trans. Signal Proc.*, Special Issue on Wavelets and Signal Processing, vol. SP-41, pp. 3579–3584, 1993.

[3] A. Faridani, E. L. Ritman, K. T. Smith, "Local tomography," *SIAM J. Appl. Math.*, vol. 52, pp. 459–484, 1992.

[4] B. Sahiner and A. E. Yagle, "Reconstruction from projections under time-frequency constraints," *IEEE Trans. Med. Imag.*, vol.14, pp. 193–204, 1995.

[5] K. C. Tam and V. Perez-Mendez, "Tomographical imaging with limited-angle input," *J. Opt. Soc. Am. A.*, vol. 71, pp. 582–592, 1981.

[6] P. Oskoui-Fard and H. Stark, "Tomographic image reconstruction using the theory of convex projections," *IEEE Trans. Med. Imaging*, vol. MI-7, pp. 45–59, 1988.

[7] M. I. Sezan and H. Stark, "Tomographic image reconstruction from incomplete data by convex projections and direct Fourier inversion," *IEEE Trans. Med. Imag.*, vol. MI-3, pp. 91–98, 1984.

[8] M. I. Sezan and H. Stark, "Image restoration by the method of convex projections: Part II," *IEEE Trans. Med. Imag.*, vol. MI-1, pp. 95–101, 1982.

[9] S. W. Rowland, "Computer implementation of image reconstruction formulas". In *Image Reconstruction from Projections, Implementation and Applications*. G. T. Herman (ed.). New York: Springer, 1978.

[10] L. A. Shepp and B. F. Logan, "The Fourier reconstruction of a head section," *IEEE Trans. Nucl. Sci.*, vol. NS-21, pp. 21–42, 1974.

[11] S. Mallat, "A theory for multiresolution signal decomposition: The wavelet representation," *IEEE Trans. Patt. Anal. Machine Intell.*, vol. PAMI-11, pp. 674–693, 1989.

[12] I. Daubechies, "Orthonormal bases of compactly supported wavelets," *Comm. Pure Appl. Math.*, vol. 41, pp. 909–996, 1988.

[13] G. Strang, "Wavelets and dilation equations: A brief introduction," *SIAM Rev.*, vol. 31, pp. 614–627, 1989.

[14] E.P. Simoncelli, W. T. Freeman, E. H. Adelson and D. J. Heeger, "Shiftable multiscale transforms," *IEEE Trans. Inform. Theory*, vol. 38, pp. 587–607, 1992.

[15] S. Mallat, "Multifrequency channel decompositions of images and wavelet models," *IEEE Trans. Acoust., Speech, Signal Proc.*, vol. ASSP-37, pp. 2091–2110, 1989.

[16] S. Mallat and S. Zhong, "Characterization of signals from multiscale edges," *IEEE Trans. Patt. Anal. Machine Intell.*, vol. PAMI-14, pp. 710–732, 1992.

[17] X.-H. Yan, R. Leahy, Z. Wu, and S. Cherry, "MAP estimation of PET images using prior anatomical information from MR scans." In *Proc. IEEE Medical Imaging Conf.*, pp. 1201–1203, Orlando, FL, October 1992.

[18] M. Bikdash and K. B. Yu, "Linear shift varying filtering of non-stationary chirp signals." In *Proc. IEEE Southeastern Symposium on System Theory*, pp. 428–432, 1988.

[19] H. L. Van Trees, *Detection, Estimation, and Modulation Theory*. New York: Wiley, 1968.

[20] A. K. Jain and S. Ansari, "Radon transform theory for random fields and optimum image reconstruction from noisy projections." In *Proc. ICASSP* pp. 12A.7.1–12A.7.4, 1984.

[21] T. Olson and J. S. Jaffe, "An explanation of the effects of squashing in limited angle tomography," *IEEE Trans. Med. Imag.*, vol. MI-9, pp. 242–246, 1990.

[22] J. A. Reeds and L. A. Shepp, "Limited angle reconstruction in tomography via squashing," *IEEE Trans. Med. Imag.*, vol. MI-6, pp. 89–97,1987.

[23] P. Oskoui and H. Stark, "A comparative study of three reconstruction methods for a limited-view computer tomography problem," *IEEE Trans. Med. Imag*, vol. MI-8, pp. 43–49, 1989.

[24] P. T. Fox, J. S. Perlmutter, and M. E. Raichle, "A stereotactic method of anatomical localization for positron emission tomography," *J. Comput. Assist. Tomogr.*, vol. 9, pp. 141–153, 1985.

[25] U. Pietrzyk, K. Herholz, and W.-D. Heiss, "Three-dimensional alignment of functional and morphological tomograms," *J. Comput. Assist. Tomogr.* vol. 14, pp. 51–59, 1990.

[26] P. A. Rattey and A. G. Lindgren, "Sampling the 2-D Radon transform," *IEEE Trans. Acoust., Speech, Signal Proc.*, vol. ASSP-29, pp. 994–1002, 1981.

[27] S. X. Pan and A. C. Kak, "A computational study of reconstruction algorithms for diffraction tomography: Interpolation versus filtered backpropagation," *IEEE Trans. Acoust., Speech, Sig. Proc.*, vol. ASSP-31, pp. 1262–1275, 1983.

Chapter 20

Wavelets Applied
to Mammograms

Walter B. Richardson, Jr.

20.1. INTRODUCTION

Wavelet analysis, a mathematical discipline which has matured rapidly during the past 15 years, represents a synthesis of ideas from classical harmonic analysis, pyramidal schemes in image processing, and subband coding methods of signal processing. Wavelets hold great promise as an integral part of a unified solution to several problems in medical imaging, and in particular, mammography, in the areas of: feature enhancement, texture discrimination, pattern recognition, data compression, de-noising, teleradiology, and image acquisition. This chapter briefly reviews the challenges faced as technology moves toward digital mammography, presents a necessarily brief overview of multiresolution analysis, and finally gives current and future applications of wavelets to several areas of mammography.

Screening mammography offers a much lower lesion size threshold and better documented record of benefit than physical examination by either the physician or patient [1]. Although mammograms are currently obtained with a film-screen combination that delivers excellent resolution, fully digital mammography will soon become common [2]. American Cancer Society guidelines for women aged 40–50 advocate screening every 1–2 years with frequency based on the patient's risk factors. Adherence to this standard would result in some 20 million mammograms per year. Archiving this data, retained for at least five years, will be expensive and difficult, requiring sophisticated data compression techniques. Even well trained radiologists misdiagnose 10%–20% of the mammograms they review [3] and computer-aided diagnosis (CAD) has the potential of reducing this rate dramatically. After retrieval, the mammograms need to be enhanced and prescreened by expert software, then compared, both for one patient over several years and multiple views, and across many patients to find similar lesions.

There are two primary signatures used by the radiologist to discriminate between normal and cancerous tissue. The first is mass, density, and shape. A benign neoplasm is smoothly marginated whereas a malignancy is characterized by an indistinct border which becomes more spiculated with time, the classic example being a stellate lesion. The second primary signature is microcalcification, although this also occurs in intramammary and intradermal ducts. Other indicators include architectural distortions in connective tissue, lucencies, nipple discharges, lymphomas, skin thickening, increased vascularity, and asymmetry of the opposite breast.

The first mammographic image by Leborgne in 1951 showed the presence of calcification in a breast tumor. Now roughly 1/2 of infraclinical tumors of the breast are revealed by microcalcifications. Calcifications less than 0.5 mm (microcalcifications) always reflect a breast pathology, benign or malignant. They evolve rapidly, either increasing or decreasing, and their morphology, distribution, location, and number must be described accurately. The Curie group described the morphology using five types in a 1976 report [4] based on 341 cases proven by histology: annular or arctiform (0% association with cancer); punctiform (20%); dust-like with a shape too small to recognize (40%); irregular punctiform (60%); and worm-like or branched (96%). Generally, the greater the number of microcalcifications, the greater the risk of malignancy. Distribution is another important factor, since greater clustering is often associated with malignancy, whereas, disseminated calcifications in both breasts suggest benignity.

Certainly the question of detecting lesions is one of pattern recognition, but the patterns involved are extremely diverse, occur at many different scales and orientations, and are characterized by discontinuous changes in intensity (microcalcifications) as well as more subtle global variations in texture (stellate lesions). Note that an essential ingredient of both primary signs is the ability of the radiologist to distinguish textures. This problem has a long history in image processing [5]; representative approaches to its solution include filtering with Gabor functions [6] and fractal image transforms [7]. Recently, Koepfler and Morel [8] describe a texture discrimination algorithm which: (1) is multiscale and pyramidal, operating in linear time; (2) has only channel weights as parameters; and (3) is a synthesis of textons, energy methods in segmentation, and wavelets.

20.2. WAVELETS AND MULTIRESOLUTION ANALYSIS

Wavelet analysis refers to a collection of methods which have found increasing use in signal processing, image analysis, and data compression. Researchers in psychophysiology of human vision, speech processing, and seismic data processing, have supplemented Fourier analysis with methods that better represent signals with near discontinuities, as often occur in speech and images. These methods replace the functions $e^{i\omega t}$ with "wavelets," which have a finite duration and a well defined average frequency. This section gives a very brief introduction to the theory of wavelets; space limitations preclude anything more than the highlighting of those features which are of particular importance to medical imaging. An extensive set of references to the wavelet literature is given at the end of this chapter.

The Fourier transform of a signal x gives its frequency content, but no time localization. An improvement is the windowed Fourier transform which first multiplies x by a windowing function g, often a square pulse or Gaussian. In contrast to this time-frequency analysis, the continuous wavelet transform

$$(Wx)(a, b) = |a|^{-1/2} \int_{-\infty}^{\infty} x(t)\psi\left(\frac{t-b}{a}\right) dt \qquad (20\text{-}1)$$

gives a time-scale description consisting of inner products of x with translated and dilated versions of a single function ψ. One may consider W as a mathematical microscope which can "zoom in" on features of interest at different scales and locations.

The analyzing wavelet ψ must satisfy a certain admissibility condition, which implies that ψ has mean zero and that $(Wx)(a, \cdot)$ results from applying a high-pass filter to x. A simple example of a wavelet with infinite support is $\psi(t) = (1 - t^2)\exp(-t^2/2)$, the so-called "Mexican hat" function. This difference-of-Gaussians function occurs in many areas of neurobiology, as described in Marr's book *Vision* [5]. For details on the deep connections between wavelet analysis and multichannel models in psychophysics and the physiology of vision see the article by Mallat [9]. Note that W involves a continuous, two-parameter family of inner products, and that there is much redundancy in the description which it gives of x. To eliminate this, one defines a discrete version of (20-1) by taking $a = 2^{-j}$, $b = 2^{-j}k$ where j, k range over the integers (2 is perhaps the most natural choice for the dilation factor, but other factors are possible). It is remarkable that for certain choices of ψ, the countable family $\psi_{j,k}(t) \equiv 2^{j/2}\psi(2^j t - k)$ forms an orthonormal basis for the class of square integrable functions, $L^2(\mathbb{R})$, and the improper integral in (20-1) can be replaced by an infinite sum. For practical purposes, a finite sum is needed so that the algorithm can be implemented with finite impulse response (FIR) filters. The simplest such function ψ was discovered by Haar in 1910 and takes the value 1 on $[0,1/2)$, -1 on $[1/2,1)$, and zero elsewhere. This wavelet serves as a "canonical" example for the theory and is useful in some applications, although lack of continuity renders it unsuitable for others. Daubechies [10] proved it is possible to construct a whole family of compactly supported wavelets with any desired degree of smoothness or differentiability. Figure 20-1 shows Daubechies wavelets with 4 and 20 nonzero taps.

Associated with the wavelet ψ is a "scaling" function ϕ, which considered as the impulse response to a filter analogous to W, is now low-pass rather than high-pass. The translates and dilates $\phi_{j,k}(t) \equiv 2^{j/2}\phi(2^j t - k)$ generate a nested sequence of subspaces $\ldots V_{-1} \subset V_0 \subset V_1 \ldots$ such that the projection of the signal x onto V_j is a coarse, blurred vision at a scale 2^{-j}. The approximate versions of x at different levels are related, because ϕ is a linear combination of scaled and translated versions of itself, as given by the dilation equation

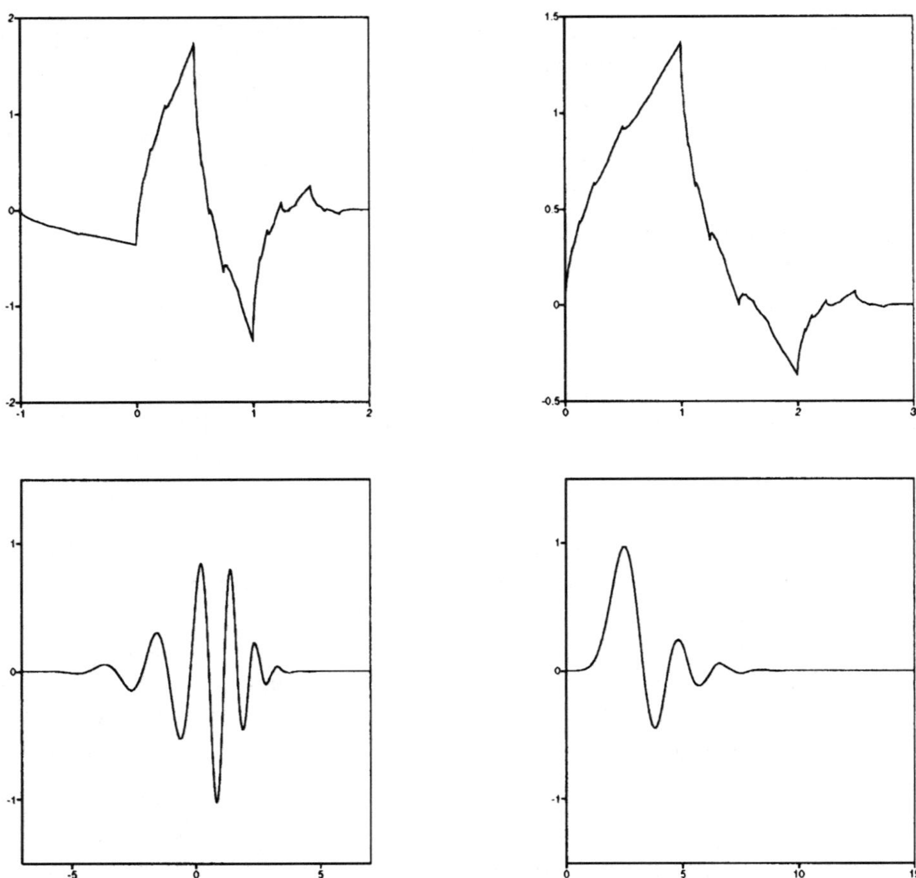

Figure 20-1 Examples of compactly supported Daubechies wavelets. The top figures
are the wavelet (left) and scaling function (right) associated with an
FIR filter with $N = 4$ "taps" or nonzero elements in the sequences $\{h_n\}$
and $\{g_n\}$. The lower two figures are $N = 20$, which results in a much
smoother wavelet having $N/2$ vanishing moments.

$$\phi(t) = \sqrt{2} \sum_n h_n \, \phi(2t - n) \tag{20-2}$$

It was Mallat and Meyer who first defined this "multiresolution analysis" of $L^2(\mathbb{R})$,
beginning now with a scaling function satisfying (20-2) and using it to generate the
corresponding wavelet.

As a consequence of (20-2), taking inner products of x with the families of
functions $\{\psi_{j,k}\}$ and $\{\phi_{j,k}\}$ can be realized using a pair of FIR filters H, G, which
are low-pass and high-pass respectively. Together these filters define an analysis-
synthesis scheme as shown in Fig. 20-2, where $\boxed{2\downarrow1}$ denotes taking every other sample
(down-sampling), $\boxed{1\uparrow2}$ denotes inserting zeroes between every other value

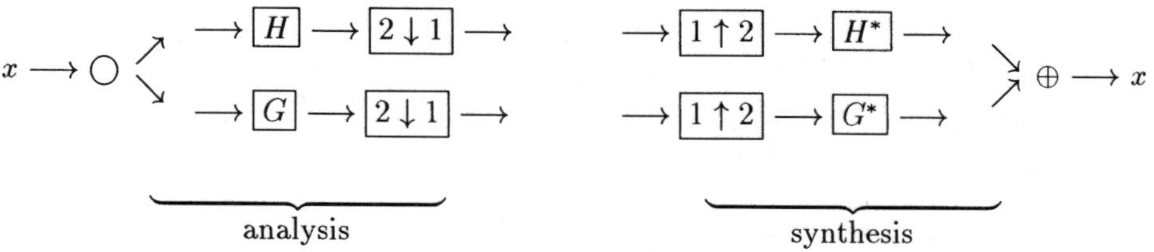

analysis   synthesis

Figure 20-2 Schematic of the forward and inverse discrete wavelet transform. A signal x is convolved with sequences $\{h_n\}$ (low-pass) and $\{g_n\}$ (high-pass) and the results down-sampled or decimated to produce a resume Hx and a detail Gx. The synthesis of x consists of up-sampling, applying the adjoints H^*, G^*, and adding the results.

(up-sampling), and A^* denotes the adjoint of the linear operator A. This convolution-decimation process has several important properties:

- In exact arithmetic, it allows for perfect reconstruction of the original signal.
- It can be applied recursively, so that the "resume" Hx is further decomposed as HHx, GHx, etc. In this way, a coarse approximation or resume of x and a series of details are generated, and from which x can be recovered exactly. If x consists of n samples then Hx and Gx would each have length $n/2$, so that (neglecting boundary effects or using circular convolutions) the transformed signal requires the same number of storage locations as the original.
- The decomposition can be carried out in linear time, i.e., $\mathcal{O}(n)$ where n is the number of samples. This is better than the $\mathcal{O}(n\log n)$ of the fast Fourier transform (FFT), although this advantage for wavelets is perhaps not as important as others involving space-time localization and the fact that wavelets form an unconditional basis for many function spaces.
- The forward/inverse wavelet transforms are elegant in their simplicity, requiring only convolutions and decimation/upsampling. They are easy to program in software or implement in hardware, resulting in special purpose chips such as are now common for the FFT. (The reader may note the connections between the wavelet algorithm and pyramidal schemes in image processing [11], as well as quadrature mirror filters.)

Figure 20-3 displays a wavelet decomposition of a mammogram "slice" using the Haar basis. The lowest graph is the signal of length 512; the panel at the left shows the resumes uniformly scaled in amplitude, each representing a coarser approximation to x than the one below. At the right the absolute value of the details (different amplitude scales) are shown. All plots are on the same horizontal scale for spatial correlation, but remember that at each level the number of samples is actually halved. The final wavelet decomposition consists of the coarsest resume and all the details, $(R_{-4}, D_{-4}, D_{-3}, D_{-2}, D_{-1})$ with lengths (32, 32, 64, 128, 256). Reconstruction requires taking R_{-4} and D_{-4} to obtain R_{-3} and then moving down the tree recursively. Note the sharp peak at position 152 in the original trace, corresponding to a microcalcification; its location is even more evident in the high-frequency details. In

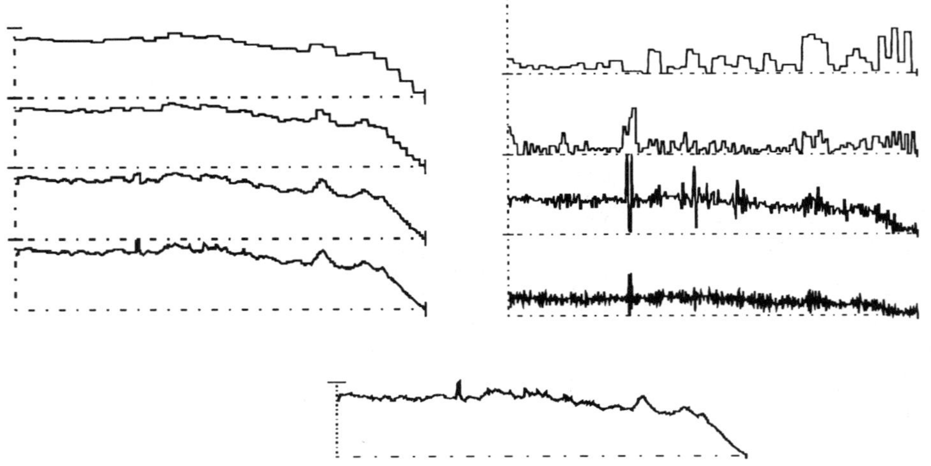

Figure 20-3   The lowest graph shows a trace extracted from a mammogram. A 1-D wavelet decomposition shows the smoothed approximations (left panel) at decreasing resolutions with the absolute value of the corresponding "details" shown at the right. A microcalcification is evident in the high-frequency detail.

two dimensions, the prescription for wavelet analysis is much the same—at least if tensor product wavelets are used. Starting with the one-dimensional (1-D) scaling function ϕ and wavelet ψ, one filters the image u with one low-pass filter $\Phi(n, m) = \phi(n)\phi(m)$ and three high-pass filters $\Psi^1(n, m) = \phi(n)\psi(m)$, $\Psi^2(n, m) = \psi(n)\phi(m)$, $\Psi^3(n, m) = \psi(n)\psi(m)$, which are directionally selective, giving horizontal, vertical, and diagonal details of an image, respectively. Figure 20-4 shows a multiscale analysis of a 512^2 mammogram snippet using linear spline wavelets. The subimage in the upper left corner represents the coarse approximation R_{-2} at a resolution $r = 2^{-2}$, i.e., the smallest detail visible is of size $1/r = 4$. The three surrounding subimages are the absolute values of the details D^1_{-2} (upper right), D^2_{-2} (lower left), and D^3_{-2} (lower right). These details and R_{-2} can be combined to yield a 256^2 resume R_{-1}, which in turn, can be used with the outer band of details D^1_{-1}, D^2_{-1}, and D^3_{-1} to reconstruct the original snippet exactly.

20.3. DATA COMPRESSION AND TELERADIOLOGY

The widespread use of digital mammography will require fast, efficient algorithms for image compression. There are two basic types of compression: lossless and lossy, which upon being uncompressed yield the original signal or an approximation, respectively. For the acquisition and storage of digitized mammograms, possible legal ramifications require lossless compression. In fact, one of the major issues facing the designers of digital systems is that of data integrity relative to a traditional

Figure 20-4 A 2-D decomposition using Haar wavelets of a 512^2 mammogram snippet. The outer band contains (clockwise order) level-one details D^1_{-1}, D^3_{-1}, D^2_{-1}, the inner band consists of the level-two details, and finally in the upper left corner is the resume R_{-2}. In two dimensions, the three high-pass filters give horizontal, vertical, and diagonal details of an image.

film-screen combination. Current proposals call for high-resolution mosaic systems consisting of 50 charge-coupled device (CCD) chips each of 1024^2 or 2048^2 pixels [12], but the question remains, will any information valuable to diagnosis be lost to the radiologist during digitization? Issues of fidelity are far more serious here than in the compact disk versus analog debate in audio recording. Of course, digital technology will ultimately prevail because of the myriad image-processing tasks that can be performed after digital data acquisition. What is required is a detailed statistical study to determine the necessary characteristics of the digital system that will replace conventional film-screen technology. It would be surprising if wavelets and their generalizations did not figure prominently in future digital systems.

In addition to its use for acquisition and storage at the hospital or imaging center, lossless compression is important in teleradiology, which involves the transmission of the X-ray from a remote site to a diagnostic center and back. In some cases this will be from a hospital workstation to a physician's home computer for a preliminary examination before further radiograms or diagnostic tests are ordered. Given the large amounts of data and limited bandwidth of conventional twisted pair telephone lines, wavelets will play an important role in improving the efficiency of telephone modems. Currently, the discrete multitone transceiver (DMT) systems [13] which lie at the heart of many modems are based on Fourier transform methods. Wavelets have been proposed as a possible method of improving classic DMT techniques by providing multiple narrowband channels with superior adjacent channel isolation and longer symbol duration [14].

There are several wavelet-based schemes for lossless compression. It is reasonable to expect that standard methods such as Huffman coding and vector quantization will yield higher compression ratios when applied to the wavelet transform rather than to the image itself. Mathematically, it is important to choose a good basis in which to represent the signals or images of interest as soon as possible after (or even during) acquisition. For representing signals with discontinuities, wavelets offer several advantages over trigonometric polynomials. On the other hand, for smooth periodic functions Fourier methods are ideal. When representing a signal, how does one choose between wavelets, sines/cosines, Walsh functions, or for that matter between different types of wavelets? The key is to find a basis which in some sense is "natural" for the class of signals being analyzed. Several approaches are possible: Tewfik and Kim [15] use optimization techniques to find the wavelet which best approximates a signal, given a constraint on the support width of the wavelet; Coifman and Wickerhauser [16] use another approach, wave packets, which are now briefly described. The standard wavelet basis decomposes a signal into a resume and detail, then decomposes the resume itself, and so on. In fact, one might choose at each step the detail or resume to decompose. This amounts to selecting a different orthonormal basis for the space of signals, one which might do a better job than the standard wavelet basis of concentrating the information contained in x into a few large coefficients. Because the corresponding orthogonal transformation is reversible, once again the original image can be reconstructed exactly from the coefficients.

More generally, both the resume and detail could be decomposed for a representation of x relative to a family of bases. Figure 20-5 shows the resulting binary tree structure: there is a one-to-one correspondence between (horizontally) nonoverlapping covers of the tree and members of a family of orthonormal bases of $L^2(\mathbb{R}^n)$. Given this family, the question becomes one of how to choose the member that is optimal for a given signal or class of signals. This is done using an information cost functional M [17] on the set of signal coefficients $\{x_i\}_{i=0}^{n-1}$ and finding the basis that minimizes M. Examples of such functionals include: (1) number of coefficients above a given threshold in magnitude; (2) l_p norm of x for $p < 2$; and (3) the Shannon–Weaver entropy of $\{x_i\}$ defined by $\mathcal{H}(x) = -\sum_i p_i \ln(p_i)$ where $p_i = |x_i|^2/\|x\|^2$ (and $p \ln p = 0$ if $p = 0$). This tree can be searched in $\mathcal{O}(n\log(n))$ time to find the "best" basis for the given signal.

For some applications, lossy compression of a signal is acceptable. For example, in high-definition television (HDTV), a controlled amount of loss during transmission can be tolerated by the user if the overall signal quality is much higher than in conventional methods. Pattern recognition techniques generally require considerable data reduction before they can be successfully applied. Wavelets have already been

x_0	x_1	x_2	x_3	x_4	x_5	x_6	x_7
s_0	**s_1**	**s_2**	**s_3**	d_0	d_1	d_2	d_3
ss_0	ss_1	ds_0	ds_1	sd_0	sd_1	**dd_0**	**dd_1**
sss_0	dss_0	sds_0	dds_0	**ssd_0**	**dsd_0**	sdd_0	ddd_0

Figure 20-5 Wave packets of Wickerhauser and Coifman. The usual wavelet decomposition applied to both resume and detail at each level generates a family of orthonormal bases of $L^2(\mathbb{R}^n)$, one of which is shown in boldface roman type. This binary tree can be searched very rapidly to find the "best" basis for a given signal.

chosen by the FBI for an automated fingerprint analysis system, which will perform quickly the task of matching a given print to one of several million others on file. The ability of multiresolution schemes to examine images at many different scales in linear time will likewise prove indispensable in processing the enormous number of mammograms taken annually.

At a very basic level, wavelet analysis offers simple lossy compression with large compression ratios, i.e., number of samples in the image divided by the number in the compressed version. One merely keeps the resume and discards all the details. This might be applicable for locating large features such as stellate lesions in a mammogram. Unfortunately, all the fine-scale features are then lost, and they often contain important information, for example the locations of microcalcifications. So-called "hard thresholding" retains at all levels only those coefficients which are above a certain cutoff level in magnitude, thereby hopefully isolating the most important features across scales. Table 20-1 [18] shows the results of compressing two 512^2 mammogram snippets with Daubechies wavelets of various lengths, using both the standard and the best-basis algorithms. The latter has higher compression ratios at the expected expense of longer computation times. Note that the longer wavelets perform substantially better than Haar (filter length = 2).

A recurring theme of the 1992 Toulouse Wavelet Conference was that the analyzing functions—be they wavelets, cosine/sine transforms, or wave packets—must be chosen to fit the specific problem. Using the approach of Coifman and Wickerhauser, one could find a best basis not just for one signal, but for a large class of signals, in this case a large database of mammograms. It is well known that if x_i, $i = 1$, M is a set of random vectors in $\mathbb{R}^n$ with $M \geq n$, then the maximum linear transform coding gain of any linear code used to transmit the set x is achieved using the basis of eigenvectors of the autocovariance matrix. Application of this Karhunen–Loeve or Hotelling transform has the effect of decorrelating the data, while the first few eigenvectors represent the principal features of the set of random

TABLE 20-1 Compression Ratios for 512^2 Mammogram Snippets using Both the Standard Wavelet and Best-Basis Algorithm of Wickerhauser with Various Length Filters and Threshold Levels. For Some Applications the Added Time for the Best-Basis Search will be Acceptable.

		Snippet 1		Snippet 2		Average CPU Time	
Filter Length	Cutoff Level	Best	Wavelet	Best	Wavelet	Best	Wavelet
2	0.001	11.4	10.5	8.3	7.9		
	0.005	36.8	35.7	22.6	22.6	197.3	128.2
	0.010	57.3	55.8	32.2	32.6		
10	0.001	20.3	18.6	11.1	11.0		
	0.005	51.2	46.2	26.4	26.9	1617.5	616.8
	0.010	72.0	68.2	37.1	36.7		
18	0.001	23.1	19.4	11.5	10.8		
	0.005	55.4	43.2	26.4	24.8	4963.6	1712.1
	0.010	75.6	58.7	36.7	33.9		

vectors. In other words, the variance of the ensemble accumulates most rapidly when the vectors are expressed in this coordinate system. In mammography, one could use many snippets as the x_i, building an orthonormal basis whose members are reminiscent of the microcalcifications, stellate lesions, and vascularity, which the trained radiologist defines as the "features" of interest. The difficulties of such a task are many; besides obtaining the required number of snippets, the resulting matrix would be so large that it would be impractical to do singular value decomposition (SVD) with standard techniques. Wickerhauser has suggested using wave packets for a very fast approximate Karhunen–Loève transform [17].

20.4. FEATURE ENHANCEMENT AND CLASSIFICATION

Wavelet analysis will prove extremely useful for feature enhancement and classification in mammography [19,20]. Figure 20-6 shows the horizontal detail D^1_{-1} of a snippet containing a well defined lesion. Note that the wavelet details correspond closely to one's perception of the detailed features of the image: the circular lesion, the fine-laced pattern of blood vessels, and microcalcifications. This "fine structure" stands out more prominently in the high-frequency detail than in the original image. At a very coarse scale stellate lesions can be isolated, because there, they too are similar mathematically to a two-dimensional (2-D) delta function. Once the larger tumors are identified, information from the fine scales could be added to identify spicules, delineate boundaries, and characterize texture.

After image enhancement has been performed, expert systems, artificial neural nets (ANNs), or fuzzy logic can be used for pattern classification. Many classifiers are based on computing the distances, in an appropriate metric, between the input

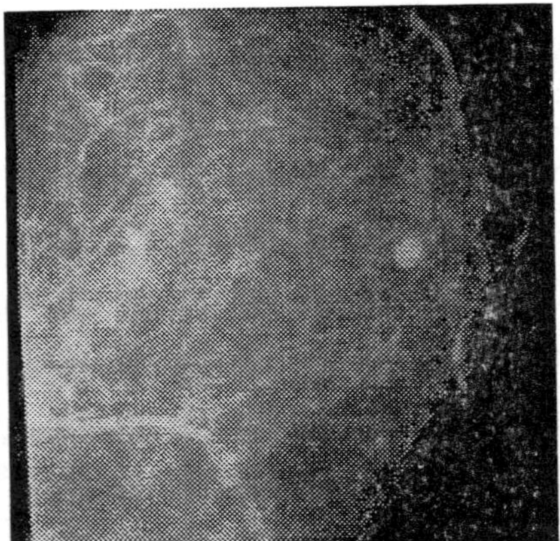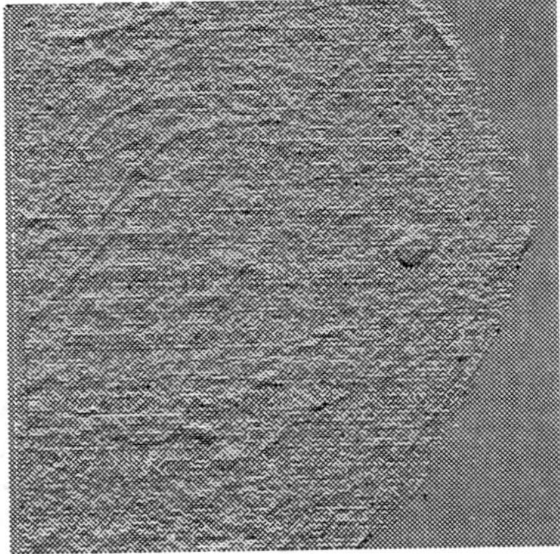

Figure 20-6   A 512^2 snippet (left) and the first level of detail (right) using Haar wavelets, in this case the horizontal, D^1_{-1}. Observe the lesion, marked vascularity, and the microcalcifications, all of which correspond to our perception of the details present in the original image.

pattern vector y and representatives $x_1, \ldots, x_m$ of M pattern classes $\omega_1, \ldots, \omega_M$. If y is "closer" to x_j than to any of the other representatives, it is said to belong to class ω_j. Neural nets extend this concept by constructing, i.e., training, a dynamical system $\dot{z}(t) = Wz(t)$ that has the representatives $\{x_j\}$ as rest points. If y belongs to the domain of attraction of x_j, as the dynamical system is run with initial data y, one has $z(t) \to x_j$ as $t \to \infty$ (see Fig. 20-7). For a survey of biosignal pattern recognition techniques see the article by Ciaccio et al. [21].

Perception psychologists are currently analyzing the visual search patterns of radiologists for clues as to which characteristics should be used to train the neural nets. Given the size of the pattern vectors (a single digitized mammogram of moderate resolution has length 1024^2), the large number of pattern classes, i.e., indicators of malignancy (see for instance [22]), and the subtle differences that distinguish benign from malignant tissue, straightforward application of neural nets is a formidable task, even for special-purpose ANN machines. Hence careful data reduction that preserves the radiologically important information is imperative before classification. As shown in Fig. 20-4, a wavelet analysis gives a compressed version of the image in the form of a high-level resume (say 64^2 pixels). The neural sets can be trained on the set of resumes to quickly recognize stellate lesions or round densities in pattern space, after which details can be added successively to further discriminate between various types of lesions. Such a hierarchical pattern recognition scheme is extremely efficient and mimics how the eye–brain analyses a scene for objects [5].

Most current work on neural nets and mammography does not work with a compressed image in the usual sense, but instead uses vectors of length 10 or 20 whose components represent various statistics of the image. For example, in [23] second-order statistics—entropy, contrast, angular second-order moment, inverse difference moment, correlation measure and five other image structure features— are used with K-means clustering and backpropagation ANNs to isolate features. It is important to note that because wavelets concentrate the energy and information into fewer large coefficients, it is quite reasonable to compute the statistics directly in the wavelet domain. The goal of using statistics is to work with a smaller, more manageable data set, which still faithfully represents the original image. Wavelets

Figure 20-7   Automated detection of microcalcifications: a 512^2 snippet (left); absolute value of a weighted average of all three details at level one using Haar wavelets (center); running a gray-scale morphological operator to "circle" the most prominent calcifications (right).

provide a mathematically precise technique for reducing data while preserving information on spatial features, structure, and smoothness.

20.5. WAVELETS, FRACTALS, AND TEXTURE

There are deep connections between wavelet analysis and the way the eye (and ear) process information. The human visual system is characterized by a multiresolution organization in that objects which appear well-defined at a fine scale are progressively lost to view as one moves to coarser scales. This reflects the fact that most natural processes are multiscale, for example, evolution of mountains and coastlines, formation of galaxy clusters, the hydrological cycle, and turbulence. Any algorithm for visual information processing, e.g., for texture discrimination, should therefore be multiscale/multiresolution. Central to the notion of scale is the mathematical concept of dimension, which tells how a quantity such as bulk scales with size: bulk $\propto$ size^D. A fractal is an object for which the scaling exponent D takes on fractional values. For a signal $1 \leq D \leq 2$, while for a gray scale image $2 \leq D \leq 3$. In both cases, a larger value of D represents a rougher, more textured surface. Fractional dimension has been used before as a feature identifier in ultrasonic images of the liver [24], X-ray images of bones [25], and mammography [26,27].

One formal definition of dimension is $D = \lim_{\epsilon \to 0^+} \ln N_\epsilon / \ln(1/\epsilon)$ where N_ϵ is the minimum number of balls of diameter ϵ needed to cover the object. This definition leads to the so-called box-counting algorithm for estimating dimension [28]; another popular method, the fractional Brownian motion or fBm algorithm, uses the expectation of the absolute intensity differences $E(|u(t + \Delta t) - u(t)|) \propto |\Delta t|^H$ to measure the Hurst exponent H, which is related to dimension by $D = 2 - H$ ($3 - H$ for an image). In fact, D may also be estimated directly from a wavelet analysis. Unlike their Fourier counterparts, wavelet coefficients give local information about the differentiability, Hölder continuity, and fractal properties of a function. For example, if a function x is Hölder continuous with exponent α, that is $|x(t) - x(s)| \leq C|t - s|^\alpha$, then its wavelet coefficients decay as $|W_{j,k}| \leq C2^{-j/2}2^{-j\alpha}$.

Fractals arise not only in many deterministic, iterative constructions, but also in random processes with long memory or statistical self-similarity. Fractional Brownian motion [29] $B_H(t)$ is a generalization of ordinary Brownian motion with a correlation structure given by

$$E(B_H(t), B_H(s)) = \frac{\sigma^2}{2}(|t|^{2H} + |s|^{2H} - |t - s|^{2H}) \tag{20-3}$$

Any portion of an fBm can be viewed as a statistically scaled version of a larger part of the same process. It is useful for describing $1/f$ noise in oscillators, classifying textures, generating images, and modeling bursts of errors in communications channels. Although nonstationary, fBm has stationary increments which are self-similar $B(at) \overset{d}{=} a^H B(t)$, a fact which is mirrored in the dilation equation for scaling functions. A realization (sample) of such a fBm process will be a fractal curve with probability one. Together nonstationarity and self-similarity suggest that fBm should be ana-

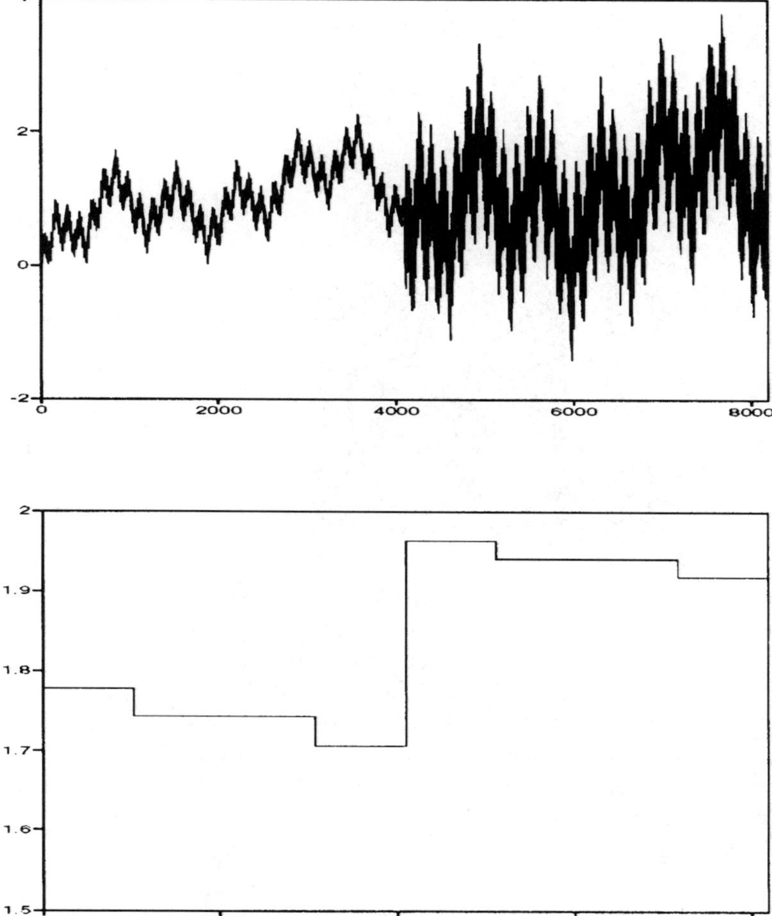

Figure 20-8   The upper graph shows a piecewise fractal curve in which the dimension is 1.7 to the left of location 4096, while to the right it is 1.9. The lower graph is a plot of "local dimension" calculated using the variance of the wavelet coefficients over a window of length 1096 samples.

lyzed using a time-scale method such as wavelets. The fBm index H (hence the dimension D) can be obtained from a log-log plot of the variance of the wavelet coefficients at the different levels plotted as a function of scale, as shown in Fig. 20-8. For more details on connections between wavelets, fractals, and stochastic processes see the articles by Flandrin [30] and Basseville et al. [31]. Some natural processes cannot be characterized by a single scaling exponent D, but rather by a continuous one-parameter family which depends on scale. Wavelets have been used with great success to measure such "multifractal" processes.

Fractal dimension, either global or local, can be used to help identify textures in mammography. Parenchymal patterns have been used for many years as an indicator for risk of breast cancer. A definitive study by Wolfe and co-workers in 1976 [32,33]

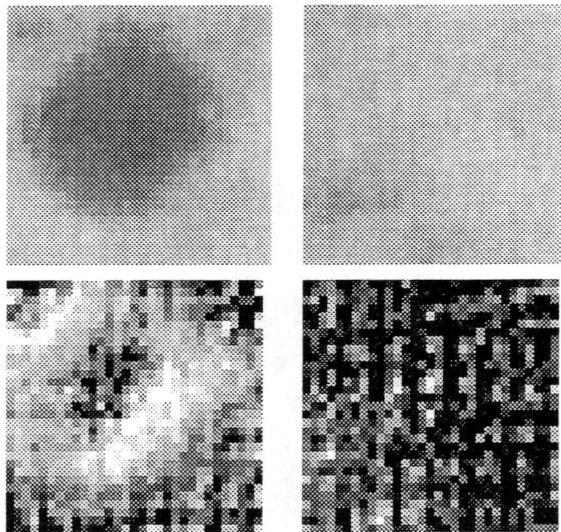

Figure 20-9 At the top left is a snippet of a suspected tumor and below it is the corresponding fractal transformation with average dimension of 2.41. The top right is a snippet of the background around the tumor; by comparison its fractal signature (bottom right) has an average dimension of 2.84. (See Moch [27].)

defined four grades of parenchyma: N1—primarily fat with no visible ducts; P1—primarily fat with less than 25% of the area covered by ducts; P2—ducts occupy more than 25%; and DY—diffuse, sheet-like regions of irregular densities. Caldwell et al. [26] use the global fractal dimension of the digitized mammograms, computed using the fBm algorithm, in order to characterize parenchyma. Such a quantitative classification will be very useful for long-term studies of patients at high risk of developing breast cancer. A simple fractal transform performed locally also provides an effective method of discriminating between a tumor and surrounding tissue, as shown in Fig. 20-9.

20.6. DE-NOISING

Any real signal is corrupted by some noise and it is generally necessary to filter or de-noise the data. The noise can be random in nature, i.e., white noise, or impulsive, in which a large number of pixels are altered during transmission to a value very much different from that of the original signal. In mammography the smallest microcalcifications are not visible on the mammogram because of noise and blur which occurs both in the X-ray system and in the digitizing camera. In the former case this is due to the geometry of the imaging process, beam scattering, and the screen-film components themselves. A detailed analysis [34], shows that to a good approximation this system can be modeled by

$$y = h * x + n_c + n_r$$

where x is the ideal image, y the observed image, h the system impulse response, n_c the camera observation noise and n_r the signal-dependent radiographic noise. The goal of a "de-noising" algorithm is to recover x given y and estimates of h, n_c, and n_r;

such inverse problems occur in many areas of medicine and engineering, including tomography, deconvolution of seismic data, and magnetic resonance imaging.

In contrast to the hard thresholding of wavelet coefficients mentioned in section 20.3, soft-thresholding described by Donoho and Johnstone [35] offers a means of removing noise from an image, signal, spectrum, or density (Fig. 20-10). Given a function $x(t)$ on [0,1] and noisy observations $y_i = x(t_i) + \sigma z_i$ where the z_i are white Gaussian noise random variables, wavelet shrinkage recovers an estimate of x by: (1) scaling ($\beta_i = y_i \sqrt{n}$) and taking the wavelet transform of β; (2) applying a soft-threshold $\eta_\tau(y) = \mathrm{sgn}(y)\,(|y| - \tau)_+$ with $\tau = \sqrt{2\log n}\,\sigma/\sqrt{n}$ to the resulting wavelet transform; and (3) performing an inverse wavelet transform to obtain $\hat{x}$, the estimate of x. This is an optimal algorithm in a minimax sense for many smoothness classes of function spaces [36]. Its effectiveness as compared with traditional Fourier techniques results from the fact that a wavelet basis is an unconditional basis for many function spaces besides L^2, for example L^p, $1 < p < \infty$. Roughly speaking, such bases do not give a strong preference to any of the coordinate directions. If more is known about the characteristics of the noise—particularly if the assumption of white Gaussian noise is suspect—other filtering operations are appropriate in the wavelet domain. Examples are the level-dependent thresholds [36] $\tau_{j,n} = \sqrt{2\log(n)}\,(2\sigma)/\sqrt{n}2^{(J-j)/2}$ or $\tau_{j,n} = \sqrt{2\log(n)}$ median absolute deviation $(\{w_{j,k}\}_k)/0.6745$, for numerical differencing and deconvolution, respectively.

For an image in which there is strong spatial correlation between the wavelet coefficients, still other nonlinear filters may be appropriate. A standard technique for eliminating noise is to convolve the input signal with a smoothing kernel such as a Gaussian. Koenderink observed that applying this low-pass filter amounts mathematically to solving the heat equation $\partial u/\partial t = \Delta u$ with $u_0 = u(n, m, 0)$ the original,

Figure 20-10   De-noising using the soft-thresholding algorithm of Donoho and Johnstone. The original image is shown on the left; in the de-noised version on the right, the microcalcifications and lesion are more clearly delineated.

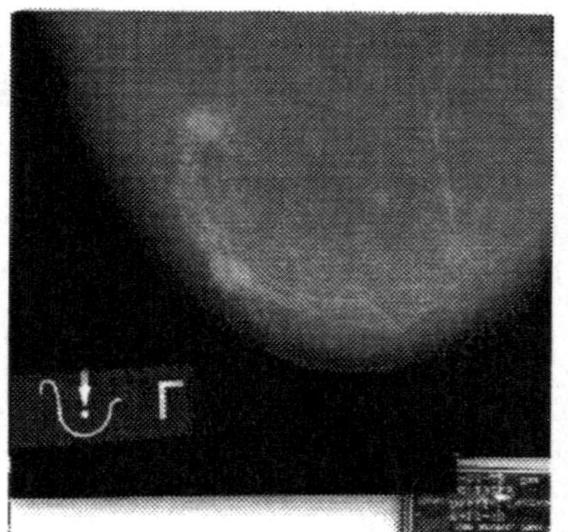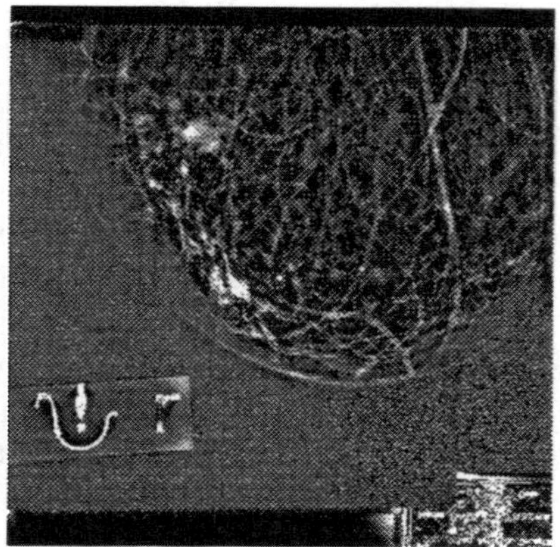

Figure 20-11   Details using mean curvature PDE filtering. The image (left) is filtered using equation (20-4). On the right is the difference between the original and the smoothed version, i.e., the details lost in going to the coarse approximation.

noisy image. The time t is a "scale" parameter which determines the minimum size of details kept in the smoothed image and the output is $G_t * u_0$ where $G_\sigma(n, m) = (C/\sigma)\exp(-(n^2 + m^2)/4\sigma)$ is a Gaussian function. Although this method does reduce impulsive noise, it can also blur true "edges," defined as a point where the norm of the gradient, $\|\nabla u\|$, is large. To smooth while maintaining edges, Lions and Morel [37] have proposed using the partial differential equation (PDE)

$$\frac{\partial u}{\partial t} = \|\nabla u\|\mathrm{div}\left(\frac{\nabla u}{\|\nabla u\|}\right) \tag{20-4}$$

The term div $(\nabla u/\|\nabla u\|)$ is the mean curvature and when multiplied by $\|\nabla u\|$ represents an operator that diffuses selectively in the direction perpendicular to the gradient, i.e., parallel to an edge, not at all across it. This filter can also be applied to the wavelet coefficients so as to preserve spatial correlation and geometric properties, while removing noise. It is also possible to combine the operations of filtering and classification by using PDEs similar to (20-4), but with added reaction terms that are a continuous analog of discrete neural nets [38].

If one has an image u_0 and a one-parameter family of smoothed, or *simplified*, versions $\{R_t\}$ of u_0, it is possible to mimic a wavelet decomposition. Much of the eloquence and efficiency of wavelets is lost, but in return the simplified versions can be obtained by nonlinear techniques, and indeed, there are both linear and nonlinear aspects to the human visual system. First, diffuse for a time t so as to smooth the image, eliminating noise and high frequencies, while preserving the edges. This gives a resume R_1 and the detail D_1 is naturally defined as $R_1 - R_0$ where $R_0 = u_0$. Figure 20-11 [19] shows the result of applying equation (20-4) to decompose a mammogram; note the excellent results in isolating the blood vessels and tumors from the large-scale features.

20.7. DISCUSSION AND CONCLUSIONS

Studies of computer-aided diagnosis (CAD) suggest that even the "primitive" CAD technology available today can improve interpretation. Groups at the University of Chicago, University of Manchester, and at GE-CGR in France have reported that CAD input to the radiologist improved his/her ability to identify subtle microcalcifications. Characterization may be of equal importance to detection since [3] "Radiologists will be more apt to take the trouble of digitizing selected cases to obtain characterization information." Several major groups funding cancer research have initiatives to develop over the next few years fully digital mammographic systems, which can be provided to the medical community at a reasonable cost. The queston then becomes: What technologies will be most effective in performing the many image processing tasks required of these complex systems? The premise of this chapter is that wavelet analysis offers significant advantages over traditional methods in meeting the needs of such a system.

The results presented here indicate that a multiresolution decomposition using wavelet analysis will offer an efficient method of data compression, texture analysis, and lesion detection for mammograms. The corresponding numerical algorithms require only convolution/decimation, can be realized using FIR filters, operate in linear time, and allow perfect reconstruction. In the areas of data compression and teleradiology, wavelets have already been chosen for the FBI's fingerprint identification system and will doubtless improve current discrete multitone transceiver technology. Having good localization in both time and frequency, wavelets offer a better means of representing functions with near discontinuities such as often occur in images, and will prove invaluable for pattern classification and image enhancement. There are strong connections between fractals and wavelets; the latter can in fact be used to analyze the fractal properties of signals and images, something not possible using classical Fourier methods. It is now known that because wavelets are an unconditional basis for many important function spaces, they are optimal in a minimax sense for solving a host of statistical problems, including de-noising and deconvolution. Finally, smoothing in the wavelet domain using soft-thresholding or mean-curvature-dependent filters offers a natural way to reduce noise that is scale or space dependent. As a mathematical discipline, wavelet analysis has matured very rapidly during the past 15 years; just as important is the fact that the applications of wavelets to technology in many different fields is taking place at an equally dizzying pace, with no signs of slowing down in the near future. In the words of Yves Meyer, "As was clear during the 1992 Toulouse conference, wavelet analysis is no longer a chapter of science but should be better viewed as a tool or a trick, like integration by parts. It will be used again and again and will play the modest but fundamental role of the furniture in a house."

One of the best general introductions to wavelets is Meyer's book, *Wavelets Algorithms and Applications* [39], in which the author does a superb job of tracing the development of wavelets and their applications to such diverse areas as computer and human vision, image and signal processing, turbulence, and the study of distant galaxies. The books by Daubechies [40] and Chui [41] are excellent also, but assume the reader has a deeper mathematical background. Several conference proceedings offer a glimpse at current wavelet research in a variety of applications [42–44]; the survey articles of Mallat [9] and Rioul and Vetterli [45] are excellent. For those interested in numerically experimenting with wavelets on their own data sets, Donoho and Johnstone's *Teach Wave* software runs under MatLab and is available via anonymous ftp from `playfair.stanford.edu`; there are wavelet toolboxes for Khoros as well. There is a wavelet news group, the *wavelet digest* (email to `wavelet@math.scarolina.edu` with subscribe as subject).

ACKNOWLEDGMENTS

The author is pleased to acknowledge support from: the Engineering Directorate, NASA Lewis Research Center (Contract C-71902-C); Motorola, Semiconductor Products Sector, Austin, Texas; and the National Science Foundation (Grant DMS-9300473). He extends special thanks to Professor Jean-Michel Morel and the image-processing group at Université Paris, Dauphine, as well as Professor David Donoho, for many stimulating discussons.

REFERENCES

[1] S. A. Feig, Decreased breast cancer mortality through mammographic screening: results of clinical trials," *Radiology*, vol. 167, pp. 659–665, 1988.

[2] F. Shtern, "Digital mammography and related technologies: a perspective from the National Cancer Institute," *Radiology*, vol. 183, pp. 629–630, 1992.

[3] H. A. Frazer, "Computerized diagnosis comes to mammography," *Diagnostic Imag.*, June, pp. 91–95, 1991.

[4] M. LeGal, J. C. Durand, M. Laurent, "Conduite à tenir devant une mammographie révélatrice de microcalcifications groupées découvertes par mammographie," *Nouv Presse Med.*, vol. 5, pp. 1623–1627, 1976.

[5] D. Marr, *Vision A Computational Investigation into the Human Representation and Processing of Visual Information*. New York: Freeman, 1982.

[6] A. C. Bovik, M. Clark, and W. S. Geisler, "Multichannel texture analysis using localized spatial filters," *IEEE Trans. PAMI*, vol. 12, pp. 55–73, 1990.

[7] S. Peleg, J. Naor, R. Hertley, and D. Avnir, "Multiple resolution texture analysis and classification," *IEEE Trans. Pattern Anal. Mach. Intell.*, vol. 6, pp. 518–523, 1970.

[8] G. Koepfler and J. Morel, Texture discrimination: a review of the theory and its algorithms. CEREMADE, Université Paris-Dauphine, 1991.

[9] S. G. Mallat, "Multifrequency channel decompositions of images and wavelet models." *IEEE Trans. Acoust., Speech, Signal Proc.*, vol. 37, pp. 2091–2110, 1989.

[10] I. Daubechies, "Orthonormal bases of compactly supported wavelets," *Comm. Pure Appl. Math.*, vol. 41, pp. 909–996, 1988.

[11] P. Burt and E. Adelson, "The Laplacian pyramid as a compact image code," *IEEE Trans. Comm.*, vol. 31, pp. 482–540, 1983.

[12] F. Shtern, "Technology transfer in digital mammography," *Report of the Joint NCI/NASA Workshop of May*, 1993.

[13] J. S. Chow, J. C. Tu, J. M. Cioffi, "A discrete multitone transceiver system for HDSL applications," *IEEE J. Select. Areas Commun.*, vol., 9, pp. 895–908, 1991.

[14] M. Tzannes, M. Tzannes, H. Resnikoff, "The DWMT: a multicarrier transceiver for ADSL using M-band wavelet transforms," *IEEE Standards Project: T1E1.4*, March 1994.

[15] A. H. Tewfik, M. Kim, "Correlation structure of the discrete wavelet coefficients of fractional Brownian motion," *IEEE Trans. Inform. Theory*, vol. 38, pp. 904–909, 1992.

[16] R. R. Coifman and M. V. Wickerhauser, "Entropy-based algorithms for best basis selection," *IEEE Trans. Inform. Theory*, vol. 38, pp. 713–718, 1992.

[17] M. V. Wickerhauser, INRIA Lectures of Wavelet Packet Algorithms. Numerical Algorithms Research Group—Yale University, 1991.

[18] W. B. Richardson, "Wavelet packets applied to mammograms," *SPIE Conf. Biomedical Image Processing and Biomedical Visualization*, pp. 504–508, 1993.

[19] W. B. Richardson, "Nonlinear Filtering and Multiscale Texture Discrimination for Mammograms," *SPIE Conf. Mathematical Methods in Medical Imaging*, pp. 293–305, 1992.

[20] W. B. Richardson, H. Longbotham, D. Gokhman, "Multiscale wavelet analysis of mammograms." *Proc. Wavelets and Applications Conf.*, Toulouse, 1992.

[21] E. J. Ciaccio, S. M. Dunn, and M. Akay, "Biosignal pattern recognition and interpretation systems," *IEEE Eng. Med. Biol.* vol. 13, pp. 129–134, 1994.

[22] A. LeTreut, and M. H. Dilhuydy, *Mammography A Guide to Interpretation*, St. Louis: Mosby, 1991.

[23] A. P. Dhawan, Y. Chitre and M. Moskowitz, "Artificial-neural-network-based classification of mammographic microcalcifications using image structure features," *SPIE Conf. Biomedical Image Processing and Biomedical Visualization*, pp. 820–831, 1993.

[24] C. C. Chen, J. S. DaPointe, and M. D. Fox, "Fractal feature analysis and classification in medical imaging," *IEEE Trans. Med. Imag.*, vol. 8, pp. 133–142, 1989.

[25] T. Lundahl, W. J. Ohley, S. M. Kay, and R. Siffert, "Fractional Brownian motion: a maximum likelihood estimator and its application to image texture," *IEEE Trans. Med. Imag.*, vol. 5, pp. 152–161, 1986.

[26] C. B. Caldwell, et al., "Characterization of mammographic parenchymal pattern by fractal dimension," *Phys. Med. Biol.*, vol. 35, pp. 235–247, 1990.

[27] A. Moch and W. B. Richardson, "Fractal signature for texture discrimination in mammography," Presented Texas Dynamics Days, Austin, 1992.

[28] J. Theiler, "Estimating fractal dimension," *J. Opt. Soc. Am. A*, vol. 7, pp. 1055–1073, 1990.

[29] B. B. Mandelbrot and B. J. Van Ness, "Fractional Brownian motion, fractional noises, and applications," *SIAM Rev.* vol. 10, pp. 422–438, 1968.

[30] P. Flandrin, "Wavelet analysis and synthesis of fractional Brownian motion," *IEEE Trans. Inform. Theory*, vol. 38, pp. 910–917, 1992.

[31] M. Basseville, A. Benveniste, K. C. Chou, S. A. Golden, R. Nikoukhah, and A. S. Willsky, "Modeling and estimation of multiresolution stochastic processes," *IEEE Trans. Inform. Theory*, vol. 38, pp. 766–784, 1992.

[32] J. N. Wolfe, "Risk for breast cancer development by mammographic parenchymal pattern," *Cancer*, vol. 37, pp. 2486–92, 1976.

[33] J. N. Wolfe, K. A. Buck, M. Salane, and N. J. Parekh, "Xeroradiography of the breast: overview of 21,057 consecutive cases," *Radiology*, vol. 165, pp. 305–311, 1987.

[34] F. Aghdasi, R. K. Ward, and B. Palcic, "Restoration of mamographic images in the presence of signal-dependent noise," *SPIE Conf. Mathematical Methods in Medical Imaging*, pp. 740–751, 1992.

[35] D. Donoho and I. Johnstone, "Minimax estimation via wavelet shrinkage," Dept. of Statistics, Stanford University, preprint 1992.

[36] D. L. Donoho, "Wavelet Shrinkage and W.V.D.: A 10-Minute Tour." *Progress in Wavelet Analysis and Applications.* Y. Meyer and S. Roques, (eds.) Gif-sur-Yvette: Editions Frontières, 1993.

[37] P. L. Lions, J. M. Morel, and L. Alvarez, *Image Selective Smoothing and Edge Detection by Nonlinear Diffusion.* CEREMADE, Université Paris-Dauphine, 1991.

[38] F. Berthommier, O. Francois, T. Herve, T. Coll, I. Marque, et al., "Asymptotic behavior of neural networks and image processing." In: A. Babloyantz (ed.) *Self-Organization, Emerging Properties, and Learning* Plenum Press, New York, pp. 219–227, 1991.

[39] Y. Meyer, *Wavelets Algorithms and Applications.* Philadelphia: SIAM, 1993.

[40] I. Daubechies, *Ten Lectures on Wavelets*, Philadelphia: SIAM, 1992.

[41] C. K. Chui, *An Introduction to Wavelets Volume I.* San Diego: Academic Press, 1992.

[42] J. M. Combes, A. Grossman, and Ph. Tchamitchian, (eds.), *Wavelets: Time-Frequency Methods and Phase Space.* 2nd ed., Berlin: Springer-Verlag, 1990.

[43] M. B. Ruskai. *Wavelets and Their Applications.* Boston: Jones and Bartlett, 1992.

[44] J. J. Benedetto, and M. W. Frazier. *Wavelets Mathematics and Applications.* Boca Raton: CRC Press, 1994.

[45] O. Rioul, and M. Vetterli, "Wavelets and signal processing," *IEEE Signal Proc. Mag.* October, pp. 14–38, 1991.

Chapter 21

Hybrid Wavelet Transform for Image Enhancement for Computer-Assisted Diagnosis and Telemedicine Applications

Laurence P. Clarke, Wei Qian, Maria Kallergi, Priya Venugopal,
Robert A. Clark

21.1. INTRODUCTION

This chapter describes the theoretical basis and application of hybrid wavelet algorithms for image enhancement as required for computer-assisted diagnosis (CAD) and telemedicine applications in medical imaging. The primary clinical application described is the enhancement of microcalcification clusters (MCCs) in digitized mammograms to improve both their visualization using a computer monitor for remote reading and their detection using CAD methods. The potential application for improved visual interpretation of digital X-ray images for telemedicine applications is also demonstrated using a preliminary set of images for other medical applications at varying resolutions [1–3]. The enhancement of MCCs is an excellent model for real-world evaluation of the wavelet transform for telemedicine applications. The detection of MCCs presents a significant challenge to the performance characteristics of X-ray imaging sensors and image display monitors since microcalcifications vary in size, shape, signal intensity, and contrast and may be located in areas of very dense parenchymal tissue making their detection difficult [1–12]. The classification of MCCs, in turn, as benign or malignant, requires their morphology and detail to be preserved particularly when using different sensors with varying noise characteristics and resolution [2].

Multiresolution methods, such as the use of the wavelet transform, originally developed in the signal-processing field [13, 14], have recently been proposed for image enhancement, segmentation and edge detection in the field of digital mammography [15, 16]. Multiresolution approaches have an inherent advantage over traditional filtering methods, which primarily focus on the coupling between image pixels on a single scale and generally fail to preserve image details of important clinical features. For example, the use of traditional single-scale CAD algorithms for enhancement or detection of MCCs has generally resulted in a sensitivity (true positive (TP) detection rate) and specificity (true negative (TN) detection rate) that does not exceed 85%, with 1–4 false positives (FPs) per image [1–3, 9–10]. Reported results, however, are very dependent on the image database used and subtlety of the MCCs evaluated [2]. One application of this work is to evaluate the potential of a hybrid wavelet transform for enhancement of microcalcifications as a preprocessing step prior to use of neural networks for their detection [2].

Multiresolution approaches based on M-channel wavelet transforms are not limited by single-scale operations and could thus improve the enhancement or classification of different mammographic features [4, 17, 18]. Application of the wavelet transform to digital mammography is still at a very early stage but has already shown significant advantages [15, 16]. The majority of reported approaches aim at edge detection, which may not be diagnostically useful since edges of parenchymal tissues, MCCs, or lesions are often mixed with image noise. Since image noise is sensor dependent, removal of noise is required to decrease the FP detection rates for MCCs [11]. Similarly, wavelet enhancement methods previously reported have included modifications to the Gabor transform that require an *a priori* knowledge of image-related parameters such as spatial frequencies, orientations, and shape of the Gaussian envelope function [17]. These enhancement methods generally are operator dependent requiring some form of interaction for visual interpretation. They may also exhibit ringing artifacts that may result in loss of image detail [19]. The clinical acceptance of CAD methods, however, ideally requires both operator-independent (automatic) algorithms that potentially reduce the inter- and intra-observer variations and adaptive noise suppression algorithms that are, ideally, image and sensor independent and preserve image detail as required for classification of clinical features [7].

The use of a novel hybrid filter architecture is proposed for MCC enhancement. First an adaptive nonlinear multistage filter (AMNF) is used for both noise suppression and image enhancement by smoothing background structures surrounding the MCCs. Second, an $M \geq 2$ channel tree structured wavelet transform (TSWT) is employed for further selective enhancement of microcalcification clusters. The hybrid filter takes advantage of the image decomposition and reconstruction processes of the TSWT, where reconstructions of specific sub-images are used to selectively enhance MCCs and separate the background structures. Finally, the hybrid filter selectively combines the filtered and reconstructed images containing MCCs and background to provide further enhancement of the microcalcifications and selective removal of parenchymal tissue structures.

The performance of the hybrid filter was evaluated using mammographic image databases at varying resolution ($35\,\mu$m and $110\,\mu$m) generated by digitization of screen/film using a high-resolution digitizer (DBA, Melbourne, FL, Image Clear 6000). Representative images of the chest and wrist are also shown using screen/film with comparison to the Sobel operator as a well known enhancement filter. The importance of image enhancement for the detection of MCCs using CAD methods is also demonstrated using a neural network.

21.2. DESIGN OF A HYBRID FILTER

21.2.1 Introduction

The design of the hybrid filter is based on our earlier work in filter design for image noise suppression and the use of the wavelet transform (two-channel TSWT), specifically for segmentation of microcalcification clusters in digitized mammograms [11, 17]. We initially developed a multistage tree structured nonlinear filter (TSF), with fixed parameters, that demonstrated significantly improved performance for noise suppression for digital mammograms, as compared to traditional single-stage filters of varying design [11]. Similarly, a two-channel TSWT was successfully used for both image decomposition and reconstruction, with selective reconstruction of different sub-images successfully used to segment the clinical feature of interest [17]. For example, the cascaded implementation of the TSF and TSWT resulted in a significant reduction in the FP detection rate for MCC analysis carried out on both simulated images with varying noise content and also on digital mammograms with biopsy proven microcalcification clusters [11]. However, the image detail of the segmented MCCs was not fully preserved, although better than in single-scale methods [3, 9, 10]. An AMNF is proposed in this work to obtain both better performance for noise suppression for application with different sensors and that includes a criterion for selective enhancement of clinical features such as MCCs while smoothing background tissue structures, i.e., on a pixel by pixel basis.

21.2.2 Hybrid Filter Architecture

A block diagram of the hybrid filter architecture is shown in Fig. 21-1. The input mammographic image $\{x(n, m)\ 1 \leq n \leq N,\ 1 \leq m \leq M\}$, expressed as $x(n, m)$, is first filtered by AMNF to enhance the MCC features while suppressing image noise and smoothing background details of parenchymal tissue structures. The output image, expressed as $x_{\text{AMNF}}(n, m)$, is processed in two different ways: (1) a weighing coefficient, α_1, is applied to the output image producing $\alpha_1\, x_{\text{AMNF}}(n, m)$; and (2) the same output image is processed by the wavelet transform. The $M \geq 2$ channel TSWT, as shown in Fig. 21-1, decomposes the output image, $x_{\text{AMNF}}(n, m)$, into a set of independent, spatially oriented frequency bands or lower resolution sub-images. These sub-images, as indicated, are then subgrouped into two categories, those that primarily contain the clinical feature of interest or the background features. The sub-images are then reconstructed by the TSWT into two images, $x_{\text{W1}}(n, m)$ and $x_{\text{W2}}(n, m)$ which contain the clinical feature of interest and background features, respectively. Finally, using coefficients α_2 and α_3, the weighted outputs of the reconstructed sub-images are combined with the original weighted output image $\alpha_1 x_{\text{AMNF}}(n, m)$, as indicated in Fig. 21-1, to yield the output image, which further enhances the clinical feature of interest, as follows:

$$\text{Output image} = \alpha_1 x_{\text{AMNF}}(n, m) + \alpha_2 x_{\text{W1}}(n, m) - \alpha_3 x_{\text{W2}}(n, m) \tag{21-1}$$

The images are then enhanced with linear gray scaling.

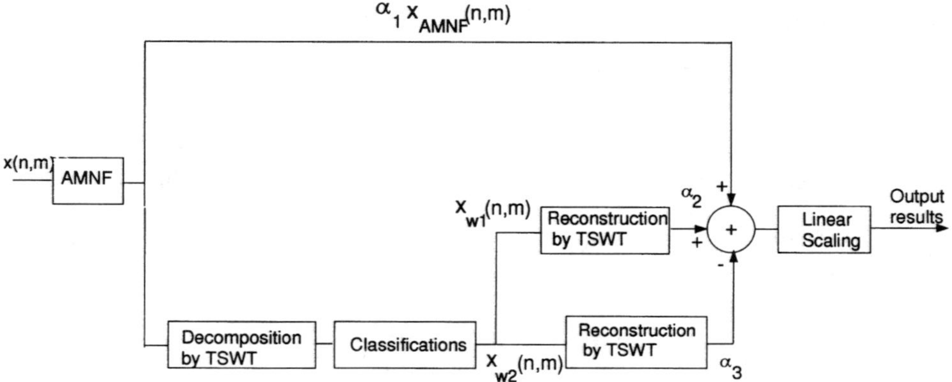

Figure 21-1   Block diagram of the hybrid filter architecture used for image enhancement that includes the AMNF, shown in Fig. 21-2, and the $M = 2$ TSWT.

21.2.3 Adaptive Multistage Nonlinear Filtering

An image, x, can be considered to consist of two parts: a low-frequency part, x_L and a high-frequent part, x_H, expressed as $x = x_L + x_H$. The low-frequency part may be dominant in the homogeneous regions, whereas the high-frequency part may be dominant in the edge regions. The two-component image model allows different treatment of its components. It can be used, therefore, for adaptive image filtering and enhancement [20]. The high-frequency part may be weighted with a signal-dependent factor. A two-component model, $x_L(n, m) = \bar{x}(n, m)$ and $x_H(n, m) = x(n, m) - \bar{x}(n, m)$, is suitable for noise suppression, and also for many other enhancement operations such as statistical differencing, contrast enhancement, shadow correction, and cloud cover removal [20]. For our application of mammographic image processing, we substitute for the low-frequency signal, $x_L(n, m)$, two-stage structured filters. In the first stage, five different kinds of filter are used, including a linear smoothing filter, nonlinear-α-trimmed mean filters with different window sizes, and a median filter. The adaptive operation is used in the second stage. The structure, called an AMNF is shown in Fig. 21-2. The output is:

$$x_{\text{AMNF}}(n, m) = x_{\text{AF}}(n, m) + b(n, m)[x(n, m) - x_{\text{AF}}(n, m)] \qquad (21\text{-}2)$$

Here, $x_{\text{AF}}(n, m)$ is a result from the second stage of the filter, called the adaptive operation in Fig. 21-2, $b(n, m)$ is a signal-dependent weighting factor, which is a measure of the local signal activity, and σ is the variance, i.e., $b(n, m) = \sigma_f^2(n, m)/(\sigma_f^2(n, m) + \sigma_n^2(n, m))$. For the case of MCC enhancement, for the flat regions of the input image, the signal-to-noise ratio is low, so $b(n, m)$ becomes small, which leads to $x_{\text{AMNF}}(n, m) \to x_{\text{AF}}(n, m)$ from Eq. (21-2). Alternatively, at image edge locations, the signal-to-noise ratio is large, so $b(n, m)$ approaches 1 and this leads to $x_{\text{AMNF}}(n, m) \to x(n, m)$. The operation of the filter, therefore, should preserve the edges in the image. The estimation of $b(n, m)$ is described in detail elsewhere [21].

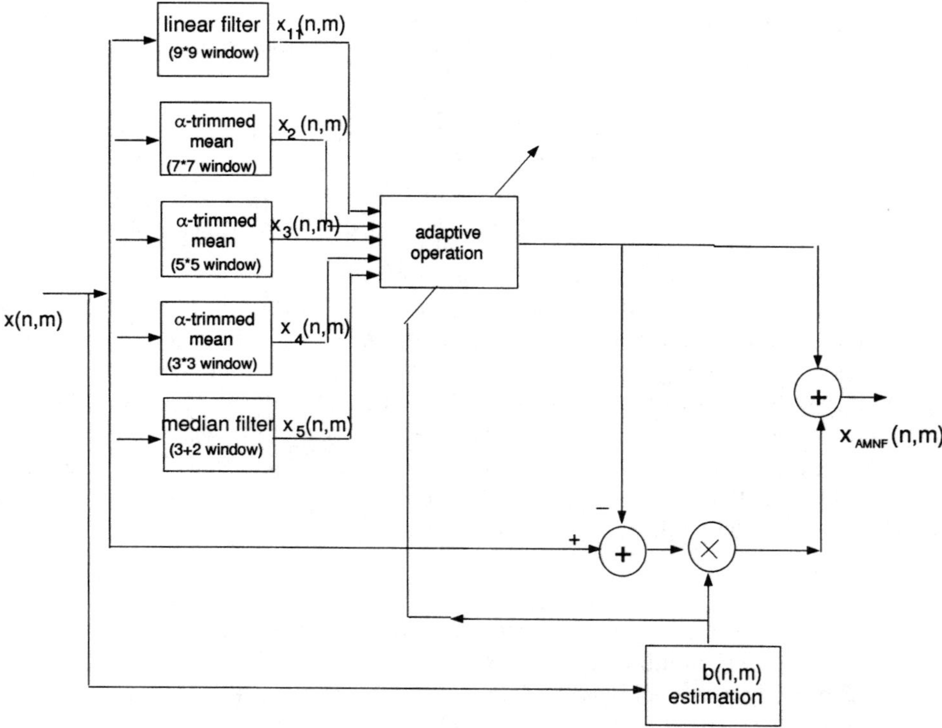

Figure 21-2   Block diagram of the AMNF used for noise suppression and image enhancement of MCCs.

In order to achieve better adaptive properties for noise suppression and enhancement for different sensors and applications, five different filters with different window sizes can be selected, according to the value of $b(n, m)$. With respect to an appropriate window size, W, two effects are taken into account. Noise suppression increases with increasing window size, and conversely, spatial resolution decreases. Linear filters will smooth the edges, average the details and noise, and decrease greatly the spatial resolution. As a consequence, a nonlinear filter with a small window (e.g., 3×3 or $3 + 2$) is used in the areas containing microcalcification; a linear filter with a large window (e.g., 7×7 or 9×9) is used in the unwanted resolution scale level areas for noise removal and background smoothing. The α-trimmed mean filter offers a compromise in performance between linear and nonlinear filters. For example, an adaptive linear operation is selected for microcalcification enhancement. In defining the linear operation, the outputs from the five filters in the first stage of Fig. 21-2 are defined as a vector of linear weighted values, expressed as:

$$x_l = c^T x_j, \qquad 1 \le j \le W, \qquad \forall j \in Z^L \tag{21-3}$$

where W is the number of the outputs from the first stage (number of pixels). The coefficients c_i determine various linear weighted operations. The weighting factor

$b(n, m)$, which is a measure of the local signal activity in Eq. (21-2), can be applied as a criterion for the adaptivity of coefficients c_i, as follows:

$$[c_1, c_2, c_3, c_4, c_5]^{\mathrm{T}} = \begin{cases} [1, 0, 0, 0, 0]^{\mathrm{T}} & b(n, m) \leq \tau_1 \\ [0, 1, 0, 0, 0]^{\mathrm{T}} & \tau_1 < b(n, m) \leq \tau_2 \\ [0, 0, 1, 0, 0]^{\mathrm{T}} & \tau_2 < b(n, m) \leq \tau_3 \\ [0, 0, 0, 1, 0]^{\mathrm{T}} & \tau_3 < b(n, m) \leq \tau_4 \\ [0, 0, 0, 0, 1]^{\mathrm{T}} & \tau_4 < b(n, m) \end{cases} \quad (21\text{-}4)$$

where τ is the adaptive thresholding of $b(n, m)$.

For enhancement of other images such as the chest or wrist, an "order statistic" (OS) operation may be better for obtaining either texture or image contrast enhancement. In defining the OS operation, the spatial or temporal distribution of the elements of x_j within the window W_j is not of interest. Instead, define the vector of order statistics (algebraically ordered versions) $x_{(j)} = \mathrm{order}[x_j] = [x_{(1)j} \; x_{(2)j} \ldots x_{(W)j}]^{\mathrm{T}}$, such that $x_{(1)j} \leq x_{(2)j} \leq \ldots \leq x_{(W)j}$. Given a vector of real-valued coefficients $a = [a_1 \; a_2 \ldots a_W]^{\mathrm{T}}$ with length m, the output of the OS filter is a matrix $\mathbf{G}$ with elements:

$$g_j = a^T y_{(j)}, \qquad 1 \leq j \leq W, \qquad \forall j \in Z^L \quad (21\text{-}5)$$

where W is the size of the selected window (number of pixels), $y_{(j)} = \mathrm{order}[y_j]$. The details of determining parameters for AMNF can be found in [21]. For enhancement of other images such as the chest or wrist an order statistic or linear operation may be used to obtain either texture or image contrast enhancement.

21.2.4 Wavelet Decomposition and Reconstruction

The theoretical basis for the $M = 2$ TSWT for the hybrid filter, as shown in Fig. 21-1, is briefly outlined below [21]. The general case for $M \geq 2$ is described elsewhere by these investigators [22] for the signal-processing field and for mammography applications. The discrete wavelet transform (DWT) utilizes two functions: the mother wavelet, Ψ, and a scaling function, Φ. Ψ is subjected to the functional operations of shifts and dyadic dilation, such as:

$$\Psi_{m,n}(x) = 2^{-m/2} \Psi_{m,n}(2^{-m}x - n) \quad (21\text{-}6)$$

where $n, m \in Z$ and Z is the set of integers.

Ψ yields an orthogonal basis of $L^2(\mathbb{R})$, which is the vector (Hilbert) space of measurable, square-integrable functions. The wavelet basis induces an orthogonal decomposition of $L^2(\mathbb{R})$, which is given by $L^2(\mathbb{R}) = \ldots \oplus W_{-2} \oplus W_{-1} \oplus W_0 \oplus W_1 \oplus W_2 \oplus \ldots$, where W_i is the wavelet subspace spanned by $\{\Psi_{m,n}(x)\}_{n=-\infty}^{n=+\infty}$. The wavelet $\Phi(x)$ is often generated from the scaling function or the "father wavelet." The scaling function $\Phi(x)$ satisfies the two-scale difference equation [13, 14] expressed as:

$$\Phi(x) = 2^{1/2} \sum_k h_1(k)\Phi(2x - k) \quad (21\text{-}7)$$

The dilations and translations of the scaling function induce a multiresolution analysis of $L^2(\mathbb{R})$, a nested chain of closed subspaces of $L^2(\mathbb{R})(\ldots \subset V_{-1} \subset V_0 \subset V_1 \subset V_2 \ldots)$ such that $V_{-\infty} = \{0\}$ and $V_\infty = L^2(\mathbb{R})$ where V_i is the subspace spanned by $\{\Phi_{m,n}(x)\}_{n=-\infty}^{n=+\infty}$. V_i and W_i are related by $V_i \oplus W_i = V_{i+1}$ which extends to $V_i \oplus W_i \oplus W_{i+1} = V_{i+2}$. Then, the wavelet kernel $\Psi(x)$ is related to the scaling function, which is:

$$\Psi(x) = 2^{1/2} \sum_k h_2(k)\Phi(2x - k) \tag{21-8}$$

where $h_2(k) = (-1)^{1-k}h_1(1 - k)$. $h_2(k)$ is the mirror filter of $h_1(k)$. In the digital signal-processing domain, $h_2(k)$ and $h_1(k)$ are called quadrature mirror filters (QMFs). The coefficients $h_1(k)$ and $h_2(k)$ play a very crucial role in a given wavelet transform. In fact, the implementation of the wavelet transform does not require the explicit forms of $\Phi(x)$ and $\Psi(x)$, but only depends on $h_1(k)$ and $h_2(k)$.

The approximation of an enhanced mammogram, x, at a resolution 2^{-m} (such as $A_m(x)$) is given by the projection of g onto the vector space V_m. The image detail features, such as MCCs, are separated when going from an approximation of x with resolution 2^{-m} to the coarser approximation $A_{m-1}(x)$ with resolution $2^{-(m-1)}$. The detail signal, $D_{m-1}(x)$ and can be obtained by projection of x onto W_{m-1}. The detail signal, $D_{m-1}(x)$, is typically a high-pass version of x, while $A_{m-1}(x)$ is a low-pass version. Given the approximation of x at a resolution of 2^{-m} [i.e, $A_m(x)$], $A_{m-1}(x)$ and $D_{m-1}(x)$ can be computed by filtering $A_m(x)$ with $h_1(k)$ and $h_2(k)$ and then keeping every other sample of the output. Approximations at lower resolutions are obtained by repeated application of this algorithm. Let Ψ and Φ be the wavelet and the scaling function necessary for reconstruction. Given $A_{m-1}(x)$ and $D_{m-1}(x)$, $A_m(x)$ can be perfectly reconstructed by interpolating $A_{m-1}(x)$ and $D_{m-1}(x)$ by a factor of two and filtering the resultant signals with $h_1(k)$ and $h_2(k)$, respectively. The synthesis filters, $h_1(k)$ and $h_2(k)$, are derived from Ψ and Φ, respectively.

It is advantageous to have wavelet bases which are orthonormal. In that case, the sub-images are orthonormal and, hence, uncorrelated. In addition, for image-processing applications, one would prefer analysis and synthesis filters to have linear-phase. Unfortunately, there exist no non-trivial, finite-length, orthogonal linear-phase filters with the perfect reconstruction property [23]. In experimental applications, this difficulty is generally overcome by using biorthogonal bases [23, 24]. Biorthogonal bases for wavelets were recently introduced, independently, by [24] and [25]. In this chapter, we have chosen to use a nine-tap spline filter as suggested by [23] and [24].

21.3. EXPERIMENTAL RESULTS

21.3.1 Influence of Preprocessing for a Hybrid Filter

The influence of the AMNF for noise suppression and enhancement prior to the use of the TSWT, as shown in the block diagrams of Figs. 21-1 and 21-2, is illustrated in Fig. 21-3 for microcalcification enhancement. This figure shows a representative section of a mammogram (512 × 512) at a resolution of $110\,\mu$m containing a microcalcification cluster as follows: (a) a raw digital mammogram, (b) an enhanced image using the hybrid filter (Fig. 21-1); (c) the enhanced image without the AMNF step in Fig. 21-1; (d) the image enhanced by replacing the AMNF by a noise suppression filter only (a TSF) as reported earlier by these investigators [11]; (e) use of a trained neural network (NN) to detect the microcalcifications using the raw data; and (f) the use of the same NN to detect microcalcifications using the enhanced data. Details of the neural network are described elsewhere [21, 26]. The image section (512 × 512) is shown to allow the microcalcification cluster to be enlarged. Analysis is automatically performed on the entire image.

The following conclusions can be formed. First, the proposed hybrid filter provided the best enhancement using visual criteria. The use of the TSWT alone or the use of the TSF generated structures that could simulate signals similar to microcalcifications that may generate FP detections. This was confirmed by the use of the NN for detection as applied to each image. Representative results for the raw and enhanced image by the hybrid filter (Fig. 21-3) reflect this problem where several FPs were detected in the raw image, but not seen in the enhanced image. The application of the NN to a large image database with ground truth files also demonstrates the importance of image enhancement [27].

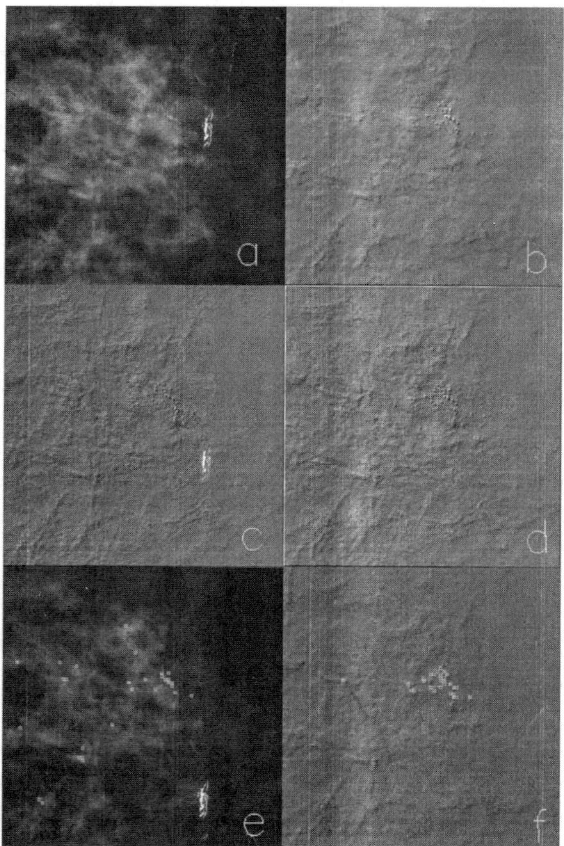

Figure 21-3 Representative mammographic images with a single biopsy proven MCC, showing: (a) the raw digital image; (b) the enhanced image using the hybrid filter (Fig. 21-1); (c) the enhanced image using the hybrid filter without the AMNF stage; (d) the enhanced image using the TSF for noise suppression; (e) the NN detection results for the raw image as input; and (f) NN detection results using the enhanced image as input. Several FP detections were observed for the raw image.

21.3.2 Influence of Sensor Resolution

The influence of sensor resolution on preservation of clinical features when using the hybrid enhancement of microcalcifications is demonstrated in Fig. 21-4. A representative mammogram sub-image (512×512) is displayed: (a) the raw image at $110\,\mu$m; (b) the raw image at $35\,\mu$m; (c) the enhancement with $M = 2$ AMNF/TSWT at $110\,\mu$m; (d) enhancement at $M = 2$ AMNF/TSWT at $35\,\mu$m; (e) enhancement at $M = 4$ AMNF/TSWT $110\,\mu$m; and (f) enhancement at $M = 4$ AMNF/TSWT at $35\,\mu$m. The $M = 2$ AMNF/TSWT performed well for image resolution at $110\,\mu$m, but did not fully delineate the microcalcification cluster at $35\,\mu$m. The

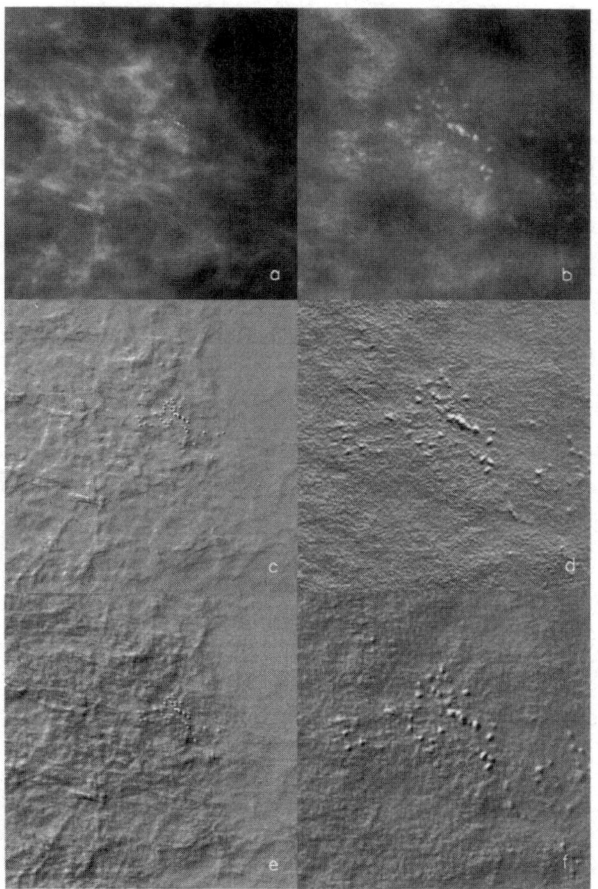

Figure 21-4   The influence of sensor resolution on preservation of clinical features. The two raw images are at 110 μm and 35 μm respectively, (a, b). (c, d) Enhancement with $M = 2$ and (e, f) enhancement with $M = 4$.

$M = 4$ AMNF/TSWT performed well at high resolution [22]. These results are to be confirmed using the NN applied to the same database at higher resolution.

21.3.3 Influence of Linear Versus Order Statistic Operator

The influence of each operator within the AMNF/TSWT is illustrated in Figs. 21-5–21-7 for images of circumscribed mass in a digitized mammogram ($210\,\mu$m

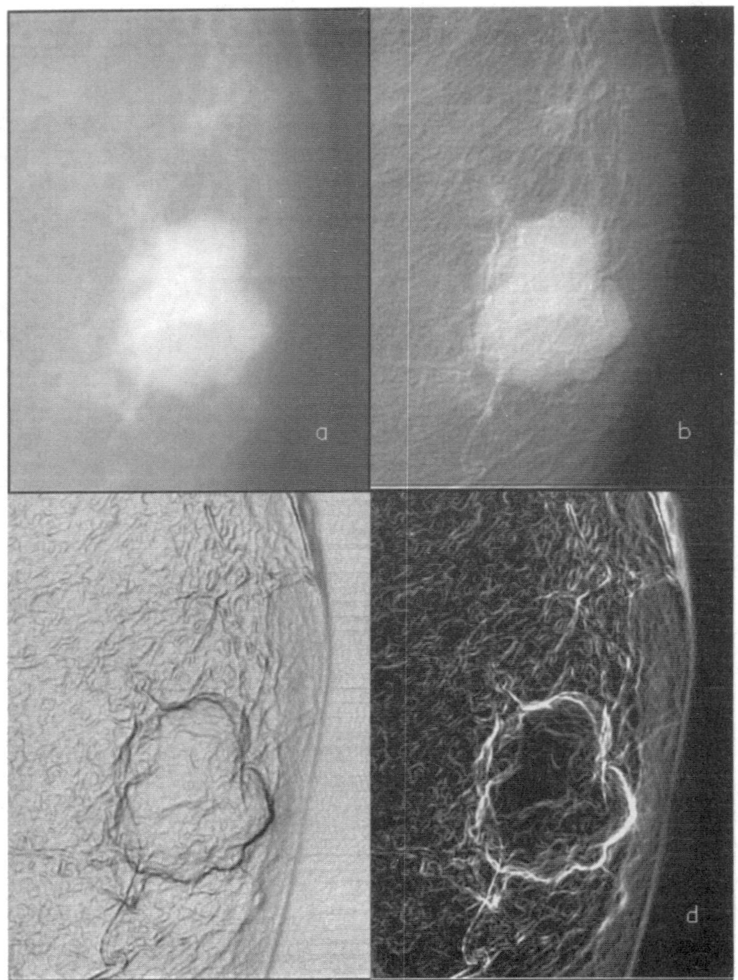

Figure 21-5 AMNF/TSWT enhancement of a circumscribed mass: (a) the original image at $200\,\mu$m; (b) the enhanced contrast image using the $M = 2$ TSWT and a linear operator (LO); (c) the enhanced texture image using the OS operator; and (d) the Sobel edge detector for comparison. The LO showing enhanced contrast is best for visual interpretation while the image generated by the OS operator may be useful for feature detection.

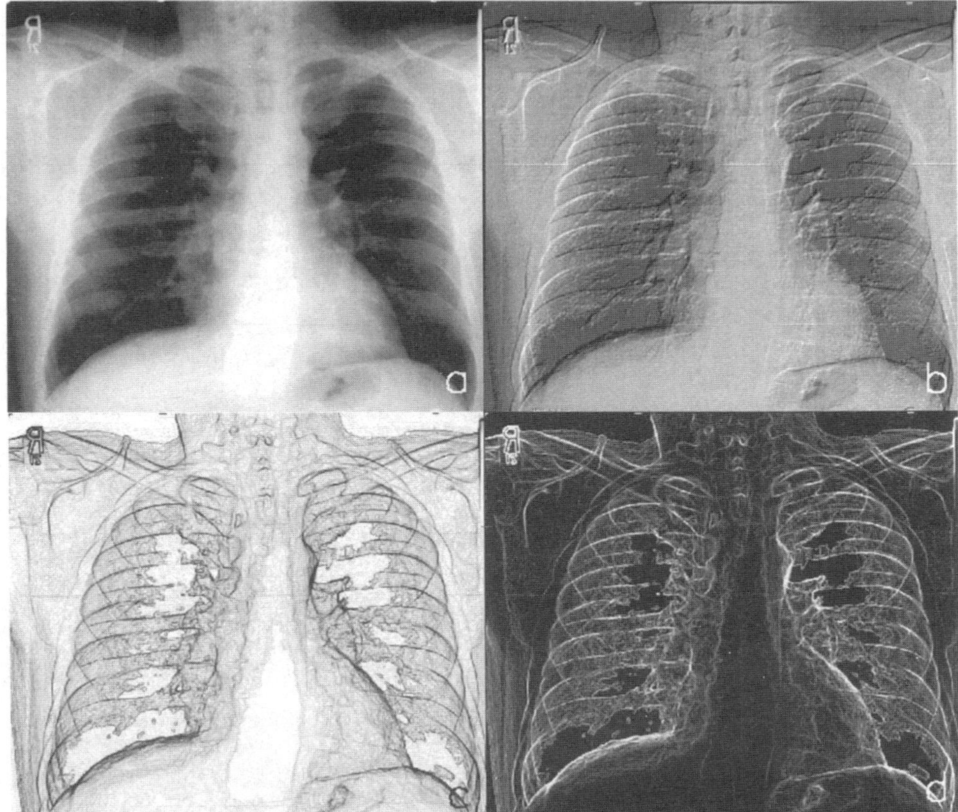

Figure 21-6   AMNF/TSWT $M = 2$ enhancement for other digital sensors: (a) the
original image of a normal chest at 200 μm using digitized film; (b) with
a linear operator; (c) the enhanced image with an OS operator; (d) the
Sobel edge detector for comparison.

resolution), digitized chest (210 μm resolution) and digitized wrist (200 μm) obtained
using different X-ray film digitizers. In each case we compare: (a) the raw image;
(b) the enhanced image with the $M = 2$ AMNF/TSWT using the linear operator;
(c) the enhanced image with the $M = 2$ AMNF/TSWT using the order statistic
operator; and (d) comparison to the well known Sobel edge detector. The parameters
of the proposed hybrid filter were unchanged for each image.

The following conclusions can be tentatively made based on this preliminary
evaluation. The linear operator provides better image contrast to the raw image
using the computer monitor for reading. The order statistic operator was very useful
for the chest and wrist image that contained many linear structures and where subtle
image details across the full gray-scale range were well observed on the computer
monitor, compared to screen/film interpretation on a light box. For the mass image
on the digitized mammogram, the enhanced image with the order statistic operator
was less useful for visual observation but may provide texture information for CAD,
particularly for parenchymal tissue distortion.

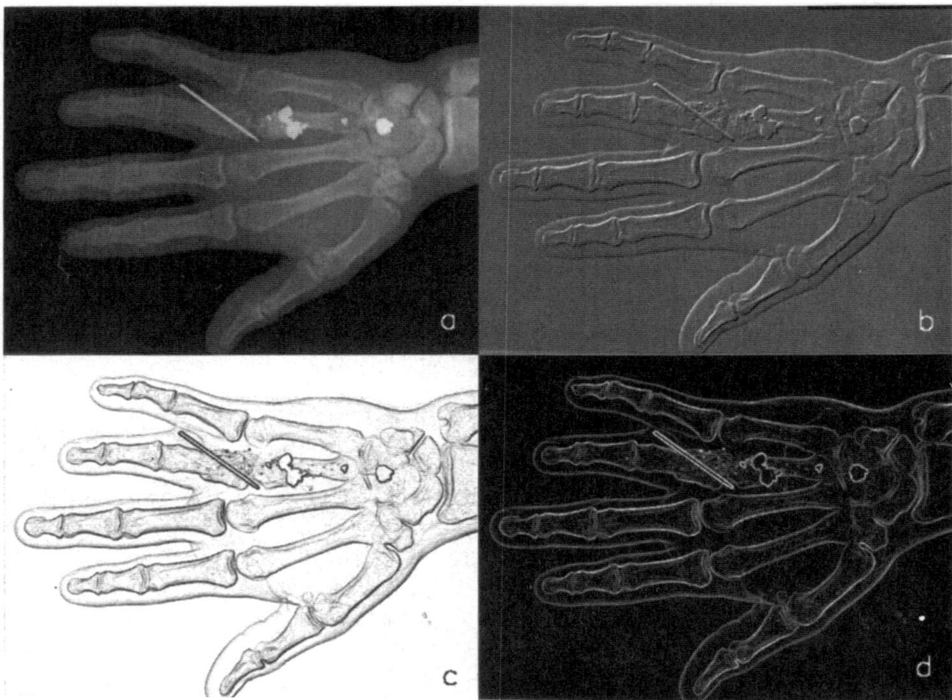

Figure 21-7   AMNF/TSWT $M = 2$ enhancement for other digital sensors: (a) the original digital X-ray image of a hand at $200\,\mu$m with a nail injury using digitized film; (b) with a linear operator; (c) the enhanced image with an OS operator; (d) the Sobel edge detector for comparison.

21.4. CONCLUSION

The use of the wavelet transform may prove to be very useful for image enhancement both for visual diagnosis and to compensate for computer monitor performance and as a preprocessing step for CAD methods. Enhancement algorithms, however, must be uniformly applied to different sensors such as X-ray film digitizers or direct X-ray detection systems that have different response characteristics (i.e., resolution, image contrast sensitivity) as required for telemedicine applications. Modifications to the wavelet transform are required, such as the hybrid filter proposed here, to allow for image noise suppression or system resolution. It is anticipated that further modifications, such as adaptive selection of sub-images to accommodate differences in contrast sensitivity for different X-ray detectors, will be needed for enhancement. The objective evaluation of enhancement algorithms is logistically difficult and will require extensive receiver operating curve (ROC) analysis by different observers. The preliminary results to date, however, suggest that image details are well preserved across the full gray-scale range of the images. These methods may allow comparable reading performance from a monitor or from a light box. These advances are particularly important in view of significant interest in telemedicine applications where remote diagnosis without loss of performance is important.

REFERENCES

[1] J. Dengler, S. Behrens, and J. F. Desaga, "Segmentation of microcalcifications in mammograms," *IEEE Trans. Med. Imag.*, vol. 12, no. 4, pp. 634–642, 1993.

[2] M. L. Giger, "Computer-aided diagnosis." In *Syllabus: A categorical course in Physics. Technical Aspects of Breast Imaging*, A. G. Haus and J. J. Yaffe (eds.). Oak Brook, IL: RSNA, pp. 257–270, 1992.

[3] W. Zhang, K. Doi, M. L. Geiger, Y. Nishikawa, Y. Wu, *et al.*, "Computerized detection of clustered microcalcifications in digital mammograms using a shift-invariant artificial neural network," *Med. Phys.*, vol. 20, no. 3, pp. 881, 1993.

[4] L. P. Clarke, M. Kallergi, W. Qian, H. D. Li, R. Velthuizen, *et al.*, "Digital mammography: review of advanced computer assisted diagnostic (CAD) methods," *President's Symposium; Proc. 35th Ann. Meeting of the AAPM (Abstract)*, Washington DC, 1993.

[5] L. P. Clarke, M. Kallergi, W. Qian, H. D. Li, R. A. Clark, and M. L. Silbiger, "Tree-structured nonlinear filter and wavelet transform for microcalcification segmentation in digital mammography," *Cancer Letts.*, vol. 77 no. 2,3, pp. 173–181, 1994.

[6] L. P. Clarke, G. J. Blaine, K. Doi, M. J. Yaffe, F. Shtern, *et al.*, "Digital mammography cancer screening: factors important for image compression," *Proc. Space and Earth Science Data Compression Workshop (NASA Conference Publication 3191)*, Invited Presentation, pp. 63–74, 1993.

[7] D. Winfield, M. Silbiger, G. S. Brown, L. P. Clarke, S. Dwyer, M. J. Yaffe, and F. Shtern, "Technology transfer in digital mammography," *Report of the joint NCI/NASA Workshop of May 19–20, 1993, Invest. Radiol.*, vol. 29, no. 4, pp. 507–515, 1994.

[8] S. J. Dwyer and B. K. Stewart, "Clinical uses of grayscale workstations," *Proc. AAPM Summer School*, Charlottesville, VA, pp. 243–264, 1993.

[9] D. H. Davies and D. R. Dance, "Automatic computer detection of clustered calcifications in digital mammograms," *Phys. Med. Biol.*, vol. 35, no.8, pp. 1111–1118, 1990.

[10] H. P. Chan, K. Doi, C. J. Vyborny, K. L. Lam and R. A. Schmidt, "Computer-aided detection of microcalcifications in mammograms, methodology and preliminary clinical study," *Invest. Radiol*, vol. 23, no. 9, pp. 664–671, 1988.

[11] W. Qian, L. P. Clarke, M. Kallergi and R. A. Clark, "Tree-structured nonlinear filters in digital mammography," *IEEE Trans. Med. Imag.* vol. 13, no. 1, pp. 25–36, 1994.

[12] R. M. Nishikawa and M. J. Yaffe, "Signal-to-noise properties of mammography film-screen systems", *Med. Phys.* vol. 12, pp. 32–39, 1985.

[13] S. Mallat, "Multifrequency channel decompositions of images and wavelet models," *IEEE Trans. Acoust., Speech, Signal Proc.*, vol. 37, pp. 2091–2110, 1989.

[14] S. Mallat, "A theory for multiresolution signal decomposition: the wavelet representation," *IEEE Trans. Pattern Anal. Machine Intell.*, vol. 11, no. 7, pp. 674–693, 1989.

[15] A. Laine and S. Song, "Wavelet processing techniques for digital Mammography," *Proc of the SPIE/IS&T Symp. on Electronic Imaging Science and Technology,* Jan. 31–Feb. 5, 1993.

[16] W. B. Richardson, "Wavelet packets applied to mammograms," *Proc. SPIE/ IS&T Symp. on Electronic Imaging Science and Technology*, San Jose, CA, Jan. 31–Feb. 5, 1993.

[17] W. Qian, L. P. Clarke, M. Kallergi, H. D. Li, R. Velthuizen, *et al.*, "Tree-structured nonlinear filter and wavelet transform for microcalcification segmentation in mammography," *Proc. of the SPIE IS&T Conf.*, San Jose, CA, Jan. 31–Feb. 4, 1993.

[18] M. Porat and Y. Y. Zeevi, "The generalized Gabor scheme of image representation in biological and machine vision," *IEEE Trans. Pattern Recog. Machine Intell.*, vol. 10, pp. 452–467, July 1983.

[19] A. Laine, S. Schuler, *et al.*, "Mammographic feature enhancement by multi-scale analysis," *IEEE Trans. on Med. Imag.*, vol. 13, no. 4, 1994.

[20] R. Bernstein, "Adaptive nonlinear filters for simultaneous removal of different kinds of noise in images," *IEEE Trans. Circ. Sys.,* vol. 34, no. 11, pp. 1275–1291, 1987.

[21] W. Qian, L. P. Clarke, B. Y. Zheng, M. Kallergi, and R. A. Clark, "Wavelet transform for computer assisted diagnosis (CAD) for digital mammography," *IEEE Eng. Med. Biol. Mag.* vol. 14, no. 5, pp. 561–569, 1995.

[22] W. Qian, L. P. Clarke, H. D. Li, M. Kallergi, R. A. Clarke, et al., "Digital mammography: M-channel quadrature mirror filters for microcalcification extraction in digital mammography," *Comput. Med. Imag. Graphics*, vol. 18, no. 5, pp. 301–314, 1994.

[23] M. Antonini, M. Barlaud, P. Mathieu, and I. Daubechies, "Image coding using wavelet transform," *IEEE Trans. Image Proc.*, vol. IP-1, pp. 205–220, 1992.

[24] A. Cohen, I. Daubechies, and J. C. Feauveau, "Biorthogonal bases of compactly supported wavelets," *Comm. Pure Appl. Math.*, vol. 41, pp. 485–560, 1992.

[25] M. Vetterli and C. Herley, "Wavelets and filter banks: theory and design," *IEEE Trans. Signal Proc.*, vol. SP-40, pp. 2207–2232, 1992.

[26] B. Zheng, W. Qian, and L. P. Clarke, "Digital mammography: mixed feature neural network with spectral entropy decision for detection of microcalcifications," *IEEE Trans. Med Imag.*, vol. 15, no. 5, pp. 589–597, 1996.

[27] B. Zheng, W. Qian and L. P. Clarke. "Artificial neural network for pattern recognition in mammography," *Proc. of World Congress on Neural Networks*, San Diego, CA, June 1994.

Chapter 22

Medical Image Enhancement Using Wavelet Transform and Arithmetic Coding

Pongskorn Saipetch, Bruce K. T. Ho, Ramesh K. Panwar, Marco Ma

22.1. INTRODUCTION

The storage requirement of a picture archiving and communication system (PACS) of a large hospital such as UCLA Medical Center exceeds two terabytes per year [1]. Network traffic is also considerably slowed down by retrieval of large images. Implementing a practical PACS thus requires a good method to compress the images for archival storage and transmission. Applications in primary-diagnosis teleradiology require that the original radiographic data are transmitted over a baud-rate-limited communication line without any degradation in quality. Affordability and practicality of viewing stations also requires that a high-speed reconstruction of the compressed images can be implemented using inexpensive, widely available hardwares.

We have developed compression algorithms that combine discrete wavelet transforms with arithmetic coding. The algorithms can be either lossy for archival applications or lossless for primary diagnosis applications. Compression is achieved by transforming the picture into wavelet coefficients and then encoding the coefficients using arithmetic coding.

Arithmetic coding is chosen over Huffman and Lempel–Ziv–Welch (LZW) for encoding the coefficients because of its optimal performance without blocking of input data and its clean separation between modeling the data and the encoding of information according to the model [2, 3]. We use the fact that arithmetic coding (AC) is a very efficient compressor and can approach the entropy bound defined by any arbitrary model in compressing a sequence of symbols. When a particular model predicts that most of the message comprises a few of the symbols (narrow histogram), the entropy bound will decrease. AC can approach this entropy bound better than any other compression methods. A more sophisticated model can give a narrower histogram at the expense of computational time and memory. We found that the computing resource of the viewing workstations is limited to an order-one model, predicting only the probabilities of symbols without using any conditional probability. Therefore, in the lossless compression case, we need to construct a reversible transform that can transform our image into another version whose histogram is extremely narrow. We construct the transform by factoring the transfer function of a maximally decimated filter bank while imposing the constraints that the filter coefficients must be rational (to get perfect reconstruction using a few bits of precision) and that the analysis filters must have second or third zero moment (to approximate the low-order polynomials well enough to make an extremely narrow histogram). In the lossy compression case, the wavelet transform is implemented using the linear-phase biorthogonal transform described by Antonini et al. because it does not need phase compensation in pyramidal implementation [5]. The edges of the images are treated according to Brislawn's descriptions [5].

22.2. WAVELET TRANSFORM

The wavelet transform is a special case of perfect-reconstruction filter banks. The main idea of the transform is to subdivide an arbitrary signal into constant-Q frequency bands using recursive filter banks generated from a small number of prototype filters. The filtering process is equivalent to decomposing the signal using a set of basis functions which are localized in both space and frequency and are scaled and shifted versions of a prototypical mother wavelet [6–11]. Figure 22-1 shows two-level analysis and synthesis of one-dimensional data. $h(n)$ and $g(n)$ are low-pass and high-pass filters for analysis. $\tilde{h}(n)$ and $\tilde{g}(n)$ are low-pass and high-pass filters for synthesis. The four filters are related in the z-domain as follows:

$$\tilde{H}(z) = G(-z)$$
$$\tilde{G}(z) = -H(-z) \tag{22-1}$$
$$H(z)G(-z) - G(z)H(-z) = 2$$

The first two conditions eliminate aliasing; the last one eliminates phase and amplitude distortion in the reconstructed signal.

22.2.1 Wavelet Transform of Images

We transform the two-dimensional images using one-dimensional row and column wavelet transforms. The process is similar to row and column fast Fourier transform (FFT) of images. The wavelet transform should have a linear phase finite impulse response (FIR) implementation to eliminate the need for phase compensation in cascade pyramidal filter structure. Details of the construction of biorthogonal bases of wavelet can be found in [7] and [8]. We choose the set of filter coefficients in

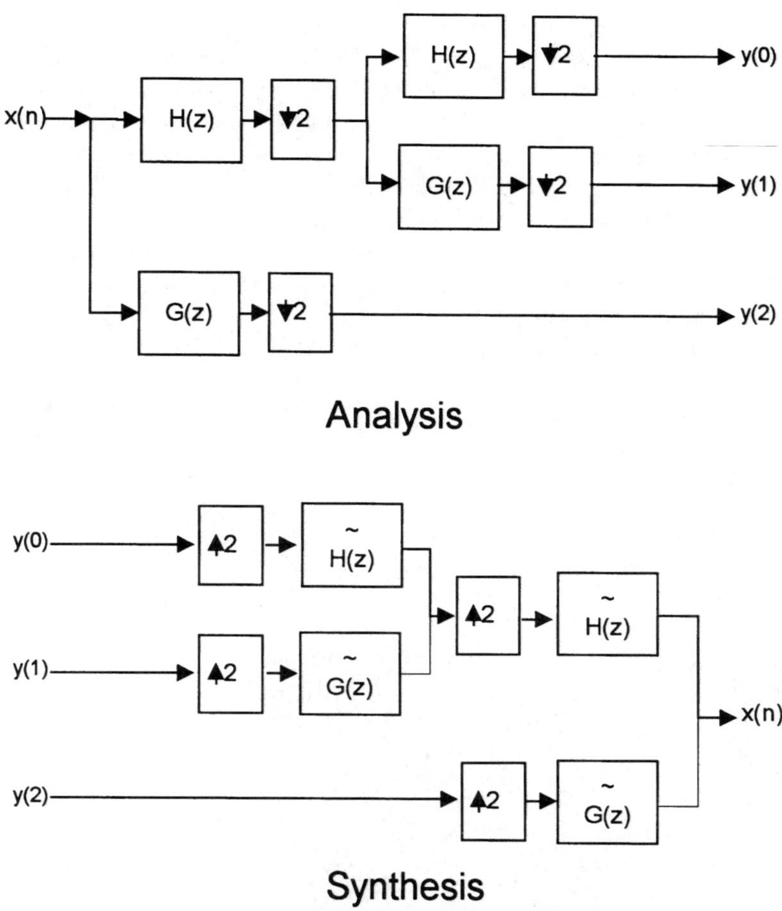

Figure 22-1   Two stages of analysis and synthesis of a wavelet transform of one-dimensional data.

Table 22-1, based on the smoothness of the linear-phased basis functions and the short length of the filters to implement the wavelet transform in our lossy compression [4, 5].

Table 22-1   Our Choice of the Filter Coefficients for Lossy Compression

n	0	±1	±2	±3	±4
$h_0(n)$	0.602 949	0.266 864	−0.078 223	−0.016 864	0.026 749
$f_0(n)$	0.557 543	0.295 636	−0.028 772	−0.045 636	0

With $h_1(n) = (-1)^n f_0(1 - n)$ and $f_1(n) = (-1)^n h_0(1 - n)$.

For lossless compression, we design the filters such that all the coefficients of all the filters are rational numbers with small denominators, using continued fraction and Diophantine techniques [8], so that all the operations are done with 32- and 64-bit integers. Furthermore, the synthesis filters, $\tilde{h}(n)$ and $\tilde{g}(n)$, are designed to have two or three consecutive zero moments to approximate low-degree polynomials which make up most images. The transform is linear-phased to eliminate the need for phase compensation in a cascade pyramidal filter structure.

The wavelet coefficients are obtained by convolution with the one-dimensional filters $h(n)$ and $g(n)$ followed by decimation (Fig. 22-1). The reconstruction is performed by up-sampling and convolution with filters $\tilde{h}(n)$ and $\tilde{g}(n)$. The resulting transformed image consists of sub-images with different resolution levels and orientations [Fig. 22-2(b)].

The edges of the image are handled by performing the convolutions on symmetric extensions of the images to avoid boundary artifacts introduced by simple periodization. The types of extensions and the associated parameters needed to perform the symmetric wavelet transform are described in detail by Brislawn in [5].

In our experiments, the images are transformed to three or four levels depending on their sizes. For 2048×2048 images, we choose to perform four levels of trans-

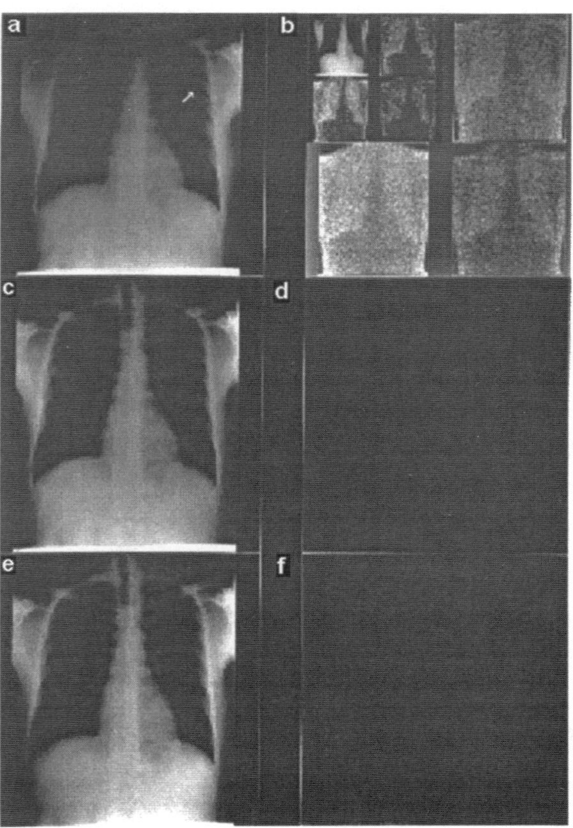

Figure 22-2 (a) An original $2048 \times 2048 \times 10$-bit chest radiograph. The arrow points to a scratch on the film prior to digitization. The scratch could serve as a benchmark on how good a compression algorithm can preserve edge sharpness. (b) a two-level wavelet transform of (a). The picture is enhanced to show the nature of wavelet coefficients. (c) 20:1 compression using the DWT. (d) 10× contrast-enhanced error image of (c). (e) 40:1 compression using the DWT. (f) 10× contrast-enhanced error image of (e). Note that the scratch mark is preserved in both (c) and (e).

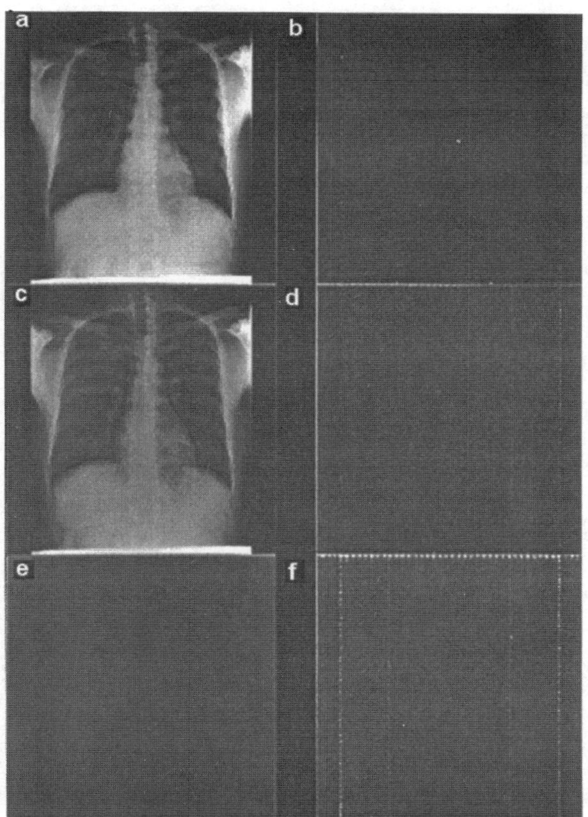

Figure 22-3 (a) 20:1 compression using the full frame discrete cosine transform (FFDCT) with bit allocation. Note the absence of the scratch mark. (b) 10× contrast-enhanced error image of (a). (c) 40:1 compression using FFDCT with bit allocation. Note the absence of the scratch mark. (d) 10× contrast-enhanced error image of (c). (e) Pixels where the error of DWT compression is larger that that of FFDCT at 20:1 compression. They distribute uniformly across the picture. (f) Pixels where the error of FFDCT is larger than that of DWT compression at 20:1 compression. They concentrate around the sharp edges.

form so that the low-resolution parts of the images are 128 × 128, a small enough size for thumb-nail picture indexing to be implemented in the future. For smaller images, a three-level transform is sufficient.

22.3. QUANTIZATION

In lossy compression, the wavelet coefficients are quantized into integral value. Since the detailed wavelet coefficients are well approximated by a Laplacian probability distribution (Fig. 22-4), we use the quantization steps $q_{i,j}$ to approximately minimize the quantization error with a constant entropy constraint:

$$q_{i,j} \propto \sqrt{n_i n_j} \qquad (22-2)$$

where n_i and n_j are the horizontal and vertical sub-sampling factors. The constant of proportionality is controlled by a quality parameter. This midtread uniform quantizer for each level of detailed coefficients is shown by Popat not to be globally optimal [12], but the deviation from the performance of the globally optimal quantizer is negligible. This choice of quantization allows us to perform quantization implicitly by scaling the filter coefficients beforehand. We do not use any perceptual weight

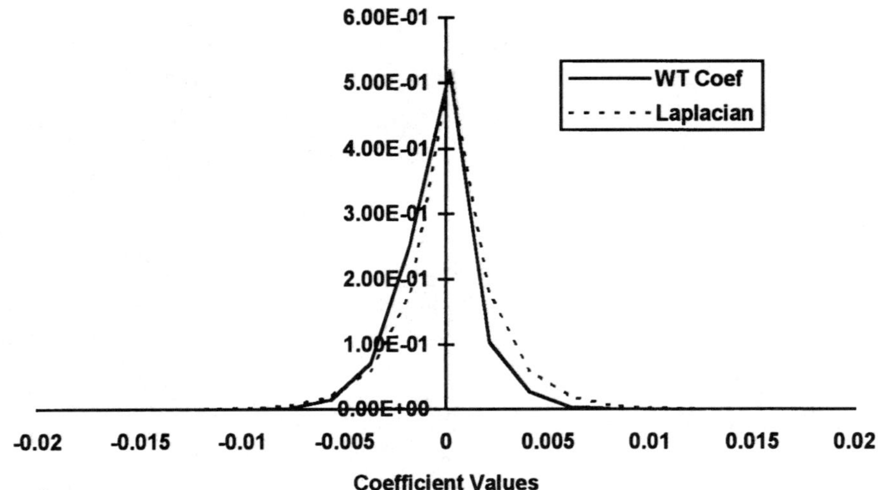

Figure 22-4 The probability distribution of the DWT coefficients compared with a Laplacian-distribution.

function in (22-2) because the coefficients will be subjected to further processing by computer in other applications. This omission of perceptual weighting modification leads to a fast overall implementation with a slight decrease in compression efficiency. However, at the compression ratios that we are interested in, slight compression inefficiency is acceptable. Scalar quantization is chosen over vector quantization because of the memory and speed requirement in affordable teleradiology workstations.

In lossless compression, quantization is unnecessary. Since the filter coefficients are rational with small denominators, the wavelet coefficients can be represented by computer integers instead of floating-point numbers.

22.4. ARITHMETIC CODING

After transforming the image using wavelet transform, the resulting coefficients are compressed by AC. To the arithmetic coder, the coefficients look just like a stream of symbols with various probabilities. The function of the coder is to represent the symbol stream using as few bits as possible.

Arithmetic coding encodes the whole stream of data with a real interval between 0 and 1. Before the arrival of the message, the interval representing the message is [0, 1). For every symbol the coder receives, the interval narrows down by a factor equal to the probability of the symbol. As the message becomes longer, the interval needed to encode it becomes smaller. The more likely symbols reduce the range by less than the less likely symbols. The number of bits per symbol after the complete encoding can approach the nonintegral entropy bound. AC is chosen over Huffman's coding because its compression performance is at least as good and sometimes, when one of the symbols' probabilities is close to 1, far exceeds it because of Huffman's requirement that the number of bits per symbols must be integral. As an extreme example, a message consisting of 999 999 zeroes followed by a 1 is represented by the interval:

$$[(0.999\,999)^{999\,999}(0.999\,999), (0.999\,999)^{999\,999}) \approx \left[e^{-1}, \frac{e^{-1}}{0.999\,999} \right)$$

while a message with one 1 followed by 999 999 zeroes is represented by the interval:

$$\left[0.999\,999, 0.999\,999 + \frac{e^{-1}}{1000\,000} \right)$$

Both messages require only 22 bits. On the other hand, typical unblocked Huffman encoding needs 1000 000 bits to represent the same messages. Analysis of the entropy of the messages shows that we need only 21.37 bits to represent them.

Arithmetic coding's independence from the modeling of data allow us to use the same coder for more sophisticated models for applications such as progressive transmission and hierarchical storage in the future.

In our experiments, the symbols are the values of the wavelet coefficients. This choice is an order-one model, using only the values of the symbols without any conditional probability among the symbols. Better compression efficiency can be obtained using a higher-order model with trade-offs in execution speed and memory requirement. The wavelet coefficients of the image are grouped according to their resolutions and orientations to skew the histogram of the coefficients. The more concentrated the histogram, the better the compression. Each group is compressed individually using AC as described in [2] and [3]. Comparison of actual compression and the entropy bound for high-frequency coefficients is shown in Fig. 22-5. The high-frequency coefficients are distributed as shown in Fig. 22-4; they aggregate densely around zero. Note that our AC implementation approaches the entropy bound to within 1%.

22.5. EXPERIMENTS

For lossy compression, a set of radiological images consisting of three computerized radiographs (CR), three chest computerized tomographs (CT), three abdominal CTs, and one head CT is used in our experiment. The CR images are $2048 \times 2048 \times 10$-bit. The CT images are $512 \times 512 \times 12$-bit. These images are transformed into either three or four levels using the coefficients from Table 22-1. The resulting coefficients are quantized. The sequence of the experiments is shown in Fig. 22-6. The error introduced in reconstruction is measured using the normalized mean square error (NMSE) and maximum difference (MD). For comparison with the joint photographic expert group (JPEG) method, we also measure the normalized nearest neighbor difference (NNND) which is defined as:

$$\text{NNND} = K \frac{\displaystyle\sum_{\text{border}} (p_i - p_{i+1})^2}{\displaystyle\sum_{\text{non-border}} (p_i - p_{i+1})^2} \tag{22-3}$$

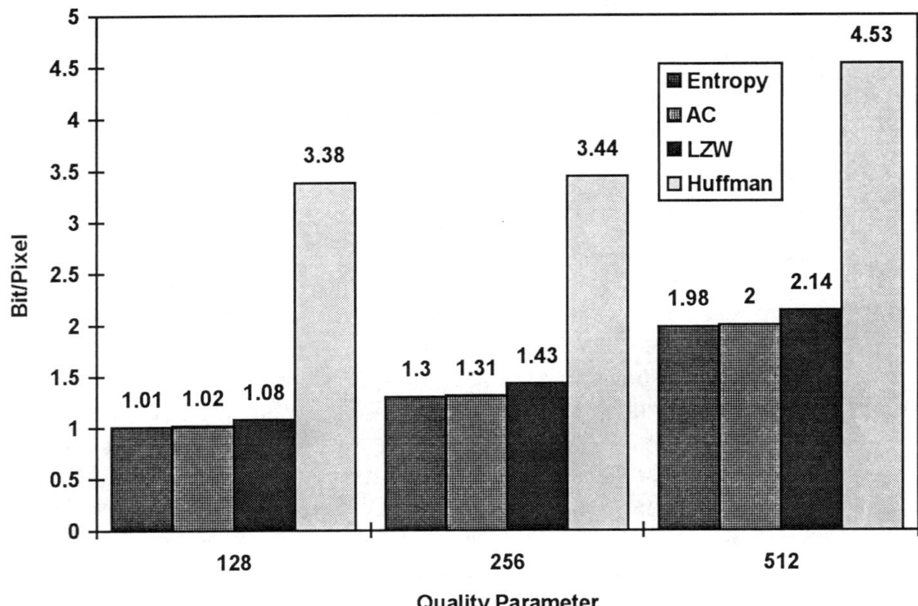

Figure 22-5 Comparison of compression performance between AC, LZW, and Huffman. Entropy is the theoretical lower bound of an order-1 model. AC is our implementation of arithmetic coding. LZW is done using Unix compress program. Huffman is an adaptive Huffman program from [3].

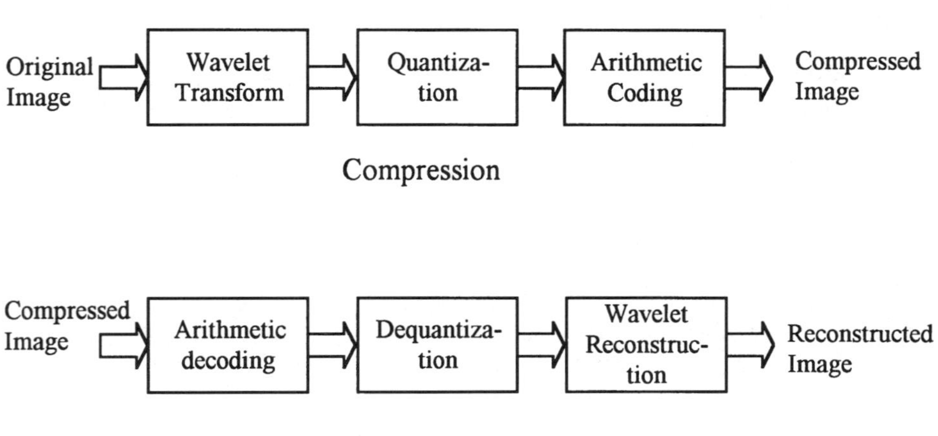

Figure 22-6 The compression and reconstruction processes.

where K is a normalization constant compensating the different number of pixel pairs summed in the numerator and denominator and p_i are the pixel values. NNND quantifies the effect of block artifacts in the reconstructed images [13]. The error images obtained from subtraction of reconstructed images from the originals are enhanced for visual inspection and are FFTed to check for any periodic artifacts.

For lossless compression, a set of radiological images consisting of five CRs, two CTs, three abdominal CTs, one head CT, three abdominal magnetic resonance images (MRI), three head MRI, and ten mammograms is used in our experiment. The radiographs are $512 \times 512 \times 8$-bit, $1024 \times 1024 \times 10$-bit, and $2048 \times 2048 \times 10$-bit. The CT images are $512 \times 512 \times 12$-bit. The MRI images are $256 \times 256 \times 12$-bit and $512 \times 512 \times 12$-bit. The mammograms are $1024 \times 1024 \times 10$-bit and $2048 \times 2048 \times 10$-bit These images are transformed into either three or four levels. The resulting coefficients are encoded with an arithmetic coder.

22.6. RESULTS

22.6.1 Lossly Compression

Quality of Reconstruction of Lossy Compression.   Comparison of NMSE and MD of the discrete wavelet transform (DWT) method with the full frame discrete cosine transform (FFDCT) method shows that both NMSE and MD from DWT increases much slower with compression ratio than those from FFDCT (Fig. 22-7 and 22-8). The subjective quality is also better since DWT's error is flatter, tending to

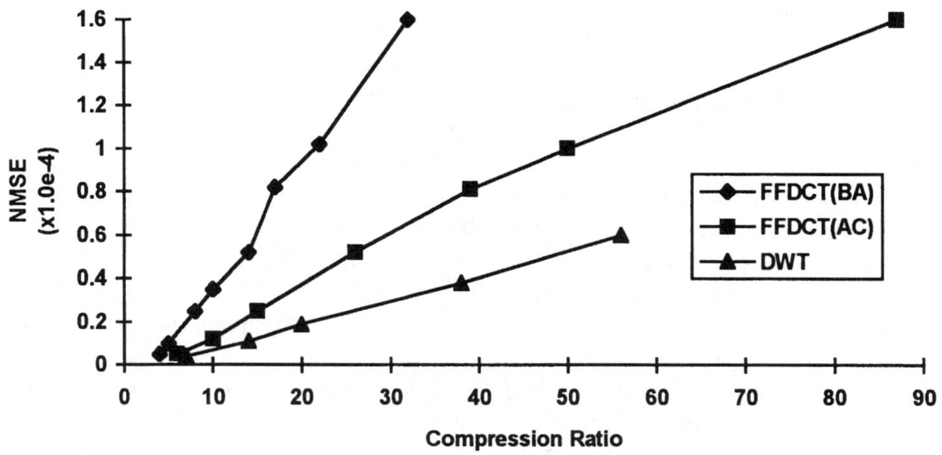

Figure 22-7   NMSE versus compression ratio. FFDCT(BA) is the FFDCT using bit allocation, FFDCT(AC) uses arithmetic coding.

Maximum Difference Vs. Compression Ratio

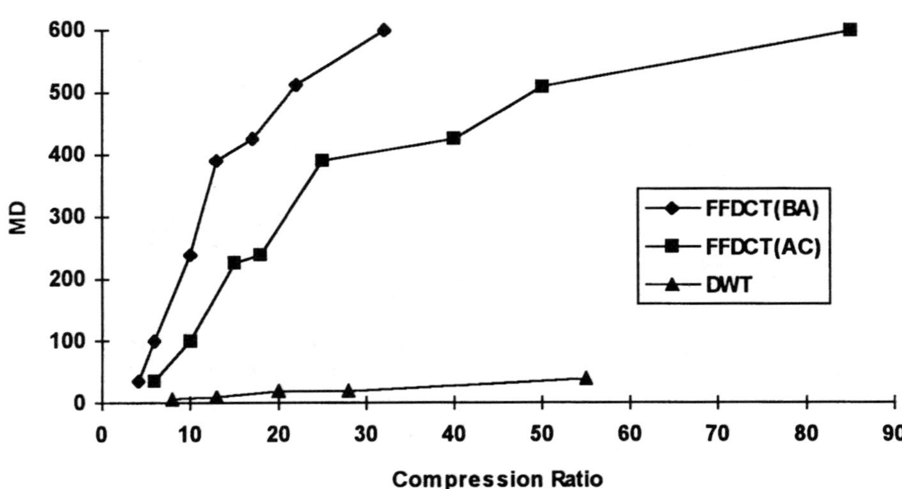

Figure 22-8   Maximum difference (MD) versus the compression ratio. FFDCT(BA) is the FFDCT using bit allocation, FFDCT(AC) uses arithmetic coding.

be small across the reconstructed pictures, while FFDCT's error tends to distribute in many local spikes (Figs. 22-2 and 22-3). When the images contain sharp edges such as patient labels, lettering, or cropping, FFDCT exhibits ringing across the reconstructed images, while DWT's error is localized. This results from the fact that the basis functions in FFDCT extend globally in the spatial domain while the basis functions in the DWT are localized both in the spatial and frequency domains. The FFT of the error images supports this subjective observation. The spectrum of the FFDCT's error shows many peaks that signify truncation of high frequency and aliasing. At a compression ratio of 20:1, DWT is more accurate than FFDCT with bit-allocation in all but 20% of the total pixels. In these 20%, a large fraction of DWT's error is larger by only 1 gray level out of 1024 (see Figs. 22-3(e), (f) and 22-9).

DWT's NNND stays approximately constant with the compression ratio, while JPEG's NNND rises rapidly as the compression ratio increases (Fig. 22-10). This is due to the fact that the JPEG is based on a block implementation of a discrete cosine transform; the errors in adjacent blocks are not correlated.

Using symmetric extension of the data, the wavelet transform can be performed on images of arbitrary sizes as shown by Brislawn in [5]. This property is especially important in radiographs of extremities, where cropping is frequently done.

Reconstruction Error.   Figure 22-11 shows the type of error introduced by this compression method. The picture was chosen because of its extreme contrast between the metal implant and surrounding tissues. Figure 22-11(c) is the unenhanced error; Fig. 22-11(d) is the error enhanced to show errors that concentrate around the edges. This type of error arises from imperfect cancellation of the coeffi-

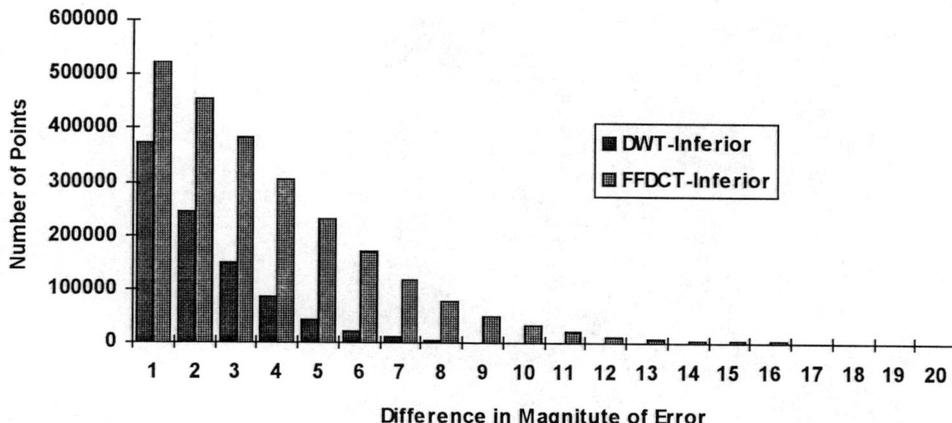

Figure 22-9   Distribution of the error magnitude comparison between the DWT and FFDCT. At a compression ratio of 20:1, DWT is more accurate than FFDCT with a bit-allocation in all but 20% of the total pixels. In more than half of those pixels where DWT is less accurate, DWT is worse by at most 2 gray-levels out of 1024.

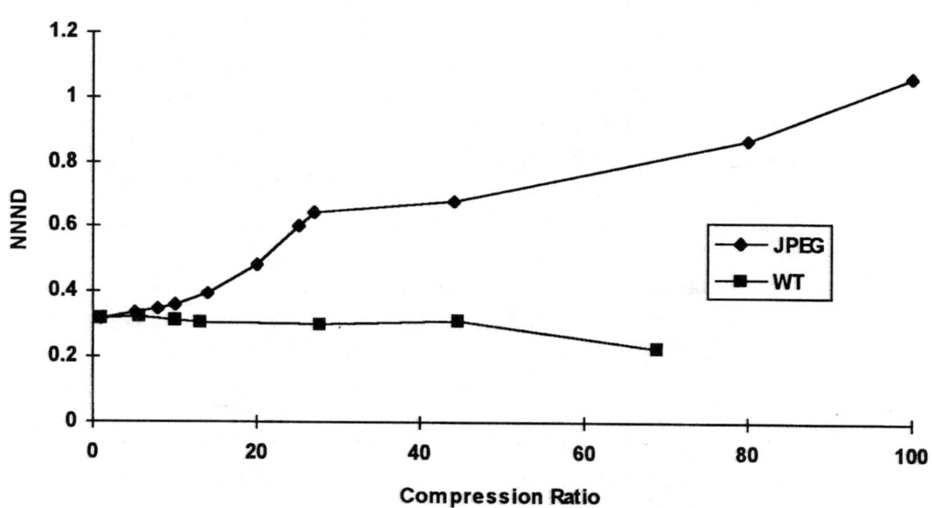

Figure 22-10   Plot of NNND as a function of compression ratio for the JPEG and DWT.

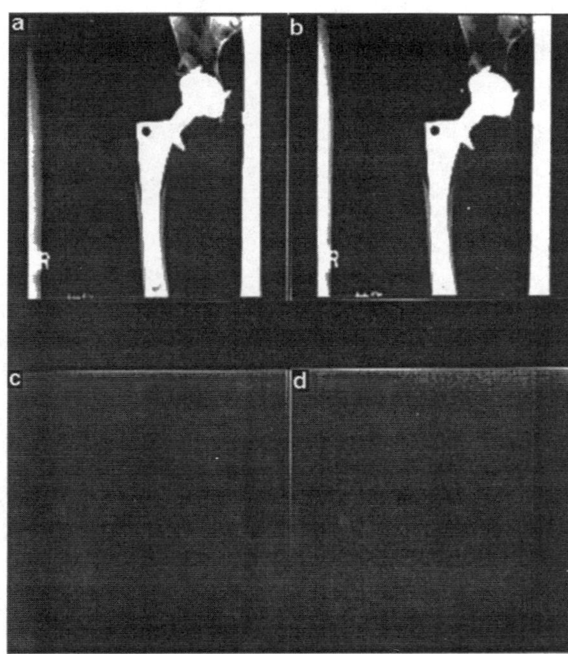

Figure 22-11 Error in DWT compression (a) The original 2048 × 2048 ×10-bit CR. This image is chosen for its extreme contrast between the metal implant and the surrounding tissues. (b) 20:1 compression using DWT. (c) Error image. (d) 10× contrast-enhanced error image

cients of different levels inherent in any quantized transform method. Note that there is no global ringing characteristic of typical global transform methods.

22.6.2 Lossless Compression

The result of the compression experiment is shown in Table 22-2, Figs. 22-4 and 22-5. The lowest compression ratio of 2.2 is obtained from a 512 × 512 × 8-bit chest CR. This is to be expected since the low-resolution image of the chest contains more high-frequency information than the high-resolution image; the distance between adjacent pixels in the low-resolution image is larger than that of the high-resolution images implying larger changes in gray level between pixels. The highest compression ratio of 4.8 is obtained from a breast biopsy at 1024 × 1024 × 10-bit. This is because the image zooms in a small physical area with relatively large constant gray-level regions.

The data show that the compression ratio increases when the original dimensions of the image increase. This is explained by the low-resolution/high-frequency-content argument above. The data also shows that large constant gray-level areas around the anatomy in CT, magnet resonance imaging (MRI), and mammograms are redundant and contribute to the high compression ratio (Figs. 22-12 and 22-13.)

The compression time is 8 s for a 1024 × 1024 × 2-byte image on a Pentium-90. The time is linear with the number of pixels. Since the computation is central processing unit (CPU)-limited and is not bound by input/output (I/O) speed, we expect linear improvement of computation time with the CPUÆs speed and the number of CPUs.

Table 22-2   Results of Lossless Compression of Test Images

Image	Type	Description	Bits/pixel	Size (pixels)	Compressed bits/pixel	Compression Ratio
CR1	CR	Chest	8	512	3.6	2.2
CR2	CR	Chest	10	2048	2.9	3.5
CR3	CR	Chest	10	2048	3.0	3.3
CR4	CR	Extremity	10	1024	3.3	3.0
CR5	CR	Extremity	10	512	4.3	2.3
CT1	CT	Chest	12	512	3.8	3.2
CT2	CT	Chest	12	512	4.0	3.0
CT3	CT	Abdomen	12	512	3.9	3.1
CT4	CT	Abdomen	12	512	3.5	3.4
CT5	CT	Abdomen	12	512	3.5	3.4
CT6	CT	Head	12	512	3.6	3.3
MR1	MRI	Abdomen	12	512	3.1	3.9
MR2	MRI	Abdomen	12	512	3.5	3.4
MR3	MRI	Abdomen	12	256	4.3	2.8
MR4	MRI	Head	12	256	3.6	3.3
MR5	MRI	Head	12	256	3.9	3.1
MR6	MRI	Head	12	256	4.3	2.8
MAM1	Mammogram	Screen	10	2048	2.8	3.6
MAM2	Mammogram	Screen	10	2048	2.7	3.7
MAM3	Mammogram	Screen	10	2048	2.6	3.9
MAM4	Mammogram	Screen	10	2048	2.9	3.4
MAM5	Mammogram	Screen	10	2048	2.6	3.9
MAM6	Mammogram	Screen	10	1024	3.4	2.9
MAM7	Mammogram	Screen	10	1024	3.3	3.0
MAM8	Mammogram	Biopsy	10	1024	2.1	4.8
MAM9	Mammogram	Biopsy	10	1024	2.4	4.2
MAM10	Mammogram	Biopsy	10	1024	2.3	4.4

22.7. CONCLUSIONS

The wavelet transform can be applied to image compression applications in radiological image archives, and primary diagnosis storage and transmission with good results. For archival application, lossy image compression based on the DWT and AC is superior to the FFDCT and JPEG. For primary diagnosis applications, lossless compression based on the DWT and AC can achieve routine 2:1 to 4:1 compression ratios which are among the best achievable by other lossless methods.

The algorithms can be implemented on inexpensive, off-the-shelf computer hardware. Our implementation is entirely software-based, meaning that we can easily take advantage of the increase in speed of the CPU. The structure of the programs eliminates the I/O bottleneck and is limited by the CPU speed. Therefore, we can expect linear improvement in compression time with the CPU's performance. The structure of the programs is also easily parallelizable; we expect the performance to scale linearly with the number of CPUs.

Compression Ratio Vs. Bits/Pixel

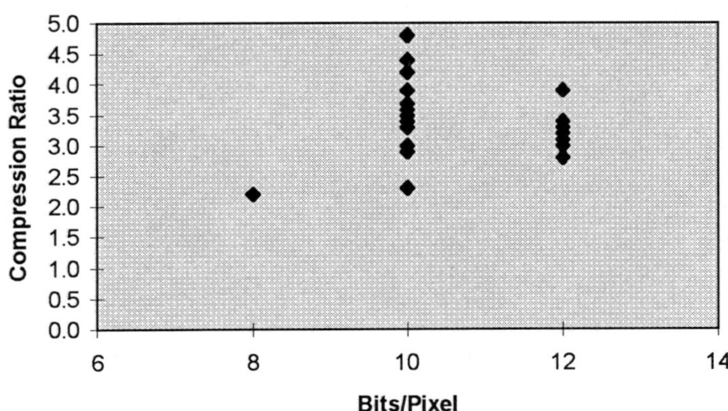

Figure 22-12 Lossless compression ratio versus bits/pixel in the images. The top three data points at 10 bits/pixel are biopsy mammograms which contain large, slowly varying regions

Compression Ratio Vs. Image Size

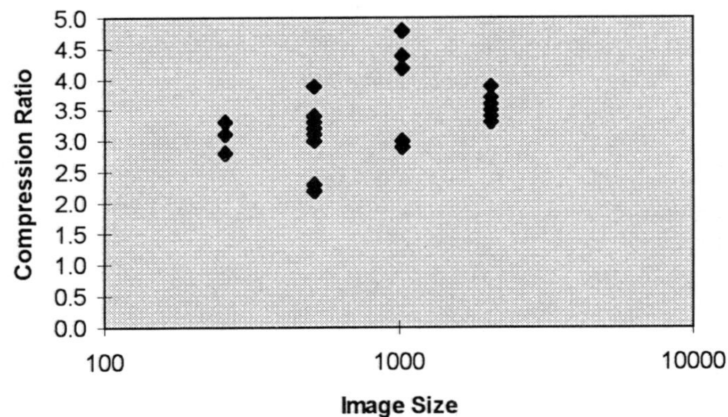

Figure 22-13 Lossless compression ratio versus image size. The top three data points at 1024 are biopsy mammograms which contain large, slowly varying region.

REFERENCES

[1] A. W. Wong, R. K. Taira, and H. K. Huang, "Implementation of a digital archive system for a radiology department," *Proc. SPIE Conf. on Medical Imaging VI: PACS Design and Evaluation* vol. 1645, pp. 182–190, 1992.

[2] I. H. Witten, R. M. Neal, and J. G. Cleary, "Arithmetic coding for data compression," *Comm. ACM*, vol. 30, no. 6, pp. 520–540, June 1987.

[3] M. Nelson, *The Data Compression Book*. M&T Books, 1992.

[4] M. Antonini et al., "Image coding using wavelet transform," *IEEE Trans. Image Proc.* vol. 1 no. 2, pp. 205–220, April 1992.

[5] C. M. Brislawn, Classification of symmetric wavelet transform. Group C-3, Computer Research, Los Alamos National Laboratory, March 1993.

[6] I. Daubechies, "Orthonormal bases of compactly supported wavelets," *Comm. Pure Appl. Math.*, vol. 41, pp. 909–996, 1988.

[7] A. Cohen, I. Daubechies, and J. C. Feauveau, "Biorthogonal bases of compactly supported wavelets," *AT&T Bell Lab., Tech. Report*, TM 11217-900529-07, 1990.

[8] S. Mallat, "A theory for multiresolution signal decomposition: The wavelet representation," *IEEE Trans. Patt. Anal. Mach. Intell.*, vol. 11, pp. 674–693, July 1989.

[9] P. P. Vaidyanathan, *Multirate Systems and Filter Bank*. Englewood Cliffs, NJ: P T R Prentice-Hall, 1993.

[10] M. Vetterli and C. Herley, "Wavelets and filter banks: Relationships and new results," *Proc. IEEE ICASSP*, Albuquerque, April 1990.

[11] M. Vetterli, "Wavelets and filter banks: Theory and design," *IEEE Trans. Signal Proc.*, vol. 40, no. 9, pp. 2207-2232, September 1992.

[12] A. C. Popat, Scalar quantization with arithmetic coding. *Master thesis* in Electrical Engineering and Computer Science, Massachusette Institute of Technology, 1990.

[13] B. K. T. Hoe et al., "A mathematical model to quantify JPEG block artifacts," *Proc. SPIE Med. Imag. Image Capture, Formatting, and Display*, vol. 1897, Newport Beach, CA, pp. 269–275, 1992.

Chapter 23

Adapted Wavelet Encoding in Functional Magnetic Resonance Imaging*

Dennis M. Healy Jr., Douglas W. Warner, John B. Weaver

From Fourier's time up to the present day, engineers and applied scientists have benefited greatly from tools and insights developed in the field of harmonic analysis. The last decade has seen an important instance of this as advances in the field of wavelets have been applied to numerous problems in signal and image processing. In this chapter we give a short sketch of how some of the recent developments in the field apply to a problem in magnetic resonance imaging (MRI).

(MRI) has become an essential tool in clinical medicine, producing exquisite contrast in images of soft tissue structures without the introduction of artificial contrast agents; see Fig. 23-1 for example. However, it can be limited in resolution and speed of image acquisition. Depending on the desired contrast in the images, acquisition time on a standard scanner can vary from a second to many minutes.

Enormous research and development effort has been devoted to the goal of reducing the imaging time. This is motivated in part by a desire to reduce the expense of MRI, often amounting to hundreds of dollars for a standard examination. Practical fast imaging techniques are also required for some of the ambitious application areas recently proposed for MRI, including heart imaging, joint and muscle motion studies, and functional imaging of the brain. The latter permits one to study evolving activity in areas of the brain during memory and motor tasks. This is an application with diagnostic potential as well as pure research interest. For example, knowing which parts of the brain are used to move fingers helps minimize damage from neurosurgery. Localizing the parts of the brain used to remember word lists

*This work supported in part by ARPA, as administered by the AFOSR under contract DOD F4960-93-1-0567.

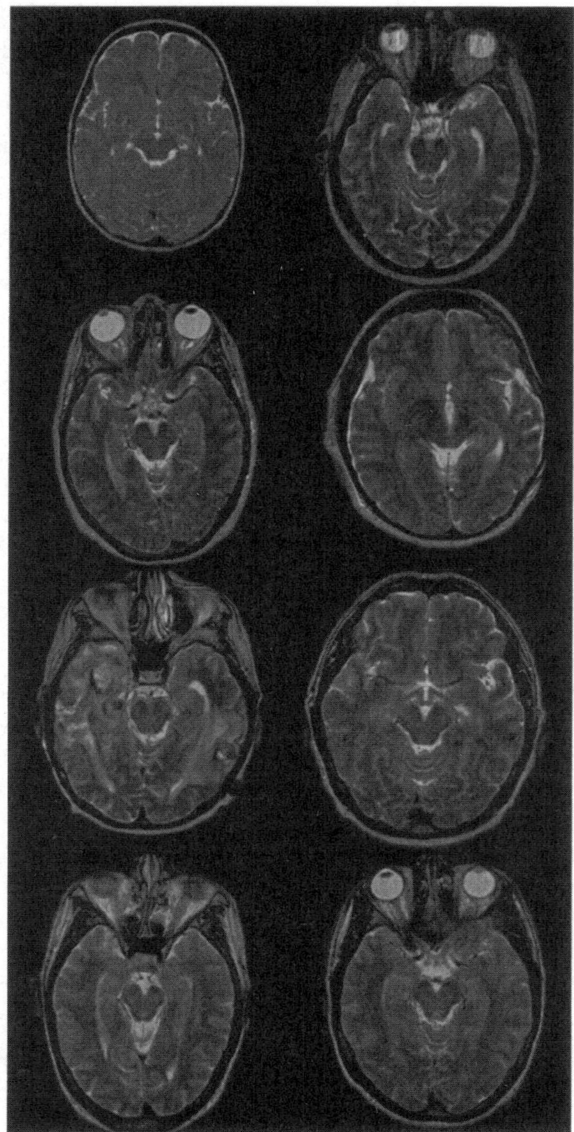

Figure 23-1 Transaxial head scans from a particular MRI image class. (Level 7 training set.)

may help identify and develop treatments for Alzheimer's patients and patients with schizophrenia.

Imaging time is largely determined by physical and engineering constraints on the rate at which the scanner takes measurements. The measurement process may be described mathematically as the projection of a function representing tissue properties onto the elements of a basis of a function space; for standard MRI, the Fourier basis is used. Each measurement requires a certain amount of time, and typically many measurements must be made. Most fast imaging methods presently under consideration are concerned with increasing the rate at which these measurements are acquired; see for example [1–8]. Unfortunately these methods can require expensive hardware modifications and can adversely affect contrast and resolution.

We take a different approach and consider the possibility of reducing the number of measurements required for certain types of images. Taking advantage of the flexibility of the MRI modality, we change the basis used for the measurements from the standard Fourier basis to a basis which incorporates prior knowledge about the imaging task at hand. The goal is to obtain the image data in as compact a form as possible. This reduces the total number of projections required and should improve imaging speed even if the measurement rate remains the same.

This notion is inspired by the field of image compression; for some interesting background see [9], for example. In this chapter, we will briefly review a simple idea from compression, the Karhunen–Loève representation, and some variants recently proposed by Wickerhauser [10]. We then indicate how this representation may be brought to bear on the problem of imaging speed in MRI. This involves only a simple modification of the bases used in standard signal-acquisition protocols. These modifications require no expensive changes to the hardware, and have little impact on contrast and resolution.

This approach to fast imaging may be used to acquire fast, preliminary "scout images." Another application of related techniques in spectroscopic imaging will be sketched. For convenience, we conclude the paper with an appendix reviewing some of the main points of MRI signal acquisition.

Our emphasis in this chapter is on the application of some of the new ideas from wavelet signal processing in the arena of signal acquisition and measurement in MRI. We should point out that there are also important research efforts devoted to the different problem of post-processing images already obtained by a scanner; some of the other chapters of this monograph discuss techniques that would be useful for this purpose. For some approaches to post-processing which have been considered by our research group, one might consult [11–13].

23.1. PARSIMONIOUS REPRESENTATIONS OF IMAGES

An important concept in image compression is the notion that efficient representations of the images in a certain class may be found using prior knowledge of statistical regularities of that class. One approach to this is given by the Karhunen–Loève (K–L) decomposition, also known as the method of principal components.

Mathematically, the K–L decomposition gives an expansion of a random process with finite second moments in a special basis of (deterministic) orthonormal functions. The expansion coefficients are random variables obtained as inner products of the process with the basis elements. The basis is chosen so that the expansion coefficients are uncorrelated random variables, see [14] for example. As it turns out, the K–L basis is also the best orthonormal basis for compressing most of the variability of the process into a few significant expansion coefficients. We now review this idea in the context of our application.

We are interested in efficient representations for collections of related MRI images. These typically consist of arrays of 256 × 256 pixels whose gray-scale values represent a weighted density of the hydrogen in water and fats of tissues in a planar slice of the subject. For our purposes, the (empirical) K–L basis is determined from a finite collection of images obtained by applying a particular type of MRI examination to many patients.

Some images from one such class, corresponding to a particular transaxial head section, are shown in Fig. 23-1. Examining a given image from this collection, we note correlations in density and texture found at various positions in the image. For example, in many neighborhoods the gray-scale values change smoothly and gradually. This suggests that although there are 256×256 pixels in each image, these pixels are not completely independent of one another.

Note that this holds for all of the images of the class. Furthermore, image features in one image bear some relationship to those in the other images. Of course there are also differences from image to image, but the most important of these tend to be localized in sub-regions of the images. This suggests that the most important variability distinguishing the various images of the class could be characterized by fewer than 256×256 parameters.

By expanding an image in the K–L basis for the class, we seek a new image representation in which a reduced set of parameters suffices to distinguish one image from another. In what sense is this possible? For the present discussion, an image is simply a vector $\mathbf{I}$ in the Euclidean space $\mathbb{R}^d$, with $d = 256 \times 256$. Suppose we have a collection of n related image vectors, $\mathcal{I} = \{\mathbf{I}_1, \mathbf{I}_2, \mathbf{I}_3, \ldots \mathbf{I}_n\}$, such as the head images shown in Fig. 23-1. Then the second-order statistical properties are indicated by the sample autocovariance matrix $R_\mathcal{I} = \frac{1}{n-1} \sum_{k=1}^{n} (\mathbf{I}_k - E(\mathbf{I}))(\mathbf{I}_k - E(\mathbf{I}))^t$, where $E(\mathbf{I})$ denotes the (sample) average or expected image. The (empirical) K–L basis is defined to be the orthonormal basis of eigenvectors of this positive matrix.

The K–L basis is characterized by many optimality criteria; the most useful ones for our purposes pertain to the decay rate of the K–L coefficients of an image from our class. Fast decay in a given representation indicates that we can omit the higher-order information from a specific image without significantly degrading its reconstruction. Specifically, if $\mathbf{I}$ is an image from our class $\mathcal{I}$, and $\mathcal{U} = \{\mathbf{u}\}_{j=1}^{d}$ is an orthonormal basis of image space, we consider the truncated estimate of $\mathbf{I}$'s $\mathcal{U}$-expansion from its first $t \leq d$ terms:

$$\widehat{\mathbf{I}}_{t,\mathcal{U}} = \sum_{j=1}^{t} \langle \mathbf{I}, \mathbf{u}_j \rangle \mathbf{u}_j + \sum_{j>t}^{d} \langle E(\mathbf{I}), \mathbf{u}_j \rangle \mathbf{u}_j$$

Note that the second sum fills in the missing higher-order terms for the particular image $\mathbf{I}$ using only *a priori* knowledge about the corresponding terms of the mean image.

To evaluate the compression of our images afforded by the basis, we form the average, or expectation, of the estimator's l^2 error over our image class: $E_\mathcal{I} \| \mathbf{I} - \widehat{\mathbf{I}}_{t,\mathcal{U}} \|$. The K–L basis for the class $\mathcal{I}$ minimizes this figure of merit over the class of all orthonormal bases for every $t \leq d$.

Alternatively and equivalently, K–L expansions minimize a variance concentration metric on orthonormal bases, as discussed in [10]. For a given orthonormal basis of image space, $\mathcal{U}$, let $\mathcal{U}(\mathcal{I})$ denote the collection of vectors comprised of the coefficients of the image vectors of class $\mathcal{I}$ with respect to the basis (their $\mathcal{U}$-coefficient vectors). The variances of these coefficients over the class are given by the diagonal elements of the covariance matrix $R_{(\mathcal{U}\mathcal{I})}$ formed from $\mathcal{U}$-coefficient vectors corresponding to the images in $\mathcal{I}$. The metric

$$H_{\mathcal{I}}(\mathcal{U}) = \sum_{j=1}^{d} \log\left[R_{\mathcal{U}(\mathcal{I})}\right]_{j,j} \tag{23-1}$$

is defined on all possible choices of the orthonormal basis $\mathcal{U}$. The K–L basis minimizes this metric.

The K–L basis does a remarkable job of concentrating the interesting information of an image class into a relatively small number of coefficients. The situation is indicated schematically in Fig. 23-2.

Finding the (empirical) K–L basis for a collection of images is computationally expensive, since it involves diagonalizing a large matrix. Wickerhauser has noted that it is often enough to find approximate K–L bases by restricting the optimization problem described above to certain nice subsets of all possible bases, such as the wavelet packet bases or local trigonometric bases [10]. This restricted optimization problem is much easier to solve, yet the sub-optimal solution is a basis which produces much the same compression results as the global optimum.

We use this approach in our MRI work, as indicated in more detail below. For now we note that the point of any best-basis search is to find a new basis or representation of the images from a class $\mathcal{I}$. With respect to this representation, images in the class can be estimated up to a small error by a number of coordinates which is small relative to the dimension of the image space.

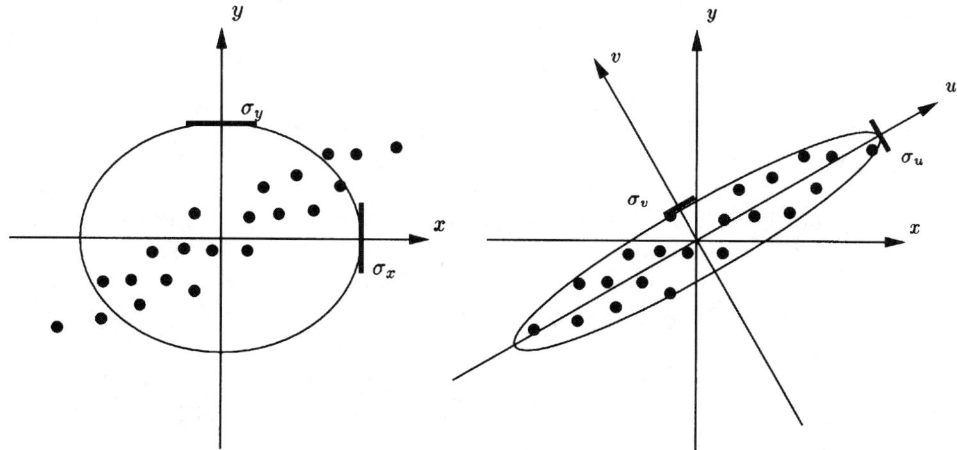

Figure 23-2 A class of signals, each represented as a two-vector. On the left, the variability of the x- and y-coordinates of the signal vectors are indicated by standard deviations, $\sigma_x \approx \sigma_y$. On the right, the K–L (u-v) axes for this class are indicated. Note that most of the variability of the class occurs along the u K–L direction. Up to a small error, we could identify a particular signal by its u-coordinate alone.

23.2. STANDARD MRI AND FOURIER TRANSFORMS

In MRI, information about an object under study is encoded in radio-frequency (RF) electromagnetic signals emitted by certain of the object's nuclei (usually hydrogen in water and fats within the tissues) as their spin state undergoes transitions in the carefully controlled magnetic field of an MRI scanner. Here we can only barely sketch how these signals are obtained and used; see Appendix A and the references therein for more information.

MRI is based on the phenomenon of "nuclear magnetic resonance" exhibited by certain atomic nuclei, such as those of hydrogen or phosphorus. When these are placed in a homogeneous external magnetic field, they absorb and radiate electromagnetic energy at a natural resonant frequency, proportional to the external magnetic field strength. This resonant frequency is called the Larmor frequency, f_{Larmor}. Because of this, it is possible to set this resonant frequency in a sample of interest by controlling its magnetic environment in the MR scanner. One may then promote the nuclei of interest to an excited state by subjecting them to an RF excitation signal at the known Larmor frequency. The excited nuclei will then respond with their own RF signal at the same resonant frequency.

By carefully controlling the scanner's field, the resonant frequency in the sample may be varied. By introducing a spatial variation of the external field strength across the sample, it is possible to make the characteristic frequency vary across the sample. It is this fact that allows one to encode spatial information by frequency, so that the measured signal can be identified with a sampling of a portion of the spatial Fourier transform of the object.

Perhaps the most straightforward approach to mapping the variation of soft tissue in a sample is to try to obtain a signal from only a localized region. For example, one may encode position by only "turning on" a portion of the sample; the resulting output signal reflects only the properties of nuclei in the activated region. This process is known as "selective excitation."

This is implemented by supplying a narrowband RF excitation while a linear magnetic gradient is applied to the sample. This makes the resonant frequency a linear function of position along the gradient. Only those nuclei whose resonant frequency coincides with a frequency component present in the excitation signal are stimulated into producing a signal. These nuclei live only at the positions along the gradient which experience magnetic field strength corresponding to the frequency components in the excitation. After excitation, only these excited nuclei produce an output signal. Therefore, any measurements reflect only the properties of the excited nuclei at this particular (known) location within the sample.

We show in Appendix A that selective excitation amounts to projection of the signal density function onto a localized amplitude profile determined from the particular signal used for the RF excitation. In the most common imaging tasks, this profile is a bump function, corresponding to a slab of excited nuclei. However, we can and will use more interesting profile shapes in the later sections of this chapter.

Selective excitation is the first step in a standard image acquisition. A carefully designed selective excitation RF pulse stimulates the relevant nuclei in a selected region of the sample, commonly a thin, planar slice. These excited nuclei output an

RF signal of their own, which is detected and measured. Parameters of this signal are controlled by manipulation of the magnetic environment in the scanner. Data collected from several iterations of this process are used to build a map of a weighted density ρ of the excited nuclei within the slice, as follows.

The signals produced in a standard imaging sequence look like:

$$S_\beta(t) = \int dx \int dy \, \rho(x, y) \, e^{-i\alpha xt} e^{-i\beta y} \qquad 0 \le t \le T \qquad (23\text{-}2)$$

where α, β are constants determined by the magnetic environment, and the integral is taken over the coordinates of the slice being imaged.

Note that each signal gives values of ρ's spatial Fourier transform sampled along a line segment parallel to the ω_x spatial frequency axis: $S_\beta(t) = \hat{\rho}(\omega_x = \alpha t, \omega_y = \beta)$, $0 \le t \le T$. A technique called "phase encoding" permits us to specify which line we measure, i.e., to set the value of β. In order to obtain enough data for image reconstruction, we must measure many of these lines. This is indicated schematically in Fig. 23-3.

Unfortunately, in many imaging situations, the nuclear spins in the sample must be allowed time to "rest," or return to the ground state, after each line is measured. The time between measurements can be long, resulting in long overall imaging time.

Given full Fourier data, the image could in principle be reconstructed by means of the inverse Fourier transform. Of course we can actually collect only finitely many phase-encoded signals, corresponding to finitely many lines of the spatial Fourier transform. Reconstruction proceeds by a numerical approximation of the inverse Fourier transform.

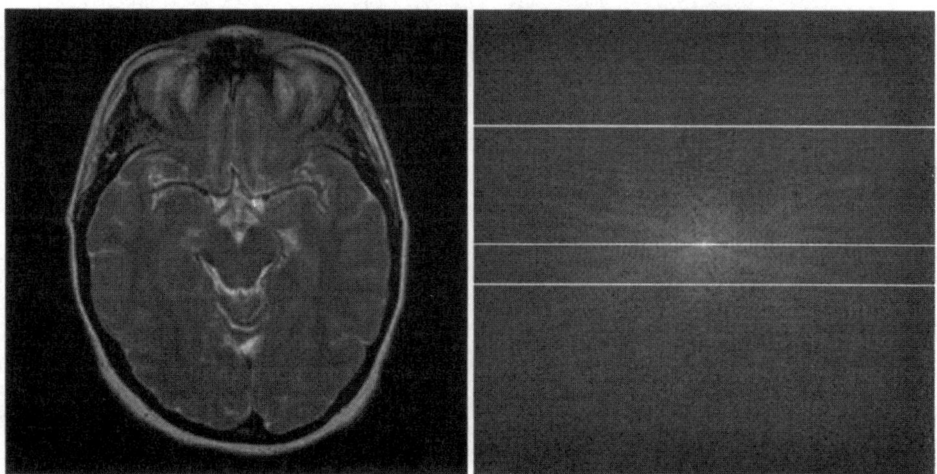

Figure 23-3   In MRI imaging, we obtain the brain image (left) by a sequence of time-domain signals which provide samples of the object's spatial Fourier transform (right). In standard imaging, each signal reads off the values of the Fourier transform along horizontal lines, such as those indicated.

In practice, the data are sampled on a finite rectangular grid in the spatial Fourier plane. It is enough to consider a finite region of the transform domain due to various practical limitations on the resolution one can expect to see in the image. Further, a discrete sampling of the Fourier transform in this finite range is sufficient because of the finite support of the actual image.

As a result, an adequate approximation of the reconstructed image is obtained by using the discrete Fourier transform (DFT), a trapezoid rule approximation to the inverse Fourier transform of the measurements taken on the rectangular grid in Fourier space,

$$\mathbf{I}(k, l) = \sum_{u,v} \hat{\rho}(u\Delta\omega_x, v\Delta\omega_x)e^{2\pi iuk/n}e^{2\pi ivl/n} \tag{23-3}$$

which gives the image as a finite rectangular pixel array.

23.3. ALTERNATIVES TO THE FOURIER BASIS

We have seen that standard MRI data must be transformed by a DFT in order to get the type of image that is useful to the radiologist. Equivalently, we can say that the object of interest is encoded by the MRI scanner as a superposition of the pure harmonics of the Fourier basis, as seen in Eqs. (23-2) and (23-3).

The Fourier basis gives a useful representation of signals and images in many situations, and has a built-in interpretation in terms of the global harmonic content of the signal. However, it is not the only basis available. In some cases, such as speech signals or music, we may want instead to have a dynamic description of the harmonic content. In terms of the music example, we want to know not only *which* frequencies (notes) are present, but also *when* the notes are to be played. The situation is similar for images where we are concerned with the variation of spatial harmonic content (texture, edges) from place to place in the image.

Many representations have been developed for this purpose over the last 15 years, including the wavelet bases, wavelet packet bases, and various time-frequency representations. A particularly useful localized harmonic representation for our purpose is provided by the local trigonometric orthonormal bases described by Coifman and Meyer, and independently by Malvar. A good description may be found in [10]; here we will just outline a picture of the ideas.

A point of departure for understanding these bases is provided by an intuitive concept of localized Fourier analysis. Suppose we are interested in studying a function supported on the line, and that we are concerned with its harmonic content in each of a sequence of intervals partitioning its support. Naively, we might simply take the portion of the function living on each interval of the partition and compute its Fourier series there. Doing this for all intervals is equivalent to expanding in a localized Fourier basis, whose typical element is the indicator or boxcar function over an interval of the partition multiplied by an appropriate pure harmonic.

The collection of all such functions may be normalized to give an orthonormal basis for $L^2(\mathbb{R})$, but the sharp localization produced by the boxcar functions introduces an extremely unpleasant blurring and lack of resolution in the coefficients describing the harmonic content. This is in accordance with a number of uncertainty principles for Fourier analysis [10]. The local trigonometric bases fix this by using smoothed versions of the boxcar functions of the intervals of the partition. An orthogonal sine or cosine basis is then employed for each interval to modulate these window functions. This construction improves frequency resolution by utilizing smooth, overlapping cutoff windows while nevertheless maintaining orthogonality by a clever parity construction. See Fig. 23-4 for an example.

We can build many bases of this type, corresponding to the different possible partitions of the domain. This permits a choice among many possible representations of a given function or class of functions, according to how we slice up the domain into sub-intervals. Local trigonometric bases can be designed which localize into many short sub-intervals, into a small number of larger sub-intervals, or into sub-intervals of various sizes. For a given function, one may determine the best representation of this type by optimizing some figure of merit over all the possibilities. Efficient algorithms have been proposed by Wickerhauser and Coifman for searching large yet manageable collections of bases to find a function's best representation with respect to a useful cost function.

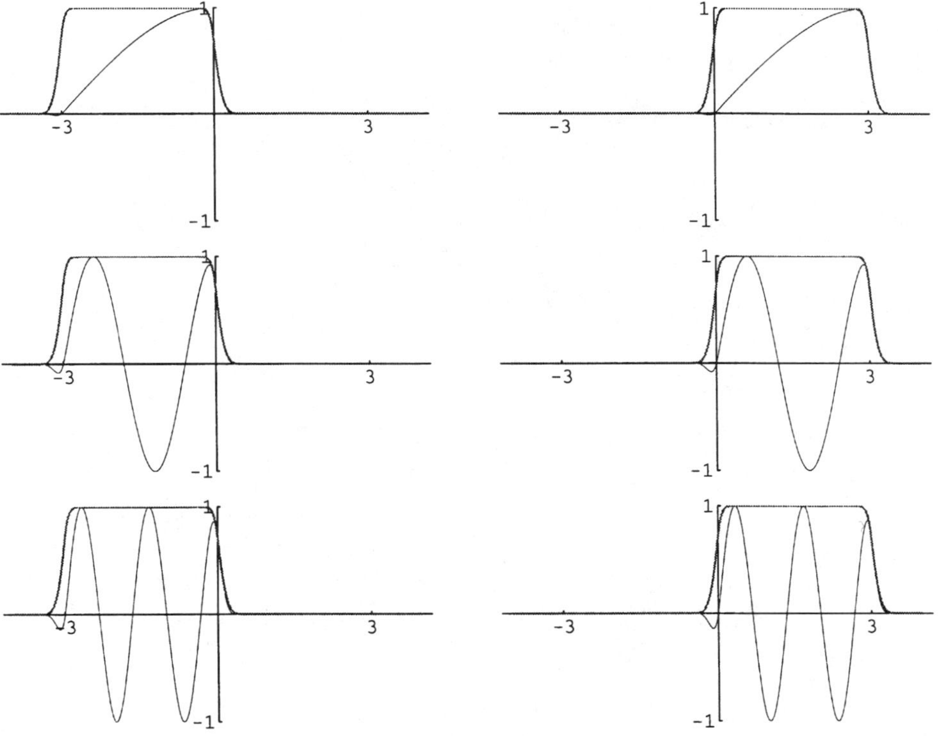

Figure 23.4 A few elements of a local trigonometric basis associated with two intervals of a partition of the line.

For example, consider the signal shown in Fig. 23-5. This signal has a transient spike and several long portions where it is essentially stationary. It seems clear that we should use a Fourier-like basis on the long intervals where the signal is stationary, and a localized basis in the vicinity of the transient. The figure shows a time-frequency representation of the signal in such a basis, obtained using a local cosine basis corresponding to the indicated partition of the time axis.

We must compare many different bases in order to find a basis like this one, which represents a given signal parsimoniously. One possibility to consider is a local cosine basis associated with the entire interval; this is essentially a Fourier basis. On the other hand, we could instead use a fine localization, in which the support of the signal is divided into 16 equal length sub-intervals, each of which has its own associated local cosine functions. Alternatively, we could merge these sub-intervals in pairs to get a partition comprised of eight double-length sub-intervals, each with its associated local cosine basis, and so forth. More generally, we should consider bases which correspond to fine localization in some regions, and coarse in others. In the example above, we ended up with a basis of this type.

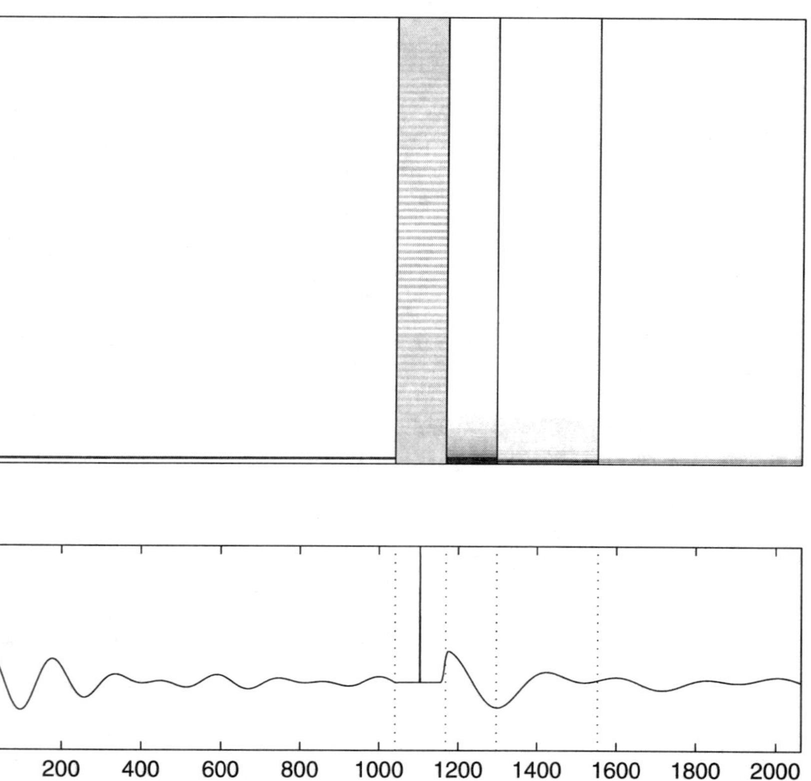

Figure 23-5   A local cosine basis time-frequency representation of a simple signal.

To expedite the search over the large collection of such bases, Coifman and his collaborators have devised an automated best-basis paradigm. To quantify what is meant by the best-basis for a given signal, we use a cost function which measures the concentration of the signal's representation in a given basis. To determine the best basis from a given collection of bases, we compare the costs of representing the given signal in the various bases of the collection; the basis with minimal cost is "best."

In order to have a fast search, Coifman and collaborators suggest comparing bases from certain highly structured collections, such as the particular collection of local trigonometric bases we now describe. We start with a partition of the signal's support interval into a power of two number of sub-intervals, $2^4 = 16$ in the case above. We then consider pair-wise mergings of these sub-intervals, merging the 16 sub-intervals into eight equal length sub-intervals as above, repeating this until all sub-intervals have been merged together to form one interval corresponding to the full support of the signal. These equal length partitions may be organized into a binary tree, as shown in Fig. 23-6, with the null partition at the root, and the 16 equal length sub-intervals as the leaves.

To find the best basis or partition of the signal, we choose a cost function and compute the cost of the signal expansion in the local cosine basis associated with the 16 sub-intervals at the bottom of the tree. That is, we compute the cost of the coefficients for the local cosine basis within each sub-interval and attach that value as the cost of that tree node. To decide whether or not such a fine localization best represents the signal, we perform the same computation for the eight sub-intervals one level up in the tree. For each of these eight parent sub-intervals, we compare its cost with the sum of its children's costs. If the cost of the parent representation wins over (is smaller than) the children, we delete the children from the tree and attach to the parent its cost. Otherwise, we do nothing and proceed to next level, comparing parent costs to children costs, repeating the procedure until we reach the top level. At this point we have a sub-tree of the original binary tree which indicates the best basis. The terminal nodes or leaves of the tree correspond to the sub-intervals whose local cosine functions make up the best basis. The darkened tree nodes indicate the terminal nodes or leaves of the sub-tree corresponding to the basis chosen for the signal in Fig. 23-5.

The binary tree search just described neglects more flexible partitions of the signal support in exchange for speed of computation. For example, we would ideally

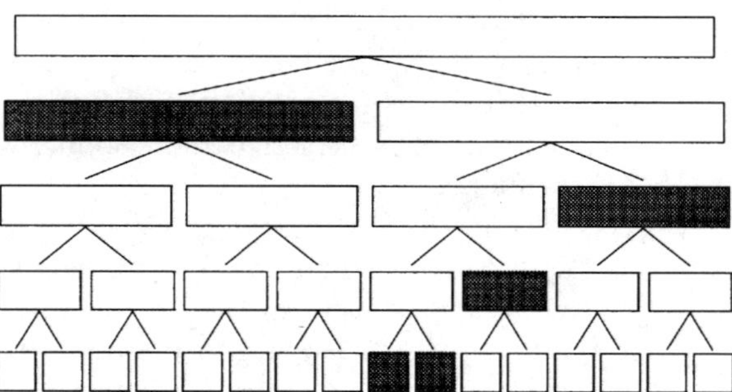

Figure 23-6 Binary tree organization of a fast best-basis search.

choose to partition the support of the signal in Fig. 5 into just three sub-intervals, i.e., a short sub-interval containing the transient and two longer sub-intervals each containing a sinusoidal waveform. The binary tree search does not include such a partition, due to the strict pair-wise merging of sub-intervals. A larger search discussed below includes such partitions at the cost of longer running time.

These same ideas can be applied to ensembles of functions to get approximate K–L bases. For instance, an approximate K–L basis for a given MRI image class can be found by minimizing the simple cost function in Eq. (23-1) over a predetermined collection of local trigonometric bases (corresponding to a collection of partitions). As we saw earlier, the cost function measures how well a given basis compresses the variability of the image class into a few significant coordinates. The resulting basis can then be used for efficient MRI encoding of new images from this class, as we will soon see.

23.4. FINDING APPROXIMATE K–L BASES

Here we review techniques for organizing large collections of local trigonometric bases into a tree data structure and efficient algorithms that may be applied to search this structure for the best basis with respect to the cost function corresponding to an ensemble of images, as given in Eq. (23-1). The result is a basis which approximates the K–L basis.

The best-basis search uses the the variance of the coefficients to evaluate the cost function of Eq. (23-1), namely the sum of the logs of the variances. This cost function is additive with respect to the partition of the interval. In particular, the cost of a basis associated with a partition of an interval into two sub-intervals equals the sum of the costs of the bases for each sub-interval separately [10].

Each local trigonometric basis corresponds to a particular partition of the interval on which the signals live. Given a specified partition (the search partition) of the domain into a power of two number of sub-intervals, we have already reviewed a fast binary tree searching strategy for finding a best partition (basis) from a certain collection of sub-partitions of the search partition. We now can find a more general tree searching strategy to find a best partition over the entire collection of sub-partitions of the search partition. In Fig. 23-7(a) we show such a tree where the search partition splits the domain into four sub-intervals.

A particular partition of the interval corresponds to a path from the root to a leaf. The first vertex below the root in one of these paths shows the position of the left-most point of the corresponding partition as a solid vertical line. Similarly the n-th vertex shows the position of the n-th partition point. The dashed lines indicate the positions of possible partition points, whose inclusion remains to be decided at a particular level of the search.

To find the best partition, we search for a path which minimizes the specified cost function. The first row below the root of the tree in Fig. 23-7(a) shows the four possible positions for the left-most partition point. The two vertices on the left of the first row correspond to the trivial partition and to a partition comprised of two sub-intervals, respectively. These partitions are completely specified at this

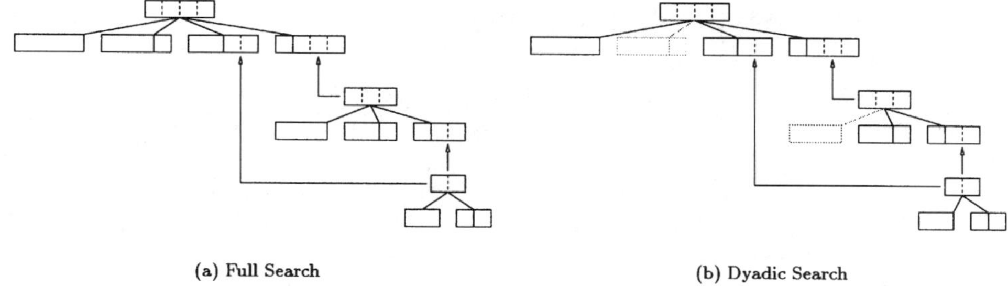

Figure 23-7   A more extensive dynamic programming search for the best sub-partition of a four-interval search partition. (a) Starting with the smallest, rightmost sub-problems we use their solutions repeatedly in solving the larger sub-problems. (b) Reducing the search tree to a set of "dyadic" sub-partitions.

level. The two vertices on the right of this row have additional possible partition points and require some further decisions about whether to include or exclude these points. In each of these cases we need only know the best splitting of the sub-interval containing those candidate partition points, because of the additivity of the cost function. In other words, we can reduce to the sub-problems of finding a best partition for the right-most sub-intervals containing two or three partition points. Then we can take those best partitions together with the indicated left-most partition point to make a list of four partitions. Now we just evaluate the cost function on the basis corresponding to each of these partitions and take the one with minimal cost as our best basis.

In practice, we solve the smallest sub-problems first and use those solutions in solving the larger sub-problems. In terms of the tree, we start at the lower right and work our way up to the root. We begin with the choice of the best partition of the interval of length 2, shown as the tree of size 2 in the lower right. There are only two choices, so we compute their costs and choose the winner. Using this result for the best partition for length 2, we need only consider three possibilities for the best partition of length 3. When this is settled, we can use the best partitions for length 2 and 3 sub-intervals to get the costs of the four remaining possibilities for partitioning the whole interval, and finally choose the best.

This applies in general, and we use the sub-problems of size $1, \ldots, n$ to solve the sub-problem of size $n + 1$. The total complexity is $O(N^3)$ where N is the maximum possible number of partition points; that is, the size of the search partition. By eliminating some of the possibilities for positions of the left-most partition points, faster searches are possible. As an example, one may restrict the search to partitions whose intervals contain a power of two of the intervals from the original search partition. In this case, the search complexity reduces to $O(N^2)$. The vertices and edges which this removes are shown ghosted out in Fig. 23-7(b). One may further reduce this to the binary tree method considered by Coifman and Wickerhauser by requiring that the midpoint of certain sub-intervals occur in any nonempty partition.

We now consider how these approximate K–L bases and others may be used for encoding in MRI.

23.5. ADAPTED WAVEFORM ENCODING IN MRI

We have studied alternatives to the standard MRI techniques we discussed in section 23.2. These new approaches [11, 15, 16] modify the encoding to permit one to apply some of the new bases discussed above. For example we can replace the phase encoding in Eq. (23-2) by an encoding with a basis of spatial amplitude profiles on the spin system. These may be chosen to suit various imaging tasks. We refer to this concept generically as "adapted waveform encoding." We will be particularly interested in choosing bases of profiles adapted to a particular imaging task.

In adapted waveform encoding, phase encoding is replaced by a second selective excitation signal which produces a specially shaped amplitude profile along the encoded axis; some details of the implementation are discussed in Appendix A of this chapter. The signal produced by this method measures the inner product of the excited profile and the hydrogen spin density along the encoded axis:

$$S_j(t) = \int dx \int dy \, \rho(x, y)\psi_j(y)e^{-i\alpha xt} \qquad (23\text{-}4)$$

We have chosen to encode the y-axis with an amplitude profile given by the j-th element of a basis $\{\psi_j\}_{j=1}^N$. This replaces the j-th harmonic of the Fourier basis which we would use in standard phase-encoded MRI. Note that frequency encoding of the x-axis permits us to obtain the complete set of Fourier measurements for that axis.

From these measurements, we can then reconstruct the projection of the spin density onto the encoded axis using the usual sort of expansion:

$$\mathbf{I}(k, l) = \sum_{u,v} S_u(v\Delta t)\psi_u(l\Delta y)e^{2\pi iuk/n}$$

which gives the image as a finite rectangular pixel array.

A Hadamard basis was first used [17] to provide profiles for encoding. Since then, wavelet packet bases have been proposed and implemented [15, 16, 18, 19]. In addition to these, one may use profiles adapted to prior knowledge of the imaging task at hand. Later in this chapter, we consider the use of K–L and approximate K–L profiles for this purpose.

In all of these situations we find some important practical considerations and constraints. For instance, we shall see that there is a practical limit to the sharpness of the profiles used for adapted waveform encoding. We are also bound by a fundamental trade-off between the signal-to-noise ratio (SNR) and imaging speed [11, 16]. SNR considerations are serious in MRI: the signals tend to be low in strength, and the primary noise source in modern scanners is thermal noise from sources other than the hardware (such as the patient), which cannot be removed.

The signal in an imaging experiment is proportional to the number of nuclei contributing to the signal. Signal strength is therefore determined in part by the size of the excited region of nuclei which provide the signal. It is also determined by the amplitude of the excitation profile, which cannot exceed a certain limit. The noise in MRI may be modeled as white noise within the sampled bandwidth. The larger the bandwidth, the greater the noise power. In adapted waveform encoding, this noise power is independent of the amplitude profile used for encoding; it is the signal strength which is strongly dependent on the profile. Since the maximum possible amplitude of signal from a given region is fixed, the signal strength from more localized profiles, such as some of those found in wavelet bases, will be smaller than from a global profile, like the harmonics in phase encoding.

In the remainder of this section, we give a brief discussion of some of the possible choices for the encoding profiles.

23.5.1 Wavelet Encoding

In [15] we described some of the benefits associated with choosing the set of excited profiles to be an orthogonal wavelet basis. This technique has since been studied by several other research groups [18, 19]. Some of the implementation consideration including analysis of SNR may be found in [11, 16]

Wavelet encoding produces an (idealized) output signal which is the wavelet transform of the spin distribution $\rho(x, y)$ measured by exciting nuclei with a sequence of RF pulses at correct dilations and translations:

$$S_{j,k}(t) = \int dx \int dy \, \rho(x, y)\psi_{j,k}(y)e^{-i\gamma G_x xt} \tag{23-5}$$

with

$$\psi_{j,k}(y) = \psi\left(\frac{x}{2^j} - k\right)$$

and $(j, k) \in \mathbb{Z}^2$. Note the lack of the usual normalizing factor for the scaled wavelets. This reflects the fact that the peak amplitude of the excitation profile is limited in MRI; we cannot exceed this limit at will. In order to obtain as much signal as possible, all scales of wavelets are normalized to this peak amplitude, and the energy normalization is applied in the reconstruction. This can lead to low SNRs [16]; intuitively, this is clear from the fact that amplitude profiles corresponding to the fine-scale wavelets have localized support, and so produce less signal than coarse-scale profiles which excite a larger group of nuclei.

Nevertheless, wavelet encoding works reasonably well in practice, and the SNR figure of merit proves to be unduly pessimistic in many circumstances. This is because the wavelet transform tends to concentrate energy in coefficients corresponding to edges. These features tend to remain above the noise floor [16]; the resulting image can be very useful, particularly for locating (oriented) edges.

In our previous work we noted the following advantages of wavelet encoding over phase encoding:

- Reduced ringing caused by partial volume effects (Gibbs ringing).
- Immunity to motion artifacts. Motion of the object over the course of standard Fourier-based image acquisition causes small inconsistencies in the measurement of the spatial Fourier transform. Traditional Fourier transform reconstruction spreads the misregistered signal all the way across the image; the basis functions of the expansion are not local, so the effect of the error is global. This is much reduced with wavelet encoding due to the localization of the basis elements.

- Novel time sharing or scheduling possibilities for certain types of examinations. Since each wavelet amplitude profile excites only a portion of the field of view, we schedule the sequence of excitations so that successive excitations involve spatially nonoverlapping bands. This means that we can use a very short resting time between excitations (TR) and yet any given spin will experience a much longer *effective* resting time, TR_{eff}, between excitations.

This last result is analogous to the multislice time sharing for volumetric imaging, discussed in Appendix A. However, the scheduling problem in this case is considerably more interesting because the excited regions have varying sizes, corresponding to the various scales. Our previous work constructed optimal schedules, so that for an $N \times N$ resolution Haar wavelet encoded image

$$TR_{\text{eff}} = \left[\frac{N-2}{\log_2 N - 1}\right] TR$$

was possible. We found that a speedup of about a factor of 3 was possible for images in which it was necessary to obtain a few slices with T_1 effects de-emphasized [15, 20].

We have used wavelet encoding schemes of various types with a standard GE Signa 1.5 T scanner. Figure 23-8 shows an example image produced with this method. This figure shows an actual wavelet encoded MR image of a calibration phantom obtained at one scale of wavelet encoding. The phantom itself is a large block of Plexiglas with various other materials (water doped with copper sulfate, air) embedded in it to produce arrays of shapes in various sizes and orientations. This is

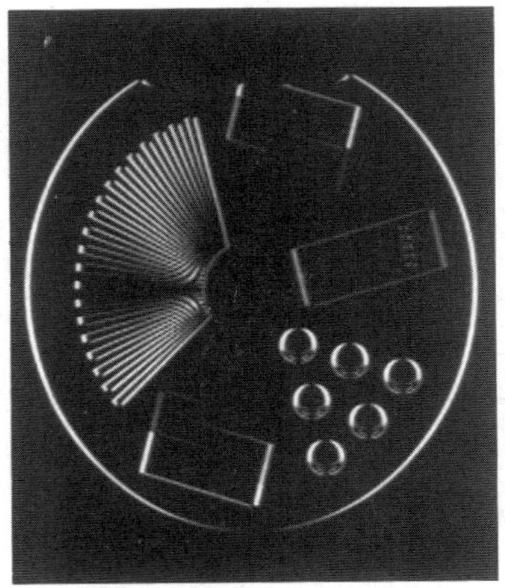

(a) Fine scale encoding only

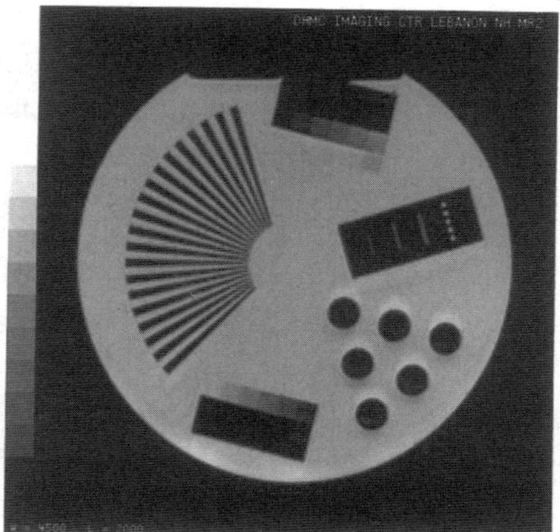

(b) Full encoding

Figure 23-8 Imaging an MR calibration phantom. On the left, the image formed from only the finest scale of wavelet encoding. This is essentially an image using an edge-sensitive RF pulse. No signal is produced from homogeneous material. The signal is produced from vertical edges. This is useful for oriented edge detection. On the right, full data reconstruction of the phantom.

placed in the scanner and imaged for testing purposes. In the case of Fig. 23-8, notice that wavelet encoding preferentially images edges oriented along the wavelet encoded direction. This feature of wavelet encoding should prove useful in several applications, such as angiography and heart wall tracking as described in [11, 16, 21].

23.5.2 More General Bases for Encoding

By utilizing more general RF pulses and gradient sequences, wavelet encoding generalizes to "adapted waveform encoding," which measures inner products of the spin density against functions from bases other than a wavelet basis. The idea here is to choose the basis to suit the given imaging task and any prior information available. In particular, we have found a great increase in flexibility by employing wavelet packet and local trigonometric libraries, as described in the papers of Coifman and Wickerhauser [22].

The term "library" in this context refers to a large collection of bases linked together by some defining common feature. In the familiar case of bases for time domain signals, a given basis from a library can be interpreted as providing a particular tiling of the time-frequency plane. Some of the bases offer better localization in time at the cost of poorer localization in frequency. Some of the bases make this trade-off in a flexible fashion across the time-frequency plane. We will be particularly concerned in this chapter with the application of the local trigonometric bases in adapted waveform encoding.

The existence of libraries of smooth functions proves to be essential for practical implementation of waveform encoding. In practice, there are limitations imposed by the hardware and physics of the spin system which make it impossible to excite accurate profiles with sharp edges. In effect, there is a limit on the bandwidth of the profiles we can use for our bases; further details may be found in the Appendix A. This limitation will play an important role in our study.

23.5.3 Choosing a Basis for Fast MRI Encoding

Recently, we have studied bases of profiles adapted to statistical regularities in a particular class of images, looking for representations which would capture most of the variability of the images within the first few coefficients. Our images can then be represented parsimoniously, in the sense that truncated image expansions corresponding to a reduced set of encodes would have minimal expected mean square error. As each encode incurs a time cost, the reduction of encodes translates directly to reduced imaging time. A natural choice is the K–L basis for the image class, as indicated in section 23.1 above. We also considered alternative approximate K–L bases, obtained by restricting the minimization of the K–L cost function to a library of local trigonometric bases associated with a family of partitions of the encoded interval, as described in section 23.4. We now turn to a detailed discussion of this approach.

23.6. K–L BASES IN MRI

In this section and the next we show how prior knowledge about an imaging task can be used to reduce the data required to form images. This work was motivated by several recent studies in the MRI literature. In one of these, Cao and Levin [23] considered the problem of finding an optimum set of phase encodes to estimate the first elements of the K–L basis from a training set of related images. This reduced set of phase encodes was used to acquire an approximate image of a new object with a reduced data set. Unfortunately, the computation of the optimal set of phase encodes could be quite computationally expensive. Cao and Levin also suggested that the direct measurement of the K–L coefficients could improve the performance of their technique. The first elements of the K–L basis have also been used by Zientara et al. to estimate changes over time in the repeated acquisitions of a given subject [24].

We have studied the feasibility of direct measurement of the K–L coefficients by selective excitation of K–L basis functions in adaptive waveform encoding. We took a typical set of images and measured the mean squared error in the reconstruction of a test image as a function of the number of K–L coefficients used. The results, given below, demonstrate that this could, in principle, greatly reduce the MRI imaging time.

This approach shares the technical limitations of any adapted waveform encoding as discussed in the last section. In particular, the accuracy of the profiles excited by the selective RF pulses is limited by bandwidth constraints. This makes excitation of the K–L basis functions difficult and of limited accuracy. In addition, the SNR of the images acquired with the K–L basis suffers compared to that in standard images. The reduction in SNR may be understood from analysis presented in previous studies, and summarized below. We will show in section 23.7 that these problems are reduced greatly through the use of approximate K–L bases constructed from local trigonometric functions.

23.6.1 K–L Waveform Encoding

The adapted waveform encoding in Eq. (23-4) shows that one axis is frequency encoded, while we must choose a basis for encoding the other axis. Assume that we are interested in imaging a planar slice of our subject. The desired result is an $m \times m$ image $\mathbf{I}$ (in matrix form) of that slice. We may model the data acquisition as measurements of the projections of the image matrix onto tensor product basis elements $e_k \otimes \psi_l = \psi_l e_k^t$ $1 \leq k, l \leq m$, where e_k is the Fourier basis vector with entries $[e_k]_l = 1/\sqrt{m}\, e^{2\pi i k l/m}$, $1 \leq l \leq m$ correspond to the frequency encoding of one direction, and the ψ_l are vectors of a yet to be determined basis of profiles for encoding of the other direction. Each excitation and signal measurement has the form of Eq. (23-4) and records the projection of the image onto one profile in y and the full Fourier basis in x. Consequently, we project onto the basis elements $e_k \otimes \psi_l$, for a fixed l and $k = 1, \ldots, m$. What is the best basis of profiles ψ_l to use for this purpose if time constraints require a limited number of acquisitions? We use the minimum expected truncation error criterion. Our truncated estimator of $\mathbf{I}$ is

$$\widehat{\mathbf{I}}_t = \sum_{l=1}^{t}\sum_{k=1}^{m} c_{k,l}\, e_k \otimes \psi_l + \sum_{l>t}^{m}\sum_{k=1}^{m} \theta_{k,l}\, e_k \otimes \psi_l$$

where the $c_{k,l} = \langle \mathbf{I}, e_k \otimes \psi_l \rangle = \psi^*_{\,l}\mathbf{I}\bar{e}_k$ are the coefficients of the image, and $\theta_{k,l}$ those of the average image $E(\mathbf{I})$. So we have for the error energy

$$\left\| \mathbf{I} - \widehat{\mathbf{I}}_t \right\|^2 = \left\| \sum_{l>t}^{m}\sum_{k=1}^{m}(c_{k,l} - \theta_{k,l})e_k \otimes \psi_l \right\|^2$$

$$= \sum_{l>t}^{m}\sum_{k=1}^{m} |c_{k,l} - \theta_{k,l}|^2$$

$$= \sum_{l>t}^{m}\sum_{k=1}^{m} |\psi^*_l\mathbf{I}\bar{e}_k - \psi^*_l E(\mathbf{I})\bar{e}_k|^2$$

$$= \sum_{l>t}^{m}\sum_{k=1}^{m} |\psi^*_l(\mathbf{I} - E(\mathbf{I}))\bar{e}_k|^2$$

$$= \sum_{l>t}^{m}\sum_{k=1}^{m} \psi^*_l(\mathbf{I} - E(\mathbf{I}))\bar{e}_k e_k^t(\mathbf{I} - E(\mathbf{I}))^*\psi_l$$

$$= \sum_{l>t}^{m} \psi^*_l(\mathbf{I} - E(\mathbf{I}))\left(\sum_{k-1}^{m} \bar{e}_k e_k^t\right)(\mathbf{I} - E(\mathbf{I}))^*\psi_l$$

$$= \sum_{l>t}^{m} \psi^*_l(\mathbf{I} - E(\mathbf{I}))(\mathbf{I} - E(\mathbf{I}))^*\psi_l$$

since for any orthonormal basis $\{e_k\}_{k=1}^m$ of m-space

$$\sum_{k=1}^{m} e_k e_k^* = \mathrm{Id}$$

the $m \times m$ identity matrix. So

$$E\left(\left\| \mathbf{I} - \widehat{\mathbf{I}}_t \right\|^2\right) = \sum_{l>t}^{m} \psi^*_l \Sigma_{\mathcal{I}} \psi_l$$

where $\Sigma_{\mathcal{I}} = E((\mathbf{I} - E(\mathbf{I}))(\mathbf{I} - E(\mathbf{I}))^*)$ is the covariance of the row vectors of the images in $\mathcal{I}$. We minimize this for each t by a clever choice of $\{\psi_l\}_{l=1}^m$. A Lagrange multiplier argument shows that such ψ_l satisfy $\Sigma_{\mathcal{I}}\psi_l = \lambda_l\psi_l$, i.e., are eigenvectors of the matrix.

23.6.2 Simulation Results

We applied the encoding described above, using excitation profiles from K–L bases associated with standard image classes. The goal was to reduce the expected number of encodes required to form a decent image, and therefore reduce image acquisition time.

We considered several image classes; each one consisted of a collection of transaxial head images taken at a given height in the skull. All images came from standard scans in daily clinical studies. Only studies with no gross pathology were used. The T_2 weighted images from eight studies were extracted and used as a training set for determining the K–L basis. See Fig. 23-1 for a look at one of the training sets. A ninth image from the class was used as a test image to measure the error in the reconstruction.

The eight images of the head used to generate the (empirical) K–L basis for each class were chosen to reflect the variation in positioning and in anatomy likely to be seen in clinical practice. Of course, a much larger data set would be required to generate a basis for the variety of pathologies seen in clinical practice.

The training matrix, X, contains the eight normalized and detrended training images as 256×256 submatrices,

$$\mathbf{X} = (\mathbf{I}_1 - E(\mathbf{I}) \mid \mathbf{I}_2 - E(\mathbf{I}) \mid \ldots \mid \mathbf{I}_8 - E(\mathbf{I}))$$

Each image comprises a set of 256 columns in the 256×2048 image matrix. X can also be viewed as a collection of 2048 vectors; each vector represents a line in the image that will be encoded with the K–L basis. The K–L basis vectors are the eigenvectors of XX^t. The eigenvectors are ordered so the associated eigenvalues decrease in size.

The K–L profiles $\{\psi_j\}_{j=1}^{256}$ obtained by this process for a given image class are used as amplitude profiles in MRI for imaging test objects from that same class. The profiles are excited by a selective RF pulse in the y-direction; the acquired data are frequency encoded in the x-direction to produce a signal as described in section 23.5. All the lines in the frequency-encoded direction are acquired at once and are all amplitude scaled with the same K–L basis vector. One frequency-encoded data string is acquired for each excitation; the inverse Fourier transform of the data string is a single K–L coefficient for all the lines in the image.

The form of the sample row covariance matrix reflects the fact that one of the two dimensions in a slice can be efficiently frequency encoded in MRI, so we are only concerned with the covariance in the second dimension. We are essentially performing a one-dimensional principal component analysis.

We applied this process in simulation to a number of the standard image classes described above. The errors in the estimates of test images for the various classes were computed as functions of the number of coefficients. Figure 23-9 plots the error energy as a percentage of the total energy in the image for the best and worst classes. For the test image of the class of head images taken at level 14, where the image plane is superior to the ventricles, the error energy is below 1% with 45 coefficients and below 0.5% with 62 coefficients. For all image planes, the error energy is below 1% with 100 coefficients and below 0.5% with 127 coefficients.

The error in the expansion of the image in the K–L basis is dominated by how closely the training set matches the image being acquired. The variation in the training set should be large enough to include all the variation likely to be encountered in practice. Therefore, the practical training set should be relatively large which can lead to poor convergence. However, better convergence can be obtained by weighting the images and recalculating another K–L basis after each acquisition. Images from the training set that are like the acquired image should have greater weight than those that are unlike it. An iterative adaptive process results. For more details, see [25]. This process yields improved results for large, diverse training sets.

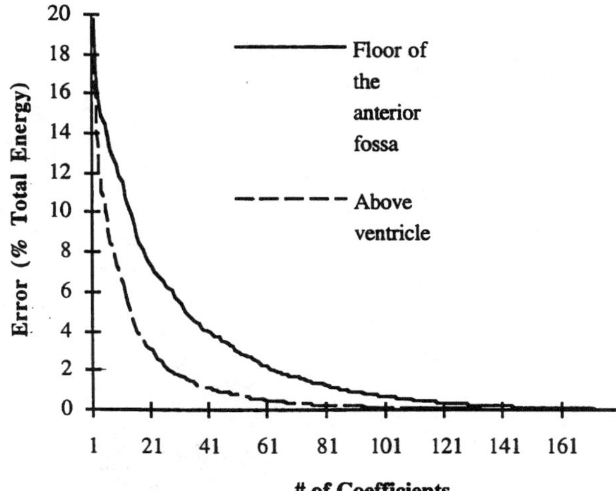

Figure 23-9 Error energy as a function of the number of coefficients used in the K–L encoding of test images from the best and worst image classes.

For the small training sets we considered, the process only yields a one to two percent decrease in root mean square (rms) error over the first eight coefficients.

23.6.3 Implementation and Practical Limitations of K–L Encoding

Standard methods [11, 16] may be applied to show that SNR in images acquired with a K–L basis is lower than in conventionally acquired images, even if a complete set of encodes is used. Basically, this reduction is caused by the fact that the encoding profiles have a fixed peak amplitude. Therefore, the use of encoding profiles in Eq. (23-4) whose amplitudes are well below peak magnitude on a large portion of the object will produce a measured signal of smaller energy than would globally supported profiles like the sinusoids used in phase encoding. This inability to excite energy-normalized basis vectors is combined with the presence of additive noise of fixed power. Because the amplitude profiles are not normalized, each projection of the spin density onto a profile must be multiplied by the appropriate factor for use in the reconstruction. The noise contribution in each measurement is multiplied by the same factor, leading to a reduction in the SNR.

Consequently, an amplitude profile which excites the magnetization at a magnitude near the peak across the entire slice has a better SNR than excitations in which the peak magnitudes are localized. The more localized the peak magnitudes of the basis function, the larger the reduction in SNR. The complex exponential basis functions used in conventional phase-encoded imaging have optimum SNR, while wavelet bases are more localized, leading to a lower average SNR performance. The K–L bases are between the two; the reduction in SNR we have seen for K–L bases averages to a factor of around 3.5.

The second implementation factor that should be considered is the length of RF pulses required for exciting basis functions with discontinuities. As indicated in

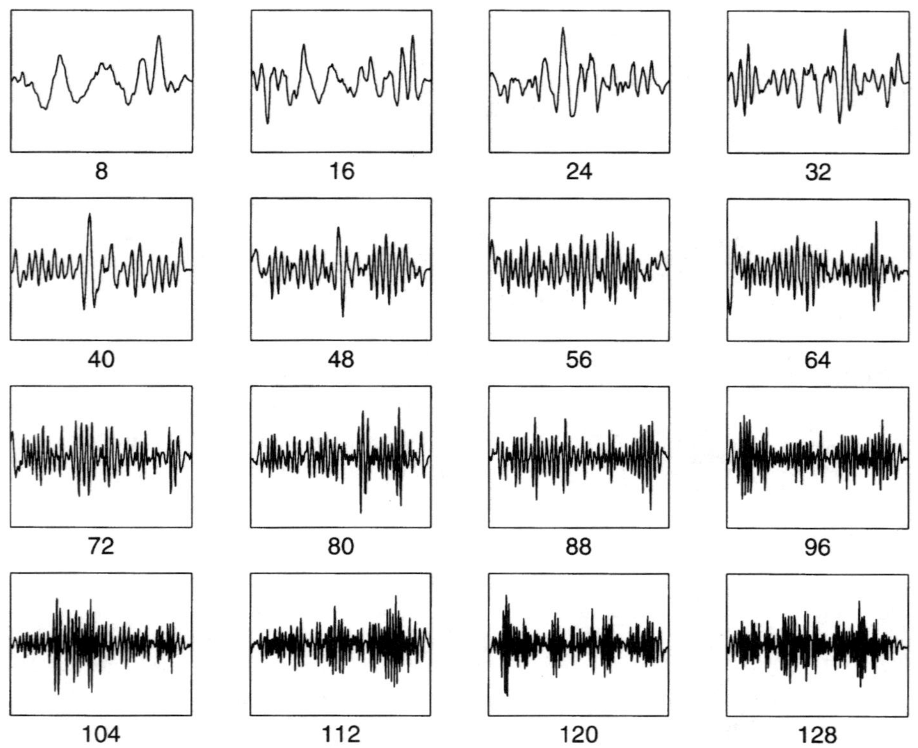

Figure 23-10 Example K–L basis elements obtained from one of the image classes.
Note the extremely rough profiles as the index increases.

Appendix A, the profile excited by an RF pulse is approximately the Fourier transform of that RF pulse. Therefore, long RF pulses with wide bandwidths are required to excite profiles with sharp edges. The K–L basis functions we have seen have many sharp edges, as seen in Fig. 23-10. In practice, the edges of these basis functions will not be present in the profile actually excited, due to limitations on RF pulse length. These band-limited profiles increase the error in the reconstructed test images significantly.

23.7. APPROXIMATE K–L BASES IN MRI

As we have seen, encoding with the K–L profiles worked quite well *in simulation*, promising to cut the number of encodes in half with very low error. Practical implementation was another matter, as it is rather difficult to excite accurate profiles from the K–L basis on the spin system. As indicated above, the K–L basis functions we have seen have many sharp edges which will not be present in the profile actually excited due to restricted bandwidth of the excitation signal. These band-limited profiles are not very close to the desired K–L profiles, which increases the error in the reconstructed test images significantly.

In light of these technical considerations, we studied the use of *approximate* K–L bases chosen from a library of localized trigonometric bases, as discussed in section 23.4. Encoding with these bases sidesteps some of the technical limitations associated with K–L encoding. In particular, the basis functions have a simple envelope with a linear phase. This offers two big advantages. First, the basis functions do not have sharp edges so they can be accurately excited in a real scanner. Secondly, the linear phase can be obtained with phase-encoding gradients. This allows the same techniques developed to reduce the number of excitations in phase encoding to be applied. In particular, several phases across the same envelope can be acquired from one excitation, leading to a further reduction of imaging time.

23.7.1 Approximate K–L Waveform Encoding

For this study, we considered the same problem with the same objectives as described in section 23.6. Once again, we use the simple adapted waveform encoding described in Eq. (23-4) with one axis frequency encoded. As in K–L encoding, we must now choose a basis for encoding the other axis so as to obtain the best image with as few encodes as possible. The difference is that we now constrain this basis to be a local trigonometric basis.

The algorithm for finding best bases discussed in section 23.4 can be used on the same training images considered in section 23.6 to obtain a localized trigonometric basis that approximates the compression performance of the K–L basis. The interval to be encoded is initially partitioned into 16 intervals of equal length; the algorithm searches the library of local cosine bases associated with the sub-partitions of this initial partition.

The resulting best local cosine basis is shown as a density plot in Fig. 23-11. The basis elements are arranged from left to right in order of decreasing importance. Figure 23-12 compares the reconstruction of the test image using the full set (256) of encodes from the basis to a sequence of reconstructions based on reduced data, each labeled by the number of encodes used. Note that we get good images with a greatly reduced data set. This translates directly into reduced imaging time.

For our purposes, the important property is that the best basis allows imaging with a reduced number of measurements without introducing a large error. Figure 23-13 compares the performances of the best basis and the K–L basis for the test image. We conclude that the approximate basis reduces encoding requirements almost as well as the K–L basis, and has the advantage of a simpler implementation.

We find these experimental results encouraging, and are studying larger data sets. The rapidity with which general image features begin to appear suggests an application of this technique to progressive imaging for scout images. We are also experimenting with different cost functions in the best-basis search to try to obtain bases that will help discriminate among tissue classes.

23.7.2 Application to Dynamical Imaging

The technique we have just outlined admits many generalizations and applications. One of these may be dynamic localization for dealing with motion in MRI. As

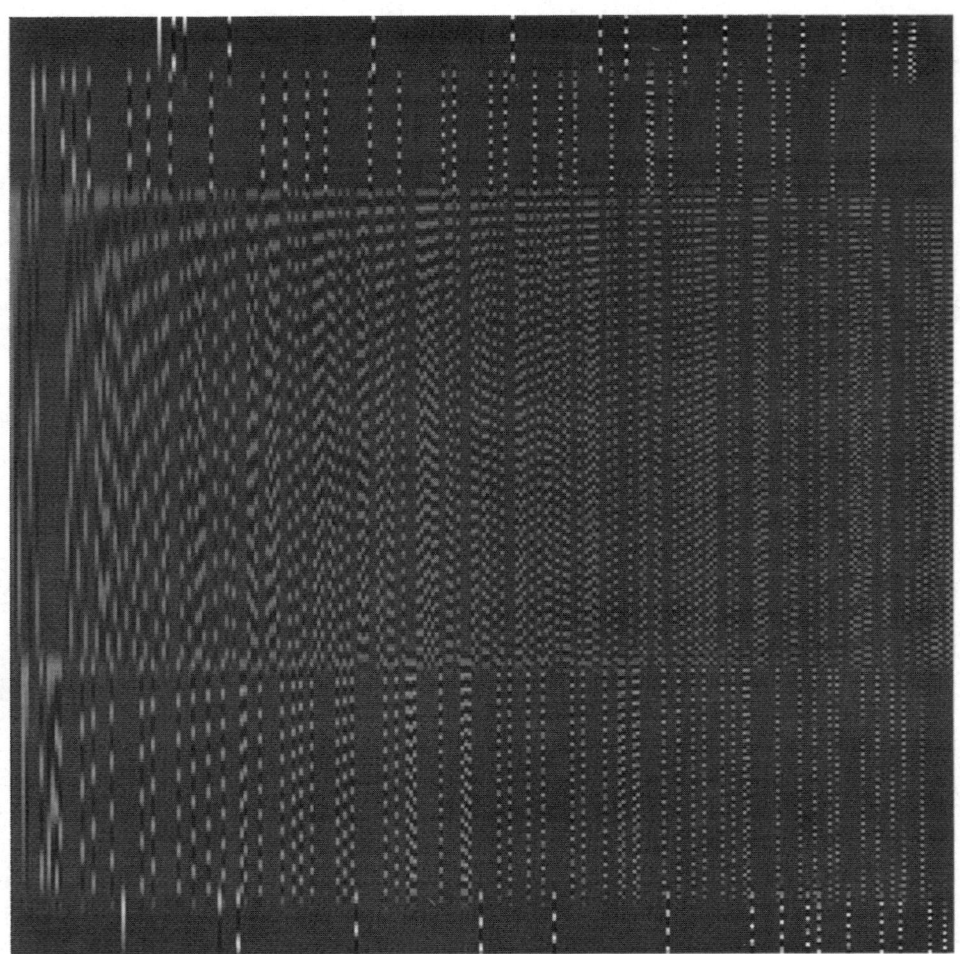

Figure 23-11 A representation of the best local trigonometric basis approximating the K–L basis for our image class. The individual basis elements are depicted in gray-scale along the vertical direction; they are seen as localized oscillations. The horizontal axis enumerates the different basis elements in order of decreasing variance, from left to right.

we have indicated earlier, motion can be a real problem in standard phase-encoded MRI, with errors due to even localized motion during the measurements being reflected in global reconstruction artifacts. We find that the techniques presented earlier may be modified to the task of adapting the encoding basis to localized changes in otherwise static anatomy.

The reduced encoding requirements attainable using approximate K–L bases enable rapid imaging of a changing object, a requirement for functional imaging of the brain or other organs. A local trigonometric approximate K–L basis adapted to a functional sequence can provide high-resolution imaging with a small number of coefficients, since the basis can localize to the region of interest. Moreover, the choice

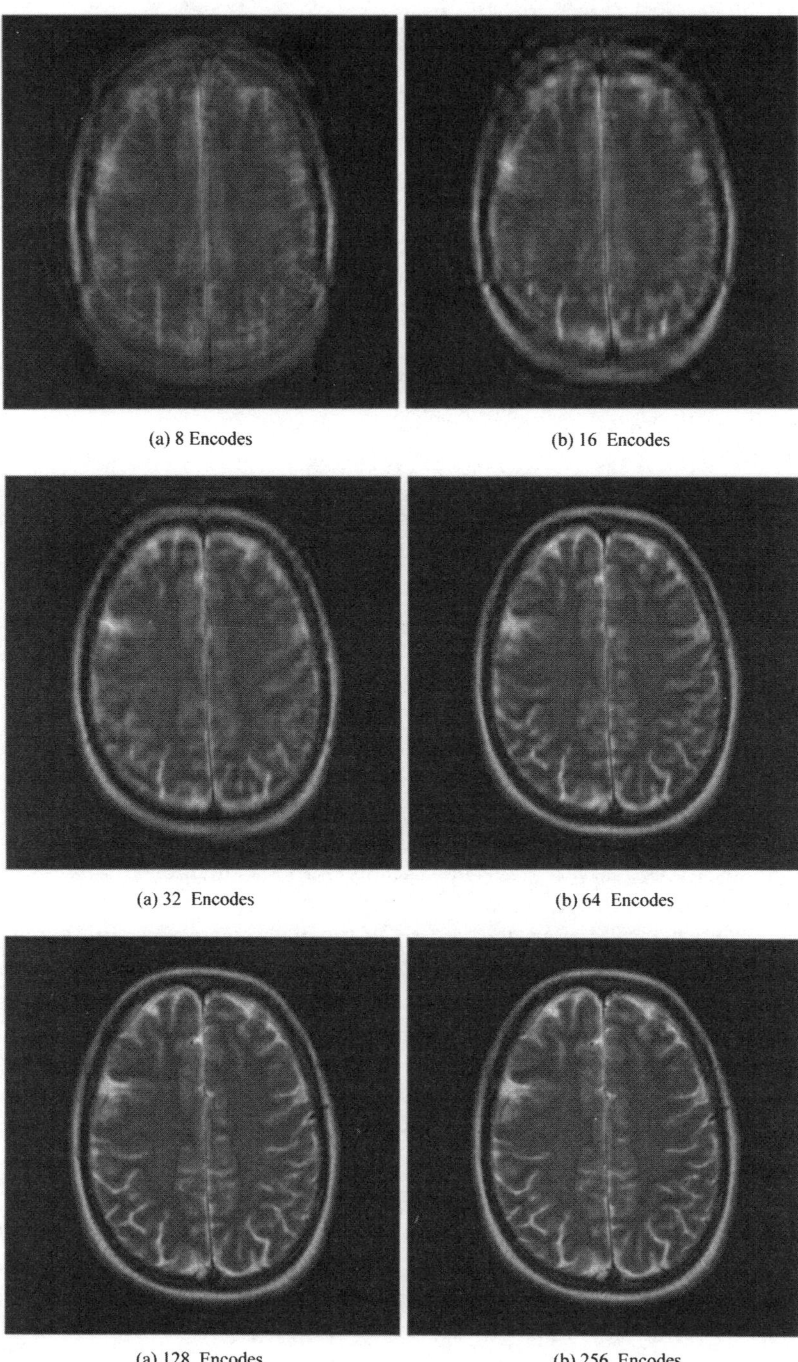

Figure 23-12   Reconstructions using partial measurement data from the approximate K–L basis. Each image is labeled with the number of encodes required for its reconstruction.

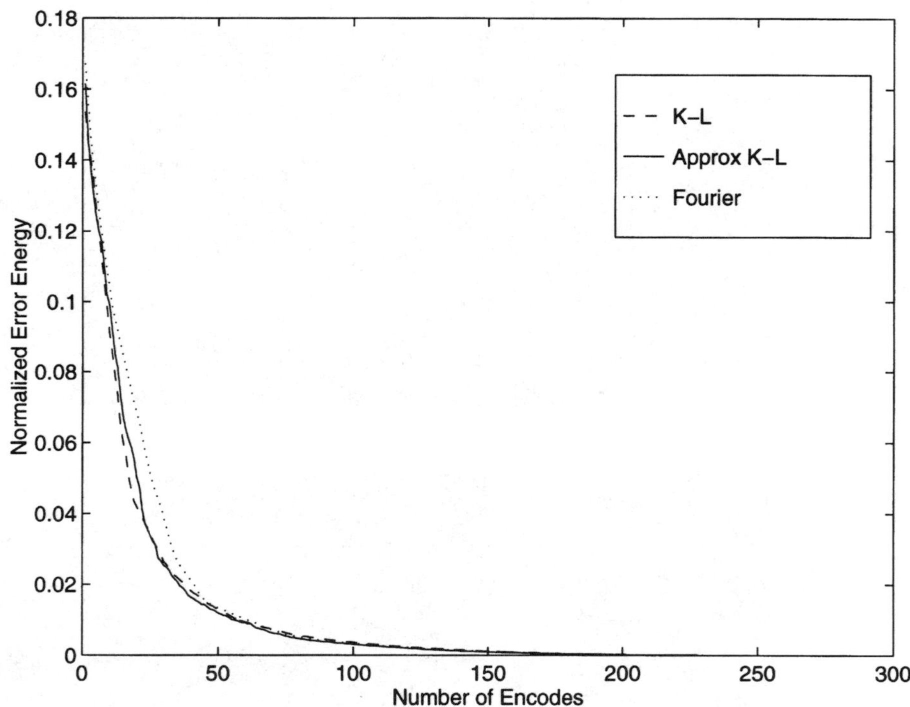

Figure 23-13   Compression performance for K–L and best-basis approximate K–L
encoding of a test image. The curves indicate the error in reconstructing the test image from a reduced number of encodes.

of basis could adaptively track changes in the image, a capability of paramount
importance as the features of interest cannot typically be known in advance.

One example of an application of this type occurs in a simple functional imaging
experiment, in which a volunteer's brain is imaged continuously while he or she is
performing certain simple cognitive tasks. It has been observed that increased blood
flow in certain portions of the brain during this task causes an increase in the
intensity in a standard image of that region. In this way researchers feel they may
be able to localize certain cognitive functions to specific areas of the brain. In order
to image the changes, the data required to build the images must be obtained fairly
rapidly.

We indicate some of the possible advantages of adapted waveform encoding
over standard Fourier encoding by the application of a local trigonometric approximate K–L basis to a simple simulated dynamical image sequence. This sequence was
obtained from a transaxial brain image by changing the intensity of the pixels in two
of the sulci in the image. The intensity of these regions was periodically intensified in
a linear manner, with the sulci alternately fading in and out.

The period consisted of 192 steps, with one frequency-encoded measurement per
step. We compared the use of phase encoding for the y-dimension to approximate
K–L encoded measurements. The 48 most important basis elements for each of these

two bases were repeatedly measured to make the estimates. The importance of a basis element was determined by its variance over the ensemble of time-varying phantoms. Image updates were made after every eight encodes.

This simulation proved very encouraging, as seen in Fig. 23-14. The approximate K–L basis estimates closely tracked the actual sequence. The cycle of fading sulci was clearly visible, with exceptional detail. In comparison, the corresponding experiment using the Fourier basis was less than satisfactory. The Fourier basis did not clearly indicate the cycle of fading and brightening sulci. At times, it was difficult to discern if any change to the image had taken place. A comparison of the error energy for the two sequence estimates underscored our observations. The Fourier basis error energy was between 3 to 15 times higher than the approximate K–L basis error energy. The

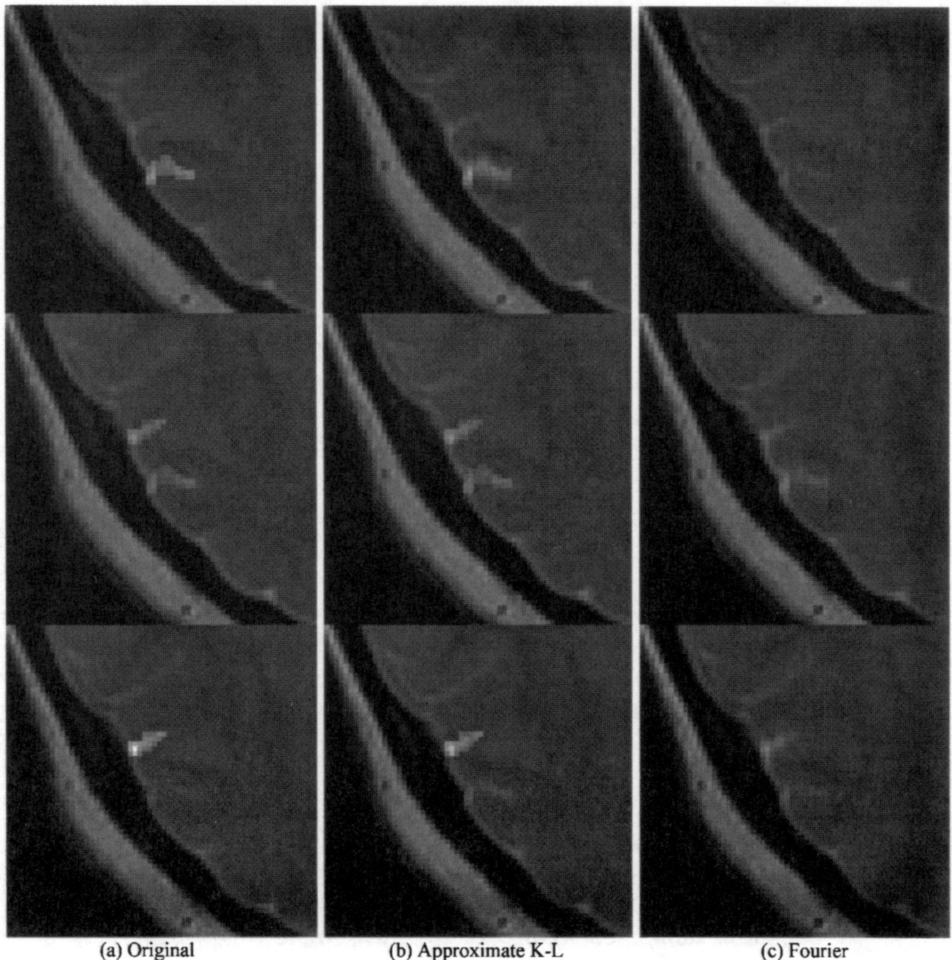

(a) Original   (b) Approximate K-L   (c) Fourier

Figure 23-14   Comparison of reconstructions based on the approximate K–L basis
and Fourier basis for the dynamic imaging sequence.

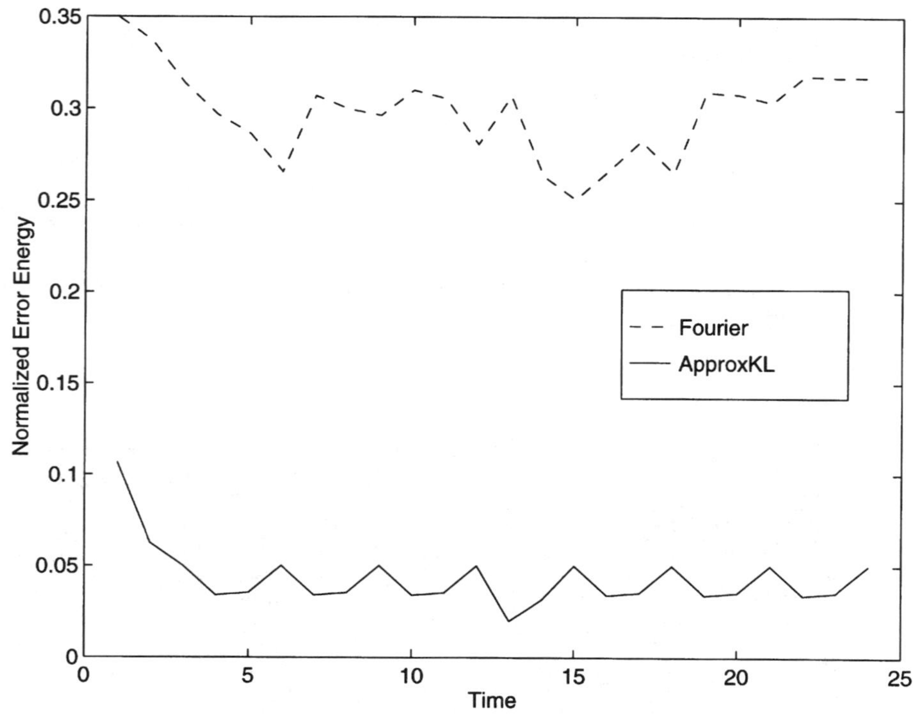

Figure 23-15   Error energy for the dynamical image sequence.

average Fourier basis error energy was about 7.5 times that of the approximate K–L basis. The error energy curve as a function of time is shown in Fig. 23-15.

23.7.3 Two-Dimensional Approximate K–L Encoding

Next, we consider an extension of our previous results to two-dimensional (2-D) approximate K–L encoding. This is useful if the frequency encoding is required for a third dimension, as in volumetric imaging, or if there is no frequency encoding at all, as in MR phosphorus spectroscopy. We present some simulation results on two-dimensional approximate K–L encoding in this section. For completeness, some of the salient issues of MR spectroscopy and chemical shift imaging may be found in Appendix A.

We first computed the one-dimensional (1-D) approximate K–L basis for both x- and y-axes as described above in section 23.6 for the level 7 training set. We then took the tensor product of the two 1-D approximate K–L bases as an approximate 2-D K–L basis. The basis elements were ranked in order of decreasing variance across the training set. Figure 23-16 shows the error energy for some truncated estimates of the test image based upon that ranking. For comparison, the figure also shows error energy for the 2-D Fourier basis and tensor product of the 1-D K–L bases, whose elements were also ranked in the same manner. The approximate

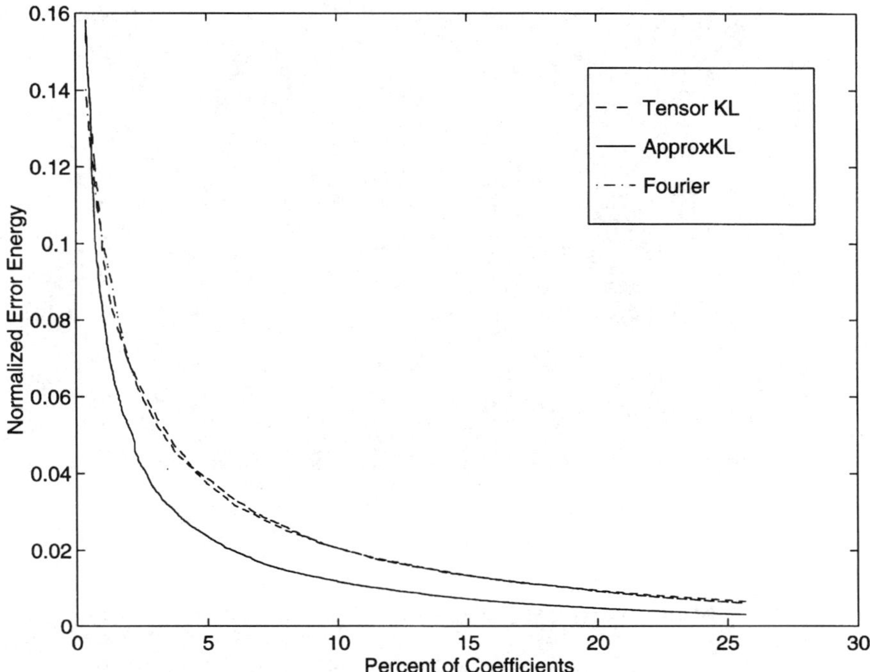

Figure 23-16   Error energy of the level 7 test image reconstructions from various 2-D encoding bases.

2-D K–L basis performed well, achieving 1% error energy using 11.8% of the coefficients and 0.5% with 19.6%. The 2-D Fourier basis required 19.4% and 30.3%, respectively.

Figures 23-17 and 23-18 compare reconstructions from the approximate 2-D K–L basis and 2-D Fourier basis using 1/8 and 1/4 of the coefficients, respectively. In both cases, the approximate 2-D K–L basis shows noticeably more detail. In particular, the regions shown in Figs. 23-19, 23-20, and 23-21, highlight some of these differences. The approximate 2-D K–L basis reconstructs the image with less pronounced ringing, noticeably so in the region of Fig. 23-21. It also resolves detailed structures quickly, allowing visual identification of small features which remain blurred when using the Fourier basis. Figures 23-19 and 23-20 show two such cases.

The chosen approximate 2-D K–L basis merely illustrates the improvements possible when using bases other than Fourier. Much more flexible 2-D local trigonometric bases exist for approximating a 2-D K–L basis. We intend to include such bases as approximate 2-D K–L basis candidates in future work.

23.8. CONCLUSION

In this chapter we have indicated an application of recent developments in adapted wavelet methods to MRI. MRI is a very flexible modality; the images produced in

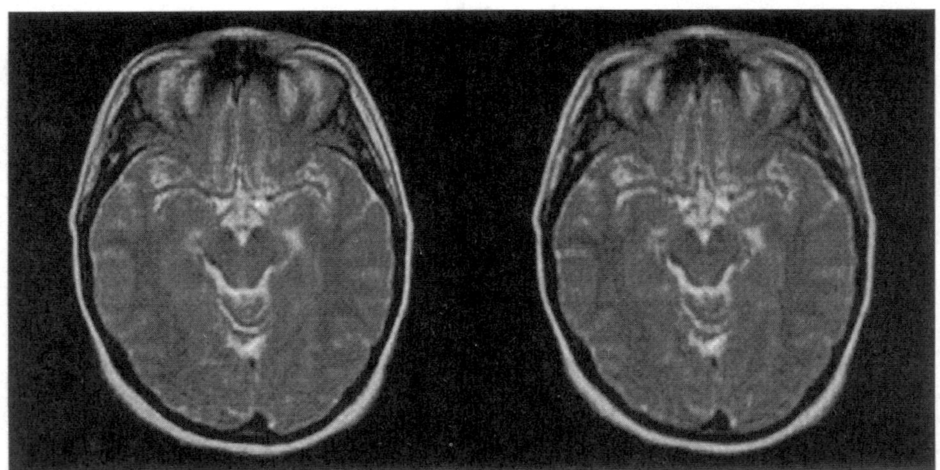

Figure 23-17   Reconstructions of the level 7 test image with approximate 2-D K–L basis (left) and Fourier basis (right) using 1/8 of the coefficients.

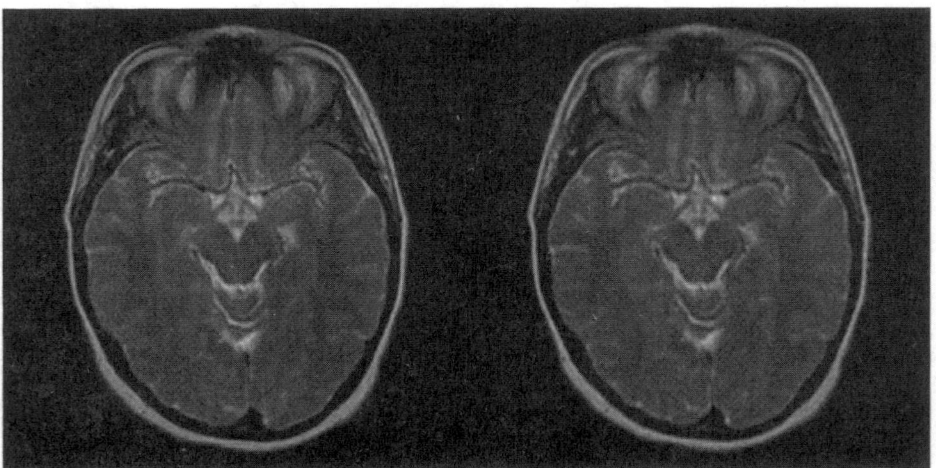

Figure 23-18   Reconstructions of the level 7 test image with approximate 2-D K–L basis (left) and Fourier basis (right) using 1/4 of the coefficients.

this modality depend critically on the values of many parameters. We have described adapted waveform encoding as a tool for the intelligent utilization of this flexibility in several specific imaging tasks.

A particular instance of the technique was presented, in which the encoding basis was chosen to reflect statistical regularities in a particular class of standard diagnostic studies. This basis captures most of the variability of the class in the first basis elements. This can be exploited for reduction of imaging times without the hardware cost of fast imaging methods like echo planar imaging. Another feature of

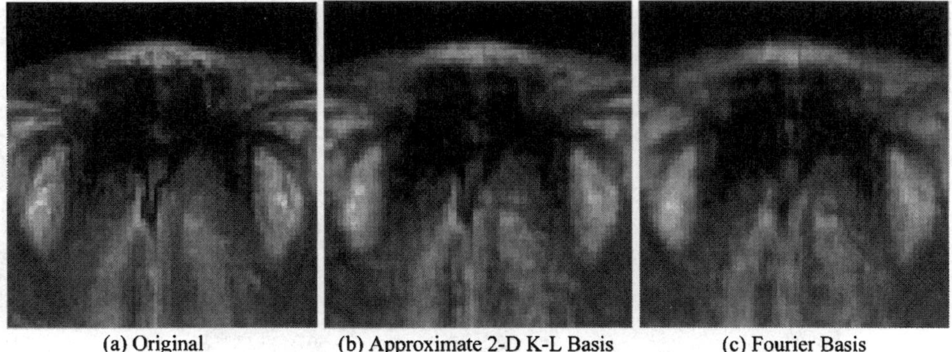

Figure 23-19   A comparison of reconstructed sinus region for the level 7 test image using an approximate 2-D K–L basis (b) and the Fourier basis (c). The actual region is shown to the left (a).

the approach we presented is that it produces a recognizable image after the first few measurements. This image is refined with each additional measurement. The operator has the option of stopping the process at any level of refinement, which may be useful for scout imaging applications. The technique may also be applied to the detection of changes in repeated scans of a single subject.

APPENDIX A: ENCODING IN MRI

This appendix presents a thumbnail sketch of some techniques and practices of MRI, which may help the reader appreciate both the unique promise and problems of this modality. We present only enough to motivate the application of wavelet and related bases for encoding diagnostically useful information about tissues. For more information, one may consult one of the many basic references describing MRI background; for example, [15, 26–28].

MRI is a noninvasive medical imaging modality which has carved out an important niche in medical imaging over the last decade or so. Some appreciation of its role in diagnostic medicine is gained by a comparison with alternatives like ultrasound and computerized tomography (CT).

Ultrasound imaging is inexpensive and is done in real time. However, ultrasound cannot be used for structures that are deep in the body or where there is air or bone around the structure of interest. Therefore, with few exceptions, ultrasound is mainly used in the abdomen and heart.

X-ray CT techniques offer excellent spatial resolution, but have very limited soft-tissue contrast. Contrast in CT is determined primarily by the electron density of the materials imaged. Bone is the only tissue in the body with dramatically different electron density; soft tissues generally have very similar electron densities and atomic numbers.

MRI contrast, on the other hand, is determined by the relaxation parameters of the water in tissue; the intrinsic contrast between soft tissues in MRI is many times

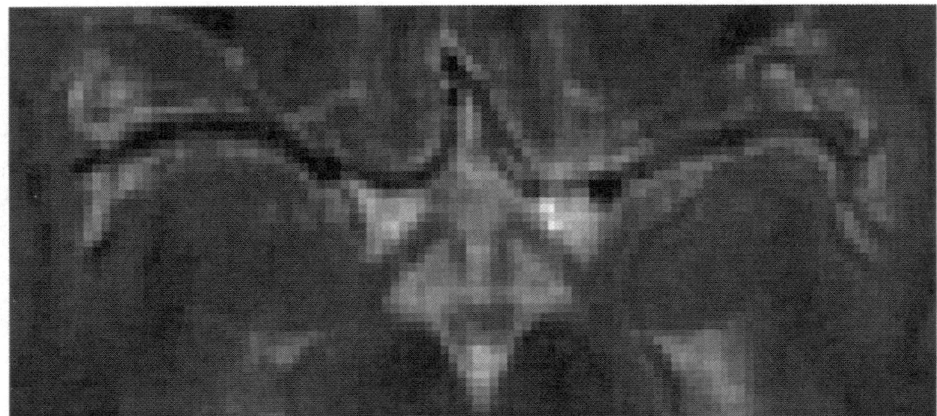

(a) Original

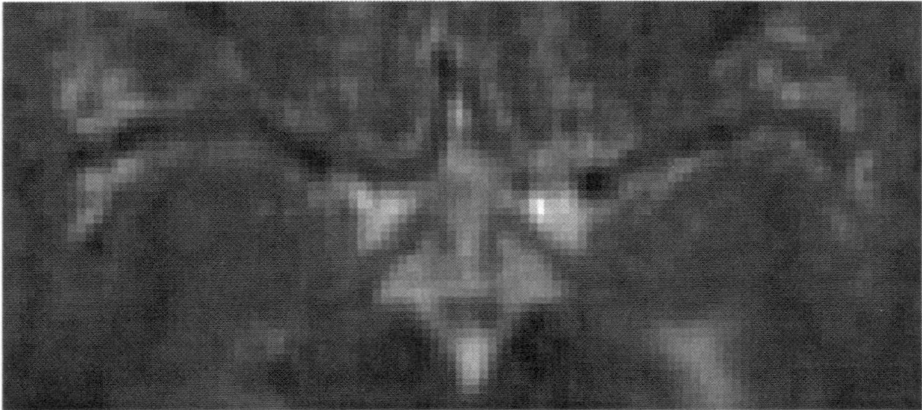

(b) Approximate 2-D K-L Basis

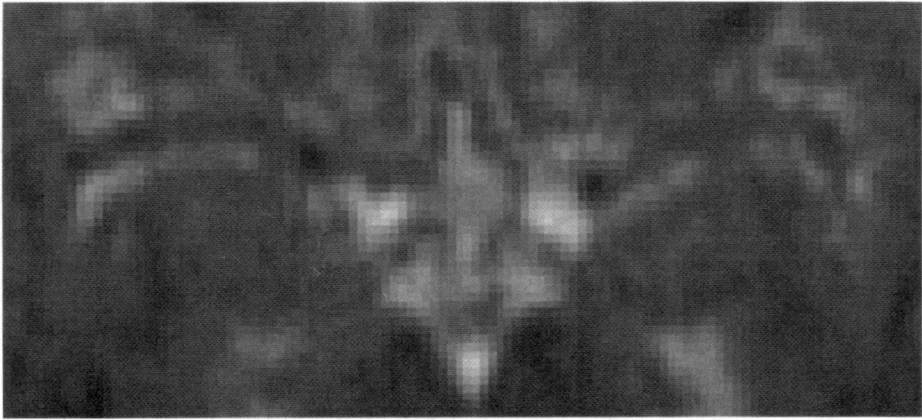

(c) Fourier Basis

Figure 23-20   A comparison of a reconstructed ventricular region for the level 7 test image using an approximate 2-D K–L basis (b) and the Fourier basis (c). The actual region is shown above (a).

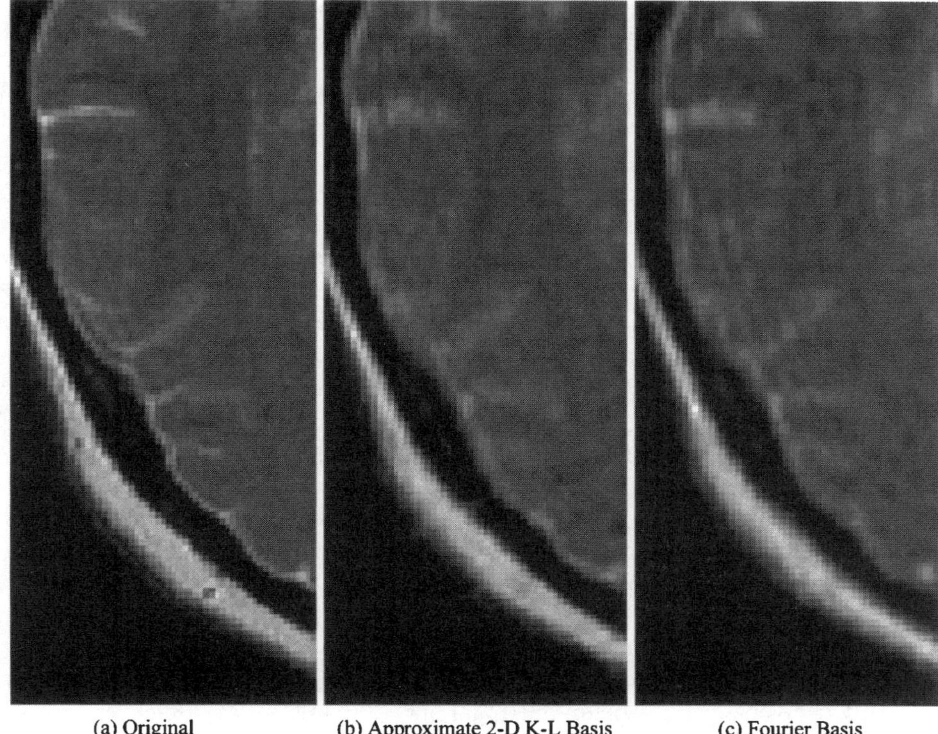

Figure 23-21   A comparison of a reconstructed cortex region for the level 7 test image using an approximate 2-D K–L basis (b) and the Fourier basis (c). The actual region is shown to the left (a).

what it is for CT. For example, the white matter of the brain has 12% lower CT number than the gray matter [29], while white matter has a 140% higher signal than gray matter in an MRI examination [30]. Another important advantage for MRI over CT is the ability to obtain views at arbitrary positions and orientations. On the other hand, MRI is sometimes limited by its spatial resolution properties and long imaging times.

MR images are obtained without the ionizing radiation of CT. Instead, MRI works with the intrinsic magnetic moments of certain types of nuclei (usually hydrogen) found in the object under investigation. It is possible to stimulate these nuclear magnetic moments within the object so that they produce an RF output signal at a characteristic resonance frequency. This signal can be used to encode the information needed for imaging. The means by which this signal is evoked from the sample is the phenomenon of nuclear magnetic resonance (NMR), which we now describe.

A.1 Nuclear Magnetic Resonance

A typical imaging session begins by placing the sample object (such as a human subject) in a strong, homogeneous, magnetic field $\mathbf{H}_0$ inside the scanner. By convention this field defines the z-direction along the scanner axis, so that $\mathbf{H}_0 = H_0\hat{\mathbf{z}}$ with $\hat{\mathbf{z}}$ the unit vector field in the z-direction.

Placing the sample in the scanner's field causes a net alignment of the magnetic moments of the hydrogen nuclei in the sample along the direction of $\mathbf{H}_0$. This allows us to work with a resultant bulk magnetization $\mathbf{M}$, created as the vector sum of the aligned nuclear magnetic moments; see Fig. 23-22.

The bulk magnetization, $\mathbf{M}$, can be knocked out of alignment, or "excited" with an RF electromagnetic pulse, resulting in its precession at a characteristic frequency around the external field $\mathbf{H}_0$. A sensitive coil detects a component of the resulting oscillating magnetic field, and produces a signal which may be used to carry information about the properties of the sample; see Fig. 23-23.

The frequency of the precession of $\mathbf{M}$, and of the resulting output signal, is proportional to the strength of the external magnetic field: $f_{\text{Larmor}} = \gamma\|\mathbf{H}_0\|$. This is the natural resonant frequency of the spin system; the term "magnetic resonance" refers to the fact that the spin system absorbs and emits radiation at the characteristic frequency f_{Larmor} determined by the magnetic environment $\mathbf{H}_0$. For hydrogen nuclei in a typical scanner $f_{\text{Larmor}} \approx 60$ MHz (megahertz).

Properties of the sample may be inferred from the parameters of the measured output signal. For instance, the amplitude of this signal is proportional to the number of nuclear magnetic moments contributing to the net magnetization $\mathbf{M}$ as it precesses. This directly reflects a property of the sample. The amplitude also depends on a user-determined parameter; namely how much energy is pumped into the sample by the RF excitation.

A typical sequence of operations is shown schematically in Fig. 23-24. One begins by stimulating the sample with an RF excitation signal at the resonant frequency f_{Larmor}, and then measures the responding output signal. The output signal generated after excitation is called the free induction decay (FID). In the absence of

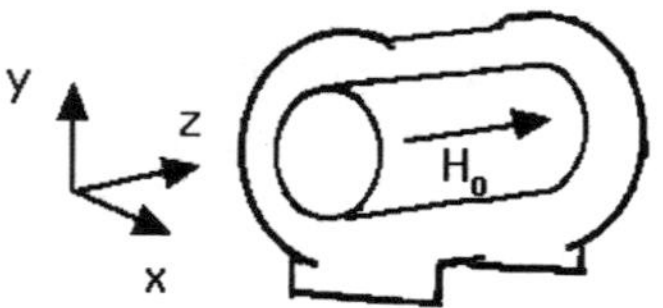

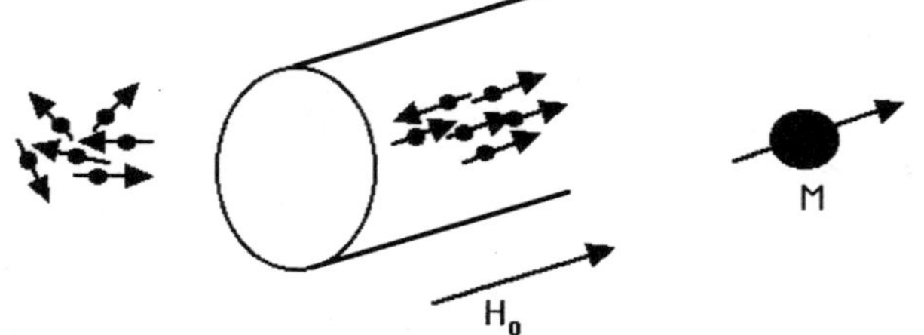

Figure 23-22 A sample placed in the strong magnetic field $\mathbf{H}_0 = H_0\hat{\mathbf{z}}$ of the MR scanner exhibits a net alignment of the nuclear magnetic moments of the hydrogen nuclei of the sample. The resulting net magnetization $\mathbf{M}$ is manipulated to produce a signal in MRI.

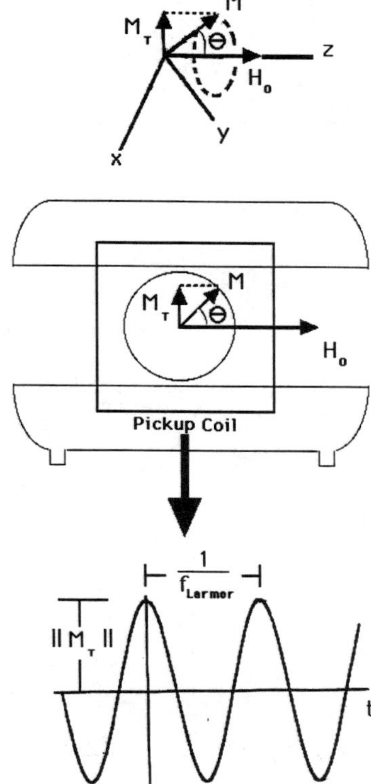

Figure 23-23 An RF excitation at the Larmor resonant frequency tips the magnetization **M** by angle θ away from **H**$_0$. It then begins to precess around at the (same) Larmor frequency. A pickup coil records an oscillation at this frequency and with amplitude proportional to the length of **M**$_T$, the transverse component of **M**.

magnetic inhomogeneities and susceptibility effects, the signal eventually dies off as a superposition of two types of exponential decay, characterized by decay constants T_1 and T_2. These time constants are properties of the materials imaged, and may vary in space throughout the sample.

- T_1 *decay*: Spin–lattice, or longitudinal relaxation is characterized by T_1. T_1 typically varies from 200 ms (milliseconds) to 800 ms. This decay reflects the magnetization realigning itself with the static magnetic field as the nuclei lose energy to the lattice.

- T_2 *decay*: Spin–spin, or transverse relaxation, occurs faster, with time constant $T_2 \approx 50$ ms in tissues. It is caused by loss of phase coherence, or dephasing, among the precessing nuclei contributing to **M**, and comes from local interactions among the nuclei.

In cases where a given group of nuclei must be measured repeatedly, the T_1 parameter helps determine a required repetition time, *TR*, between successive excitations. This resting period is mandated by the necessity of waiting for the spin system to relax back towards alignment with the external magnetic field before the next excitation. *TR* can be quite long when imaging material with a long T_1 relaxation time, resulting in long imaging times.

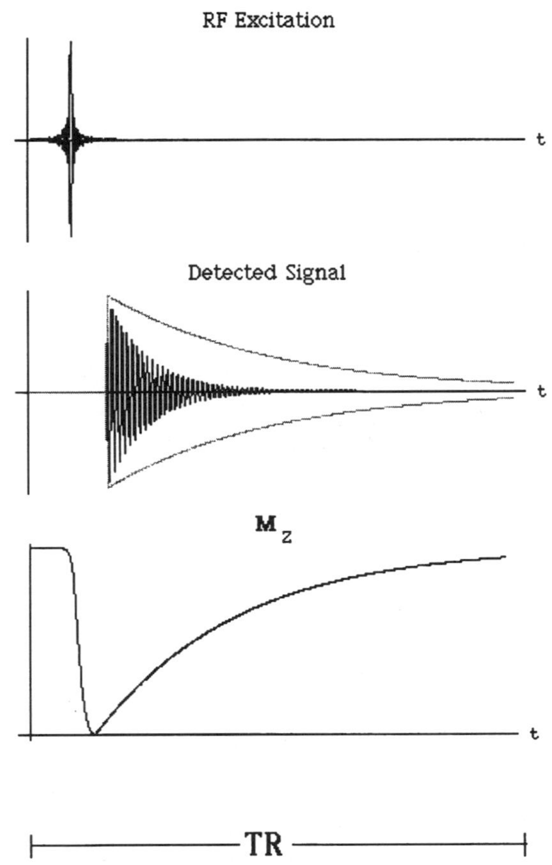

Figure 23-24 The sequence of events in a typical MR measurement, indicated on three time lines. The sequence begins when an RF excitation (top line) at the Larmor frequency is input to excite the nuclei in the subject. These produce a free induction decay (FID) signal (middle line) at the Larmor frequency. The decay is partly caused by the eventual realignment of the nuclear magnetizations with the static field, as indicated by the recovery of the longitudinal magnetization (bottom line). This recovery determines the repetition time, *TR*, which should be allowed before the spin system is queried again. Unfortunately, signal measurements are usually unavailable for much of this time, due to another, faster decay effect stemming from the loss of coherence among the nuclear magnetizations which contribute to the bulk magnetization.

Contributing to the difficulty of MR measurements is the fact that the usable signal generally dies away by fast T_2 decay (or faster) long before the longitudinal magnetization has recovered by the slower T_1 process. This can have the effect that much of the *TR* resting period is wasted as far as obtaining useful measurements is concerned. This is the cause of the long imaging times in standard MRI.

So far, the output signal reflects properties of the sample as a whole. Imaging requires localization; i.e., the ability to determine what is contributed to the signal by a given region of the sample, and therefore permits inference about the local properties of each portion of the sample.

A.2 Imaging

For a sample in the uniform field $\mathbf{H}_0$, all the nuclei produce signals at the corresponding Larmor frequency after excitation. The output signal thus reflects the properties of the sample as a whole. In order to perform imaging, we must somehow map the variation of some property of interest within the sample. In most examinations, we are trying to determine a map of the weighted spin density over the extent of the sample. Typically, this is accomplished by perturbing the external magnetic field so as to vary its strength across the sample. Nuclei at different positions now experience a slightly different field strength and therefore have a Larmor frequency that depends on their position.

This may be used in several ways to encode information on the RF signal so that one may map tissue properties across a region of the sample. The principal techniques are:

- Selective excitation
- Frequency, phase encoding

The measurements made with these methods may be interpreted as projections against elements of various bases. We now present the standard MRI technique from this perspective; later we will see how this may be generalized.

Selective Excitation.   Perhaps the most straightforward approach to obtaining localized information is to collect signals from only a small region of the sample. For example, one may study sample properties near a given position by only "turning on" the nuclei in that portion of the sample; the resulting output signal necessarily reflects only the properties of nuclei in the activated region.

This "selective excitation" is obtained by supplying a narrow-band RF excitation, which stimulates only those nuclei whose resonant frequency is near that of the excitation signal. The nuclei so selected live only in the region of the sample which experiences the corresponding magnetic field strength. Their output signal is associated only with that location. Selective excitation amounts to projection of the signal density onto a localized amplitude profile, as we now see.

The requisite variation of Larmor frequency across the sample is provided by special gradient coils in the scanner, which add to the background field $\mathbf{H}_0$ a linear variation in one direction across the sample, say the z direction, while the narrow-band RF excitation signal is supplied. The natural frequencies of the nuclei then vary linearly with z, $f_{\text{Larmor}} = f_{\text{Larmor}}(z) = \gamma(H_0 + G_z z)$, where G_z is a constant reflecting the gradient of magnetic field strength in the z-direction. Only nuclei in those regions corresponding to frequency components present in the RF pulse will be excited into contributing to the output signal. To first order (low tip angle approximation), the amplitude and phase of the nuclear excitations in a given plane $z = z_0$ will be proportional to the amplitude and phase of the frequency component of the RF excitation at the corresponding frequency $f_{\text{Larmor}}(z_0)$; see Fig. 23-25.

For example, an idealized RF excitation pulse could be a modulated sinc function comprised of pure harmonics with amplitude 1 in a certain band and amplitude 0 outside that band. Nuclei at positions with Larmor frequencies inside the band are excited; nuclei with Larmor frequencies outside are not excited. This results in a slab of excited nuclei transverse to the z-axis, with the depth of the slab corresponding to the support of the Fourier transform of the RF excitation. The thickness of the slab can be changed by changing either the time constant of the sinc or the gradient strength. The z-position of the slab can be changed by changing the center frequency

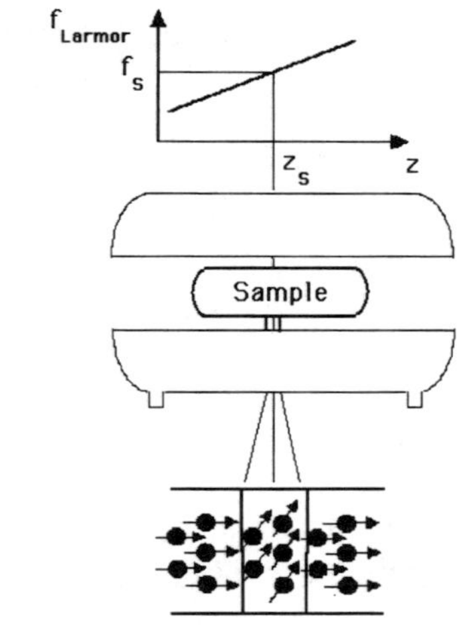

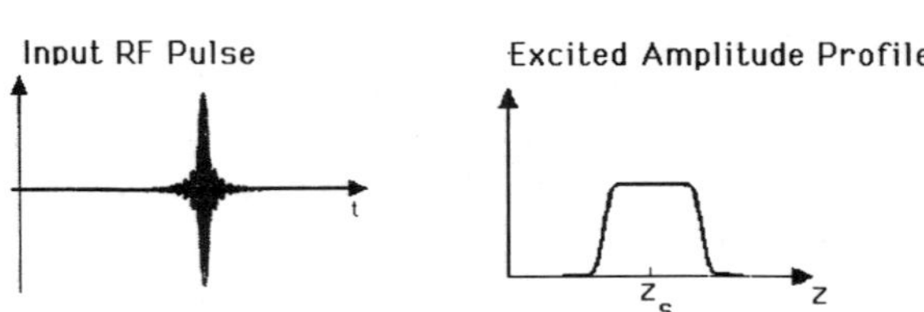

Figure 23–25 Selective excitation of nuclear spins in a sample. A magnetic gradient imposed along the z-direction of the sample, makes the resonant frequency vary linearly with position. A narrow-band RF excitation tips only those nuclei whose resonant frequencies correspond to frequency components in the excitation; only these nuclei precess and produce a signal. The resulting amplitude profile along the z-axis is approximated by the Fourier transform of the RF pulse.

of the RF pulse. More general amplitude profiles would require more complicated RF pulses with a different balance of Fourier components.

Obtaining a precisely shaped excited amplitude profile along the z-axis can be nontrivial due to nonlinearities in the physical system; to first order one makes the RF pulse equal to the Fourier transform of the desired amplitude profile as a function of z. See [31–35] for some discussions of the general excitation problem.

At the end of the RF excitation, the z-gradient is turned off and the output signal is measured. The result is a sinusoidal signal obtained by summing the contributions of all the excited nuclei now precessing again at the common $f_{\text{Larmor}} = \gamma \|\mathbf{H}_0\|$. The amplitude contributed from nuclei near a given plane $z = z_0$ depends on the following.

- The number of nuclei there, $\rho(z_0)\Delta z$, where $\rho(z)$ denotes the density of nuclei along the z-axis.
- The excitation amplitude at $z = z_0$, $A(z_0)$, determined by how much RF excitation was provided to that position [$\approx$ the Fourier component of the RF pulse at frequency $\gamma(H_0 + G_z z)$].

Essentially, the output signal looks like a weighted sum of the spin density:

$$S_A(t) = \int A(z)\rho(z)e^{i\gamma H_0 t}\, dz = \langle A, \rho \rangle e^{i\gamma H_0 t} \tag{23-6}$$

If we choose a *basis* of amplitude profiles, we can reconstruct the density in z. As a simple example of the use of selective excitation encoding, suppose we were interested in mapping the variation of spin density along the z-axis of a sample. This could be done by selectively exciting a succession of slabs along the z-axis, and recording the signal strength from each slab. Assume the profiles are ideal boxcar functions partitioning the z-axis into slabs of width Δz:

$$\phi_k(z) = \begin{cases} 1 & \text{if } k\Delta z < z < (k+1)\Delta z \\ 0 & \text{otherwise} \end{cases}$$

These correspond to sinc-like RF excitations applied with gradient G_z turned on; this gradient is turned off at the end of the RF pulse; we would then measure signals proportional to

$$S_k(t) = \langle \phi_k, \rho \rangle e^{i\gamma H_0 t} \tag{23-7}$$

successively, for $k = 1, \ldots, N$. This imaging process is represented schematically in Fig. 23-26. After demodulating the Larmor frequency carrier, we have the projections of the spin density onto the elements of a boxcar basis, or equivalently, the averages of the spin density over the voxels determined by the partition of the z-axis. From this, we could reconstruct ρ up to the resolution Δz.

One could obtain the same information by selective excitation of different amplitude profiles, which can be done by using an appropriate RF excitation along a selection gradient as discussed above. In principle, any of a number of bases could be used as amplitude profiles for selective excitation encoding; one which has been considered for MRI is the Hadamard basis [17, 26, 36]. See Fig. 23-27 for a depiction of the Hadamard basis as well as some other bases used for encoding. Orthogonality is obtained in the Hadamard encoding by weighting some portions of the spin system with amplitude $+1$ and other portions with amplitude -1. The physical meaning of weighting some spin regions with a negative amplitude is simply that the phase of their precession is π radians ahead of those nuclei weighted with a positive amplitude.

The Hadamard function basis is very different from the boxcar basis in that all of the Hadamard functions are globally supported; that is to say, they are non-vanishing on the entire interval being encoded. This has important implications in MRI, as we shall see when we discuss the considerations which make one basis preferable to another.

The "stack of boxcars" basis described above is commonly used in volumetric imaging to resolve the sample into slabs transverse to the z-axis, but the variation of density with x and y in each slab is usually not constant, and remains to be mapped. This is usually done with the techniques of "frequency and phase encoding." As we shall see, these correspond to techniques for projecting onto a Fourier basis.

Frequency Encoding.   If we are interested in mapping the spin density in all three dimensions, $\rho(x, y, z)$, we must go beyond encoding only the z-axis. This may be done as follows. After selective excitation along the z-axis, as described above, the magnetic field strength is now made to vary linearly in one direction *across* the excited region, say the x direction, during the time at which output signal is being measured. The Larmor frequency in the excited region of the sample now varies linearly with x; $f_{\text{Larmor}} = f_{\text{Larmor}}(x) = \gamma(H_0 + G_x x)$, where G_x is a constant reflecting the gradient of magnetic field strength along the x direction; see Fig. 23-28.

The demodulated signal output then amounts to

$$S(t) = \int dx \int dy \int dz \; \phi(z)\rho(x, y, z)e^{i\gamma G_x x t}$$

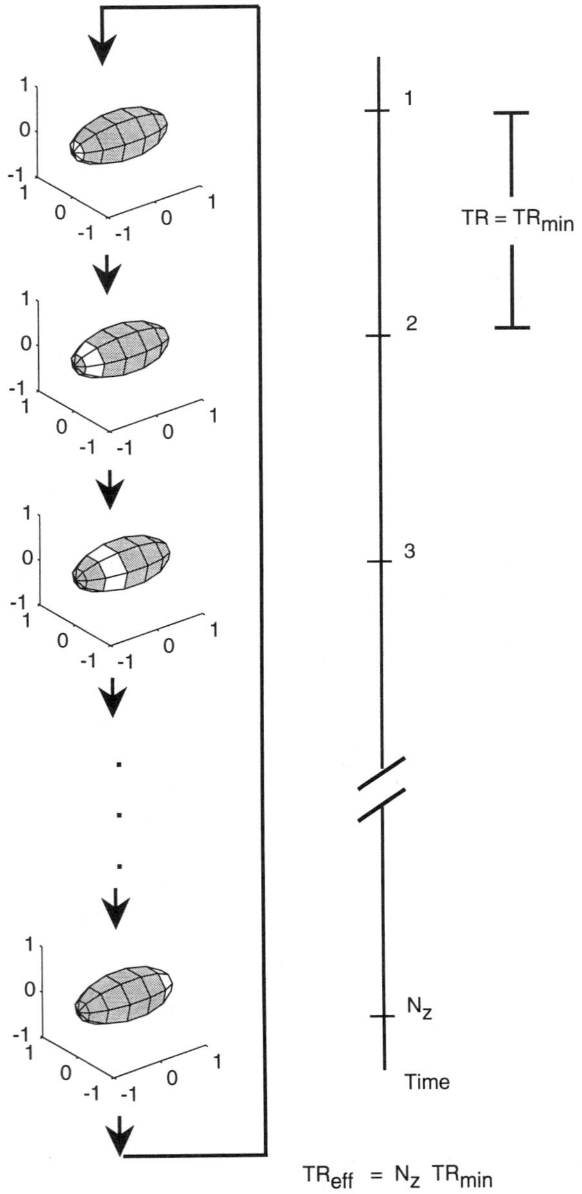

Figure 23-26 Imaging a volume by successively exciting slices along the z-axis. White indicates excited regions, gray regions do not produce a signal. In many situations, each slice must be measured more than once. By looping through the excitations in the sequence given here, we assure that a long time passes between the consecutive excitations of a given slice.

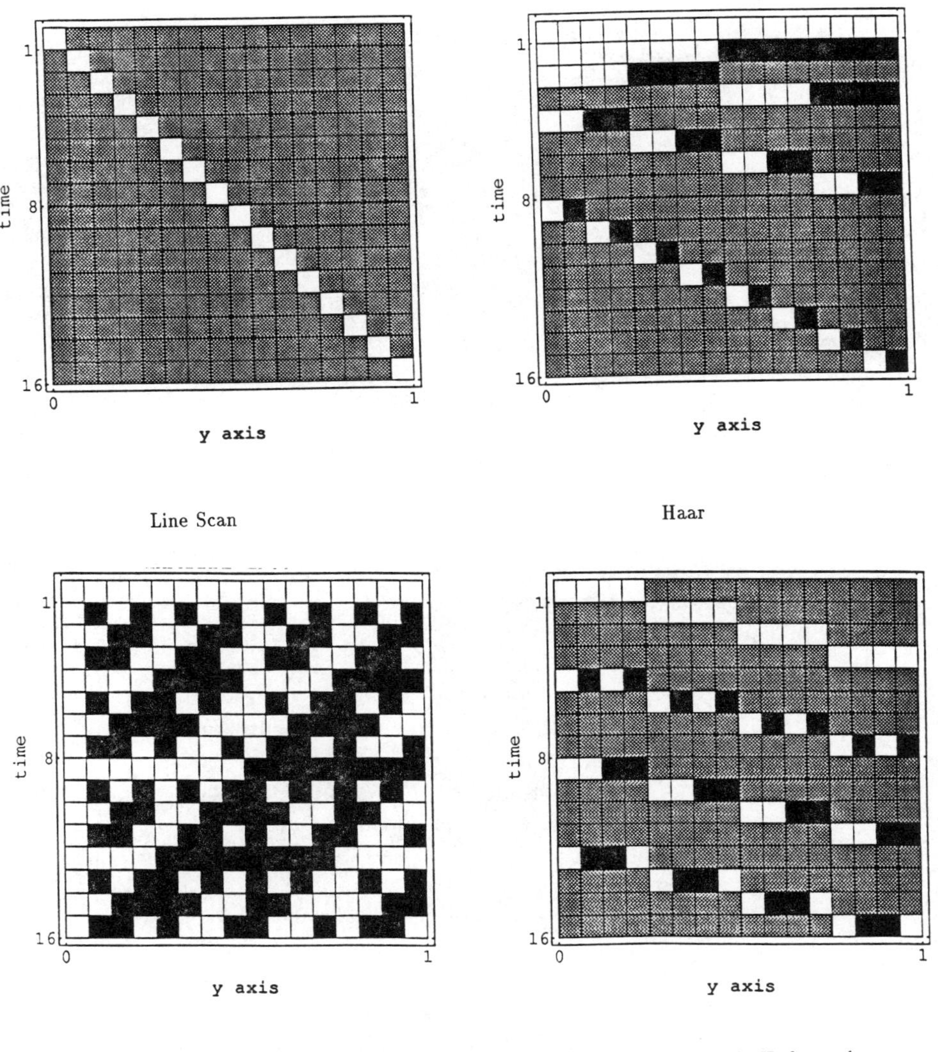

Figure 23-27   Representations of four orthonormal bases for encoding an interval to a resolution of 16 pixels. Each row represents one encoding profile; white represents a +1 value, black represents −1, and gray represents 0.

where ϕ is the amplitude profile of the selective excitation. The time samples of this signal give projections of the spin density onto the elements of a basis which look like the Fourier basis elements in the x-coordinate, constant in the y-coordinate, and have the slice profile in z:

$$S(k\Delta t) = \langle \rho, e_k \rangle$$

with $e_k(x, y, z) = e^{-ikx(\Delta t \gamma G_x)}\, \phi(z).$

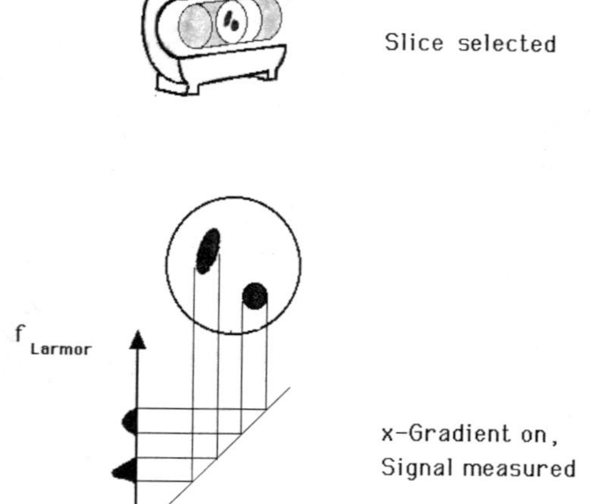

Figure 23-28 Frequency encoding the x-axis. After selective excitation of a slice, a magnetic gradient is imposed across it during signal readout and measurement. This gradient causes the Larmor frequency to vary linearly across the slice. Consequently, a given frequency component of the output signal corresponds to a particular x position; its amplitude reflects the amount of precessing nuclei at that position.

Notice this output signal gives no information about the variation of density in the y direction. It is common current practice to acquire this additional information by phase encoding.

Phase Encoding.   In this technique one uses a magnetic gradient to build up a linear phase variation in the y-direction of the excited slice, prior to frequency encoding and signal measurement. The signal output is then a modification of what we saw in the case of frequency encoding alone:

$$S_\theta(t) = \int dx \int dy \int dz \, \phi(z) \, \rho(x, y, z) \, e^{i\gamma G_x x t} \, e^{i\theta y} \tag{23-8}$$

measured for t over a small interval of time, and θ is the constant phase gradient.

For example, in a common imaging application the profile ϕ corresponds to a thin slice near $z = z_0$. Then our signal gives us the values of the 2-D spatial Fourier transform of the spin density of that slice, $\widehat{\rho_{z_0}}(\omega_x, \omega_y)$, along one line segment through the 2-D spatial frequency plane: $\omega_x = \gamma G_x t$, $\omega_y = \theta$ for t varying over the measurement time interval.

Of course, this one signal only gives part of the information required to build up a map of the spin density within the slice. In practice, one gets more data by repeating the steps of excitation and encoding according to a specific schedule. This consists of many repetitions of the basic sequence: slice excitation, followed by phase encoding, and then finally signal measurement while frequency encoding. With each repetition, the MR scanner outputs a signal corresponding to the values along a new line of the 2-D spatial Fourier transform of the proton spin density in the slice being imaged. This is done by varying the phase gradient in the y-direction from repetition to repetition. In this way, a sufficiently dense sampling of the Fourier transform of the spin density may be acquired at a rate of one line per repetition. When this is done, the spin density in the slice may be reconstructed by inverse discrete Fourier transform, up to a given resolution.

Note an important distinction between phase and frequency encoding. Frequency encoding implements the projection onto a whole range of Fourier basis elements in one signal acquisition. In a standard scanner, phase encoding usually only computes the projection onto one Fourier basis element in a given signal measurement.

A.3 Imaging Time and SNR

All imaging techniques involve a trade-off among resolution, imaging speed, and SNR. SNR considerations are serious in MRI: the signals tend to be low in strength, and the primary noise source in modern scanners is thermal noise from sources other than the hardware (such as the patient) and cannot be removed.

Most standard scanners can also provide only limited imaging speed. Hardware modifications can increase this, but often at the cost of resolution and contrast. This limitation can make imaging moving structures difficult, and also contributes to the rather high cost of MRI examinations.

Let us illustrate these considerations in the example of volumetric imaging which we considered earlier. We may choose to encode a volume with a stack of N_z boxcar excitations as a basis to break the z-axis into nonoverlapping slices, as in Eq. (23-2) above.

The time required to generate an image of this volume with a resolution of N_z pixels in the encoded z direction is:

$$T_I = M_s \, (N_z \, TR)$$

because N_z different slices must be successively excited and measured, each in a time TR. M_s is the number of times each slice must be measured; this will be more than one if the density within the slice is to be resolved with standard phase encoding; repetitions for signal averaging can also take into account if the SNRs in the images are too low and one is forced to average images to reduce noise. In practice, M_s is almost always greater than one, i.e., each slice must be called on more than once to give a signal. The overall imaging time now depends critically on the resting time between successive excitations of a given group of nuclei, and on the ordering of the excitations.

The resting time is an important parameter in determining signal strength and image contrast. It is typically much larger than the time for which usable signal can be measured from a given group of nuclei. This resting time is required to allow the precessing magnetization to realign via (slow) T_1 relaxation with the static magnetic field $\mathbf{H}_0$; otherwise, there is little signal produced by the next excitation of these nuclei. This is the primary obstruction to fast imaging.

The minimum possible time interval between successive signal measurements, $TR_{\min}$ is determined by hardware considerations and the particular type of contrast weighting; for a T_2 weighted image it might be around 100 ms whereas the resting time for a given group of nuclei needs to be about 2500 ms.

In this situation, the ordering of the signal measurements has a critical impact on the total imaging time. We know we must obtain M_s signals from each slice, and that a relatively long resting time must be allowed to elapse between the successive excitation and signal measurements from this slice. In this case, the time difference between $TR_{\min}$ and the (order of magnitude longer) resting time need not be wasted. Other slices can be imaged during that time, one after the other, while allowing the first slice a sufficient recuperation interval.

This idea is indicated schematically in Fig. 23-26. In this approach, the slices in the stack along the z-axis are excited and measured sequentially, one every $TR_{\min}$ seconds. This sequence of measurements obtains from each slice the first of the M_s signals it must contribute. When the last slice is finished, we return to the top of the stack and start over, obtaining the second signal required of each slice. The loop is repeated M_s times to obtain all the required measurements from each slice.

Suppose the number of slices along the z-axis to be imaged is N_z. Although we conduct a measurement every $TR_{\min}$ seconds, the effective resting time for a *given* slice of nuclei is much longer: $TR_{\mathrm{eff}} = N_z\, TR_{\min}$. The *effective* time cost per slice is then:

$$\frac{T_I}{N_z} = TR_{\min}\, M_s$$

Note, however, that a single slice could not be imaged in T_I/N_z; T_I must be used to get a batch of N_z slices.

The effective repetition time, $N_z TR_{\min}$, can be made very large by dividing the sample into a large number of slices. However, this has its cost in the SNR as we shall now see. The signal in an imaging experiment is proportional to the number of nuclei contributing to the signal. The signal is therefore proportional to voxel size, and so resolution is purchased at the cost of lowered signal. The noise may be modeled as white noise within the sampled bandwidth. The larger the gradients used, the larger the bandwidth and so the noise power. Signal strength is also determined by the timing parameters. For example, a longer TR buys more signal.

Let us examine these SNR considerations in our example. We encoded with a stack of N_z boxcar excitations as a basis to break the z-axis into nonoverlapping slices, as in Fig. 23-26. Applying the time sharing described above, the effective repetition time experienced by any slice is $N_z TR_{\min}$, which will be large if there are many slices. However, as the number of slices goes up, their width decreases, and so does the signal strength obtained from each one. This eventually becomes prohibitive.

On the other hand, we might instead encode the z-axis with amplitude profiles from the Hadamard basis. Again we wish to resolve the z-axis into N_z pixels of equal width, but now encode by exciting all of these pixels at once with profiles corresponding to the elements of a Hadamard basis on these N_z pixels; see Fig. 23-29. Each Hadamard encode yields a signal from the entire sample at once, which leads to a good SNR. In fact, in some sense the Hadamard encoding is optimal in this respect, as may be seen from standard considerations of statistical weighting designs [37]. However, any time-sharing possibility is precluded; since all of the nuclei in the

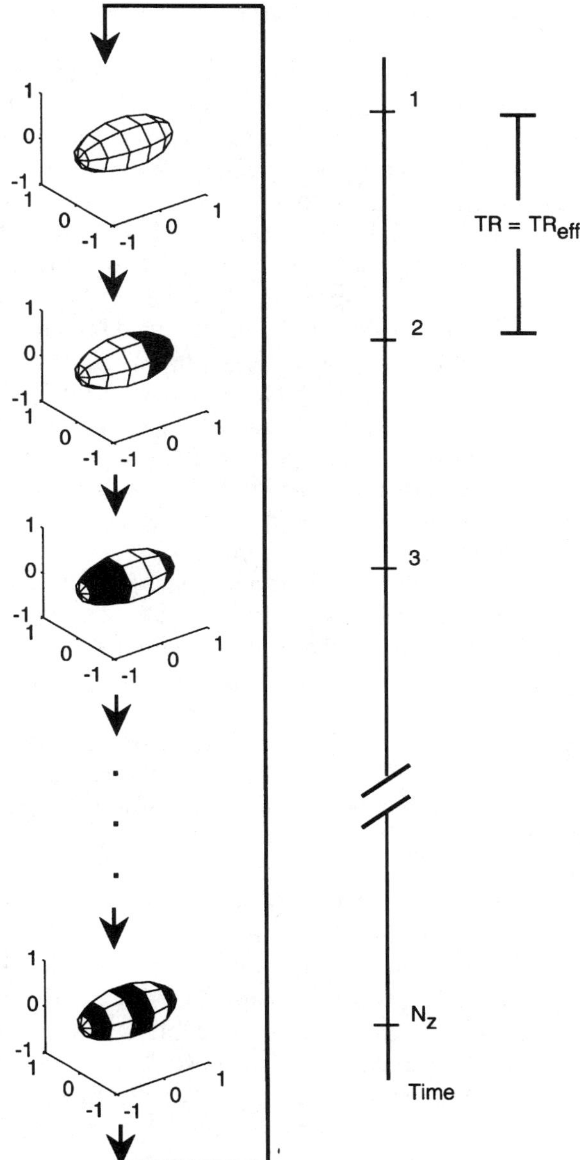

Figure 23-29 Volumetric imaging by Hadamard encoding of the z-axis. Each encode excites the entire sample with an amplitude profile drawn from the Hadamard basis. White corresponds to an amplitude of $+1$, black to -1. The global excitation produces a good SNR, but longer imaging times than the line scanning approach in Fig. 23-26.

sample are excited on every encode, we must wait a long TR time between successive excitations in order to get decent signal strength. In general this will force a big increase in imaging time.

These approaches represent two extremes with respect to the critical figures of merit of the SNR and imaging time. The question naturally arises as to whether there is any practical way to interpolate between these two extremes. We take up this issue now.

A.4 Adapted Waveform Encoding in MRI

One may greatly expand the number of possible bases of amplitude profiles used in MR encoding by utilizing more general RF pulses and gradient sequences. Here we consider some particular bases for encoding in this way, and give some criteria for choosing the basis for a given imaging situation. We will use the term "adapted waveform encoding" to refer to the measurement of projections of the spin density onto bases tailored to the imaging requirements of a particular task. In this section we consider the advantages and disadvantages these new schemes offer over more conventional approaches to MRI, such as the volumetric slicing by encoding with the boxcar basis or Hadamard encoding described previously.

In order to make this comparison, we must describe some of the considerations of wavelet and adapted waveform encoding. We offer a discussion of some important implementation details, such as the need for bases comprised of smooth profiles. We present some important properties and figures of merit to be applied in choosing a basis suited to a particular imaging task. In particular, we summarize the results of a quantitative analysis of the signal-to-noise issues associated with adapted wave-form encoding, and compare this to SNR computations for more conventional schemes. We present results concerning the trade-off of SNR against effective repetition time, TR_{eff}, a parameter related to imaging speed and image contrast. This gives us a quantitative measure to employ in the evaluation of encoding schemes, and which we use in our efforts to balance the various requirements of a particular imaging task.

The tailored sequences we design through these considerations allow us to obtain useful images by accommodating *a priori* information about the type of scan and the particular information we are seeking.

A.5 MRI Encoding with a Basis

Previously, we discussed standard MRI encoding schemes which employed boxcar or Hadamard bases for the excited amplitude profiles in selective excitation. Adapted waveform encoded MRI is a simple variant of these techniques which addresses the imaging time problem and several other issues.

We may encode any given direction by selective excitation along a magnetic gradient in that direction. Each RF pulse excites the spin density with a wavelet or more general waveform-shaped excitation profile along the encoded axis. The signal produced is the inner product of the excited profile and the spin density along the encoded axis:

$$S_j(t) = \left(\int dx \int dy \int dz \, \rho(x, y, z) \, \psi_j(y) \right) e^{i\gamma H_0 t} \qquad (23\text{-}9)$$

Here we have chosen to encode the y-axis with an amplitude profile given by the j-th element of a basis. Any direction could be encoded in the same way. We can then reconstruct the projection of the spin density onto the encoded axis from these measurements in the usual fashion.

Note that we can frequency encode another axis with no cost in time by turning on a readout gradient in, say, the x-direction during signal measurement. There are several slice selection techniques that can be used [18, 38, 39] to additionally localize to a slice perpendicular to the z-direction. This gives rise to the adapted waveform encoding signal described in Eq. (23-4). This type of encoding could also be applied in three-dimensional (3-D) imaging, using phase encoding to localize the third direction.

In [15] we described some of the benefits associated with choosing the set of excited profiles to be an orthogonal wavelet basis. This technique has since been studied by several other research groups [18, 19]. Many other bases can be used with this technique, as we have seen in this chapter. Some of the implementation considerations which inform the choice of basis for a given task, including SNR may be found in [11, 16].

Our first work on wavelet encoding used the well known Haar wavelets for simplicity of presentation. Various other workers in MRI [17, 26, 36] have considered conceptually similar encoding techniques in which excited profiles from the Hadamard basis were used instead of Haar wavelets. In contrast to the Haar basis, all of the Hadamard basis functions are globally supported on the image, and so offer a different set of advantages and disadvantages. Nevertheless, both bases are drawn from a single library of bases, the Haar–Hadamard wavelet packet library [22]. This library contains many other bases, all comprised of piecewise constant waveforms. Another basis from this collection has been used in MRI in the technique of "line scanning." Here, one divides the axis to be encoded into narrow strips, and excites one strip on each repetition. This approach was described earlier in the context of volumetric imaging as boxcar encoding. It too may be viewed as adapted waveform encoding using the basis of unit impulses. A representation of these three bases for the unit interval may be found in Fig. 23-37.

In line scanning, the excited profiles are zero over most of the encoded interval, so that only a small band of nuclei is excited on each repetition. This can be exploited to increase the effective repetition time experienced by any group of nuclei, which can in turn be used to reduce imaging time in certain situations. Also, the localized excitation profiles can reduce motion artifacts. However, as discussed earlier, the local profiles used in line scanning cause a much reduced SNR compared to Hadamard or phase encoding. This comparison is indicated in more detail in [11]. As originally proposed, wavelet encoding with Haar wavelets steers a middle ground between Hadamard encoding and line scanning, using basis elements more localized (on average) than Hadamard functions, but less so than line scanning. More generally, we can encode with any basis from the Haar–Hadamard wavelet packet library, such as the dilated multiple Hadamard basis represented in Fig. 23-27.

This flexibility allows us to modify our encoding to fit the task at hand. For example, the variety of adapted waveform bases allows us to construct customized encoding schemes that trade the SNR for the advantages of local bases: a longer TR_{eff}, reduced motion artifacts, and reduced Gibbs ringing. We may choose a basis of localized elements to increase the TR_{eff}, or a less localized basis to increase the SNR. We may also consider compromise bases, comprised of local elements in some regions of the field of view (FOV), and of more globally supported profiles in other regions.

All of the bases we have mentioned up to now are drawn from the Haar–Hadamard library, and are all comprised of piecewise constant functions. We have briefly described the freedom to choose bases from within this library to meet certain design criteria. Just as important is the freedom to consider other libraries of bases, and in particular libraries with smooth waveforms.

From the implementation point of view, a disadvantage of the Haar–Hadamard library is the fact that all of its functions have discontinuities, or equivalently, poor localization in the frequency domain. This greatly increases the duration of the RF pulse required to produce these shaped excited profiles, which can cause great difficulties. In fact, our simulations for Haar wavelet encoding required RF pulses as long as 20 ms [21]; RF pulses commonly used in MRI are 3 ms to 8 ms long. Longer RF pulses can cause significant difficulties in MRI. The RF power amplifier and the gradient amplifiers are usually not built to withstand long pulses. The echo time is necessarily lengthened by long RF pulses; long echo times produce necessarily T_2 weighted images which is often undesirable, and long echo times eliminate signals from nuclei that decay swiftly. More importantly, the excited profile is degraded by the T_2^* decay during the RF pulse; the decay could be compensated for if all the nuclei experienced the same decay but images will always have nuclei with many decay rates. This implies that the excited profiles obtained with long RF pulses are always degraded, which is unpleasant since wavelet encoding depends on accurate excited profiles. Many techniques have been employed to reduce the RF pulse length required to obtain a given profile [40, 41] and to generate more accurate excited profiles [31, 32, 34, 35, 42]. Hadamard encoding faces the same problems [17, 36, 42]: long RF pulses and the resulting inaccurate excited profiles. Some of the pulse shaping techniques developed for Hadamard encoding may mitigate the long RF problem.

However, there is a simpler alternative to Haar–Hadamard encoding available in the smooth wavelet packet libraries associated with the wide variety of smooth wavelet bases that have been discovered [22, 43–46]. Alternatively, we may use the local trigonometric libraries of Coifman and Wickerhauser [22]. These smooth libraries offer many of the same useful features as the Haar–Hadamard library, such as localization, and obviate the RF problems, as they provide good localization simultaneously in the space and spatial frequency domains. As indicated above, the localization in the frequency domain corresponds to shorter RF pulses.

Given the desired excited profile for use in waveform encoding, we can estimate the length of the RF pulse required to make it. A common approach is to linearize the actual procedure which requires solving the Bloch equations [47] governing the dynamics of the magnetization. The linearization, corresponding to low tip angles, gives a Fourier transform relationship between the excited profile and the RF pulse. For example, with the first-order approximation, the RF pulse used to generate a boxcar excited profile is given by a sinc function in time. This first approximation can then be improved by various iterative schemes if desired.

In [11], we compared the RF pulse length required by wavelet encoding with several smoother wavelet bases with the requirements of the Haar basis. We presented a study of one of the Daubechies family of wavelets [43, 44], the Battle-Lemarié wavelet [45], and the Meyer wavelet [46]. We began with the low tip angle approximation of the Bloch equations described earlier. This approximation is used to estimate the length of the RF pulse required to excite an accurate wavelet profile. The time-scale is that required to obtain 1 mm resolution with magnetic gradients of 10 mT/m; other resolutions and times can be obtained by simple scaling. The estimated RF pulse was cut off at different lengths to produce a set of shorter RF pulses. These shorter RF pulses were put through an inverse Fourier transform to estimate the resulting solutions of the Bloch equations, and the error in the estimate of the excited profile was calculated. The results, presented in more detail in [11], allow us to conclude that adapted waveform encoding can be implemented with very short RF pulses. With 10 mT/m magnetic gradients, 1 mm resolution can be obtained with RF pulses approximately 3 ms long [48]. Wavelet encoding, and by extension, adapted waveform encoding is therefore not limited by long RF pulses and may be a good replacement for phase encoding.

A.6 MR Phosphorus Spectroscopy

MR spectroscopy provides a means of studying physiology by providing information about the proportions of certain important chemical species and their distribution in the tissue of interest. For instance, phosphorus and hydrogen spectroscopy can be used to provide quantitative measurements of the distribution of certain chemical byproducts of metabolism. This may be used to study the energy metabolism of certain tissues over time. This is an important example of the developing area of functional imaging, whereby one obtains localized information about the function of various tissues and organs in the body.

In addition to basic science applications, the use of spectroscopy to study the energy metabolism of tissue provides clues about many important pathologic states. Measurements of the relative amounts of the important chemical byproducts of metabolic activity made with phosphorus and hydrogen spectroscopy can be used to provide information about some important diseases. For example, pH can be measured from the chemical shift of the inorganic phosphate in the tissue. The pH has shown promise in predicting how permanent the damage produced by stroke will be. Another important instance is lactate. Because lactate is produced by aerobic glycolysis in many cancers, it provides a marker for cancer.

Spectroscopy tends to be slow because of an intrinsically low signal from phosphorus as compared to hydrogen. Phosphorus also has a long T_1, which mandates a long resting time between successive signal acquisitions. This can be a difficulty when it comes to applying spectroscopic techniques to functional imaging, as the phenomena under study in this case are changing. In this setting we have incompatible requirements of long acquisition times for the individual measurements, low spectral leakage, and reasonably short overall imaging time. This is a situation where it would be desirable to use a reduced number of measurements or to pack many long measurements into a short time. Both of these approaches may be implemented with our techniques.

Chemical Shift Imaging. In spectroscopy, the "chemical shift" of each signal provides information about the various molecules contributing to the signal. In particular, phosphorus spectroscopy makes use of the different chemical shifts of different phosphorus compounds to detect the relative amounts of these compounds in a given region. This has shown great promise in studying the metabolism of both cancer and heart disease. The physiology of tissue can be studied by observing the phosphorus compounds such as ADP, ATP and inorganic phosphate [49, 50]. Spectroscopic determination of the pH of a tissue may permit inference about tissue insult. Spectroscopic study of cardiac function is particularly promising, as it is very difficult to predict how permanent the damage to cardiac tissue will be from the temporary loss of perfusion [51, 52].

The chemical shift is due to the slight variation of Larmor frequency among the various phosphorus compounds. This is a consequence of the fact that the magnetic field each phosphorus nucleus feels is the sum of the static magnetic field of the scanner and the fields produced by the surrounding atoms. Consequently, the local Larmor frequency at the nucleus of a phosphorus atom is affected by the particular atoms it is bonded to. For example, the phosphorus in ATP produces three spectral peaks in the signal that represents the three phosphates in ATP. ADP has two of the peaks but lacks the third because one phosphate is missing. As a result, a phosphorus spectrum allows the relative amounts of ATP, ADP, phosphocreatin, and inorganic phosphorus to be estimated from the relative sizes of the peaks. Changes in the relative concentrations can provide information about the damage produced by ischemia or hypoxia. The relative amount of lactate that is present can be estimated from the hydrogen spectra. Lactate is produced during glycolysis, so it reflects the oxygen supply to the tissue and is thought to cause a drop in pH.

Spectra are acquired with no gradient fields because the Larmor frequency is being used to measure small chemical shifts instead of position; i.e., frequency encoding cannot be used to localize the various contributions to the spectra. This is unfortunate; localization is important because physiology changes with blood supply and tissue type. Therefore, several different methods have been devised for localizing the volume from which each spectrum is obtained [53, 54]. The signal from a coil can be used when it is small enough to be sensitive to a limited volume. Alternately, a single volume can be localized by exciting the nuclear spins with a series of slice-selective RF pulses all in orthogonal planes. Another common technique is 1-D or 2-D chemical shift imaging (CSI). CSI uses phase encoding to localize the spectra. The problem here is always the balance between SNR and localization. The signal in spectroscopy is small because the amount of each chemical is relatively small and the sensitivity of phosphorus is small compared with that of hydrogen. As a result spatial resolution is always poor and the acquisition times are long. Hadamard encoding was developed to localize the signal while maintaining the SNR ratio in the spectra [17, 36], and still is actively pursued [35, 36].

We have pointed out [11] that wavelet encoding promises improvement over other localization methods in spectroscopy. First, the effective TR can be made very long. Phosphorus spectroscopy requires T_1 effects to be relatively small so a long effective TR would be very helpful, especially considering the long T_1 of phosphorus [57]. Wavelet encoding schemes can be designed to have a long effective TR and to offer low spectral leakage. Reduced spectral leakage is important for improving the quantitation of the spectra. Quantitation is important in applications like finding the ratios of peak sizes which are used in many applications: e.g., for measurements of the pH of the tissue. Spectral leakage is often a major source of error in the ratios. For example, when a phosphorus signal from skeletal muscle in the chest wall leaks into the spectra from the myocardium, the ratios become contaminated.

The quantitative localization of the signal is generally more important in spectroscopy than in conventional MRI. In conventional imaging, trained observers can generally "read through" localization errors such as the Gibbs artifact. It is rare that a small lesion will be masked by the bright bands of Gibbs ringing.

In spectroscopy, on the other hand, quantitative information is generally needed. Usually, the desired results are the amounts of metabolites which are estimated from ratios of peaks in the spectra. Also, because the pixels are larger than most anatomical structures, it is difficult to "read through" the under-sampling artifacts in spectroscopy. Hadamard encoding has been studied as a method of reducing the under-sampling artifacts in spectroscopic localization. The treatment was basically correct but qualitative. Adapted waveform techniques permit a more general and quantitative treatment.

The Hadamard basis, can be formed as a linear combination of a set of boxcars centered on each pixel in the reconstructed image. Therefore, the linear reconstruction process can isolate any response to one of those boxcars. The "minimum resolution" in the reconstructed spin profile is the size of the boxcar; i.e., the point spread function is the boxcar. The same argument is valid for the Haar wavelets. This argument generalizes easily to all wavelet packet bases which includes the Haar–Hadamard library, as well as the libraries obtained from the Daubechies, Battle-Lemarié, and Meyer wavelets. The boxcar is the scaling function at the smallest scale for the Haar–Hadamard wavelet packet library. All wavelet packet libraries can be formed from linear combinations of the scaling functions for those libraries. Therefore, the point spread functions for those libraries are the scaling functions for those libraries at the smallest scale sampled.

The energy that is displaced by undersampling is roughly the same for all these bases. However, the concentration of that energy is very different. The sinc function, which is the point spread function of the Fourier basis, is much less concentrated than the boxcar. Similarly, the point spread functions for various smooth wavelet packet bases show better localization than Fourier imaging. This suggests a definite advantage for adapted waveform encoding which we will pursue in detail in a later publication.

REFERENCES

[1] L. Axel and L. Dougherty, "Heart wall motion: improved method of spatial modulation of magnetization for MR imaging," *Radiology*, vol. 172, pp. 349–350, 1989.

[2] J. W. Belliveau, B. R. Rosen, H. L. Kantor, et al., "Functional cerebral imaging by susceptibility-contrast NMR," *Magn. Res. Med.*, vol. 14, p. 538, 1990.

[3] V. A. Ferrari, J. A. C. Lima, L. Axel, C. M. Kramer, M. R. Llaneras, L. C. Palmon, B. Tallant, and N. Reichek, "Regional changes in segmental rigid body motion and deformation after myocardial infarction by magnetic resonance tissue tagging," *Proc. Soc. of Magnetic Resonance in Medicine*, p. 16, Berlin, Aug. 1992.

[4] F. Shellock, J. Mink, and J. Fox, "Patellofemoral joint: kinematic MR imaging to assess tracking abnormalities," *Radiology*, vol. 168, p. 551, 1988.

[5] R. Turner, D. LeBihan, C. T. W. Moonen, D. Despres, and J. Frank, "Echo-planar time course MRI of cat brain oxygenation changes," *Mag. Res. Med.*, vol. 22, pp. 159, 1991.

[6] A. Villringer, B. R. Rosen, J. W. Belliveau, et al., "Dynamic imaging with lanthanide chelates in normal brain: contrast due to magnetic susceptibility effects," *Mag. Res. Med.*, vol. 6, p. 164, 1988.

[7] V. Wedeen, "Magnetic resonance imaging of myocardial kinematics: a technique to detect, localize and quantify the strain rates of the active human myocardium," *Mag. Res. Med.*, vol. 27, p. 52, 1992.

[8] E. A. Zerhouni, D. M. Parish, W. J. Rodgers, A. Yang, and E. P. Shapiro, "Human heart tagging with MR imaging—a method for noninvasive assessment of myocardial motion," *Radiology*, vol. 169, pp. 59–63, 1988.

[9] M. Vetterli and J. Kovacevic, *Wavelets and Subband Coding*. Englewood Cliffs, NJ: Prentice Hall PTR, pp. 369–389, 1995.

[10] M. Wickerhauser, *Adapted Wavelet Analysis from Theory to Software*. A. K. Peters, Wellesley. 1994.

[11] D. M. Healy, J. Lu, and J. B. Weaver, "Two applications of wavelets and related techniques in medical imaging," *Ann. Biomed. Eng.*, vol. 23, pp. 637–655, 1995.

[12] J. Lu, D. M. Healy Jr., and J. B. Weaver, "Contrast enhancement of medical images using multiscale edge representation," *Opt. Eng.*, vol. 33, pp. 2151–2161.

[13] J. Lu, J. B. Weaver, and D. M. Healy Jr., "Noise reduction with multiscale edge representation and perceptual criteria," *Proc. IEEE-SP Int. Symp. on Time-Frequency and Time-Scale Analysis*, pp. 555–558, Victoria, B.C., Oct. 1992.

[14] M. Loève, *Probability Theory*, Princeton, NJ: Van Nostrand, p. 478, 1955.

[15] D. M. Healy and J. B. Weaver, "Two applications of wavelet transforms in MR imaging," *IEEE Trans. Inform. Theory*, vol. 38, pp. 840–860, 1992.

[16] J. B. Weaver and D. M. Healy, Jr., "Signal to noise ratios and effective repetition times for wavelet and adapted wavelet encoding," *J. Mag. Res.*, Series A, vol. 113, pp. 1–10, 1995.

[17] G. Goelman, V. H. Subramanian, and J. S. Leigh, "Transverse Hadamard spectroscopic imaging technique," *J. Mag. Res.*, vol. 89, pp. 437–454, 1990.

[18] L. P. Panych, P. D. Jakab, and F. A. Jolesz, "An implementation of wavelet encoded magnetic resonance imaging," *J. Mag. Res. Imag.*, vol. 3, pp. 649–655, 1993.

[19] R. D. Peters and M. L. Wood, "Practical considerations for implementation of wavelet encoding," *Proc. Soc. of Magnetic Resonance in Medicine*, p, 1212, New York, Aug. 1993.

[20] J. R. Driscoll, D. M. Healy, and G. Isaak, "Scheduling dyadic intervals," *Discrete Appl. Math.*, (submitted for publication) 1995.

[21] J. B. Weaver, Y. Xu, D. M. Healy and J. R. Driscoll, "Wavelet encoded MR imaging," *Mag. Res. Med.*, vol. 24, pp. 275–287, 1992.

[22] R. R. Coifman and M. V. Wickerhauser, "Entropy based algorithms for best basis selection," *IEEE Trans. Inf. Theory*, vol. 38, pp. 713–718, 1992.

[23] Y. Cao and D. Levin, "Feature recognizing MRI," *Magn. Res. Med.*, vol. 30, pp. 305–317, 1993.

[24] G. Zientara, L. Panych, and F. A. Jolesz, *Mag. Res. Med.*, vol. 32, p. 268, 1994.

[25] J. B. Weaver and D. M. Healy, *Proc. IEEE Int. Conf. Image Proc.*, III-35. Austin, TX, 1994.

[26] Z. H. Cho, O. Nalcioglu, and H. W. Park, "Methods and algorithms for Fourier-transform nuclear magnetic resonance tomography," *J. Opt. Soc. Am.*, vol. 4, pp. 923–932, 1987.

[27] P. Mansfield and P. G. Morris, *NMR Imaging in Biomedicine*. San Diego: Academic Press, p. 354, 1982.

[28] C. P. Slichter, *Principles of Magnetic Resonance*. New York: Springer Verlag, p. 397, 1980.

[29] R. A. Brooks, G. DiChiro, and M. R. Keller, "Explanation of cerebral white-gray contrast in computed tomography," *J. Comput. Assisted Tomog.*, vol. 4, pp. 489–491, 1980.

[30] R. T. Droege, S. N. Wiencer, M. S. Rzeszotarski, G. N. Holland, and I. R. Young, "Nuclear magnetic resonance: a gray scale model for head images," *Radiology*, vol. 148, pp. 763–771, 1983.

[31] S. Conolly, D. Nishimura, and A. Macovski, "Optimal control solutions to the magnetic resonance selective excitation problem," *IEEE Trans. Med. Imag.*, vol. MI-5, no. 2, pp. 106–115, 1986.

[32] F. A. Grunbaum and A. Hasenfeld, "An exploration of the invertibility of the Bloch transform," *Inverse Problems*, vol. 2, pp. 75–81, 1986.

[33] H. W. Korin, J. P. Felmlee, R. L. Ehman, and S. J. Riederer, "Adaptive technique for three-dimensional MR imaging of moving structures," *Radiology*, vol. 177, pp. 217–221, 1990.

[34] D. J. Lurie, "A systematic design procedure for selective pulses in NMR imaging," *Mag. Res. Imag.*, vol. 3, pp. 235–243, 1985.

[35] A. E. Yagle, "Inversion of the Bloch transform in magnetic resonance imaging using asymmetric two-component inverse scattering," *Inverse Problems*, vol. 6, pp. 131–151, 1990.

[36] G. Goelman and J. S. Leigh, "B1-insensitive Hadamard spectroscopic imaging technique." *J. Mag. Res.*, vol. 91, pp. 93–101, 1991.

[37] N. J. A. Sloane, "Hadamard and other discrete transforms in spectroscopy." In *Fourier, Hadamard, and Hilbert transforms in Chemistry*, A. G. Marshall (ed.) New York: Plenum, pp. 45–67, 1982.

[38] W. P. Aue, S. Muller, T. A. Cross, and Seelig, "Volume selective excitation: a novel approach to topical NMR," *J. Mag. Res.*, vol. 56, pp. 350–354, 1984.

[39] P. Mansfield, A. A. Maudsley, and T. Baines, "Fast scan proton density by NMR," *J. Phys. E. Sci. Instrum.*, vol. 9, pp. 271–278, 1976.

[40] H. Kooijman, "Short low power self-refocussing pulse," *Proc. Soc. of Magnetic Resonance in Medicine*, p. 685, San Franciso, Aug. 1991.

[41] J. Pauly, S. Conolly, D. Nishimura, and A. Macovski, "Slice-selective excitation for very short T2 species," *Proc. Soc. of Magnetic Resonance in Medicine*, p. 28, San Francisco, Aug. 1991.

[42] M. Shinnar and J. S. Leigh, 'Frequency response of soft pulses," *J. Mag. Res.*, vol. 75, pp. 502–505, 1987.

[43] I. Daubechies, "The wavelet transform, time-frequency localization, and signal analysis," *IEEE Trans. Inform. Theory*, vol. 36, pp. 961–1005, 1990.

[44] I. Daubechies, "Orthonormal bases of compactly supported wavelets," *Comm. Pure Appl. Math.*, vol. 41, pp. 909–996, 1988.

[45] S. Mallat, "Multiresolution approximation and wavelets," *Trans. AMS*, vol. 135, pp. 69–88, 1989.

[46] Y. Meyer, "Principe d'incertitude, bases hilbertiennes, et algèbres d'opérateurs." In *Séminaire Bourbaki, 1985/86, exposés 651–668*. Paris: Société Mathématique de France, pp. 209–223, 1987.

[47] F. Bloch, "Nuclear induction," *Phys. Rev.*, vol. 70, p. 460, 1946.

[48] J. B. Weaver, D. M. Healy Jr., D. Crean, and Y. Xu, "Wavelet encoding with smooth wavelets: short RF pulses," *Proc. Soc. of Magnetic Resonance in Medicine*, p. 4264, Berlin, Aug. 1992.

[49] G. Radda, P. J. Bore, and B. J. Rajagopalan, "Clinical aspects of 31-P NMR spectroscopy," *Br. Med. Bull.*, vol. 40, pp. 155–159, 1984.

[50] P. J. Bore, "Principles and applications of phosphorus magnetic resonance spectroscopy." In *Magnetic Resonance Annual*, H. Y. Kressel (ed.) New York: Raven Press, pp. 45–69, 1985.

[51] B. Massie, G. G. Schwartz, J. Garcia, S. Steinman, M. W. Weiner, and J. Wisneski, "Changes in energy phosphates with increased left ventricular workload represent "demand ischemia"," *Proc. Soc. of Magnetic Resonance in Medicine*, p. 343, Berlin, Aug. 1992.

[52] R. S. Menon, K. Hendrich, X. Hu, and K. Ugurbil, "31P NMR spectroscopy of the human heart at 4 T: detection of uncontaminated cardiac spectra and differentiation of subepicardium and subendocardium," *Proc. Soc. of Magnetic Resonance in Medicine*, p. 342, Berlin, Aug. 1992.

[53] W. P. Aue, "Localization methods for in vivo nuclear magnetic resonance spectroscopy," *Rev. Mag. Res. Med.*, vol. 1, pp. 21–72, 1986.

[54] R. E. Lenkinski, *Clinical MR Spectroscopy*, Syllabus: Special Course. RNSA Publications, p. 279, 1990.

[55] G. Goelman and J. S. Leigh, "The use of high order Hadamard matrices for metabolic maps," *Proc. Soc. of Magnetic Resonance in Medicine*, p. 3823, Berlin, Aug. 1992.

[56] P. Chen and O. Nalcioglu, "3-D multi-voxel NMR spectroscopy by Hadamard encoding," *Proc. Soc. Magnetic Resonance in Medicine*, p. 3824, Berlin, Aug. 1992.

[57] S. J. Li, G. Y. Jin, and J. E. Mooulder, "Effects of local irradiation on spin-lattice relaxation time of phosphate metabolites in mouse tumors monitored by 31-P magnetic resonance spectroscopy," *Mag. Reson. Med.*, vol. 23, no. 2, pp. 302–310, 1992.

Chapter 24

A Tutorial Overview
of a Stabilization Algorithm
for Limited-Angle Tomography*

Tom Olson

24.1. INTRODUCTION

Many problems in applied mathematics involve recovering a function f from measurements of Lf, where L is a known operator. If L is a linear operator with a bounded inverse, then f can be recovered from noisy data $Lf + n$ via standard tech-niques with little difficulty. The recovery of f from noisy data $Lf + n$ is much more difficult if L is an operator whose inverse is unbounded.

In this chapter we will study the recovery of a function $f\colon \mathbb{R}^2 \to \mathbb{R}$, from limited knowledge of its Fourier transform $\hat{f}$. Thus we are interested in reconstructing f from Lf where L is an operator which reflects the limited knowledge of the Fourier transform of f. This problem is motivated by the limited-angle tomography problem. This recovery problem is ill-posed and unsolvable without *a priori* assumptions about f. With proper *a priori* knowledge, however, this problem is solvable. For example, if f is compactly supported, then $\hat{f}$ will be an analytic function, and therefore $\hat{f}$ will be uniquely determined by its values on any region containing a limit point [1].

Classical analytic continuation techniques are not numerically feasible [2] since the operator L will generally not have a bounded inverse. A technique for doing analytic continuation was introduced by Papoulis in [3]. This algorithm depends upon the eigenfunctions of the continuous time and band-limiting operators, which were extensively studied in the classical works of Slepian and Pollack, [4]. The convergence of the algorithm is dependent upon the eigenvalues of the joint time- and band-limiting operators, and except in extreme cases, this algorithm is not feasible numerically [5].

*This work was supported in part by DARPA as administered by the AFOSR under contract AFOSR-90-0292.

We will model these data recovery problems as the inversion of a compact operator L (for limited data); where

$$Lf = \mathcal{X}_S(\widehat{\mathcal{X}_C \hat{f}})$$

S is the common support of the functions under consideration, C is the set where we can measure the Fourier transform of f, and $\mathcal{X}_S$ is a characteristic, or indicator function on the set S.

We will study the inversion of L, from a discrete signal-processing perspective. In this context, the continuous operator L is naturally viewed as the limiting case of a series of discrete operators. Since L will generally be compact and self-adjoint, its spectrum can been analyzed via standard techniques. Real-world problems are generated from discretely sampled data sets. Therefore we believe that it is more informative to study the spectra of the discrete approximations to L rather than the spectrum of L itself.

Our main tool for analyzing the spectra of these discrete approximations to L will be the theory of finite Toeplitz forms, originally introduced by Szegö [6]. We will show that the study of these finite Toeplitz forms will give us some clues concerning the construction of an accurate, stable inversion for L, even when the continuous spectrum of L suggests that it is not invertible.

24.2. BACKGROUND AND DEFINITIONS

24.2.1 The Radon Transform

The fundamental theoretical tool behind tomography has always been the Radon transform [2,7]; see Fig. 24-1. The Radon transform, which is sometimes referred to as the X-ray transform or the projection transform of a function $f : \mathbb{R}^2 \to \mathbb{R}$ is defined by

$$P_f(\theta, \theta \cdot \vec{x}) = \int_{\mathbb{R}} f(x + t\theta^\perp) dt \tag{24-1}$$

It follows that $P_f(\theta, \theta \cdot \vec{x})$ is the line integral of f through $x \in \mathbb{R}^2$ and perpendicular to θ. We will often abuse this notation by associating the vector θ with the angle θ, between θ and the positive x-axis.

An essential property of the Radon transform is that the one-dimensional Fourier transform of $P_f(\theta, \theta \cdot x)$ with respect to the variable $t \equiv \theta \cdot x$ is a line through the two-dimensional Fourier transform of f, i.e.,

$$F_1[P_f(\theta, \theta \cdot x)] = F_2(f)(s\theta) \equiv \hat{f}(s\theta). \tag{24-2}$$

Using this fact and a polar Fourier inversion formula, one can easily relate the Radon transform $P_f(\theta, \theta \cdot x)$ to $f(x)$ through the reconstruction formula

$$f(x) = \int_{S^1} \int_{\mathbb{R}} \hat{f}(s\theta)|s|e^{i(x \cdot s\theta)} ds d\theta = \int_{S^1} \int_{\mathbb{R}} F_1\big(P_f(\theta, \theta \cdot x)\big)(s)|s|e^{i(x \cdot s\theta)} ds d\theta \tag{24-3}$$

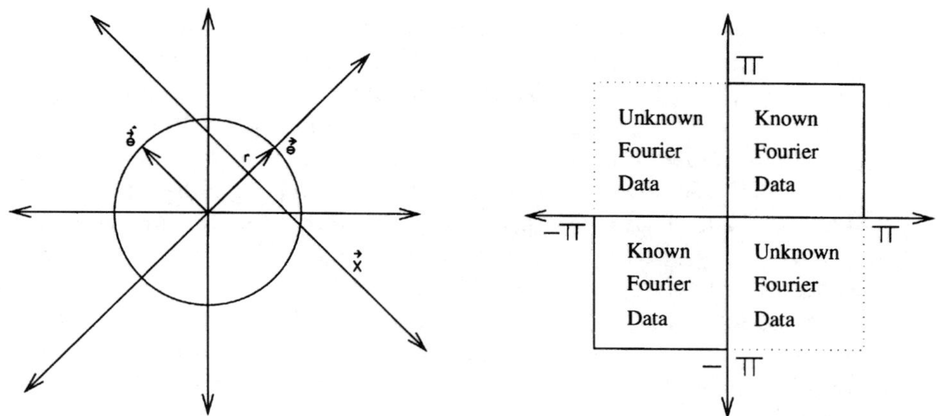

Figure 24-1   The Radon transform is illustrated on the left. The notation $\vec{\theta}^{\perp}$ stands for the vector which is perpendicular to $\vec{\theta}$. The line integral $R_f(\theta, r) = \int f(\vec{\theta} + t\vec{\theta}^{\perp})dt$ is taken along the illustrated line at a distance r from the origin. On the right, above, is an illustration of the 90 degree limited-angle problem.

The inner integral in (24-3) is simply a one-dimensional inverse Fourier transform. Insertion of an appropriate band-limiting window into (24-3) yields the standard filtered backprojection formula [2]

$$f_w(x) = \int_{S^1} \int_{\mathbb{R}} F_1\left(P_f(\theta, \theta \cdot x)\right)(w(s)|s|)e^{i(x \cdot \theta)s} ds d\theta = \int_{S^1} P_f(\theta, \theta \cdot x) * F_1^{-1}(w(s)|s|)d\theta$$

$$(24\text{-}4)$$

24.2.2 Tomography and Limited-Angle Tomography

In this chapter, tomography refers to the recovery of a two-dimensional density $f: \mathbb{R}^2 \to \mathbb{R}$ from one-dimensional line integrals via an inversion formula such as (24-4). Tomography will generally refer to the recovery of such a two-dimensional density from a full complement of one-dimensional line integrals, i.e., $P_f(\theta, \theta \cdot x)$ for all $\theta \in S^1$, or an evenly subsampled version of these line integrals. Typical sampling conditions can be found in Natterer [2].

Limited-angle tomography refers to the recovery of a function f from a proper subset of its line integrals. More specifically, in limited-angle tomography only the line integrals $P_f(\theta, \theta \cdot x)$ where $|\theta| < \Phi < \pi/2$ are available for the inversion. From (24-2) we know that the one-dimensional Fourier transform of each projection corresponds to a line through the origin of the two-dimensional Fourier transform of the original image.

The difficulty of recovering a function from limited-angle tomographic data can be understood through the reconstruction formulas (24-2), (24-3), and (24-4). We can see via (24-2) that if the projection data $P_f(\theta, \theta \cdot x)$ is only available for a restricted angular range, $-\Phi < \theta < \Phi$, where $\Phi < \pi/2$ then the Fourier data

$F_2(f)(s\theta) \equiv \hat{f}(s\theta)$ will only be available for the restricted angular range $-\Phi < \theta < \Phi$. We know from (24-3), and (24-4) that all of the Fourier, or projection data is needed to accurately reconstruct f. Thus we are forced to try to recover an unknown portion of the Fourier transform of f, as is illustrated in Fig. 24-2.

If we know nothing about the objective function $f\colon \mathbb{R}^2 \to \mathbb{R}$, then this reconstruction would certainly be impossible. Usually, though, the support of the object is finite and known. It is also generally the case that the function is positive and real. Therefore, we assume throughout that we are interested in reconstructing a bounded function $f\colon S \cap \mathbb{R}^2 \to \mathbb{R}$, from the values of its Fourier transform, where S is a compact set, containing the origin. This will imply that the Fourier transform of f will be an entire function of finite exponential type, and therefore the continuous problem is solvable. Classical analytic continuation methods, however, are not numerically feasible [2, 5].

The creation of a stable numerical inversion for this problem is the subject of this chapter.

OVERVIEW:

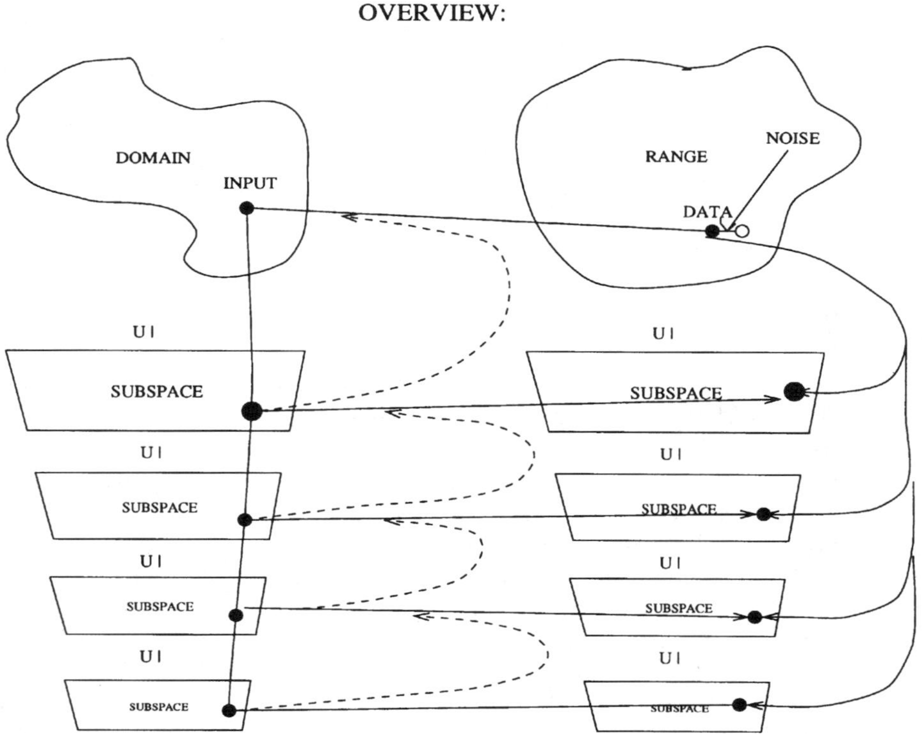

Figure 24-2 The nested subspace approach. Although the SVD also generates nested subspaces, these subspaces will not be generated by orthogonal sets of functions. We need to correlate the difference subspaces, in order to mollify the inversion.

24.2.3 Physical Motivation and Prior Work

The limited-angle tomography problem arises in a number of contexts. In biology, transmission electron microscopy is used to image the structure of chromatin fibers in tissue. Because of the rapid absorption of the electrons an extremely thin slice of tissue is prepared. This slice becomes prohibitively thick when the angle of incidence of the rays goes beyond a certain threshold. Therefore only projections at angles within approximately 60 degrees of the normal to the slice can be utilized.

Limited angle tomography has been extensively studied. It has been proven [8–10] that when the singular values of the associated operator are viewed as a function of the angular limit Φ, they converge to zero extremely quickly and as a result the problem is highly ill-conditioned.

One approach to the problem of reconstructing a density function from limited-angle data is to compute the projection of the known data onto the the range of the Radon transform [2]. The result is a completed data set, which can be processed via typical reconstruction algorithms to produce an approximation to the original density function. The range theorems for the Radon transform make it possible to recover the low-frequency components of the image through this process. The crucial step in this process is the determination of the coefficients of a trigonometric polynomial, from nonuniform samples. The high-frequency components of the image require the recovery of high order trigonometric polynomials from nonuniform samples, however. This process is notoriously ill-conditioned, and as a result the algorithm will not perform well when the angular limit Φ is not close to $\pi/2$ and a high-resolution reconstruction is desired.

In [11] a alternative reconstruction algorithm was proposed. It was speculated that it might be close to optimal. It was shown in [12], however, that this algorithm formed a commutative diagram with filtered back projection, and as a result was of little use in limited-angle tomography.

Finally, a thorough review of other approaches to the limited-angle tomography problem can be found in Natterer [2].

24.3. LIMITATIONS OF THE SINGULAR VALUE DECOMPOSITION

24.3.1 Unbounded Inverses and Approximate Identities

The standard numerical method for inverting a linear operator is the singular value decomposition (SVD). The singular value decomposition of a linear operator L consists of the normalized eigenfunctions of L^*L, $\{\psi_n\}$, and the functions $\{\phi_n\} = \{1/\sqrt{\lambda_n} L\psi_n\}$ where λ_n is the eigenvalue corresponding to ψ_n. The vectors ψ_n and ϕ_n are the singular vectors, and the values $\sigma_n = \sqrt{\lambda_n}$ are the singular values. It follows that both $\{\psi_n\}$ and $\{1/\sqrt{\lambda_n} L\psi_n\}$ are orthonormal sets for the domain and range of L. Most important is the fact that these vectors allow a decomposition of L given by

$$Lf = \sum_n \sigma_n \langle f, \psi_n \rangle \phi_n. \tag{24-5}$$

Moreover, the singular value decomposition also yields an inversion formula

$$f = \sum_n \sigma_n^{-1} \langle Lf, \phi_n \rangle \psi_n. \tag{24-6}$$

A great number of problems in applied mathematics can be reduced to the inversion of a linear operator, and as a result, the SVD has been extremely important in applied and numerical mathematics.

Unfortunately, since the operator L is usually compact, the values $\sigma_n \to 0$ as $n \to \infty$ [13]. Thus when only finite precision is available in both the measurements and computer facilities, the usefulness of (24-6) is somewhat limited for two reasons. The first is that the storage of the data Lf with finite precision will cause the higher-order terms in the sum (24-5) to be beyond the wordsize of the machine, and thus they will be rounded to zero. In addition, the multipliers σ_n^{-1} in (24-6) will approach ∞ as $n \to \infty$. This makes the computation of the higher-order terms in the sum in (24-6) very unstable or impossible and therefore an altered, truncated version of (24-6) is often used.

These mollified methods, such as the truncated SVD, and Tikhonov–Phillips [2] regularization can usually be represented in the form

$$Mf = \sum_n w(n)\sigma_n^{-1} \langle Lf, \phi_n \rangle \psi_n$$

where the windowing function $w(n)$ is chosen so that $w(n)\sigma_n^{-1} \in l^2$. For instance the truncated SVD would define $w(n) = 1$ for $n < N$ and zero for $n \geq N$. Thus an approximate inversion is obtained, which is computationally stable, but not exact.

The second problem with the SVD is that if there are two distinct classes of singular values: those which are close to a fixed value $b > 0$ and those which are nearly 0, then the inversion formulas will all reduce to an approximation of a truncated SVD. If in addition, the operator is self-adjoint, the functions $\{\psi_n\}$ will be equal to the functions $\{\phi_n\}$, and will be eigenvectors. In this case the truncated SVD will act as an approximate identity, since division by the singular values which are all in the vicinity of b will only scale the function Lf, and the singular values which are nearly 0 will be truncated from the expansion. Thus the effort of constructing and inverting the SVD will not cause a noticeable improvement in the final result.

This is the case in limited-angle tomography. The operator is compact and self-adjoint, and the singular values, or eigenvalues, of the operator separate distinctly into two classes, those which are nearly 2 and those which are nearly 0.

24.3.2. Uncorrelated, Exact Bases versus Induced Correlations and Redundant Bases

The SVD produces exact orthonormal bases for the range and domain of any compact linear operator. Although there are many instances when an exact representation is useful, we believe that an exact representation is not desirable in this situation.

When reconstruction problems or inversion problems are ill-conditioned, we believe that an over complete basis, which is correlated and redundant, is more useful than an exact basis. It is well known that in limited-angle tomography, and many other imaging problems, low-resolution information can easily be reconstructed. With an exact, uncorrelated basis, this low-resolution information will not help to reconstruct the high-frequency details which are difficult to recover. If a correlated, redundant basis, is used, however, then the low-resolution information can aid in recovering the high-frequency details.

We will utilize nonlinear or linear constraints to induce correlations between the coefficients of an orthonormal basis. In our case, the initial nonlinear constraints are positivity, compact support, and the known data. After we reconstruct the low-resolution information in the image, we will use the low-resolution information as an additional nonlinear constraint.

24.3.3 Decreasing Signal-To-Noise Ratio

The linear projection onto convex sets (POCS) algorithm, as outlined in [3], uses the singular functions of the time- and band-limiting operators to construct an approximate inverse to the problem. More specifically, if f is the desired function, with the eigenfunction expansion for $f = \sum c_n e_n$, and we observe a blurred version of f, namely Lf, then the K-th step of POCS yields the result

$$P^K(Lf) = \sum (1 - (1 - \lambda_n)^K) c_n e_n$$

We observe that as $K \to \infty$ this converges to f. In the presence of white noise, however, we will observe $Lf = \sum (\lambda_n c_n + g_n) e_n$, where g_n are i.i.d. observations of white noise. Now the reconstruction procedure on the K-th step will yield

$$P^K(f + g) = \sum (1 - (1 - \lambda_n)^K)\left(c_n + \frac{1}{\lambda_n} g_n\right) e_n$$

As $K \to \infty$, it follows that the signal-to-noise ratio of the solution will go to zero (assuming that L was an infinite dimensional compact operator). Moreover, the solution will converge to

$$\sum \left(c_n + \frac{1}{\lambda_n} g_n\right) e_n$$

Since it is usually the case that $\lambda_n \to 0$, this solution is not desirable.

24.4. MOLLIFICATION METHODS

24.4.1 Szegö's Theory for Finite Toeplitz Operators

In this section we will study the spectral properties of discrete Toeplitz operators generated from compact operators of the type

$$Lf = \mathcal{X}_S F_2^{-1}(\hat{f} \mathcal{X}_C) = \mathcal{X}_S(\mathcal{K}*\})$$

where $*$ denotes convolution. We will study the spectral properties of these discrete operators, as a function of the discretization N. We will then use this study to introduce an altered inversion formula which will prove to be stable.

Szegö's one-dimensional theory of Toeplitz operators extends, essentially with a change of notation, to n-dimensions. For completeness, we will present a small portion of this here. What follows can be found in the one-dimensional case in [14].

Much is known about the eigenvalues of Toeplitz forms [14]. As the discretization, or size of the pixel array, $N \to \infty$, the eigenvalues become distributionally equivalent to $\hat{K}$ in the following sense. The distribution of the eigenvalues $\{\lambda_\mu^N\}_{\mu \in S_N}$ of finite Toeplitz operators, converges to the distribution of the values of $K(s)$, i.e.,

$$\lim_{N \to \infty} \frac{|\{\lambda_\mu^N | a < \lambda < b\}|}{|S_N|} = \frac{1}{|\Omega|} |\{s \,|\, a < \hat{K}(s) < b\}|, \qquad (24\text{-}7)$$

as long as the measure of the sets where $F(s) = a$ or $F(s) = b$ is zero. In words, the proportion of eigenvalues in a given interval, converges to the measure of the proportion of the set where F is in the corresponding interval.

To gain some insight into the ill-conditioned nature of inverting any limited-angle problem, we will apply (24-7) to the finite Toeplitz forms associate with the Fourier transform of our kernel $\hat{K} = 2\mathcal{X}_C$. Since all of the eigenvalues are positive, it follows that

$$\lim_{N \to \infty} \frac{|\{\lambda_\mu^N | 0 < \lambda < \epsilon\}|}{|S_N|} = \lim_{N \to \infty} \frac{|\{\lambda_\mu^N | -\epsilon < \lambda < \epsilon\}|}{|S_N|}$$

$$= \frac{1}{|\Omega|} |\{s \,|\, -\epsilon < \hat{K}(s) < \epsilon\}| = \frac{|\Omega| - |C|}{|\Omega|} \qquad (24\text{-}8)$$

Thus the proportion of eigenvalues which eventually tend toward 0 is exactly the same as the proportion of Ω where K is 0 or the measure of Ω/C. More explicitly,

$$\lim_{N \to \infty} \frac{|\{\lambda_\mu^N | 0 < \lambda < \epsilon\}|}{|S_N|} = \frac{|\text{Unknown data}|}{|\text{Necessary data}|} \qquad (24\text{-}9)$$

In the case which we have been studying, i.e. the 90 degree case, half of the eigenvectors must approach 0 and half must approach 2.

This discussion highlights the problems associated with using a truncated or mollified SVD to invert this type of operator, when the discretization N is large. There will be eigenvalues which are essentially zero, which will represent the unknown data, and eigenvalues which are near 2, which represent the known data. If N is large, the truncated SVD will only rescale the final image, by dividing the known data by 2. The unknown coefficients will be truncated to zero, and no new information will be gathered.

24.4.2 Limited-Angle Spectra

We have noted above that the spectra of limited data operators separate into two classes. There are eigenvalues which are essentially one and eigenvalues which are essentially zero. Because of the above discussion of the geometry of the eigenvectors, it is not at all surprising that half of the eigenvalues are greater than 1/2, and nearly 1, and half of the eigenvalues are less than 1/2, and nearly 0.

For the 90 degree limited-angle problem, 8944 out of the 16 384 eigenvalues are greater than 0.01, or 54%. Thus a truncated SVD with a tolerance of 0.01 will recover only 4% more information than was present in the original 50% of the known data. The difference in image quality due to this recovered information is negligible. Thus a truncated SVD, used at a discretization of 128 $\times$ 128 will act essentially as an identity operator on the blurred image, rather than an effective inversion tool.

We now consider the spectra of the limited-angle operators from smaller discretizations. In Fig. 24-3, we have plotted the spectra of the limited-angle operators at various discretizations (note that normalizing factors cause them to tend toward either 0 or 2. From the figure, we see that multiresolution analysis can be used to mollify the inversion. The low-resolution operators are not massively ill-conditioned, and we should be able to reverse the effect of the blurring at low resolutions. We want to use this information to help reverse the blurring at higher resolutions.

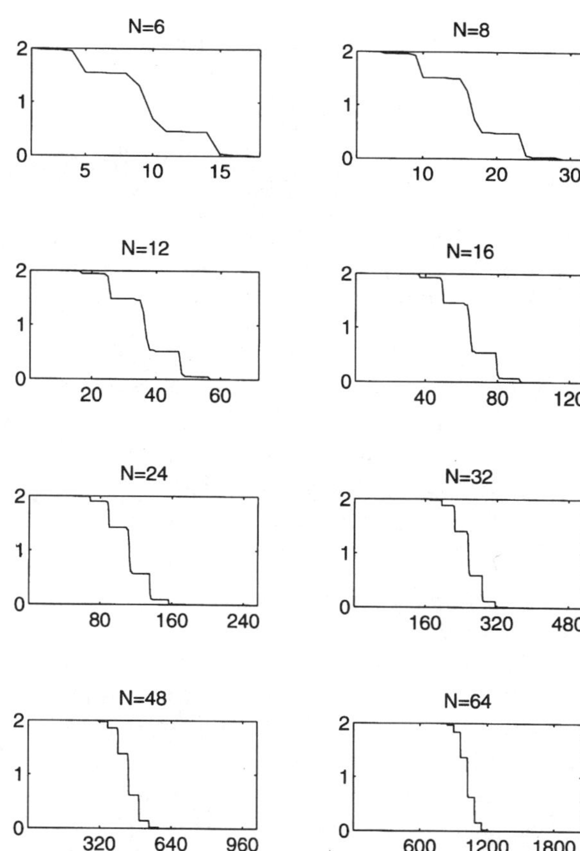

Figure 24-3   The spectra of the 90 degree limited-angle operators. Notice that as the discretization becomes large, the eigenvalues converge to either 0 or 2. This is consistent with the Szegö theory of finite Toeplitz forms.

24.4.3 Uncertainty Principles and Signal Recovery

Another way to understand the difficulties of signal recovery, and the advantages of our approach, is via the work of Donoho and Stark on uncertainty principles [5]. The classical uncertainty principle states that if a function $f(t)$ is essentially zero outside an interval of length Δt and its Fourier transform $\hat{f}(w)$ is essentially zero outside an interval of length Δw, then

$$\Delta t \Delta w \geq 1$$

A basic tenet of [5] is the following. Suppose that you are trying to recover a signal $f(t)$ which is nonzero in a region of size Δt, from knowledge of the Fourier transform $\hat{f}(\omega)$. Suppose further that for some reason you do not know the values of $\hat{f}(\omega)$ on some interval I where

$$|I| < \Delta w \equiv \frac{1}{\Delta t}$$

Then by the uncertainty principle $\hat{f}(\omega)$ cannot be essentially zero outside of I, and thus you can observe $f(t)$ from $\hat{f}(\omega)$ where $\omega \notin I$. Since this problem is linear, this is equivalent to stating that the problem is invertible if and only if the kernel of the operator is 0.

Donoho and Stark constructed discrete analogs of the uncertainty principle which are of use here. The first of these is the following

Lemma 1. *Let $\{x_t\}$ be a discrete sequence with at most N_t nonzero elements, and let $\{\hat{x}_\omega\}$ be its discrete Fourier transform. Then $\{\hat{x}_\omega\}$ cannot have N_t consecutive zeros.*

This Lemma can be generalized to an n-dimensional discrete Fourier transform, with a proof essentially identical to that in [5], except that the Vandermonde matrix of [5] will be replaced with a tensor product of Vandermonde matrices. We state this below.

Lemma 2. *Let x_t be a two-dimensional sequence on $S = \mathbb{Z}_N \times \mathbb{Z}_N$, and let $\hat{x}_\omega$ be its two-dimensional discrete Fourier transform. If x_t is nonzero only on a block of the type $R = [r_1, r_1 + 1, ..., r_1 + n] \times [r_2, r_2 + 1, ..., r_2 + n] \subset S$ then $\hat{x}_\omega$ cannot vanish on a block $Q = [q_1, q_1 + 1, ..., q_1 + n] \times [q_2, q_2 + 1, ..., q_2 + n] \subset S$.*

We use this to construct the following two-dimensional result, which applies directly to our problem; see Fig. 24-4.

Suppose that a function f_t, where $t \in S$ is nonzero only on the interior $N/2 \times N/2$ region of S. Then we can recover it from the values of its discrete Fourier transform on the region $C = [0, 1, ..., N/2 - 1] \times [0, 1, ..., N/2 - 1] \cup [N/2, N/2 + 1, ..., N - 1] \times [N/2, N/2 + 1, ..., N - 1]$. This result follows directly from the above Lemma, since the problem is linear and $\hat{f}(\omega)$ cannot vanish on either $[0, 1, ..., N/2 - 1] \times [N/2, N/2 + 1, ..., N - 1]$ or $[N/2, N/2 + 1, ..., N - 1] \times [0, 1, ..., N/2 - 1]$, thus the problem must be nonsingular. This is of interest in

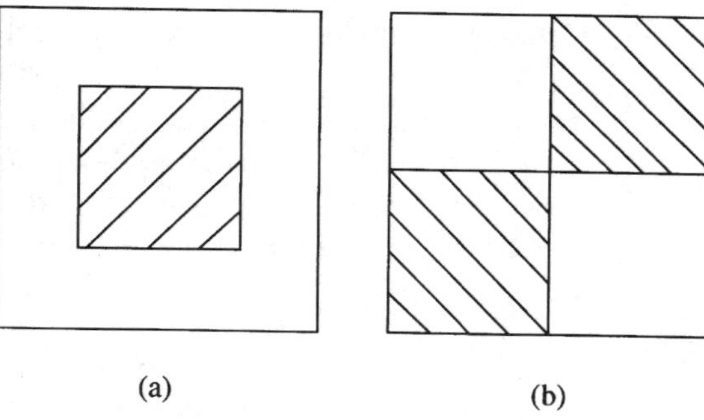

Figure 24-4 The discrete invertibility result. In (a), we illustrate the support of f, on the discrete lattice $S_N = \mathbb{Z}_N \times \mathbb{Z}_N$. In (b), the known portions of the discrete Fourier transform $\hat{f}$, are illustrated. These known regions correspond to the known regions of the Fourier transform in limited-angle tomography. The discrete invertibility result implies that $\hat{f}$ cannot vanish in these known regions, which implies invertibility.

(a)

(b)

our case, because the set C is the discrete analog of the first and third quadrants in the continuous case.

Thus we find that if we over-sample the discrete Fourier transform of a function f_t by a factor of 2, the recovery of the original function from limited-angle Fourier data is nonsingular. This corresponds with having a "true" rather than a modulo convolution on the support of the image. The problem is very ill-conditioned, however, if the discretization is large.

We once again return to the "moral" argument that a signal recovery problem is invertible, if the data to be retrieved cannot vanish in the unknown region. Although the data cannot vanish in the unknown region, they can essentially vanish, if the discretization is large. Thus, if we try to recover fine details from the beginning, the problem will be very ill-conditioned. If, however, we can recover a low-resolution reconstruction first, and use this low resolution reconstruction to mollify the fine-resolution reconstructions, then we might have some hope.

24.4.4 Nonlinear Constraints, Induced Correlations, and POCS

We assume throughout that the functions which we are recovering are compactly supported, and have finite support, which is known to some degree. Moreover, we know a portion of the Fourier transform of these functions. Finally, we assume that there is some maximum density which can be observed, i.e., gold markers in electron microscopy.

Thus our algorithm will assume that the image f obeys the following constraints: (1) $0 \leq f \leq M$, where M is known; (2) $\hat{f}\mathcal{X}_C = g$ where g is the given data; (3) $\mathrm{supp} f \subseteq \Omega$.

The techniques which we will use to stabilize the algorithm are highlighted below.

- *Multiresolution analysis*: It is well known that a low-resolution version of f can be reconstructed accurately. This low-resolution version of f is then added to the constraints. If we utilize projection operators P_k for the projections onto these resolution subspaces, it then follows that we are interested in using our ability to recover $P_1 f$, or a low-resolution version of f, to aid us in the recovery of f on higher subspaces.

- *Preservation of nonlinear constraints*: We would like the nonlinear constraints outlined earlier to hold on these subspaces. In other words, we would like $P_n f$ to be positive, and compactly supported. This is the case if we use the Haar functions for our reconstruction.

- *Induced correlations*: We want to preserve the nonlinear constraints on the low-resolution subspaces, so that the difference subspaces $D_n = P_n / P_{n-1}$ will be correlated to P_{n-1}. In the case of the Haar subspaces, positivity implies that if the scaling coefficient at level $n-1$ is given by $s_{k,n-1}$, then the difference coefficients on this interval will be bounded above and below by $\pm s_{k,n-1}/2$.
  Similarly, the support constraint induces correlations between coefficients at different scales. This is of utmost importance to this algorithm. If we think only of the linear problem, then one would have to retrieve $c_n e_n$ when one has only observed the noisy coefficient $(\lambda_n c_n + g_n)e_n$. If λ_n is small, this is a very ill-conditioned problem, but if there is a great deal of correlation between scales, then it implies a great deal of correlation between singular functions. Thus we can observe $c_n e_n$ from the coefficients $c_k e_k$ where $k < n$. The "low-resolution" information will allow us to recover the "high-resolution" information.

- *Signal to noise recovery*: As outlined previously, the linear POCS algorithm will necessarily produce a very noisy image, if the input data are noisy. More specifically, the signal-to-noise ratios will be essentially driven to zero as $n \to \infty$ (although this assumes the existence of truly white noise, which would require infinite noise power, the approximation in finite dimensions is essentially true).
  To combat this we want to de-noise the reconstruction periodically, in order to restore the dominant features and remove the noise. We do this with the nonlinear thresholding techniques of Donoho and Johnstone [15, 16]. We utilize this type of technique because images are generally dominated by their edges. Nonlinear thresholding is designed to allow edges to remain sharp, while polynomially smooth regions can be smoothed or de-noised. Since this is incorporated into the more general POCS algorithm, we will use the "hard" nonlinear thresholding algorithm, rather than the slightly more stylish "soft" nonlinear thresholding.

24.5 THE ALGORITHM

Our algorithm consists of three different procedures.

- *Stable low-resolution reconstruction*: We have shown that when the resolution of the problem, or the size of the numerical grid S_N is sufficiently large, the spectrum of the problem will make a truncated or mollified SVD essentially useless. Thus we will not be able to effectively invert an operator of the type

$$Lf = \mathcal{X}_B F_2^{-1}(\hat{f}\mathcal{X}_C)$$

unless the discretization is sufficiently coarse or the frequency resolution is very low.

If the discretization is sufficiently coarse, however, the spectra of the discrete operators is quite different, as can be seen in Figure 24-3. Therefore we will begin by recovering the image on a very low-resolution Haar subspace, i.e., we begin by recovering local averages f_1. This will serve as the first approximation to the desired object function f. We reiterate that f_1 will be the image of f in a low-resolution Haar subspace.

- *Stable iterative improvement*: Once we have a first approximation f_1 to f, the natural question to ask is whether we can use f_1 and our original data Lf to reconstruct a higher resolution reconstruction of f, which we will call f_2.

  As outlined before, the Haar subspaces preserve the nonlinear constraint of positivity, as well as the linear constraint of compact support. These constraints induce correlations between the scales, and allow the low-resolution subspace to mollify the reconstruction on the higher-resolution subspace.

  Thus we use the induced nonlinear constraint given by the solution on the former subspace, as well as positivity, which induces a bound on the size of the coefficients in the difference subspace. We also utilize the compact support for further stabilization.

- *Iterative error reduction*: As outlined before, any type of algorithm which yields an actual inverse to a noisy ill-conditioned problem, will drive the signal-to-noise ratio in the final image to zero. Therefore we need to remove noise from the image along the way.

  We utilize both POCS and nonlinear shrinkage to accomplish this goal.

  POCS [17, 18] is a standard iterative algorithm for utilizing *a priori* knowledge in a reconstruction algorithm. We assume throughout that $0 \leq f \leq M$, supp$f \subset B_N$, and that we know $\hat{f}\mathcal{X}_C$. All of these assumptions are consistent with tomography, where positive, compactly supported densities are recovered. Moreover, the maximum density of the image is also well known, (i.e., bone when doing medical imaging, gold tracers when doing electron microscopy, etc.). Thus we have three convex sets, corresponding to these three assumptions, and we want our answer to lie in the intersection of these convex sets.

  This technique allows us to reduce the residual computational error during the algorithm, and stabilize the final outcome.

24.6. NUMERICAL RESULTS

We will illustrate our algorithm with the following three numerical experiments. We do not feel that this experiment is conclusive, but we do feel that it demonstrates the potential of this type of algorithm. We begin with limited-angle projection data and use the second Fourier reconstruction algorithm, outlined in Natterer [2], to construct the "known" portion of the Fourier data. The algorithm is then run as described above. The "full" data set consists of 256 evenly spaced angles, and 256 samples per angle. The limited-angle data sets consist of correspondingly fewer angles.

In the first experiment, we use 145 degrees of projection data. The only "noise" which the algorithm will be subject to is the "noise" created by the Fourier reconstruction algorithm, which is significant. We assumed that knowledge concerning the support of the object is not exact. We assumed only that the support of the object is embedded in a circle inscribed in the reconstruction square. Nevertheless, the algorithm recovered the edges correctly, although their magnitudes were not exactly reconstructed. The results of this experiment can be seen in Fig. 24-5. Note that there is one irritating portion of the reconstruction, namely the widened edge at the top right.

In the second experiment, we use 145 degrees of noisy projection data. We also assumed that we knew the support of the object fairly exactly. The results of this experiment can be seen in Fig. 24-6.

The third experiment utilizes only 90 degrees of projection data, and once again assumes knowledge of the support of the object. The reconstruction can be seen in Fig. 24-7.

We should note at this time, that it seems that utilizing the complete knowledge of the support of the object is useful, but also can cause artifacts. If the data are not

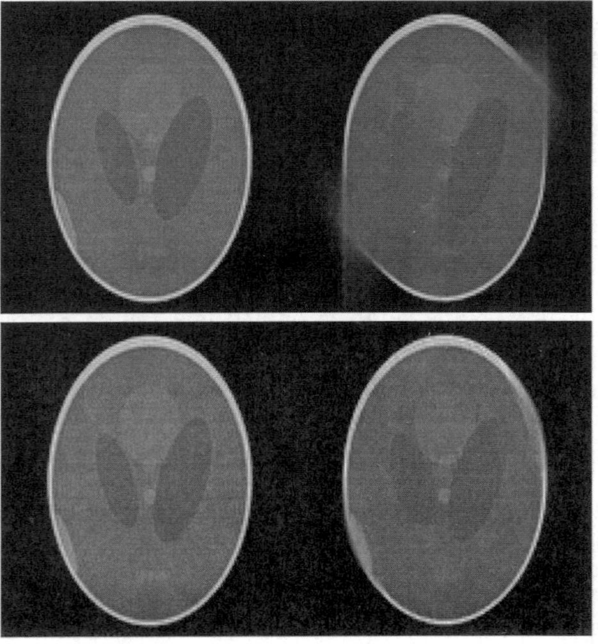

Figure 24-5 The reconstruction of the Shepp–Logan phantom from 145 degrees of data. The full data reconstruction is shown top left. The reconstruction using only the limited data is shown top right. The full data reconstruction is shown again, for reference, bottom left, and the reconstruction using our limited-angle algorithm is shown bottom right. Notice that the edges are recovered in their proper places, although their magnitudes are somewhat lower. The only irritating artifact is the edge at the top right, which is somewhat enlarged.

Figure 24-6 The reconstruction of the Shepp–Logan phantom from 145 degrees of noisy data. The full data reconstruction from these noisy data is shown top left. The reconstruction using only the limited data is shown top right. The full data reconstruction is shown again, for reference, bottom left, and the reconstruction using our limited-angle algorithm is shown bottom right. Notice that the noise level is somewhat elevated, but that most of the structure is recovered. The noise is white Gaussian and the signal-to-noise ratio is 20:1.

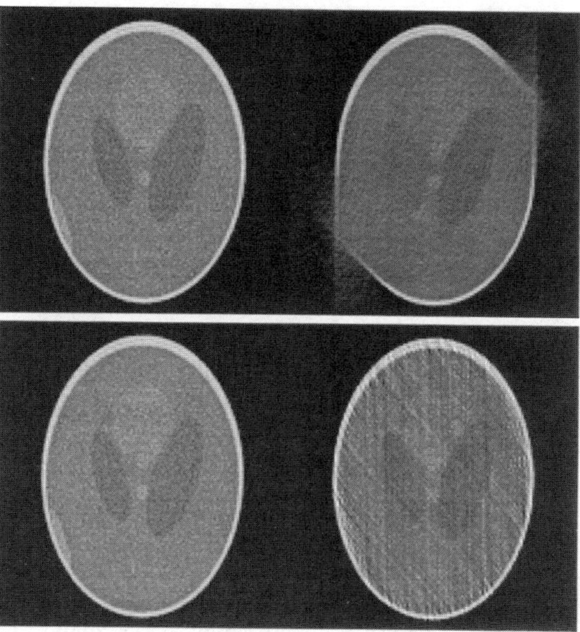

Figure 24-7 The reconstruction of the Shepp-Logan phantom from 90 degrees of data. The full data reconstruction from these noisy data is shown top left. The reconstruction using only the limited data is shown top right. The full data reconstruction is shown again, for reference, bottom left, and the reconstruction using our limited angle algorithm is shown bottom right. Athough the reconstruction is not perfect, this is not surprising given that only half of the angular data were utilized.

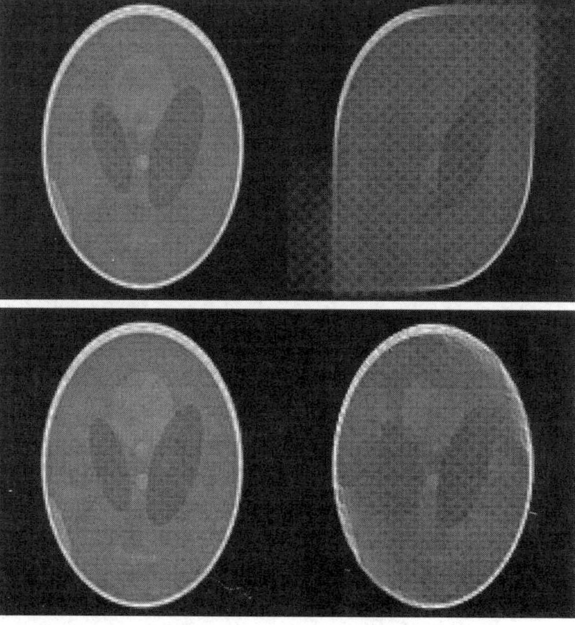

exact, a full data reconstruction will not produce an object with exactly the support of the original image. As a result, insisting on this constraint can cause artifacts. This constraint seemed necessary, however, if one were to recover the image from the extreme limited-angle case of 90 degrees.

24.7. CONCLUSION

The explanation for why we are able to extrapolate into the missing region of the Fourier transform is simple. In order to get the true convolution of the image against our kernel, we have to zero pad, and therefore over-sample the Fourier transform of the function $f(x, y)$. Since f has compact support this is equivalent to interpolating with a sinc series which is very nonlocal. Therefore we are implicitly getting nonlocal information about our Fourier transform. This is the information which we are utilizing in order to recover the missing region of the Fourier transform. The uniqueness result, which was constructed from the results of Donoho and Stark, confirms the validity of this sampling scheme.

There are a number of issues which we have not addressed. We have utilized empirical timing schedules to decide when to impose the low-resolution constraints on our algorithm. We believe, however, that these timing schedules can either be replaced with fixed schedules, or with adaptive schedules as in simulated annealing. We have also utilized empirical results to fine tune the nonlinear filtering. The nonlinear filtering has been extensively studied and this does not seem to be a concern.

A recent paper by Quinto, [19], states that "if a singularity is not stably visible from limited data, no algorithm can reconstruct it stably." We certainly agree that the edges which are tangential to the view directions will be much easier to reconstruct. Our results are quite a bit more optimistic than those of [19], however. The results of [14] are generated from Sobolev norms and microlocal analysis, and are very elegant. In practical numerical analysis, however, decay at infinity is not really an issue. It is well known that the low frequencies can be reconstructed [2]. We use the low frequencies to build outward and recover edge detail. In a certain sense, we are solving a series of problems which are much easier than the limited-angle tomography problem.

Another issue which should be addressed is that in most applications where limited-angle tomography is needed, the object is not effectively compactly supported. The wall of a nuclear reactor is better modeled as infinitely long for all practical purposes. A weld is generally a local nonuniformity in a uniform substance, however, so the projections can be changed synthetically, after they are gathered, in order to make them correspond to the local, compactly supported region of interest. In electron microscopy, however, the solution to this problem is not obvious.

We are currently working on a rigorous and complete analytical and statistical study of techniques of this type. We believe that minor alterations of the above-outlined algorithm will be able to produce reliable limited-angle reconstructions.

ACKNOWLEDGMENTS

Thanks are due to Dennis Healy, Reese Prosser, Dan Rockmore, and Richard Zalik for their advice and encouragement.

REFERENCES

[1] I. M. Gelfand and G. E. Šilov, "Fourier transforms of rapidly increasing functions and questions of the uniqueness of the solution of Cauchy's Problem," *Am. Math. Soc. Trans.*, vol. 15, pp. 221–283.

[2] F. Natterer, *The Mathematics of Computerized Tomography*, New York: John Wiley, 1986.

[3] A. Papoulis, "A new algorithm in spectral analysis and band-limited extrapolation," *IEEE Trans. Circul. Syst.*, vol. 9, September 1975.

[4] D. Slepian and H. O. Pollack, "Prolate spheroidal wave functions, Fourier analysis and uncertainty, I and II," *Bell Syst. Tech. J.*, vol. 40, pp. 43–84, 1961.

[5] D. L. Donoho and P. B. Stark, "Uncertainty Principles and Signal Recovery," *SIAM J. Appl. Math.*, vol. 49, no. 3, pp. 906–931, June 1989.

[6] G. Szego, "Ein Grenzwertsatz über die Toeplitzschen Determinanten einer reellen positiven Funktion," *Math. Ann.* vol. 76, pp. 490–503, 1915.

[7] J. Radon, "Über die Bestimmung von Funktionen durch ihre Integralwerte langs gewisser Mannigfaltigkeiten," *Berichte Sächsische Akademie der Wissenschaften, Leipzig: Math.-Phys.Kl*, vol. 69, pp. 262–267.

[8] F. A. Grünbaum, "A study of Fourier space methods for 'Limited Angle' image reconstruction," *Num. Func. Anal. Opt.*, vol. 2, no. 1, pp. 31–42, 1980.

[9] M. E. Davison, "The ill-conditioned nature of the limit angle tomography problem," *SIAM J. Appl. Math.*, vol. 43, April, pp. 428–448, 1983.

[10] A. K. Louis, "Incomplete data problems in x-ray computerized tomography I," *Num. Mathematik*, vol. 48, pp. 251–262, 1986.

[11] J. A. Reeds and L. A. Shepp, "Limited angle reconstruction in tomography via squashing," *Trans. Med. Imag.*, vol. MI-6, no. 2, June, pp. 89–97, 1987.

[12] T. Olson and J. Jaffe, "An explanation of the effects of squashing in limited angle tomography," *Trans. Med. Imag.*, vol. 9, no. 3, September 1990.

[13] M. Reed and B. Simon, *Functional Analysis I*, New York: Academic Press, 1980.

[14] U. Grenander and G. Szego, *Toeplitz Forms and Their Applications*, Berkeley, CA: University California Press, 1958.

[15] D. L. Donoho and I. M. Johnstone, "Minimax estimation via wavelet shrinkage," *Tech. Rep. 402*, Dept. of Statistics, Stanford University, Stanford, CA.

[16] D. L. Donoho and I. M. Johnstone, "Ideal spatial adaptation by wavelet shrinkage," *Tech. Rep. 400*, Dept. of Statistics, Stanford University, Stanford, CA.

[17] M. I. Sezan and H. Stark, "Applications of convex projections theory to image recovery in tomography and related areas." In *Image Recovery: Theory and Application*, New York: H. Stark, (ed.). New York: Academic Press, 1987.

[18] D. C. Youla, "Mathematical theory of image restoration by the method of convex projections." In *Image Recovery, Theory and Applications*, Academic Press: San Diego, 1987.

[19] E. T. Quinto, "Singularities of the X-ray transform and limited data tomography in $\mathbb{R}^2$ and $\mathbb{R}^3$," *SIAM J. Math. Anal.* submitted 1993.

Chapter 25

Wavelet Compression of Medical Images

Armando Manduca

25.1. INTRODUCTION

Despite rapid gains in storage and data transmission technology, there is an increasing need for medical image compression. This need is due to the development of new imaging modalities and the digitization of conventional modalities, which generate vast amounts of data. There is also increasing demand for transmission of these data across computer networks. There are two basic kinds of image compression: lossless and lossy. Many types of algorithms exist within each of these categories [1]. Lossless compression is fully reversible, with no loss of information, but is typically limited to compression ratios on the order of 3:1. Lossy compression is irreversible and loses information; however, quite high compression ratios can sometimes be achieved with small or imperceptible differences between the original and reconstructed images. Physicians are historically very reluctant to consider a technique that would discard even a small amount of information from a medical image. However, the high compression ratios offered by lossy techniques are often required for the efficient storage, manipulation, and transfer of large amounts of digital image data. In addition, there is much evidence that many medical images may be compressed by ratios of 10:1 or more with no significant loss of diagnostic accuracy, as discussed below.

Wavelet transforms have recently gained wide application in many areas of signal and image processing, and are particularly well-suited to image compression. Wavelet transform coefficients are partially localized in both space and frequency, and form a multiscale representation of the image with a constant scale factor, leading to localized frequency subbands with equal widths on a logarithmic scale.

They also have some orientation specificity. Because of these properties, wavelet transforms do an excellent job of efficiently encoding real-world images. This transform is theoretically better-suited to image compression than other common transforms, including the discrete cosine transform (DCT), on which the joint photographic experts group (JPEG) standard [2] is based. The localized nature of the wavelet transform also lends itself to allowing the user to specify areas of interest that can be preserved with maximum fidelity, while the remainder of the image, which provides mostly context, is compressed. We have developed software modules (both stand-alone and in the biomedical image analysis and display package ANALYZE [3]) that perform wavelet-based compression on both two-dimensional (2-D) and three-dimensional (3-D) gray-scale images. We present examples of such compression on a variety of medical images and comparisons with JPEG and other compression schemes. We demonstrate the improvements gained by true 3-D compression of a 3-D image (as opposed to 2-D compression of each slice), and discuss issues such as the treatment of edge effects and human visual system response in the context of a wavelet-based approach. Finally, we discuss extensions of the current approach to still more efficient compression schemes.

25.2. DISCRETE WAVELET TRANSFORMS

Discrete wavelet transforms have recently gained wide application in many areas; some examples in the imaging domain are spatial filtering, edge detection, feature extraction and texture analysis. Excellent review papers on these subjects are available [4–6]. The work presented here is based on the development and algorithms presented by Simoncelli and Adelson [7] and Antonini et al. [8]. Like the fast Fourier transform (FFT), the discrete wavelet transform (DWT) is a fast, linear operation that operates on a data vector, transforming it into a numerically different vector of the same length. Also like the FFT, the DWT is invertible and orthonormal, but represents image information in a different way. In the FFT, the basis functions are sines and cosines. In the DWT, they are a hierarchical set of "wavelet functions" that satisfy certain mathematical criteria [6, 9] and are all translations and scalings of each other. Such a wavelet basis is, in a sense, intermediate between image data and the frequency spectrum. While image data are fully localized in space but totally unlocalized in frequency, and frequency data are fully localized in frequency but unlocalized in space, data expressed in a wavelet basis are partially localized both in space and in frequency.

To compute a one-dimensional (1-D) DWT, a pair of quadrature mirror filters are defined from the underlying wavelet function [6–8], and both are applied to the signal and down-sampled by a factor of two. This process splits the signal into two components, each of half the original length, with one containing the low-frequency or "smooth" information and the other the high-frequency or "difference" information. The process is performed again on the smooth component, breaking it up into "high–low" and "low–low" components, in turn, and this is repeated several times. A multilevel hierarchy is thereby generated, with the initial "difference" component having half the original number of points and carrying the information in the upper half of the frequency range. Thus, these coefficients are well-localized in space but only slightly localized in frequency. Each higher level has half as many points as the previous level and only half the spread in frequency. Thus, each higher level is progressively more localized in frequency and less localized in space. The final "smooth" signal has only a relatively few points and carries the information in the lowest frequency band. The inverse transform is essentially identical to the forward transform, but with up-sampling instead of down-sampling. There is a good deal of overlap between this type of wavelet transform and subband coding [10, 11]. In fact, the particular scheme described here is identical to hierarchical subband coding with octave width subbands and exact reconstruction filters [7, 8]. Wavelet theory, however, imposes an extra regularity requirement on the filters, which is usually not satisfied by the exact reconstruction filters in the subband coding literature [8].

There are infinitely many possible wavelet functions, with various classes more or less suited to particular applications. For image compression, desirable properties are that the wavelet function be smooth, that the filters be short (for fast computation, however, shorter filters are generally less smooth, so this is a trade-off), and that the filters be linear phase (i.e., symmetric), so that they can be cascaded as above without phase compensation. However, there are no nontrivial, orthonormal, finite impulse response, linear-phase filters [8, 9]. The well-known Daubechies [9] filters, which maximize the number of vanishing moments, are sometimes used, but they are neither very smooth nor symmetric and in our experience give only fair compression results. The filters traditionally used in subband coding tend to be quite long and therefore slow to calculate, such as the 32-tap filter of Johnston [12]. Simoncelli and Adelson [7] derive fairly small, odd-length, symmetric filters, which are almost orthonormal and give quite good results, particularly the nine-tap filter. Perhaps the best solution is to relax the orthonormality requirement slightly, and allow the use of biorthogonal bases that involve two related sets of complementary filters. Simoncelli and Adelson [7] derive various sets of such biorthogonal filters, in which one set (for the initial compression) is long and relatively slow to compute, while the second set (for the decompression) is very small and fast to compute. Antonini et al. [8] also experiment with various sets of such filters. We have specifically adopted the nine-tap/seven-tap biorthogonal filters given in their Table II as our default filter set, since they appear to give very good results for general use, both in our experience and in a recent comparative evaluation [13]. All the figures below were generated with this filter set.

Like the FFT, the DWT is separable, and to apply it to a 2-D image one applies it to each dimension in turn. Usually, the filtering is performed once for each dimension, creating four sub-images (low–low, low–high, high–low, high–high in x and y respectively), and then recursively performed only on the low–low sub-image. The DWT of the classic Lena image is shown in Fig. 25-1. Note the hierarchical scales in each dimension, and how vertical features (high frequency in x, low in y) are captured in the lower right sub-images. Horizontal features (low in x, high in y) are captured in the upper left sub-images, and diagonal features (high in both x and y) in the upper right sub-images at each level. This process partitions frequency space into octave-spaced oriented subbands [7]. A 3-D DWT can be defined as the obvious

Figure 25-1   The classic Lena image and the logarithms of the amplitudes of its DWT.

extension of the 2-D case, and a similar process with recursion only on the "low–low–low" sub-image can be used to calculate a 3-D DWT of a volume image.

25.3. IMAGE COMPRESSION WITH WAVELETS

Real-world images, since they tend to have internal morphological consistency, tend to have first-order correlations (locally similar luminance values), second-order or dipole correlations (e.g., oriented edge continuation), and higher-order correlations (e.g., texture) [14]. It is precisely these correlations that distinguish real-world images from random noise, but this distinction is not exploited in the standard pixel-by-pixel image representation [14]. Wavelet transforms, by exploiting these correlations, do a good job of efficiently encoding image structure, so most of the information in the image is carried in a relatively small number of coefficients. There are interesting analogies between wavelet transforms and the way the visual cortex processes incoming visual data in higher animals and humans [14].

Figure 25-2 shows the histograms of the Lena image (left) and the amplitudes of the coefficients of its wavelet transform (right). The image histogram is broad and multimodal, and has high entropy. Conversely, most of the transform coefficients have zero or near-zero values (83% of the coefficients fall in the lowest of the 256 bins, whose amplitude is off the scale in the figure), and there is a small tail of coefficients with significant amplitudes. It is this latter set that is carrying most of the information in the image, and the large number of coefficients with small or zero values can be ignored or approximated with little effect on the image quality.

The above description is actually the central idea underlying a large class of lossy image compression techniques, called transform-based methods [1, 15]. The first step with such a method is to perform an invertible transformation on the

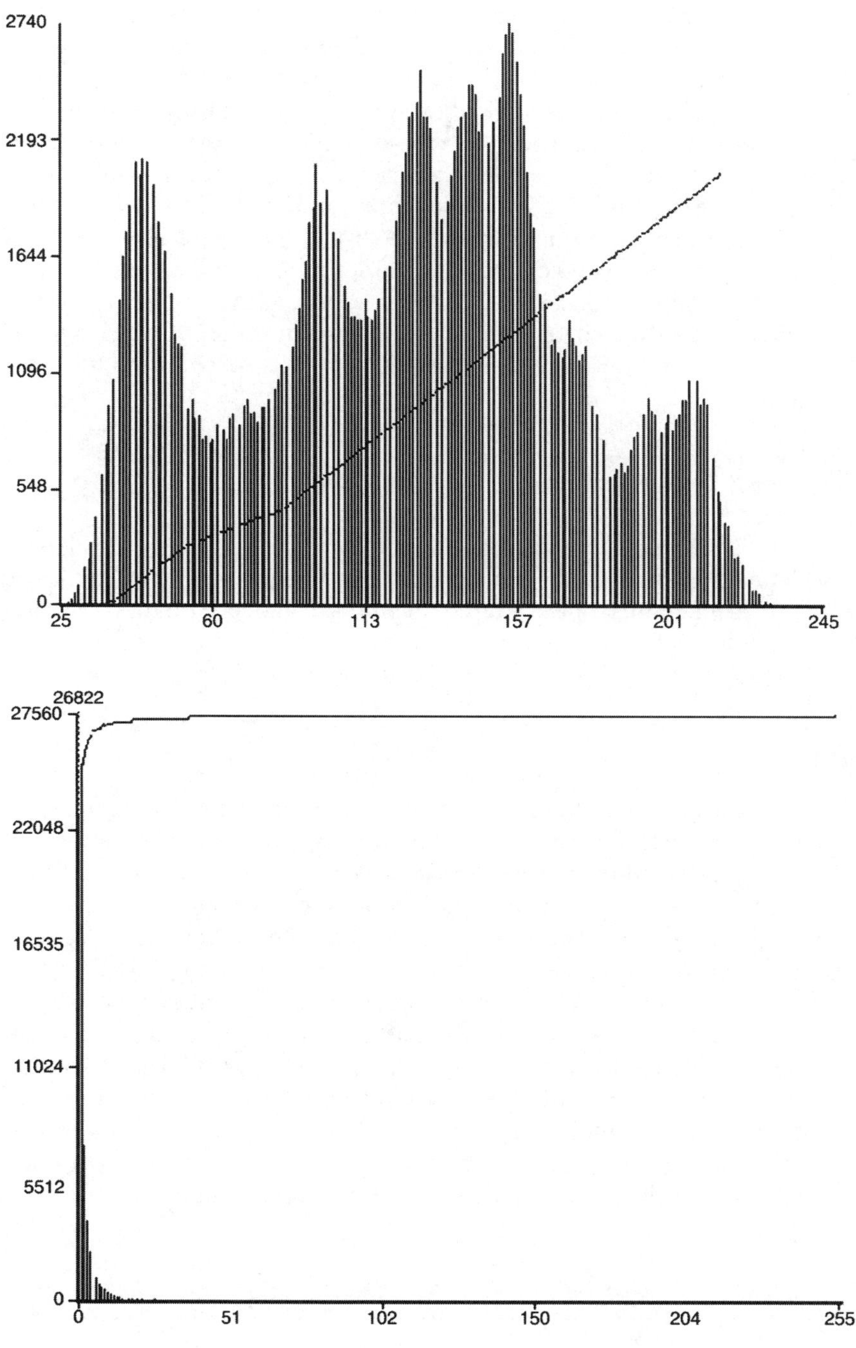

Figure 25-2   The histograms and cumulative histograms of the Lena image (top) and
the amplitudes of its DWT (bottom). In the latter, the bin with value 0 is
off the scale and contains 216 822 or 83% of the total values.

image, converting it to a basis in which the coefficients are much less correlated than the pixels in the original image. This step removes much of the redundancy in the representation, and typically many or most of the coefficients are very small. This transformation step is lossless except for numerical precision errors. Typically, the transform coefficients are then quantized—the real coefficients are replaced by lower-precision approximations, which can be represented by scaled integer values and are easily encoded. It is here that information is lost. The output of this step is a stream of (usually small) integers, many of which are zero. Finally, this stream of integers is losslessly compressed. To decompress the image, the steps are reversed. However, the reconstructed image does not exactly match the original, due to the alteration of the coefficients in the quantization step. One version of wavelet compression described here uses essentially this approach, using the DWT as the transform.

25.3.1 Implementation

The specific compression scheme currently implemented is based on first calculating the DWT of the image to four or five levels of resolution, with the biorthogonal nine-tap/seven-tap filters mentioned above [8] as the default. The low-pass analysis (h_0) and synthesis (g_0) filter coefficients are given in Table 25-1, and the high-pass analysis (h_1) and synthesis (g_1) coefficients are given by:

$$h_1 = (-1)^n g_0(1 - n)$$
$$g_1 = (-1)^n h_0(1 - n)$$

Images of arbitrary size may need to be padded to the next higher multiple of 16 or 32 in each dimension to enable calculation of the DWTs to four or five levels, respectively. At the borders, the image is extended virtually by reflection, rather than using cyclic wraparound (which will introduce artificial edges into the scene). Technically, a more correct approach would be to use wavelets defined on an interval [16], which naturally and explicitly handles the borders of real images. However, it is not clear that much is gained in practice from this added complexity.

We now describe two alternative ways to proceed. The standard approach, which we have used previously [17], proceeds by uniformly quantizing the wavelet coefficients by dividing by a user-specified quantization parameter and rounding off to the nearest integer. Typically, a large majority of coefficients with very small values are quantized to zero by this step. If the quantization parameter is increased, more coefficients are quantized to zero, the remaining coefficients are quantized

Table 25-1 Coefficients for the Default Filter Set

	0	±1	±2	±3	±4
$2^{-1/2}h_0(n)$	0.602 949	0.266 864	−0.078 223	−0.016 864	0.026 749
$2^{-1/2}g_0(n)$	0.557 543	0.295 636	−0.028 772	−0.045 636	0.0

more coarsely, the representation accuracy decreases, and the compression ratio increases. Conversely, if the quantization parameter is decreased, the reverse changes occur.

Once the coefficients are quantized, the zeroes in the resulting sequence are run-length encoded, and Huffman [1] and/or arithmetic coding [18] is performed on the resulting sequence. Arithmetic coding is more efficient but significantly slower; sometimes a useful approach is to first Huffman encode and then arithmetic encode the resulting (much shorter) sequence. Coding the various subband blocks of coefficients separately, rather than as one long sequence, improves the overall compression slightly [19].

This process represents the simplest, most common implementation of wavelet coding, and can be regarded as a baseline method to which variations can be added to improve performance. For example, the quantization process can be modified to take advantage of known properties of the human visual system, quantizing more carefully the frequencies to which the viewer is more sensitive. This is discussed in more detail below. As another example, vector quantization can be used to encode the coefficients [8]. This can yield better results but usually requires restricting oneself to a certain class of images and first analyzing a "training set" of such images, and deriving an appropriate vector codebook.

25.3.2 Set Partitioning in Hierarchical Trees

An alternative scheme for encoding the wavelet coefficients was initially proposed by Shapiro [20] and later modified and enhanced by Said and Pearlman [21, 22]. Their approach, termed set partitioning in hierarchical trees (SPIHT), yields significantly better compression than the scheme above with similar computational complexity, and represents the state-of-the-art in general-purpose image compression.

The SPIHT encoding technique is based on three principles: (1) exploitation of the hierarchical structure of the wavelet transform, by using a quadtree organization of the coefficients; (2) partial ordering of the transformed coefficients by magnitude, with the ordering data not explicitly transmitted but recalculated by the decoder; and (3) ordered bit plane transmission of refinement bits.

The partial ordering is a result of comparing coefficient magnitudes to a set of octavely decreasing thresholds. Coefficients which are smaller in magnitude than the current threshold are deemed insignificant, and the decoder considers their value (for the moment) to be zero. It is often the case that entire quadtree subsets are insignificant, and this can be expressed very efficiently by the encoder. Coefficients which have just become significant with the last threshold decrease have their location and sign transmitted and are approximated crudely by the decoder. Finally, coefficients which are already significant have their values refined by an additional bit of information. Very efficient methods for implicitly transmitting the location and ordering information and the current states of coefficients and quadtrees are described in [21, 22].

The scheme above leads to a compressed bitstream in which the most important coefficients (regardless of location) are transmitted first, the values of all coefficients are progressively refined, and the relationship between coefficients representing the same location at different scales is fully exploited for compression efficiency. Remarkably, this bitstream is fully embedded, which means that it can be truncated (or the compression process stopped) at any point and the image decompressed and reconstructed. At any time, the transmission of another byte results in simply further refining the values of one or more coefficients. The desired compression ratio or bit rate can thus be fully specified in advance, in contrast to JPEG or to the more standard approach discussed above. With these, one only roughly knows the compression ratio that will result from a given quantization parameter, and achieving a specific compression ratio may require several iterations. The reader is referred to [21, 22] for further details on the SPIHT algorithm.

25.3.3 Sample Compressions

Figure 25-3 shows the Lena image (8-bit data) along with a 16:1 compressed version, which corresponds to 0.50 bits per pixel (bpp). Slight artifacts are visible in various areas, but the fidelity of the image is quite good. The root mean square (rms) difference between the original and compressed images is 3.68, which corresponds to a power signal-to-noise ratio (PSNR) of 36.81 dB. This performance is significantly better than the JPEG standard algorithm [2] in terms of both rms error (4.70 for JPEG, PSNR = 34.69 dB) and subjective image quality (especially when magnified). Indeed, the results are comparable to the best reported in the literature by any method, including much more computationally complex or less general methods.

We also present sample compressions of a variety of medical images. Figure 25-4 shows a portion of a chest X-ray and a 50:1 compression (rms error = 10.16, 0.32 bpp). The original image 12-bit data, scaled from 0 to 4096; note that the rms error corresponds to less than one gray level if the dynamic range of the image is

Figure 25-3   The Lena image (left) and a 16:1 compressed version (right).

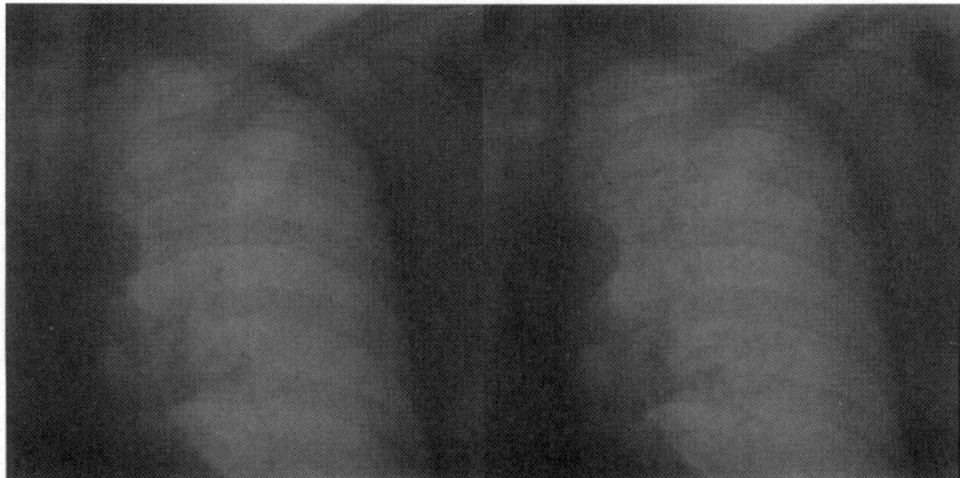

Figure 25-4   A portion of a chest X-ray (left) and a 50:1 compressed version (right).

reduced to eight bits for display. The compressed image is remarkably good for such a high compression ratio, although degradations and artifacts are obvious if the image is printed onto full-size radiographic film and examined on a light box. This is too high a compression ratio for diagnostic use, but it may be adequate for less critical applications. Figure 25-5 shows a portion of a cervical spine radiograph and a 50:1 compression (rms = 14.80, 0.32 bpp). Again, the compressed image is remarkably good, and while not of diagnostic quality, may be adequate for other uses. Figure 25-6 shows a portion of a hand radiograph and a 20:1 compression (rms error = 10.58, 0.40 bpp). Figure 25-7 shows a computerized tomography (CT) image

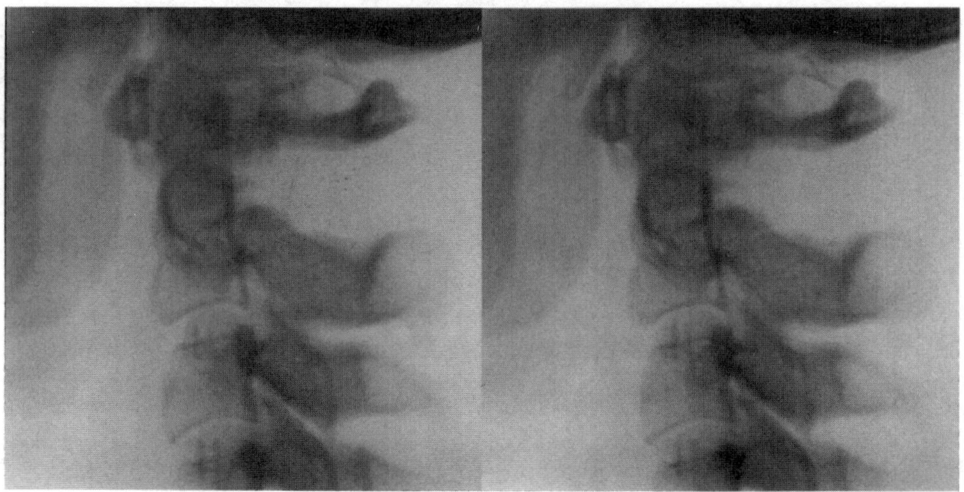

Figure 25-5   A portion of a cervical spine X-ray (left) and a 50:1 compressed version (right).

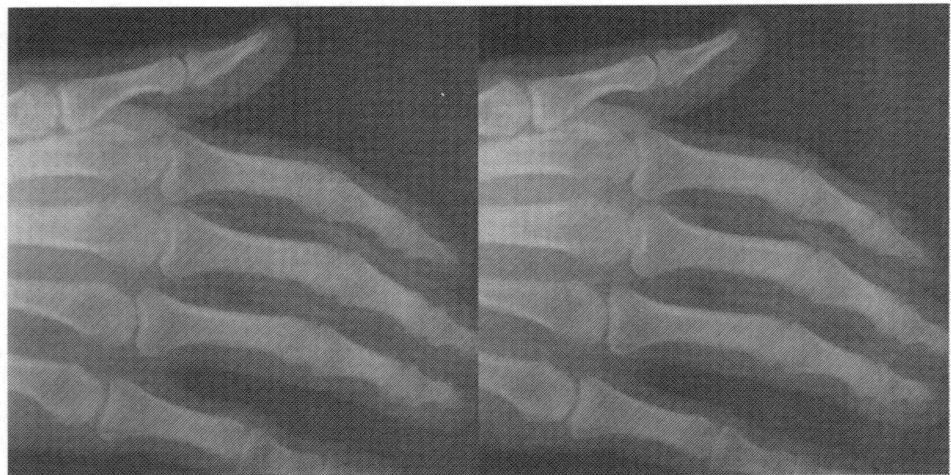

Figure 25-6 A portion of a hand X-ray (left) and a 20:1 compressed version (right).

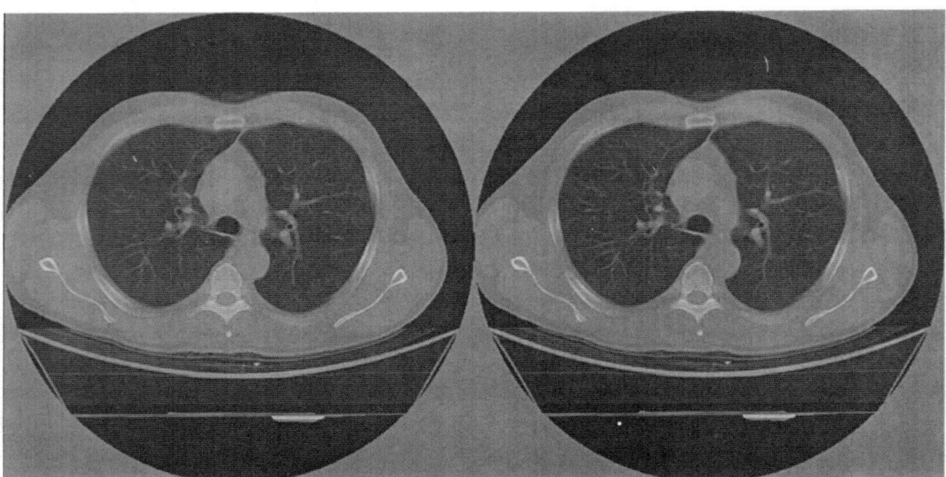

Figure 25-7 A CT image (left) and a 20:1 compressed version (right).

and a 20:1 compression (rms error = 11.59, 0.8 bpp). The compressed image is quite good and may be diagnostically useful. Figure 25-8 shows a magnetic resonance imaging (MRI) image (8-bit data) and a 10:1 compression (rms error = 3.23, 0.80 bpp). Finally, Fig. 25-9 shows a transmission light microscope image of pitting in bone tissue and a 25:1 compression (rms error = 3.37, 0.32 bpp).

25.3.4 Compression of 3-D Images

As mentioned above, a 3-D DWT can be defined as the obvious extension of the 2-D case. The procedure described here can thus be used to perform true 3-D

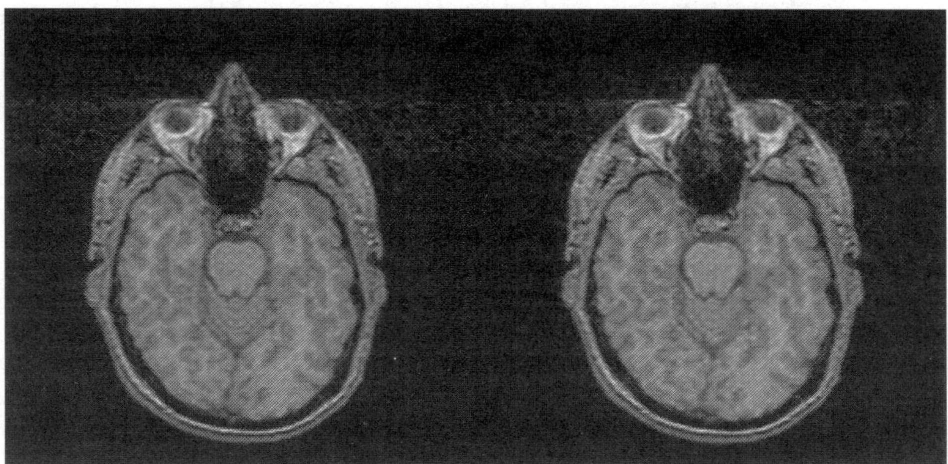

Figure 25-8   An MRI image (left) and a 10:1 compressed version (right).

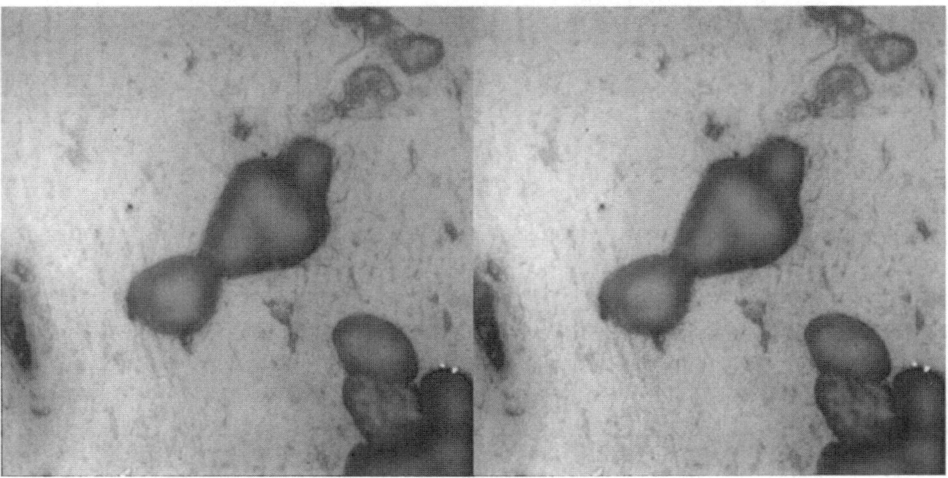

Figure 25-9   A transmission light microscope image of pitting in bone tissue (left)
and a 25:1 compressed version (right).

compression of a volume image, taking full advantage of structure along the third
dimension, as illustrated in Fig. 25-10. Two sample slices of a $256 \times 256 \times 32$ MRI
volume image are shown in the left column, and the results of a full 3-D 20:1
compression and individual 2-D 20:1 compressions are shown in the middle and
right columns, respectively. Again, a high compression ratio has been chosen to
better show the artifacts introduced. The 3-D compression is significantly better,
with many fewer serious artifacts, as one would expect, since the redundancy in
the third dimension can be exploited. The rms errors are 1.77 for the 3-D compression and 2.35 for each of the 2-D compressions. The obvious drawback here is that

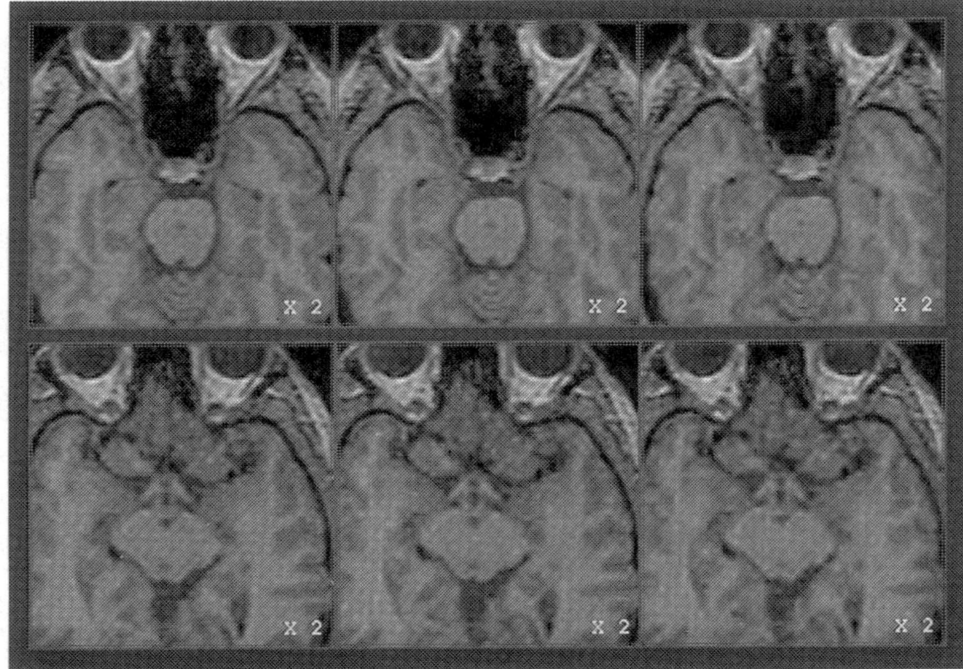

Figure 25-10   Two sample slices of a 3-D MRI data set (left column), the slices after
3-D compression by a 20:1 ratio (middle column), and the slices after
individual 2-D compression by a 20:1 ratio (right column).

operating with the full 3-D transform of large volume images requires large amounts
of computer memory and processing time (see below).

25.3.5 Preserving Arbitrary Regions

The partially localized nature of the wavelet transform means that the value of
any one pixel in the image depends on only a small number of wavelet coefficients.
Thus, it is possible to specify an arbitrary region of the image and prevent that region
from being badly degraded during the compression process by simply representing
more accurately (i.e., quantizing more finely) the wavelet coefficients that map to
that region. This can be accomplished by simply multiplying those values by an
arbitrary factor (here chosen to be 64) before quantization, and then dividing by
that factor when decompressing the image. It is necessary, of course, to record which
coefficients are treated in this way. In the current implementation, the user is allowed
to define any number of such regions, either as rectangular areas or by tracing out
arbitrary shapes on the image. If the image is 3-D, the user also specifies to which
slices each such region is applicable. This allows the user to specify areas of interest
that can be preserved with maximum fidelity, while the rest of the image, which
provides mostly context, is compressed. Such compression may be useful in applications such as teleradiology, or in any situation where data storage or transmission
bandwidth is limited and only certain portions of the image must be viewed at the

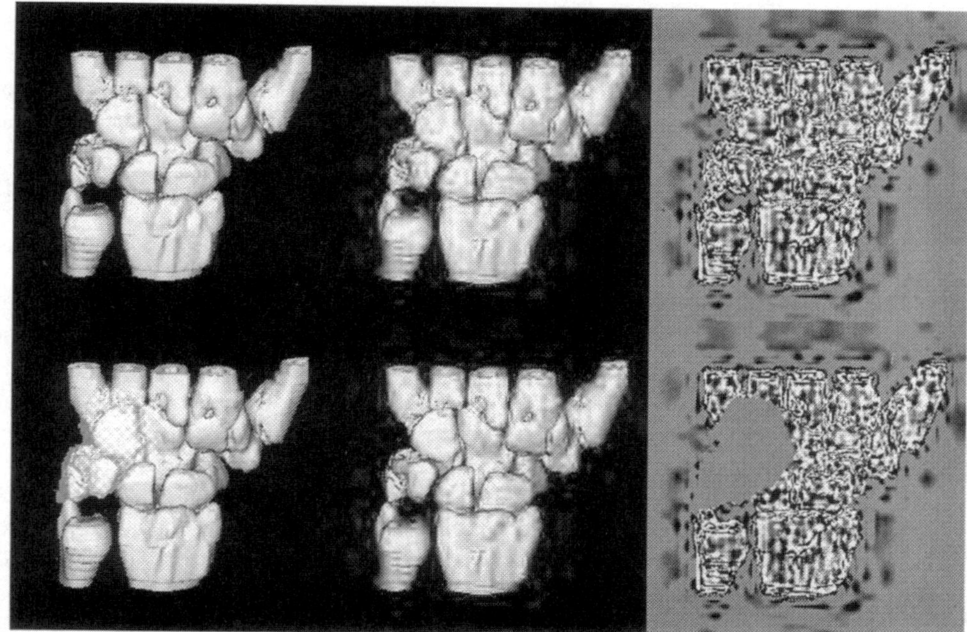

Figure 25-11   Top row from left to right: the rendered wrist image, a 40:1 compression, and the difference between the two. Bottom row from left to right: the wrist image showing a region designated to be preserved, the resulting compressed image (now requiring 10.2% of the original size), and the difference between the two.

highest resolution. The effects of preserving a region in this manner are shown in Fig. 25-11. The top row shows a volume rendered image of a wrist, a badly degraded 40:1 compression, and the difference between the two. In the bottom row, a region has been selected to be preserved. The compressed image, which now totals 10.2% of its original size, preserves the original detail in this area. The difference image shows clearly that there are no differences between the compressed image and the original in the designated area, but that large differences remain elsewhere.

25.4. DISCUSSION

The DWT as described above is useful for image compression since the coefficients are localized both in space and frequency, form a multiscale representation of the image (with a constant scale factor, leading to frequency subbands with equal widths on a logarithmic scale), and have some orientation specificity. These are precisely the properties that are most desirable for encoding real-world images [7]. Let us contrast this with the most commonly used transform, the discrete cosine transform (DCT), applied either to the full frame image or to individual blocks of pixels as in JPEG. The coefficients in a full-frame DCT have no spatial localization, so the transform cannot take advantage of any local structure—a small feature anywhere affects all the coefficients. Also, the frequency subbands are of equal width rather than equal logarithmic width. The coefficients in a block-based DCT have an abrupt spatial localization at a single scale (the block size), and the equal-width frequency subbands are not well-localized (due to the abrupt block edges) [7]. These characteristics lead to aliasing and artifacts at the block edges, and failure to take advantage of features at scales other than the block size.

25.4.1 Comparisons with JPEG

The international standard for still image compression is the JPEG algorithm [2], which is based on a 2-D DCT of 8×8 blocks of pixels. While this algorithm is very good for general purposes, and is in common use, it suffers from some drawbacks. It degrades ungracefully at high compression ratios, with prominent artifacts at block boundaries. These block artifacts are still present at low compression ratios, and even if they are not objectionable to a human viewer, they can adversely affect algorithms that detect and treat sharp edges [15, 19]. As noted above, the algorithm also cannot take advantage of correlations larger than the 8×8 block size. Finally, the basic assumptions of the JPEG algorithm treatment of the human visual system are violated by simply zooming in and out, as well as by other simple operations (as discussed in more detail below).

In all cases tried, wavelet-based compression achieves significantly better results than the JPEG algorithm in terms of rms error (which is not a good measure of image quality), maximum difference, and subjective quality at a given compression ratio. Figure 25-12 shows corresponding wavelet and JPEG compressions for the Lena image at a 33:1 compression ratio. It is apparent that the JPEG performs

Figure 25-12   The Lena image compressed by 33:1 with the wavelet algorithm (left) and the standard JPEG algorithm (right).

poorly at this high ratio; the block artifacts are very noticeable. The wavelet-based compression degrades more gracefully and is far superior. Figure 25-13 shows a magnified portion of the image at 20:1 compression. Here again, the JPEG block artifacts are very obvious, and the wavelet compression is far superior. These are purposely high compression ratios in order to show better the artifacts introduced by the two methods; at lower compression ratios it becomes more difficult to discern differences. Saipetch et al. [19] have made quantitative studies of the relative merits of DWT, 8×8 DCT, and full-frame DCT compression with respect to blocking artifacts and overall distortion, and find the DWT to be superior. A recent comparison of the JPEG versus a wavelet-based method on a variety of CT, magnetic resonance (MR) and X-ray images also reported the wavelet compression to be uniformly superior [23].

The speed of transform-based compression schemes is generally determined by the speed of the transform, which is usually the slowest step of the process. The speed of wavelet compression depends on the size of the filter set chosen, but with the default filter set above, it is slightly slower than the JPEG for small images. The complexity of the algorithm is $O(n)$, where n is the number of pixels. This is the same order of complexity as the JPEG: since the JPEG always performs 8×8 DCTs, its complexity simply depends on the number of blocks, which is proportional to the number of pixels. However, the memory requirements for the DWT are more severe, since the entire image must be transformed at once. Therefore, on a given platform, wavelet compression speed will remain comparable to JPEG only as long as there is sufficient memory to perform the DWT efficiently, and will become slower than JPEG for images larger than this size. Actual speeds on an SGI Indigo 2 XZ are 0.5 s for the DWT (and 0.9 s for the entire compression or decompression process) for a 512×512 image. In three dimensions, a $256 \times 256 \times 32$ transform takes 19 s. A

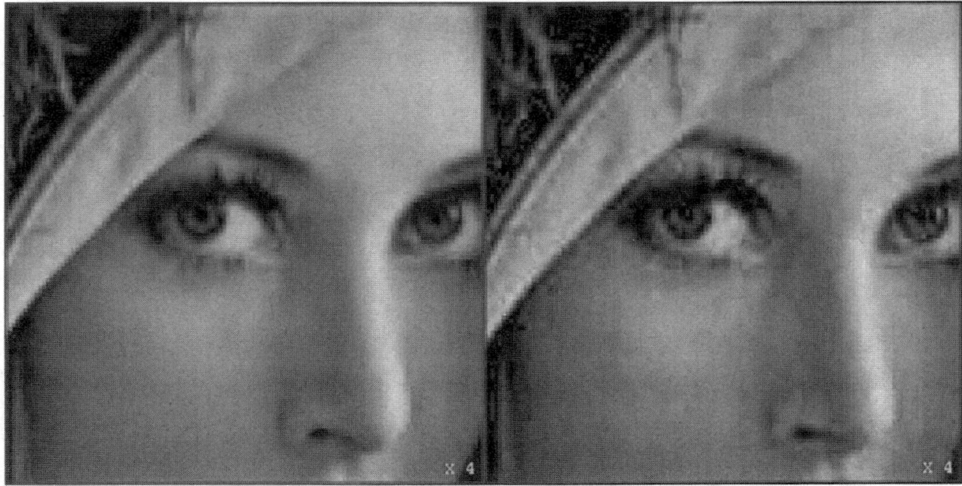

Figure 25-13   A magnified view of a section of the Lena image compressed by 20:1
with the wavelet algorithm (left) and the standard JPEG algorithm
(right).

full-frame DCT has higher complexity, $O(n \log n)$, and the same memory requirements as the DWT. There are chips available to perform the wavelet transform in hardware (e.g., from Aware, Inc., Cambridge, MA).

25.4.2 Human Visual System Response

Most image compression schemes take advantage of known properties of the human visual system (HVS) and attempt to suppress information that a human viewer would not perceive anyway. For example, the JPEG (and DCT-based compression schemes, in general) use a "quantization matrix," which quantizes the higher frequencies more coarsely, in accordance with the HVS spatial frequency sensitivity [1, 2]. Sometimes, more sophisticated schemes are used; e.g., Lewis and Knowles [24] adjust the quantization parameter for individual transform coefficients, based not only on spatial frequency but also on background luminance, edge proximity, and texture masking—all modeling well-documented HVS effects.

Such attempts to obtain maximum compression without perceptually altering the image are quite appropriate when the image is meant only to be viewed, especially at the same size and viewing distance as the original. However, simple zooming in and out (or moving one's head closer to or further from the display) alters the effective visual response, rendering such a specific treatment incorrect. The correct treatment becomes more problematical if the compressed image is to be later manipulated, processed, or automatically interpreted in some way—since information that is not visible to the human in the original image may still be quite relevant. In fact, extracting such information may be the point of the processing. For some medical applications, in particular, the high-frequency components that are suppressed may contain crucial diagnostic detail. If the data are 3-D, further questions arise, since the data may be viewed from different orientations, or viewed obliquely, or be surface or volume rendered. Also, in many 3-D medical data sets, the resolution is commonly different in the third dimension, further complicating matters. For these reasons, we prefer to perform uniform quantization and not take advantage of HVS response effects in the general case, although we do offer, as a user-selectable option, HVS weighting of the coefficients prior to quantization.

25.4.3 Medical Acceptance of Lossy Compression

The extent to which images can be usefully compressed by wavelet-based or other lossy techniques depends on the specific application and how much image degradation can be tolerated. Perhaps the most stringent application is medical diagnosis, where subtle observations on images can have life-or-death importance. Historically, physicians have resisted any lossy compression technique, on the principle that no information whatsoever should be thrown out of the original image. However, a broader look at the entire imaging process often makes it clear that information is in fact discarded or altered at many steps in the processing chain, such as image acquisition or digitization. In some modalities, such as computed radiography, image-processing steps are routinely used which alter the information significantly.

There is much evidence based on receiver operating curves (ROC) and other studies that a significant amount of lossy compression is possible with no perceptible change in the image and, more importantly, no loss of diagnostic information [23, 25–32]. For instance, chest X-rays can almost certainly be compressed by factors of 10:1 [23, 25–30]. One recent preliminary study concluded that no clinically relevant degradation was found at ratios of 20:1, although the presence of degradation in the compressed image was sometimes observable [30]. Mammograms can probably also be compressed by 20:1 with no loss of diagnostic accuracy [31, 32]. Such compression ratios can enable the practical implementation of teleradiology and digital picture archiving and communications systems (PACS).

The exact amount of compression that is acceptable in a medical image depends on the type of image and the diagnosis to be made. Such a determination requires detailed and carefully designed ROC studies. We are currently carrying out such studies, as are many other groups using a wide variety of compression algorithms. Also, in areas other than primary diagnosis, significantly higher compression ratios may be acceptable.

25.4.4 Related Advanced Techniques

Wavelet packets are a generalization of wavelet transforms which generate a large library of different orthonormal bases rather than a single one as does the DWT. Briefly, in the DWT, the filtering is performed recursively only on the low-frequency or "smooth" result from a previous iteration. With wavelet packets, filtering is performed on the high-frequency or "difference" result at each level as well. This leads to a large set of related bases, of which the DWT basis is a special case. These bases have a natural quadtree organization, and can be calculated and searched efficiently, with complexity $O(n \log n)$. The idea is to search for the single basis that best represents a given image [33], and then encode the coefficients in that basis in a standard manner as above. The identity of the specific basis chosen must be transmitted or stored with the compressed coefficients, and the calculation and search processes, while efficient, add a significant amount of computational complexity. However, better compression results than with the DWT are almost guaranteed, since the DWT is itself one of the many bases in the library.

Simoncelli and Adelson [7] describe an advanced approach based on hexagonal nonseparable filters, which provides improved orientation selectivity and may achieve better performance than the separable filters used here. They also describe a 3-D extension using rhombic dodecahedral filters. Mallat and Zhong [34] describe a more complicated approach, based on the evolution of wavelet local maxima across scales and an iterative reconstruction of images from these maxima alone, which may yield even higher compression ratios. Methods which combine techniques, such as first identifying significant edges in an image, coding those very efficiently, and then wavelet coding the residual information, are also being researched [35]. The entire area of wavelet compression is a very active field of research, and new advances are reported continually.

REFERENCES

[1] M. Rabbini and P. W. Jones, *Digital Image Compression Techniques.* Bellingham, WA: SPIE Press, 1991.

[2] G. K. Wallace, "The JPEG still picture compression standard," *Comm. ACM* vol. 34, pp. 30–44, 1991.

[3] R. A. Robb, "A software system for interactive and quantitative analysis of biomedical images." In *3D Imaging in Medicine (NATO ASI Series Vol. F60),* K. H. Hohne, H. Fuchs, and S. M. Pizer (eds.). Berlin: Springer-Verlag, pp. 333–361, 1990.

[4] G. Strang, "Wavelets and dilation equations: a brief introduction," *SIAM Rev.* vol. 31, pp. 614–627, 1989.

[5] O. Rioul and M. Vetterli, "Wavelets and signal processing," *IEEE Signal Proc. Mag.*, October pp. 14–38, 1991.

[6] S. G. Mallat, "A theory for multiresolution signal decomposition," *IEEE Trans. Patt. Anal. Mach. Intell.*, vol. 11, pp. 674–693, 1989.

[7] E. Simoncelli and E. Adelson, "Subband transforms." In *Subband Image Coding*, J. W. Woods (ed.). Norwell, MA: *Kluwer Academic Publishers*, pp. 143–192, 1991.

[8] M. Antonini, M. Barlaud, P. Mathieu, and I. Daubechies, "Image coding using wavelet transform," *IEEE Trans. Image Proc.*, vol. 1, pp. 205–220, 1992.

[9] I. Daubechies, "Orthonormal bases of compactly supported wavelets," *Comm. Pure Appl. Math.*, vol. 41, pp. 909–996, 1988.

[10] J. W. Woods and S. D. O'Neil, "Subband coding of images," *IEEE Trans. Acoust., Speech, Signal Proc.*, vol. 34, pp. 1278–1288, 1986.

[11] M. J. Smith and D. P. Barnwell, "Exact reconstruction for tree-structured subband coders," *IEEE Trans. Acoust., Speech, Signal Proc.*, vol. 34, pp. 434–441, 1986.

[12] J. D. Johnston. "A filter family designed for use in quadrature filter banks," *Proc. ICASSP*, pp. 291–294, 1980.

[13] J. D. Villasensor, B. Belzer, and J. Liao, "Wavelet filter evaluation for image compression," *IEEE Trans. Image Proc.*, vol. 4, pp. 1053–1060, 1995.

[14] J. Daugman, "Complete discrete 2-D Gabor transforms by neural networks for image analysis and compression," *IEEE Trans. Acoust., Speech, Signal Proc.*, vol. 36, pp. 1169–1179, 1988.

[15] M. V. Wickerhauser, "High-resolution still picture compression," *Digital Signal Proc. Rev. J.*, vol. 2, pp. 204–226, 1992.

[16] A. Cohen, I. Daubechies, B. Jawerth, and P. Vial, "Multiresolution analysis, wavelets and fast algorithms on the interval," *Comptes Rendus, Acad. Sci. Paris*, vol. 316, pp. 417–421, 1992.

[17] A. Manduca, "Interactive wavelet-based 2-D and 3-D image compression," *Medical Imaging 1993: Image Capture, Formatting and Display, Proc. SPIE*, vol. 1897, pp. 307–318, 1993.

[18] I. H. Witten, R. M. Neal, J. G. Cleary, "Arithmetic coding for data compression," *Comm. ACM*, vol. 30, pp. 520–540, 1987.

[19] P. Saipetch, B. K. Ho, M. Ma, K. S. Chuang, and J. Wei, "Applying wavelet transforms with arithmetic coding to radiological image compression", *IEEE Eng. Med. Bio.*, vol. 14, pp. 587–593, 1995.

[20] J. M. Shapiro, "Embedded image coding using zerotrees of wavelet coefficients," *IEEE Trans. Signal Proc.*, vol. 41, pp. 3445–3462, 1993.

[21] A. Said and W. A. Pearlman, "Image compression using the spatial-orientation tree," *IEEE Int. Symp. on Circuits and Systems*, May 1993, pp. 279–282, 1993.

[22] A. Said and W. A. Pearlman, "A new fast and efficient image codec based on set partioning in hierarchical trees," *IEEE Trans. Circ. Syst. Video Tech.*, vol. 6, pp. 243–250, 1996.

[23] P. Kotsas, D. W. Piraino, M. P. Recht, and B. J. Richmond, "Comparison of adaptive wavelet-based and discrete cosine transform algorithms in image compression," *Radiology*, vol. 193 (P), p. 331, 1994.

[24] A. Lewis and G. Knowles, "Image compression using the 2-D wavelet transform," *IEEE Trans. Image Proc.*, vol. 1, pp. 244–250, 1992.

[25] J. Sayre, D. R. Aberle, I. Boechat, T. R. Hall, H. K. Huang et al., "Effect of data compression on diagnostic accuracy in digital hand and chest radiography," *Medical Imaging 1992: Image Capture, Formatting and Display, Proc. SPIE* vol. 1653, pp. 232–240, 1992.

[26] H. MacMahon, K. Doi, S. Sanada, S. Montner, M. Giger et al., "Data compression: effect on diagnostic accuracy in digital chest radiography," *Radiology*, vol. 178, pp. 175–179, 1991.

[27] T. Ishigaki, S. Sakuma, M. Ikeda, Y. Itoh, M. Suzuki et al., "Clinical evaluation of irreversible image compression: Analysis of chest imaging with computed radiography," *Radiology*, vol. 175, pp. 739–743, 1990.

[28] P. C. Cosman, C. Tseng, R. M. Gray, R. A. Olshen, L. E. Moses et al., "Tree-structured vector quantization of CT chest scans: Image quality and diagnostic accuracy," *IEEE Trans. Med. Imag.*, vol. 12, pp. 727–739, 1993.

[29] H. Lee, Y. Kim, A. H. Rowberg, M. S. Frank, W. Lee, "Lossy compression of medical images using prediction and classification," *Medical Imaging 1993: Image Capture, Formatting and Display, Proc. SPIE*, vol. 1897, pp. 282–287, 1993.

[30] M. Goldberg, M. Pivovarov, W. M. Mayo-Smith, M. P. Bhalla, J. G. Blickman et al., "Application of wavelet compression to digitized radiographs," *Am. J. Radiol.*, vol. 163, pp. 463–468, 1994.

[31] A. Baskurt and I. Magnin, "Image coding for archiving mammograms," *Medical Imaging 1992: Image Capture, Formatting and Display, Proc. SPIE*, vol. 1653, pp. 219–227, 1992.

[32] B. Lucier, M. Kallergi, W. Qian, R. A. DeVore, R. A. Clark et al., "Wavelet compression and segmentation of digital mammograms," *J. Digital Imag.*, vol. 7, pp. 27–38, 1994.

[33] R. R. Coifman and M. V. Wickerhauser, "Entropy-based algorithms for best basis selection," *IEEE Trans. Inform. Theory*, vol. 38, pp. 713–718, 1992.

[34] S. Mallat and S. Zhong, "Characterization of signals from multiscale edges," *IEEE Trans. Patt. Anal. Mach. Intel.*, vol. 14, pp. 710–732, 1992.

[35] B. Zhu, A. H. Twefik, M. A. Colestock, Ö. N. Gerek, and A. E. Çetin, "Image coding with wavelet representations, edge information, and visual masking," *Proc. ICIP* vol. 95(I), pp. 582–585, 1995.

PART IV

Wavelets, Neural Networks and Fractals

In this part, we will focus on the hybrid signal processing methods such as wavelet neural networks and wavelet-based fractal estimation methods.

Chapter 26 by Zhang gives a summary of artificial neural networks, wavelet frames and wavelet networks.

Chapter 27 by Heinrich and Dickhaus discusses the analysis of evoked potentials using wavelet networks to diagnose children with attention deficit disorder with hyperactivity.

Chapter 28 by Kobayashi presents the self-organizing wavelet networks and shows the performance of this network in function approximation.

Chapter 29 by Flandrin summarizes the fractal processes and emphasizes the estimation of fractal dimension using the wavelet transform method.

Chapter 30 by Fischer and Akay presents the fractal analysis of heart rate variability signals using Fourier transform and wavelet transform methods.

Chapter 26

Single Side Scaling Wavelet Frame and Neural Network

Qinghua Zhang

In this chapter we first give a short introduction to artificial neural networks, then we present some results on wavelet frames and their application to a particular class of neural networks, the wavelet networks.

26.1. A SHORT INTRODUCTION TO NEURAL NETWORKS

Artificial neural networks, often shortened to *neural networks*, are computational architectures composed of interconnected units (neurons). Its name reflects its initial inspiration from biological neural systems, though the functioning of today's artificial neural networks may be quite different from that of the biological ones. Sometimes the term neural network also refers to the corresponding mathematical model, but properly speaking a network is an architecture. It is difficult to give a clear definition of artificial neural networks, due to their variety. However, at least the following two particularities distinguish them from other computational architectures or mathematical models.

- *Neural networks are naturally massively parallel*: This is the structural similarity of artificial neural networks to biological ones. Though in some cases neural network models are implemented in software on ordinary digital computers, they are naturally suitable for parallel implementations.

"

- *Neural networks are adaptive*: A neural network is composed of "living" units or neurons. It can "learn" or memorize information from data. "Learning" is the most fascinating feature of neural networks.

The beginning of studies on neural networks can be traced back to ancient times; however, it is only since the middle of the 1980s that artificial neural networks have attracted the attention of researchers from many fields, from image processing to financial management. Recently the number of publications and conferences on neural networks has increased explosively. Let us just mention two review papers [1, 2] and several books among many others [3–6]. When looking into these publications, one finds that the purposes of studies on neural networks are various. Some researchers try to model and to understand better the functioning of biological neural systems, some others try to design new computer architectures. The most remarkable successes of neural networks, however, have been in solving engineering problems for which traditional mathematic tools fail to work.

The neural networks so far studied in the literature can be roughly classified as feedforward networks, recurrent networks, associative memory networks, and self-organizing networks. In the following we give a short description for each type of network.

1. *Feedforward networks*: The units or neurons are usually organized in layers. They are typically used as classifiers or used to model (nonlinear) static mappings.
2. *Recurrent networks*: The outputs of neurons are fed back to the inputs of some neurons. They are usually used to model (nonlinear) dynamic systems.
3. *Associative memory networks*: They are a particular type of recurrent network whose equilibrium states are used to memorize information. The memorized information can be extracted from a possibly corrupted addressing key.
4. *Self-organizing networks*: The neurons of a self-organizing network are organized on a "map" and evolve according to the data presented to the network. The network partitions the input data into groups or clusters. This type of clustering or vector quantization is used to extract useful information from data.

In this chapter our attention is focused on feedforward networks. Indeed the main success of neural networks has been in the application of multilayer feedforward networks together with the "back-propagation algorithm" [7, 8]. It has been proved that, with at least one hidden layer, a feedforward network composed of the most popular neurons (sigmoid or radial) is capable of approximating all continuous functions in a finite domain, and that the approximation can be made arbitrarily accurate if the number of neurons is not limited [9–11]. In spite of their popularity, feedforward networks still lack mathematical analysis tools and efficient construction methods. For this reason, in this chapter we apply results of the wavelet theory to feedforward neural networks.

In order to uncover the sometimes mythical aspect of artificial neural networks, let us make the following comments. Today's artificial neural networks can only imitate a small portion of the behaviors of biological neural systems. The most powerful computer so far constructed is far less complicated than the human brain. Moreover, the main functionalities of neural networks, i.e., data classification and regression, have also been research topics of some older fields of applied mathematics, for instance, regression analysis, pattern recognition and system identification. It turns out that some results and techniques developed in these fields can and should be borrowed by researchers of neural networks.

26.2. WAVELET SERIES AND WAVELET NETWORK

The wavelet transform and wavelet series are becoming popular in signal processing and numerical analysis. Informally speaking, a function $f(x)$ can be decomposed into

$$f(x) = \sum_m \sum_n w_{m,n} \psi_{m,n}(x) \tag{26-1}$$

where $\psi_{m,n}(x)$ are wavelet functions, usually obtained by dilating and translating a mother wavelet function $\psi(x)$, m and n denote the dilation and translation indexes respectively. Compared with Fourier series, an important advantage of wavelet series is that decomposition (26-1) provides a powerful tool for spatio-frequency (or time-frequency, if x represents the time) analysis.

The most popular wavelet series are built with "orthonormal wavelet bases" [12, 13] for which very fast and elegant algorithms exist for computing the decomposition coefficients $w_{m,n}$ in (26-1). Besides these, the less often used "wavelet frames" [13–15], for which the computation of the decomposition coefficients is more complicated, have the advantage of leaving more freedom to the choice of the mother wavelet $\psi(x)$. Loosely speaking, a frame is a redundant basis ensuring stable decompositions.

Wavelet decompositions are useful for signal representation, function approximation, data compression, and some other purposes in signal processing and numerical analysis [16, 17]. The wavelet approaches are particularly efficient when $f(x)$ has localized irregularities. However, studies and applications of wavelets have concentrated on one- or two-dimensional wavelets, i.e. $x \in \mathbb{R}$ or $x \in \mathbb{R}^2$ in (26-1). The reason is that the implementation of wavelet bases or frames of large dimension is of prohibitive cost [18]. One practically reasonable use of large dimensional wavelets is in neural networks. Some studies on this subject have been reported [19–21]. The basic idea is to consider (26-1) as a "one hidden layer neural network," and to avoid the implementation of "complete" wavelet bases or frames. This yields a class of neural networks, named "wavelet networks" [19]. Moreover, techniques of regression analysis can be used to build wavelet networks in a more constructive way [22]. It consists in selecting, within an appropriate wavelet frame, the wavelet functions which best fit the observed data. Wavelet frames are used rather than orthonormal wavelet bases, since a wavelet network only implements a portion of a (truncated) wavelet frame (or basis), and the observed data on x are usually sparse in large dimensional cases, the fast algorithms associated with orthonormal wavelet bases are therefore not suitable.

An important question is thus how to construct multidimensional wavelet frames. Since single-dimensional wavelets are much more studied, it is natural to construct multidimensional wavelets and wavelet frames from single-dimensional ones. Tensor product construction and radial construction of multidimensional wavelet frames are discussed in [15]. The computational complexity of a tensor product wavelet is approximately proportional to its dimension, whereas that of a radial wavelet only weakly depends on its dimension. Therefore, radial wavelets are more suitable for use in large dimensions. On the other hand, as discussed in [15], tensor product construction is related to vectorial scaling, i.e., tensor product wavelets should use independent dilation parameters in different dimensions in order to form wavelet frames (in (26-1) the dilation index m has the same dimension as x); in contrast, radial construction is related to scalar scaling wavelet frames, i.e., for each wavelet function a common dilation parameter is used in all the dimensions (m is scalar). Scalar scaling wavelet frames are structurally less complex than vectorial scaling ones. For these reasons, we mainly consider scalar scaling radial wavelet frames.

The usual wavelet frames are composed of wavelets from an infinitely coarse scale to an infinitely fine scale, i.e., m ranges from $-\infty$ to $+\infty$. We refer to these kinds of wavelet frames as "double side scaling" wavelet frames. Similarly, there also exist double side scaling orthonormal wavelet bases. However, the widely used orthonormal wavelet bases are single side scaling: in addition to the single side scaled wavelet functions (with m ranging from 1 to ∞), a "scaling function" is used to cover all the coarse scales. It is also well known that $2^d - 1$ different mother wavelets and one scaling function are required to generate a scalar scaling orthonormal wavelet basis of $L_2(\mathbb{R}^d)$. This fact is one of the difficulties for implementing large dimensional orthonormal wavelet bases.

In the following sections, we consider the construction of single side scaling wavelet frames, i.e., frames with m ranging from 1 to $+\infty$. We shall show that, unlike orthonormal wavelet bases, it is possible to generate a single side scalar scaling wavelet frame of $L_2(\mathbb{R}^d)$ with only *one* mother wavelet and one scaling function. The complexity of the single side scaling wavelet frame is considerably lower than that of the double side scaling wavelet frame. It is therefore preferable to use single side scaling wavelet frames in wavelet networks.

26.3. DOUBLE SIDE SCALING WAVELET FRAMES

In this section we review some results on multidimensional wavelet frames in $L_2(\mathbb{R}^d)$. We only consider scalar scaling wavelet frames. For vectorial scaling wavelet frames the reader is referred to [15].

Before discussing wavelet frames, let us recall the definition of frame in a general Hilbert space.

Definition 1.   *A family of functions $\{\phi_j : j \in J\}$ in a Hilbert space $\mathcal{H}$ is called a frame of $\mathcal{H}$ if there exist $A > 0$, $B < \infty$ so that, for all $f \in \mathcal{H}$,*

$$A\|f\|^2 \leq \sum_{j \in J} |\langle f, \phi_j \rangle|^2 \leq B\|f\|^2 \tag{26-2}$$

A and B are called the frame bounds.

The right inequality in (26-2) ensures that any function $f \in \mathcal{H}$ can be fully "characterized" by its "frame coefficients" $\langle f, \phi_j \rangle$, and the left inequality guarantees the numerical stability of the reconstruction of f from $\langle f, \phi_j \rangle$; see [13] section 3.1.

Let $\psi \in L_2(\mathbb{R}^2)$ be a multidimensional wavelet function. We consider families of wavelet functions generated by dilating and translating ψ in the following form:

$$\psi_{m,n}(x) = a^{\frac{1}{2}dm} \psi(a^m x - nb), \qquad m \in \mathbb{Z}, n \in \mathbb{Z}^d$$

where m and n are respectively dilation and translation indexes, $a \in \mathbb{R}$ and $b \in \mathbb{R}$ specify respectively the dilation step size and the translation step size. We assume that $a > 1$ and $b > 0$ in the following. Note that the dilation index m is scalar, and so is the dilation parameter a^m. The dilation parameter a^m is thus common in all the dimensions. Such wavelet families are called "scalar scaling" wavelet families.

26.3.1 A Sufficient Condition

An important question is how to check the frame property of a wavelet function ψ and its associated dilation and translation step sizes a, b. The following theorem gives such a sufficient condition in $L_2(\mathbb{R}^d)$.

Theorem 1.   *Let $\psi(x) \in L_2(\mathbb{R}^d)$ be a multidimensional wavelet function and $\hat{\psi}(\omega)$ be its Fourier transform. If for some scalar $a > 1$ the following three conditions are satisfied ($\|\omega\|$ and $\|\eta\|$ denote the Euclidean norms of ω and η in the following formulae)*

Condition 1.1

$$m(\psi, a) \triangleq \operatorname*{ess\,inf}_{\|\omega\| \in [1,a]} \sum_{m \in \mathbb{Z}} |\hat{\psi}(a^{-m}\omega)|^2 > 0 \tag{26-3}$$

Condition 1.2

$$M(\psi, a) \triangleq \operatorname*{ess\,sup}_{\|\omega\| \in [1,a]} \sum_{m \in \mathbb{Z}}^{\infty} |\hat{\psi}(a^{-m}\omega)|^2 < \infty \tag{26-4}$$

Condition 1.3

$$\sup_{\eta \in \mathbb{R}^d} \left[(1 + \|\eta\|^2)^{d(1+\epsilon)/2} \beta(\eta) \right] = C_\epsilon < \infty \qquad \textit{for some } \epsilon > 0 \tag{26-5}$$

where

$$\beta(\eta) \triangleq \sup_{\|\omega\| \in [1,a]} \sum_{m \in \mathbb{Z}} |\hat{\psi}(a^{-m}\omega)| \cdot |\hat{\psi}(a^{-m}\omega + \eta)| \tag{26-6}$$

then there exists a scalar $b_0 > 0$, such that $\forall b \in (0, b_0)$, the wavelet family

$$\Psi(a, b) = \{\psi_{m,n}(x) = a^{\frac{1}{2}dm}\psi(a^m x - nb) : m \in \mathbb{Z}, \ n \in \mathbb{Z}^d\} \tag{26-7}$$

constitutes a frame of $L_2(\mathbb{R}^d)$.

This theorem is a generalization of Daubechies' theorem for one-dimensional wavelet frames [13, 14]. Its proof can be found in [15]. In the conditions of the above theorem, the sup values are defined with the restriction $\|\omega\| \in [1, a]$, since other values of $\|\omega\|$ can be shifted to this range by multiplication with a suitable a^m, except $\omega = 0$ which constitutes a set of measure zero, and therefore does not matter.

Note that in Theorem 1, ψ is a general multidimensional wavelet function. Given any function $\psi \in L_2(\mathbb{R}^d)$, Theorem 1 gives a way to check if ψ can generate wavelet frames of the form (26-7). However, Theorem 1 does not tell us how to construct such wavelets. On the other hand, computations involving the multidimensional Fourier transform of ψ required by the theorem may be complicated. It is desirable to find methods with which multidimensional wavelets can be constructed from single-dimensional ones, and to check the frame property of a thus constructed multidimensional wavelet family in a more convenient way. As mentioned earlier, radial construction is of particular interest for large-dimensional wavelet frames. So we consider radial wavelet frames in the next subsection.

26.3.2 Radial Case

In this subsection, we consider radial wavelet functions. A function $\psi : \mathbb{R}^d \to \mathbb{R}$ is radial if there exists $\rho : \mathbb{R} \to \mathbb{R}$ such that for all $x \in \mathbb{R}^d$, $\psi(x) = \rho(\|x\|)$, where $\| \cdot \|$ denotes the Euclidean norm. The following theorem gives a convenient way to check if a radial wavelet can generate frames.

Theorem 2. *Let $\mu(t) \in L_2(\mathbb{R})$ be a symmetric function, i.e. for all $t \in \mathbb{R}$, $\mu(-t) = \mu(t)$. Denote by $\hat{\mu}$ the Fourier transform of μ. Assume that for some scalar $a > 1$ the following two conditions are satisfied*

Condition 2.1

$$m(\mu, a) \triangleq \operatorname*{ess\,inf}_{\|\xi\| \in [1,a]} \sum_{m \in \mathbb{Z}} |\hat{\mu}(a^{-m}\xi)|^2 > 0 \tag{26-8}$$

Condition 2.2. There exist constants $C > 0$, $\alpha > 0$, and $\gamma > \alpha + d$ such that for all $\xi \in \mathbb{R}$

$$|\hat{\mu}(\xi)| \leq C|\xi|^\alpha (1 + |\xi|^2)^{-\frac{\gamma}{2}}$$

Take $\hat{\psi}(\omega) = \hat{\mu}(\|\omega\|)$, $\omega \in \mathbb{R}^d$. Let $\psi(x)$, $x \in \mathbb{R}^d$, be the inverse Fourier transform of $\hat{\psi}(\omega)$. Then, there exists a scalar $b_0 > 0$, such that $\forall b \in (0, b_0)$, the wavelet family

$$\Psi(a, b) = \{\psi_{m,n}(x) = a^{\frac{1}{2}dm}\psi(a^m x - nb) : m \in \mathbb{Z},\ n \in \mathbb{Z}^d\} \qquad (26\text{-}9)$$

constitutes a frame of $L_2(\mathbb{R}^d)$.   □

(See [15] for a proof of this theorem.)

Theorem 2 is practically much easier to use than Theorem 1, since it only considers a single variable function $\hat{\mu}$ instead of a multidimensional wavelet function. Note that since $\hat{\psi}(\omega) = \hat{\mu}(\|\omega\|)$ is radial, $\psi(x)$ is necessarily radial.

Given a radial wavelet function ψ, we find its Fourier transform $\hat{\psi}$, derive the scalar function $\hat{\mu}$ such that $\hat{\psi}(\omega) = \hat{\mu}(\|\omega\|)$ for all $\omega \in \mathbb{R}^d$, then, applying Theorem 2 to $\hat{\mu}$ allows us to check the frame property of ψ. On the other hand, given a scalar function $\hat{\mu}$ satisfying Theorem 2, then a radial wavelet function and the associated wavelet frames can be easily constructed.

26.4. SINGLE SIDE SCALING WAVELET FRAME

In a wavelet frame as described in the previous section, the scale index m ranges from $-\infty$ to $+\infty$. We refer to such frames as "double side scaling" wavelet frames. For many applications, only a finite domain in the x-space is of interest. Inside a finite domain, if the mother wavelet is sufficiently regular, wavelets of very large scale (i.e., with $m \ll 0$) are very flat and therefore redundant. In order to reduce this redundancy we propose to construct single side scaling wavelet frames for which the scale index m ranges from 1 to $+\infty$. The construction proposed in this section is an analogy of the widely used single side scaling orthonormal wavelet bases in which an additional scaling function is used to cover all the coarse scales [12, 13]. It is well known that $2^d - 1$ different mother wavelets and one scaling function are required to generate a single side scalar scaling orthonormal wavelet basis of $L_2(\mathbb{R}^d)$. However, as we are going to show, it is possible to generate a single side scalar scaling wavelet frame of $L_2(\mathbb{R}^d)$ with only *one* mother wavelet and one scaling function.

26.4.1 A Sufficient Condition for Single Side Scaling Wavelet Frame

Typical wavelet functions are bandpass filters, i.e., their Fourier transforms are concentrated in some finite domain between zero and infinity in the ω-space. Loosely speaking, the first condition of Theorem 1 requires that the zeros of $\hat{\psi}(a^{-m}\omega)$ do not "conspire" so that the series $\sum |\hat{\psi}(a^{-m}\omega)|^2$ covers all the ω-space. The second and third conditions of Theorem 1 require some sufficient decay speed of $|\hat{\psi}(\omega)|$.

Due to the bandpass property of $\hat{\psi}(\omega)$, when $m \to -\infty$, the support of $|\hat{\psi}(a^{-m}\omega)|^2$ approaches $\omega = 0$. Therefore, one can expect to use a single function $\hat{\varphi}(\omega)$, concentrated around $\omega = 0$, to replace all the wavelets of large scales. Since the corresponding function $\varphi(x)$ plays a role similar to that of the scaling function in an orthonormal wavelet basis, we also adopt the name of "scaling function" for $\varphi(x)$. Typically, $\varphi(x)$ is a low-pass filter.

From the above heuristic, we propose the following theorem which gives a sufficient condition for constructing mixed wavelet and scaling function frames.

Theorem 3.   *Let $\varphi, \psi \in L_2(\mathbb{R}^d)$ be a scaling function and a wavelet function, respectively. If for some scalar $a > 1$ the following three conditions are satisfied:*

Condition 3.1

$$m(\varphi, \psi, a) \triangleq \operatorname*{ess\,inf}_{\omega} \left(|\hat{\varphi}(\omega)|^2 + \sum_{m=1}^{\infty} |\hat{\psi}(a^{-m}\omega)|^2 \right) > 0 \tag{26-10}$$

Condition 3.2

$$M(\varphi, \psi, a) \triangleq \operatorname*{ess\,sup}_{\omega} \left(|\hat{\varphi}(\omega)|^2 + \sum_{m=1}^{\infty} |\hat{\psi}(a^{-m}\omega)|^2 \right) < \infty \tag{26-11}$$

Condition 3.3

$$\sup_{\eta \in \mathbb{R}^d} \left[(1 + \|\eta\|^2)^{d(1+\epsilon)/2} \beta_{\varphi}(\eta) \right] = C_{\epsilon} < \infty \qquad \textit{for some } \epsilon > 0 \tag{26-12}$$

$$\sup_{\eta \in \mathbb{R}^d} \left[(1 + \|\eta\|^2)^{d(1+\varepsilon)/2} \beta_{\psi}(\eta) \right] = C_{\varepsilon} < \infty \qquad \textit{for some } \varepsilon > 0 \tag{26-13}$$

where

$$\beta_{\varphi}(\eta) \triangleq \sup_{\omega} |\hat{\varphi}(\omega)| \cdot |\hat{\varphi}(\omega + \eta)| \tag{26-14}$$

$$\beta_{\psi}(\eta) \triangleq \sup_{\omega} \sum_{m=1}^{\infty} |\hat{\psi}(a^{-m}\omega)| \cdot |\hat{\psi}(a^{-m}\omega + \eta)| \tag{26-15}$$

then there exists $b_0 > 0$, such that $\forall b \in (0, b_0)$, the union of the two families

$$\Phi(b) = \{\varphi_n(x) = \varphi(x - nb) : n \in \mathbb{Z}^d\} \tag{26-16}$$

$$\Psi(a, b) = \{\psi_{m,n}(x) = a^{\frac{1}{2}dm}\psi(a^m x - nb) : m \in \mathbb{N},\ n \in \mathbb{Z}^d\} \tag{26-17}$$

constitutes a frame of $L_2(\mathbb{R}^d)$, in other words, there exist two constants $A > 0$ and $B < +\infty$, such that $\forall f \in L_2(\mathbb{R}^d)$, the following inequalities hold

$$A\|f\|^2 \leq \sum_{n \in \mathbb{Z}^d} |\langle \varphi_n, f \rangle|^2 + \sum_{m \in \mathbb{N}} \sum_{n \in \mathbb{Z}^d} |\langle \psi_{m,n}, f \rangle|^2 \leq B\|f\|^2$$

$\square$

The proof of this theorem is given in Appendix A.

Note that in Theorem 1 the sup values were defined with $\|\omega\| \in [1, a]$, whereas in Theorem 3 they are defined over all $\omega \in \mathbb{R}^d$. It is therefore more difficult to check the conditions of Theorem 3.

26.4.2 Radial Case

As for the double side scaling wavelet frames, in the radial case some practically more convenient result can be obtained. The following theorem is an analogy of Theorem 2 in this sense.

Theorem 4.   *Let $\phi, \mu \in L_2(\mathbb{R})$ be two symmetric functions, i.e., for all $t \in \mathbb{R}$, $\phi(-t) = \phi(t)$ and $\mu(-t) = \mu(t)$. $\hat{\phi}$ and $\hat{\mu}$ denote the Fourier transforms of ϕ and μ, respectively. Assume that for some scalar $a > 1$ the following two conditions are satisfied*

Condition 4.1

$$m(\phi, \mu, a) \triangleq \operatorname*{ess\,inf}_{\xi}\left(|\hat{\phi}(\xi)|^2 + \sum_{m=1}^{\infty} |\hat{\mu}(a^{-m}\xi)|^2 \right) > 0 \qquad (26\text{-}18)$$

Condition 4.2. There exist constants $C_1 > 0$, $C_2 > 0$, $\tau > d$, $\alpha > 0$ and $\gamma > \alpha + d$, such that for all $\xi \in \mathbb{R}$

$$|\hat{\phi}(\xi)| \le C_1(1 + |\xi|^2)^{-\frac{\tau}{2}}$$

and

$$|\hat{\mu}(\xi)| \le C_2|\xi|^{\alpha}(1 + |\xi|^2)^{-\frac{\gamma}{2}}$$

Take $\hat{\varphi}(\omega) = \hat{\phi}(\|\omega\|)$ and $\hat{\psi}(\omega) = \hat{\mu}(\|\omega\|)$, $\omega \in \mathbb{R}^d$. Let $\varphi(x)$ and $\psi(x)$, $x \in \mathbb{R}^d$, be the inverse Fourier transforms of $\hat{\varphi}(\omega)$ and $\hat{\psi}(\omega)$ respectively. Then, there exists $b_0 > 0$, such that $\forall b \in (0, b_0)$, the union of the two families

$$\Phi(b) = \{\varphi_n(x) = \varphi(x - nb) : n \in \mathbb{Z}^d\} \qquad (26\text{-}19)$$

$$\Psi(a, b) = \{\psi_{m,n}(x) = a^{\frac{1}{2}dm}\psi(a^m x - nb) : m \in \mathbb{N}, \ n \in \mathbb{Z}^d\} \qquad (26\text{-}20)$$

constitutes a frame of $L_2(\mathbb{R}^d)$. $\square$

The proof of this theorem is given in Appendix B.

Given two radial wavelet functions φ and ψ, we find their Fourier transforms $\hat{\varphi}$ and $\hat{\psi}$, derive the scalar functions $\hat{\phi}$ and $\hat{\mu}$ such that $\hat{\varphi}(\omega) = \hat{\phi}(\|\omega\|)$ and $\hat{\psi}(\omega) = \hat{\mu}(\|\omega\|)$ for all $\omega \in \mathbb{R}^d$, then, applying Theorem 4 to $\hat{\phi}$ and $\hat{\mu}$ allows us to check the frame property of φ and ψ. On the other hand, given two scalar functions $\hat{\phi}$ and $\hat{\mu}$ satisfying Theorem 4, the corresponding radial scaling and wavelet functions can be easily constructed as well as the associated single side scaling wavelet frames.

26.4.3 Some Practical Considerations

The radial case of the single side scaling wavelet frame is of particular interest, not only because the use of radial functions limits the computational complexity of large-dimensional wavelets and scaling functions, but also the frame property of φ and ψ can be checked via the corresponding single-variable functions $\hat{\phi}$ and $\hat{\mu}$. It can be shown that this checking only weakly depends on the dimension d, mainly through the mild assumptions $\tau > d$ and $\gamma > a + d$ in Condition 4.2. More specifically, Theorem 4 can be used in a very convenient manner in practice: if $\hat{\phi}$ and $\hat{\mu}$ satisfy Condition 4.1 with a chosen value of a and Condition 4.2 for a sufficiently large value of d, say d_M, find a value of $b \leq \frac{\pi}{3}$ so that the union of (26-19) and (26-20) constitutes a frame in the case of $d = 1$, then for all $d \leq d_M$, a frame can be constructed by using the same value of a, b and the corresponding d-dimensional radial functions φ and ψ. See Appendix C for more explanation.

26.5. COMBINING WAVELET AND NEURAL NETWORK

In this section we consider the use of wavelet frames in neural networks, or more specifically, in "wavelet networks."

26.5.1 Modeling Nonlinear Systems

Informally speaking, wavelet expansion (26-1) is structurally similar to one hidden layer neural networks. This fact is better illustrated in Fig. 26-1. We will explain later why in Fig. 26-1 the wavelets are indexed by $i = 1, \ldots, s$ instead of the double index (m, n) as used in Eq. (26-1). We call a wavelet expansion a wavelet network when it is considered as a feedforward neural network. In principle, such wavelet expansions (or wavelet networks), like other feedforward neural networks, can be constructed by standard procedures (typically the back-propagation procedure). However, since wavelets are used, we expect some more efficient construction methods. Of course, the most efficient methods would be obtained if orthonormal wavelet bases were used. Unfortunately, the implementation of orthonormal wavelet bases is practically feasible only in small-dimensional spaces. Moreover, for large-dimensional applications, the available data are usually sparse and irregularly sampled, the fast algorithms of orthonormal wavelet bases are not designed for

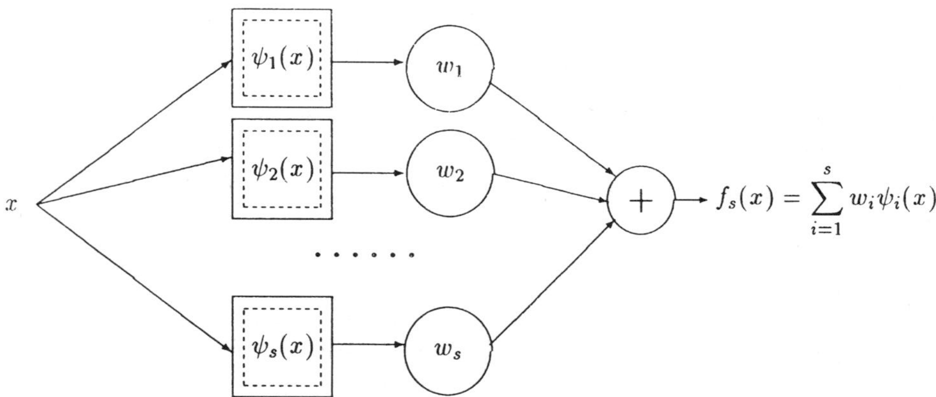

Figure 26-1   An illustration of wavelet expansion as a network.

these situations. For these reasons, it is more reasonable to use wavelet frames in the construction of wavelet networks.

Like other feedforward neural networks, wavelet networks can be used to approximate nonlinear functions. This property is particularly useful for nonlinear system modeling. Here we only consider static nonlinear system modeling. For dynamic nonlinear system modeling, the reader is referred to [23, 24]. More formally, let x and y be respectively the input and the output of a system, assume that there exists a function f such that

$$y = f(x) + e \tag{26-21}$$

with some noise e. The problem to be solved is to estimate the function f from observations on x and y. Let us consider the case of multi-input and single-output systems, i.e., $x \in \mathbb{R}^d$, $y \in \mathbb{R}$. Note that a multioutput system can always be decomposed into single-output systems. Assume that we are given a sample of N couples of observations on (x, y) denoted by

$$\mathcal{X} = \{x_1, \ldots, x_N\}$$
$$\mathcal{Y} = \{y_1, \ldots, y_N\}$$
$$(\mathcal{X}, \mathcal{Y}) = \{(x_1, y_1), \ldots, (x_N, y_N)\}$$

Some assumption on the regularity of f must be made if we want to estimate f with an observation sample of finite length. We shall assume $f \in L_2(\mathbb{R}^d)$ in the following. Under this assumption, results of the previous sections tell us that wavelet frames are suitable tools for estimating f. In practical implementations, infinite wavelet frames are always truncated into finite sets. The truncation of a wavelet frame may be considered as some additional assumption on the regularity of f. In practice, the truncation follows two practical considerations. First, only a finite sample of observations is available, so f can only be estimated up to a certain "scale" of details. Second, observed data usually stay inside a finite domain in the x-space, and wavelet functions (and scaling functions) used in practice are compactly supported or almost compactly supported (typically with exponential decay), so the terms in a wavelet frame whose supports do not overlap the domain of the observed data can be truncated. These considerations usually result in regular truncations, preserving the "pyramidal" structure of wavelet frames. For large-dimensional applications, the implementation of regularly truncated wavelet frames is often too expensive in terms of data storage and computation. The key point for wavelet network construction is thus to avoid the implementation of "complete" regularly truncated wavelet frames.

26.5.2 Sparse Data and Thinned Wavelet Frame

In most practical situations of large dimension, the observed data are sparse in the x-space, consequently many terms in the truncated wavelet frame do not contain any data point in their supports. Removing these "empty" terms often considerably reduces the size of the truncated frame. We refer to the thus reduced set as the "thinned wavelet frame" and denote it by W. In the following we give more details about the thinning procedure.

It is preferable to use single side scaling wavelet frames in order to limit the number of terms in the truncated and thinned frames. The truncated frame is then a subset of $\Phi(b) \cup \Psi(a, b)$ as defined in (26-19) and (26-20). For the convenience of notation, let $\psi_{0,n}(x) = \varphi_n(x)$, then we can write

$$\Phi(b) \cup \Psi(a, b) = \{\psi_{m,n}(x) : m \in \{0\} \cup \mathbb{N}, \ n \in \mathbb{Z}^d\}$$

If $\psi_{m,n}(x)$ is compactly supported, denote by $S_{m,n}$ its support, i.e.,

$$S_{m,n} = \{x \in \mathbb{R}^d : \psi_{m,n}(x) \neq 0\}$$

If $\psi_{m,n}(x)$ is almost compactly supported, $S_{m,n}$ is defined by

$$S_{m,n} = \{x \in \mathbb{R}^d : |\psi_{m,n}(x)| > \epsilon \max_x |\psi_{m,n}(x)|\}$$

where ϵ is a chosen small positive number. With these notations, the truncated wavelet frame is thinned in the following way. For each $x_k \in \mathcal{X}$, find M_k, the index set of wavelets whose supports contain x_k. Then the union of M_k, $k = 1, \ldots, N$, gives the indexes of the wavelets whose supports contain at least one data point. The thinned wavelet frame W is then determined. The algorithm is summarized as follows.

Algorithm 1

Begin-loop For $k = 1 : N$

$$M_k = \{(m, n) : x_k \in S_{m,n} \text{ and } x_k \in \mathcal{X}\}$$

End-loop

$$W = \{\psi_{m,n} : (m, n) \in M_1 \cup M_2 \cup \cdots \cup M_N\}.$$

In order to facilitate the determination of M_k, the wavelet supports $S_{m,n}$ can be approximated by hyper-cubes in the x-space. With this approximation, M_k can be determined by separately considering the components of $x_k \in \mathcal{X}$.

Let L be the number of terms in W. For convenience of notation, let us replace the double index (m, n) by a single index $j = 1, \ldots, L$, thus

$$W = \{\psi_1, \ldots, \psi_L\}$$

For the same reason the single index has been used for the wavelets in Fig. 26.1. Though ψ_j may be a scaling function or a wavelet function, we call it a wavelet function in order to facilitate the following discussions.

26.5.3 Regression Analysis Applied to Wavelets

Wavelet networks could be constructed as expansions over the thinned frame W. It turns out that the terms of W are often redundant for estimating f. The reason is that the wavelet frame has only been thinned regarding the input data $\mathcal{X}$. The output data $\mathcal{Y}$ should also be considered. For this purpose, consider a wavelet network as a regression model, and consider the terms $\psi_j(x) \in W$ as regressors, then techniques for regressor selection in "regression analysis" [25–27] can be applied.

Let us define the following vectorial notations:

$$\boldsymbol{\psi}_j = \zeta_j[\psi_j(x_1) \cdots \psi_j(x_N)]^{\mathrm{T}}, \qquad \psi_j \in W, \qquad j = 1, \ldots, L,$$
$$\boldsymbol{y} = [y_1 \ldots y_N]^{\mathrm{T}}, \qquad (x_k, y_k) \in (\mathcal{X}, \mathcal{Y}), \qquad k = 1, \ldots, N$$

where ζ_j is chosen so that $\boldsymbol{\psi}_j^{\mathrm{T}}\boldsymbol{\psi}_j = 1$. With these notations, selecting the "best" subset of W for the estimation of f amounts to selecting a subset of $\{\boldsymbol{\psi}_1, \ldots \boldsymbol{\psi}_L$ that spans the space the "closest" to the vector $\boldsymbol{y}$. Two questions arise for solving this problem. First, how to choose the size of this subset? Second, how to select each term of the subset? We will answer the first question later and for the moment assume that this size s is known. The second question could be answered by searching through all the possible combinations of the terms of W. The number of possible combinations is usually, however, too large for a reasonable search. There exist some heuristic algorithms for solving this problem with reasonable computational costs, see [22] for more details. Here we only recall one of the heuristic algorithms which makes a good trade-off between computational cost and performance.

The algorithm first selects the wavelet in W that best fits the observed data, then it repeatedly selects the wavelet in the remainder of W that best fits the data while combining with the previously selected wavelets. For computational efficiency, later selected wavelets are orthonormalized to earlier selected ones.

Denote by l_i the index of the wavelet selected at iteration i. If i is the current iteration number, then $\boldsymbol{\psi}_{l_1}, \boldsymbol{\psi}_{l_2}, \ldots, \boldsymbol{\psi}_{l_{i-1}}$ have been selected in the previous iterations. Define

$$\boldsymbol{q}_{l_1} = \boldsymbol{\psi}_{l_1}$$
$$\boldsymbol{p}_{l_j} = \boldsymbol{\psi}_{l_j} - \left[\left(\boldsymbol{\psi}_{l_j}^{\mathrm{T}}\boldsymbol{q}_{l_1}\right)\boldsymbol{q}_{l_1} + \cdots + \left(\boldsymbol{\psi}_{l_j}^{\mathrm{T}}\boldsymbol{q}_{l_{j-1}}\right)\boldsymbol{q}_{l_{j-1}}\right], \qquad j = 2, 3, \ldots, i - 1$$
$$\boldsymbol{q}_{l_j} = \left(\boldsymbol{p}_{l_j}^{\mathrm{T}}\boldsymbol{p}_{l_j}\right)^{-1/2}\boldsymbol{p}_{l_j}, \qquad j = 2, 3, \ldots, i - 1$$

then $\boldsymbol{q}_{l_j}$ is the orthonormalized version of $\boldsymbol{\psi}_{l_j}$, $j = 1, \ldots, i - 1$. Now, orthogonalize the remaining vectors $\boldsymbol{\psi}_j$ to $\boldsymbol{q}_{l_1}, \ldots, \boldsymbol{q}_{l_{i-1}}$:

$$\boldsymbol{p}_j = \boldsymbol{\psi}_j - [(\boldsymbol{\psi}_j^{\mathrm{T}}\boldsymbol{q}_{l_1})\boldsymbol{q}_{l_1} + \cdots + (\boldsymbol{\psi}_j^{\mathrm{T}}\boldsymbol{g}_{l_{i-1}})\boldsymbol{q}_{l_{i-1}}]$$

The vectors $\boldsymbol{\psi}_{l_1}, \ldots, \boldsymbol{\psi}_{l_{i-1}}$ and $\boldsymbol{\psi}_j$ span the same space as $\boldsymbol{q}_{l_1}, \cdots, \boldsymbol{q}_{l_{i-1}}$ and $\boldsymbol{p}_j$. Since $\boldsymbol{q}_{l_1}, \cdots, \boldsymbol{q}_{l_{i-1}}$ and $\boldsymbol{p}_j$ are orthogonal, the best $\boldsymbol{\psi}_j$ to be chosen for the current iteration corresponds to the $\boldsymbol{p}_j$ the "closest" to $\boldsymbol{y}$. This results in the following algorithm.

Algorithm 2

Initialization:

$$I = \{1, 2, \ldots, L\}; \quad \boldsymbol{p}_j = \boldsymbol{\psi}_j \text{ for all } j \in I$$
$$l_0 = 0, \quad \boldsymbol{q}_{l_0} = 0$$

Begin-loop For $i = 1 : s$

$$p_j = p_j - \left(\psi_j^{\mathrm{T}} q_{l_{i-1}}\right) q_{l_{i-1}} \text{ for all } j \in I$$

$$I = I - \{j : p_j = 0\}$$

$$l_i = \arg\max_{j \in I} \frac{\left(p_j^{\mathrm{T}} y\right)^2}{p_j^{\mathrm{T}} p_j}$$

$$\alpha_{ji} = \psi_{l_i}^{\mathrm{T}} q_{l_j} \text{ for } j = 1, \ldots, i - 1$$

$$\alpha_{ii} = \sqrt{p_{l_i}^{\mathrm{T}} p_{l_i}}$$

$$q_{l_i} = \alpha_{ii}^{-1} p_{l_i}$$

$$\tilde{w}_{l_i} = q_{l_i}^{\mathrm{T}} y$$

End-loop

The wavelet network estimating $f(x)$ is given by

$$f_s(x) = \sum_{i=1}^{s} w_{l_i} \psi_{l_i}(x) \tag{26-22}$$

where the coefficients w_{l_i} are determined by

$$
\begin{bmatrix}
\alpha_{11} & \alpha_{12} & \alpha_{13} & \cdots & & \alpha_{1s} \\
0 & \alpha_{22} & \alpha_{23} & \cdots & & \alpha_{2s} \\
0 & 0 & \alpha_{33} & \cdots & & \alpha_{3s} \\
\vdots & \vdots & \ddots & \ddots & & \vdots \\
0 & 0 & \cdots & 0 & \alpha_{s-1\,s-1} & \alpha_{s-1\,s} \\
0 & 0 & \cdots & & 0 & \alpha_{ss}
\end{bmatrix}
\begin{bmatrix}
w_{l_1} \\
\vdots \\
w_{l_s}
\end{bmatrix}
=
\begin{bmatrix}
\tilde{w}_{l_1} \\
\vdots \\
\tilde{w}_{l_s}
\end{bmatrix}
$$

26.5.4 The Network Size

In the previous subsection we have assumed that the network size s (the number of wavelets used in the network) is known. The choice for the value of s is not very easy in practice. We would choose s so that the performance of the resulting wavelet network is optimized. The problem is how to evaluate the performance of the network. A natural evaluation is the mean squared error (MSE) of the network, defined as follows:

$$\mathrm{MSE} = \frac{1}{N} \sum_{k=1}^{N} [y_k - f_s(x_k)]^2$$

where $f_s(x)$ is the wavelet network composed of s wavelets as defined in (26-22). If the MSE is evaluated on the same data used for estimating the parameters of $f_s(x)$, the larger s is, the smaller the MSE is, so s cannot be determined in this way. In fact, a better performance evaluation is the generalization ability of the network, i.e., how the network behaves on data which are not used in the estimation of $f_s(x)$. This evaluation method is called "cross-validation" (CV). It consists in splitting the available data into two sets, an "estimation set" and a "validation set". The first set is used to estimate $f_s(x)$ and the second set is used to evaluate the MSE. The value of s is chosen so that the MSE evaluated on the validation set is minimized.

However, splitting the data into two sets reduces the information used in the estimation of $f_s(x)$. This effect is not desirable especially when the available data set is relatively small. In this case, an alternative method, the "generalized cross-validation" (GCV), can be considered. It consists in estimating the expected value of the MSE evaluated on validation data [28]. Under certain assumptions, it turns out that this expectation is approximately given by

$$\text{GCV} = \text{MSE} + \frac{2s}{N}\sigma_e^2$$

where the MSE is evaluated on a data sample of length N, s is the size of the wavelet network, and σ_e^2 is the variance of the noise e in (26-21). s is then determined so that the GCV is minimized.

Now we must have a method for estimating σ_e^2. A simple method was proposed by Moody in [28]:

$$\sigma_e^2 \approx \frac{N}{N - s}\text{MSE}$$

The GCV criterion can be viewed as the MSE plus a penalty on the complexity of the network: the larger s is, the more complex $f_s(x)$ is. The GCV is thus closely related to the famous Akaike criterion.

26.5.5 Additional Optimization

In the above procedure, the construction of a wavelet network for estimating $f(x)$ was done by choosing scaling and wavelet functions from a wavelet frame. The dilation and translation parameters of these scaling and wavelet functions issuing from the wavelet frame were fixed. Since a wavelet network is a feedforward network, all of its parameters, including the dilation and translation parameters can be trained by some optimization algorithm, for example, the popular "back-propagation procedure" [7, 8] minimizing the MSE of the network. The above-proposed procedure then constitutes an initialization of the optimization algorithm. A quasi-Newton algorithm is recommended for the optimization, due to its good initialization. The optimization algorithms are usually very time consuming. They should be considered as an optional choice.

26.5.6 Implementation of the Wavelet Network

For the implementation of the wavelet network, the first question one may ask is how to choose the scaling function $\varphi(x)$, the wavelet function $\psi(x)$, and the related frame parameters a, b as in (26-19) and (26-20). For practical considerations, functions of simple analytical form are preferred. Radial functions are suitable for large-dimensional applications since the computation needed to evaluate a radial function only weakly depends on the dimension. For usual double side scaling wavelet frames, it is relatively easy to check the sufficient frame conditions (Conditions 1.1–1.3 and 2.1–2.2) in order to choose the values of the frame parameters a, b, since only $\|\omega\| \in [1, a]$ needs to be considered in Conditions 1.1–1.3 or $|\xi| \in [1, a]$ in Condition 2.1. It is more difficult to check the frame conditions for single side scaling frames, since $\omega \in \mathbb{R}^d$ or $\xi \in \mathbb{R}$ must be considered. In practice, the user may try various functions and values of a, b by experiments instead of checking the theoretical conditions. When doing this, one should keep in mind that $\varphi(x)$ is a "low-pass filter" and $\psi(x)$ is a "bandpass filter," so that the $\hat{\varphi}(\omega)$ covers low frequencies and the dilated $\hat{\psi}(\omega)$ covers higher frequencies. The choice of frame parameters a, b may be suggested by the corresponding double side scaling wavelet frames. For example, for some typical wavelet frames, values of a, b and the corresponding frame bounds A, B are provided in [13, 14].

Some preprocessing of the data may improve the numerical property of the estimation algorithms. The simplest processing consists in re-scaling the data so that they have zero mean and that the variance is component-wisely normalized. Sometimes it is helpful to remove outliers in the data set and to perform a principal component analysis in order to eventually reduce the data dimension.

Finally, the implementation of the wavelet network is summarized in Fig. 26-2.

26.5.7 Numerical Example

In order to illustrate the performance of the wavelet network, we present an example on the estimation of a function of two variables from noise corrupted observations. The wavelet network is suitable for solving problems of larger dimensions with sparse data, but this example has been chosen for the convenience of graphical visualization.

In our example $x \in \mathbb{R}^2$ and the used scaling function and wavelet function are

$$\varphi(x) = e^{-\frac{1}{2}\|x\|^2}, \qquad \psi(x) = (2 - \|x\|^2)e^{-\frac{1}{2}\|x\|^2}, \qquad \|x\|^2 = x^T x$$

which are illustrated in Fig. 26-3.

Let us denote by x_1 and x_2 the two components of x, i.e., $x = (x_1, x_2)^T$. The function to be estimated is

$$f(x) = 20e^{x_1 - 2x_1^2 - 2x_2^2} \sin[3(x_1 - 0.6)^2] \sin(2x_2)$$

and is illustrated in Fig. 26-4(a).

A sample $\mathcal{X}$ of 2000 points is randomly drawn with the uniform law for $x \in [-1, 1]^2$, then the corresponding $\mathcal{Y}$ sample is generated with $y = f(x) + e$, where e is the Gaussian noise with variance $\sigma_e^2 = 1$. The wavelet network is then used to estimate f from the data sample $(\mathcal{X}, \mathcal{Y})$. The GCV criterion is used to choose the size of the network, resulting in a network of three scaling functions and 26 wavelet functions. The estimated function obtained with the network before the final optimization is illustrated in Fig. 26-4(b) and the corresponding MSE is 1.483. All the parameters of the network are then optimized by 12 iterations of the Levenberg–Marquardt algorithm. The result given by the optimized network is illustrated in Fig.

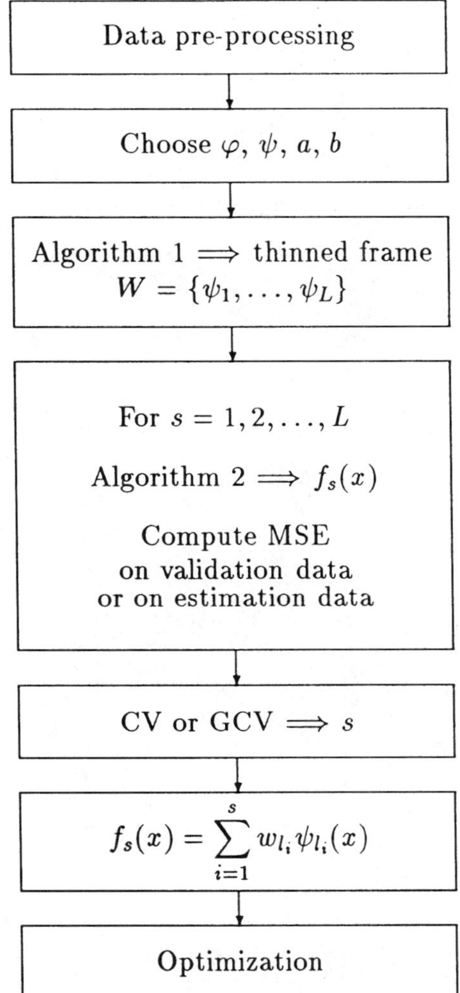

Figure 26-2   Implementation schema of the wavelet network.

26-4(c) and the corresponding MSE is 0.923. The difference between the function $f(x)$ and its estimation given by the optimized network is shown in Fig. 26-4(d).

The optimization with the Levenberg–Marquardt algorithm slightly improves the quality of the estimation, but it takes about half an hour on a Sun Sparc 2 workstation, whereas only about 40 s are needed for the estimation without the final optimization. Note that MSE = 0.923 is close to the variance of the noise e, indicating a good choice of the network size.

26.6. CONCLUSION

Large-dimensional wavelets are much less often studied than one- or two-dimensional wavelets, due to practical difficulties for the implementation of

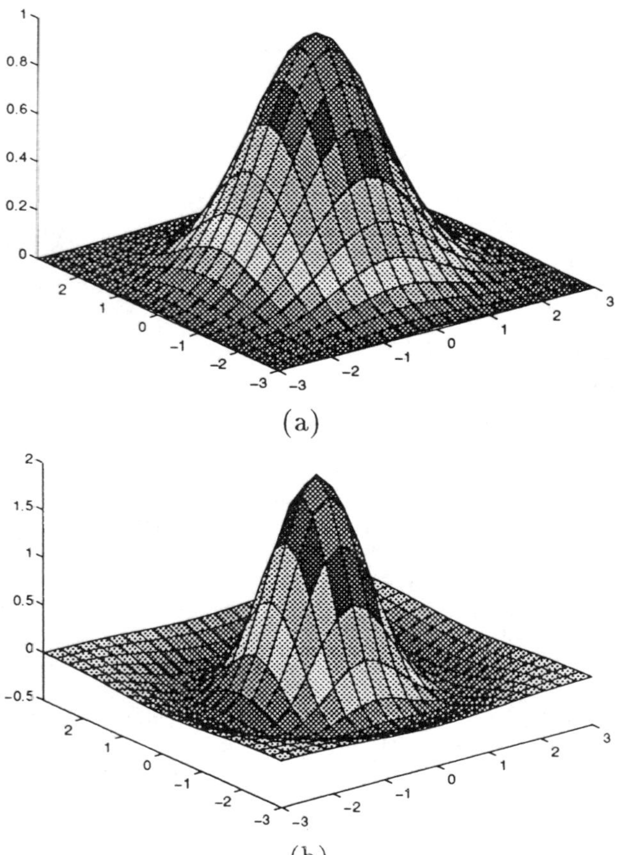

Figure 26-3   The scaling function (a) and the wavelet function (b) used in the example.

large-dimensional wavelet bases or frames. A method for constructing large-dimensional wavelet frames is presented in this paper. In the considered frame, the scalar dilation index is shared by all the dimensions (scalar scaling) and ranges from 1 to ∞ (single side scaling). A radial construction of such frames is proposed. The use of large-dimensional wavelet frames in wavelet networks is also considered, as a practically reasonable application.

APPENDIX A: PROOF OF THEOREM 3

The proof closely follows Daubechies' scheme for one dimension (see [14] and [13] section 3.3.2) and that of [15].

First, we need the following generalized Poisson formula:

$$\sum_{k \in \mathbb{Z}^d} e^{iCk^{\mathrm{T}}x} = \left(\frac{2\pi}{C}\right)^d \prod_{j=1}^{d} \sum_{k_j \in \mathbb{Z}} \delta\left(x_j - \frac{2\pi}{C}k_j\right)$$

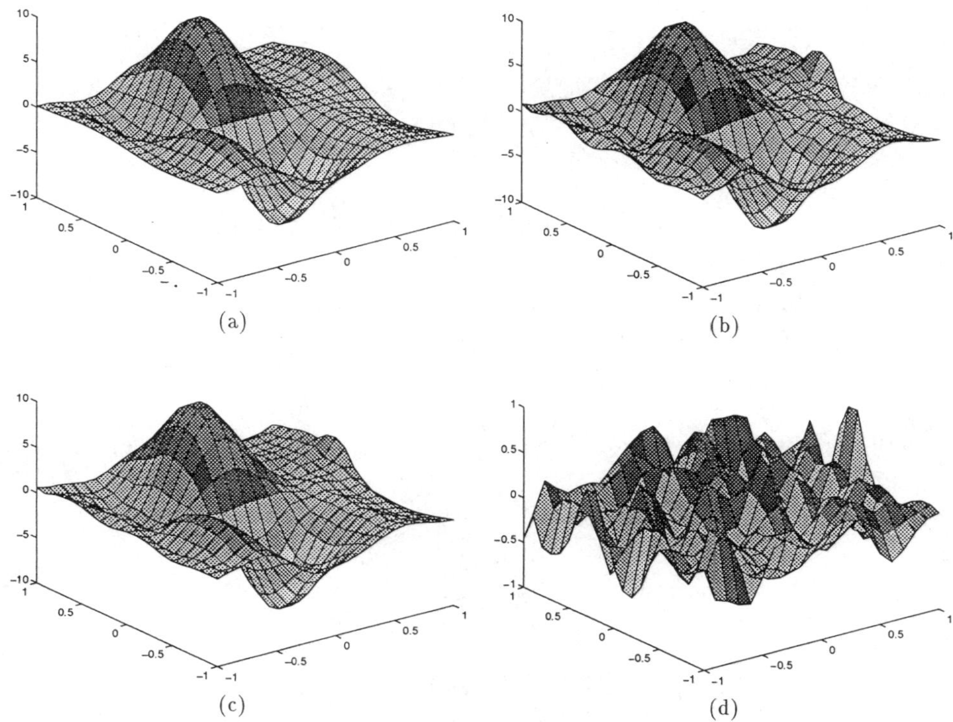

Figure 26-4 Estimation of a function of two variables. (a): the original (noiseless) function $f(x)$, (b): the estimation before optimization, (c): the estimation after optimization, (d): the difference between $f(x)$ and the estimation after optimization.

where i is the imaginary unity, C is any real non zero constant, and $k = (k_1, \ldots, k_d)^{\mathrm{T}}$, $x = (x_1, \ldots, x_d)^{\mathrm{T}}$.

By applying this generalized Poisson formula and the Parseval's theorem, straightforward computations give, for all $f \in L_2(\mathbb{R}^d)$,

$$\sum_{n \in \mathbb{Z}^d} |\langle \varphi_n, f \rangle|^2 + \sum_{m \in \mathbb{N}} \sum_{n \in \mathbb{Z}^d} |\langle \psi_{m,n}, f \rangle|^2$$

$$= \left(\frac{2\pi}{b} \right)^d \left\{ \int_{\mathbb{R}^d} \left(|\hat{\varphi}(\omega)|^2 + \sum_{m \in \mathbb{N}} |\hat{\psi}(a^{-m}\omega)|^2 \right) |\hat{f}(\omega)|^2 d\omega + \Lambda_\varphi + \Lambda_\psi \right\}$$

where

$$\Lambda_\varphi = \sum_{n \neq 0} \int_{\mathbb{R}^d} \hat{\varphi}(\omega) \overline{\hat{\varphi}\left(\omega - \frac{2\pi}{b} n \right)} \overline{\hat{f}(\omega)} \hat{f}\left(\omega - \frac{2\pi}{b} n \right) d\omega$$

$$\Lambda_\psi = \sum_{m \in \mathbb{N}} \sum_{n \neq 0} \int_{\mathbb{R}^d} \hat{\psi}(a^{-m}\omega) \overline{\hat{\psi}\left(a^{-m}\omega - \frac{2\pi}{b} n \right)} \overline{\hat{f}(\omega)} \hat{f}\left(\omega - \frac{2\pi}{a^{-m}b} n \right) d\omega$$

By applying the Cauchy–Schwarz inequality, we get ($k \neq 0$ means $k \in \mathbb{Z}^d$ but $\|k\| \neq 0$):

$$|\Lambda_\varphi| \leq \sum_{k \neq 0}\left[\beta_\varphi\left(\frac{2\pi}{b}k\right)\beta_\varphi\left(-\frac{2\pi}{b}k\right)\right]^{\frac{1}{2}}\|f\|^2 = \Upsilon_\varphi\|f\|^2$$

$$|\Lambda_\psi| \leq \sum_{k \neq 0}\left[\beta_\psi\left(\frac{2\pi}{b}k\right)\beta_\psi\left(-\frac{2\pi}{b}k\right)\right]^{\frac{1}{2}}\|f\|^2 = \Upsilon_\psi\|f\|^2$$

where $\beta_\varphi(\cdot)$ and $\beta_\psi(\cdot)$ are as in (26-14) and (26-15), Υ_φ and Υ_ψ are used to denote the two corresponding multi-indexed series.

These two inequalities together with Conditions 3.1 and 3.2 result in

$$\left(\frac{2\pi}{b}\right)^d[m(\varphi,\psi,a) - \Upsilon_\varphi - \Upsilon_\psi]\|f\|^2 \leq \sum_{n \in \mathbf{Z}^d}|\langle\varphi_n,f\rangle|^2 + \sum_{m \in \mathbb{N}}\sum_{n \in \mathbb{Z}^d}|\langle\psi_{m,n},f\rangle|^2$$

$$\leq \left(\frac{2\pi}{b}\right)^d[M(\varphi,\psi,a) + \Upsilon_\varphi + \Upsilon_\psi]\|f\|^2$$

We still need to show that Condition 3.3 ensures the convergence of the multi-indexed series Υ_φ and Υ_ψ. From (26-12) we get

$$\Upsilon_\varphi \leq C_\epsilon\left(\frac{b}{2\pi}\right)^{d(1+\epsilon)}\sum_{k \neq 0}\left[\left(\frac{b}{2\pi}\right)^2 + \|k\|^2\right]^{-\frac{d(1+\epsilon)}{2}}$$

$$\leq C_\epsilon\left(\frac{b}{2\pi}\right)^{d(1+\epsilon)}\sum_{l=1}^{\infty}2d(2l+1)^{d-1}\left[\left(\frac{b}{2\pi}\right)^2 + l^2\right]^{-\frac{d(1+\epsilon)}{2}}$$

$$\leq C_\epsilon\left(\frac{b}{2\pi}\right)^{d(1+\epsilon)}\sum_{l=1}^{\infty}2d(3l)^{d-1}\left[\left(\frac{b}{2\pi}\right)^2 + l^2\right]^{-\frac{d(1+\epsilon)}{2}}$$

$$\leq C_\epsilon 2d3^{d-1}\left(\frac{b}{2\pi}\right)^{d(1+\epsilon)}\sum_{l=1}^{\infty}\left[\left(\frac{b}{2\pi}\right)^2 + l^2\right]^{-\frac{1+d\epsilon}{2}} \tag{26-23}$$

So, Υ_φ converges and tends to zero when $b \to 0$. The same result can be similarly obtained for Υ_ψ. The proof of Theorem 3 is thus established.

APPENDIX B: PROOF OF THEOREM 4

Take $\hat{\varphi}(x) = \hat{\phi}(\|x\|)$ and $\hat{\psi}(x) = \hat{\mu}(\|x\|)$. Then, it is trivial to verify that Condition 4.1 is equivalent to Condition 3.1. It is also straightforward to show that Condition 4.2 implies Condition 3.2.

It is more complicated to show that Condition 4.2 also implies Condition 3.3. We shall need the following inequality that holds for all $x, y \in \mathbb{R}^d$:

$$(1 + \|x+y\|^2)(1 + \|x-y\|^2) \geq 1 + 4\|y\|^2 \tag{26-24}$$

From Condition 4.2,

$$\beta_\varphi(\eta) = \sup_\omega |\hat{\phi}(\|\omega\|)| \cdot |\hat{\phi}(\|\omega+\eta\|)|$$

$$\leq C_1^2 \sup_\omega \left[(1+\|\omega\|^2)(1+\|\omega+\eta\|^2)\right]^{-\frac{\tau}{2}}$$

Applying (26-24) with $x = \omega + \frac{1}{2}\eta$ and $y = \frac{1}{2}\eta$ yields

$$\beta_\varphi(\eta) \leq C_1^2 (1+\|\eta\|^2)^{-\frac{\tau}{2}}$$

Note that $\tau > d$, so (26-12) holds

$$\beta_\psi(\eta) = \sup_\omega \sum_{m\in\mathbb{N}} |\hat{\mu}(\|a^{-m}\omega\|)| \cdot |\hat{\mu}(\|a^{-m}\omega+\eta\|)|$$

$$\leq C_2^2 \sup_\omega \sum_{m\in\mathbb{N}} \|a^{-m}\omega\|^\alpha (1+\|a^{-m}\omega\|^2)^{-\frac{\gamma}{2}} \|a^{-m}\omega+\eta\|^\alpha (1+\|a^{-m}\omega+\eta\|^2)^{-\frac{\gamma}{2}}$$

We distinguish the two cases $\|\omega\| \leq a$ and $\|\omega\| > a$.
If $\|\omega\| \leq a$,

$$\beta_\psi(\eta) \leq C_2^2 \sup_{\|\omega\|\leq a} \sum_{m\in\mathbb{N}} \|a^{-m}\omega\|^\alpha (1+\|a^{-m}\omega+\eta\|^2)^{-\frac{\gamma-\alpha}{2}}$$

Note that for $\|\omega\| \leq a$, we have $\|a^{-m}\omega\| \leq 1$, then

$$3(1 + \|a^{-m}\omega+\eta\|^2) \geq 1 + 2(1 + \|a^{-m}\omega+\eta\|^2)$$

$$\geq 1 + (1 + \|a^{-m}\omega+\eta\|)^2$$

$$\geq 1 + (\| - a^{-m}\omega\| + \|a^{-m}\omega+\eta\|)^2$$

$$\geq 1 + \|\eta\|^2$$

therefore,

$$(1 + \|a^{-m}\omega+\eta\|^2)^{-1} \leq 3(1+\|\eta\|^2)^{-1}$$

So,

$$\beta_\psi(\eta) \leq C_2^2 3^{\frac{\gamma-\alpha}{2}} \sup_{\|\omega\|\leq a} \sum_{m\in\mathbb{N}} \|a^{-m}\omega\|^\alpha a^{-(m-1)\alpha} (1+\|\eta\|^2)^{-\frac{\gamma-\alpha}{2}}$$

$$\leq C_2^2 3^{\frac{\gamma-\alpha}{2}} (1+\|\eta\|^2)^{-\frac{\gamma-\alpha}{2}} \sup_{\|\omega\|\leq a} \sum_{m\in\mathbb{N}} \|a^{-m}\omega\|^\alpha a^{-(m-1)\alpha}$$

If $\|\omega\| > a$, for any $\delta \in [0, 1]$,

$$\beta_\psi(\eta) \leq C_2^2 \sup_{\|\omega\|>a} \sum_{m\in\mathbb{N}} \left[(1+\|a^{-m}\omega\|^2)(1+\|a^{-m}\omega+\eta\|^2)\right]^{-\frac{\gamma-\alpha}{2}}$$

$$\leq C_2^2 \sup_{\|\omega\|>a} \sum_{m\in\mathbb{N}} \left[(1+\|a^{-m}\omega\|^2)(1+\|a^{-m}\omega+\eta\|^2)\right]^{-(1-\delta)\frac{\gamma-\alpha}{2}}$$

$$\cdot \left[(1+\|a^{-m}\omega\|^2)(1+\|a^{-m}\omega+\eta\|^2)\right]^{-\delta\frac{\gamma-\alpha}{2}}$$

Applying (26-24) with $x = \omega + \frac{1}{2}\eta$ and $y = \frac{1}{2}\eta$ yields

$$\beta_\psi(\eta) \leq C_2^2 (1 + \|\eta\|^2)^{-(1-\delta)\frac{\gamma-\alpha}{2}} \sum_{m \in \mathbb{N}} (a^{-m+1})^{-\delta\frac{\gamma-\alpha}{2}}$$

Since $\gamma > \alpha + d$, there exist $\delta > 0$ and ε, both small enough, so that $(1 - \delta)(\gamma - \alpha) \geq d(1 + \varepsilon)$. So, in all the cases, $\beta_\psi(\eta)$ is bounded by $(1 + \|\eta\|^2)^{-d(1+\varepsilon)/2}$ times a constant, with some $\varepsilon > 0$. Thus $\beta_\psi(\eta)$ satisfies (26-13).

APPENDIX C: SOME COMMENTS ON THEOREM 4

It is trivial to see that Condition 4.1 does not involve the dimension d. Condition 4.2 involves the dimension d through the assumptions $\tau > d$ and $\gamma > \alpha + d$. They are quite mild assumptions on the decay of $\hat{\phi}$ and $\hat{\mu}$. For instance, all functions with exponential decay satisfy these assumptions for any finite d. Obviously, if Condition 4.2 is satisfied for $d = d_M$, then for all $d \leq d_M$ Condition 4.2 is also satisfied. Therefore, it suffices to check Condition 4.2 for a sufficiently large dimension in order to ensure its validity for smaller dimensions.

The choice of the value of a can be considered as independent of the dimension d. The value of b, however, does depend on d and a. This dependency is through the two multi-indexed series Υ_φ and Υ_ψ defined in Appendix A. It is easy to show, through (26-23), that if $b < 2\pi$ then Υ_φ is bounded by $d(3b/2\pi)^d$ times a constant. If $b \leq \pi/3$, we have $(d + 1)(3b/2\pi)^{d+1} \leq d(3b/2\pi)^d$ for all $d \geq 1$. So, when $b \leq \pi/3$, the larger d is, the smaller is the bound of the series Υ_φ. The same result can be shown for Υ_ψ in the same way. Therefore, if $\hat{\phi}$, $\hat{\mu}$ and a satisfy Theorem 4 for $d = d_M$, and a value $b \leq \pi/3$ is checked for $d = 1$, then for all $d \leq d_M$, with the same values of a and b, the corresponding d-dimensional radial functions generate a frame of $L_2(\mathbb{R}^d)$.

REFERENCES

[1] R. P. Lippmann, "An introduction to computing with neural nets," *IEEE ASSP Mag.*, vol. 4, no. 2, pp. 4–22, 1987.

[2] D. R. Hush, and B. G. Horne, "Progress in supervised neural networks, what's new since Lippmann?" *IEEE Signal Proc. Mag.*, vol. 10, no. 1, pp. 8–39, 1993.

[3] S. Kung, *Digital Neural Networks.* NJ: Prentice-Hall, Englewood Cliffs, 1993.

[4] S. Haykin, *Neural Networks: A Comprehensive Foundation.* New York: Macmillan College Publishing Company, 1994.

[5] C. Lau, (ed.), *Neural Networks, Theoretical Foundations and Analysis.* New York: IEEE Press, 1992.

[6] E. Sanchez-Sinencio, and C. Lau, (eds.), *Artificial Neural Networks: Paradigms, Applications, and Hardware Implementations.* New York: 1992.

[7] D. Rumelhart, G. Hinton, and R. Williams, "Learning representations by backpropagating errors," *Nature*, vol. 323, no. 9, pp. 533–536, 1986.

[8] S. Saarinen, R. Bramley, and G. Cybenko, "Ill-conditioning in neural network training problems," *SIAM J. Sci. Comput.*, vol. 14, no. 3, pp. 693–714, 1993.

[9] G. Cybenko, "Approximation by superpositions of a sigmoidal function," *Math. Control, Signals, Syst.*, vol. 2, pp. 303–314, 1989.

[10] K. Hornik, "Multilayer feedforward networks are universal approximators." *Neural Networks*, vol. 2, pp. 359–366, 1989.

[11] A. Barron, "Universal approximation bounds for superpositions of a sigmoidal function," *IEEE Trans. Inform. Theory*, vol. 39, no. 3, 1993.

[12] S. Mallat, "Multiresolution approximation and wavelets orthonormal bases of $L^2(\mathbb{R})$," *Trans. Am. Math. Soc.*, vol. 315, no. 1, pp. 69–87, 1989.

[13] I. Daubechies, *Ten Lectures on Wavelets*. CBMS-NSF regional series in applied mathematics, Society for Industrial and Applied Mathematics, Philadelphia, 1992.

[14] I. Daubechies, "The wavelet transform, time-frequency," *IEEE Trans. Inform. Theory*, vol. 36, no. 5, pp. 961–1005, 1990.

[15] T. Kugarajah, and Q. Zhang, "Multi-dimensional wavelet frames," *IEEE Trans. Neural Networks*, vol. 6, no. 6, pp. 1552–1556, 1995.

[16] C. Chui, (ed.), *Wavelets: A Tutorial in Theory and Applications*. Boston, San Diego: Academic Press, Inc., 1992.

[17] M. Ruskai, G. Beylkin, R. Coifman, I. Daubechies, S. Mallat, Y. Meyer, and L. Raphael, (eds.), *Wavelets and their Applications*. Boston: Jones and Bartlett, 1992.

[18] A. Juditsky, Q. Zhang, B. Delyon, P.-Y. Glorennec, and A. Benveniste, "Wavelets in identification," *Technical report*, IRISA, 1994.

[19] Q. Zhang, and A. Benveniste, "Wavelet networks," *IEEE Trans. Neural Networks*, vol. 3, no. 6, pp. 889–898, 1992.

[20] Y. Pati and P. Krishnaprasad, "Analysis and synthesis of feedforward neural networks using discrete affine wavelet transformations," *IEEE Trans. Neural Networks*, vol. 4, no. 1, pp. 73–85, 1993.

[21] B. Bakshi and G. Stephanopoulos, "Wave-net: a multiresolution, hierarchical neural network with localized learning," *Am. Inst. Chem. Eng. J.*, vol. 39, no. 1, pp. 57–81, 1993.

[22] Q. Zhang, "Using wavelet network in nonparametric estimation," *IEEE Trans. Neural Networks,* vol. 8, no. 2, pp. 227–236, 1997.

[23] J. Sjöberg, Q. Zhang, L. Ljung, A. Benveniste, B. Deylon, P.-Y. Glorennec, H. Hjalmarsson, and A. Juditsky, "Non-linear black-box modeling in system identification: A unified overview," *Automatica*, vol. 31, no. 12, pp. 1691–1724, 1995.

[24] A. Juditsky, H. Hjalmarsson, A. Benveniste, B. Deylon, L. Ljung, J. Sjöberg, and Q. Zhang, "Nonlinear black-box models in system identification: Mathematical foundations," *Automatica*, vol. 31, no. 12, pp. 1725–1750, 1995.

[25] N. Draper, and H. Smith, *Applied Regression Analysis*. 2nd Ed., Series in Probability and Mathematical Statistics. New York: Wiley, 1981.

[26] W. Dillon, and M. Goldstein, *Multivariate Analysis: Method and Applications.* New York: John Wiley & Sons, Inc., 1984.

[27] A. Sen, and M. Srivastava, *Regression Analysis: Theory, Methods, and Applications.* New York: Springer-Verlag, 1990.

[28] J. E. Moody, "Note on generalization, regularization and architecture selection in nonlinear learning systems." In *Neural Networks for Signal Processing, Proc. 1991 IEEE Workshop*, B. H. Juang, S. Y. Kung, and C. A. Kamm (eds.), New York: IEEE Press, pp. 1–10, 1991.

Chapter 27

Analysis of Evoked Potentials Using Wavelet Networks

Hartmut Heinrich, Hartmut Dickhaus

27.1. INTRODUCTION

Time-frequency representations describe the characteristics of time-varying signals by means of two-dimensional energy distributions in an excellent manner. Many examples in this book demonstrate the advantages of time-frequency transformations in different medical and biological applications. However, normally these methods do not automatically characterize signals with a set of meaningful parameters as is necessary for typical classification or discrimination tasks.

For example, in our application, evoked potentials are used to discriminate children with attention deficit disorder with hyperactivity (ADDH) from a control group. To verify the clinical hypothesis of frontal brain lesions, the central nervous system is tested by auditory evoked potentials (AEPs). This activity is particularly useful because it is independent of an intact behavioral response system [1]. However, the main problem of our study is to extract parameters from evoked potentials (EPs) which can be easily used to find group differences between normal and ADDH children. Frequency, EP peak parameters, i.e., amplitudes and latencies, are considered as relevant features [2]. Besides this, specific parts of the signal represented by its sample points are used for the classification task. In any case additional selection procedures are required to reduce the number of features.

Because a strong systematical parameter optimization is usually very time consuming, simple heuristic strategies are frequently applied. Of course, the results depend on the underlying model and sometimes the evaluated features seem more or less selected by chance. Therefore an approach would be desirable which automatically provides a representation of particular time-varying signal characteristics with only a small set of parameter values. Furthermore, these parameters should be easy to interpret in the context of the time-varying signal structure or in terms of the underlying physiological process.

Since wavelet functions provide an adequate, clear formal description of a signal's time-frequency behavior, it seems reasonable to represent a given signal by these functions. Moreover, the typical wavelet parameters can be regarded as sensitive features for the signal pattern. Depending on the signal's time-frequency behavior a composition of several wavelets, all of the same type but with different values, is useful.

The important question of which parameter configuration of a set of wavelet functions is optimally tuned for a given signal should be answered by applying training strategies of neural networks (NNs). Experience in this field [3, 4] is ideally suited to constructing a powerful algorithm with high performance and good approximation or representation quality.

This chapter gives a short introduction on how wavelet functions can be used for an optimal formal description of a time-varying signal. This description is achieved by only a small number of representative parameters. Particular attention is paid to the learning algorithm in order to get reliable parameter estimates. Furthermore, the derived procedure is applied to our clinical study concerning the discrimination of ADDH children by AEPs. The results are discussed in comparison with conventional heuristic strategies.

27.2. WAVELET NETWORKS

Wavelet nets (WNs) can be regarded as a signal model based on a particular network structure with wavelet functions describing the nodes. Depending on their intended purpose, different kinds of wavelet networks can be constructed [5, 6]:

- WN for representing signals with a nonstationary, time-varying character by continuous wavelet functions or their parameters respectively and
- WN for classification with automatically optimized features

This chapter stresses only the first topic, WNs for representation.

27.2.1 Basic Method

The structure of a WN for signal representation is shown in Fig. 27-1: a signal function $x(t)$ is approximated by a linear combination of K modified wavelet functions $h(t)$, which represent the wavelet nodes.

$$\hat{x}(t) = \sum_{k=1}^{K} w_k h\left(\frac{t - b_k}{a_k}\right) \tag{27-1}$$

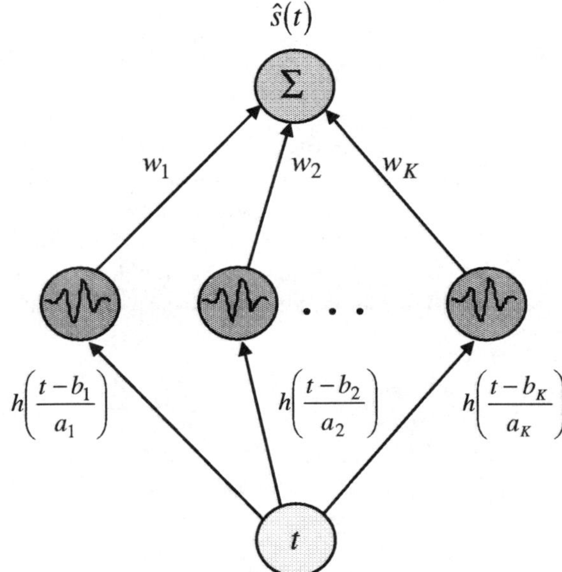

Figure 27-1 The structure of a WN for representation and parameterization of signals.

Each wavelet node is characterized by

- A shift parameter b_k, which describes the center time of the wavelet function, respectively the corresponding node
- A scale parameter a_k, which is responsible for the node's frequency as well as its temporal and spectral spread, and
- A weight factor w_k, representing a node's contribution to a signal $x(t)$

The shift and the scale parameter have a similar meaning to the corresponding variables of the continuous wavelet transform.

As in many studies concerning the continuous wavelet transform [7, 8] the complex Morlet wavelet is used as a basis wavelet for the WN in our application:

$$h(t) = \exp\left(-\frac{t^2}{2} + j\omega_0 t\right), \qquad \text{with} \quad \omega_0 = 5.33 \tag{27-2}$$

Inserting Eq. (27-2) into Eq. (27-1) and taking into account that $h(t)$ is a complex function, we get:

$$\hat{x}(t) = \sum_{k=-K, k\neq 0}^{K} w_k \exp\left[-j\omega_k\left(\frac{t-b_k}{a_k}\right)\right] \exp\left[-0.5\left(\frac{t-b_k}{a_k}\right)^2\right] \tag{27-3}$$

As in a Fourier series, the complex notation can be converted into a real one:

$$\hat{x}(t) = \sum_{k=1}^{K}\left[w_{\cos,k}\cos\left(\omega_k\frac{t-b_k}{a_k}\right) + w_{\sin,k}\sin\left(\omega_k\frac{t-b_k}{a_k}\right)\right]\exp\left[-0.5\left(\frac{t-b_k}{a_k}\right)^2\right]$$

$$(27\text{-}4)$$

In contrast to the wavelet transformation, a variable frequency parameter ω_k is explicitly used in our approach because it results in a better approximation with a smaller number of parameters [9].

A WN for representation decomposes a signal in the time-frequency plane. For this task the wavelet net parameters have to be adapted for a given signal. This is usually done with a least-square-error (LSQ) approach. The error E is defined as:

$$E = \sum_{i=1}^{N}[x(t_i) - \hat{x}(t_i)]^2 \overset{!}{=} \min \tag{27-5}$$

Since the minimization task is a nonlinear problem, no closed-form solution exists. Therefore, the following iterative procedure is frequently applied.

- Assume a number of WN nodes.
- Initialize several wavelet functions for each node with different starting values (= a pool of candidates), which are spread over the time-frequency plane covered by the signal; the weights may have zero initial values.
- Use a gradient technique for the optimization of the parameters $w_{\cos,k}$, $w_{\sin,k}$, ω_k, a_k and b_k. In detail this is performed as follows.
  - Calculate the WN output $\hat{x}(t)$ and the error signal $e(t) = x(t) - \hat{x}(t)$.
  - Calculate the five partial derivatives for each candidate in each iteration

$$\frac{\partial E}{\partial w_{\cos,k}} = -\sum_{i=1}^{N}[x(t_i) - \hat{x}(t_i)]\cos\left(\omega_k\frac{t_i-b_k}{a_k}\right)\exp\left[-0.5\left(\frac{t_i-b_k}{a_k}\right)^2\right]$$

$$\frac{\partial E}{\partial w_{\sin,k}} = -\sum_{i=1}^{N}[x(t_i) - \hat{x}(t_i)]\sin\left(\omega_k\frac{t_i-b_k}{a_k}\right)\exp\left[-0.5\left(\frac{t_i-b_k}{a_k}\right)^2\right]$$

$$\frac{\partial E}{\partial b_k} = -\sum_{i=1}^{N}[x(t_i) - \hat{x}(t_i)]\exp\left[-0.5\left(\frac{t_i-b_k}{a_k}\right)^2\right]\frac{1}{a_k}$$

$$\left(w_{\cos,k}\left[\frac{t_i-b_k}{a_k}\cos\left(\omega\frac{t_i-b_k}{a_k}\right) + \omega_k\sin\left(\omega_k\frac{t_i-b_k}{a_k}\right)\right] \right.$$

$$\left. + w_{\sin,k}\left[\frac{t_i-b_k}{a_k}\sin\left(\omega_k\frac{t_i-b_k}{a_k}\right) - \omega_k\cos\left(\omega_k\frac{t_i-b_k}{a_k}\right)\right]\right)$$

$$\frac{\partial E}{\partial \omega_k} = -\sum_{i=1}^{N}[x(t_i) - \hat{x}(t_i)]\exp\left[-0.5\left(\frac{t_i-b_k}{a_k}\right)^2\right]\left(\frac{t-b_k}{a_k}\right)$$

$$\left(-w_{\cos,k}\sin\left(\omega_k\frac{t_i-b_k}{a_k}\right) + w_{\sin,k}\cos\left(\omega_k\frac{t_i-b_k}{a_k}\right)\right)$$

$$(27\text{-}6)$$

$$\frac{\partial E}{\partial a_k^{-1}} = -\sum_{i=1}^{N}[x(t_i) - \hat{x}(t_i)]\exp\left[-0.5\left(\frac{t_i - b_k}{a_k}\right)^2\right](t_i - b_k)$$

$$\left\{-w_{\cos,k}\left[\frac{t_i - b_k}{a_k}\cos\left(\omega_k\frac{t_i - b_k}{a_k}\right) - \omega_k\sin\left(\omega_k\frac{t_i - b_k}{a_k}\right)\right]\right.$$

$$\left.-w_{\sin,k}\left[\frac{t_i - b_k}{a_k}\sin\left(\omega_k\frac{t_i - b_k}{a_k}\right) + \omega_k\cos\left(\omega_k\frac{t_i - b_k}{a_k}\right)\right]\right\}$$

    It should be noted the inverse scale parameter a_k^{-1} is trained instead of a_k itself. Thus a faster convergence is achieved [9].

    – Update the parameters until a (local) minimum has been reached.

• Choose the candidate with the minimal error E.

    This algorithm results in a good approximation for a given signal; however, if we want to use the evaluated WN parameters in a classification task, they have to fulfill some constraints, besides resulting in a good approximation. In the next sub-section these additional conditions are explained. The resulting WN learning algorithm is presented in section 27.2.3.

27.2.2 Constraints for a Uniform WN Parameterization

    For an appropriate algorithm the resulting WN parameters should be located at specific regions in the time-frequency plane for all trials of an experiment, e.g., if a WN node represents higher frequencies in the beginning of a trial x_i, this node must represent similar time-frequency characteristics for another trial x_j, but if the LSQ algorithm mentioned is used, often several solutions with different WN parameters and a comparable error E exist.

    There are several reasons for this phenomenon, for example in most cases biomedical signals are embedded in noise which leads to an indefinite parameter estimate. Furthermore, the underlying WN model cannot be regarded as an orthogonal decomposition.

    This effect is clearly demonstrated by the simulated test signal in Fig. 27-2. The given WN components [Fig. 27-2(b)] sum up to a signal that resembles a visual EP [Fig. 27-2(a)]. If a WN is trained to represent noisy instances of this signal (signal-to-noise ratio (SNR) = 10 dB), not only the true underlying components are obtained [Fig. 27-2(b)], but also signal components as shown in Fig. 27-2(c) and (d). So, the basic method does not yield definite solutions, i.e., the WN parameters, obtained for nearly identical signals by the basic method, cannot be compared with each other. Therefore, the learning algorithm has to be modified in order to use WNs for parameterization tasks in a useful manner.

    For example, the WN parameters of a node should be restricted to a specific region of the time-frequency plane by some constraints during the optimization process. As is explained in more detail in section 27.2.3, we have realized these constraints by filtering and windowing the signal.

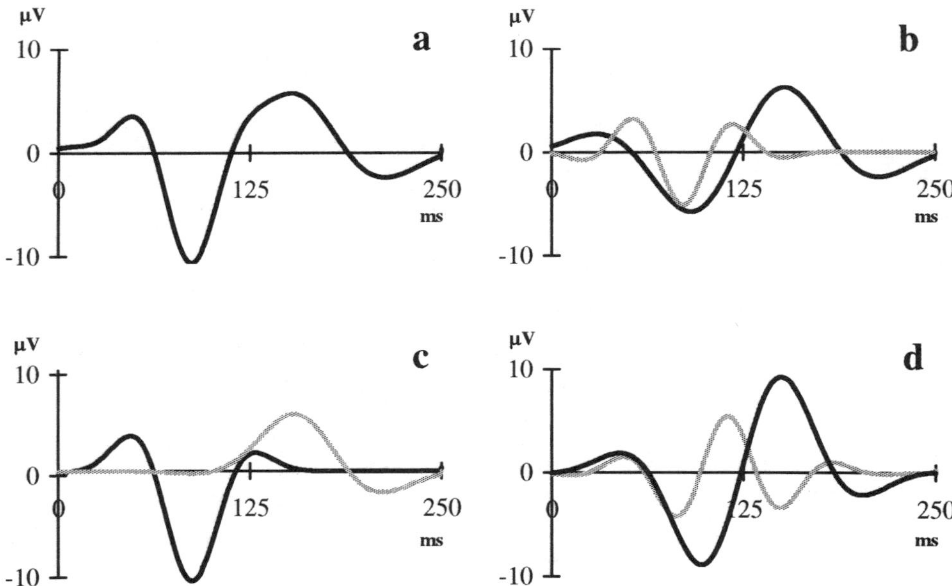

Figure 27-2 The signal $x(t)$ presented in (a) is the sum of two given WN components as shown in (b). Because the basic LSQ method leads to indefinite solutions, various WNs with comparable error E are achieved. For example, besides the "true" WN components in (b), solutions as shown in (c) and (d) result.

Another problem is given by the so-called "herd effect." This means, if we train all the nodes simultaneously, different nodes compete for the high-energy signal components since the nodes cannot communicate. As a result low-energy components, particularly the higher-frequency signal parts, are neglected. This holds especially for the WN because the WN nodes may overlap in the time-frequency plane. Fahlman, who first introduced this "herd effect" [10], recommends using a recursive strategy to avoid this phenomenon.

Finally, the required number of WN nodes is often not known *a priori*. Therefore an objective procedure which determines the optimal number of nodes would be desirable. A recursive strategy may offer a solution to this problem too.

27.2.3 Advanced WN Learning Algorithm

In order to fulfill the mentioned constraints and criteria, we derived the following recursive learning algorithm. The whole procedure is divided into two consecutive steps. In an initial phase, specific regions of the time-frequency plane are defined for a representative signal of the phenomenon, e.g., the grand mean of all trials. This procedure is schematically presented in the flow chart in Fig. 27-3. During the second step, WN parameters for all trials are optimally tuned within the predefined time-frequency regions; see Fig. 27-4.

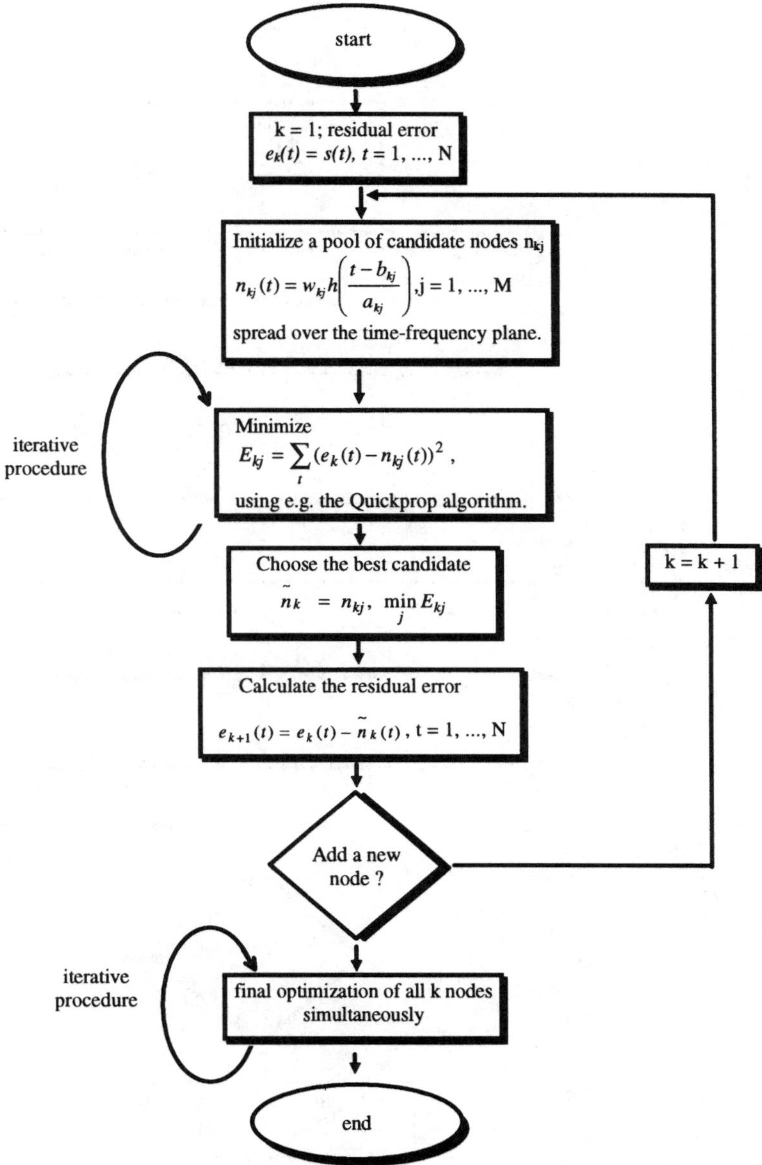

Figure 27-3   A flow chart of the algorithm for training a representative signal of an experiment (phase 1 of the recursive strategy). The resulting WN parameters are a prerequisite to achieve a uniform parameterization of all trials of the experiment.

The following paragraph explains the initial phase in more detail. The principal task of the first phase is to define an adequate region of the time-frequency plane for each WN node. These regions can be described by the WN parameters estimated for a noise-free representative signal using a recursive strategy. This recursive procedure

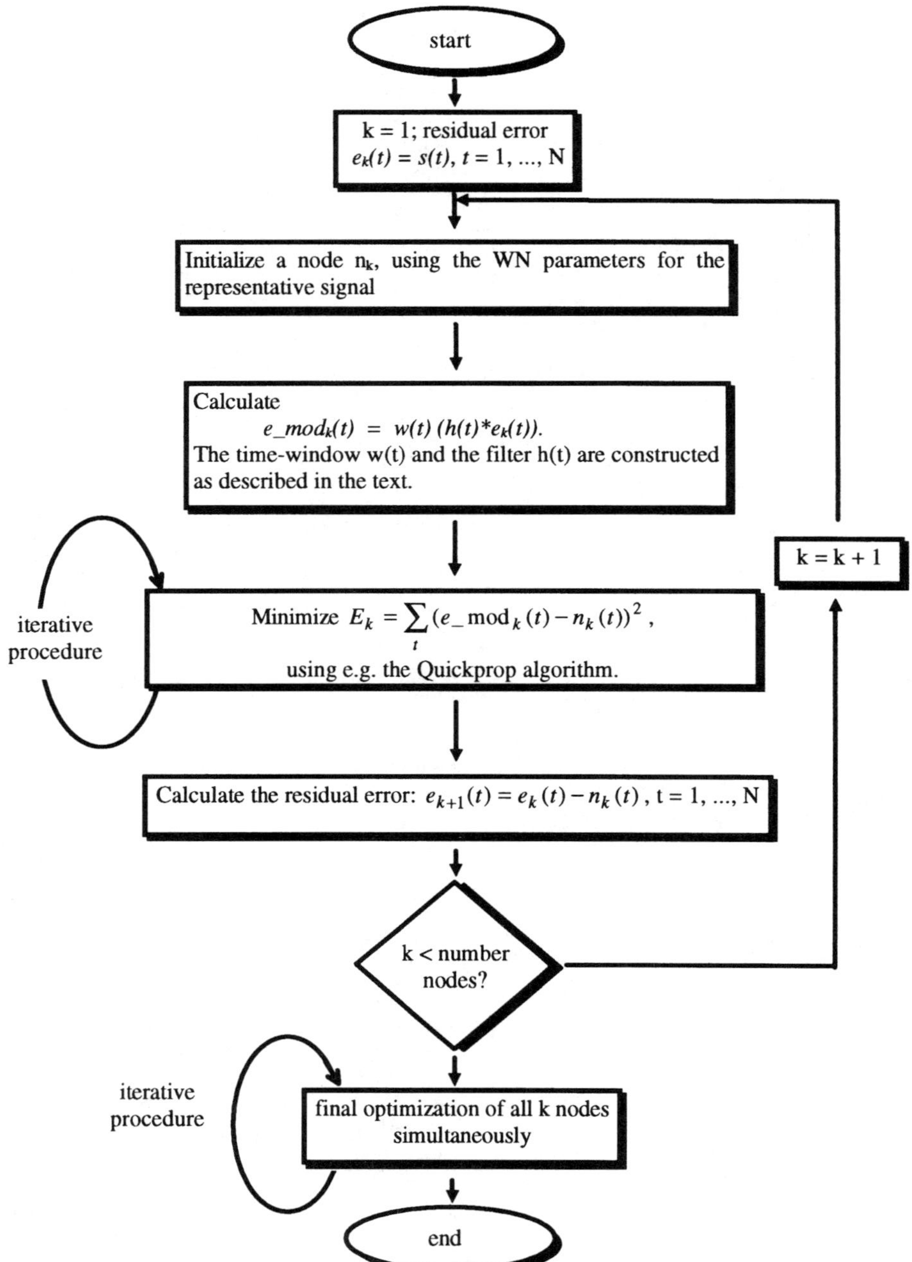

Figure 27-4   A flow chart of the algorithm for training all the trials of an experiment (phase 2 of the recursive strategy). The WN parameters of the representative signal (resulting from phase 1) are used for the preprocessing steps (time-windowing, filtering). Thus, a node's "view" is concentrated on a specific region of the time-frequency plane in order to get a uniform WN parameterization for all trials of an experiment.

works in the following manner. The first WN node $n_1(t)$ "looks" at the desired signal $x(t)$ and tries to compensate for as much of the error E as possible. When the error is no longer significantly reduced, a second WN node $n_2(t)$ is created. The residual error $e_2(t)$, the difference between the signal $x(t)$ and the first node's output $n_1(t)$, is used to train this second WN node $n_2(t)$. This procedure is repeated until a given criterion has been fulfilled, e.g., a high percentage of the signal's energy is represented by the wavelet net.

But we do not use an LSQ criterion for all nodes. In order to emphasize higher-frequency components with low energy, nodes which represent these components are trained by a criterion called "error-of-the-gradient" [11]:

$$E_{\text{grad}} = \sum_{i=1}^{N} \left[\left. \frac{\partial e_k(t)}{\partial t} \right|_{t_i} - \left. \frac{\partial n_k(t)}{\partial t} \right|_{t_i} \right]^2 \overset{!}{=} \min \tag{27-7}$$

This means, instead of the function itself its first derivative is used to train the WN node. The residual error's gradient $\partial e_k(t)/\partial t$ can be estimated using a formula like

$$\left. \frac{\partial e_k(t)}{\partial t} \right|_{t_i} = \frac{1}{12h} \left[e_k(t_{i-2}) - 8e_k(t_{i-1}) + 8e_k(t_{i+1}) - e_k(t_{i+2}) \right] \tag{27-8}$$

where h is the sampling interval [12].

The gradient of a wavelet node $n_k(t)$ is calculated via

$$\begin{aligned}
\frac{\partial n_k(t)}{\partial t} = {} & \exp\left[-0.5 \left(\frac{t - b_k}{a_k} \right)^2 \right] \frac{1}{a_k} \\
& \left\{ w_{\cos,k} \left[-\frac{t - b_k}{a_k} \cos\left(\omega_k \frac{t - b_k}{a_k} \right) - \omega_k \sin\left(\omega_k \frac{t - b_k}{a_k} \right) \right] \right. \\
& + w_{\sin,k} \left[-\frac{t - b_k}{a_k} \sin\left(\omega_k \frac{t - b_k}{a_k} \right) + \omega_k \cos\left(\omega_k \frac{t - b_k}{a_k} \right) \right] \left. \right\}
\end{aligned} \tag{27-9}$$

In order to find the optimal parameter values for the representative signal it is recommended to train a *pool* of candidate nodes, i.e., nodes with different random initial parameters spread over the time-frequency plane. The candidate node with the miminal error is chosen.

For a fast convergence of the WN parameters a gradient technique like Fahlman's quickprop algorithm [13] is used. The final update of all the WN nodes performs fine tuning of the parameters.

In the second phase of the learning algorithm, WNs for all the individual trials of the experiment are trained. Therefore, the presented recursive algorithm is extended by some preprocessing steps: before a WN node is trained, the residual error is filtered and multiplied by a time window (preprocessing). The corresponding window and filter function are defined by means of the WN parameters resulting from the representative signal in the initial phase. Thus a wavelet node is focused on its relevant part of the time-frequency plane. Figure 27-5 illustrates how the time-window and the filter are constructed from a node which has been trained to approximate the representative signal. The shift parameter b_k of this node determines the center of the time-window, the inverse scale parameter a_k^{-1} its width. The parameters

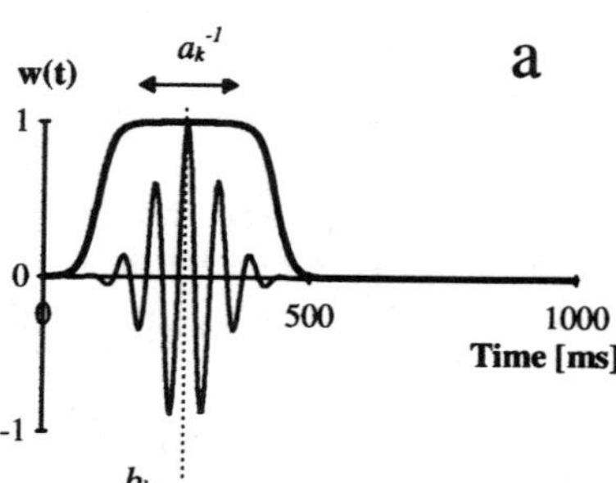

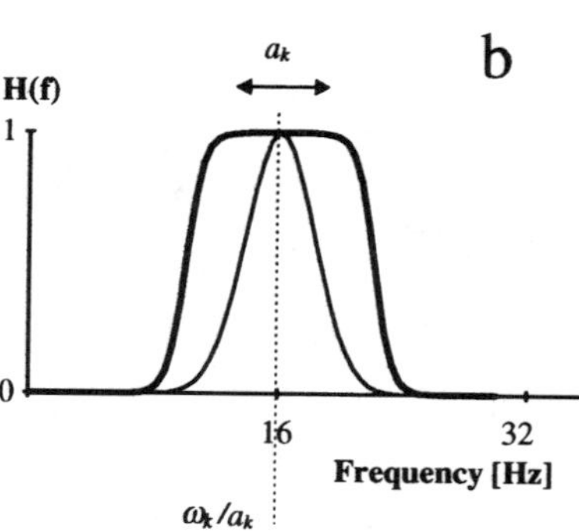

Figure 27-5 (a) Design of a window function $w(t)$ (thick line) derived from the time course of a wavelet function (node) which was adapted to a representative signal (thin line). The shift parameter b_k of this node determines the center of the time-window, the inverse scale parameter a_k^{-1} is its width. (b) The design of a filter function $H(f)$ (thick line) derived from a node's spectrum (thin line). The parameters ω_k and a_k define the filter's passband.

ω_k and a_k define the filter's passband. Note that the width of the time-window is the reciprocal of the filter's bandwidth.

Besides these preprocessing steps, a node is trained in a similar way to that in the algorithm's first phase. However, it is recommended to train only one candidate node during a recursive step of the second phase. This node should be initialized according to the corresponding node of the representative signal.

In Fig. 27-4 the complete algorithm for training the trials of an experiment is presented schematically. For the example in Fig. 27-2 the algorithm works perfectly well, even under worse SNRs. The clinical study in section 27.3 will demonstrate the efficiency of this modified recursive algorithm for a reliable EP parameterization.

27.3. WAVELET NETS APPLIED TO EP SIGNALS

27.3.1 Clinical and Methodical Background

Frontal lobes of the brain have been proposed as playing a major role in child psychiatric disorders. There is some evidence for a deficit in the nervous activity of frontal lobes in children with ADDH [14]. For instance in [15] a lower blood perfusion, i.e., a lower metabolic activity, at the level of frontal lobes was found in all examined hyperactive children. These children, mainly boys, suffer from inadequate attention, improper control of impulsiveness, and poor control of response sequences.

To investigate these abnormalities EPs are particularly useful since they reflect the dynamic behavior of central nervous system activity. Thus, in [1] EP elicited by a selective attention paradigm (SELA) are used to discriminate ADDH children from healthy children. The N2 and the Nd amplitudes of the EP were found to be significantly smaller in children with ADDH.

In the clinical study described in this chapter, SELA EP are approximated and parameterized by wavelet networks. The appropriate WN parameters are used to discriminate ADDH children from normal children. The results yielded by the WN parameters are compared with conventional EP features, namely latencies and amplitudes of prominent peaks.

27.3.2 Data Acquisition and Preprocessing

EPs were recorded from 25 healthy children and 25 children with ADDH, using an auditory SELA paradigm. The groups are matched with regard to age and IQ. 240 stimuli were delivered, consisting of a random sequence of 96 acoustic events of 1500 Hz and 144 acoustic events of 1000 Hz. The presentation of the high- and low-frequency tones was equally distributed to the right and left ear. A prompt motor response was required to the high-frequency tone on the attended side. In series I the right-hand side was the attended side, in series II the attended side changed. The stimuli characteristics in detail are:

– Duration :	120 ms	– Rise and fall time : 10 ms
– Intensity :	85 dB	– Interstimulus interval : 1300 ± 200 ms

Frontal (F3, Fz, F4), central (C3, Cz, C4) and parietal leads (P3, P4) were recorded, according to the 10–20-system. The EP signals were sampled with 500 Hz. The response is available for 150 ms pre-stimulus and 1000 ms post-stimulus time. In order to remove artifacts from the EP trials, the following preprocessing steps were necessary. A regression-based electrooculogram (EOG) correction algorithm in the time domain was used to reduce artifacts caused by eye movements and blinking [16]. Trials containing muscle artifacts and overmodulated segments were excluded from the averaging process.

27.3.3 Parameterization and Discrimination by Means of WN Parameters

As an example the averaged target response of the frontocentral lead (Fz) is investigated. However, the characteristics (number and shape of the WN nodes, order of the optimization criterion) also hold true for all the frontal and central leads.

Usually, EPs are parameterized by maximal and mean amplitudes of prominent peaks, like the N1, P2, or N2 and their corresponding latencies. The grand-mean in Fig. 27-6(a) (thin line) clearly shows these peaks. But these features do not provide a satisfactory parameterization for all the EPs. As an example, in Fig. 27-7 parts (a) the time courses of five EPs recorded from healthy children are shown (thin line). It is

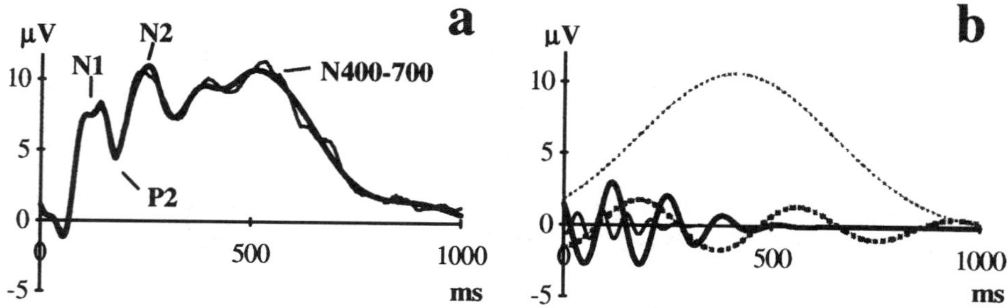

Figure 27-6　Grand mean SELA EP (target response, lead Fz) recorded in a clinical study concerning ADDH children. (a) The grand mean (thin line) and the WN output (thick line) resulting from four wavelet nodes. The prominent peaks are marked. (b) The time course of the four wavelet functions and nodes: nodes 1 (dashed, thin line) and 4 (dashed, thick line) are trained by the LSQ criterion, nodes 2 (solid, thick line) and 3 (solid, thin line) by the "error-of-the-gradient" criterion. The order of the nodes corresponds to the order in which the nodes are created in the recursive learning algorithm.

obvious that the various peaks of the different trials are not so clearly pronounced as one would expect. For example, the N1 is sometimes two-fold. In the extreme case of example (1a) the N1-P2-N2 complex even consists of four negative half waves. Furthermore, the N400-700 complex is often disturbed by noisy components [example (3a), (4a)]. Therefore it is nearly impossible to give a reliable estimate for the peak parameters in view of the demonstrated interindividual variabilities.

In order to overcome these peak identification problems, we tried to represent the averaged EP responses by adequate WN parameters. The WN learning algorithm described in section 27.2.3 was applied to the grand mean first. The approximation yielded by a linear combination of four wavelet nodes perfectly fits the averaged EP. These wavelet nodes are represented by their characteristic time courses in Fig. 27-6(b). The first node (dashed, thin line) trained by the traditional LSQ criterion represents the component with the highest energy. It lasts for the whole EP segment. Next, two nodes are trained according to the "error-of-the-gradient" criterion. These nodes are concentrated in the early phase of the EP (node 2—solid, thick line; node 3—solid, thin line). As a matter of fact they represent the higher-frequency components. The fourth node (dashed, thick line) is trained using the LSQ criterion again. Using this empirical order of the optimization criterion, the best results have been found for the frontal and central leads. It is interesting to notice that the WN nodes can be related to the conventional EEG bands, e.g., the third node's spectrum nearly coincides with the β-band, which is usually defined between 13–30 Hz.

How WNs are adapted to each individual signal of the study under consideration, is described next. As already mentioned, in Fig. 27-7 parts (a) five SELA EPs of healthy children are shown. The diagrams represent the WN output $\hat{x}(t)$ (thick line) together with the averaged EP signal (thin line). The corresponding diagrams on the right side [Fig. 27-7 parts (b)] display the four different wavelet nodes. Two conclusions can be drawn from these plots. First, the WNs fit the recorded EP quite well,

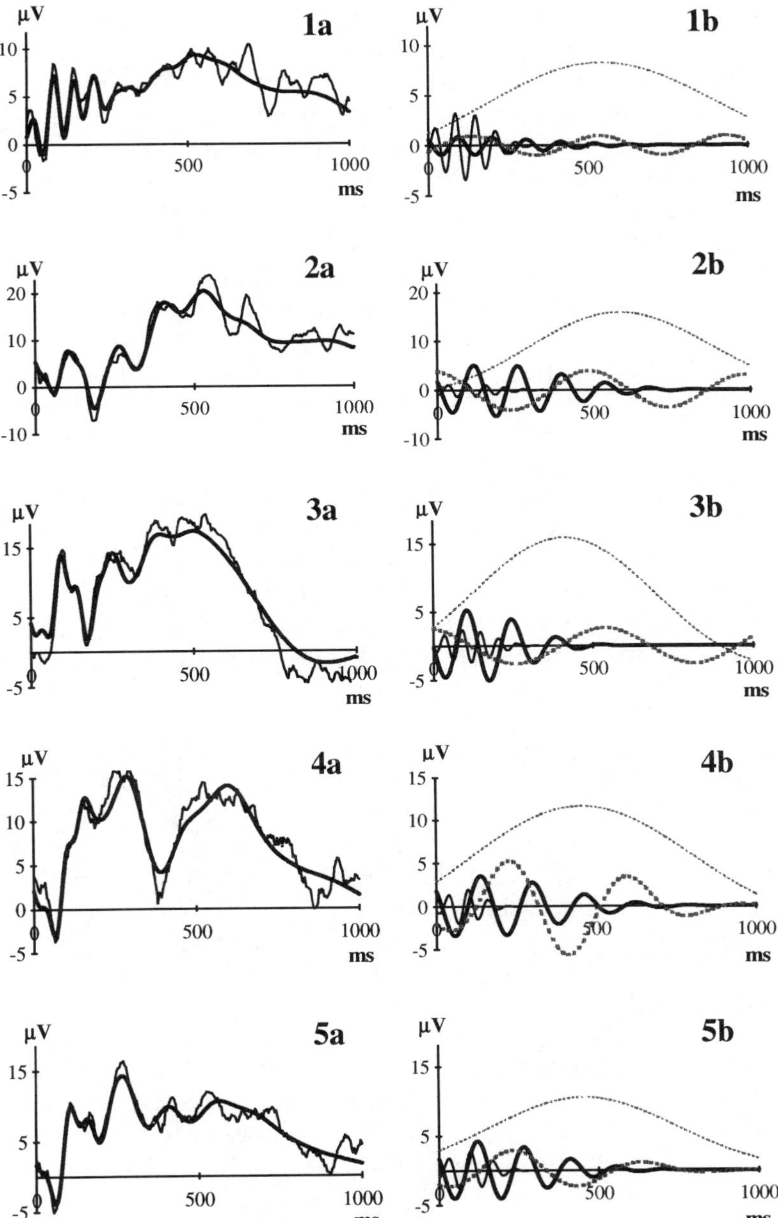

Figure 27-7   Averaged SELA EPs (target response, lead Fz) of five healthy children. Parts (a) EP signal (thin line) and the WN output (thick line), resulting from four wavelet nodes. Parts (b) time course of the four wavelet functions and nodes: nodes 1 (dashed, thin line) and 4 (dashed, thick line) are trained by the LSQ criterion, nodes 2 (solid, thick line) and 3 (solid, thin line) by the "error-of-the-gradient" criterion. The order of the nodes corresponds to the order in which the nodes are created in the recursive learning algorithm.

especially the early components [see examples (1a)–(5a)]. Second, the WN nodes, resulting for different signals, show a similar time course [see examples (1b)–(5b)]. Thus, a uniform parameterization of the EP is attained by using WNs which are trained according to the algorithm introduced in section 27.2.3.

But the WN parameters are not only suited to represent EP signals, they also permit discrimination between healthy and ADDH children. Using a quadratic polynomial classifier, which is trained by four WN parameters, results in an 88% correct classification in SELA series I (right side being the attended side), and even 90% in series II (left side being the attended side). The weight parameters of the WN nodes 2 and 3 mainly account for the group differences. If conventional EP parameters (latencies as well as maximum and mean amplitudes of prominent peaks) are used to train the polynomial classifier, no more than 72% correct classification in series I and 74% in series II can be achieved. So the WN parameters seem to be an attractive alternative, or are even superior, to the traditionally used EP parameters.

In order to interpret the wavelet parameters with regard to the group effects, we have studied single WN parameters in a nonparametric test. Differences are considered as significant for $p < 0.05$. As already mentioned, the third node's spectrum nearly coincides with the β-band. Comparing these 'β-nodes' for the frontal leads, the weight coefficients of the ADDH children show larger values ipsilateral to the stimulus than contralateral. For the normal children this difference is of opposite sign. This interesting phenomenon, which is evident in both SELA series, may be interpreted as an orienting deficit of the ADDH children.

Besides this, mean differences for the weight coefficients of the first WN node ($w_{\mathrm{NORM}} > w_{\mathrm{ADDH}}$) can be found in the frontal leads. This first node represents a kind of basic evoked activity. If a selective averaging depending on the previous stimulus is performed, a significant effect results. This phenomenon is not evident, using only the mean differences without regarding the stimulus sequence. The weights of the first node are significantly smaller in the F3 lead of the ADDH children in both series, if a nontarget stimulus on the nonattended side precedes the target stimulus.

27.4. CONCLUSION

WNs can be regarded as an interesting extension or application of the continuous wavelet transform. Adapting a WN to a given signal with the powerful learning algorithms of NNs, WNs allow a signal representation that is easy to understand. In particular, the time-varying characteristics of the signal are taken into account by the nodes' wavelet parameters (shift, scale, frequency, weight coefficient). The number of nodes introduced depends primary on the complexity of the signal's time-frequency characteristics. Since the decomposition is linear, each individual node's contribution is clear and easy to interpret.

It has been shown that the usually applied LSQ algorithm for training NNs requires some additional modifications to achieve an appropriate EP parameterization by WNs. Mainly the subdivision of the learning strategy into two consecutive steps and the preprocessing procedure (tapering, filtering) enable reliable and meaningful parameters for describing EPs.

The WN parameters, which have been estimated for the SELA EP to discriminate between normal and ADDH children, yielded better results than the empirical latency and amplitude values of prominent peaks. It has been shown that the optimized WN parameters are appropriate features for a multivariate classification. Besides this, precise evaluation and interpretation of WN parameters enable a better understanding of the phenomenon and stimulates attempts for physiological explanations. Although this clinical study is only of a preliminary nature, the encouraging results demonstrate the efficiency and usefulness of the WN approach which can be easily adapted to other applications.

REFERENCES

[1] J. H. Satterfield, A. M. Schell, T. Nicholas, and R. W. Backs, "Topographic study of auditory event-related potentials in normal boys and boys with attention deficit disorder with hyperactivity," *Psychophysiology*, vol. 25, no. 5, pp. 591–606, September 1988.

[2] B. Rockstroh, T. H. Elbert, N. Birbaumer, and W. Lutzenberger, *Slow Brain Potentials and Behaviour*. Munich: Urban & Schwarzenberg, 1982.

[3] R. P. Lippmann, "An introduction to computing with neural nets," *IEEE Acoust., Speech, Signal Proc. Mag.*, vol. 4, pp. 4–22, April 1987.

[4] R. Battiti, "First- and second-order methods for learning: Between steepest descent and Newton's method," *Neural Comput.*, vol. 4, pp. 141–166, 1992.

[5] H. H. Szu, B. Telfer, and S. Kadambe, "Neural network adaptive wavelets for signal representation and classification," *Opt. Eng.*, vol. 31, no. 9, pp. 1907–1916, September 1992.

[6] Q. Zhang and A. Benveniste, "Wavelet Networks," *IEEE Trans. Neural Networks*, vol. 3, no. 6, pp. 889–898, 1992.

[7] H. Dickhaus, L. Khadra, A. Lipp, and M. Schweizer, "Ventricular late potentials studied by nonstationary signal analysis," In *Proc. Int. Conf. IEEE EMBS*, J. P. Morucci et al. (eds.), vol. 14, pp. 490–491, 1992.

[8] L. M. Khadra, M. Matalgah, B. El-Asir, and S. Mawagdeh, "The wavelet transform in phonocardiogram signal analysis," *Med. Inf. (Lond.)*, vol. 16, pp. 271–277, 1991.

[9] H. Heinrich, H. Dickhaus, and U. Klauck, "Klassifikation von Biosignalen am Beispiel visuell evozierter Potentiale mit Hilfe von Wavelet-Netzen." In *15. DAGM-Symp. Mustererkennung*, S. J. Pöppl, H. Handels, (eds.). Berlin: Springer, pp. 208–213, 1993.

[10] S. E. Fahlman and C. Lebiere, "The Cascade-correlation architecture," *Tech. Report CMU-CS-90-100*, School of Computer Science, Carnegie Mellon University, Pittsburgh, February 1990.

[11] M. Sri-Jayantha and R. F. Stengel, "Determination of nonlinear aerodynamic coefficients using the estimation-before-modeling method," *J. Aircraft*, vol. 25, no. 9, pp. 796–804, September 1988.

[12] I. N. Bronstein and K. A. Semendjajew, *Taschenbuch der Mathematik*. Thun und Frankfurt/Main: Deutsch, 1981.

[13] S. E. Fahlman, "An empirical study of learning speed in back-propagation networks," *Tech. Report CMU-CS-88-162*, School of Computer Science, Carnegie Mellon University, Pittsburgh, PA, 1988.

[14] A. Rothenberger, "The role of frontal lobes in child psychiatric disorders." In *Brain and Behavior in Child Psychiatry*, A. Rothenberger, (ed.). Berlin: Springer, pp. 34–58, 1990.

[15] H. C. Lou, L. Henriksen, and P. Bruhn, "Focal cerebral hypoperfusion in children with dysphasia and/or attentional deficit disorder," *Arch. Neurol.*, vol. 41, pp. 825–829, 1984.

[16] C. H. M. Brunia, J. Möcks, and M. M. C. van den Berg-Lenssen, "Correcting ocular artifacts in the EEG: a comparison of several methods," *J. Psychophysiol.*, vol. 3, pp. 1–50, 1989.

Chapter 28

Self-Organizing Wavelet-Based Neural Networks

Kunikazu Kobayashi

28.1. INTRODUCTION

It has been shown that neural networks (NNs) can realize any mappings [1–4]. These are important for the theoretical exploration of the potential of NNs.

Poggio and coworkers proposed a regularization network based on classical regularization theory [5]. They insisted that the network has the property of "best approximation" [6]. Lee and Kil [7] proposed a mapping network, called the GPFN (Gaussian potential function network), which has interesting properties.

Backpropagation (BP) networks are now the most popular mapping networks [8]. Recently, the theoretical support was provided [9]. It is, however, well known that BP networks have some problems such as trapping into local minima and slow convergence. In addition, the network structures are determined by trial and error because the algorithm cannot provide any optimization schemes. Therefore, networks with fewer hidden units than the optimal number cannot realize the desired approximation and the networks with excess hidden units result in poor interpolation due to over-fitting.

Recently, many researchers have proposed various network optimization schemes in order to solve the above problems. There are two approaches: the reproducing (adding) method and the pruning (deleting) method. In the former, starting from a small-size network, hidden units are gradually added to satisfy a criterion as the network learns [7, 10, 11]. On the other hand, in the latter, a large-size network gradually removes redundant hidden units as it learns [12–14]. In addition, other

685

approaches have been presented: a method using a genetic algorithm (GA) [15] and modular structured networks [16].

It is expected that wavelets will be a new powerful tool for signal analysis [17]. Wavelets can approximately realize the time-frequency analysis using a kernel function, which is called a mother wavelet. The mother wavelet has a squared window in the time-frequency space. The size of the window can be almost freely variable by two parameters. Thus, wavelets can identify the localization of unknown signals at any level. From the uncertainty principle, however, the resolution for frequency will be lowered.

In this chapter, wavelets and NNs are combined and a "self-reproducing wavelet neural network" (SERWANN) is proposed. Wavelet-based NNs have been proposed by Zhang and Benveniste [18] and Pati and Krishnaprasad [19]. Zhang and Benveniste's network consists of processing elements named "wavelon" not biological neuron model and the number of wavelons is determined by trial and error. Pati and Krishnaprasad's network determines its structure using discrete affine wavelet transforms. That is, in both networks the number of hidden units is determined before learning. On the other hand, SERWANN has a concept of network optimization and self-organization. According to a given target, SERWANN gradually adds hidden units in order to self-organize an optimal-size network.

SERWANN has four merits: self-organization of networks, partial retrieval of approximated function, fast convergence and escaping local minima. Incorporating the idea of wavelets, the output function is localized in both the time and frequency domains. Therefore, each hidden unit has a square window in the time-frequency plane. Thus, SERWANN can capture function approximation problems as two tasks: (1) to effectively cover the time-frequency region of a given target by the windows of hidden units; (2) to minimize training and generation errors. In this connection, the proposed algorithm comprises two processes: self-organization of networks and minimization of errors. In the first process, the network gradually reproduces hidden units to effectively and sufficiently cover the time-frequency region occupied by a given target, i.e., to optimize the network structure. Simultaneously, the network parameters are updated to preserve the network topology and take advantage of the later process. In the second process, the parameters of the initialized network are updated using the δ-rule [8] in order to minimize the errors of approximation. This rule is only applied to the hidden units where the selected point falls into their windows. Therefore, the learning cost can be reduced. In addition, the localization of the output function results in the partial retrieval of the approximated function.

This chapter is organized as follows. In section 28.2, three bases regarding wavelets—the wavelet transform, the inversion formula and the window of a wavelet—are reviewed. Section 28.3 describes the network expression of inversion formula. In section 28.4, the interpretation of function-approximating problems in SERWANN is outlined and the proposed algorithm is explained. Section 28.5 confirms the capabilities of approximation and interpolation of SERWANN through the comparison with a standard BP network. Section 28.6 contains conclusions and describes future problems.

28.2. PRELIMINARIES

This section explains three fundamental concepts regarding wavelets: the wavelet transform (28.2.1), the inversion formula (28.2.2) and windows (28.2.3) [17]. In this chapter, we will use the underlying Hilbert space $L^2(R)$, whose inner product and norm are defined as follows.

$$\langle f, g \rangle \triangleq \int_R f(x)\,\overline{g(x)}\,dx$$

$$\|f\| \triangleq \langle f, f \rangle^{1/2}$$

where $\overline{\psi}$ denotes the dual of ψ.

28.2.1 Wavelet Transform

The integral wavelet transform of function $f \in L^2(R)$, $T(a, b)$, is defined as:

$$T(a, b) \triangleq \langle f, \psi^{(a,b)} \rangle \tag{28-1}$$

where

$$\psi^{(a,b)}(x) = \frac{1}{\sqrt{a}}\,\psi\left(\frac{x - b}{a}\right)$$

This describes the correlation between function f and $\psi^{(a,b)}$, which is obtained from a basic function by dilation of a and translation of b.

In wavelet analysis, the basic function ψ is called the mother wavelet but we call it the "fitting wavelet" (FW) because our goal is function approximation. This FW must satisfy the following admissibility condition.

$$c_\psi = \int_R \frac{|\Psi(w)|^2}{w}\,dw < \infty \tag{28-2}$$

where Ψ is the Fourier transform of ψ:

$$\Psi(w) = \int_R \psi(x)e^{-jwx}\,dx$$

For the localized case, the above condition simply means:

$$\int_R \psi(x)\,dx = 0$$

That is, the FW has no bias.

28.2.2 Inversion Formula

The inversion formula is used to reconstruct function f from $T(a, b)$. The recovery from a, $b \in R$ is defined as:

$$f(x) \triangle = \frac{1}{c_\psi} \iint_{R^2} T(a, b)\,\psi^{(a,b)}(x)\,\frac{da\,db}{a^2} \tag{28-3}$$

where c_ψ is a constant defined by Eq. (28-2).

However, using NNs, such recovery is impossible because it requires continuum or infinite units. Therefore, NNs must reconstruct the target function from partial information of $T(a, b)$. In this chapter, the dyadic partition $a = 2^j$, $b = kb_0 2^j$ $(j, k \in Z)$ is used (b_0 is a sampling rate). This inversion formula is defined as:

$$f(x) \triangleq \sum_{j,k \in Z} \langle f, \psi_{j,k} \rangle \, \psi^{j,k}(x) \tag{28-4}$$

where

$$\psi_{j,k}(x) = \psi^{(2^j, kb_0 2^j)}(x)$$
$$= 2^{-j/2} \psi(2^{-j} x - kb_0) \tag{28-5}$$

and $\{\psi^{j,k}\}$ is the dual basis of $\{\psi_{j,k}\}$.

The FW should satisfy the following stability condition because f must be reconstructed from partial information of $T(a, b)$:

$$A \, \|f\|^2 \leq \sum_{j,k \in Z} |\langle f, \psi_{j,k} \rangle|^2 \leq B \, \|f\|^2 \tag{28-6}$$

where A and B are real constants, $0 \leq A \leq B < \infty$.

28.2.3 Windows

In wavelets, one of the most important concepts is "windows." This refers to a rectangular region in the time-frequency plane defined for a FW. That is, the FW can *see* such a region but not other regions. This property results in the identification of localization.

For FW ψ, the support in the time domain, $\mathrm{supp}_\epsilon(\psi)$, is defined as:

$$\mathrm{supp}_\epsilon(\psi) \triangleq [x_{\min}^{FW}, x_{\max}^{FW}] \tag{28-7}$$

where $x_{\min}^{FW}$ and $x_{\max}^{FW}$ satisfies the following inequality.

$$1 - \epsilon < \frac{\int_{x_{\min}^{FW}}^{x_{\max}^{FW}} |\psi(x)|^2 \, dx}{\int_R |\psi(x)|^2 \, dx} \tag{28-8}$$

where ϵ is a constant, $0 \leq \epsilon \leq 1$. It is implied that the power of ψ in $[x_{\min}^{FW}, x_{\max}^{FW}]$ is at least concentrated in the time domain at rate $1 - \epsilon$. Similarly, for the Fourier transform of the FW, $\Psi(w)$, the support in the frequency domain is defined as:

$$\mathrm{supp}_\epsilon(\Psi) \triangleq [w_{\min}^{FW}, w_{\max}^{FW}] \tag{28-9}$$

where

$$1 - \epsilon < \frac{\int_{w_{\min}^{FW}}^{w_{\max}^{FW}} |\Psi(w)|^2 \, dw}{\int_R |\Psi(w)|^2 \, dw} \tag{28-10}$$

Therefore, ψ has a time-frequency window, $[x_{\min}^{FW}, x_{\max}^{FW}] \times [w_{\min}^{FW}, w_{\max}^{FW}]$ (see Fig. 28-1).

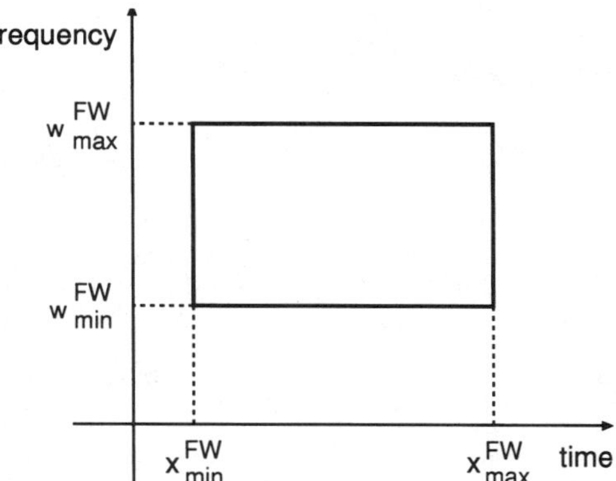

Figure 28-1   A window in the time-frequency plane.

In general, $\psi^{(a,b)}(x)$ has a window, $[b + ax_{min}^{FW}, b + ax_{max}^{FW}] \times [w_{min}^{FW}/a, w_{max}^{FW}/a]$. The size of the window is constant for any translation or dilation.

$$(ax_{max} - ax_{min})(w_{max}/a - w_{min}/a) = (x_{max} - x_{min})(w_{max} - w_{min})$$
$$= \text{constant}$$

28.3. NETWORK EXPRESSION

In this section, let us consider the network expression of the inversion formula, Eq. (28-4). The inversion formula cannot be expressed by finite NNs. Actually, however, most targets are restricted in both the time and frequency domains. Thus, the inversion formula can be approximately realized using finite NNs.

Consider a $1 \times N \times 1$ network; see Fig. 28-2. The input and output units are linear elements and the output function of the hidden units satisfies both admissibility and stability conditions, i.e., Eq. (28-2) and Eq. (28-6). It is assumed that the network sufficiently approximates the target. Intuitively, this means that the time-frequency region is effectively covered by their N windows (e.g., Fig. 28-3). The estimate of the network, $\hat{f}$, is represented by:

$$\hat{f}(x) = \sum_{i=1}^{N} c_i \, \psi(a_i x - b_i) \tag{28-11}$$

where a_i, b_i and c_i are called dilation, translation, and amplification parameters, respectively (see Fig. 28-4). a_i, b_i, and c_i correspond to the weight between the input unit and unit i, the threshold of unit i and the weight between unit i and the output unit, respectively.

28.4. FUNCTION APPROXIMATION AND NETWORK OPTIMIZATION

Firstly, this section gives the interpretation of function approximation problems in SERWANN (section 28.4.1). Secondly, the proposed algorithms are explained in sections 28.4.2 and 28.4.3.

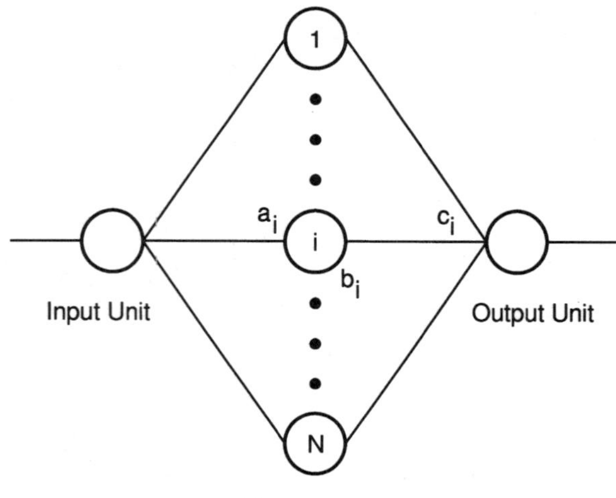

Figure 28-2　A network expression of the inversion formula.

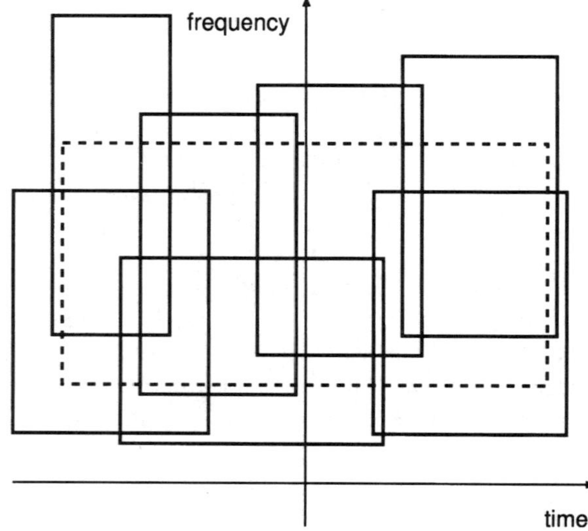

Figure 28-3　An example of covering by the windows of FWs. The square drawn by the broken line is the window of the target and the other squares are the windows of FWs.

28.4.1 Function Approximation Problem

Given sparse examples, NNs will internally construct a desired mapping through learning. SERWANN captures function approximation problems as two tasks.

1. To effectively cover the time-frequency region of a given target by the windows of FWs.
2. To approximate training data as precisely as possible and to interpolate test data as plausibly as possible.

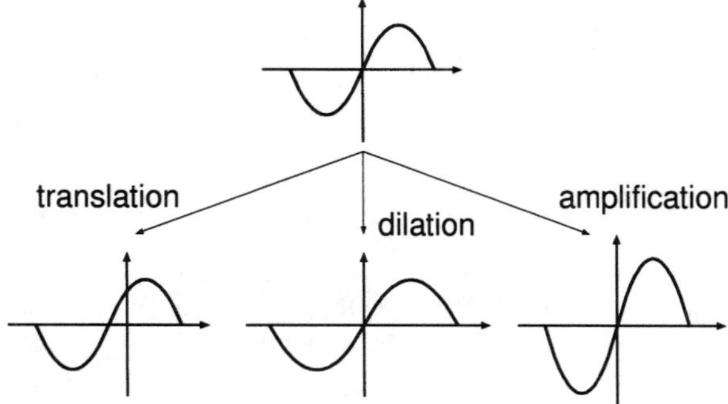

Figure 28-4   Wavelet operators (translation, dilation, and amplification).

The former refers to optimizing the network structure and the latter to realizing the best approximation and best interpolation. Thus, SERWANN must solve both tasks.

28.4.2 Self-Organization of Networks

In the first process, the number of hidden units is determined to effectively and sufficiently approximate a given target. Simultaneously, the parameters are updated by Kohonen's rule [20].

Consider that the power of a target f is localized in $[x_{\min}, x_{\max}]$, i.e., $\mathrm{supp}_\epsilon(f) = [x_{\min}, x_{\max}]$ and f is sampled at sampling rate w_s. The sampled set, T, is:

$$T = \{(x_\alpha, f_\alpha) \mid x_\alpha \in X,\ f_\alpha \in F,\ \alpha = 1, 2, \ldots, M\} \tag{28-12}$$

where

$$X = \{x_1 = x_{\min}, \ldots, x_\alpha, \ldots, x_M = x_{\max}\}$$
$$F = \{f(x_1), \ldots, f(x_\alpha), \ldots, f(x_M)\}$$
$$M = w_s(x_{\max} - x_{\min})$$

The algorithm is described as follows.

1. At first, we estimate the bandwidth of T by the discrete Fourier transform (DFT)

$$supp_\epsilon(T) = [w_{\min}, w_{\max}]$$

Of course, since the DFT requires discrete frequency, taking the inequality $a > 0$ into consideration,

$$w_{\min},\ w_{\max} \in \Omega$$
$$= \{w_1, \cdots, w_\alpha, \cdots, w_{M/2}\}$$

where $w_{M/2} \leq w_s/2$ because of sampling theorem.

2. Using the following equations, $x \in X$ and $w \in \Omega$ are transformed to dilation and translation, respectively

$$a = \frac{w_{\min}^{\text{FW}} + w_{\max}^{\text{FW}}}{2w}$$

$$b = x - \frac{a(x_{\min}^{\text{FW}} + x_{\max}^{\text{FW}})}{2}$$

$$= x - \frac{(x_{\min}^{\text{FW}} + x_{\max}^{\text{FW}})(w_{\min}^{\text{FW}} + w_{\max}^{\text{FW}})}{4w}$$

The above equations imply that the center of the window is in the time-frequency region occupied by the target.

As a result, we take the dilation and translation sets:

$$S_a = \{a_1, \ldots, a_\alpha, \ldots, a_{M/2}\}$$
$$S_b = \{b_1, \ldots, b_\alpha, \ldots, b_M\}$$

3. Wavelet spectra of the target are calculated for all the combinations of S_a and S_b.

4. We create the training set S_{CL}

$$S_{\text{CL}} = \{(x_\alpha, w_\alpha) \mid x_\alpha \in X, \ w_\alpha \in \Omega, \ \alpha = 1, 2, \ldots, M\} \tag{28-13}$$

where w_α refers to the frequency at the maximum value of wavelet spectra with respect to x_α.

5. We determine the initial number of hidden units, N^0 and initialize their parameters, a_i, b_i, and c_i $(i = 1 \sim N^0)$.

6. A training point (x_α, w_α) is selected from S_{CL} at probability p_α defined as:

$$p_\alpha = |T'(a_\alpha, b_\alpha)|$$

where T' refers to normalized T, i.e., $\sum_{i=1}^{M} |T'(a_i, b_i)| = 1$. This results in effective covering.

7. We determine the nearest neighbor c of the selected point

$$c = \arg \min_i \|(x_\alpha, w_\alpha) - (b_i, a_i)\|$$

8. Its neighbor set N_c is defined

$$N_c = \{i \mid |i - c| \leq l\}$$

where l is a positive integer.

9. The parameters of unit i are updated if and only if i belongs to N_c and its window contains the selected point

$$\begin{aligned}
a_i^{t+1} &= a_i^t + \alpha_{\text{CL}}(w_\alpha - a_i^t) \\
b_i^{t+1} &= b_i^t + \alpha_{\text{CL}}(x_\alpha - b_i^t) \qquad \text{for } i \in N_c \quad \text{and } (x_\alpha, w_\alpha) \in W_i
\end{aligned} \tag{28-14}$$

where α_{CL} is a learning constant and W_i is the window of unit i.

Otherwise, a new unit is created at rate ρ and its parameters are initialized as follows:

$$a^0_{N^{l+1}} = w_\alpha + \delta$$

$$b^0_{N^{l+1}} = x_\alpha + \delta$$

$$c^0_{N^{l+1}} = T(a^0_{N^{l+1}}, b^0_{N^{l+1}}) + \delta$$

where $N^{l+1} = N^l + 1$ and δ denotes small fluctuations. The fluctuations will reduce the effects of "ghost," which is caused by the nonorthogonality of FWs in NNs and is involved in $T(a, b)$. The new unit is reproduced so as to preserve the topology between the input unit and the hidden layer. Actually, the distance between units i and j is defined as:

$$d_{i,j} = \|(b_i, a_i) - (b_j, a_j)\|$$

The adding location is between the nearest unit c^l to the parameter vector of the new unit and unit c'^l defined below

$$c^l = \arg \min_{i \in N^l} d_{i, N^{l+1}}$$

$$c'^l = \arg \min_{i \in \{c^l - 1, c^l + 1\}} d_{i, N^{l+1}}$$

The reproducing rate is then defined by the zero-crossing rate. The reason for this is that more fluctuating targets will require more hidden units.

10. Repeat steps 6 to 9 until the windows of the FWs fully cover the time-frequency region of the target and the network settles down to a stable state.

28.4.3 Minimization of Errors

In this process, the parameters of the initialized network are updated using the δ rule [8]. They are updated if and only if the selected point falls into the windows of hidden units. We call it a localized backpropagation (LBP) algorithm.

The LBP algorithm is outlined as follows.

1. At first, training set S_{LBP} is created

$$S_{\text{LBP}} = \{(x_\alpha, w_\alpha; f_\alpha) \mid x_\alpha \in X, \ w_\alpha \in \Omega, \ f_\alpha \in F, \ \alpha = 1 \sim M\} \tag{28-15}$$

2. A training point from S_{LBP} is selected at random.

3. We calculate the estimate of the network using Eq. (28-11).

4. We calculate the squared error

$$E_\alpha = \tfrac{1}{2} |f(x_\alpha) - \hat{f}(x_\alpha)|^2 \tag{28-16}$$

5. The parameters are updated if and only if the selected point falls into the windows of their units.

$$a_i^{t+1} = a_i^t - \alpha_{\mathrm{LBP}}\, \delta_\alpha\, \frac{c_i^t(x_\alpha - b_i^t)}{(a_i^t)^2}\, \psi'$$

$$b_i^{t+1} = b_i^t - \alpha_{\mathrm{LBP}}\, \delta_\alpha\, \frac{c_i^t}{a_i^t}\, \psi' \qquad\qquad (28\text{-}17)$$

$$c_i^{t+1} = c_i^t + \alpha_{\mathrm{LBP}}\, \delta_\alpha\, \psi \qquad \text{for } (x_\alpha, w_\alpha) \in W_i$$

where $\delta_\alpha = f(x_\alpha) - \hat{f}(x_\alpha)$ and α_{LBP} represents a learning coefficient.

6. Repeat steps 2 to 5 until the RMSE-A (root mean squared error for approximation) satisfies a predetermined convergence criterion.

28.5. COMPUTER SIMULATIONS

This section confirms the capabilities of the function approximation and the interpolation of SERWANN. It is evaluated through the comparison with a BP network.

The mapping networks are evaluated by two factors: the capabilities of approximation and interpolation. The former shows how precisely the network approximates training data. The latter shows how plausibly the network interpolates test data.

We used the following function as an FW

$$\psi(x) = \frac{x}{x_0}\exp\left(\frac{x^2}{2x_0^2}\right)$$

This satisfies both the admissibility and stability conditions [Eqs. (28-2) and (28-6)] and can consist of three sigmoidal neurons [19].

The value of x_0 was set to 0.07 (see Fig. 28-5). The time and frequency supports of FW are calculated as:

$$\mathrm{supp}_{0.05}(\psi) = [-0.1, 0.1]$$

$$\mathrm{supp}_{0.05}(\Psi) = [1.0, 7.0]$$

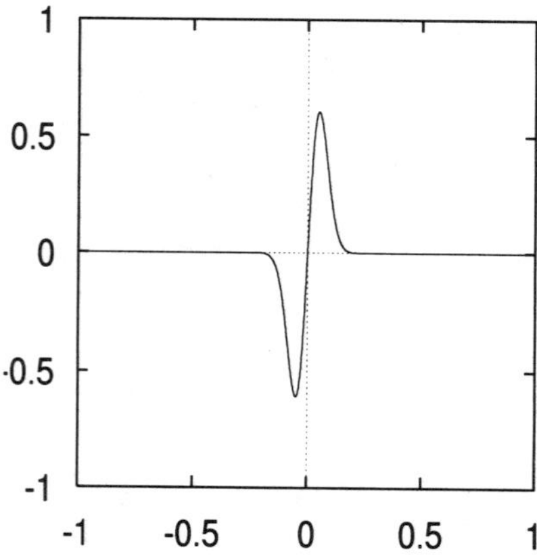

Figure 28-5   The shape of the basic FW used in the simulations, $\psi(x) = (x/x_0)\exp(x^2/2x_0^2)$, $x_0 = 0.07$.

We experimented under the following conditions. The learning steps in the self-organization of networks and the minimization of errors were 5000 and 50 000, respectively. The number of initial hidden units was zero, i.e., $N^0 = 0$. In this section, the neighbors have not been considered, i.e., $l = 0$. Furthermore, the learning constant and learning coefficient were both set to 0.03, i.e., $\alpha_{CL} = \alpha_{LBP} = 0.03$. The fluctuations in reproducing hidden units were determined from the interval $[-0.1, 0.1]$ at random.

28.5.1 Simulation I

First, we used the function

$$f_1(x) = \sin(3x)\cos[5(x - 0.5)]$$

as a target. f_1 was sampled at a sampling rate of 8 Hz for training. In Fig. 28-6, the solid line represents the target function f_1 and the symbol $\diamond$ is a sampling point.

The supports of the target are:

$$\mathrm{supp}_{0.05}(f_1) = [-1.0, 1.0]$$
$$\mathrm{supp}_{0.05}(F_1) = [0.5, 2.0]$$

The ranges of dilation and translation are $a \in [2.0, 8.0]$ and $b \in [-1.0, 1.0]$, respectively. The wavelet spectrum of f_1 is illustrated in Fig. 28-7.

SERWANN converged to 5–9 hidden units, 6.6 on average in 100 trials (see Table 28-1). The best result by SERWANN with 7 hidden units is illustrated in Fig. 28-8. In this figure, the solid line denotes the target function and the broken line is the approximated function.

For comparison, the result by a BP network with 7 hidden units is shown in Fig. 28-9. This is the best result in 100 trials. The values of learning and inertial coefficients were 0.05 and 0.5, respectively, and the gradient of the sigmoid function was

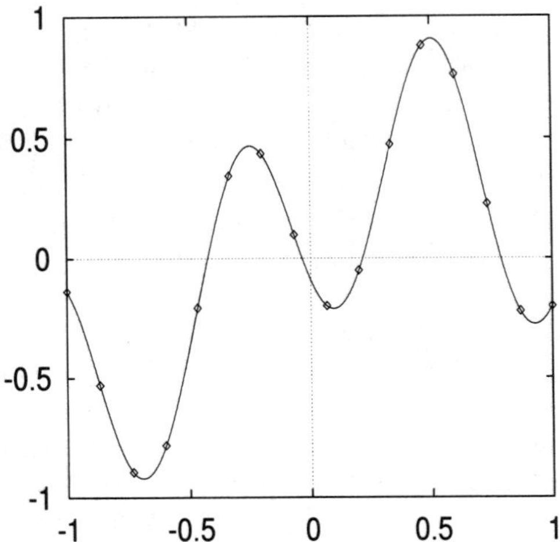

Figure 28-6   The target function $f_1(x) = \sin(3x)\ \cos[5(x - 0.5)]$ (solid line) and sampling points (symbol $\diamond$).

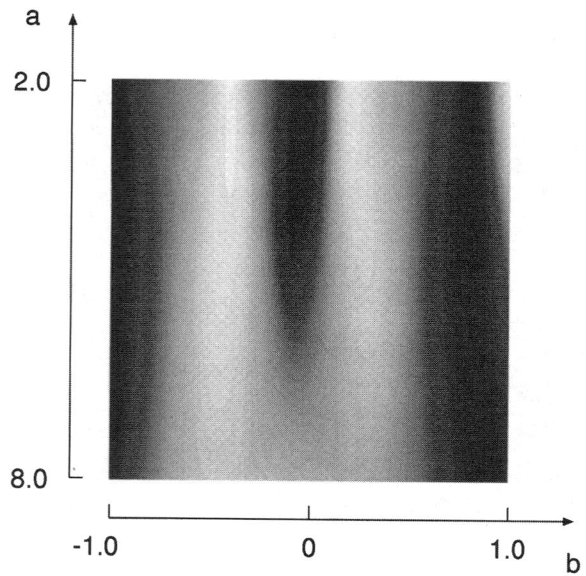

Figure 28-7   The wavelet spectrum of f_1. The gray levels represent the intensity of the spectrum (intensity high $\rightarrow$ low corresponds to white $\rightarrow$ black).

Table 28-1   The Number of Successful and Failed Trials in Simulation I. N Denotes the Number of Hidden Units After Learning

N	Success (failure)
5	9 (0)
6	34 (4)
7	39 (0)
8	11 (1)
9	2 (0)
Total	95 (5)

0.2. The initial values of weights and thresholds were determined from the interval $[-1.0, 1.0]$ at random. Furthermore, on-line algorithms were adapted as the updating method because we focus on the convergence speed.

The learning curves of both networks are illustrated in Fig. 28-10. Table 28-2 shows the figure of merit of each network. In this table, RMSE-A and RMSE-I represent the RMSE for training data and test data, respectively.

SERWANN has a better result than the BP network. As seen in Fig. 28-10, SERWANN realizes fast convergence. Of the total learning time*, the computational cost of SERWANN was 70% that of the BP network. This implies the merit of the LBP algorithm. The learning steps for a convergence criterion RMSE-$A < 1.0 \times 10^{-2}$ in SERWANN and the BP network were about 2000 and 27 000 steps, respectively. The rate of convergence was 95% and 67% in SERWANN and the BP network, respectively.

*This refers to the sum of both processes in SERWANN.

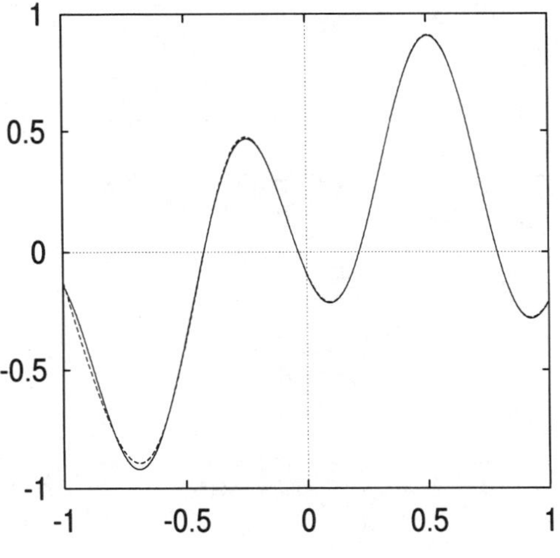

Figure 28-8 The best result obtained by SERWANN (solid line, target function; broken line, approximated function).

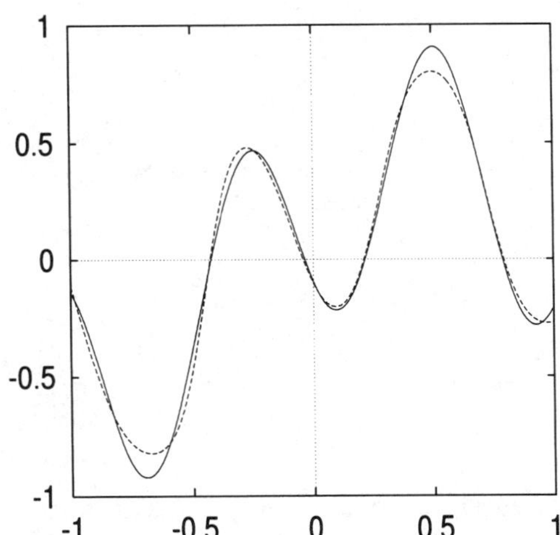

Figure 28-9 The best result obtained by the BP network with 7 hidden units (solid line, target function; broken line, approximated function).

The partial retrieval by SERWANN is shown in Fig. 28-11, where the approximated function is reconstructed in the interval [0, 1]. This required 5 out of 7 hidden units. Therefore, SERWANN can quickly retrieve a part of the target.

28.5.2 Simulation II

In this simulation, the function f_2 was used as a target:

$$f_2(x) = \sin(3\pi x)\cos[2(x - 1)]\exp[-(x - 1)^2]$$

It was sampled at a sampling rate of 16 Hz for training.

The supports of the target are:

$$\text{supp}_{0.05}(f) = [-1.0, 1.0]$$
$$\text{supp}_{0.05}(F) = [0.5, 4.0]$$

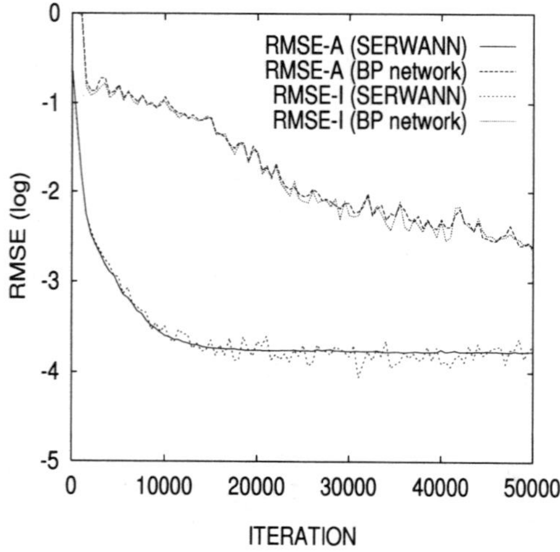

Figure 28-10 Learning curves (RMSE-A, RMSE for approximation; RMSE-I, RMSE for interpolation).

Table 28-2 The RMSEs of a SERWANN or BP Network in Simulation I

Network	RMSE	RMSE-A	RMSE-I
SERWANN		1.70×10^{-4}	2.02×10^{-4}
BP network		2.45×10^{-3}	2.26×10^{-3}

The ranges of dilation and translation are $a \in [1.0, 8.0]$ and $b \in [-1.0, 1.0]$, respectively.

SERWANN converged to 14–18 hidden units in 100 trials (see Table 28-3). The best result by SERWANN with 16 hidden units is illustrated in Fig. 28-12. The BP network with 16 hidden units did not converge at all. Therefore, the result is not given. Table 28-4 shows RMSEs of SERWANN.

28.5.3 Simulation III

The Gaussian function f_3 was used as a target:

$$f_3(x) = \exp[-(x - 0.5)^2]$$

It was sampled at a sampling rate of 16 Hz for training.

The supports of the target are:

$$\text{supp}_{0.05}(f) = [-1.0, 1.0]$$
$$\text{supp}_{0.05}(F) = [0.5, 3.5]$$

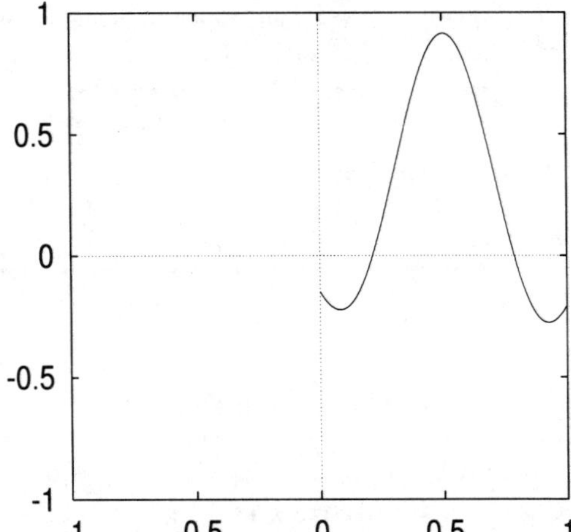

Figure 28-11   Partial retrieval of an approximated function by SERWANN in the positive range of the input domain, i.e., $x \in [0, 1]$.

Table 28-3   The Number of Successful and Failed Trials in Simulation II

N	Success (failure)
14	5 (0)
15	26 (3)
16	37 (5)
17	21 (0)
18	2 (1)
Total	91 (9)

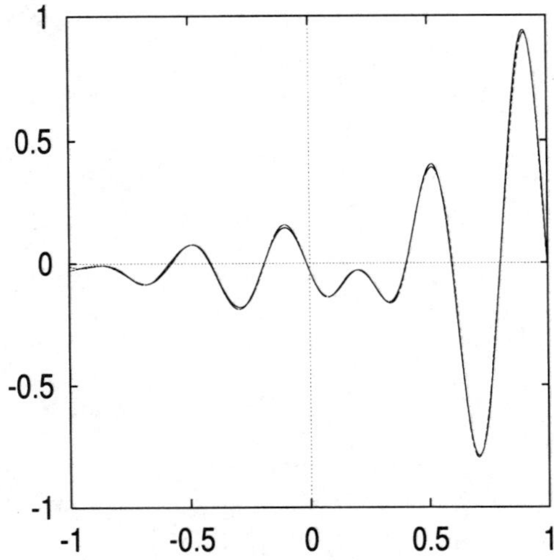

Figure 28-12   Result obtained by SERWANN (solid line, target function; broken line, approximated function).

Table 28-4   The RMSEs of a SERWANN or BP Network in Simulation II.
The Convergence Criterion is RMSE-A $< 5.0 \times 10^{-2}$.

Network	RMSE	RMSE-A	RMSE-I
SERWANN		5.91×10^{-5}	5.55×10^{-5}
BP network		—	—

The ranges of dilation and translation are $a \in [1.1, 8.0]$ and $b \in [-1.0, 1.0]$, respectively.

SERWANN converged to 5–7 hidden units in 100 trials (see Table 28-5). The best result by SERWANN with 6 hidden units is illustrated in Fig. 28-13. In this simulation, the BP network with 6 hidden units did not converge at all. Thus, the result is not given. Table 28-6 shows RMSEs of SERWANN.

Table 28-5   The Number of Successful and Failed Trials in Simulation III

N	Success (failure)
5	45 (0)
6	52 (0)
7	2 (1)
Total	99 (1)

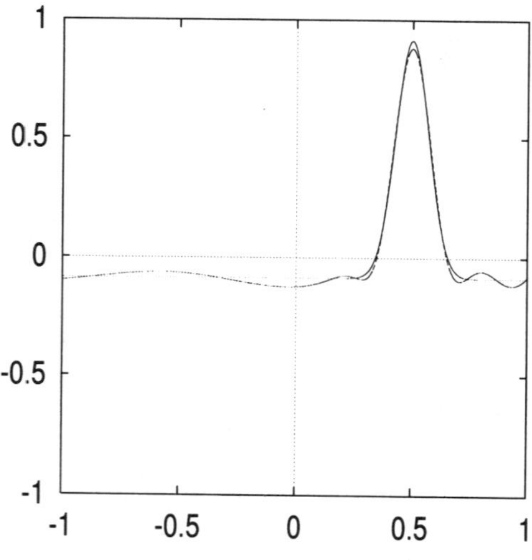

Figure 28-13   Result obtained by SER-WANN (solid line, target function; broken line, approximated function).

Table 28-6   The RMSEs of a SERWANN or BP Network in Simulation III.
The Convergence Criterion is RMSE-A $< 1.0 \times 10^{-2}$.

Network	RMSE	RMSE-A	RMSE-1
SERWANN		3.94×10^{-4}	3.94×10^{-4}
BP network		—	—

28.6. CONCLUSIONS

This chapter has proposed a mapping network called SERWANN, which incorporates wavelets into NNs. SERWANN handled function-approximating problems as two tasks: optimizing the network structure and minimizing training and generation errors.

The algorithm consisted of two processes. First, the training set was created by the time-frequency analysis. Since each hidden unit has a window in the time-frequency plane, hidden units were gradually reproduced to cover the time-frequency region of the target by the windows of FWs. Simultaneously, the network parameters were updated by Kohonen's learning law in order to preserve the network topology and take advantage of the later process. Next, the network parameters were updated by the LBP algorithm to minimize the errors. The LBP algorithm could dramatically reduce the learning cost.

The performance of SERWANN was confirmed by computer simulations through the comparison with the BP network. In addition, it was shown that SERWANN realizes a fast convergence and high convergence rate. Furthermore, it was shown that SERWANN can partially retrieve the approximated function using some hidden units.

REFERENCES

[1] R. Hecht-Nielsen, "Theory of the back propagation neural network," *Proc. IJCNN*, pp. 593–608, 1988.

[2] B. Irie and S. Miyake, "Capabilities of three-layered perceptrons," *Proc. ICNN*, pp. 641–648, 1988.

[3] K. Funahashi, "On the approximate realization of continuous mappings by neural networks," *Neural Networks*, vol. 2, pp. 183–192, 1989.

[4] K. Hornik, M. Stinchcombe, and H. White, "Multilayer feedforward networks are universal approximators," *Neural Networks*, vol. 2, pp. 359–366, 1989.

[5] T. Poggio and F. Girosi, "Networks for approximation and learning," *Proc. IEEE*, vol. 78, pp. 1481–1497, 1990.

[6] F. Girosi and T. Poggio, "Networks and the best approximation property," *Biol. Cybern.*, vol. 63, pp. 169–176, 1990.

[7] S. Lee and R. M. Kil, "A gaussian potential function network with hierarchically self-organizing learning," *Neural Networks*, vol. 4, pp. 207–224, 1991.

[8] D. E. Rumelhart, G. E. Hinton, and R. J. Williams, "Learning internal representations by error propagation." In *PDP: Explorations in the Microstructures of Cognition*, vol. 1, Cambridge, MA: MIT Press, pp. 318–362, 1985.

[9] R. Hecht-Nielsen, "Kolmogorov's mapping neural network existence theorem," *Proc. ICNN*, pp. 11–13, 1987.

[10] O. Fujita, "Optimization of the hidden unit function in feedforward neural networks," *Neural Networks*, vol. 5, pp. 755–764, 1992.

[11] M. R. Azimi-Sadjadi, S. Sheedvash and F. O. Trujillo, "Recursive dynamic node creation in multilayer neural networks," *IEEE Trans. on Neural Networks*, vol. 4, pp. 207–220, 1993.

[12] M. Hagiwara, "Novel back propagation algorithm for reduction of hidden units and acceleration of convergence using artificial selection," *Proc. ICNN*, pp. 625–630, 1990.

[13] Y. Matsunaga, Y. Nakade, O. Yamakawa, and K. Murase, "A backpropagation algorithm with automatic reduction of association units in multi-layered neural network," *Trans. IEICE (in Japanese)*, vol. J74-D-II, pp. 1118–1121, 1991.

[14] T. Oshino, J. Ojima, and S. Yamamoto, "Method for gradually reducing a number of hidden units on back propagation learning algorithm," *Trans. IEICE (in Japanese)*, vol. J76-D-II, pp. 1414–1424, 1993.

[15] S. Bornholdt and D. Graudenz, "General asymmetric neural networks and structure design by genetic algorithms," *Neural Networks*, vol. 5, pp. 327–334, 1992.

[16] M. Ishikawa, "A new approach to problem solving by modular structured networks," *Proc. ICFL & NN*, pp. 855–858, 1992.

[17] C. K. Chui, *An Introduction to Wavelets, Wavelet Analysis and Its Applications* vol. 1. New York: Academic Press, 1992.

[18] Q. Zhang and A. Benveniste, "Wavelet networks," *IEEE Trans. Neural Networks*, vol. 3, pp. 889–898, 1992.

[19] Y. C. Pati and P. S. Krishnaprasad, "Analysis and synthesis of feedforward neural networks using discrete affine wavelet transformations," *IEEE Trans. Neural Networks*, vol. 4, pp. 73–85, 1993.

[20] T. Kohonen, *Self-Organization and Associative Memory*. Berlin: Springer Verlag, 1989.

Chapter 29

On Wavelets and Fractal Processes

Patrick Flandrin

29.1. INTRODUCTION

In many areas (solid-state physics, hydrology, biology, turbulence, telecommunication networks, etc.), one is faced with processes which have specific and challenging properties: correlations decay very slowly, empirical power spectra exhibit power-law divergences at low frequencies, sample paths are extremely irregular with no characteristic scale. In fact, in many cases, such processes evidence "fractal" characteristics, thus requiring the use of specific tools for their analysis.

In recent times, fractal processes have received much attention, especially from the point of view of wavelet analysis because of its ability to reveal self-similarity features in signals and processes. Comprehensive studies devoted to wavelet analysis of fractal (or related) processes have already been given elsewhere [1–4]. It is the purpose of this chapter to summarize key results on the subject, with the hope that it will provide the reader with a general perspective which encompasses in the same framework situations corresponding to continuous processes, point processes, and filtered point processes.

Notation. For the sake of simplicity, discussion will be mostly limited in the following to orthonormal wavelet decompositions on a dyadic grid. At a given scale 2^j ($j \in \mathbf{Z}$) and at a given time instant $2^j n$ ($n \in \mathbf{Z}$), *approximation* coefficients will be given by

$$r_x[j, n] = 2^{-j/2} \int_{-\infty}^{+\infty} x(t)\, \varphi(2^{-j}t - n)\, dt$$

where $\varphi(t)$ is the "scaling function," whereas "detail" coefficients will correspond to

$$d_x[j, n] = 2^{-j/2} \int_{-\infty}^{+\infty} x(t)\, \psi(2^{-j}t - n)\, dt$$

where $\psi(t)$ is the associated "wavelet" such that the collection

$$\{\psi_{jn}(t) \equiv 2^{-j/2}\, \psi(2^{-j}t - n), (j, n) \in \mathbf{Z}^2\}$$

is an orthonormal basis of $L^2(\mathbb{R})$. (For basics on wavelets, the reader is invited to consult classical textbooks such as, e.g., [5].)

29.2. FRACTAL PROCESSES

Fractal processes are characterized by irregular sample paths at all scales, so that they do not possess any characteristic scale [6]. In other words, any magnification of a fractal trajectory reveals finer detail and ends up with a new trajectory which undergoes a similar behavior. More precisely, this idea is formalized by the concept of (statistical) *self-similarity* according to which a process $\{x(t), t \in \mathbb{R}\}$ is said to be H-self-similar if, for any $k > 0$,

$$\{x(kt)\} \stackrel{d}{=} \{k^H x(t)\}$$

where $\stackrel{d}{=}$ stands for an equality of all finite-dimensional laws.

The irregularity of a fractal curve can be measured by its "fractal dimension," a noninteger number which lies between 1 and 2 (the more irregular the curve, the closer to 2 its dimension). Different definitions can be adopted for a fractal dimension, among which the "Hausdorff dimension" plays a prominent role [6]. From the point of view of estimation, simpler definitions may, however, be preferred. The key idea underlying most approaches is that of a measurement at a scale a and the evaluation of its power-law (limit) behavior when a goes to zero. Assuming for instance that a curve $x(t)$ has "length" $L_a(x)$ when measured by a ruler of elementary length a, its fractal character can be evidenced by the fact that $L_a(x) \sim a^{-D}$ when a goes to zero, thus defining a fractal dimension as

$$D \equiv -\lim_{a \to 0} \frac{\log L_a(x)}{\log a}$$

It is important to note that fractality corresponds primarily to scaling laws defined in the time domain although, as we will see, it may induce typical behaviors in other domains, such as frequency or wavelet domains.

29.3. WAVELETS AND FRACTIONAL BROWNIAN MOTION

29.3.1 The Fractional Brownian Motion Model

Fractional Brownian motion (fBm) is the most useful model of a statistically self-similar process. By definition [7], fBm of the Hurst exponent H $(0 < H < 1)$ is a zero-mean Gaussian process $\{B_H(t), t \in \mathbb{R}\}$ which

1. Is nonstationary (although it has stationary increments) since

$$\mathbf{E}B_H(t)B_H(s) = \frac{\sigma^2}{2}\left(|t|^{2H} + |s|^{2H} - |t - s|^{2H}\right)$$

2. Has a "$1/f$" average spectrum given by

$$S_{B_H}(f) = \frac{\sigma^2}{|f|^{2H+1}}$$

the (power-law) divergence of which at the zero frequency induces a slowly-decaying correlation and, hence, long-range dependence

3. Is H-self-similar since

$$\{B_H(kt)\} \stackrel{d}{=} \{k^H B_H(t)\}, k \in \mathbb{R}^+$$

As a by-product of this property, sample paths of fBm are fractal curves of Hausdorff dimension $D_H = 2 - H$.

Fractional Brownian motion can be viewed as a natural generalization of ordinary Brownian motion (oBm), which corresponds to $H = 1/2$. One of the main differences between oBm and fBm is that, whereas increments of oBm are uncorrelated, those of fBm are correlated, either positively when $H > 1/2$ (persistence, $1 < D_H < 3/2$) or negatively when $H < 1/2$ (antipersistence, $3/2 < D_H < 2$). Moreover, those correlations can be shown to decay slowly since we have

$$|\tau| \gg \delta \Longrightarrow \mathbf{E}\, G_{H,\delta}(t + \tau)G_{H,\delta}(t) \sim \sigma^2 H(2H - 1)|\tau|^{2H-2}$$

with

$$G_{H,\delta}(t) = \frac{1}{\delta}[B_H(t + \delta) - B_H(t)]$$

Of course, true fBms are unlikely to be observed in natural phenomena, but the fBm model offers an extremely powerful idealization and a very convenient starting point for studying statistically self-similar processes, as does, for example, white noise in the case of stationary processes.

29.3.2 Wavelet Analysis of fBm

Because of the properties of "nonstationarity" (which require us to take into account time in the description the process) and "self-similarity" (which is naturally connected to scaling properties), it is appealing to consider a mixed time-scale description of fBm. This is precisely what can be given by the wavelet transform. In fact, and in parallel with the above-mentioned characteristic properties, the following points can be shown.

1. Although fBm is nonstationary, its wavelet coefficients are stationary at each scale [8, 9]. This means that there exists a function

$$f_H[k] \equiv -\iint_{-\infty}^{+\infty} |\tau|^{2H}\, \psi(s)\, \psi(s - \tau - k)\, ds\, d\tau$$

such that

$$\mathbf{E}d_{B_H}[j, n]d_{B_H}[j, m] = \frac{\sigma^2}{2} f_H[n - m]\,(2^j)^{2H+1}$$

This is merely due to the fact that fBm has stationary increments, the stationarizing property of the wavelet transform holding for any process with stationary increments [10].

2. Although fBm exhibits positive long-range correlations in the range $1/2 < H < 1$, its wavelet coefficients have a correlation which can be arbitrarily reduced. More precisely, it can be shown that, for a wavelet with R vanishing moments, we have asymptotically [9, 11]

$$f_H[k] \sim O\!\left(|k|^{2(H-R)}\right), k \longrightarrow +\infty$$

provided that the mild condition $R > H + 1/2$ is satisfied.

3. The self-similarity of fBm is directly reflected in its wavelet coefficients, whose variance varies as a power law as a function of scale [8, 9, 12]

$$\log_2 \mathbf{E}d_{B_H}^2[j, n] = (2H + 1)j + \log_2\!\left(\frac{\sigma^2}{2}f_H[0]\right)$$

This property can also be given a frequency interpretation which reveals the "$1/f$" character of fBm since, denoting by $\Gamma_{d_j}(f)$ the power spectral density of the (stationary) sequence $d_{B_H}[j, .]$ at a given scale j, we have

$$\Gamma_{d_j}(f) = \frac{\sigma^2}{|f|^{2H+1}}\, 2^j \left|\Psi(2^j f)\right|^2$$

a relation which can formally be interpreted as the input–output relationship of the "wavelet filter" at scale j—whose transfer function is $2^{j/2}\,\Psi(2^j f)$—excited by a process of power spectrum $\sigma^2/|f|^{2H+1}$. (For a more detailed study of wavelet-based spectrum analysis of fBm, the reader is referred to [2, 13].)

29.3.3 Wavelet Estimation of the Hurst Exponent

All the above properties evidence that wavelet analysis is naturally well-suited to fBm. Each of them provides in fact a key ingredient for a problem of major importance when analyzing fBm, i.e., the estimation of the Hurst parameter H or of the related spectral exponent $\alpha = 2H + 1$. Starting from the above-mentioned observation that—in the wavelet transform—variance progression follows a power law across scales

$$v[j] \equiv \mathbf{E}d^2_{B_H}[j, n] = C.2^{j\alpha}$$

one can simply make use of the empirical variance estimator at scale j (based on $N_j = 2^{-j}N_0$ coefficients for a sample of total length N_0)

$$\widehat{v}[j] \equiv \frac{1}{N_j}\sum_{n=1}^{N_j} d^2_{B_H}[j, n]$$

This is made possible because, when decomposed via the wavelet transform, fBm becomes stationary at each scale. Moreover, for a sample of finite length, the efficiency of this variance estimator can be improved by choosing for the analysis a wavelet with a number of vanishing moments high enough so as to get almost uncorrelated wavelet coefficients. Within this assumption, and making use of the Gaussian properties of fBm, it can be shown [2, 3] that the quantity

$$\eta[j] \equiv \log_2\left(\frac{\widehat{v}[j]}{\mathbf{E}d^2_{B_H}[j, n]}\right)$$

(a quantity which can be referred to as a normalized log-scalogram) has as its probability density function

$$p(\eta[j]) = \frac{(N_j/2)^{N_j/2}\ln 2}{\Gamma(N_j/2)}\exp\left\{\frac{N_j}{2}\left(\eta[j]\ln 2 - e^{\eta[j]\ln 2}\right)\right\}$$

with the asymptotic property

$$\eta[j] \overset{j\to\infty}{\longrightarrow} \mathcal{N}\left(0, \frac{2^{j+1}}{N_0\ln^2 2}\right)$$

where $\mathcal{N}(m, \sigma^2)$ denotes a Gaussian density with mean m and variance σ^2.

One can then use this result to find that the estimator $\widehat{\alpha}[J]$ of α based on a (weighted) linear regression in a log–log plot (variance versus scale, with J consecutive scales) is unbiased and that its variance reads [2, 13]

$$var(\widehat{\alpha}[J]) = \frac{1}{N_0\ln^2 2}\frac{2^J - 1}{2^J + 2^{-J} - J^2/2 - 2} \overset{J\to\infty}{\longrightarrow} \frac{1}{N_0\ln^2 2}$$

Such an estimator is furthermore efficient in the sense that its variance attains the Cramér–Rao lower bound [4]. It is worthwhile to remark that the asymptotic lower bound (which theoretically requires an infinite number of consecutive scales in the estimation) is almost reached for fairly small J values (typically, $J \sim 5$ [2]), a result which is very important from a practical point of view, since finite length data always impose a natural limit on the number of available dyadic scales.

29.3.4 Some Further Remarks on Wavelets and fBm

Allan Variance. Wavelet-based estimators dedicated to fBm can be viewed as versatile generalizations of previous techniques. An important feature of fBm is that its increments are stationary and such that

$$\mathbf{E}\left(B_H(t+\tau) - B_H(t)\right)^2 = \sigma^2 |\tau|^{2H}$$

thus suggesting that this variance should be estimated in order to find H. Although feasible, this approach is faced with a difficulty due to long-range dependence. Classical (empirical) variance estimators are obviously poor estimators in such a context and specific estimators have to be designed [14]. An example of a modified variance coping with long-range dependence is the so-called "Allan variance," defined as [15]

$$V_X(T) = \frac{1}{2T^2}\, \mathbf{E}\left[\int_{t-T}^{t} X(s)\,ds - \int_{t}^{t+T} X(s)\,ds\right]^2$$

and such that

$$V_{B_H}(T) \sim O(T^{2H}), \qquad T \longrightarrow +\infty$$

By construction, Allan variance is built upon differences between averages over increasing time-scales and it can be shown [9] that its discrete-time counterpart can be recast in wavelet terms, with the Haar wavelet as the basis function, since

$$V_{B_H}(2^{j-1}) = C(j).f_H^{(\text{Haar})}[0]$$

This interpretation is most useful for interpreting reported results on estimates based on the Allan variance and suggesting possible improvements by using more sophisticated wavelets. In fact, the Haar wavelet has only one vanishing moment ($R = 1$) and, within the range $1/2 < H < 1$, which corresponds to positive long-range correlations, this is not enough to satisfy the relation $R > H + 1/2$ which guarantees almost uncorrelated coefficients. This drawback is easily overcome by replacing the Haar wavelet by some more regular wavelet with at least two vanishing moments [16].

Synthesis. Another remark concerns the structure of the wavelet transform which, by construction, builds a signal by successive refinements, starting from a coarse approximation and adding finer and finer details at each step. Such a procedure is of course reminiscent of basic fractal constructions, thus suggesting we should make use of wavelets to synthesize a fractal process. In the case of orthonormal constructions, it can be shown [9] that wavelets offer a framework generalizing classical procedures such as "random midpoint displacement" [17], but that iterations based on uncorrelated wavelet coefficients at each scale, whose variance progression follows the expected power law indicated above, can only end up with approximate fBm processes [18]. It has, however, been recently shown that exact synthesis can nevertheless be achieved, provided that the constraint of orthogonality is relaxed and replaced by a biorthogonal scheme [19], making effective the construction of fBm via a fractional integration of white noise.

29.4. WAVELETS AND POINT PROCESSES

Fractional Brownian motion is a most useful model for continuous stochastic processes which evidence self-similarity properties and/or $1/f$ spectral behaviors. However, there are many instances in which—while spectra are of the $1/f$ type—a modeling of the data via a continuous process is not relevant. This is for instance the case in all the situations where observations are made of many individual and isolated contributions. In those situations too, where the introduction of point processes is necessary [20, 21], wavelet-based tools can be introduced [1].

29.4.1 Some Models

In order to cope with situations related to point processes, it is useful to introduce the very general model

$$P(t) = \sum_{k=-\infty}^{+\infty} g(t - t_k)$$

where the t_k are Poisson distributed with a (stationary) density $\lambda(t)$ and where long-range dependence may be induced by the input and/or the filter of impulse response $g(t)$.

Fractal Point Processes.   A first example, referred to as a fractal point process (FPP) [22], consists in assuming the following.

1. The filter $g(t)$ is all-pass: $g(t) = \delta(t)$.
2. The input Poisson density is non-zero fractional Gaussian noise ("derivative of fBm"): $\lambda(t) = \epsilon_\lambda + \delta B_H(t)$.

In such a case, long-range dependence arises from a complicated law of the occurrence of simply shaped events, the power spectral density of the resulting point process being given by

$$S_P(f) - \epsilon_\lambda - \epsilon_\lambda^2 \delta(f) \sim |f|^{1-2H}$$

at low frequencies.

Power-Law Shot Noise.   A second example, referred to as power-law shot noise (PSN) [23], assumes the following.

1. Given $0 < \beta < 1$, the filter $g(t)$ has a power-law impulse response of the form: $g(t) = t^{-\beta}$ if $0 < A \leq t < B < +\infty$ and 0 elsewhere.

2. The input Poisson density is constant: $\lambda(t) = \epsilon_\lambda$.

In this case, long-range dependence arises from a simple law of the occurrence of complexly shaped events, and the power spectral density of the corresponding filtered point process is given by

$$S_P(f) - \epsilon_\lambda^2 |G(0)|^2 \sim |f|^{2(\beta-1)}$$

at low frequencies.

29.4.2 A Wavelet-Based Fano Factor

As for fBm, some methods have been specifically designed for estimating parameters of a fractal point process [22]. One of them stems from a remark concerning standard Poisson processes (SPP), characterized by $\partial\lambda(t)/\partial t = 0$. In this case, it is well known that the associated counting process $N(.)$ is such that $\mathrm{var}N(T) = \mathbf{E}N(T)$ for any T, thus motivating the introduction of the so-called "Fano factor" [24]

$$F(T) \equiv \frac{\mathrm{var}N(T)}{\mathbf{E}N(T)}$$

as a test statistics for measuring a deviation from SPP.

Indeed, in the case of a fractal point process, we have [22]

$$F(T) = 1 + \frac{\sigma^2}{\epsilon_\lambda} T^{2H-1}$$

when $H > 1/2$. However, for the very same reasons as mentioned in the fBm case, estimation of the Fano factor is very difficult in the case of long-range dependence. If not performed with special care, this can even lead to largely erroneous results.

In order to improve this situation, one can first revisit the Fano factor and the way it is classically constructed. Basically, the Fano factor compares fluctuations and averages over larger and larger observation scales. This has clearly the flavor of quantities that can be captured by a multiresolution analysis and it is therefore natural to propose a definition of a wavelet-based Fano factor according to [1, 3, 25]

$$WF(j) = (2^j)^{\frac{1}{2}} \frac{\mathbf{E}d_P^2[j, n]}{\mathbf{E}r_P[j, n]}$$

It is easy to show that, in the FPP case, one has

$$WF(j) \sim 1 + C(2^j)^{2H-1}$$

for large j values. Moreover, an explicit connection between this new definition and the classical one can be given, which reads

$$WF^{(\mathrm{Haar})}(j) = F^{(\mathrm{Allan})}(2^j)$$

In words, this means that, when variance is estimated according to the definition of Allan, the classical Fano factor can be identified exactly with the wavelet-based factor, with Haar as the basis. The proposed definition is therefore generalizable beyond Haar, with an increased efficiency in the estimation when $R > 1$.

One further advantage of the wavelet-based Fano factor is that it offers the possibility of dealing directly with filtered point processes, which the usual factor does not. More precisely, in the PSN case, one gets

$$WF(j) \sim (2^j)^{2(1-\beta)}$$

within the range $1/B \ll 2^{-j} \ll 1/A$, provided that $R > \frac{1}{2} - \beta$.

By choosing a wavelet with enough vanishing moments, which can therefore be "blind" to polynomials of any desired degree, it also permits us to get rid of possible trends in the data, a situation known to heavily corrupt estimations in long-range-dependent situations.

29.5. FURTHER COMMENTS AND EXTENSIONS

29.5.1 On Implementation

All the above characterizations of fractal processes have been based on dyadic wavelet transforms and—provided that elementary precautions are taken, concerning, for example, the adaptation of the number R of vanishing moments to the Hurst exponent H—this poses no problem and any standard pyramidal algorithm [12] can be applied, as long as the same scaling law exists over the whole range of scales under study. Unfortunately, this is very often not the case and, in practice, scaling laws are generally observed over finite ranges of scales only, which obviously have no reason to correspond to those arbitrarily fixed by the dyadic analysis. In such, more realistic, situations, classical pyramidal algorithms cannot be used without some extra care and the results can be heavily affected by a too sparse sampling of the scale axis [2].

A first solution to this problem could be to make use of a continuous wavelet transform, but this is at the expense of a computational cost which is considerably increased. An intermediate and more satisfactory solution is to compute an approximate but quasi-continuous transform via a pyramidal scheme, as discussed, for example, in [26] or [27]. This solution, which amounts essentially to making use of a battery of wavelets associated with any wanted scale sampling in between any two consecutive dyadic scales, combines the advantages of the computational efficiency attached to dyadic wavelet transforms (recursive filter-bank structure) and of the versatility attached to continuous wavelet transforms (arbitrary choice of the "mother" wavelet and of the scale sampling).

More precisely, the procedure requires three steps.

1. In the first step, it is necessary to choose a multiresolution analysis (MRA), which amounts to selecting a scaling function $\varphi_0(.)$ and its associated two-scale sequence $u[.]$, such that

$$\varphi_0(t) = \sqrt{2} \sum_k u[k]\, \varphi_0(2t - k)$$

2. In a second step, a family of M mother wavelets is generated according to

$$\psi_m(t) \equiv \frac{1}{\sqrt{\alpha_m}}\, \psi\!\left(\frac{t}{\alpha_m}\right)$$

with $1 \le \alpha_m < 2$ and $m = 0, 1, \ldots, M - 1$

3. Finally, in a third step, all of these different wavelets are replaced by their MRA approximations, i.e., by the functions

$$\tilde{\psi}_m(t) = \sum_k v_m[k]\, \varphi_0(t - k)$$

with $v_m[k]$ the orthogonal projections

$$v_m[k] = \langle \psi_m(\cdot), \tilde{\varphi}_0(\cdot - k)\rangle, \quad k \in \mathbf{Z},\ m = 0, 1, \ldots, M - 1,$$

and where $\{\tilde{\varphi}_0(t - k)\}_{k\in\mathbf{Z}}$ is the dual Riesz basis of $\{\varphi_0(t - k)\}_{k\in\mathbf{Z}}$ in V_0, the space spanned by $\tilde{\varphi}_0(.)$ and its integer translates.

The three steps described above can therefore be implemented via a filter-bank pyramidal algorithm according to which approximations are given by the recursion

$$r_x[j, n] = \sum_k u[-k]\, r_x[2^{j-1}n - k, j - 1]$$

whereas the associated (approximate) wavelet coefficients at scales $2^{jm} = \alpha_m 2^j$ can be expressed as

$$\tilde{d}_x[j_m, n] = \sum_k v_m[-k]\, r_x[2^{j-1}n - k, j - 1]$$

For N data points, J octaves and M subbands per octave, the computational cost of this algorithm is $O(JMN)$, making it very efficient when—as it is commonly the case for correctly evidencing scaling laws—large amounts of data are to be processed.

29.5.2 On Time-Dependent Fractal Processes

In the fBm case, sample paths are everywhere singular and only one type of singularity exists at any time, characterized by the only Hurst exponent H. A useful generalization consists in allowing the singularity exponent to become time-dependent $[H \longrightarrow H(t)]$, thus generating a new process $B_{H(t)}(t)$ such that, e.g.,

$$\mathbf{E}\left[B_{H(t)}(t + \tau) - B_{H(t)}(t)\right]^2 = \sigma^2 |\tau|^{2H(t)}$$

In such a situation, which has been heuristically introduced in [28, 29] and, more recently, formalized in [30], the increment process is no longer stationary, which forbids us to globally apply the techniques mentioned above for estimating $H(t)$. Provided that variations of $H(t)$ are smooth enough, similar techniques, or variants based on time-scale energy distributions [29, 31], can, however, be applied locally, with a local estimation of variance at a given scale being obtained via time averages over an observation window which is itself scale-dependent.

29.5.3 On Multifractal Processes

The assumption that only one singularity exponent fully characterizes a fractal process (homogeneous case) can be reasonable—this is for instance the case in DNA sequences [32]—but it is often an oversimplification. Apart from the class of time-dependent fractal processes, other situations exist that correspond to the existence of a whole range of singularity exponents and of different values of those exponents at different times; this is observed, for example, in turbulence [33]. Multifractal processes are a model for such situations.

Considering again the fBm case, its homogeneous fractal properties can be viewed as a result of its Gaussian properties since, for any $q > 0$, the power-law variance progression of its wavelet coefficients can be generalized to

$$\log_2 \mathbf{E} d^q_{B_H}[j, n] = q\left(H + \frac{1}{2}\right)j + C$$

In fact, in many cases and because they are not Gaussian, fractal processes are heterogeneous and such a relation, in which the scaling exponent of order q is some affine function of q, is not observed. Such processes are referred to as multifractal processes, in the sense that the existence of q-dependent scaling exponents accounts for the existence of a whole hierarchy of singularities in the process, occurring not everywhere but on some subsets only of the real line. In contrast with homogeneous fractal (or monofractal) processes which are characterized by one number only—the Hurst exponent H in the fBm case—multifractal processes are best characterized by a function $D(H)$, referred to as their "singularity spectrum," which corresponds for any possible H to the Hausdorff dimension of the set of time instants t where the singularity exponent $H(t)$ is equal to H.

Arnéodo and coworkers have introduced a powerful wavelet-based method for estimating the singularity spectrum of a multifractal process [33–35]. The method is based on the continuous wavelet transform, defined in their case as

$$T_x(t, a) = \frac{1}{a}\int_{-\infty}^{+\infty} x(s)\, \psi\!\left(\frac{s - t}{a}\right) ds$$

It exploits the fact that, when a singularity exists in a signal at some time t_0, it is revealed by large coefficients in its wavelet transform and that, furthermore, the strength of the singularity [the scaling exponent $H(t)$] can be estimated from the decay across scales of the wavelet transform modulus at the time instant where the singularity occurs [36]:

$$\left|T_x(t_0, a)\right| \sim a^{H(t_0)}, \qquad a \longrightarrow 0$$

In order to be concerned with singularities only, the complete wavelet transform is first reduced to a skeleton formed by the set $\mathcal{L}(a)$ of all the lines l connecting maxima in the wavelet transform modulus up to scale a and converging towards a singularity when a goes to zero. A partition function

$$\mathcal{Z}(q, a) = \sum_{l \in \mathcal{L}(a)} \left\{ \sup_{(t, a') \in l} |T_x(t, a')| \right\}^q$$

is then introduced and it can be shown [35] that, for a fractal process,

$$\mathcal{Z}(q, a) \sim a^{\tau(q)}$$

when a goes to zero. In the multifractal case, $\tau(q)$ is a nonlinear function of q, from which the singularity spectrum $D(H)$ can be directly obtained via a Legendre transformation according to

$$D(H) = \min_q \left\{ qH - \tau(q) \right\}$$

One can check that, in the monofractal case, where $\tau(q)$ is a linear function of q, the singularity spectrum reduces to a unique point. In contrast, singularity spectra not reduced to a point are an indication of the plausability of a multifractal model and the necessity of not characterizing the process with a second-order fractal property only.

29.6. CONCLUSION

Wavelet-based tools have been briefly shown to be well-suited to (multi)fractal processes and their analysis. This is mainly due to the fact that the wavelet transform incorporates in its definition two basic features—time and scale—which are of primary importance for fractal processes: time, because fractal processes are characterized by singularities in sample paths, which eventually can be different at different instants; scale, because no characteristic scale can be attached to a fractal process and that characterization of a fractal process relies on the estimation of scaling laws and scaling exponents.

Let us, however, conclude with a remark useful in data analysis: when a process is fractal, a consequence of the nonexistence of a characteristic scale is that its spectral behavior is of the $1/f$ type, and that a (wavelet-based) spectrum estimation can be used to reveal some of the fractal characteristics of the process. The converse is, however, not true. In general, evidencing a $1/f$ spectrum does not suffice to guarantee that the analyzed process is fractal [37], and it is in general illegitimate to deduce a (meaningful) fractal dimension from a spectral exponent only. Indeed, given, for example, a fBm of Hurst exponent H and of fractal dimension $2 - H$, its power spectrum is proportional to $1/|f|^{2H+1}$ but, more importantly, the phase of its Fourier transform is uniformly distributed over $[0, 2\pi]$. Changing the way this phase is distributed therefore allows us to build as many surrogate processes as desired, with exactly the same power spectrum but with completely different time-domain characteristics and, hence, different fractal dimensions (with no reason for them to remain equal to $2 - H$, since fractal dimension is basically a time-domain feature and the time structure of sample paths is directly governed by phase relationships). As a consequence, it is in general necessary to decouple "fractal" from "$1/f$" and to draw conclusions about a parameter from the domain in which it finds its prominent significance.

Nevertheless, it is clear that, whereas they can be considered as either "fractal" or "$1/f$," processes which can be characterized by scaling and/or spectral exponents certainly can be analyzed with wavelets, and that such exponents can be most useful parameters for detection or classification purposes. In fact, first results obtained by such approaches have already been reported for, for example, seizure detection in EEG records [38] or the study of neural activity in auditory nerves [1, 21, 22]. More generally, it is believed that wavelet-based tools aimed at fractal or $1/f$ signals offer new perspectives from which new insights should be gained in other specific areas of biomedical signal processing.

ACKNOWLEDGMENTS

Several results summarized in this brief overview have been obtained in joint work with Patrice Abry and Paulo Gonçalvès, who are both gratefully acknowledged for numerous and stimulating discussions.

REFERENCES

[1] P. Abry and P. Flandrin, "Point processes, long-range dependence and wavelets." In *Wavelets in Biology and Medicine*, A. Aldroubi and M. Unser (eds.). Boca Raton, FL: CRC Press, 1995.

[2] P. Abry, P. Gonçalvès, and P. Flandrin, "Wavelets, spectrum analysis and $1/f$ processes." In *Wavelets and Statistics (Lecture Notes in Statistics)*, A. Antoniadis (ed.). Berlin: Springer, 1995.

[3] P. Flandrin, "Time-scale analyses and self-similar stochastic processes." In *Wavelets and Their Applications*, J. Byrnes et. al., (eds.). Dordrecht: Kluwer, pp. 121–142, 1994.

[4] G. W. Wornell, "Wavelet-based representations for the $1/f$ family of fractal processes," *Proc. IEEE*, vol. 81, pp. 1428–1450, 1993.

[5] I. Daubechies, *Ten Lectures on Wavelets*. Philadelphia: SIAM, 1992.

[6] K. Falconer, *Fractal Geometry*, New York: J. Wiley and Sons, 1990.

[7] B. B. Mandelbrot and J. W. van Ness, "Fractional Brownian motions, fractional noises and applications," *SIAM Rev.*, vol. 10, no. 4, pp. 422–437, 1968.

[8] P. Flandrin, "On the spectrum of fractional Brownian motions," *IEEE Trans. Inform. Theory*, vol. IT-35, no. 1, pp. 197–199, 1989.

[9] P. Flandrin, "Wavelet analysis and synthesis of fractional Brownian motion," *IEEE Trans. Inform. Theory*, vol. IT-38, no. 2, pp. 910–917, 1992.

[10] E. Masry, "The wavelet transform of stochastic processes with stationary increments and its application to fractional Brownian motion," *IEEE Trans. Inform. Theory*, vol. IT-39, no. 1, pp. 260–264, 1993.

[11] A. H. Tewfik and M. Kim, "Correlation structure of the discrete wavelet coefficients of fractional Brownian motion," *IEEE Trans. on Inform. Theory*, vol. 38, no. 2, pp. 904–909, 1992.

[12] S. G. Mallat, "A theory for multiresolution signal decomposition: the wavelet representation," *IEEE Trans. Pattern Anal. Machine Intell.*, vol. PAMI-11, no. 7, pp. 674–693, 1989.

[13] P. Abry, P. Gonçalvès, and P. Flandrin, "Wavelet-based spectral analysis of $1/f$ processes," *IEEE-ICASSP-93*, pp. III.237–III.240, Minneapolis, MN, 1993.

[14] J. Beran, "Statistical methods for data with long-range dependence," *Stat. Sci.*, vol. 7, no. 4, pp. 404–427, 1992.

[15] D. W. Allan, "Statistics of atomic frequency standards," *Proc. IEEE*, vol. 54, pp. 221–230, 1966.

[16] P. Flandrin, "Fractional Brownian motion and wavelets." In *Wavelets, Fractals and Fourier Transforms—New Developments and New Application*, M. Farge et al. (eds.). Oxford: Oxford University Press, 1992.

[17] H. O. Peitgen and D. Saupe, *The Science of Fractal Images*. New York: Springer Verlag, 1988.

[18] G. W. Wornell, "A Karhunen-Loève-like expansion for $1/f$ processes via wavelets," *IEEE Trans. Inform. Theory*, vol. IT-36, no. 4, pp. 859–861, 1990.

[19] F. Sellan, "Synthèse de mouvements browniens fractionnaires à l'aide de la transformation en ondelettes," *C. R. Acad. Sci. Paris*, t. 321, Série I, pp. 351–358, 1995.

[20] S. B. Lowen and M. C. Teich, "Fractal renewal processes generate $1/f$ noise," *Phys. Rev. E*, vol. 47, pp. 992–1001, 1993.

[21] S. B. Lowen and M. C. Teich, "Estimation and simulation of fractal point processes," *Fractals*, vol. 3, no. 1, pp. 183–210, 1995.

[22] D. H. Johnson and A. R. Kumar, "Modeling and analyzing fractal point processes," *IEEE-ICASSP-90*, Albuquerque, NM, pp. 1353–1356, 1990.

[23] S. B. Lowen and M. C. Teich, "Power-law shot noise," *IEEE Trans. Inform. Theory*, vol. IT-36, no. 6, pp. 1302–1318, 1990.

[24] U. Fano, "Ionization yield of radiations II. The fluctuations of the numbers of ions," *Phys. Rev.*, vol. 72, no. 1, pp. 26–29, 1947.

[25] P. Abry and P. Flandrin, "Wavelet-based Fano factor for long-range dependent point processes," *IEEE-EMBS-94*, Baltimore, MD, 1994

[26] P. Flandrin, E. Chassande-Mottin, and P. Abry, "Reassigned scalograms and their fast algorithms," San Diego, CA: SPIE, 1995.

[27] S. Maes, "A fast quasi-continuous wavelet transform algorithm," *Colloque TOM-94*, Lyon, pp. 31.1–31.4, 1994.

[28] P. Gonçalvès and P. Flandrin, "Scaling exponents estimation from time-scale energy distributions," *IEEE-ICASSP-92*, San Francisco, CA: IEEE, pp. V.157–V.160, 1992.

[29] P. Flandrin and P. Gonçalvès, "From wavelets to time-scale energy distributions." In *Recent Advances in Wavelet Analysis*, L. L. Schumaker and G. Webb (eds.). Academic Press, pp. 309–334, 1994.

[30] R. F. Peltier and J. Lévy-Véhel, "Multifractional Brownian motion," *INRIA Research Report* No. RR-2645, 1995.

[31] P. Flandrin, *Temps-Fréquence*, Paris: Hermès, 1993.

[32] A. Arnéodo, E. Bacry, P. V. Graves, and J. F. Muzy, "Characterizing long-range correlations in DNA sequences from wavelet analysis," *Phys. Rev. Lett.*, vol. 74, no. 16, pp. 3293–3296, 1995.

[33] J. F. Muzy, E. Bacry, and A. Arnéodo, "Wavelets and multifractal formalism for singular signals: Application to turbulence data," *Phys. Rev. Lett.*, vol. 67, no. 25, pp. 3515–3518, 1991.

[34] J. F. Muzy, E. Bacry, and A. Arnéodo, "Multifractal formalism for fractal signals: The structure function approach versus the wavelet transform modulus maxima method," *Phys. Rev. E*, vol. 47, no. 2, 1993.

[35] E. Bacry, J. F. Muzy, and A. Arnéodo, "Singularity spectrum of fractal signals from wavelet analysis: exact results," *J. Stat. Phys.*, vol. 70, p. 635, 1993.

[36] S. G. Mallat and W. L. Hwang. "Singularity detection and processing with wavelets," *IEEE Trans. Inform. Theory*, vol. IT-38, no. 2, pp. 617–643, 1992.

[37] T. Higuchi, "Relationship between the fractal dimension and the power law index for a time series: a numerical investigation," *Physica D*, vol. 46, pp. 254–264, 1990.

[38] S. V. Mehta, R. W. Koser, and P. J. Venziale, "Wavelet analysis as a potential tool for seizure detection," *IEEE-SP Int. Symp. on Time-Frequency and Time-Scale Analysis*, Philadelphia, PA, pp. 584–587, 1994.

Chapter 30

Fractal Analysis of Heart Rate Variability

Russell Fischer, Metin Akay

30.1. INTRODUCTION

Fractal models have found wide acceptance in many fields of science. They provide a powerful tool for the study of systems that demonstrate long-term correlations and $1/f$-type spectral behavior. The same properties that make fractals powerful models are those that also complicate analysis, however. The extended correlation structure is difficult to capture with standard techniques (e.g., finite-order autoregressive moving average (ARMA) models), and frequency-domain analysis is complicated by the nonstationary character of many fractal processes [1].

Nevertheless, a model known as fractional Brownian motion (fBm) has provided a mathematical framework for the development of analytical methods appropriate for such processes. Fractional Brownian motion is a nonstationary random process that has infinitely long-run correlations and demonstrates $1/f$-type spectral behavior. Fractional Brownian motion also has the interesting property of statistical self-similarity, obeying the relationship:

$$\mathrm{var}[B_H(t_2) - B_H(t_1)] \propto (t_2 - t_1)^{2H}$$

where $B_H(t)$ is the fBm process, and H is the Hurst exponent, a single parameter that characterizes the scaling relationship. In this sense, fBm is fractal; it possesses no characteristic time-scale, and any selected segment properly rescaled is statistically equivalent to the original process.

The fBm model has been successfully applied in many fields of science, including physiology. Fractional Brownian motion models have been proposed for signals such as the electroencephalogram (EEG), heart rate variability (HRV), and regional coronary flow distributions [2, 3]. Typically an fBm model will be tentatively proposed for such processes based on demonstrated $1/f$-type spectral behavior, and the problem is then to estimate H for the process. In this chapter, we discuss a number of techniques that take advantage of the scaling nature of fBm in different ways to estimate H.

We begin by reviewing the fBm model and the process obtained by calculating the increments of fBm, discrete fractional Gaussian noise. We then review three estimators for H. These estimators, based upon the maximum likelihood estimation (MLE), the power spectral density (PSD), and the wavelet transform, exploit the scaling behavior in different ways to estimate H.

30.2. THE fBm MODEL

The fBm model can be understood as a generalization of Brownian motion. In fact in [4], Mandelbrot and Van Ness define fBm as a moving average of the increments of Brownian motion. Although fBm is a nonstationary random function, the increments of fBm are stationary and possess the following properties.

1. The increments have zero mean: $E\{B_H(k) - B_H(k - 1)\} = 0$.
2. The increments are in a strict sense stationary.
3. The variance obeys the scaling law: $E\{B_H(k + T) - B_H(k)\}^2 = T^{2H} V_H$ where V_H is a constant.

In an experimental situation, we will typically obtain samples of a fractional Brownian process:

$$B_H[k] = B_H(kT_S) \tag{30-1}$$

where T_s is the sampling period. The sampled fBm is called the discrete fractional Brownian motion (DfBm). The increments of DfBm are called discrete fractional Gaussian noise (DFGN):

$$X_H = B_H[k] - B_H[k - 1] \tag{30-2}$$

30.3. THE AUTOCORRELATION FUNCTION FOR DFGN

As DFGN is stationary and Gaussian, only the mean and autocorrelation are necessary to completely describe the process. From property 1 for fBm, the DFGN process is zero mean. The autocorrelation function for DFGN was derived by Lundahl et al. [1] as follows. Define the autocorrelation function for DFGN:

$$r[k] = E\{X_H[n + k]X_H[n]\} \tag{30-3}$$

substitute by (30-2):

$$r[k] = E\{(B_H[n+k] - B_H[n+k-1])(B_H[n] - B_H[n-1])\} \tag{30-4}$$

expanding, we have:

$$\begin{aligned} r[k] =&E\{B_H(n+k)B_H(n) - B_H(n+k)B_H(n-1) \\ &- B_H(n+k-1)B_H(n) - B_H(n+k-1)B_H(n-1)\} \end{aligned} \tag{30-5}$$

Then using the relation $E\{xy\} = -\frac{1}{2}[E\{x-y\}^2 - E\{x\}^2 - E\{y\}^2]$,

$$\begin{aligned} r[k] = -\frac{1}{2}[&E\{B_H[n+k] - B_H[n]\}^2 - E\{B_H[n+k]B_H[n-1]\}^2 \\ &- E\{B_H[n+k-1]B_H[n]\}^2 - E\{B_H[n+k-1]B_H[n-1]\}^2] \end{aligned} \tag{30-6}$$

Using property (30-3) for fBm we finally obtain:

$$r[k] = \frac{V_H}{2}\left[|k+1|^{2H} - 2|k|^{2H} + |k-1|^{2H}\right] \tag{30-7}$$

30.4. THE PROBABILITY DENSITY FUNCTION FOR DFGN

With the autocorrelation function for DFGN in hand, we may now directly write the probability density function (PDF) for DFGN:

$$p(\mathbf{x}, H) = 2\pi^{-\frac{N}{2}}|\mathbf{R}|^{\frac{-1}{2}}\exp\left\{-\frac{1}{2}\mathbf{x}^{\mathrm{T}}\mathbf{R}^{-1}\mathbf{x}\right\} \tag{30-8}$$

where $\mathbf{x} = \{x_0, x_1, \ldots, x_{N-1}\}^{\mathrm{T}}$ is the dataset and R is the covariance matrix $[\mathbf{R}]_{ij} = r[|i-j|]$, and $|\mathbf{R}|$ is the determinant of R.

30.5. A MAXIMUM LIKELIHOOD ESTIMATOR FOR DFGN

A maximum likelihood estimate of a parameter is often considered the best possible estimate, having minimum variance and being asymptotically unbiased and asymptotically Gaussian in distribution. Lundahl [1] developed a maximum likelihood estimator for H with the following approach. We consider the measured dataset to be a realization of a random variable whose pdf is parameterized by an unknown parameter. In our case, the parameter is H. We then seek a unique value of H for which the observed dataset is more probably observed than for any other. We call this value of H for which the dataset is most probable the maximum likelihood estimate of H.

The liklihood function for DFGN is equivalent to (30-8) with the interpretation that it expresses the likelihood of dataset $\mathbf{x}$ occurring given a value H. To simplify the likelihood function, the logarithm of the likelihood function may be used. The logarithm is a monotonic function, so maximizing the log-likelihood function is equivalent to maximizing the likelihood. From (30-8),

$$\ln p(\mathbf{x}, H) = -\frac{N}{2}\ln 2\pi - \frac{1}{2}\ln|\mathbf{R}| - \frac{1}{2}\mathbf{x}^{\mathrm{T}}\mathbf{R}^{-1}\mathbf{x} \tag{30-9}$$

The autocorrelation function is a function of H and V_H. It will simplify the calculations to remove V_H as follows. Define

$$\mathbf{R} = V_H \bar{\mathbf{R}} \tag{30-10}$$

The likelihood function then becomes

$$\ln p(\mathbf{x}, H) = -\frac{N}{2}\ln 2\pi - \frac{N}{2}\ln V_H - \frac{1}{2}\ln |\bar{\mathbf{R}}| - \frac{1}{2V_H}\mathbf{x}^T\bar{\mathbf{R}}^{-1}\mathbf{x} \tag{30-11}$$

The likelihood function can then be minimized with respect to V_H by taking the derivative of (30-11) with respect to V_H and setting it equal to zero:

$$V_H = \frac{1}{N}\mathbf{x}^T\bar{\mathbf{R}}^{-1}\mathbf{x} \tag{30-12}$$

This expression for V_H is then inserted into (30-11) to form the final likelihood function:

$$\ln p(\mathbf{x}, H) = -\frac{N}{2}\ln 2\pi - \frac{N}{2}\ln\left(\frac{1}{N}\mathbf{x}^T\bar{\mathbf{R}}^{-1}\mathbf{x}\right) - \frac{1}{2}\ln |\bar{\mathbf{R}}| - \frac{N}{2} \tag{30-13}$$

A search algorithm may now be used to find H as a function of the dataset.

30.6. PSD ESTIMATORS FOR fBm AND DFGN

The nonstationary aspect of the fBm model implies that the variance and PSD will be time dependent. In Mandelbrot and van Ness's work [4], it is proposed that the nonstationarity can be dealt with by calculating the increments of the process (DFGN) to create a stationary process that is better suited to analysis. However, many researchers have measured the PSD of fBm processes to find that the slope of the PSD is quite consistent regardless of the position of the observation window. The apparent contradiction was explained by Keshner [5], who showed that when the observation window for such a process is short compared to the total elapsed time since the start of the process, the PSD will maintain the $1/f$ shape although the amplitude of the PSD may vary.

Assuming that we are analyzing a trace of fBm that is very short relative to the total elapsed time of the process (the typical case), there are then two viable options for the determination of H via the PSD: application of the PSD to the fBm directly, or application of the PSD to the increments of the process. In the first case the PSD will be proportional to $|f|^{-2H-1}$, in the second case proportional to $|f|^{1-2H}$. Studies using simulated fBm datasets have indicated that the preferred approach is to analyze fBm directly; a PSD estimator for H utilizing DFGN can be biased in a nonlinear manner [6].

Typically in an experimental situation, the acquired data are discrete and care must be taken to avoid aliasing of the sampled data. Assuming the aliasing problem has been minimized, an estimate of the PSD can then be obtained by the periodogram. The fast Fourier transform (FFT) is used to calculate the discrete Fourier transform:

$$X[k] = \frac{1}{N} \sum_{n=0}^{N-1} x[n] e^{-j\frac{2\pi kn}{N}} \tag{30-14}$$

where $x[n] = \{x_0, x_1, \ldots, x_{N-1}\}$ is the dataset. The periodogram estimate of the power spectrum is defined at $\frac{N}{2} + 1$ frequencies as:

$$P[0] = P[f_0] = |X[0]|^2$$
$$P[f_k] = |X[k]|^2 + |X[N-k]|^2 \qquad k = 1, 2, \ldots, \left(\frac{N}{2} - 1\right) \tag{30-15}$$
$$P[f_c] = P\left[f_{\frac{N}{2}}\right] = \left|X\left[\frac{N}{2}\right]\right|$$

The estimates of power generated by the periodogram are typically high in variance. It is usually beneficial to implement a strategy for reducing the variance. One approach is to partition the original time-domain data evenly into segments. The periodograms of the individual segments are then calculated, and averaged to form the final estimate of spectral power [7].

To calculate an estimate of H from the periodogram, the slope of the PSD is calculated from a log–log plot using standard regression techniques. The estimate of H is then:

$$\frac{\text{slope of PSD} + 1}{-2} \tag{30-16}$$

30.7. A WAVELET ESTIMATOR FOR DFGN

An essential characteristic of the wavelet transform is the ability to analyze a process relative to scale or resolution. Thus the wavelet transform is uniquely well suited to the analysis of fBm, which is essentially a scale-invariant process. Flandrin demonstrated how fBm, nonstationary in the time domain, could be rendered stationary in the wavelet domain and more amenable to analysis [8].

The wavelet transform can be understood by drawing some analogies to the Fourier transform. Both transforms map the input data from the time domain to a new domain in which the data are represented as a weighted sum of characteristic basis functions. For the Fourier transform, the basis functions are sines and cosines. For the wavelet transform the basis is composed of scaled and shifted versions of a single function, the analyzing or "mother" wavelet:

$$\Psi_n^m(t) = 2^{m/2} \Psi(2^m t - n) \tag{30-17}$$

In this continuous-time representation, $\Psi(t)$ is the mother wavelet, and m and n are the dilation and translation indices, respectively. Typically, we require that the wavelet system provide an orthonormal basis for transformation of our signal

$$x(t) \leftrightarrow x_n^m \tag{30-18}$$

and the transformation is defined throughout the synthesis and analysis equations:

$$x(t) = \sum_m \sum_n x_n^m \Psi_n^m(t) \tag{30-19a}$$

$$x_n^m = \int_{-\infty}^{+\infty} x(t)\Psi_n^m(t)dt \tag{30-19b}$$

Note that in analogy to the Fourier transform, the wavelet transform analysis is performed by projecting the input data onto our basis functions.

The work of Mallat [9] led to an efficient algorithm for wavelet decomposition of a discrete signal, and a framework for understanding the wavelet transform in terms of a multiresolution analysis. The algorithm, called the discrete wavelet transform (DWT) utilizes a pyramidal algorithm to recursively decompose the signal into "approximation" and "detail" signals at successive resolutions.

Let a_n^0 denote the input signal, where the superscript 0 denotes the resolution of the original acquired data and n is an index to the vector elements. We then recursively calculate approximation coefficients a_n^m and detail coefficients d_n^m at lower (coarser) resolutions $m > 0$ via

$$a_n^{m+1} = \sum_k h[k - 2n]a_k^m \qquad \text{"smooth"} \tag{30-20a}$$

$$d_n^{m+1} = \sum_k g[k - 2n]a_k^m \qquad \text{"detail"} \tag{30-20b}$$

For each decrease in resolution $m + 1$, the representation provides a lower resolution approximation of the signal a_n^{m+1} and a detail signal d_n^{m+1} equivalent to the difference in information between two successive resolutions. The filters $h[\ ]$ and $g[\ ]$ are quadrature mirror filters, respectively a low-pass and high-pass, that possess impulse responses with the relationship:

$$g[k] = (-1)^k h[1 - k] \tag{30-21}$$

The quadrature filters are derived for a given wavelet basis, and provide the linkage between the multiresolution analysis and the wavelet transform. In this study, the class of compactly supported wavelets proposed by Daubechies [10] was used, satisfying an approximation condition of order 4.

For an input signal of length $N = 2^r$, the DWT produces a series of $r - 1$ detail signals of length $N/(2^m)$. The detail signals have been shown to be weakly correlated in scale and time for fBm [11], and obey the following scaling relationship for their variance [12]:

$$\text{var}[x_n^m] = \sigma^2 2^{-(2H+1)m} \tag{30-22}$$

Essentially the DWT acts as a whitening filter for fBm such that the d_n^m coefficients are rendered weakly correlated in scale and time. Kaplan and Kuo [13] improved the analytical approach by recognizing that the nonstationarity present in fBm can propagate through the recursive calculation used by the DWT to calculate the detail coefficients and bias the variance progression. They demonstrated that a lower bias result could be obtained by applying the DWT to DFGN instead of fBm, and our work supports that conclusion. For analysis of DFGN, the scaling relationship is changed to:

$$\mathrm{var}[d_n^m] = 2^{(2H-1)(m-1)}\sigma^2\left(2 - 2^{2H-1}\right) \tag{30-23}$$

which can be simplified as follows:

$$\log_2[\mathrm{var}] = (2H - 1)m + f(H, \sigma) \tag{30-24}$$

In Kaplan and Kuo's work a maximum likelihood estimator developed by Wornell and Oppenheim [12] was used to find H based on the variance progression (30-11). An estimate-maximize (EM) algorithm was then used to search for the maximum likelihood (ML) value of H. To improve the speed of calculation we used the formulation (30-12) directly, and found H by performing a linear regression on a $\log_2$ plot of the detail variance versus the resolution level. By calculating H in this manner, we are no longer assured of the desirable properties attributed to the ML method, i.e., asymptotically unbiased and efficient. However, the method is more directly comparable to the PSD method which employs regression analysis, and is better suited to real-time processing.

An additional implementational note concerns the exclusion of detail signals. It was empirically found in this study that the accuracy and precision of the wavelet method significantly improved upon exclusion of the first detail signal from analysis. Accordingly, the first detail signal was omitted from the regression calculation used to find H for the results presented in this study.

30.8. THE HEART RATE VARIABILITY SIGNAL

The HRV signal is derived by measurement of the time interval between successive R–R events [Fig. 30-1(a)]. In this way, it is fundamentally different from a signal derived by the sampling of a continuous process, and requires special consideration before analysis. Although the HRV signal is a sequence of events (R-waves), one strategy is to assume that there is an underlying continuous process behind the events. With this approach the R-wave events are considered "snapshots" of the continuous process, and reconstruction techniques are used to estimate them. The advantage of this method is that the continuous process, once derived, may be sampled equidistantly in time and analyzed by standard techniques. The drawback is the risk of adding spurious information to the HRV measurements via the reconstruction algorithm. It is also unclear whether there is a sound physiological basis for a heart rate signal defined for all points in time.

In this study we adopt the approach outlined by DeBoer [14], in which the HRV signal is defined in terms of R–R intervals rather than sampled time. When defined in this way, the HRV signal is equidistantly sampled in terms of R–R intervals, not time [Fig. 30-1(b)], and may be processed by methods that require equidistant samples. It is important to note that when we speak of the spectral content of such a signal, we are speaking of the *interval* power spectra, and not the conventional power spectra of a time-sampled process. If necessary, it is possible to map the interval spectra to time or frequency by mapping the R–R events to the time axis, separated by time intervals equivalent to the average heart rate [Fig. 30-1(c)]. With this approximation, the frequencies of other physiological processes (i.e., breathing) can be located in the interval spectrum.

To illustrate the scaling behavior of the HRV signal in the frequency and wavelet domains, we will investigate a segment of real HRV data acquired by a Holter monitor (Fig. 30-2(a)]. In this figure, the abscissa is the interval count, and the ordinate is the corresponding measure of time between the R–R waves that define the interval. The duration of the segment shown is approximately three hours. The complete segment was arbitrarily divided into ten analysis windows of equal length. In each analysis window, an estimate of the wavelet or power spectra was calculated for five 256-point data segments, and the five estimates were then averaged in the spectral or wavelet domains to form the final estimate for the analysis window. In Fig. 30.2(b), the ten sequential estimates of the PSD are shown.

In Fig. 30-2(c) the estimates of wavelet variance as a function of detail level are shown for the same data set. In order to directly compare the wavelet variance spectra with the PSD, we take the convention of plotting successive detail variance estimates from right to left. This reflects the progression of DWT analysis, which begins as a

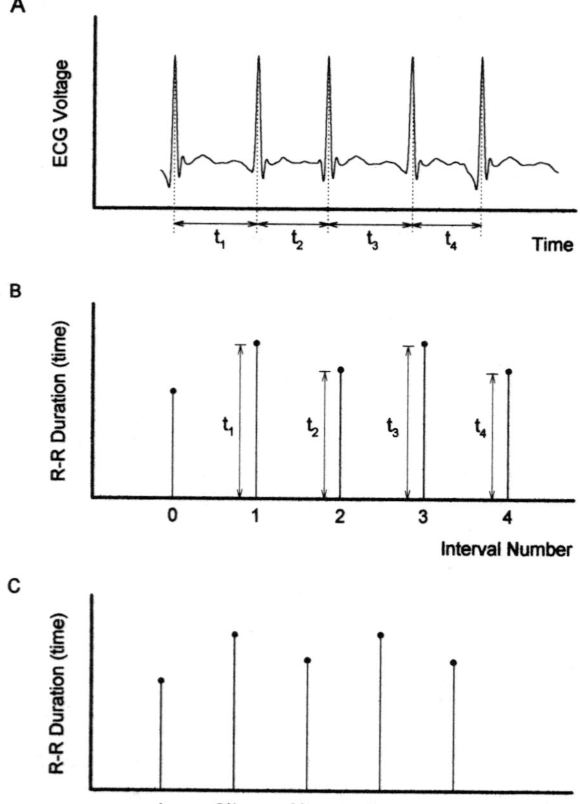

Figure 30-1 The HRV signal: (a) Measurement of R–R intervals; (b) HRV represented as R–R duration versus interval number; (c) mapping intervals back to the time axis by using the average interval duration.

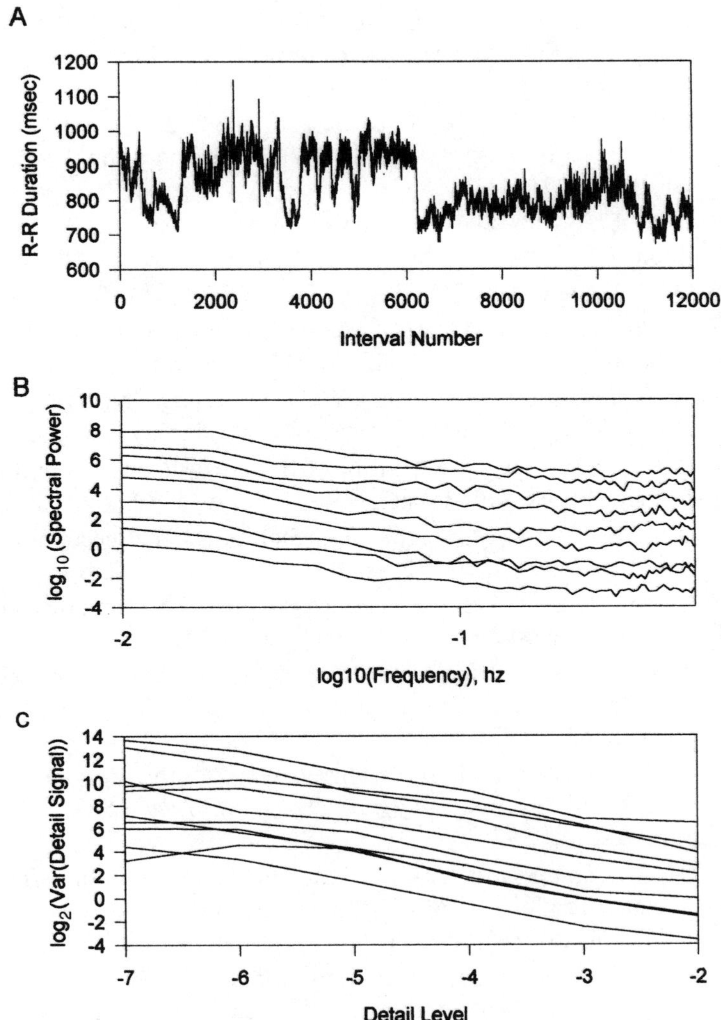

Figure 30-2   (a) A real HRV record acquired by a Holter monitor. (b) The power spectral density for ten successive windows of the HRV record. (c) Wavelet detail variance for the same record.

high-resolution analysis, and proceeds to successively lower resolutions. In both the wavelet and frequency domains, the characteristic $1/f$ behavior is present.

REFERENCES

[1] T. Lundahl, W. J. Ohley, S. M. Kay, and R. Siffert, "Fractional Brownian motion: A maximum likelihood estimator and its application to image texture," *IEEE Trans. Med. Imag.*, vol. MI-5, no. 3, pp. 152–161, 1986.

[2] J. B. Bassingthwaighte, and G. M. Redmond, "Evaluation of the dispersional analysis method for fractal time series," *Ann. Biomed. Eng.*, vol. 23, pp. 491–505, 1995.

[3] J. P. Saul, P. Albrecht, R. D. Berger, and R. J. Cohen, "Analysis of long term heart rate variability: Methods, $1/f$ scaling and implications," *Comp. Cardiol.*, vol. 14, pp. 419–422, 1987.

[4] B. B. Mandelbrot and J. W. van Ness, "Fractional Brownian motions, fractional noises and applications," *SIAM Rev.*, vol. 10, no. 4, pp. 422–437, 1968.

[5] M. S. Keshner, "$1/f$ Noise," *Proc. IEEE*, vol. 70, no. 3, pp. 212–218, 1982.

[6] R. Fischer and M. Akay, "A comparison of analytical methods for the study of fractional Brownian motion," *Ann. Biomed. Eng.*, vol. 24, pp. 537–543, 1996.

[7] W. H. Press, S. A. Teukolsky, W. T. Vetterling, and B. P. Flannery, *Numerical Recipes in C*, 2nd Ed. Cambridge: Cambridge University Press, p. 994, 1992.

[8] P. Flandrin, "On the spectrum of fractional Brownian motion," *IEEE Trans. Inform. Theory*, vol. 35, no. 1, pp. 197–199, 1989.

[9] S. G. Mallat, "A theory for multiresolution signal decomposition: The wavelet representation," *IEEE Trans. PAMI*, vol. 11, no. 7, pp. 674–693, 1989.

[10] I. Daubechies, "Orthonormal bases of compactly supported wavelets," *Comm. Pure Appl. Math.*, vol. 44, pp. 909–996, 1988.

[11] A. H. Tewfik and M. Kim, "Correlation structure of discrete wavelet coefficients of fractional Brownian motion," *IEEE Trans. Inform. Theory*, vol. 38, pp. 904–909, 1992.

[12] G. W. Wornell, and A. V. Oppenheim, "Estimation of fractal signals from noisy measurements using wavelets,' *IEEE Trans. Signal Proc.*, vol. 40, no. 3, pp. 611–623, 1992.

[13] L. M. Kaplan, and C. C. J. Kuo, "Fractal estimation from noisy data via discrete fractional Gaussian noise (DFGN) and the Haar basis," *IEEE Trans. Signal Proc.*, vol. 41, no. 12, pp. 3554–3562, 1993.

[14] R. W. DeBoer, J. M. Karemaker, and J. Strackee, "Comparing spectra of a series of point events particularly for heart rate variability data," *IEEE Trans. Biomed. Eng.*, vol. BME-31, no. 4, pp. 384–387, 1984.

Index

Editor's Biography

Metin Akay is IEEE Press Series Editor for the IEEE Press Series in Biomedical Engineering, and a member of the IEEE Engineering in Medicine and Biology Society Publication Committee. Dr. Akay has authored *Biomedical Signal Processing* (Academic Press, 1994); *Detection and Estimation of Biomedical Signals* (Academic Press, 1996); and coauthored the most recent edition of *Theory and Design of Biomedical Instruments* (Academic Press, 1991). He has published a number of technical papers in the areas of noninvasive detection of coronary artery disease, early human development, and control of breathing. In addition, Dr. Akay holds two U.S. patents and has given several keynote/plenary and invited talks at internationl conferences, workshops, and symposiums in these areas.